Algebraic Structures and Applications

CONTEMPORARY MATHEMATICS

826

Algebraic Structures and Applications

Conference
NonCommutative Rings and their Applications, VIII
August 28–31, 2023
Lens, France

Conference
Quadratic Forms, Rings and Codes
August 31, 2023
Lens, France

Ahmed Laghribi
André Leroy
Editors

AMS American Mathematical Society

2020 *Mathematics Subject Classification.* Primary 16D10, 16D25, 16D70, 16G10, 16Z10, 11T71, 94Bxx, 11E04, 11E81.

Library of Congress Cataloging-in-Publication Data

Names: Leroy, Andre (Andre Gerard), 1955– editor | Laghribi, Ahmed, 1967– editor | NonCommutative Rings and Their Applications (NCRA ,VIII) (2023 : Lens, France) | Quadratic Forms, Rings and Codes (QFRC) (2023 : Lens, France)

Title: Algebraic structures and applications / NonCommutative Rings and Their Applications (NCRA ,VIII), August 28–31, 2023, Lens, France ; Quadratic Forms, Rings and Codes (QFRC), August 31, 2023, Lens, France ; André Leroy, Ahmed Laghribi, editors.

Description: Providence, Rhode Island : American Mathematical Society, 2025. | Series: Contemporary mathematics, 02714132 ; vol 826 | "NonCommutative Rings and Their Applications (NCRA ,VIII), August 28–31, 2023, Lens, France"

Identifiers: LCCN 2025013720 | ISBN 9781470477639 paperback | ISBN 9781470481032 ebook

Subjects: LCSH: Coding theory | Graph theory | Number theory | Linear algebraic groups | AMS: Associative rings and algebras – Modules, bimodules and ideals – General module theory | Associative rings and algebras – Modules, bimodules and ideals – Ideals | Associative rings and algebras – Modules, bimodules and ideals – Structure and classification | Associative rings and algebras – Representation theory of rings and algebras – Representations of Artinian rings | Number theory – Finite fields and commutative rings (number-theoretic aspects) – Algebraic coding theory; cryptography | Information and communication, circuits – Theory of error-correcting codes and error-detecting codes | Number theory – Forms and linear algebraic groups – Quadratic forms over general fields | Number theory – Forms and linear algebraic groups – Algebraic theory of quadratic forms; Witt groups and rings

Classification: LCC QA268 .A334 2025

LC record available at https://lccn.loc.gov/2025013720

DOI: https://doi.org/10.1090/conm/826

Color graphic policy. Any graphics created in color will be rendered in grayscale for the printed version unless color printing is authorized by the Publisher. In general, color graphics will appear in color in the online version.

♾ The paper used in this book is acid-free and falls within the guidelines
established to ensure permanence and durability.
Visit the AMS home page at `https://www.ams.org/`

10 9 8 7 6 5 4 3 2 1   30 29 28 27 26 25

Contents

Preface

This volume covers many different aspects of research in algebra and considers their applications, in particular towards coding theory and graph theory. Classical rings and modules are present of course as attested by papers dealing with Baer rings, group algebras, essential extensions, and the separativity question for regular rings. Noncommutative and nonassociative structures also appear in this volume. In coding theory, various classical topics such as orthogonal codes or MacWilliams identities are naturally present. Topics related to quadratic forms are also treated along the pages. Both survey papers and articles paving the way of future research enrich this book.

Many of the articles in this proceeding are related to talks presented at the international meetings "NonCommutative Rings and their Applications VIII" and "Quadratic Forms Codes and Rings" that were held in an hybrid format in Lens in August 2023. The course given by Michela Ceria was welcomed by all the participants both present in person and online. We would like to thank her warmly for the efforts she made. She carefully prepared every lecture and vividly presented the course. The international aspect and the friendly atmosphere of both meetings also contributed to make these gatherings extremely pleasant to remember.

All the papers were subject to a strict process of refereeing and we would like to thank warmly all the colleagues who served as anonymous referees for the articles in this volume and provided their valuable comments and recommendations. Their expertise and professionalism greatly contributed to the quality of this volume. Without their help this proceeding would not exist.

The meetings were supported by the Laboratory of Mathematics in Lens (LML), by the BQR (Bonus Qualité Recherche) of the Artois University. and the program ValoRIS of the région Hauts-de-France.

We would also like to thank Aline Lorthios, secretary of the Laboratory in Lens, for all her work and for making the administrative tasks as light as possible for all of us.

Thanks are due to the editorial staff of the AMS especially to Christine Thivierge and Bridget Manzella. Their kindness, patience, and professionalism helped us a lot in preparing the volume.

The editors

List of Participants

Mehdi Aaghabali
University of Edinburg

Evrim Akalan
Hacettepe University, Ankara.

Mustafa Alkan
Akademiz University, Antalya.

Yusuf Alagöz
Siirt University (Turkey).

Lulwah Al-Essa
Imam Abdulrahman bin Faisal University, Dammam (Saudi Arabia).

Dilshad Alghazzawi
King Abdulaziz University, Rabigh (Saudi Arabia).

Shakir Ali
Department of Mathematics Aligarh Muslim University (India).

Meltem Altun Özarslan
Hacettepe University (Turkey).

Maria José Arroyo
Paniagua Universidad Autonoma Metropolitana, México (Mexico).

Samir Assuena
Centro Universitário da FEI (Brazil).

Masood Aryapoor
Mälardalen University (Sweeden).

Per Bäck
Mälardalen University (Sweden).

Demba Barry
Sciences University, Bamako (Mali).

Volodymyr Bavula
University of Sheffield (UK).

Nabil Bennenni
USTHB, Alger (Algeria).

Nathan Blacher
University of Sheffield (UK).

Tomasz Brzezinski
Swansea University and University of Bialystok (G.B. and Poland).

Engin Büyükasik
Izmir Institute of Technology (Turkey).

Stefan Catoiu
DePaul University, Chicago (U.S.A.).

Adam Chapman
Academic College of Tel-Aviv-Yaffo (Israel).

Anna Cichocka
University of Technology (Poland).

Gabriella D'Este
University of Milano (Italy).

Max Dickmann
Sorbonne University, Paris (France).

Steven Dougherty
Scranton University (USA).

André Duarte
Sao Paulo University (Brazil).

Nguyen Viet Dung
Ohio University (USA).

Yilmaz Durgun
Çukurova University (Turkey).

Michela Ceria
University Politecnico di Bari (Italy).

Fatma Ebrahim
Al-Azhar University, Cairo (Egypt).

Mahdieh Ebrahipour
Vali-e-Asr University of Rafsanjan (Iran).

Susan El-Deken
Helwan University, Cairo (Egypt)

Lhoussain El Fadil
Faculty of Sciences, Sidi Mohamed Ben Abdellah University (Morocco).

Alberto Facchini
University of Padova (Italy).

Raul Ferraz
Sao Paulo University (Brazil).

Walter Ferrer, Dma
Universidad de la República. (Uruguay)

José Gómez Torrecillas
Universidad de Granada (Spain).

Huda Hamdan
Damietta University (Egypt).

Leila Heidari Zadeh
Islamic Azad University, Shoushtar (Iran).

Malgorzata Hryniewicka
Faculty of Mathematics, University of Bialystok (Poland).

Surender Jain
University of Ohio, Athens (USA).

Malgorzata Jastrzebska
Siedlce University (Poland).

Yeliz Kara
Bursa Uludag University(Turkey).

Mahmoud Khaled Ahmed
Elgazeera Higher Institute (Egypt).

Ayten Koc
Gebze Technical University (Turkey).

Emrah Korkmaz
Çukurova University (Turkey).

Ahmed Laghribi
Université d'Artois (France).

André Leroy
Université d'Artois (France).

Sergio Lopez-Permouth
Ohio University, Athens (USA).

Guanglin Ma
Nanjing University of Information Science and Technology (China).

Praveen Manju
Indian Institute of Technology Delhi (India).

Jerzy Matczuk
Warsaw University (Poland).

Eli Matzri
Bar-Ilan University (Israel).

Zeinab Mohamed
South Valley University (Egypt).

Teo Mora
University of Genoa (Italy).

Azam Mozaffarikhah
Ohio University (USA).

Diksha Mukhija
University of Alberta (Canada).

Mehrdad Nasernejad
Université d'Artois (France).

ahra Nazemian
Graz University (Austria).

Cornelia Ott
Ulm University (Germany).

Murad Ozaydin
University of Oklahoma (USA).

André Perreira
Sao Paulo University (Brazil).

Dmitri Piontkovski
HSE University, Moscow (Russia).

Cesar Polcino Milies
Sao Paulo University (Brazil).

Rakhi Pratihar
Inria Saclay Centre (France).

Louis Rowen
Bar-Ilan University (Israel).

Esengül Saltürk
Istanbul Atlas University (Turkey).

Martha Lizbeth Sandoval
Universidad Autónoma Metropolitana
(UAM-I) (Mexico)

Serap Sahinkaya
Tarsus University (Turkey).

Steven
University of Sheffield (UK).

Virgilio Sison
University of the Philippines Los Banos
(Philippines).

Daniel Smertnig
Graz University (Austria).

Patrick Solé
I2M (France).

Ashish Srivastava
University of Saint Louis, Missouri
(USA).

Steve Szabo
Eastern Kentucky University (USA).

Yaser Toloei, Razi University,
Kermanshah (Iran).

Felix Ulmer
University of Rennes (France).

Gulsen Ulucak
Gebze Technical University (Turkey).

Jay Wood
Western Michigan University (USA).

Tulay Yildirim
Karabuk University (Turkey).

Mohamed F. Yousif
The Ohio State University (USA).

Yiqiang Zhou
Memorial University of Newfoundland
(Canada).

Contemporary Mathematics
Volume **826**, 2025
https://doi.org/10.1090/conm/826/16569

On matrix Lie affgebras

Tomasz Brzeziński and Krzysztof Radziszewski

ABSTRACT. Lie brackets or Lie affgebra structures on several classes of affine spaces of matrices are studied. These include general normalised affine matrices, special normalised affine matrices, anti-symmetric and anti-hermitian normalised affine matrices and special anti-hermitian normalised affine matrices. It is shown that, when retracted to the underlying vector spaces, they correspond to classical matrix Lie algebras: general and special linear, anti-symmetric, anti-hermitian and special anti-hermitian Lie algebras respectively.

1. Introduction

The study of Lie brackets on affine spaces or *Lie affgebras* was initiated in [**6**] as a part of the geometric programme of developing a frame-independent formulation of classical mechanics [**12**] in which vector bundles are replaced by bundles with affine fibres [**7**–**9**]. The definition of a Lie bracket in [**6**] relies on the linear structure of the vector space underlying an affine space. This reliance has been shown to be not necessary, leading to the intrinsic or vector-space independent formulation of Lie affgebras [**5**]. From this viewpoint a vector space is an artefact of the definition of an affine space [**3**, **10**], while any Lie bracket on an affine space can be retracted to the (bilinear) Lie bracket on this vector space. In the present paper we analyse and extend the class of matrix Lie affgebras introduced in [**4**]. In particular we present equivalent formulation of resulting classes, which allows one to identify quickly classical matrix Lie algebras to which matrix Lie affgebras are retracted by fixing an element in accordance with [**5**, Theorem 3.15].

The paper is organised as follows. In Section 2 we recall the intrinsic definitions of an affine space and a Lie affgebra. Next we describe the process of retracting of a bi-affine Lie bracket into a bi-linear Lie bracket on an underlying vector space. Section 3 contains main results. First several classes of matrix Lie affgebras are introduced. These are then shown to be isomorphic with Lie affgebras which in a natural way relate to classical matrix Lie algebras. As a corollary we thus obtain Lie algebras to which the matrix Lie affgebras are retracted.

We use the following notation for classical (matrix) Lie algebras. The Lie algebra of $n \times n$-matrices over a field $\mathbb{F}$ is denoted by $\mathrm{gl}(n, \mathbb{F})$ with $\mathrm{sl}(n, \mathbb{F})$ standing for

2020 *Mathematics Subject Classification.* Primary 17B99, 15B30, 20N10.

Key words and phrases. Heap, affine space, Lie bracket, matrix Lie affgebra.

The research was supported by the National Science Centre, Poland, through the WEAVE-UNISONO grant no. 2023/05/Y/ST1/00046. In addition the research of the first author was partially supported by the National Science Centre, Poland, grant no. 2019/35/B/ST1/01115.

traceless matrices with entries from $\mathbb{F}$; $\mathrm{o}(n)$ denotes real anti-symmetric matrices, while $\mathrm{u}(n)$ stands for complex anti-hermitian matrices. Finally, $\mathrm{su}(n)$ is the Lie algebra of all traceless anti-hermitian matrices.

2. Lie brackets on affine spaces

In a vector-free formulation (see e.g. [**3**]) an affine space over a field $\mathbb{F}$ can be realised as an algebraic system consisting of a set A together with ternary operations

$$\langle -,-,-\rangle : A\times A\times A\to A, \qquad (a,b,c)\mapsto \langle a,b,c\rangle, \tag{2.1a}$$

$$-\triangleright_{-} - : \mathbb{F}\times A\times A\to A, \qquad (\alpha,a,b)\mapsto \alpha\triangleright_a b, \tag{2.1b}$$

satisfying the following conditions, for all $a,b,c,d,e\in A$ and $\alpha,\beta,\gamma\in\mathbb{F}$,
(2.2a)

$$\langle\langle a,b,c\rangle,d,e\rangle = \langle a,b,\langle c,d,e\rangle\rangle, \quad \langle a,b,b\rangle = a = \langle b,b,a\rangle, \quad \langle a,b,c\rangle = \langle c,b,a\rangle,$$

$$(\alpha-\beta+\gamma)\triangleright_a b = \langle \alpha\triangleright_a b, \beta\triangleright_a b, \gamma\triangleright_a b\rangle, \tag{2.2b}$$

$$\alpha\triangleright_a \langle b,c,d\rangle = \langle \alpha\triangleright_a b, \alpha\triangleright_a c, \alpha\triangleright_a d\rangle, \tag{2.2c}$$

$$(\alpha\beta)\triangleright_a b = \alpha\triangleright_a(\beta\triangleright_a b), \quad 1\triangleright_a b = b, \quad 0\triangleright_a b = a, \tag{2.2d}$$

$$\alpha\triangleright_a b = \langle \alpha\triangleright_c b, \alpha\triangleright_c a, a\rangle. \tag{2.2e}$$

A homomorphism of affine spaces or *affine transformation* is a function $f : A\to B$ that preserves both operations, that is, for all $a,b,c\in A$ and $\alpha\in\mathbb{F}$,

$$f(\langle a,b,c\rangle) = \langle f(a),f(b),f(c)\rangle, \quad \text{and} \quad f(\alpha\triangleright_a b) = \alpha\triangleright_{f(a)} f(b). \tag{2.3}$$

A few explanations and forays into terminology are in order now. The conditions (2.1a) mean that A together with $\langle -,-,-\rangle$ is an *abelian heap*, the notion introduced by Prüfer in [**11**]. We will refer to $\langle -,-,-\rangle$ as the *heap operation.* The first condition in (2.2a) might be referred to as the *quasi-* or *heap associativity*, the second is known as the *Mal'cev identities*, and the last one is the commutativity of the heap operation (required originally in [**11**], but not needed for general heaps [**2**]). Altogether, the laws that heap operation of an abelian heap satisfies imply that the placement of brackets $\langle , \rangle$ does not play a role, hence we can write succinctly $\langle a,b,c,d,e\rangle$ for $\langle\langle a,b,c\rangle,d,e\rangle$ or any other distribution of heap operations.

A homomorphism of heaps is a function that preserves the ternary operations, i.e. satisfies the first of condition (2.3). Every abelian group A is a heap in the canonical way with the heap operation, for all $a,b,c\in A$,

$$\langle a,b,c\rangle = a-b+c. \tag{2.4}$$

Any additive map is automatically a heap homomorphism, but there are more heap homomorphisms than group homomorphisms, as any additive map composed with the translation by an element is also a heap homomorphism.

Conversely, every (non-empty) abelian heap can be made into an abelian group or, to be precise, the family of isomorphic groups, by retracting the heap operation at any element $o\in A$. As a set, the group $G(A;o)$ is the same as A, with o as the neutral element and the addition and inverse

$$a+b = \langle a,o,b\rangle, \quad -a = \langle o,a,o\rangle. \tag{2.5}$$

We refer to $G(A;o)$ as the *retract* of A at o. The isomorphism between retracts of the same heap at different elements is given by the *translation automorphism*,

$$\tau_o^{\bar{o}} : G(A;o) \to G(A;\bar{o}), \qquad a \mapsto \langle a, o, \bar{o}\rangle. \tag{2.6}$$

In view of all this a commutative heap operation can always be understood as a combination of binary operations as in (2.4). Although the binary operations are defined relative to a chosen element, their combination is independent of the choice made. This observation immediately justifies various equivalences such as

$$\langle a,b,c\rangle = d \iff \langle a,d,c\rangle = b \iff \langle b,c,d\rangle = a$$

and

$$a = b \iff \langle a,b,c\rangle = c, \qquad \text{for all } c \in A.$$

The map $-\triangleright_{-}-$ is called the *action* of $\mathbb{F}$ on A and a in $-\triangleright_a-$ is known as the *base element*. Equation (2.2b) means that $-\triangleright_a b : \mathbb{F} \to A$ is a heap homomorphism, and similarly, (2.2c) states the heap homomorpism property of $\alpha \triangleright_a - : A \to A$. The first two conditions in (2.2d) state the associativity and unitality of the actions at a fixed base; the last normalisation property when combined with (2.2b) means that $-\triangleright_a b : \mathbb{F} \to G(A;a)$ is a group homomorphism. Finally, the *base change property* (2.2e) ensures that all elements of A play exactly the same role; every element can be a base for the action and the definition as a whole is independent of the choices made. This fits perfectly into the observer-independent framework alluded to in Introduction.

In the standard definition of an affine space as a set A with a free and transitive action of a vector space $\vec{A}$, the heap operation and action are defined as, for all $a,b,c \in A$ and $\alpha \in \mathbb{F}$,

$$\langle a,b,c\rangle = a + \overrightarrow{bc}\,, \qquad \alpha \triangleright_a b = a + \alpha\,\overrightarrow{ab}\,,$$

where $\overrightarrow{ab}$ denotes the vector from a to b etc.

Conversely, fixing any $o \in A$, the retract $G(A;o)$ becomes a vector space with multiplication by scalars

$$\alpha a = \alpha \triangleright_o b.$$

We denote this vector space by $V(A;o)$. The vector from a to b is then given by the formula $\langle o,a,b\rangle = b - a$, where the difference is written in $G(A;o)$.

The major difficulty in defining Lie brackets on an affine space is to make all conditions independent from a specific element. A way to overcome this obstacle is presented in the following

DEFINITION 2.1 ([**5**, Definition 3.1]). Let A be an affine space over $\mathbb{F}$. A *Lie bracket* on A is a binary operation $[-,-] : A \times A \to A$ that satisfies the following conditions:

(a) for all $a \in A$, both $[a,-]$ and $[-,a]$ are affine transformations $A \to A$;
(b) affine antisymmetry, that is, for all $a,b \in A$,

$$\langle [a,b],[a,a],[b,a]\rangle = [b,b]; \tag{2.7}$$

(c) the affine Jacobi identity, that is, for all $a,b,c \in A$,

$$\langle [a,[b,c]],[a,a],[b,[c,a]],[b,b],[c,[a,b]]\rangle = [c,c]. \tag{2.8}$$

An affine space together with a Lie bracket is called a *Lie affgebra.*

A homomorphism of Lie affgebras is an affine map that preserves the Lie bracket.

Condition (a) means that the Lie bracket is a *bi-affine transformation.*

Basic examples of Lie affgebras include an affine space A itself with the bracket

$$[a,b] = \zeta \triangleright_a b, \tag{2.9}$$

where ζ is any fixed scalar. As shown in [**5**, Proposition 3.4], if dimension of A is at least one, brackets corresponding to different scalars equip A with non-isomorphic Lie affgebra structures. Furthermore, for any associative bi-affine multiplication on A, or more generally, for a bi-affine multiplication that satisfies conditions similar to that of a pre-Lie algebra one can define the *affine commutator* Lie bracket

$$[a,b] = \langle ab, ba, b\rangle; \tag{2.10}$$

see [**5**, Definition 3.8 & Proposition 3.9].

It has been observed in [**5**, Theorem 3.15] that with every Lie affgebra $(A, [-,-])$ one can associate a family of isomorphic Lie algebras. Each member of this family has $V(A;o)$ as the underlying vector space and the Lie bracket

$$[a,b]_o := \langle [a,b],[a,o],[o,o],[o,b],o\rangle = [a,b]-[a,o]+[o,o]-[o,b]. \tag{2.11}$$

We denote this Lie algebra by $L(A;o)$ and call it the *Lie algebra retract* of A at o. Different Lie affgebra structures might and often do collapse to the same Lie algebra. For example, for all $\zeta \in \mathbb{F}$ and $o \in A$, the Lie algebra associated to the bracket (2.9) is trivial, that is $[a,b]_o = o$. On the other hand, as explained in [**1**, Theorem 4.3], any associative bi-affine multiplication on A makes $V(A;o)$ into an associative algebra with the product

$$a \bullet b = ab - ao + o^2 - ob.$$

The affine commutator bracket (2.10) on A retracts at o to the bilinear commutator bracket $[a,b]_o = a\bullet b - b\bullet a$.

Any Lie algebra $\mathfrak{g}$ with the bracket $[-,-]_\mathfrak{g}$ is a Lie affgebra with the affine structure

$$\langle a,b,c\rangle = a-b+c, \qquad \alpha \triangleright_a b = (1-\alpha)a + \alpha b,$$

and the bi-affine bracket $[a,b] = [a,b]_\mathfrak{g} + b$. The retract of this Lie affgebra at 0 is precisely $\mathfrak{g}$, i.e. $L(\mathfrak{g};0) = \mathfrak{g}$ and $[-,-]_0 = [-,-]_\mathfrak{g}$.

Finally, let us point out that if a Lie bracket on an affine space A is an idempotent operation, that is $[a,a]=a$, for all $a\in A$, then, for all $o \in A$, $[a,b]_v := [a,b]-b$ is a vector-valued Lie bracket on the affine space viewed traditionally as a pair $(A, V(A;o))$, as introduced in [**6**]. This is the case for both (2.9) and (2.10) structures.

3. Matrix Lie affgebras and corresponding Lie algebras

In this paper we consider the following sets of matrices, which are defined for all positive integers n.

General normalised affine matrices:

$$\mathrm{gna}(n,\mathbb{F}) = \{\mathbf{a} = (a_{kl})_{k,l=1}^{n+1} \in \mathrm{gl}(n+1,\mathbb{F}) \mid \sum_k a_{kl} = \sum_k a_{lk} = 1,\ \forall l\}.$$

Special normalised affine matrices:

$$\mathrm{sna}(n,\mathbb{F}) = \{\mathbf{a} \in \mathrm{gna}(n,\mathbb{F}) \mid \mathrm{tr}\,\mathbf{a} = 0\}.$$

Anti-symmetric normalised affine matrices:

$$\mathrm{ona}(n) = \{\mathbf{a} \in \mathrm{gna}(n,\mathbb{R}) \mid a_{kk} = 1,\ a_{kl} = -a_{lk},\ \forall k \neq l\}.$$

Anti-hermitian normalised affine matrices:

$$\mathrm{una}(n) = \{\mathbf{a} = (a_{kl})_{k,l=1}^{n+1} \in \mathrm{gl}(n+1,\mathbb{C}) \mid \sum_k a_{kl} = \sum_k a_{lk} = i,\ a_{lm} = -\bar{a}_{ml}, \forall l,m\}.$$

Special anti-hermitian normalised affine matrices:

$$\mathrm{suna}(n) = \{\mathbf{a} \in \mathrm{una}(n) \mid \mathrm{tr}\,\mathbf{a} = 0\}.$$

LEMMA 3.1. *Let* $\mathfrak{a}$ *denote any of the sets* $\mathrm{gna}(n,\mathbb{F})$, $\mathrm{sna}(n,\mathbb{F})$, $\mathrm{ona}(n)$, $\mathrm{una}(n)$ *or* $\mathrm{suna}(n)$. *Then* $\mathfrak{a}$ *is a Lie affgebra with the following operations:*

(a) *the heap operation:*

$$\langle \mathbf{a}, \mathbf{b}, \mathbf{c}\rangle = \mathbf{a} - \mathbf{b} + \mathbf{c},$$

(b) *the affine* $\mathbb{F}$*-action*

$$\alpha \triangleright_{\mathbf{a}} \mathbf{b} = \alpha\mathbf{b} - \alpha\mathbf{a} + \mathbf{a},$$

(c) *the affine commutator Lie bracket*

$$[\mathbf{a}, \mathbf{b}] = \mathbf{a}\mathbf{b} - \mathbf{b}\mathbf{a} + \mathbf{b},$$

for all $\mathbf{a}, \mathbf{b}, \mathbf{c} \in \mathfrak{a}$ *and* $\alpha \in \mathbb{F}$.

PROOF. This is a straightforward calculation no different from the method of the proof that $\mathrm{sna}(n,\mathbb{C})$ is a Lie affgebra used in [**4**, Proposition 4]. □

The main result of this paper provides one with alternative descriptions of matrix Lie affgebras in terms of classical (matrix) Lie algebras.

THEOREM 3.2. *For all* n *such that* $\mathrm{char}(\mathbb{F}) \nmid n, n+1$, *there are Lie affgebra isomorphisms with the following Lie affgebras (all with the affine structure and the brackets as in Lemma 3.1):*

$$\mathrm{gna}(n,\mathbb{F}) \cong \begin{pmatrix} 0 & 0 \\ 0 & 1 \end{pmatrix} + \begin{pmatrix} \mathrm{gl}(n,\mathbb{F}) & 0 \\ 0 & 0 \end{pmatrix} \cong \mathrm{gl}(n,\mathbb{F});$$

$$\mathrm{sna}(n,\mathbb{F}) \cong \begin{pmatrix} -\frac{1}{n}\mathrm{I}_n & 0 \\ 0 & 1 \end{pmatrix} + \begin{pmatrix} \mathrm{sl}(n,\mathbb{F}) & 0 \\ 0 & 0 \end{pmatrix} \cong \mathrm{sl}(n,\mathbb{F});$$

$$\mathrm{ona}(n) \cong \mathrm{I}_{n+1} + \begin{pmatrix} \mathrm{o}(n) & 0 \\ 0 & 0 \end{pmatrix} \cong \mathrm{o}(n);$$

$$\mathrm{una}(n) \cong \begin{pmatrix} 0 & 0 \\ 0 & i \end{pmatrix} + \begin{pmatrix} \mathrm{u}(n) & 0 \\ 0 & 0 \end{pmatrix} \cong \mathrm{u}(n);$$

$$\mathrm{suna}(n) \cong \begin{pmatrix} -\frac{i}{n}\mathrm{I}_n & 0 \\ 0 & i \end{pmatrix} + \begin{pmatrix} \mathrm{su}(n) & 0 \\ 0 & 0 \end{pmatrix} \cong \mathrm{su}(n),$$

where I_m *denotes the* $m \times m$ *identity matrix.*

PROOF. We start by proving the first in each of the five chains of isomorphisms. In order to develop a universal argument that covers all cases listed above, let us fix an element $c \in \mathbb{F}$ and define

$$\mathrm{ga}_c(n, \mathbb{F}) = \{\mathbf{a} = (a_{kl})_{k,l=1}^{n+1} \in \mathrm{gl}(n+1, \mathbb{F}) \mid \sum_k a_{kl} = \sum_k a_{lk} = c, \ \forall l\}.$$

For all $c \in \mathbb{F}$, $\mathrm{ga}_c(n, \mathbb{F})$ is a Lie affgebra with the same affine and Lie algebra structures as in Lemma 3.1. Furthermore, the first three Lie affgebras in the theorem are Lie sub-affgebras of $\mathrm{gna}(n, \mathbb{F}) = \mathrm{ga}_1(n, \mathbb{F})$, while the last two are Lie sub-affgebras of $\mathrm{ga}_i(n, \mathbb{C})$. In fact, for all $c, c' \neq 0$, $\mathrm{ga}_c(n, \mathbb{F}) \cong \mathrm{ga}_{c'}(n, \mathbb{F})$ as Lie affgebras. Indeed, consider the following function

$$f : \mathrm{ga}_c(n, \mathbb{F}) \to \mathrm{ga}_{c'}(n, \mathbb{F}), \qquad \mathbf{a} \mapsto \mathbf{a} + (c' - c)\mathrm{I}_{n+1}. \tag{3.1}$$

Being a combination of a linear and constant terms this is an affine isomorphism, which changes the normalisation of rows and columns in the desired way. Since the matrix that is a multiple of identity is central, one easily checks that, for all $\mathbf{a}, \mathbf{b} \in \mathrm{ga}_c(n, \mathbb{F})$

$$f(\mathbf{ab} - \mathbf{ba} + \mathbf{b}) = f(\mathbf{a})f(\mathbf{b}) - f(\mathbf{b})f(\mathbf{a}) + f(\mathbf{b}),$$

i.e. that f preserves the affine commutator Lie brackets.

Clearly, as affine spaces

$$\mathrm{ga}_c(n, \mathbb{F}) = \mathbf{o} + \mathrm{ga}_0(n, \mathbb{F}), \tag{3.2}$$

where $\mathbf{o} \in \mathrm{ga}_c(n, \mathbb{F})$ is freely fixed. One easily checks that this identification is an isomorphism of Lie affgebras, provided that $\mathbf{o}$ commutes with all elements of $\mathrm{ga}_0(n, \mathbb{F})$.

First we show the existence of a similarity transformation which implements an isomorphism of affine spaces

$$\mathrm{ga}_c(n, \mathbb{F}) \cong \begin{pmatrix} \mathrm{gl}(n, \mathbb{F}) & 0 \\ 0 & c \end{pmatrix}. \tag{3.3}$$

To this end, consider the $(n+1) \times (n+1)$-matrix

$$\mathbf{P} = \begin{pmatrix} 1 & 1 & 1 & \cdots & 1 & 1 \\ 0 & 0 & 0 & \cdots & -1 & 1 \\ \cdots & \cdots & \cdots & \cdots & \cdots & \cdots \\ 0 & 0 & -1 & \cdots & 0 & 1 \\ 0 & -1 & 0 & \cdots & 0 & 1 \\ -1 & 0 & 0 & \cdots & 0 & 1 \end{pmatrix}. \tag{3.4}$$

If $\mathrm{char}(\mathbb{F}) \nmid n+1$, $\mathbf{P}$ is invertible with the inverse

$$\mathbf{P}^{-1} = \frac{1}{n+1} \begin{pmatrix} 1 & 1 & 1 & \cdots & 1 & -n \\ 1 & 1 & 1 & \cdots & -n & 1 \\ \cdots & \cdots & \cdots & \cdots & \cdots & \cdots \\ 1 & 1 & -n & \cdots & 1 & 1 \\ 1 & -n & 1 & \cdots & 1 & 1 \\ 1 & 1 & 1 & \cdots & 1 & 1 \end{pmatrix}.$$

Since the last column of $\mathbf{P}$ and the last row of $\mathbf{P}^{-1}$ contain only identities (up to an overall normalisation factor in the latter case),

$$\mathbf{P}\begin{pmatrix}0 & 0\\ 0 & c\end{pmatrix}\mathbf{P}^{-1} = \frac{1}{n+1}\begin{pmatrix} c & c & \cdots & c\\ c & c & \cdots & c\\ \cdots & \cdots & \cdots & \cdots\\ c & c & \cdots & c\end{pmatrix} \in \mathrm{ga}_c(n,\mathbb{F}). \tag{3.5}$$

Second, note that the sums of the elements in each of the first n columns of $\mathbf{P}$ and in each of the first n rows of $\mathbf{P}^{-1}$ vanish. This, combined with the observation that in any matrix $\mathbf{a} \in \begin{pmatrix}\mathrm{gl}(n,\mathbb{F}) & 0\\ 0 & 0\end{pmatrix}$ the last row and column contain zeros only, we immediately conclude that

$$\mathbf{P}\mathbf{a}\mathbf{P}^{-1} \in \mathrm{ga}_0(n,\mathbb{F}). \tag{3.6}$$

Since both $\mathrm{ga}_0(n,\mathbb{F})$ and $\begin{pmatrix}\mathrm{gl}(n,\mathbb{F}) & 0\\ 0 & 0\end{pmatrix} \cong \mathrm{gl}(n,\mathbb{F})$ are $n \times n$ dimensional vector spaces, the similarity transformation establishes an isomorphism of vector spaces. In view of the identification (3.2) we obtain the required isomorphism (3.3) of affine spaces:

$$f : \mathrm{ga}_c(n,\mathbb{F}) \to \begin{pmatrix}\mathrm{gl}(n,\mathbb{F}) & 0\\ 0 & c\end{pmatrix}, \qquad \mathbf{a} \mapsto \mathbf{P}^{-1}\mathbf{a}\mathbf{P}. \tag{3.7}$$

The map f is given by conjugation by an invertible matrix, hence it preserves the Lie brackets in Lemma 3.1. Finally, $\begin{pmatrix}0 & 0\\ 0 & c\end{pmatrix}$ commutes with all elements of $\begin{pmatrix}\mathrm{gl}(n,\mathbb{F}) & 0\\ 0 & 0\end{pmatrix}$, its similarity transform commutes with all elements of $\mathrm{ga}_0(n,\mathbb{F})$ and, consequently, the map f is a Lie affgebra isomorphism. Therefore, the first assertion follows.

As the similarity transformation f preserves traces, when restricted to $\mathrm{sna}(n,\mathbb{F})$ the map f gives the isomorphism

$$\begin{aligned}\begin{pmatrix}-\frac{1}{n}\mathrm{I}_n & 0\\ 0 & 1\end{pmatrix} + \begin{pmatrix}\mathrm{sl}(n,\mathbb{F}) & 0\\ 0 & 0\end{pmatrix} &\longrightarrow \mathbf{P}\begin{pmatrix}-\frac{1}{n}\mathrm{I}_n & 0\\ 0 & 1\end{pmatrix}\mathbf{P}^{-1} + \mathrm{sa}_0(n,\mathbb{F})\\ &= \frac{1}{n}\begin{pmatrix} 0 & 1 & 1 & \cdots & 1\\ 1 & 0 & 1 & \cdots & 1\\ \cdots & \cdots & \cdots & \cdots & \cdots\\ 1 & 1 & 1 & \cdots & 0\end{pmatrix} + \mathrm{sa}_0(n,\mathbb{F})\\ &= \mathrm{sna}(n,\mathbb{F}),\end{aligned}$$

where $\mathrm{sa}_0(n,\mathbb{F})$ denotes the Lie subalgebra of traceless matrices in $\mathrm{ga}_0(n,\mathbb{F})$. Since $\begin{pmatrix}-\frac{1}{n}\mathrm{I}_n & 0\\ 0 & 1\end{pmatrix}$ commutes with all elements of $\begin{pmatrix}\mathrm{sl}(n,\mathbb{F}) & 0\\ 0 & 0\end{pmatrix}$, its similarity transform commutes with all elements of $\mathrm{sa}_0(n,\mathbb{F})$ and hence we obtain the required isomorphism of Lie affgebras.

For the remaining three isomorphisms, $\mathbb{F}$ is either $\mathbb{R}$ or $\mathbb{C}$. Since $\mathbf{P}$ is a real matrix, the Gram-Schmidt orthonormalisation of its columns with respect to the Euclidean or dot product, starting with the last column $(1,1,\ldots,1)^T$, produces the real unitary, hence orthogonal matrix $\mathbf{U}$ in which the sums of the elements in each column except the last one are zero. Explicitly,

$$\mathbf{U}=\begin{pmatrix}\frac{1}{\sqrt{(n+1)n}} & \frac{1}{\sqrt{n(n-1)}} & \frac{1}{\sqrt{(n-1)(n-2)}} & \cdots & \frac{1}{\sqrt{6}} & \frac{1}{\sqrt{2}} & \frac{1}{\sqrt{n+1}}\\ \frac{1}{\sqrt{(n+1)n}} & \frac{1}{\sqrt{n(n-1)}} & \frac{1}{\sqrt{(n-1)(n-2)}} & \cdots & \frac{1}{\sqrt{6}} & -\frac{1}{\sqrt{2}} & \frac{1}{\sqrt{n+1}}\\ \frac{1}{\sqrt{(n+1)n}} & \frac{1}{\sqrt{n(n-1)}} & \frac{1}{\sqrt{(n-1)(n-2)}} & \cdots & -\sqrt{\frac{2}{3}} & 0 & \frac{1}{\sqrt{n+1}}\\ \frac{1}{\sqrt{(n+1)n}} & \frac{1}{\sqrt{n(n-1)}} & \frac{1}{\sqrt{(n-1)(n-2)}} & \cdots & 0 & 0 & \frac{1}{\sqrt{n+1}}\\ \cdots & \cdots & \cdots & \cdots & \cdots & \cdots & \cdots\\ \frac{1}{\sqrt{(n+1)n}} & \frac{1}{\sqrt{n(n-1)}} & -\sqrt{\frac{n-2}{n-1}} & \cdots & 0 & 0 & \frac{1}{\sqrt{n+1}}\\ \frac{1}{\sqrt{(n+1)n}} & -\sqrt{\frac{n-1}{n}} & 0 & \cdots & 0 & 0 & \frac{1}{\sqrt{n+1}}\\ -\sqrt{\frac{n}{n+1}} & 0 & 0 & \cdots & 0 & 0 & \frac{1}{\sqrt{n+1}}\end{pmatrix}.$$

The last column in $\mathbf{U}$ and hence the last row of $\mathbf{U}^{-1}=\mathbf{U}^{\dagger}$ are multiples of $(1,1,\dots,1)^T$. Therefore, $\mathbf{U}\begin{pmatrix}0&0\\0&c\end{pmatrix}\mathbf{U}^{\dagger}$ is the same matrix as in (3.5). Similarly, the vanishing of the sums of the first n columns of $\mathbf{U}$ and correspondingly first n rows of $\mathbf{U}^{\dagger}$ allows one to conclude as in (3.6). In consequence one obtains the unitary transformation

$$g:\mathrm{ga}_c(n,\mathbb{F})\to\begin{pmatrix}\mathrm{gl}(n,\mathbb{F})&0\\0&c\end{pmatrix},\qquad \mathbf{a}\mapsto\mathbf{U}^{\dagger}\mathbf{a}\mathbf{U},\tag{3.8}$$

which covers all the remaining of the first isomorphisms in each chain.

To prove the second set of isomorphisms we note that each of the Lie affgebras in the middle is of the form

$$\mathfrak{a}=\mathbf{o}+\begin{pmatrix}\mathfrak{g}&0\\0&0\end{pmatrix},$$

where $\mathfrak{g}$ is a classical matrix Lie algebra. Since $\mathbf{o}$ commutes with all elements of $\begin{pmatrix}\mathfrak{g}&0\\0&0\end{pmatrix}$, the map

$$\mathfrak{a}\to\mathfrak{g},\qquad \mathbf{o}+\begin{pmatrix}\mathbf{a}&0\\0&0\end{pmatrix}\mapsto\mathbf{a},$$

is an isomorphism of Lie affgebras, provided $\mathfrak{g}$ is understood as a Lie affgebra with the structure as in Lemma 3.1. □

The second isomorphisms in Theorem 3.2 enable one immediately to identify (up to isomorphism) Lie algebra retracts of all the matrix Lie affgebras listed there.

COROLLARY 3.3. *Provided that* $\mathrm{char}(\mathbb{F})\nmid n,n+1$, *the following table lists Lie algebra retracts of matrix Lie affgebras:*

matrix Lie affgebra with the affine Lie bracket $[\mathbf{a},\mathbf{b}]=\mathbf{ab}-\mathbf{ba}+\mathbf{b}$	*Lie algebra retract with the Lie bracket* $[\mathbf{a},\mathbf{b}]=\mathbf{ab}-\mathbf{ba}$
$\mathrm{gna}(n,\mathbb{F})\cong\mathrm{gl}(n,\mathbb{F})$	$\mathrm{gl}(n,\mathbb{F})$
$\mathrm{sna}(n,\mathbb{F})\cong\mathrm{sl}(n,\mathbb{F})$	$\mathrm{sl}(n,\mathbb{F})$
$\mathrm{ona}(n)\cong\mathrm{o}(n)$	$\mathrm{o}(n)$
$\mathrm{una}(n)\cong\mathrm{u}(n)$	$\mathrm{u}(n)$
$\mathrm{suna}(n)\cong\mathrm{su}(n)$	$\mathrm{su}(n)$

Acknowledgments

The authors are grateful to Ryszard R. Andruszkiewicz for comments that led to the second set of equivalences in Theorem 3.2.

References

[1] Ryszard R. Andruszkiewicz, Tomasz Brzeziński, and Bernard Rybołowicz, *Ideal ring extensions and trusses*, J. Algebra **600** (2022), 237–278, DOI 10.1016/j.jalgebra.2022.01.038. MR4396016

[2] Reinhold Baer, *Zur Einführung des Scharbegriffs* (German), J. Reine Angew. Math. **160** (1929), 199–207, DOI 10.1515/crll.1929.160.199. MR1581184

[3] Simion Breaz, Tomasz Brzeziński, Bernard Rybołowicz, and Paolo Saracco, *Heaps of modules and affine spaces*, Ann. Mat. Pura Appl. (4) **203** (2024), no. 1, 403–445, DOI 10.1007/s10231-023-01369-0. MR4685735

[4] Tomasz Brzeziński, *Special normalised affine matrices: an example of a Lie affgebra*, Geometric methods in physics XL, Trends Math., Birkhäuser/Springer, Cham, [2024] ©2024, pp. 115–125, DOI 10.1007/978-3-031-62407-0_9. MR4807453

[5] Tomasz Brzeziński and James Papworth, *Lie and Nijenhuis brackets on affine spaces*, Bull. Belg. Math. Soc. Simon Stevin **30** (2023), no. 5, 683–704, DOI 10.36045/j.bbms.231013. MR4705165

[6] Katarzyna Grabowska, Janusz Grabowski, and Paweł Urbański, *Lie brackets on affine bundles*, Ann. Global Anal. Geom. **24** (2003), no. 2, 101–130, DOI 10.1023/A:1024457728027. MR1990111

[7] Katarzyna Grabowska, Janusz Grabowski, and Paweł Urbański, *AV-differential geometry: Poisson and Jacobi structures*, J. Geom. Phys. **52** (2004), no. 4, 398–446, DOI 10.1016/j.geomphys.2004.04.004. MR2098833

[8] Katarzyna Grabowska, Janusz Grabowski, and Paweł Urbański, *Frame-independent mechanics: geometry on affine bundles*, Travaux mathématiques. Fasc. XVI, Trav. Math., vol. 16, Univ. Luxemb., Luxembourg, 2005, pp. 107–120. MR2223154

[9] Katarzyna Grabowska, Janusz Grabowski, and Paweł Urbański, *AV-differential geometry: Euler-Lagrange equations*, J. Geom. Phys. **57** (2007), no. 10, 1984–1998, DOI 10.1016/j.geomphys.2007.04.003. MR2348274

[10] Fritz Ostermann and Jürgen Schmidt, *Der baryzentrische Kalkül als axiomatische Grundlage der affinen Geometrie* (German), J. Reine Angew. Math. **224** (1966), 44–57, DOI 10.1515/crll.1966.224.44. MR206783

[11] Heinz Prüfer, *Theorie der Abelschen Gruppen* (German), Math. Z. **20** (1924), no. 1, 165–187, DOI 10.1007/BF01188079. MR1544670

[12] W. M. Tulczyjew, *Frame independence of analytical mechanics* (English, with Italian summary), Atti Accad. Sci. Torino Cl. Sci. Fis. Mat. Natur. **119** (1985), no. 5-6, 273–279 (1986). MR929906

Department of Mathematics, Swansea University, Fabian Way, Swansea SA1 8EN, United Kingdom; and Faculty of Mathematics, University of Białystok, K. Ciołkowskiego 1M, 15-245 Białystok, Poland

Email address: T.Brzezinski@swansea.ac.uk

Doctoral School & Faculty of Mathematics, University of Białystok, K. Ciołkowskiego, 15-245 Białystok, Poland

Email address: k.radziszewski@uwb.edu.pl

Contemporary Mathematics
Volume **826**, 2025
https://doi.org/10.1090/conm/826/16570

Gröbnerian and Gröbner free techniques on non-associative algebras

Michela Ceria and Teo Mora

ABSTRACT. We are investigating to what extent Gröbnerian technologies and the Gröbner-free approach can allow to make deductions on the structure of the non-associative magmas.

1. Introduction

In this paper we study, both via Gröbner and Gröbner-free techniques, the structure of the ideals and algebras *representing the internal relations defining the operations in magmas.*

This paper is one (but probably not the last) in a long series [**15**–**21**, **52**] of investigations on several interlacing threads dedicated to the different computational algebraic aspects of (not necessarily 0-dimensional) ideal theory.

In his Ph.D Thesis Buchberger [**9**] introduced in 1965 the notion of Gröbner basis, which labelled such also in the 1976's papers [**10**, **11**], and his algorithm for computing these bases; later he gave a completely different algorithm for Gröbner bases [**13**], adapting the original construction by Gordan [**32**, **33**]. On the basis of this construction, Buchberger's school [**72**] was able to determine the upper bound of the maximal degree of a Gröbner basis. Such algorithm was dealing with (not necessarily 0-dimensional) ideals of polynomial rings over a field but soon the coefficient setting was extended by a series of papers which culminated with Pan's [**56**] formulation of weak and strong Gröbner basis over a principal ideal domain (PID). In the meantime, Gebauer and Möller improved the criteria introduced by Buchberger [**12**] in order to avoid the bottleneck of computing *useless pairs*, giving a new and more efficient reformulation [**30**, **31**]; on the basis of that and of the results of [**67**, **73**], Möller introduced a stronger characterization (see the analysis in [**21**]), the *Lifting Theorem*, applicable over each (commutative) ring over a principal ideal ring (PIR).

2020 *Mathematics Subject Classification.* Primary 17-08; Secondary 13P10.
Key words and phrases. Non-associative algebras, Buchberger reduction, degröbnerization.
This work has been done within Italian National Group for Algebraic and Geometric Structures and their Applications (GNSAGA– INdAM). This work was supported by the Italian Ministry of University and Research under the Programme "Department of Excellence" Legge 232/2016 (Grant No. CUP - D93C23000100001).

In 1982 Buchberger and Möller gave *an algorithm for constructing a basis of the ideal of all polynomials, which vanish at a preassigned set of points* $\{y_1, \dots, y_m\} \subset K^n, K$ *a field.* $[\cdots]$ *These polynomials admit an immediate construction of interpolating polynomials and allow to shorten the algorithm, if it is applied to an enlarged set* $\{y_1, \dots, y_{m_1}\} \subset K^n, m_1 > m$ [**48**, p.24]. Such algorithm was later studied in [**45,46**], extending it to all 0-dimensional ideals defined by functionals and applying to this setting the complexity analysis of [**28**]; there they proposed four different algorithms, one of which (the *Möller Algorithm*) is much more efficient (see the analysis in [**27**] and its evaluation in [**52**, pgg.40-42]) than the original algorithm (the *Buchberger-Möller Algorithm* [**51**]) of [**48**].

In [**23**] (see also [**22**]) Cerlienco and Mureddu proposed an iterative algorithm defining a correspondence $y_i \leftrightarrow t_i$ between a finite ordered set of points $\{y_1, \dots, y_m\} \subset K^n, K$ a field, and an ordered set of terms $\{t_1, \dots, t_m\} \subset K[x_1, \dots, x_n]$ in a such way that, for each $i, 1 \leq i \leq m$, $\{t_1, \dots, t_i\}$ is an order ideal of terms, generating the same vector space as the one generated by the interpolating polynomials produced by the Buchberger-Möller Algorithm for the set of points $\{y_1, \dots, y_i\}$. The curious and positive aspect of their approach is that, when the terms are ordered according a lexicographical ordering, each term t_i is the leading term of the interpolating polynomial associated to the point y_i by the Buchberger-Möller Algorithm. The negative aspect of their approach, however, is that it required recursion on the variables, giving a complexity $\mathcal{O}\left(n^2N^2\right)$ but combinatorial tools as the *lex game* [**29**] and the *Bar Code* [**15,16**] allowed to reduce the complexity to $\mathcal{O}\left(nN^2 \log(N)\right)$.

The *Tag-Variable Techniques* introduced in [**67**] were applied in [**66**] to characterize subalgebras of $K[x_1, \dots, x_n]$ and to compute their SAGBI bases [**65**].

All the techniques discussed here have been often directly introduced for the case of the polynomial rings $K[x_1, \dots, x_n]$ over a field K and extended to the case of the polynomial ring $R[x_1, \dots, x_n]$, R being a PIR, but, if not, they can anyway be trivially extended. The same holds for most of the extensions discussed here. The first was the extension of the notion of (left-right) Gröbner Basis [**6**] and the Gebauer Möller Algorithm [**8, 50, 62**] to compute them in the non-commutative polynomial ring $R\langle x_1, \dots, x_n\rangle$. The bilateral case was solved by Pritchard's reduction to the left case [**63**] and by Weispfenning via both its completion procedure [**42**] and the introduction of the notion of *restricted Gröbner bases* [**71**]. A version of the Möller Algorithm and the Cerlienco-Mureddu algorithm in a more general setting than $R\langle x_1, \dots, x_n\rangle$ is available in [**19**]. Tag-variable techniques in $K\langle x_1, \dots, x_n\rangle$ are discussed in [**40, 54**].

The notion of Gröbner Basis and the related algorithm were extended to twisted algebras by [**1, 2**] which study Lie Algebras, but soon the Passau group successfully extended the original approach introduced by Ore [**58, 59**] in a series of papers culminated with [**60**]. The extension of the results of the Lifting Theorem and the notion of restricted Gröbner bases allowed to cover all *effective associative rings* [**17, 18, 57**]. For a completely different approach consider the reduction of the non-commutative case to the commutative one discussed in [**43, 44**]. The Möller Algorithm and the Cerlienco-Mureddu Correspondence into effective associative rings are discussed in [**19**], tag-variable techniques in [**55, 70**].

Thus we have reached the most general setting for 0-dimensional ideals into an *associative* ring. What suggested us to drop associativity in our research was, on one side, the queries suggested by [**5**], on the other side, the research of suitable settings for satisfying the request of [**61**] of new models for *design of experiments*.

Since the most natural examples were octonions, we decided to place ourselves in the natural setting suggested by [**34**] for *real algebras*; more so, since it has a strong connection with Mourrain's notion [**53**] of *connectedness to* 1.

Thus, we consider a $\mathbb{F}$−vector space $V, \dim_{\mathbb{F}}(V) = n$, label and enumerate as $\mathbf{S} = \{x_1, \dots, x_n\}$ an independent basis of it, and consider on V a non-associative binary operation $\circ : V \times V \mapsto V$ and the magma $\overline{\mathbf{S}}$ generated by it. In this setting, Gröbner bases have been studied in [**24, 35, 36**] which defined as Buchberger reduction the natural reduction and proved that any interreduced set is a reduced Gröbner basis G of the ideal $\mathbb{I}(G)$ it generates, so that neither S-pair completions nor criteria are needed.

In order to describe such ideal $\mathbb{I}(G)$ we make use of the alternative description proposed by Buchberger and Möller in [**48**] in terms of functionals; this requires to define, for each $P := (a_1, \dots, a_n) \in \mathbb{F}^n$, the functional $L_P : \mathbb{F}\{\overline{\mathbf{S}}'\} \to \mathbb{F}$, to give an adaptation of the Möller Algorithm which we are sketching in Examples 7.12 and 7.18 and state a proper (and trivial) Cerlienco–Mureddu Correspondence, which allows us to prove that $\mathbb{I}(G) = \{f \in \mathbb{F}\{\overline{\mathbf{S}}'\} : L_P(f) = 0, P \in \mathbb{F}^n\}$.

Finally we simply adapt the original constructions for SAGBI bases and illustrate it in Example 7.4.

After introducing the natural notation (Section 2), we discuss in our setting Gröbner bases, Buchberger reduction and the related results of [**24, 35, 36**] (Section 3), the Möller Algorithm and the Cerlienco–Mureddu Correspondence (Section 4) and subalgebras (Section 5). In Section 6 we adapt the notion and notation of Kuzmin and Shestakov [**41**] with the aim of investigating in our magmas the state of the main laws related to alternative algebras.

Finally (Section 7) we illustrate our theory on the most natural non-associative binary operations: *subtraction* (Section 7.1), with its relation with the associative addition, and *average* (Section 7.2), both with a behaviour strictly connected with char($\mathbb{F}$), *division* (Section 7.3), *set difference* (Section 7.4), which for the case $\overline{\mathbf{S}} = \{x, y, z\}$ (see Examples 7.18) gives a natural instance of octonions. This orients us to consider *cross product* (Section 7.5), the similar but weaker *Rock-Paper-Scissors* (Section 7.7), *quaternions* (Section 7.6), *octonions* (Section 7.8) and their representation as a *Steiner Triple System* (Section 7.9).

As we stated above this is definitely not the last paper of ours in this series.

Our aim to answer the challenging queries of [**61**] on new models for design of experiments, is orienting us toward functionals which do not satisfy left/right-distributivity as assumed by our last version of the Möller Algorithm; the request of [**61**] for a symmetric Gröbnerian solution for the symmetric model

$$\{(0,0), (1,-1), (-1,1), (1,0), (0,1)\} \subset \mathbb{Q}^2$$

suggests to investigate toward a termordering free ordering [**7**].

Such termordering free orderings are central in Mourrain's [**53**] *connectedness to* 1 and, as pointed out in the *divertissment* [**3**], could allow to drop the finiteness assumption and extend our theory to each $\mathbb{F}$−vector space V, forcing on it a proper Gröbnerian setting for its non-associative binary operation $\circ : V \times V \mapsto V$.

2. Notation

Adapting the definitions introduced in [**34**, pg.183] in relation to real algebras, we consider the set $\mathbf{S} = \{x_1, \ldots, x_n\}$ and, on the vector space V generated by $\mathbf{S}$ over the field $\mathbb{F}$

$$V := \mathrm{Span}_{\mathbb{F}}(\mathbf{S}) = \left\{ \sum_{i=1}^{n} a_i x_i : a_i \in \mathbb{F} \right\} =: \mathbb{F}\{\mathbf{S}\},$$

a *product mapping* or (*multiplication*)

$$\begin{aligned} \circ :& V \times V \mapsto V \\ &(x, y) \mapsto x \circ y := xy \end{aligned}$$

which satisfies the two distributive laws

$$(ax + by) \circ z = a(x \circ z) + b(y \circ z), \quad x \circ (ay + bz) = a(x \circ y) + b(x \circ z) \tag{2.1}$$

for all $a, b \in \mathbb{F}$ and all $x, y, z \in V$. We label such vector space with such multiplication a $\mathbb{F}$-*algebra* and denote it as $A = (\mathbb{F}\{\mathbf{S}\}, \circ) = (V, \circ)$.

Note that Eq. (2.1) implies the relations $a(x \circ y) = (ax \circ y) = x \circ ay$.

An element $e \in \mathbb{F}\{\mathbf{S}\}$ is called an *identity element* (or *unit*) if

$$ex = xe = x, \text{ for each } x \in \mathbb{F}\{\mathbf{S}\}.$$

Note that every algebra has at most a single identity:

$$\text{for each } e, \eta \text{ holds } ex = xe = x, \eta x = x\eta = x, \text{ for each } x \in V \implies \eta = e\eta = e.$$

In case the $\mathbb{F}$-algebra $A = (\mathbb{F}\{\mathbf{S}\}, \circ) = (V, \circ)$ does not have a unit we add it, denoting it by e, and consider the set $\mathbf{S}' := \mathbf{S} \cup \{e\}$ and the vector space $V' := \mathrm{Span}_{\mathbb{F}}(\mathbf{S}')$. We make then $\mathbb{F}\{\mathbf{S}'\} \equiv \mathbb{F}e \oplus \mathbb{F}\{\mathbf{S}\}$ a $\mathbb{F}$-algebra with the identity e, by defining on $\mathbb{F}\{\mathbf{S}'\}$ a product $\cdot$, extending $\circ$ in such a way that

$$e^2 = 1, \text{ and } ew = we, \text{ for each } w \in \mathbb{F}\{\mathbf{S}\},$$

for $z := ae + c = a_0 e + \sum_{i=1}^{n} a_i x_i$ and $w := be + d = b_0 e + \sum_{i=1}^{n} b_i x_i$

$$\begin{aligned} z \cdot w &= (ae + c) \cdot (be + d) = aebe + cbe + ade + cd = ab + cbe + ade - c \circ d \\ &= (ab - c \circ d) + (ad + cb)e \end{aligned}$$

and

$$\begin{aligned} z \cdot w &= \left(a_0 e + \sum_{i=1}^{n} a_i x_i \right) \cdot \left(b_0 e + \sum_{i=1}^{n} b_i x_i \right) \\ &= \left(a_0 b_0 + \sum_{i,j=1}^{n} a_i b_j \gamma_{ij0} \right) e + \sum_{u=1}^{n} \left(a_u b_0 + a_0 b_u + \sum_{i,j=1}^{n} a_i b_j \gamma_{iju} \right) x_u \\ &= \left(a_0 b_0 - \sum_{i=1}^{n} a_i b_i \right) e + \sum_{u=1}^{n} \left(a_u b_0 + a_0 b_u + \sum_{i,j=1}^{n} a_i b_j \gamma_{iju} \right) x_u \end{aligned}$$

where we specialized the generic expression $x_i \circ x_j = \gamma_{ij0} e + \sum_{u=1}^{n} \gamma_{iju} x_u, 1 \leq i \neq j \leq n$ by setting $\gamma_{ij0} = 0, 1 \leq i \neq j \leq n$, and, for each $1 \leq i \leq n, \gamma_{ii0} = -1$ and $\gamma_{iiu} = 0, 1 \leq u \leq n$ obtaining

$$x_i \circ x_j = \sum_{u=1}^{n} \gamma_{iju} x_u, 1 \leq i \neq j \leq n \text{ and } x_i \circ x_i = -e, 1 \leq i \leq n.$$

This classical construction allows and forces us to define at the same time

- a *scalar product* $\langle -,-\rangle : \mathbb{F}\{\mathbf{S}\} \times \mathbb{F}\{\mathbf{S}\} \to \mathbb{F}$

$$(z,w) \mapsto \langle z,w\rangle = \left\langle \sum_{i=1}^{n} a_i x_i, \sum_{i=1}^{n} b_i x_i \right\rangle := \sum_{i=1}^{n} a_i b_i$$

  id est a bilinar form, symmetric and positive definite

$$\langle x,y\rangle = \langle y,x\rangle, \quad \langle x,x\rangle > 0 \text{ for each } x \neq 0$$

- a *cross product* $\times : \mathbb{F}\{\mathbf{S}\} \times \mathbb{F}\{\mathbf{S}\} \to \mathbb{F}\{\mathbf{S}\}$

$$u \times v := u \circ v = \sum_{i,j=1}^{n} a_i b_j x_i \circ x_j = \sum_{u=1}^{n} \left(\sum_{i,j=1}^{n} a_i b_j \gamma_{iju} \right) x_u$$

- and a product $\cdot : \mathbb{F}\{\mathbf{S}'\} \times \mathbb{F}\{\mathbf{S}'\} \to \mathbb{F}\{\mathbf{S}'\}$

$$u \cdot v = a_0 b_0 - \langle u,v\rangle + u \times v = a_0 b_0 - \sum_{i=1}^{n} a_i b_i + \sum_{u=1}^{n} \left(\sum_{i,j=1}^{n} a_i b_j \gamma_{iju} \right) x_u;$$

- for each element $z := a_0 e + \sum_{i=1}^{n} a_i x_i \in \mathbb{F}\{\mathbf{S}'\}$, the *real part* $\Re(z) = a_0 \in \mathbb{F}$ and *imaginary part* $\Im(z) = \sum_{i=1}^{n} a_i x_i \in \mathbb{F}\{\mathbf{S}\}$ and
- the related *real linear form* $\Re$ and *imaginary map* $\Im$

$$\Re : \mathbb{F}\{\mathbf{S}'\} \to \mathbb{F} : z = \Re(z)e + \Im(z) \mapsto \Re(z) = a_0 \in \mathbb{F},$$

$$\Im : \mathbb{F}\{\mathbf{S}'\} \to \mathbb{F}\{\mathbf{S}\} : z = \Re(z)e + \Im(z) \mapsto \Im(z) = \sum_{i=1}^{n} a_i x_i \in \mathbb{F}\{\mathbf{S}\}$$

- the set

$$\mathrm{Im}(\mathbb{F}\{\mathbf{S}'\}) := \{x \in \mathbb{F}\{\mathbf{S}'\} : x^2 \in \Re(x)e \text{ and } x \notin \Re(x)e \setminus \{0\}\}$$

  of the *purely imaginary* elements.

The *dimension* of A is by definition its dimension as a vector space, so we have $\dim(\mathbb{F}\{\mathbf{S}\}) = n$ and $\dim(\mathbb{F}\{\mathbf{S}'\}) = n+1$.

A $\mathbb{F}$-algebra A is said *associative* if $\circ$ is associative, i.e.

$$x \circ (y \circ z) = (x \circ y) \circ z \text{ for each } x, y, z \in A,$$

and it is said *commutative* if

$$x \circ y = y \circ x \text{ for each } x, y \in A.$$

Note that in general a $\mathbb{F}$-algebra is neither associative nor commutative.

Also in $\mathbb{F}$-algebras, powers are usually defined inductively by $x^m = x \circ x^{m-1}$ but it does not necessarily hold $x \circ x^2 \neq x^2 \circ x$ and, even in the commutative case, it can be $(x^2)^2 \neq x^4$. Thus we say that an algebra is *power-associative* if

$$x^m \circ x^n = x^{m+n} \text{ for each } x \in A \text{ and } m, n \in \mathbb{N}^*$$

LEMMA 2.1. *Every associative algebra A is power associative.*

PROOF. For $x \in A$ by induction we have:

$$\forall n : x^{m-1}x^n = x^{m+n-1} \implies x^m x^n = (xx^{m-1})x^n = x(x^{m-1}x^n) = xx^{m+n-1} = x^{m+n}$$

and

$$\forall m : x^m x^{n-1} = x^{m+n-1} \implies x^m x^n = x^m(x^{n-1}x) = (x^m x^{n-1})x = x^{m+n-1}x = x^{m+n}$$

starting with the seed $x^m x^n = x^{m+n}$ with $m = n = 1$. □

An element $x \in A, x \neq 0$, of the algebra A is said to be a *divisor of zero* if there is $y \in A, y \neq 0$ such that $xy = 0$ or $yx = 0$.

An algebra A is said to have no zero divisors if it contains no divisors of zero, which means that

$$\forall x, y \in A, xy = 0 \implies \text{ either } x = 0 \text{ or } y = 0.$$

A good model for describing non-associative algebras is *magmas*: we start from the set $\mathbf{S}$ and define the magma $\overline{\mathbf{S}}$ freely generated by $\mathbf{S}$, setting $\mathbf{S}_0 = \mathbf{S}$, $\mathbf{S}_1 = \mathbf{S} \times \mathbf{S}$ and, in general, $\mathbf{S}_n = \sqcup_{a+b=n-1}\mathbf{S}_a \times \mathbf{S}_b$. The magma $\overline{\mathbf{S}}$ we are looking for is $\overline{\mathbf{S}} = \sqcup_{n=0}^{\infty}\mathbf{S}_n$.

The non-associative binary operation

$$\circ : \overline{\mathbf{S}} \times \overline{\mathbf{S}} \mapsto \overline{\mathbf{S}},$$

corresponding to the bracketing that is defined in $\overline{\mathbf{S}}$ by the use of the Cartesian product, is the same used in defining the product mapping of the $\mathbb{F}$-algebra $A = (\mathbb{F}\{\mathbf{S}\}, \circ) = (V, \circ)$ so that, for instance, $(x_i, (x_j, x_l))$ corresponds to $x_i \circ (x_j \circ x_l)$ while $((x_i, x_j), x_l) = (x_i \circ x_j) \circ x_l$.

In order to describe as a magma $\overline{\mathbf{S}}'$, the $\mathbb{F}$-algebra

$$\mathbb{F}\{\overline{\mathbf{S}}'\} := \left\{ a_0 e + \sum_{i=1}^{n} a_i x_i : a_i \in \mathbb{F} \right\},$$

with unit e over the field $\mathbb{F}$, we associate to each element m of $\overline{\mathbf{S}}'$ a tree $T(m)$, whose leaves are always elements of $\mathbf{S}$. In particular:

- for the element e we associate the tree consisting of the root only;
- for $x_i \in S$, $T(x_i)$ is a tree with only one branch, which connects the root with a leaf, labelled by x_i;
- the operation $m = m_1 \circ m_2$ is translated, in the trees' context as *join* $T(m_1)$ and $T(m_2)$ making their roots being the children of a newly built root.

In this way the non-associative nature of $\circ$ is mimicked in the shape of the trees' branches; we need only to extend it by linearity, so that a linear combination of elements of $\overline{\mathbf{S}}'$ can be seen as a linear combination of trees, giving a map $\mathbb{F}\{\overline{\mathbf{S}}'\}^{\delta} \to \mathbb{F}\{\overline{\mathbf{S}}'\}$.

We also define the *degree* of an element $m \in \overline{\mathbf{S}}'$, setting $\deg(e) = 0$, $\deg(m) = 1$ for $m \in \mathbf{S}$ and $\deg(m) = \deg(m_1) + \deg(m_2)$ if $m = m_1 \circ m_2$.

3. Gröbnerian technology

As we saw in the Introduction, Gröbner bases' Theory can be defined also in this very generic context; in this section we decline to our scenery the main concepts and tools related to Gröbner bases computation.

Definition 3.1 (de Graaf–Wisliceny [**24**]). A 1-*appliance* on $\overline{\mathbf{S}}'$ is the assignment

$$\alpha := (T, j, m_1, \ldots, m_{j-1}, m_{j+1}, \ldots, m_\mu)$$

of

- a finite binary rooted tree T, $\mu := \deg(T)$;
- an integer $j, 1 \leq j \leq \mu$;
- $\mu - 1$ elements $m_1, \ldots, m_{j-1}, m_{j+1}, \ldots, m_\mu$ of $\overline{\mathbf{S}}'$.

We denote by $A_1(\overline{\mathbf{S}}')$, the set of the 1-appliances on $\overline{\mathbf{S}}'$.

For each 1-appliance α, we denote by P_α the $\mathbb{F}$-linear map

$$P_\alpha : \mathbb{F}\{\overline{\mathbf{S}}'\} \to \mathbb{F}\{\overline{\mathbf{S}}'\} : f \mapsto P_\alpha(f) := T(m_1, \ldots, m_{j-1}, f, m_{j+1}, \ldots, m_n).$$

This means that we are performing the operation $\circ$ in the order and with the bracketing induced by T and we distribute the j-th node among the terms in $\overline{\mathbf{S}}'$ and, for the particular case $\alpha := (T, 1)$, T being the root only, and $m \in \overline{\mathbf{S}}'$, $P_\alpha(m)$ is the root to which the tree m is directly glued.

In the following example we show that for different 1-appliances the linear map can be the same.

Example 3.2. Let us consider the tree T

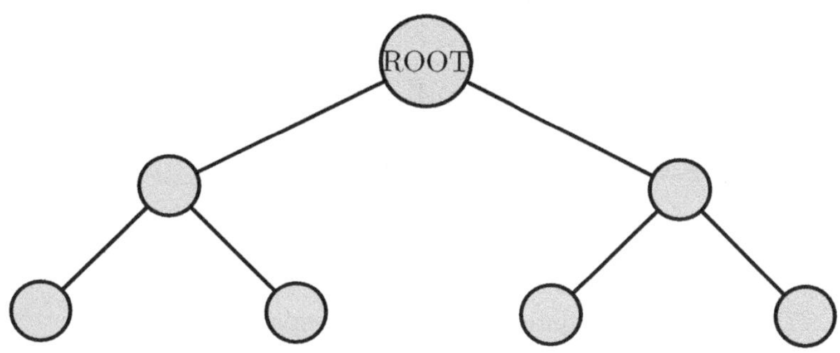

and the two 1-appliances $\alpha = (T, 2, xy, x)$ and $\alpha' = (T, 3, x, y, x)$. In this case, for each $f \in \overline{\mathbf{S}}$, $P_\alpha(f) = (T, xy, f, x)$ and $P_{\alpha'}(f) = (T, x, y, f, x)$, both meaning

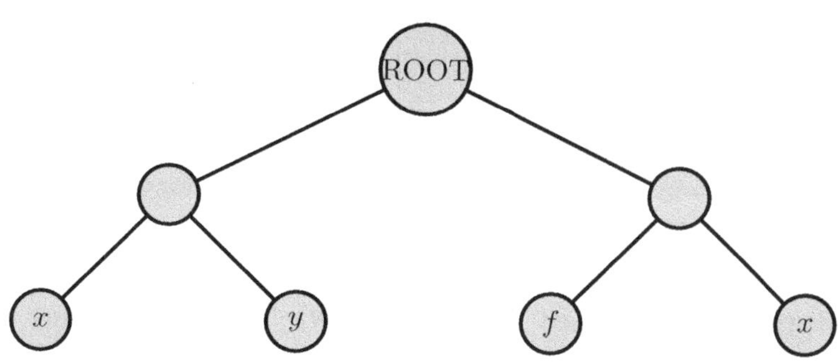

namely $(x \circ y) \circ (f \circ x)$.

Of course for an $f \in \mathbb{F}\{\overline{\mathbf{S}}'\}$ we would get two identical linear combinations of trees. For example if $f = x + y$ then we have the sum of

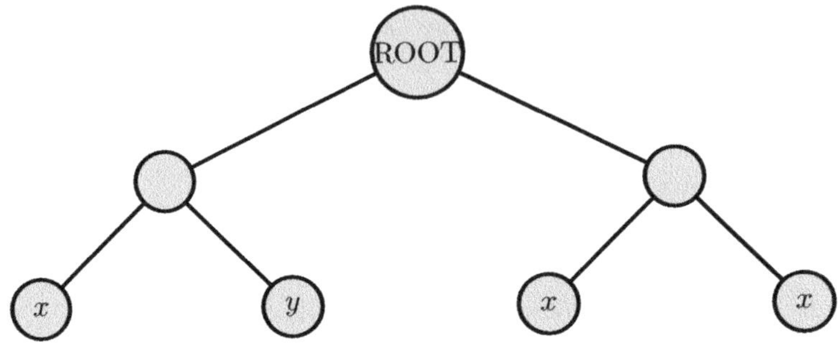

and

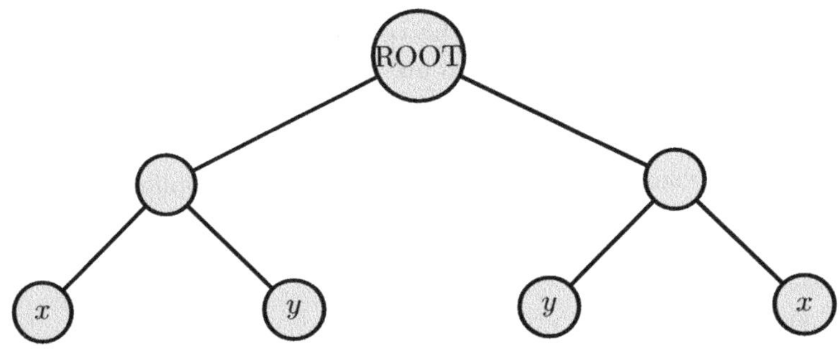

An *ideal* $\mathsf{I} \subset \mathbb{F}\{\overline{\mathbf{S}}'\}$ in the $\mathbb{F}$-algebra $A' = (\mathbb{F}\{\mathbf{S}'\}, \cdot) = (V', \cdot) = \mathbb{F}\{\overline{\mathbf{S}}'\}$, being a $\mathbb{F}$-vector space closed under left and right multiplication by elements in $\mathbb{F}\{\overline{\mathbf{S}}'\}$, can be described as any $\mathbb{F}$-vector space which satisfies

$$P_\alpha(f) \in \mathsf{I} \text{ for each } f \in \mathsf{I}, \alpha \in A_1(\overline{\mathbf{S}}').$$

DEFINITION 3.3. For any set $F \subset \mathbb{F}\{\overline{\mathbf{S}}'\}$ the *ideal generated by* F is the ideal $\mathbb{I}(F) \subset A' = (\mathbb{F}\{\overline{\mathbf{S}}'\}, \cdot)$ whose underlying $\mathbb{F}$-vector-space has

$$\mathcal{B}(F) := \{P_\alpha(f) : f \in F, \alpha \in A_1(\overline{\mathbf{S}}')\}$$

as generating set.

For $m_1, m_2 \in \overline{\mathbf{S}}'$ we say that m_1 *divides* (or *is a factor of*) m_2 if and only if $m_2 = P_\alpha(m_1)$ for some $\alpha \in A_1(\overline{\mathbf{S}}')$ and we denote it as $m_1 \mid m_2$.

A *term ordering* on $\mathbb{F}\{\overline{\mathbf{S}}'\}$ is a total well-ordering $<$ on $\overline{\mathbf{S}}' := \overline{\mathbf{S}} \cup \{e\}$ satisfying

(1) $e < m$ for each $m \in \overline{\mathbf{S}}$,
(2) $m_1 < m_2 \implies (n, m_1) = n \cdot m_1 < n \cdot m_2 = (n, m_2)$ for each $n, m_1, m_2 \in \overline{\mathbf{S}}'$,
(3) $m_1 < m_2 \implies (m_1, n) = m_1 \cdot n < m_2 \cdot n = (m_2, n)$ for each $n, m_1, m_2 \in \overline{\mathbf{S}}'$.

□

These definitions allow to introduce on the magma algebra $\mathbb{F}\{\overline{\mathbf{S}}'\}$ the Gröbnerian technology and terminology. In particular, w.r.t. a term ordering $<$ we can speak, for each $f \in \mathbb{F}\{\overline{\mathbf{S}}'\}$, of its *maximal term*[1] $\mathbf{T}(f)$, *leading coefficient* $\mathrm{lc}(f)$ and *maximal monomial* $\mathbf{M}(f)$.

[1]Remark that, since $<$ is a term ordering, we have $P_\alpha(\mathbf{T}(f)) = \mathbf{T}(P_\alpha(f))$ for each $f \in \mathbb{F}\{\overline{\mathbf{S}}'\}$ and each $\alpha \in A_1(\overline{\mathbf{S}}')$.

For each set F we can denote $\mathbf{T}\{F\} := \{\mathbf{T}(f) : f \in F\} \subset \overline{\mathbf{S}}'$ and[2]

$$\mathbf{T}(F) := \{\mathbf{T}(P_\alpha(f)) : \alpha \in A_1(\overline{\mathbf{S}}'), f \in F\} = \{P_\alpha(\mathbf{T}(f)) : \alpha \in A_1(\overline{\mathbf{S}}'), f \in F\}.$$

Finally we can denote

$$\mathbf{N}(F) := \overline{\mathbf{S}}' \setminus \mathbf{T}(F) \text{ and } \mathbb{F}[\mathbf{N}(F)] := \mathrm{Span}_{\mathbb{F}}(\mathbf{N}(F))).$$

As remarked in [**49**, IV. pg.187] one can then, for each ideal $\mathsf{I} \subset \mathbb{F}\{\overline{\mathbf{S}}'\}$ and each set $G \subset \mathsf{I}$,

- prove the relations $\mathbb{F}\{\overline{\mathbf{S}}'\} \cong \mathsf{I} \oplus \mathbb{F}[\mathbf{N}(\mathsf{I})]$, and $\mathbb{F}\{\overline{\mathbf{S}}'\}/\mathsf{I} \cong \mathbb{F}[\mathbf{N}(\mathsf{I})]$;
- define a *Gröbner basis* of an ideal $\mathsf{I} \subset \mathbb{F}\{\overline{\mathbf{S}}'\}$ any set $G \subset \mathsf{I}$ such that
$$\mathbf{T}(G) = \mathbf{T}(\mathsf{I}) = \mathbf{T}\{\mathsf{I}\};$$
- introduce the notions of (strong) *Gröbner representation*, *normal form* and *canonical form* in terms of G;
- compute both normal and canonical forms by the obvious adaptations of Buchberger reduction.
- We can also introduce the classical Buchberger reduction by iteratively appling on
$$f = \sum_{i=1}^{\nu} c_i t_i, c_i \in \mathbb{F}, t_i \in \overline{\mathbf{S}}', t_1 > \ldots > t_\nu$$
for which exists $I, 1 \le I \le \nu$ and $g = \mathrm{lc}(g)\mathbf{T}(g) - h \in G$ such that $\mathbf{T}(g)$ divides t_I
$$t_I = P_\alpha(\mathbf{T}(g)), \alpha := (T, j, m_1, \ldots, m_{j-1}, m_{j+1}, \ldots, m_m)$$
the rewriting rule
$$f = \sum_{i=1}^{I-1} c_i t_i + c_I T(m_1, \ldots, m_{j-1}, t_I, m_{j+1}, \ldots, m_n) + \sum_{i>I}^{\nu} c_i t_i \mapsto$$
$$\sum_{i=1}^{I-1} c_i t_i + c_I \,\mathrm{lc}(g)^{-1} T(m_1, \ldots, m_{j-1}, h, m_{j+1}, \ldots, m_n) + \sum_{i>I}^{\nu} c_i t_i$$
until $f \in \mathbb{F}[\mathbf{N}(\mathsf{I})]$.

DEFINITION 3.4. A set $G \subset \mathbb{F}\{\overline{\mathbf{S}}'\}$ such that

- for each $g_1, g_2 \in G$, $\mathbf{T}(g_1) \nmid \mathbf{T}(g_2)$,
- for each $g \in G, \mathrm{lc}(g) = 1$,
- for each $g \in G, g - \mathbf{T}(g) \in \mathbb{F}[\mathbf{N}(G)]$,

is called an *interreduced set*.

DEFINITION 3.5. A set $G \subset \mathsf{I} \subset \mathbb{F}\{\overline{\mathbf{S}}'\}$ such that

- $\mathbf{T}(G) = \mathbf{T}(\mathsf{I}) = \mathbf{T}\{\mathsf{I}\}$,
- for each $g_1, g_2 \in G$, $\mathbf{T}(g_1) \nmid \mathbf{T}(g_2)$,
- for each $g \in G, \mathrm{lc}(g) = 1$,
- for each $g \in G, g - \mathbf{T}(g) \in \mathbb{F}[\mathbf{N}(G)]$,

is called the *reduced Gröbner basis* of I.

---

[2]Remark that for an ideal $\mathsf{I} \subset \mathbb{F}\{\overline{\mathbf{S}}'\}$ we have $\mathbf{T}(\mathsf{I}) = \mathbf{T}\{\mathsf{I}\}$.

Finally we recall the main result introduced in [**24, 35, 36**] as reported in [**49**, IV.47.12]:

COROLLARY 3.6. *Any interreduced set is a reduced Gröbner basis of the ideal it generates.* □

On the basis of these remarks we can conclude

THEOREM 3.7. *For each* $i, j, 1 \leq i, j \leq n$, *let* $\gamma_{iju} \in \mathbb{F}, \forall u, 0 \leq u \leq n$, *such that*

$$x_i \circ x_j = \gamma_{ij0} + \sum_{u=1}^{n} \gamma_{iju} x_u$$

and denote $G := \{x_i \circ x_j - \gamma_{ij0} - \sum_{u=1}^{n} \gamma_{iju} x_u, 1 \leq i, j \leq n\}$. *Then* G *is the reduced Gröbner basis of the ideal it generates.*

In order to give a combinatorial description of the ideal generated by G, we need to introduce our main Gröbner-Free technique, the Möller Algorithm.

4. The Möller Algorithm

Let us now consider the functionals $L_P : \mathbb{F}\{\overline{\mathbf{S}}'\} \to \mathbb{F}, P := (a_1, \ldots, a_n) \in \mathbb{F}^n$ defined inductively by

$$\begin{array}{lclcll} L_P(1) & = & & = & 1, & \\ L_P(x_i) & = & & = & a_i, & 1 \leq i \leq n, \\ L_P(t_i \circ t_j) & = & L_P(t_i t_j) & = & L_P(t_i) \circ L_P(t_j), & t_i, t_j \in \overline{\mathbf{S}}, \\ L_P(\sum_{i=1}^{u} c_i t_i) & = & & = & \sum_{i=1}^{u} c_i L_P(t_i), & \sum_{i=1}^{u} c_i t_i \in \mathbb{F}\{\overline{\mathbf{S}}'\}. \end{array}$$

Given any set of points $\mathbf{X} \subset \mathbb{F}^n$ it is clear that the ideal $I(\mathbf{X}) := \{f \in \mathbb{F}\{\overline{\mathbf{S}}'\} : L_P(f) = 0, P \in \mathbf{X}\}$ contains the ideal $\mathbb{I}(G)$:

$$\begin{aligned} \mathbb{I}(G) \subset \mathsf{I} := \{f \in \mathbb{F}\{\overline{\mathbf{S}}'\} : L_P(f) = 0, P \in \mathbb{F}^n\} \subset \mathbb{F}\{\overline{\mathbf{S}}'\} \\ \subset \{f \in \mathbb{F}\{\overline{\mathbf{S}}'\} : L_P(f) = 0, P \in \mathbf{X}\}. \end{aligned}$$

Thus, in principle the techniques related to the Möller Algorithm and the Cerlienco-Mureddu Correspondence can be easily adapted in this setting.

EXAMPLE 4.1. Let us consider the case $\overline{\mathbf{S}} = \{x\}, \overline{\mathbf{S}}' = \{x, e\}$ so that by the previous results $x^2 - 1 \in \mathsf{I}$ and $\mathbf{N}(I(\mathbf{X})) \subset \mathbf{N}(\mathbb{I}(G)) = \{e, x\}$.

Thus for $\mathbf{X} := \left\{ P_i := i \in \{1, 2, 3, -1\} \subset \mathbb{F} := \mathbb{Q} \right\}$ a direct application of the Möller Algorithm gives

	L_1	L_2	L_3	L_{-1}
1	1	1	1	1
x	1	2	3	-1

	L_1	L_2	L_3	L_{-1}
1	1	1	1	1
$x-1$	0	1	2	-2

	L_1	L_2	L_3	L_{-1}
$2-x$	1	0	-1	3
$x-1$	0	1	2	-2

Thus, as we expected, $\dim_{\mathbb{F}}(L_1, L_2, L_3, L_{-1}) = 2$ and $\mathrm{Span}_{\mathbb{F}}(L_1, L_2, L_3, L_{-1}) = \mathrm{Span}_{\mathbb{F}}(L_1, L_2)$. The last step of reduction allows to deduce an orthogonal basis $\{q_1, q_2\}, q_1 = 2 - x, q_2 = x - 1$ of $\{L_1, L_2\}$ and, by column reduction, the dependencies of L_3 and L_{-1} in terms of $\{L_1, L_2\}$:

$$L_3 = -L_1 + 2L_2, \text{ and } L_{-1} = 3L_1 - 2L_2.$$

In the same mood

	L_{-1}	L_3	L_1	L_2
1	1	1	1	1
x	-1	3	1	2

	L_{-1}	L_3	L_1	L_2
1	1	1	1	1
$\frac{x+1}{4}$	0	1	$\frac{1}{2}$	$\frac{3}{4}$

	L_{-1}	L_3	L_1	L_2
$\frac{3-x}{4}$	1	0	$\frac{1}{2}$	$\frac{1}{4}$
$\frac{x+1}{4}$	0	1	$\frac{1}{2}$	$\frac{3}{4}$

The last step of reduction allows to deduce an orthogonal basis $\{q_{-1}, q_3\}, q_{-1} = \frac{3-x}{4}, q_3 = \frac{x+1}{4}$ of $\{L_{-1}, L_3\}$ and, by column reduction the dependencies of L_1 and L_2 in terms of $\{L_{-1}, L_3\}$, namely $2L_1 = L_{-1} + L_3$ and $4L_2 = L_{-1} + 3L_3$.

It is then sufficient to choose the functionals $L_i := L_{P_i}, 0 \le i \le n$ with $P_0 = (0, \dots, 0)$ and[3] $P_i := (\delta_{i1}, \dots, \delta_{in}), 1 \le i \le n$ and set the Cerlienco-Mureddu Correspondence $1 \leftrightarrow P_0, x_i \leftrightarrow P_i, 1 \le i \le n$, to deduce that

LEMMA 4.2. *Setting* $\mathbf{X} := \{P_i : 0 \le i \le n\}$, *we have*

- $\dim_{\mathbb{F}}(L_i, 0 \le i \le n) = n+1$
- $\mathbf{N}(I(\mathbf{X})) = \mathbf{S}'$.

COROLLARY 4.3. *The set*

$$G := \{x_i \circ x_j - \gamma_{ij0} - \sum_{u=1}^{n} \gamma_{iju} x_u, 1 \le i, j \le n\}$$

is the reduced Gröbner basis of the ideal associated to the points in $\mathbf{X}$*:*

$$\mathbb{I}(G) = I(\mathbf{X}) = \{f \in \mathbb{F}\{\overline{\mathbf{S}}'\} : L_P(f) = 0, P \in \mathbf{X}\} \subset \mathbb{F}\{\overline{\mathbf{S}}'\}.$$

□

5. Subalgebras

We plan to investigate subalgebras (and SAGBI bases) of Magmas but the first remarks seem to be even more trivial then our considerations on Gröbner bases and the Möller algorithm; in fact, if we set

$$\mathbf{F} := \{f_1, \dots, f_n\} \subset \mathbb{F}\{\overline{\mathbf{S}}'\},$$

we consider the subalgebra $\mathbb{F}\{\overline{\mathbf{F}}\} \subset \mathbb{F}\{\overline{\mathbf{S}}'\}$ and apply the techniques discussed in [**49**, II.26.5] to the magma $\mathbb{F}\{\overline{\mathbf{T}'}\}, \mathbf{T} := \{Y_1, \dots, Y_n, x_1, \dots, x_n\}$ and to the map

$$\gamma : \mathbb{F}\{\overline{\mathbf{T}'}\} \to \mathbb{F}\{\overline{\mathbf{S}}'\} : \gamma(x_i) = x_i, \gamma(Y_j) = f_j$$

the computation of the proper Gröbner basis G of the ideal $\ker(\gamma) = \mathbb{I}(f_j - Y_j, 1 \le j \le n)$, trivially gives $G = \{f_j - Y_j\}$ thus granting (via [**49**, II. Theorem 26.5] $\mathbb{F}\{\overline{\mathbf{F}}\} = \mathbb{F}\{\overline{\mathbf{S}}'\}/\bar{\mathbf{I}}(G)$.

6. Freely generated algebras

We recall that Kuzmin and Shestakov [**41**] (see also [**64**]) consider the alternative laws **A1**, **A2** (see them below), consider the ideal

$$I_{\text{alt}} := \left\{(xx)y - x(xy), (xy)y - x(yy) : x, y \in \mathbf{S}'\right\}$$

and the related *alternative algebra freely generated by* $\mathbf{S}$

$$A_{\text{alt}} := \mathbb{F}\{\overline{\mathbf{S}}'\}/I_{\text{alt}}.$$

[3] δ_{ij} being the Kronecker's δ.

In a similar mood we can consider for each of the formulas below and each non-associative binary operation $\circ$ the ideal I°_F generated by all instances of the formula F and the related *algebra freely generated by* $\mathbf{S}$

$$A^\circ_F := \mathbb{F}\{\overline{\mathbf{S}}'\}/I^\circ_F.$$

REMARK 6.1. Note that in general $\circ$ is **not** left/right distributive, *i.e.* it does not satisfy the distributive laws **L**, **R** (see them in Definition 6.2) For instance for subtraction

L: $(x+y)z-xz-yz = (x+y)-z-(x-z)-(y-z) = x+y-z-x+z-y+z = z$
R: $z(x+y)-zx-zy = z-(x+y)-(z-x)-(z-y) = z-x-y-z+x-z+y = z-z-z = -z$

which informs that only 0 distributes with respect to the sum. □

DEFINITION 6.2. Let us consider the magma $\mathbb{F}\{\overline{\mathbf{S}}'\} = \mathbb{F}\langle x_1, x_2, \cdots, x_n\rangle$ freely generated by the set $\mathbf{S} = \{x_1, x_2, \cdots, x_n\}$ and define for each $x, y, z \in \mathbb{F}\{\overline{\mathbf{S}}'\}$ the following relations:

A1: (left alternative law): $(xx)y - x(xy)$
A2: (right alternative law): $(xy)y - x(yy)$
F: (flexible law): $(xy)x - x(yx)$
L: (left distributive law): $(x+y)z - xz - yz$
M1: (left Moufang identity): $(x(zx))y = x(z(xy))$
M2: (right Moufang identity): $((yx)z)x = y(x(zx))$
M3: (central Moufang identity): $((xy)(zx)) = x((yz)x)$
R: (right distributive law): $z(x+y) - zx - zy$.

REMARK 6.3. Note the relations[4]

$$\mathbf{M1} \iff \mathbf{M2} \iff \mathbf{M3} \implies \mathbf{F} \Leftarrow \mathbf{A1} \iff \mathbf{A2}$$

holding in each associative algebra [**68**].

7. Magmas

Let us now discuss what these techniques give on specific magmas $A' = (\mathbb{F}\{\mathbf{S}'\}, \cdot) = (V', \cdot) = \mathbb{F}\{\overline{\mathbf{S}}'\}$ with identity e related to magmas $A = (\mathbb{F}\{\mathbf{S}\}, \circ) = (V, \circ)$ with a non-associative binary operation $\circ$. With a natural abuse of notation we will denote still $\circ$ its extension $\cdot$ to A'. We will compactly express it by $A' = (\mathbb{F}\{\mathbf{S}'\}, \circ) = (V', \circ) = \mathbb{F}\{\overline{\mathbf{S}}'\}$.

7.1. Subtraction. Let us begin with the more natural example considering the non-associative[5] but anticommutative[6] binary subtraction $- : \overline{\mathbf{S}'} \times \overline{\mathbf{S}'} \mapsto \overline{\mathbf{S}'}$.

The results discussed in the above sections give, in this context:

COROLLARY 7.1. *Denoting* $\mathbf{S} = \{x_1, \ldots, x_n\}$, $A' = (\mathbb{F}\{\mathbf{S}'\}, -) = (V', -) = \mathbb{F}\{\overline{\mathbf{S}}'\}$, $\mathsf{I} := \{f \in \mathbb{F}\{\overline{\mathbf{S}}'\} : L_P(f) = 0, P \in \mathbb{F}^n\} \subset \mathbb{F}\{\overline{\mathbf{S}}'\}$ *the ideal of relations for the operation of subtraction,* $\mathsf{A} := \mathbb{F}\langle x_1, \ldots, x_n\rangle/\mathsf{I}$, $\mathbf{N}(\mathsf{I}) = \{1, x_1, \ldots, x_n\}$, $P_0 = (0, \ldots, 0) \in \mathbb{F}^n$, $P_i := (\delta_{i1}, \ldots, \delta_{in}) \in \mathbb{F}^n, 1 \le i \le n$, $\mathbf{X} := \{P_i : 0 \le i \le n\} \subset \mathbb{F}^n$, $I(\mathbf{X}) := \{f \in \mathbb{F}\{\overline{\mathbf{S}}'\} : L_P(f) = 0, P \in \mathbf{X}\}$ *it holds that*

---

[4]To go from **M3** to **F** set $z = 1$.
[5]$a - (b - c) = a - b + c \neq a - b - c = (a - b) - c$ for each $a, b \in \mathbb{F}, c \in \mathbb{F}^*$.
[6]$(a \circ b) = -(b \circ a)$.

(1) $\mathsf{A} \cong \mathbb{K}[\mathbf{N}(\mathsf{I})] = \mathrm{Span}_{\mathbb{F}}\{1, x_1, \dots, x_n\}$

(2) *The Auzinger-Stetter matrices* [**4**] *related to such representation are*

$$A_{x_i} = \begin{array}{r|cccccccc} & v_0 & v_1 & \dots & v_i & \dots & v_j & \dots & v_n \\ \hline v_0 = 1 & 0 & 0 & \dots & 1 & \dots & 0 & \dots & 0 \\ v_1 = x_1 & 0 & -1 & \dots & 1 & \dots & 0 & \dots & 0 \\ \vdots & \vdots & \vdots & \ddots & \vdots & \ddots & \vdots & \ddots & \vdots \\ v_i = x_i & 0 & 0 & \dots & 0 & \dots & 0 & \dots & 0 \\ \vdots & \vdots & \vdots & \ddots & \vdots & \ddots & \vdots & \ddots & \vdots \\ v_j = x_j & 0 & 0 & \dots & 1 & \dots & -1 & \dots & 0 \\ \vdots & \vdots & \vdots & \ddots & \vdots & \ddots & \vdots & \ddots & \vdots \\ v_n = x_n & 0 & 0 & \dots & 1 & \dots & 0 & \dots & -1 \end{array},$$

$$_{x_i}A = \begin{array}{r|cccccccc} & v_0 & v_1 & \dots & v_i & \dots & v_j & \dots & v_n \\ \hline v_0 = 1 & 0 & 0 & \dots & 1 & \dots & 0 & \dots & 0 \\ v_1 = x_1 & 0 & 1 & \dots & -1 & \dots & 0 & \dots & 0 \\ \vdots & \vdots & \vdots & \ddots & \vdots & \ddots & \vdots & \ddots & \vdots \\ v_i = x_i & 0 & 0 & \dots & 0 & \dots & 0 & \dots & 0 \\ \vdots & \vdots & \vdots & \ddots & \vdots & \ddots & \vdots & \ddots & \vdots \\ v_j = x_j & 0 & 0 & \dots & -1 & \dots & 1 & \dots & 0 \\ \vdots & \vdots & \vdots & \ddots & \vdots & \ddots & \vdots & \ddots & \vdots \\ v_n = x_n & 0 & 0 & \dots & -1 & \dots & 0 & \dots & 1 \end{array}$$

(3) *The reduced Gröbner basis of* I *is* $G = \{x_i \circ x_i, 1 \leq i \leq n\} \cup \{x_i \circ x_j + x_j - x_i, x_j \circ x_i - x_j + x_i, 1 \leq i < j \leq n\}$.

(4) *There is the following Cerlienco-Mureddu Correspondence*

$$1 \leftrightarrow P_0, x_i \leftrightarrow P_i, 1 \leq i \leq n$$

between $\mathbf{N}(\mathsf{I})$ *and* $\mathbf{X}$.

(5) $\dim_{\mathbb{F}}(L_i, 0 \leq i \leq n) = n + 1$.

(6) $\mathbf{N}(I(\mathbf{X})) = \mathbf{S} \cup \{1\}$.

(7) $I(\mathbf{X}) = \mathsf{I}$.

LEMMA 7.2. *It holds for* $\mathcal{F} \in \Big\{\mathbf{M1}\Big\}$ $I^{\circ}_{\mathcal{F}} = 0$ *and* $A^{\circ}_{\mathcal{F}} = \mathbb{F}\{\overline{\mathbf{S}}'\}$.

PROOF. $(x(yx))z - x(y(xz)) = (x - (y - x)) - z - \Big(x - (y - (x - z))\Big) = x - (y - x) - z - \Big(x - y + (x - z)\Big) = x - y + x - z - \Big(x - y + x - z\Big) = x - y + x - z - x + y - x + z = 0.$ □

LEMMA 7.3. *For* $\mathcal{F} \in \Big\{\mathbf{A1}, \mathbf{A2}, \mathbf{F}, \mathbf{M2}, \mathbf{M3}\Big\}$ *we have that*

(1) *if* $\mathrm{char}(\mathbb{F}) = 2$ *then* $A^{\circ}_{\mathrm{alt}} = A^{\circ}_{\mathcal{F}} = \mathbb{F}\{\overline{\mathbf{S}}'\}$ *and* $I^{\circ}_{\mathcal{F}} = 0$.

(2) $2 \in I^{\circ}_{\mathrm{alt}}$ *and* $2 \in I^{\circ}_{\mathcal{F}}$

(3) *if* $\mathrm{char}(\mathbb{F}) = 0$ *then* $A^{\circ}_{\mathrm{alt}} = A^{\circ}_{\mathcal{F}} = \mathbb{Z}_2 \otimes \mathbb{F}\{\overline{\mathbf{S}}'\}$;

(4) *if* $\mathrm{char}(\mathbb{F}) = p, \gcd(p, 2) = 1$ *then* $I^{\circ}_{\mathrm{alt}} = I^{\circ}_{\mathcal{F}} = \mathbb{F}\{\overline{\mathbf{S}}'\}$.

PROOF. (1) If $\mathrm{char}(\mathbb{F}) = 2$, subtraction coincides with addition;

(2) We have

A1: $(xx)y - x(xy) = (x-x) - y - (x - (x-y)) = -2y = 2$, for $y = -1$ and any $x, z \in \mathbb{F}\{\overline{\mathbf{S}}'\}$

A2: $(xy)y - x(yy) = (x-y) - y - (x - (y-y)) = -2y = 2$, for $y = -1$ and any $x, z \in \mathbb{F}\{\overline{\mathbf{S}}'\}$

F: $(xy)x - x(yx) = (x-y) - x - \Big(x - (y-x)\Big) = x - y - x - x + y - x = -2x = 2$, for $x = -1$ and any $y, z \in \mathbb{F}\{\overline{\mathbf{S}}'\}$

M2: $-\Big(((yx)z)x\Big) + \Big(y(x(zx))\Big) = -\Big(((y-x)-z)-x\Big) + \Big(y-(x-(z-x))\Big) = -\Big(y-x-z-x\Big) + \Big(y-x+z-x\Big) = -y+x+z+x+y-x+z-x = 2z = 2$ for $z = 1$ and any $x, y \in \mathbb{F}\{\overline{\mathbf{S}}'\}$

M3: $((xy)(zx)) - \Big(x((yz)x)\Big) = ((x-y)-(z-x)) - \Big(x-((y-z)-x)\Big) = (x-y-z+x) - \Big(x-y+z+x\Big) = (x-y-z+x) - x + y - z - x = -z - z = -2z = 2$ for $z = -1$ and any $x, y \in \mathbb{F}\{\overline{\mathbf{S}}'\}$

(3) obvious

(4) Let $c, d \in \mathbb{Z} : c2 + dp = 1$ so that $2^{-1} = \frac{1}{2} = c \in \mathbb{F}$ whence $1 = c2 \in \mathbb{I}(2) \subset I^{\circ}_{F}$.

□

EXAMPLE 7.4. We intend to prove that $I^{\circ}_{\text{alt}} \subsetneq \mathbb{Z}\{\overline{\mathbf{S}}'\}$ showing that, for each $s \in \mathbf{S}$, $\mathbb{Z}\{\overline{\mathbf{S}}'\} \ni s \notin I^{\circ}_{\text{alt}}$ simply remarking that $s \neq -2y = (xx)y - x(xy)$ for each $y, x \in \mathbb{Z}\{\overline{\mathbf{S}}'\}$.

Consider instead the case $\mathbb{F} := \mathbb{Z}_5$ where we intend to proof that $I^{\circ}_{\text{alt}} = \mathbb{F}\{\overline{\mathbf{S}}'\}$; to do so it is sufficient to show that, for each $s \in \mathbf{S}$, it holds $s \in I^{\circ}_{\text{alt}}$ by considering $x := y := 2s$ getting

$$(xx)y - x(xy) = x - x - y - (x - (x-y)) = 2s - 2s - 2s - (2s - 2s + 2s) = -4s = s$$

and

$$(xy)y - x(yy) = x - y - y - (x - (y-y)) = 2s - 2s - 2s - (2s - 2s + 2s)) = -4s = s.$$

EXAMPLE 7.5. We intend to illustrate here on a trivial example the tag-variables techniques discussed in [**49**, II.26.5] and mainly [**49**, II. Theorem 26.5]; thus let us consider the magma $\mathbb{F}\{\overline{\mathbf{S}'}\}, \mathbf{S} := \{x_1, x_2, x_3\}$, its related reduced Gröbner basis

$$G = \{x_i \circ x_i, 1 \leq i \leq 3\} \cup \{x_i \circ x_j + x_j - x_i, x_j \circ x_i - x_j + x_i, 1 \leq i < j \leq 3\}$$

and the polynomial $f := x_1 - x_3$. According to [**49**, II.26.5] we consider $\mathbf{T} := \{x_1, x_2, x_3, Y\}$ and define the map

$$\gamma : \mathbb{F}\{\overline{\mathbf{T}'}\} \to \mathbb{F}\{\overline{\mathbf{S}}'\} : \gamma(x_i) = x_i, \gamma(Y) = f$$

and we compute the proper Gröbner basis H of the ideal $\ker(\gamma) = \mathbf{I}(Y - f)$ which is

$$\begin{aligned} H =& \{x_i \circ x_i, 1 \leq i \leq 2\} \cup \{x_i \circ x_j + x_j - x_i, x_j \circ x_i - x_j + x_i, 1 \leq i < j \leq 2\} \\ & \cup \{Y - x_1 + \mathbf{x_3}\} \cup \{Y \circ x_i - x_i + Y, x_i \circ Y - Y + x_i, 1 \leq i \leq 3\}. \end{aligned}$$

If we intend to verify under which conditions $g := (x_1 \circ x_3) \circ x_3 + ax_3 \in \mathbb{F}\{\overline{\mathbf{F}}'\} \subset \mathbb{F}\{\overline{\mathbf{S}}'\}$ we perform Buchberger reduction $\mathbf{x_3} \mapsto x_1 - Y$ obtaining

$$\begin{aligned} g &\to (x_1 - x_3) \circ x_3 + ax_3 \to (x_1 - x_3) \circ (x_1 - Y) + ax_3 \\ &= x_1 - x_3 - x_1 + Y + ax_3 = (a-1)x_3 + Y \to (a-1)(x_1 - Y) + Y \\ &= (a-1)x_1 + (2-a)Y \end{aligned}$$

whence the answer is that $g \in \mathbb{F}\{\overline{\mathbf{F}}'\} \iff a = 1$ in which case $g = f$.

7.2. Average. Considering the commutative non-associative[7] binary *average* function

$$\circ : \overline{\mathbf{S}} \times \overline{\mathbf{S}} \mapsto \overline{\mathbf{S}} : a \circ b := \frac{a+b}{2} = \frac{a}{2} + \frac{b}{2}$$

which is defined for each field with $\mathrm{char}(\mathbb{F}) \neq 2$. For a field $\mathbb{F}$ with prime characteristic $\mathrm{char}(\mathbb{F}) = p$, we denote with the symbols $\boxed{\frac{1}{2}}$ and $\boxed{\frac{1}{4}}$ the unique elements such that $\boxed{\frac{1}{2}}2 \equiv \boxed{\frac{1}{4}}4 \equiv 1 \pmod p$ and, in general, with $\boxed{{}^{\mathrm{x}}/_{\mathrm{y}}}$ the unique element such that $\boxed{{}^{\mathrm{x}}/_{\mathrm{y}}}y \equiv x \pmod p$; with this notation we have $a \circ b = \boxed{\frac{1}{2}}a + \boxed{\frac{1}{2}}b$.

COROLLARY 7.6. *Denoting* $\mathbf{S} = \{x_1, \dots, x_n\}$, $A' = (\mathbb{F}\{\mathbf{S}'\}, \cdot) = (V', \cdot) = \mathbb{F}\{\overline{\mathbf{S}}'\}$, $\mathsf{I} := \{f \in \mathbb{F}\{\overline{\mathbf{S}}'\} : L_P(f) = 0, P \in \mathbb{F}^n\} \subset \mathbb{F}\{\overline{\mathbf{S}}'\}$ *the ideal of relation for the average operation,* $\mathsf{A} := \mathbb{F}\langle x_1, \dots, x_n\rangle/\mathsf{I}$, $\mathbf{N}(\mathsf{I}) = \{1, x_1, \dots, x_n\}$, $P_0 = (0, \dots, 0) \in \mathbb{F}^n$, $P_i := (\delta_{i1}, \dots, \delta_{in}) \in \mathbb{F}^n, 1 \leq i \leq n$, $\mathbf{X} := \{P_i : 0 \leq i \leq n\} \subset \mathbb{F}^n$, $I(\mathbf{X}) := \{f \in \mathbb{F}\{\overline{\mathbf{S}}'\} : L_P(f) = 0, P \in \mathbf{X}\}$ *it holds that*

(1) $\mathsf{A} \cong \mathbb{K}[\mathbf{N}(\mathsf{I})] = \mathrm{Span}_{\mathbb{F}}\{1, x_1, \dots, x_n\}$

(2) *The Auzinger-Stetter matrices* [**4**] *related to such representation are*

$$A_{x_i} = {}_{x_i}A = \begin{array}{r|cccccccc} & v_0 & v_1 & \dots & v_i & \dots & v_j & \dots & v_n \\ \hline v_0 = 1 & 0 & 0 & \dots & 1 & \dots & 0 & \dots & 0 \\ v_1 = x_1 & 0 & \boxed{\frac{1}{2}} & \dots & \boxed{\frac{1}{2}} & \dots & 0 & \dots & 0 \\ \vdots & \vdots & \vdots & \ddots & \vdots & \ddots & \vdots & \ddots & \vdots \\ v_i = x_i & 0 & 0 & \dots & 1 & \dots & 0 & \dots & 0 \\ \vdots & \vdots & \vdots & \ddots & \vdots & \ddots & \vdots & \ddots & \vdots \\ v_j = x_j & 0 & 0 & \dots & \boxed{\frac{1}{2}} & \dots & \boxed{\frac{1}{2}} & \dots & 0 \\ \vdots & \vdots & \vdots & \ddots & \vdots & \ddots & \vdots & \ddots & \vdots \\ v_n = x_n & 0 & 0 & \dots & \boxed{\frac{1}{2}} & \dots & 0 & \dots & \boxed{\frac{1}{2}} \end{array}$$

(3) *The reduced Gröbner basis of* I *is* $G = \{x_i \circ x_i - x_i, 1 \leq i \leq n\} \cup \{x_i \circ x_j - \boxed{\frac{1}{2}}x_j - \boxed{\frac{1}{2}}x_i, x_j \circ x_i - \boxed{\frac{1}{2}}x_j - \boxed{\frac{1}{2}}x_i, 1 \leq i < j \leq n\}$.

(4) *There is the following Cerlienco-Mureddu Correspondence*

$$1 \leftrightarrow P_0, x_i \leftrightarrow P_i, 1 \leq i \leq n$$

between $\mathbf{N}(\mathsf{I})$ *and* $\mathbf{X}$

(5) $\dim_{\mathbb{F}}(L_i, 0 \leq i \leq n) = n + 1$.

(6) $\mathbf{N}(I(\mathbf{X})) = \mathbf{S} \cup \{1\}$

(7) $I(\mathbf{X}) = \mathsf{I}$.

---

[7] $a \circ (b \circ c) = \frac{a}{2} + \frac{b}{4} + \frac{c}{4} \neq \frac{a}{4} + \frac{b}{4} + \frac{c}{2} = (a \circ b) \circ c$ for each $a, b, c \in \mathbb{F}, a \neq c$.

REMARK 7.7. We illustrate using average the correctness of the relation, satisfied by each commutative magma,

X: $(xy)(zw) = (xz)(yw)$

by developing each pass

$$\begin{aligned}
(xy)(zw) &= \boxed{\tfrac{1}{2}}(x \circ y) \circ \boxed{\tfrac{1}{2}}(z \circ w) \\
&= \boxed{\tfrac{1}{4}}x + \boxed{\tfrac{1}{4}}y + \boxed{\tfrac{1}{4}}z + \boxed{\tfrac{1}{4}}w \\
&= \boxed{\tfrac{1}{4}}x + \left(\boxed{\tfrac{1}{4}}y + \boxed{\tfrac{1}{4}}z\right) + \boxed{\tfrac{1}{4}}w \\
&= \boxed{\tfrac{1}{4}}x + \left(\boxed{\tfrac{1}{4}}z + \boxed{\tfrac{1}{4}}y\right) + \boxed{\tfrac{1}{4}}w \\
&= \boxed{\tfrac{1}{4}}x + \boxed{\tfrac{1}{4}}z + \boxed{\tfrac{1}{4}}y + \boxed{\tfrac{1}{4}}w \\
&= \boxed{\tfrac{1}{2}}(x \circ z) \circ \boxed{\tfrac{1}{2}}(y \circ w) \\
&= (xz)(yw).
\end{aligned}$$

LEMMA 7.8. *For* $\mathsf{F} \in \{\mathbf{F}\}$, *we have* $I^{\circ}_{\mathsf{F}} = 0$ *and* $A^{\circ}_{\mathsf{F}} = \mathbb{F}\{\overline{\mathbf{S}}'\}$.

PROOF. **F:** $(xy)x - x(yx) = \boxed{\tfrac{1}{2}}(x \circ y) + \boxed{\tfrac{1}{2}}x - \left(\boxed{\tfrac{1}{2}}x + \boxed{\tfrac{1}{2}}(y \circ x)\right) = \boxed{\tfrac{1}{4}}x + \boxed{\tfrac{1}{4}}y + \boxed{\tfrac{1}{2}}x - \left(\boxed{\tfrac{1}{2}}x + \boxed{\tfrac{1}{4}}y + \boxed{\tfrac{1}{4}}x)\right) = 0.$

□

LEMMA 7.9. *For* $\mathsf{F} \in \{\mathbf{A1}, \mathbf{A2}, \mathbf{M1}, \mathbf{M2}, \mathbf{M3}\}$ *we have*

$$I^{\circ}_{\text{alt}} = I^{\circ}_{\mathsf{F}} = \mathbb{I}(x_i - x_1, 1 < i \le n) \text{ and } A^{\circ}_{\text{alt}} = A^{\circ}_{\mathsf{F}} = \mathbb{F}\{\overline{\{\mathbf{x_1}\}}\}.$$

PROOF. It holds

A1: $(xx)y - x(xy) = \boxed{\tfrac{1}{2}}(x \circ x) + \boxed{\tfrac{1}{2}}y - \left(\boxed{\tfrac{1}{2}}x + \boxed{\tfrac{1}{2}}(x \circ y)\right) = \boxed{\tfrac{1}{2}}x + \boxed{\tfrac{1}{2}}y - \left(\boxed{\tfrac{1}{2}}x + \boxed{\tfrac{1}{4}}x + \boxed{\tfrac{1}{4}}y\right) =$
$= \boxed{\tfrac{1}{2}}x + \boxed{\tfrac{1}{2}}y - \boxed{\tfrac{1}{2}}x - \boxed{\tfrac{1}{4}}x - \boxed{\tfrac{1}{4}}y = \boxed{\tfrac{1}{4}}(y - x),$

A2: $(xy)y - x(yy) = \boxed{\tfrac{1}{4}}x + \boxed{\tfrac{1}{4}}y + \boxed{\tfrac{1}{2}}y - \left(\boxed{\tfrac{1}{2}}x + \boxed{\tfrac{1}{2}}y\right) = (\boxed{\tfrac{1}{4}} - \boxed{\tfrac{1}{2}})x + (\boxed{\tfrac{1}{4}} + \boxed{\tfrac{1}{2}} - \boxed{\tfrac{1}{2}})y = \boxed{\tfrac{1}{4}}(y - x),$

M1: $(x(zx))y - x(z(xy)) = \boxed{\tfrac{1}{2}}(x \circ (z \circ x)) + \boxed{\tfrac{1}{2}}y - \left(\boxed{\tfrac{1}{2}}x + \boxed{\tfrac{1}{2}}(z \circ (x \circ y))\right) = \boxed{\tfrac{1}{4}}x + \boxed{\tfrac{1}{4}}(z \circ x) + \boxed{\tfrac{1}{2}}y - \left(\boxed{\tfrac{1}{2}}x + \boxed{\tfrac{1}{4}}z + \boxed{\tfrac{1}{4}}(x \circ y)\right) = \boxed{\tfrac{1}{4}}x + \boxed{1/8}z + \boxed{1/8}x + \boxed{\tfrac{1}{2}}y - \left(\boxed{\tfrac{1}{2}}x + \boxed{\tfrac{1}{4}}z + \boxed{1/8}x + \boxed{1/8}y\right) = \left(\boxed{\tfrac{1}{4}} + \boxed{1/8} - \boxed{\tfrac{1}{2}} - \boxed{1/8}\right)x + \left(\boxed{\tfrac{1}{2}} - \boxed{1/8}\right)y + \left(\boxed{1/8} - \boxed{\tfrac{1}{4}}\right)z = -\boxed{\tfrac{1}{4}}x + \boxed{3/8}y - \boxed{1/8}z,$

M2: $\left(((yx)z)x\right) - \left(y(x(zx))\right) = \left(\boxed{\tfrac{1}{2}}((y \circ x) \circ z) + \boxed{\tfrac{1}{2}}x\right) - \left(\boxed{\tfrac{1}{2}}y + \boxed{\tfrac{1}{2}}(x \circ (z \circ x))\right) = \left(\boxed{1/8}y + \boxed{1/8}x + \boxed{\tfrac{1}{4}}z + \boxed{\tfrac{1}{2}}x\right) - \left(\boxed{\tfrac{1}{2}}y + \boxed{\tfrac{1}{4}}x + \boxed{1/8}z + \boxed{1/8}x\right) = \left(\boxed{1/8} + \boxed{\tfrac{1}{2}} - \boxed{\tfrac{1}{4}} - \boxed{1/8}\right)x + \left(\boxed{1/8} - \boxed{\tfrac{1}{2}}\right)y + \left(\boxed{\tfrac{1}{4}} - \boxed{1/8}\right)z = \boxed{\tfrac{1}{4}}x - \boxed{3/8}y + \boxed{1/8}z,$

M3: $((xy)(zx)) - \Big(x((yz)x)\Big) = \Big(\boxed{\frac{1}{4}}x + \boxed{\frac{1}{4}}y + \boxed{\frac{1}{4}}z + \boxed{\frac{1}{4}}x\Big) - \Big(\boxed{\frac{1}{2}}x + \boxed{1/8}y + \boxed{1/8}z + \boxed{\frac{1}{4}}x\Big) = \Big(\boxed{\frac{1}{4}} + \boxed{\frac{1}{4}} - \boxed{\frac{1}{2}} - \boxed{\frac{1}{4}}\Big)x + \Big(\boxed{\frac{1}{4}} - \boxed{1/8}\Big)y + \Big(\boxed{\frac{1}{4}} - \boxed{1/8}\Big)z = -\boxed{\frac{1}{4}}x + \boxed{1/8}y + \boxed{1/8}z.$

□

7.3. Division. For the non-associative[8] binary *division* function

$$\circ : \overline{\mathbf{S}} \times \overline{\mathbf{S}} \mapsto \overline{\mathbf{S}} : a \circ b := \frac{a}{b}$$

it is sufficient to remark that $x \circ x = \frac{x}{x} = 1$ to obtain an elementary Gröbner and Möller solution.

But, to do so, we need to define a proper notation for a correspondence between the set of all points $\mathbf{X} = \mathbb{Z}_2^n$, the set of integers $\{i \in \mathbb{N}, 1 \leq i \leq 2^n\}$ and the set of monomials $\mathbf{N} := \{x_1^{\alpha_1}, \dots, x_n^{\alpha_n} \in \mathbb{F}\{\overline{\mathbf{S}}\}, \alpha_i \in \mathbb{Z}_2\}$. To do so we express each integer $i : 0 \leq i < 2^n$ as

$$i = \sum_{j=1}^{n} \alpha_j 2^{j-1}, \text{ with } 0 \leq \alpha_j \leq 1 \text{ for each } j, 1 \leq j \leq n$$

and set the correspondence

$$\begin{aligned}\sigma : \{i : i \in \mathbb{N}, 1 \leq i \leq 2^n\} \mapsto \mathbb{Z}_2^n = \mathbf{X} \leftrightarrow \mathbf{N} : i& \\ = 1 + \sum_{j=1}^{n} \alpha_j 2^{j-1} \leftrightarrow \sigma(i) = (\alpha_1, \dots, \alpha_n) &\leftrightarrow x_1^{\alpha_1}, \dots, x_n^{\alpha_n}.\end{aligned}$$

Denoting, for each $i \in \mathbb{N}, 1 \leq i \leq 2^n$,

- $P_i := \sigma(i) = (\alpha_1, \dots, \alpha_n) \in \mathbf{X} = \mathbb{Z}_2^n$,
- $m_i := x_1^{\alpha_1}, \dots, x_n^{\alpha_n} \in \mathbf{N} \subset \mathbb{F}\{\overline{\mathbf{S}}\}$,
- $L_i := L_{P_i}$ the related functional,

where we defined inductively the functionals $L_P : \mathbb{F}\{\overline{\mathbf{S}}\} \to \mathbb{F}, P := (a_1, \dots, a_n) \in \mathbb{F}^{*n}$ as

$$\begin{array}{lcccll} L_P(1) & = & & = & 1, & \\ L_P(x_i) & = & & = & a_i, & 1 \leq i \leq n, \\ L_P(t_i \circ t_j) & = & L_P(t_i t_j) & = & \frac{L_P(t_i)}{L_P(t_j)}, & t_i, t_j \in \overline{\mathbf{S}}, \\ L_P(\sum_{i=1}^{u} c_i t_i) & = & & = & \sum_{i=1}^{u} c_i L_P(t_i), & \sum_{i=1}^{u} c_i t_i \in \mathbb{F}\{\overline{\mathbf{S}'}\}. \end{array}$$

we obtain

COROLLARY 7.10. *Denoting* $\mathbf{S} = \{x_1, \dots, x_n\}$, $A' = (\mathbb{F}\{\mathbf{S}'\}, \cdot/\cdot) = (V', \cdot/\cdot) = \mathbb{F}\{\overline{\mathbf{S}}'\}$, $\mathsf{I} := \{f \in \mathbb{F}\{\overline{\mathbf{S}}'\} : L_P(f) = 0, P \in \mathbb{F}^n\} \subset \mathbb{F}\{\overline{\mathbf{S}}'\}$ *the ideal of relation for the operation of division,* $\mathsf{A} := \mathbb{F}\langle x_1, \dots, x_n\rangle/\mathsf{I}$, $\mathbf{N}(\mathsf{I}) = \mathbf{N} = \{x_1^{\alpha_1}, \dots, x_n^{\alpha_n} \in \mathbb{F}\{\overline{\mathbf{S}}\}, \alpha_i \in \mathbb{Z}_2\}$, $\mathbf{X} := \mathbb{Z}_2^n := \{P_i, 1 \leq i \leq 2^n\}$, $\Lambda = \{L_i, 1 \leq i < 2^n\}$, $I(\mathbf{X}) := \{f \in \mathbb{F}\{\overline{\mathbf{S}}\} : L_P(f) = 0, P \in \mathbf{X}\}$ *and for each* $i, j, 1 \leq i \leq n, 1 \leq j \leq 2^n$, $\gamma_{iju} \in \mathbb{F}, \forall u, 1 \leq u \leq 2^n$, *such that*

$$x_i \circ x_j = \gamma_{ij0} + \sum_{u=1}^{n} \gamma_{iju} x_u$$

it holds that

[8] $a \circ (b \circ c) = a \circ \frac{b}{c} = \frac{ac}{b} \neq \frac{a}{bc} = \frac{a}{b} \circ c = (a \circ b) \circ c$ for $a \in \mathbb{F}, b, c \in \mathbb{F}^*$.

(1) $\mathsf{A} \cong \mathbb{F}[\mathbf{N}(\mathsf{I})] = \mathrm{Span}_{\mathbb{F}}\{x_1^{\alpha_1} \cdots x_n^{\alpha_n} : \alpha_j \in \{0,1\}\}$

(2) *The Auzinger-Stetter matrices* [**4**] *related to such representation are*

$$A_{x_i} = \begin{array}{r|cccccccccccc}
 & v_0 & v_1 & \dots & v_i & \dots & v_j & \dots & v_n & \dots & v_\ell & \dots & v_{2^n} \\ \hline
v_0 = 1 & 0 & 0 & \dots & 1 & \dots & 0 & \dots & 0 & \dots & 0 & \dots & 0 \\
v_1 = x_1 & \gamma_{i10} & \gamma_{i11} & \cdots & \gamma_{i1i} & \cdots & \gamma_{i1j} & \cdots & \gamma_{i1n} & \cdots & \gamma_{i1\ell} & \cdots & \gamma_{i12^n} \\
\vdots & \vdots & \vdots & \ddots & \vdots & \ddots & \vdots & \ddots & \vdots & \dots & \vdots & \dots & \vdots \\
v_i = x_i & 1 & 0 & \dots & 0 & \dots & 0 & \dots & 0\dots & 0 & \dots & 0 & \\
\vdots & \vdots & \vdots & \ddots & \vdots & \ddots & \vdots & \ddots & \vdots & \dots & \vdots & \dots & \vdots \\
v_j = x_j & \gamma_{ij0} & \gamma_{ij1} & \cdots & \gamma_{iji} & \cdots & \gamma_{ijj} & \cdots & \gamma_{ijn} & \cdots & \gamma_{ij\ell} & \cdots & \gamma_{ij2^n} \\
\vdots & \vdots & \vdots & \ddots & \vdots & \ddots & \vdots & \ddots & \vdots & \dots & \vdots & \dots & \vdots \\
v_n = x_n & \gamma_{in0} & \gamma_{in1} & \cdots & \gamma_{ini} & \cdots & \gamma_{inj} & \cdots & \gamma_{inn} & \cdots & \gamma_{in\ell} & \cdots & \gamma_{in2^n} \\
\vdots & \vdots & \vdots & \ddots & \vdots & \ddots & \vdots & \ddots & \vdots & \dots & \vdots & \dots & \vdots \\
v_\ell = m_\ell = x_1^{\alpha_1} \cdots x_n^{\alpha_n} & \gamma_{i\ell 0} & \gamma_{i\ell 1} & \cdots & \gamma_{i\ell i} & \cdots & \gamma_{i\ell j} & \cdots & \gamma_{i\ell n} & \cdots & \gamma_{i\ell\ell} & \cdots & \gamma_{i\ell 2^n} \\
\vdots & \vdots & \vdots & \ddots & \vdots & \ddots & \vdots & \ddots & \vdots & \dots & \vdots & \dots & \vdots \\
v_{2^n} = x_1 \dots x_n & \gamma_{i2^n0} & \gamma_{i2^n1} & \cdots & \gamma_{i2^ni} & \cdots & \gamma_{i2^nj} & \cdots & \gamma_{i2^nn} & \cdots & \gamma_{i2^n\ell} & \cdots & \gamma_{i2^n2^n}
\end{array},$$

$$_{x_i}A = \begin{array}{r|cccccccccccc}
 & v_0 & v_1 & \dots & v_i & \dots & v_j & \dots & v_n & \dots & v_\ell & \dots & v_{2^n} \\ \hline
v_0 = 1 & 0 & 0 & \dots & 1 & \dots & 0 & \dots & 0 & \dots & 0 & \dots & 0 \\
v_1 = x_1 & \gamma_{1i0} & \gamma_{1i1} & \cdots & \gamma_{1ii} & \cdots & \gamma_{1ij} & \cdots & \gamma_{1in} & \cdots & \gamma_{1i\ell} & \cdots & \gamma_{1i2^n} \\
\vdots & \vdots & \vdots & \ddots & \vdots & \ddots & \vdots & \ddots & \vdots & \dots & \vdots & \dots & \vdots \\
v_i = x_i & 1 & 0 & \dots & 0 & \dots & 0 & \dots & 0\dots & 0 & \dots & 0 & \\
\vdots & \vdots & \vdots & \ddots & \vdots & \ddots & \vdots & \ddots & \vdots & \dots & \vdots & \dots & \vdots \\
v_j = x_j & \gamma_{ji0} & \gamma_{ji1} & \cdots & \gamma_{jii} & \cdots & \gamma_{jij} & \cdots & \gamma_{jin} & \cdots & \gamma_{ji\ell} & \cdots & \gamma_{ji2^n} \\
\vdots & \vdots & \vdots & \ddots & \vdots & \ddots & \vdots & \ddots & \vdots & \dots & \vdots & \dots & \vdots \\
v_n = x_n & \gamma_{ni0} & \gamma_{ni1} & \cdots & \gamma_{nii} & \cdots & \gamma_{nij} & \cdots & \gamma_{nin} & \cdots & \gamma_{ni\ell} & \cdots & \gamma_{ni2^n} \\
\vdots & \vdots & \vdots & \ddots & \vdots & \ddots & \vdots & \ddots & \vdots & \dots & \vdots & \dots & \vdots \\
v_\ell = m_\ell = x_1^{\alpha_1} \cdots x_n^{\alpha_n} & \gamma_{\ell i0} & \gamma_{\ell i1} & \cdots & \gamma_{\ell ii} & \cdots & \gamma_{\ell ij} & \cdots & \gamma_{\ell in} & \cdots & \gamma_{\ell i\ell} & \cdots & \gamma_{\ell i2^n} \\
\vdots & \vdots & \vdots & \ddots & \vdots & \ddots & \vdots & \ddots & \vdots & \dots & \vdots & \dots & \vdots \\
v_{2^n} = x_1 \dots x_n & \gamma_{2^ni0} & \gamma_{2^ni1} & \cdots & \gamma_{2^nii} & \cdots & \gamma_{2^nij} & \cdots & \gamma_{2^nin} & \cdots & \gamma_{2^ni\ell} & \cdots & \gamma_{2^ni2^n}
\end{array}.$$

(3) *The reduced Gröbner basis of* I *is* $G = \{x_i \circ x_i - 1, 1 \le i \le n\} \cup \{x_i \circ x_j - \gamma_{ij0} - \sum_{u=1}^n \gamma_{iju} x_u, 1 \le i, j \le n\}$.

(4) *There is the Cerlienco-Mureddu Correspondence* $m_i \leftrightarrow P_i, 1 \le i \le 2^n$ *between* $\mathbf{N}(\mathsf{I})$ *and* $\mathbf{X}$.

(5) $\dim_{\mathbb{F}}(L_i, 1 \le i \le 2^n) = 2^n$.

(6) $I(\mathbf{X}) = \mathsf{I}$.

LEMMA 7.11. *For* $F \in \left\{\mathbf{A1}, \mathbf{A2}, \mathbf{F}, \mathbf{M1}, \mathbf{M2}, \mathbf{M3}\right\}$, *it holds* $I^\circ_{\text{alt}} = I^\circ_F = 0$ *and* $A^\circ_{\text{alt}} = A^\circ_F = \mathbb{F}\{\overline{\mathbf{S}}'\}$.

PROOF. **A1:** $(xx)y - x(xy) = \frac{x}{yx} - \frac{xy}{x} = \frac{1}{y} - y = \frac{1-y^2}{y} = 0$,

A2: $(xy)y - x(yy) = \frac{x}{y^2} - \frac{xy}{y} = \frac{x}{y^2} - \frac{x}{1} = x\left(\frac{1}{y^2} - 1\right) = x\left(\frac{1-y^2}{y^2}\right) = \frac{x(1-y^2)}{y^2} = 0$,

F: $(xy)x - x(yx) = \frac{x}{yx} - \frac{x^2}{y} = \frac{1}{y} - \frac{x^2}{y} = \frac{1-x^2}{y} = 0$,

M1: $(x(zx))y - x(z(xy)) = \frac{x^2}{zy} - \frac{x^2}{zy} = 0$,

M2: $\Big(((yx)z)x\Big) - \Big(y(x(zx)\Big) = \frac{y}{zx^2} - \frac{yz}{x^2} = \frac{y(1-z^2)}{zx^2} = 0$,

M3: $((xy)(zx)) - \Big(x((yz)x)\Big) = \frac{x^2}{yz} - \frac{x^2 z}{y} = \frac{x^2(1-z^2)}{yz} = 0$.

□

EXAMPLE 7.12. Let us consider the case $\overline{\mathbf{S}} = \{x, y\}$ and $\mathbb{F} := \mathbb{F}_3$ so we have 4 points

$$\mathbf{X} := \{(a,b) = P_{a+2b-2}, a, b \in \{1,2\}\} \subset {\mathbb{F}_3}^* \times {\mathbb{F}_3}^*$$

and the related functionals $\Lambda = \{L_1, \dots, L_4\}$, $L_{(a,b)} = L_{a+2b-2} = L_{P_{a+2b-2}} = L_{P_{(a,b)}}$.

$$P_1 = (1,1)$$

	$(1,1)$	$(2,1)$	$(1,2)$	$(2,2)$
1	**1**	1	1	1
$x-1$	0	1	0	1
$y-1$	0	0	1	1

,

$$\mathcal{B} = \{x-1, y-1\} \quad \mathcal{A} = \{v_1 = 1\}$$

$$P_2 = (2,1)$$

	$(1,1)$	$(2,1)$	$(1,2)$	$(2,2)$
1	**1**	1	1	1
$x-1$	0	**1**	0	1
$y-1$	0	0	1	1
x^2-1	0	0	0	0
xy	1	-1	-1	1
$xy-1$	0	1	1	0
$xy-x$	0	0	1	-1
yx	1	-1	-1	1
$yx-1$	0	1	1	0
$yx-x$	0	0	1	-1

,

$$\mathcal{B} = \{y-1, x^2-1, xy-x, yx-x\}$$

$$\mathcal{A} = \{v_1 = 1, v_2 = x-1; x^2-1, , xy-x, yx-x\}$$

$P_3 = (1,2)$

	$(1,1)$	$(2,1)$	$(1,2)$	$(2,2)$
1	**1**	1	1	1
$x-1$	0	**1**	0	1
$y-1$	0	0	**1**	1
x^2-1	0	0	0	0
$xy-x$	0	0	1	-1
$xy-x-y+1$	0	0	0	1
$yx-x$	0	0	1	-1
$yx-x-y+1$	0	0	0	1
y^2	1	1	1	1
y^2-1	0	0	0	0

,

$$\mathcal{B} = \{x^2-1, xy-x-y+1, yx-x-y+1, y^2-1\}$$

$$\mathcal{A} = \{v_1 = 1, v_2 = x-1, v_3 = y-1; x^2-1, y^2-1, xy-x-y+1, yx-x-y+1\}$$

$P_4 = (2,2)$

	$(1,1)$	$(2,1)$	$(1,2)$	$(2,2)$
1	$\mathbf{1}$	1	1	1
$x-1$	0	$\mathbf{1}$	0	1
$y-1$	0	0	$\mathbf{1}$	1
$xy-x-y+1$	0	0	0	$\mathbf{1}$
x^2-1	0	0	0	0
$yx-x-y+1$	0	0	0	1
$yx-xy$	0	0	0	0
y^2-1	0	0	0	0
$x(xy)$	1	1	-1	-1
$x(xy)-y$	0	0	0	0
$(xy)x$	1	1	-1	-1
$(xy)x-y$	0	0	0	0
$y(xy)$	1	-1	1	-1
$y(xy)-x$	0	0	0	0
$(xy)y$	1	-1	1	-1
$(xy)y-x$	0	0	0	0
$(xy)(xy)$	1	1	1	1
$(xy)(xy)-1$	0	0	0	0

$$\mathcal{B} = \{x^2-1, yx-xy, y^2-1, x(xy)-y, (xy)x-y, y(xy)-x, (xy)y-x, (xy)(xy)-1\}$$

$$\mathcal{A} = \{v_1 = 1, v_2 = x-1, v_3 = y-1, v_4 = xy-x-y+1; x^2-1, yx-xy, y^2-1,$$
$$x(xy)-y, (xy)x-y, y(xy)-x, (xy)y-x, (xy)(xy)-1x^2-1, y^2-1x^2-1,$$
$$yx-xy, y^2-1\}$$

Let us perform Buchberger reduction on

$$\begin{aligned}
&\Big(((\underline{x(xy)})y)((xy)(yx))\Big)\Big(((y(xy))y)((yy)(xx))\Big)\\
&\to \Big(((\mathbf{y})y)((xy)\underline{(yx)})\Big)\Big(((y(xy))y)((yy)(xx))\Big)\\
&\to \Big((y^2)((xy)(\mathbf{xy}))\Big)\Big(((\underline{y(xy)}y)((yy)(xx))\Big)\\
&\to \Big((y^2)((xy)xy)\Big)\Big(((\mathbf{x}y)(\underline{(yy)}(xx))\Big)\\
&\to \Big((y^2)((xy)(xy))\Big)\Big((xy)(\mathbf{1}\underline{(xx)})\Big)\\
&\to \Big(\underline{(y^2)}((xy)(xy))\Big)\Big((xy)\mathbf{1}\Big)\\
&\to \Big(\mathbf{1}\underline{((xy)(xy))}\Big)\Big((xy)1\Big)\\
&\to \Big(\mathbf{1}\Big)\Big(\underline{(xy)}\Big)
\end{aligned}$$

and then we rewrite it with respect to the basis $\{v_1, \dots, v_4\}$:

$$\begin{aligned}
xy &\to \mathbf{v_4} + \mathbf{x} + \mathbf{y} - \mathbf{1}\\
&\to v_4 + \underline{y} + x - 1\\
&\to v_4 + \mathbf{v_3} + \mathbf{1} + \underline{x} - 1\\
&\to v_4 + v_3 + \mathbf{v_2} + \mathbf{1}\\
&\to v_4 + v_3 + v_2 + \underline{1}
\end{aligned}$$

$$\to v_4 + v_3 + v_2 + v_1.$$

DEFINITION 7.13. In an algebra A with unit element e any three elements $u, v, w \in A$ which satisfy the Hamiltonian conditions **[38, 39]** $u^2 = v^2 = w^2 = -e, w = uv = -vu, u = vw = -wv, v = wu = -uw$ are called a *Hamiltonian triple*

REMARK 7.14. If we set $n = 3$, we have $\#\mathbf{X} = 8$ and since we have $x_i \circ x_i = \frac{x_i}{x_i} = 1, 1 \leq i \leq 3$, and for each $x, y \in \overline{\mathbf{S}} = \{x_1, x_2, x_3\}$ it holds

$$x \circ y = \frac{x}{y} \implies (x \circ y) \circ \frac{x}{y} = \frac{x}{y} \circ \frac{x}{y} = 1$$

we can conclude that they generate a Hamiltonian triple and thus $\#\mathbf{X}$ is a model of the octonions (Section 7.8); the relation is described below.

x_2^2			
x_2	j	k	
1	1	i	
1	1	x_1	x_1^2

x_2^2			
x_2	jl	kl	
1	l	il	
x_3	1	x_1	x_1^2

7.4. Set difference. For the non-associative[9] nor commutative[10] binary *set difference* function

$$\circ : \overline{\mathbf{S}} \times \overline{\mathbf{S}} \mapsto \overline{\mathbf{S}} : A \circ B := A \setminus B = \{x \in A : x \notin B\}, A, B \subset \mathbf{S} = \{x_1, \ldots, x_n\},$$

we can adapt the notation settled for division (Section 7.3); to do so we associate each subset

$$A := \{x_{i_1}, \cdots, x_{i_m} : 1 \leq i_1 < \cdots < i_m \leq n\} \subset \mathbf{S} = \{x_1, \ldots, x_n\}$$

to the monomial $m := x_{i_1} \cdots x_{i_m} \in \mathbf{N} := \{x_1^{\alpha_1} \cdots x_n^{\alpha_n} \in \mathbb{F}\{\overline{\mathbf{S}}\}, \alpha_i \in \mathbb{Z}_2\}$ and enumerate them using the correspondence

$$\begin{aligned}\sigma :& \{i : i \in \mathbb{N}, 1 \leq i \leq 2^n\} \mapsto \mathbb{Z}_2^n \mapsto \mathbf{N} \overset{\rho}{\leftarrow} \{A \subset \mathbf{S}\} : \\ & i = 1 + \sum_{j=1}^{n} \alpha_j 2^{j-1} \to \sigma(i) \\ & = (\alpha_1, \ldots, \alpha_n) \leftrightarrow x_1^{\alpha_1}, \ldots, x_n^{\alpha_n} \overset{\rho}{\leftarrow} \{x_i, 1 \leq i \leq n : \alpha_i = 1\}.\end{aligned}$$

If we specialize to the case $n = 3$ we have

	v_0	v_1	v_2	v_3	v_4	v_5	v_6	v_7
$v_0 = \emptyset$	v_0	v_0	v_0	v_0	v_0	v_0	v_0	v_0
$v_1 = \{x_1\}$	v_1	v_0	v_1	v_0	v_1	v_0	v_1	v_0
$v_2 = \{x_2\}$	v_2	v_2	v_0	v_0	v_2	v_2	v_0	v_0
$v_3 = \{x_1, x_2\}$	v_3	v_2	v_1	v_0	v_3	v_2	v_1	v_0
$v_4 = \{x_3\}$	v_4	v_4	v_4	v_4	v_0	v_0	v_0	v_0
$v_5 = \{x_1, x_3\}$	v_5	v_4	v_5	v_4	v_1	v_0	v_1	v_0
$v_6 = \{x_2, x_3\}$	v_6	v_6	v_4	v_4	v_2	v_2	v_0	v_0
$v_7 = \{x_1, x_2, x_3\}$	v_7	v_6	v_5	v_4	v_3	v_2	v_1	v_0

Denoting, for each $i \in \mathbb{N}, 1 \leq i \leq 2^n$,

- $m_i := x_1^{\alpha_1} \cdots x_n^{\alpha_n} \in \mathbf{N} \subset \mathbb{F}\{\overline{\mathbf{S}}\}$,

[9] $\{x_1, x_2\} \circ (\{x_2, x_3\} \circ \{x_3, x_1\}) = \{x_1, x_2\} \circ \{x_2\} = \{x_1\} \neq (\{x_1, x_2\} \circ \{x_2, x_3\}) \circ \{x_3, x_1\} = \emptyset$.

[10] $\{x_1, x_2, x_3\} \circ \{x_1\} = \{x_2, x_3\} \neq \emptyset = \{x_1\} \circ \{x_1, x_2, x_3\}$.

- $P_i := \sigma(i) = (\alpha_1, \ldots, \alpha_n) \in \mathbf{X} = \mathbb{Z}_2^n$,
- $L_i := L_{P_i}$ the related functional,

where we defined inductively the functionals $L_P : \mathbb{F}\{\overline{\mathbf{S}}\} \to \mathbb{F}, P := (a_1, \ldots, a_n) \in \mathbb{F}^n$ as

$$\begin{array}{llll} L_P(1) & = & = 1, & \\ L_P(x_i) & = & = a_i, & 1 \le i \le n, \\ L_P(t_i \circ t_j) & = L_P(t_i t_j) & = L_P\left(\rho\left(\rho^{-1}(t_i) \setminus \rho^{-1}(t_j)\right)\right), & t_i, t_j \in \overline{\mathbf{S}}, \\ L_P(\sum_{i=1}^u c_i t_i) & = & = \sum_{i=1}^u c_i L_P(t_i), & \sum_{i=1}^u c_i t_i \in \mathbb{F}\{\overline{\mathbf{S}}\}. \end{array}$$

we obtain

COROLLARY 7.15. *Denoting* $\mathbf{S} = \{x_1, \ldots, x_n\}$, $A' = (\mathbb{F}\{\mathbf{S}'\}, \setminus) = (V', \setminus) = \mathbb{F}\{\overline{\mathbf{S}}'\}$, $\mathsf{I} := \{f \in \mathbb{F}\{\overline{\mathbf{S}}'\} : L_P(f) = 0, P \in \mathbb{F}^n\} \subset \mathbb{F}\{\overline{\mathbf{S}}'\}$ *the ideal of relation for the operation of set difference,* $\mathsf{A} := \mathbb{F}\langle x_1, \ldots, x_n\rangle/\mathsf{I}$, $\mathbf{N}(\mathsf{I}) = \{1, x_1, \ldots, x_n\}$, $P_0 = (0, \ldots, 0) \in \mathbb{F}^n$, $P_i := (\delta_{i1}, \ldots, \delta_{in}) \in \mathbb{F}^n, 1 \le i \le n$, $\mathbf{X} := \{P_i : 0 \le i \le n\} \subset \mathbb{F}^n$, $I(\mathbf{X}) := \{f \in \mathbb{F}\{\overline{\mathbf{S}}'\} : L_P(f) = 0, P \in \mathbf{X}\}$ *it holds that*

(1) $\mathsf{A} \cong \mathbb{F}[\mathbf{N}(\mathsf{I})] = \mathrm{Span}_{\mathbb{F}}\{x_1^{\alpha_1} \cdots x_n^{\alpha_n} : \alpha_j \in \{0, 1\}\}$
(2) *The Auzinger-Stetter matrices* [4] *related to such representation are*

$$A_{x_i} = \begin{array}{c|ccccccccc} & v_0 & v_1 & \ldots & v_i & \ldots & v_j & \ldots & v_n \\ \hline v_0 = 1 & 0 & 0 & \ldots & 1 & \ldots & 0 & \ldots & 0 \\ v_1 = x_1 & 0 & 0 & \ldots & 1 & \ldots & 0 & \ldots & 0 \\ \vdots & \vdots & \vdots & \ddots & \vdots & \ddots & \vdots & \ddots & \vdots \\ v_i = x_i & 1 & 0 & \ldots & 0 & \ldots & 0 & \ldots & 0 \\ \vdots & \vdots & \vdots & \ddots & \vdots & \ddots & \vdots & \ddots & \vdots \\ v_j = x_j & 0 & 0 & \ldots & 1 & \ldots & 0 & \ldots & 0 \\ \vdots & \vdots & \vdots & \ddots & \vdots & \ddots & \vdots & \ddots & \vdots \\ v_n = x_n & 0 & 0 & \ldots & 1 & \ldots & 0 & \ldots & 0 \end{array},$$

$$_{x_i}A = \begin{array}{c|ccccccccc} & v_0 & v_1 & \ldots & v_i & \ldots & v_j & \ldots & v_n \\ \hline v_0 = 1 & 1 & 0 & \ldots & 0 & \ldots & 0 & \ldots & 0 \\ v_1 = x_1 & 0 & 1 & \ldots & 0 & \ldots & 0 & \ldots & 0 \\ \vdots & \vdots & \vdots & \ddots & \vdots & \ddots & \vdots & \ddots & \vdots \\ v_i = x_i & 1 & 0 & \ldots & 0 & \ldots & 0 & \ldots & 0 \\ \vdots & \vdots & \vdots & \ddots & \vdots & \ddots & \vdots & \ddots & \vdots \\ v_j = x_j & 0 & 0 & \ldots & 0 & \ldots & 1 & \ldots & 0 \\ \vdots & \vdots & \vdots & \ddots & \vdots & \ddots & \vdots & \ddots & \vdots \\ v_n = x_n & 0 & 0 & \ldots & 0 & \ldots & 0 & \ldots & 1 \end{array}.$$

(3) *The reduced Gröbner basis of* I *is* $G = \{x_i \circ x_i - 1, 1 \le i \le n\} \cup \{x_i \circ x_j - x_i, 1 \le i, j \le n\}$.
(4) *There is the Cerlienco-Mureddu Correspondence* $m_i \leftrightarrow P_i, 0 \le i \le n$ *between* $\mathbf{N}(\mathsf{I})$ *and* $\mathbf{X}$.
(5) $\dim_{\mathbb{F}}(L_i, 1 \le i \le 2^n) = n + 1$.
(6) $I(\mathbf{X}) = \mathsf{I}$.

LEMMA 7.16. *For $F \in \Big\{\mathbf{F}\Big\}$ we have $I^\circ_F = \mathbb{I}(x_1, \dots, x_n)$ and $A^\circ_F = \mathbb{F}$*

PROOF. **F:** $(xy)x - x(yx) = \Big(\{x\} \setminus \{y\}\Big) \setminus \{x\} - \{x\} \setminus \Big(\{y\} \setminus \{x\}\Big) = \{x\} \setminus \{x\} - \{x\} \setminus \{y\} = \emptyset - \{x\} = -\{x\}$. $\square$

LEMMA 7.17. *For $F \in \Big\{\mathbf{A1}, \mathbf{A2}, \mathbf{M1}, \mathbf{M2}, \mathbf{M3}\Big\}$ we have $I^\circ_{\text{alt}} = I^\circ_F = 0$ and $A^\circ_{\text{alt}} = A^\circ_F = \mathbb{F}\{\overline{\mathbf{S}}'\}$.*

PROOF. **A1:** $(xx)y - x(xy) = \Big(\{x\} \setminus \{x\}\Big) \setminus \{y\} - \{x\} \setminus \Big(\{x\} \setminus \{y\}\Big) = \emptyset \setminus \{y\} - \{x\} \setminus \{x\} = \emptyset$,

A2: $(xy)y - x(yy) = \Big(\{x\}\setminus\{y\}\Big)\setminus\{y\} - \{x\}\setminus\Big(\{y\}\setminus\{y\}\Big) = \{x\}\setminus\{y\} - \{x\}\setminus\emptyset = \{x\} - \{x\} = 0$,

M1: $(x(zx))y - x(z(xy)) = \Big(\{x\}\setminus(\{z\} \setminus \{x\})\Big)\setminus\{y\} - \{x\}\setminus\Big(\{z\}\setminus(\{x\} \setminus \{y\})\Big) = \Big(\{x\}\setminus\{z\}\Big)\setminus\{y\} - \{x\}\setminus\Big(\{z\}\setminus\{x\}\Big) = \{x\}\setminus\{y\} - \{x\}\setminus\{z\} = \{x\} - \{x\} = 0$,

M2: $\Big(((yx)z)x\Big) - \Big(y(x(zx))\Big) = \Big((\{y\} \setminus \{x\}) \setminus \{z\}\Big) \setminus \{x\} - \Big(\{y\}\setminus((\{x\} \setminus (\{z\}) \setminus \{x\})\Big) = \Big(\{y\}\setminus\{z\}\Big)\setminus\{x\} - \Big(\{y\}\setminus(\{x\} \setminus \{z\})\Big) = \{y\} \setminus \{x\} - \{y\} \setminus \{x\} = \{y\} - \{y\} = 0$,

M3: $\Big((xy)(zx)\Big) - \Big(x\,((yz)x)\Big) = \Big((\{x\} \setminus \{y\}) \setminus (\{z\} \setminus \{x\}\Big) - \Big(\{x\} \setminus ((\{y\} \setminus (\{z\}) \setminus \{x\})\Big) = \Big(\{x\} \setminus \{z\}\Big) - \Big(\{x\} \setminus (\{y\} \setminus \{x\})\Big) = \{x\} - \Big(\{x\} \setminus \{y\}\Big) = \{x\} - \{x\} = 0$. $\square$

EXAMPLE 7.18. Let us consider the case $\overline{\mathbf{S}} = \{x, y, z\}$ and $\mathbb{F} := \mathbb{F}_2$ so we have 8 points

$$\mathbf{X} := \{(a, b, c) = P_{a+2b+4c+1}, a, b, c \in \{0, 1\}\} \subset (\mathbb{F}_2)^3.$$

and the related functionals $\Lambda = \{L_1, \dots, L_8\}$, $L_{(a,b,c)} = L_{a+2b+4c+1} = L_{P_{a+2b+4c+1}} = L_{P_{(a,b,c)}}$.

		$(0,0,0)$	$(1,0,0)$	$(0,1,0)$	$(1,1,0)$	$(0,0,1)$	$(1,0,1)$	$(0,1,1)$	$(1,1,1)$
		L_1	L_2	L_3	L_4	L_5	L_6	L_7	L_8
$P_1 = (0,0,0)$	1	**1**	1	1	1	1	1	1	1
	x	0	1	0	1	0	1	0	1
	y	0	0	1	1	0	0	1	1
	z	0	0	0	0	1	1	1	1

,

$$\mathcal{B} = \{x, y, z\}$$
$$\mathcal{A} = \{v_1 = 1\}$$

		$(0,0,0)$	$(1,0,0)$	$(0,1,0)$	$(1,1,0)$	$(0,0,1)$	$(1,0,1)$	$(0,1,1)$	$(1,1,1)$
		L_1	L_2	L_3	L_4	L_5	L_6	L_7	L_8
	1	**1**	1	1	1	1	1	1	1
	x	0	**1**	0	1	0	1	0	1
	$x \circ x$	0	0	0	0	0	0	0	0
$P_2 = (1,0,0)$	$x \circ y - x$	0	0	0	0	0	0	0	0
	$x \circ z - x$	0	0	0	0	0	0	0	0
	y	0	0	1	1	0	0	1	1
	$y \circ x = y$	0	0	1	1	0	0	1	1
	z	0	0	0	0	1	1	1	1
	$z \circ x = z$	0	0	0	0	1	1	1	1

,

$$\mathcal{B} = \{x^2, xy - x, xz - x, y, z\}$$

$$\mathcal{A} = \{v_1 = 1, v_2 = x\}$$

		$(0,0,0)$	$(1,0,0)$	$(0,1,0)$	$(1,1,0)$	$(0,0,1)$	$(1,0,1)$	$(0,1,1)$	$(1,1,1)$
		L_1	L_2	L_3	L_4	L_5	L_6	L_7	L_8
	1	**1**	1	1	1	1	1	1	1
	x	0	**1**	0	1	0	1	0	1
	y	0	0	**1**	1	0	0	1	1
	$x \circ x$	0	0	0	0	0	0	0	0
	$x \circ y - x$	0	0	0	0	0	0	0	0
$P_3 = (0,1,0)$	$x \circ z - x$	0	0	0	0	0	0	0	0
	$y \circ x - y$	0	0	0	0	0	0	0	0
	$y \circ y$	0	0	0	0	0	0	0	0
	$y \circ z - y$	0	0	0	0	0	0	0	0
	z	0	0	0	0	1	1	1	1
	$z \circ x = z$	0	0	0	0	1	1	1	1
	$z \circ y = z$	0	0	0	0	1	1	1	1

,

$$\mathcal{B} = \{x^2, xy - x, xz - x, yx - y, y^2, yz - y, z\}$$

$$\mathcal{A} = \{v_1 = 1, v_2 = x, v_3 = y\}$$

		$(0,0,0)$	$(1,0,0)$	$(0,1,0)$	$(1,1,0)$	$(0,0,1)$	$(1,0,1)$	$(0,1,1)$	$(1,1,1)$
		L_1	L_2	L_3	L_4	L_5	L_6	L_7	L_8
	1	**1**	1	1	1	1	1	1	1
	x	0	**1**	0	1	0	1	0	1
	y	0	0	**1**	1	0	0	1	1
	z	0	0	0	0	**1**	1	1	1
	$x \circ x$	0	0	0	0	0	0	0	0
$P_4 = (0,0,1)$	$x \circ y - x$	0	0	0	0	0	0	0	0
	$x \circ z - x$	0	0	0	0	0	0	0	0
	$y \circ x - y$	0	0	0	0	0	0	0	0
	$y \circ y$	0	0	0	0	0	0	0	0
	$y \circ z - y$	0	0	0	0	0	0	0	0
	$z \circ x - z$	0	0	0	0	0	0	0	0
	$z \circ y - z$	0	0	0	0	0	0	0	0
	$z \circ z$	0	0	0	0	0	0	0	0

,

$$\mathcal{B} = \{x^2, xy - x, xz - x, yx - y, y^2, yz - y, zx - z, zy - z, z^2\}$$

$$\mathcal{A} = \{v_1 = 1, v_2 = x, v_3 = y, v_4 = z\}.$$

We can conclude that $\dim_{\mathbb{F}}(\Lambda) = 4$ and $\mathrm{Span}_{\mathbb{F}}(\Lambda) = \mathrm{Span}_{\mathbb{F}}\{L_i : i \in \{1,2,3,5\}\}$. Left column reduction allows to find the following dependencies of $L_i, i \in \{4,6,7,8\}$ in terms of $\mathbb{L} := \{L_1, L_2, L_3, L_5\}$

$$L_4 = L_3 + L_2 - L_1, L_6 = L_5 + L_2 - L_1, L_7 = L_5 + L_3 - L_1, L_8 = L_5 + L_3 + L_2 - 2L_1$$

	L_1	L_2	L_3	L_4	L_5	L_6	L_7	L_8
1	**1**	1	1	1	1	1	1	1
x	0	**1**	0	1	0	1	0	1
y	0	0	**1**	1	0	0	1	1
z	0	0	0	0	**1**	1	1	1

Finally up row reduction allows to deduce the obvious separator polynomials $\mathbf{q} := \{q_1, q_2, q_3, q_5\}$ for $\mathbb{L}$, namely $q_1 = 1 - x - y - z, q_2 = x, q_3 = y, q_4 = z$:

	L_1	L_2	L_3	L_5
$1 - x - y - z$	**1**	0	0	0
x	0	**1**	0	0
y	0	0	**1**	0
z	0	0	0	**1**

7.5. Cross Product. For the non-associative[11] anticommutative binary *cross product* function $\circ : \overline{\mathbf{S}} \times \overline{\mathbf{S}} \mapsto \overline{\mathbf{S}}$ defined on $\mathbf{S} = \{i, j, k\}$, as $i \times j = k, j \times k = i, k \times i = j, j \times i = -k, k \times j = -i, i \times k = -j, i \times i = j \times j = k \times k = 0$ the relations can be summarized in the following table:

	1	i	j	k
$v_0 = 1$	1	i	j	k
$v_1 = i$	i	0	k	$-j$
$v_2 = j$	j	$-k$	0	i
$v_3 = k$	k	j	$-i$	0

COROLLARY 7.19. *Denoting by* $\mathbf{S} = \{i, j, k\}$, $\mathsf{I} \subset \mathbb{F}\{\overline{\mathbf{S}}'\}$ *the ideal of relations for the cross product operation,* $\mathsf{A} := \mathbb{F}\langle i, j, k\rangle/\mathsf{I}$, $\mathbf{N}(\mathsf{I}) = \{1, i, j, k\}$, $P_0 = (0,0,0) \in \mathbb{F}^3$, $P_i := (\delta_{i1}, \delta_{i2}, \delta_{i3}) \in \mathbb{F}^3, 1 \le i \le 3$, *where* δ *is the Kronecker's delta,* $\mathbf{X} := \{P_i : 0 \le i \le 3\} \subset \mathbb{F}^3$, $I(\mathbf{X}) := \{f \in \mathbb{F}\{\overline{\mathbf{S}}'\} : L_P(f) = 0, P \in \mathbf{X}\}$ *it holds that*

(1) $\mathsf{A} \cong \mathbb{F}[\mathbf{N}(\mathsf{I})] = \mathrm{Span}_{\mathbb{F}}\{1, i, j, k\}$

(2) *The Auzinger-Stetter matrices* [4] *related to such representation are*

$A_i =$

	1	i	j	k
$v_0 = 1$	0	1	0	0
$v_1 = i$	0	0	0	0
$v_2 = j$	0	0	0	1
$v_3 = k$	0	0	-1	0

$A_j =$

	1	i	j	k
$v_0 = 1$	0	0	1	0
$v_1 = i$	0	0	0	-1
$v_2 = j$	0	0	0	0
$v_3 = k$	0	1	0	0

$A_k =$

	1	i	j	k
$v_0 = 1$	0	0	0	1
$v_1 = i$	0	0	1	0
$v_2 = j$	0	-1	0	0
$v_3 = k$	0	0	0	0

${}_iA =$

	1	i	j	k
$v_0 = 1$	0	1	0	0
$v_1 = i$	0	0	0	0
$v_2 = j$	0	0	0	-1
$v_3 = k$	0	0	1	0

${}_jA =$

	1	i	j	k
$v_0 = 1$	0	0	1	0
$v_1 = i$	0	0	0	1
$v_2 = j$	0	0	0	0
$v_3 = k$	0	-1	0	0

${}_kA =$

	1	i	j	k
$v_0 = 1$	0	0	0	1
$v_1 = i$	0	0	-1	0
$v_2 = j$	0	1	0	0
$v_3 = k$	0	0	0	0

[11] $i \circ (i \circ j) = i \circ k = -j \neq 0 = 0 \circ j = (i \circ i) \circ j$.

(3) *The reduced Gröbner basis of* I *is* $G = \{i \times i, i \times j - k, i \times k + j, j \times i + k, j \times j, j \times k - i, k \times i - j, k \times j + i, k \times k\}$.
(4) *There is the following Cerlienco-Mureddu Correspondence*

$$1 \leftrightarrow P_0, i \leftrightarrow P_1, j \leftrightarrow P_2, k \leftrightarrow P_3$$

between $\mathbf{N}(\mathsf{I})$ *and* $\mathbf{X}$
(5) $\dim_{\mathbb{F}}(L_i, 0 \le i \le 3) = 4$.
(6) $\mathbf{N}(I(\mathbf{X})) = \mathbf{S} \cup \{1\}$
(7) $I(\mathbf{X}) = \mathsf{I}$.

Note that, denoting π the permutation acting on $\{i, j, k\} : \pi(i) = j, \pi(j) = k, \pi(k) = i$ so that $\pi^2(i) = k, \pi^2(j) = i, \pi^2(k) = j$ and $\pi^3 = \text{Id}$,

$$G = \{i \times j - k, j \times k - i, k \times i - j, j \times i + k, k \times j + i, i \times k + j, i \times i, j \times j, k \times k\}$$

satisfies $G = \pi(G) = \pi^2(G)$. On the basis of this remark for each formula $\mathcal{F}$ it is sufficient to prove that $i \in I^\circ_{\mathcal{F}}$ and remark that each formula are naturally satisfied by identity.

LEMMA 7.20. *For* $\mathcal{F} \in \Big\{\mathbf{F}\Big\}$ *we have* $I^\circ_{\mathcal{F}} = 0$ *and* $A^\circ_{\mathcal{F}} = \mathbb{F}\{\overline{\mathbf{S}}'\}$.

PROOF. $\mathbf{F}$: $(i \times j) \times i - i \times (j \times i) = k \times i - i \times (-k) = j - j = 0$. □

LEMMA 7.21. *For* $\mathcal{F} \in \Big\{\mathbf{A1}, \mathbf{A2}, \mathbf{M1}, \mathbf{M2}, \mathbf{M3}\Big\}$ *we have that*

$$I^\circ_{\text{alt}} = I^\circ_{\mathcal{F}} = \mathbb{I}(i, j, k) \text{ and } A^\circ_{\text{alt}} = A^\circ_{\mathcal{F}} = \mathbb{F}.$$

PROOF. $\mathbf{A1}$: $(k \times k) \times i - k \times (k \times i) = 0 - k \times j = i$,
$\mathbf{A2}$: $(i \times j) \times j - i \times (j \times j) = k \times j - 0 = -i$,
$\mathbf{M1}$: $(i \times (j \times i)) \times k - i \times (j \times (i \times k)) = (i \times (-k)) \times k - i \times (j \times (-j)) = j \times k - 0 = i$,
$\mathbf{M2}$: $\Big(((k \times i) \times j) \times i\Big) - \Big(k \times (i \times (j \times i))\Big) = \Big((j \times j) \times i\Big) - \Big(k \times (i \times (-k))\Big) = 0 - \Big(k \times j\Big) = i$,
$\mathbf{M3}$: $-\Big((i \times j) \times (k \times i)\Big) + \Big((i \times ((j \times k) \times i))\Big) = -\Big(k \times j\Big) - \Big((i \times ((i) \times i))\Big) = i + 0 = i$. □

7.6. Quaternion product. We have just discussed the behaviour of the cross product $\times$ defined on $\mathbf{S} = \{i, j, k\}$, over the magma algebra $\mathbb{F}\{\overline{\mathbf{S}}'\} \cong \text{Span}_{\mathbb{F}}\{1, i, j, k\}$; here we discuss how the hamiltomian product[12] $\cdot$ imposes on the same module the structure of the quaternions $\mathbb{H}$.

Technically it is sufficient to identify the unit element with a new symbol e getting

$$\mathbb{H} = \text{Span}_{\mathbb{F}}\{e, i, j, k\} \cong \text{Span}_{\mathbb{F}}\{1, i, j, k\},$$

[12] $u \cdot v = - < u, v > + u \times v$.

define $\cdot : \overline{\mathbf{S}}' \times \overline{\mathbf{S}}' \mapsto \overline{\mathbf{S}}'$ defined on $\mathbf{S}' = \{e, i, j, k\}$, according the following relations

$$H := \begin{array}{c|cccc} & e & i & j & k \\ \hline v_0 = e & e & i & j & k \\ v_1 = i & i & -e & k & -j \\ v_2 = j & j & -k & -e & i \\ v_3 = k & k & j & -i & -e \end{array}$$

whence for $u = -\sum_{i=0}^{3} a_i v_i$ and $v = -\sum_{i=0}^{3} b_i v_i$ we have

$$\begin{aligned} u \cdot v &= (a_0, a_1, a_2, a_3) H (b_0, b_1, b_2, b_3)^T \\ &= (a_0, a_1, a_2, a_3)(b_0 e + b_1 i + b_2 j + b_3 k, b_0 i - b_1 e + b_2 k - b_3 j, b_0 j - b_1 k \\ &\quad - b_2 e + b_3 i, b_0 k + b_1 j - b_2 i - b_3 e) \\ &= (a_0 b_0 - a_1 b_1 - a_2 b_2 - a_3 b_3) e + (a_0 b_1 + a_1 b_0 + a_2 b_3 - a_3 b_2) i \\ &\quad + (a_0 b_2 - a_1 b_3 + a_2 b_0 + a_3 b_1) j + (a_0 b_3 + a_1 b_2 - a_2 b_1 + a_3 b_0) k. \end{aligned}$$

Hamilton compacted his result under the fundamental formula

$$i^2 = j^2 = k^2 = ijk = -1.$$

Remark that the permutations of the three symbols $\{i, j, k\}$ are generated by the two permutations π acting on $\{i, j, k\}$ and ρ acting on $\{i, j\}$ so that the formula above then can be expressed compactly as $i^2 = 1, ij = -ji = k$.[13]

COROLLARY 7.22. *Denoting* $\mathbf{S} = \{e, i, j, k\}$, $\mathsf{I} \subset \mathbb{F}\{\overline{\mathbf{S}}\}$ *the ideal of the relations of the quaternion's product,* $\mathsf{A} := \mathbb{F}\langle e, i, j, k\rangle/\mathsf{I}$, $\mathbf{N}(\mathsf{I}) = \{e, i, j, k\}$, $P_i := (\delta_{i1}, \delta_{i2}, \delta_{i3}, \delta_{i4}) \in \mathbb{F}^4, 1 \le i \le 4$, $\mathbf{X} := \{P_i : 1 \le i \le 4\} \subset \mathbb{F}^4$, $I(\mathbf{X}) := \{f \in \mathbb{F}\{\overline{\mathbf{S}}'\} : L_P(f) = 0, P \in \mathbf{X}\}$ *it holds that*

(1) $\mathsf{A} \cong \mathbb{K}[\mathbf{N}(\mathsf{I})] = \mathrm{Span}_{\mathbb{F}}\{e, i, j, k\}$

(2) *The Auzinger-Stetter matrices* [4] *related to such representation are*

$$A_i = \begin{array}{c|cccc} & e & i & j & k \\ \hline v_1 = e & 0 & 1 & 0 & 0 \\ v_2 = i & -1 & 0 & 0 & 0 \\ v_3 = j & 0 & 0 & 0 & 1 \\ v_4 = k & 0 & 0 & -1 & 0 \end{array} \qquad A_j = \begin{array}{c|cccc} & e & i & j & k \\ \hline v_1 = e & 0 & 0 & 1 & 0 \\ v_2 = i & 0 & 0 & 0 & -1 \\ v_3 = j & -1 & 0 & 0 & 0 \\ v_4 = k & 0 & 1 & 0 & 0 \end{array}$$

$$A_k = \begin{array}{c|cccc} & e & i & j & k \\ \hline v_1 = e & 0 & 0 & 0 & 1 \\ v_2 = i & 0 & 0 & 1 & 0 \\ v_3 = j & 0 & -1 & 0 & 0 \\ v_4 = k & -1 & 0 & 0 & 0 \end{array} \qquad {}_iA = \begin{array}{c|cccc} & e & i & j & k \\ \hline v_1 = e & 0 & 1 & 0 & 0 \\ v_2 = i & 0 & -1 & 0 & 0 \\ v_3 = j & 0 & 0 & 0 & -1 \\ v_4 = k & 0 & 0 & 1 & 0 \end{array}$$

[13]Note that

$$i^2 = j^2 = k^2 = ijk = -1 \implies ij = ij(-k^2) = (-ijk)k = k$$
$$i^2 = j^2 = k^2 = ijk = -1 \implies -ji = ji(ijk) = ji^2jk = -j^2k = k.$$

$$
{}_jA = \begin{array}{c|cccc} & e & i & j & k \\ \hline v_1 = e & 0 & 0 & 1 & 0 \\ v_2 = i & 0 & 0 & 0 & 1 \\ v_3 = j & -1 & 0 & 0 & 0 \\ v_4 = k & 0 & -1 & 0 & 0 \end{array} \qquad {}_kA = \begin{array}{c|cccc} & e & i & j & k \\ \hline v_1 = e & 0 & 0 & 0 & 1 \\ v_2 = i & 0 & 0 & -1 & 0 \\ v_3 = j & 0 & 1 & 0 & 0 \\ v_4 = k & -1 & 0 & 0 & 0 \end{array}
$$

$$
\textit{and } {}_eA = A_e = \begin{array}{c|cccc} & e & i & j & k \\ \hline v_1 = e & 1 & 0 & 0 & 0 \\ v_2 = i & 0 & 1 & 0 & 0 \\ v_3 = j & 0 & 0 & 1 & 0 \\ v_4 = k & 0 & 0 & 0 & 1 \end{array}
$$

(3) *The reduced Gröbner basis of* I *is* $G = \{i \times j - k, j \times k - i, k \times i - j, j \times i + k, k \times j + i, i \times k + j, e \times e - e\} \cup \{x \times x + e, x \times e - x, e \times x - x : x \in \{i, j, k\}\}$.

(4) *There is the following Cerlienco-Mureddu Correspondence*

$$e \leftrightarrow P_1, i \leftrightarrow P_2, j \leftrightarrow P_3, k \leftrightarrow P_4$$

between $\mathbf{N}(\mathsf{I})$ *and* $\mathbf{X}$.

LEMMA 7.23. *For* $\mathcal{F} \in \Big\{\mathbf{A1}, \mathbf{A2}, \mathbf{F}, \mathbf{M1}, \mathbf{M2}, \mathbf{M3}\Big\}$ *it holds* $I^\circ_{\text{alt}} = I^\circ_{\mathcal{F}} = 0$ *and* $A^\circ_{\text{alt}} = A^\circ_{\mathcal{F}} = \mathbb{F}\{\overline{\mathbf{S}}'\}$.

PROOF. We have

A1: $(k \times k) \times i - k \times (k \times i) = -e \times i - k \times j = -i + i = 0$,

A2: $(i \times j) \times j - i \times (j \times j) = k \times j + i = -i + i = 0$,

F: $(i \times j) \times i - i \times (j \times i) = k \times i - i \times (-k) = j - j = 0$,

M1: $(i \times (j \times i)) \times k - i \times (j \times (i \times k)) = (i \times (-k)) \times k - i \times (j \times (-j)) = j \times k - i = i - i = 0$,

M2: $\Big(((k \times i) \times j) \times i\Big) - \Big(k \times (i \times (j \times i))\Big) = \Big((j \times j) \times i\Big) - \Big(k \times (i \times (-k))\Big) = -i - \Big(k \times j\Big) = -i + i = 0$,

M3: $\Big((i \times j) \times (k \times i)\Big) - \Big((i \times ((j \times k) \times i))\Big) = \Big(k \times j\Big) - \Big((i \times (i \times i))\Big) = 0$.

□

7.7. Rock-Paper-Scissors. We intend to investigate a different binary relation on a set $\mathbf{S} = \{r, p, s\}$ of three objects, which is commutative but not associative,[14] the game usually known as *Rock-Paper-Scissors*:

$$
\begin{array}{c|cccc} & 1 & r & p & s \\ \hline v_0 = 1 & 1 & r & p & s \\ v_1 = r & r & r & p & r \\ v_2 = p & p & p & p & s \\ v_3 = s & s & r & s & s \end{array}
$$

Note that such algebra has non-trivial zero divisors, independently from the way we define $x \circ x$ for each $x \in \mathbf{S}$: for instance $r(1 - s) = 0$. Therefore we can represent such algebra as a direct sum of fields.

[14] $r \circ (p \circ s) = r \circ s = r \neq s = p \circ s = (r \circ p) \circ s$.

To find the potential idempotents $i = a + br + cp + ds$ consider the equation $i^2 - i = 0$ obtaining

$$\begin{aligned} 0 = {} & a^2 - a \\ & + (2ab + b^2 + 2bd - b)r \\ & + (2ac + 2bc + c^2 - c)p \\ & + (2ad + 2cd + d^2 - d)s, \end{aligned}$$

the related equations

$$\begin{aligned} 0 &= a(a-1) \\ 0 &= b(2a + b + 2d - 1) \\ 0 &= c(2a + 2b + c - 1) \\ 0 &= d(2a + 2c + d - 1), \end{aligned}$$

and the following solution

- $a = b = c = d = 0; i_{0000} = 0$
- $a = b = c = 0, d = 1; i_{0001} = s$
- $a = b = 0, 0 = 2a + 2b + c - 1 = c - 1, d = 0; i_{0010} = p$
- $a = b = 0, 0 = 2a + 2b + c - 1 = c - 1, 0 = 2a + 2c + d - 1 = d + 1; i_{0011} = p - s = i_{0010} - i_{0001}$
- $a = c = d = 0, 0 = 2a + b + 2d - 1 = b - 1; i_{0100} = r$
- $a = c = 0, 0 = 2a + b + 2d - 1 = b + 2d - 1 = b + 1, 0 = 2a + 2c + d - 1 = d - 1; i_{0101} = -r + s$.
- $a = 0, 0 = 2a + b + 2d - 1 = b - 1, 0 = 2a + 2b + c - 1 = 2 + c - 1 = c + 1, d = 0; i_{0110} = r - p.$
- $a = 0, 0 = 2a+b+2d-1 = b+2d-1, 0 = 2a+2b+c-1 = -4d+2+c-1 = 1-4d+c, 0 = 2a+2c+d-1 = -2+8d+d-1 = 9d-3 \implies d = \frac{1}{3}, c = -1+4d = -1+\frac{4}{3} = \frac{1}{3}, b = 1-2d = 1-\frac{2}{3} = \frac{1}{3} - 2 + d - 1; i_{0111} = \frac{1}{3}r + \frac{1}{3}p + \frac{1}{3}s$[15]
- $a = 1, b = c = d = 0; i_{1000} = 1.$
- $a = 1, b = c = 0, 0 = 2a + 2c + d - 1 = 2 + d - 1 = d + 1; i_{1001} = 1 - s.$
- $a = 1, b = 0, 0 = 2a + 2b + c - 1 = 2 + c - 1 = 1 + c, d = 0; i_{1010} = 1 - p$
- $a = 1, b = 0, 0 = 2a + 2b + c - 1 = 2 + c - 1 = c + 1, 0 = 2a + 2c + d - 1 = 2 - 2 + d - 1 = d - 1; i_{1011} = 1 - p + s.$
- $a = 1, 0 = 2a + b + 2d - 1 = 2 + b - 1 = b + 1, c = d = 0; i_{0100} = 1 - r.$
- $a = 1, 0 = 2a + b + 2d - 1 = 2 + b + 2d - 1 = 2 + b - 2 - 1 = b - 1, c = 0, 0 = 2a + 2c + d - 1 = 2 + d - 1 = d + 1; i_{1101} = 1 + r - s.$
- $a = 1, 0 = 2a + b + 2d - 1 = 2 + b - 1 = b + 1, 0 = 2 - 2 + c - 1 = c - 1, d = 0; i_{1110} = 1 - r + p.$
- $a = 1, 0 = 2a+b+2d-1 = b+2d+1, 0 = 2a+2b+c-1 = 2-4d-2+c-1 = -1 - 4d + c, 0 = 2a + 2c + d - 1 = 2 + 2 + 8d + d - 1 = 9d + 3 \implies d = -\frac{1}{3}, c = 1 + 4d = -1 - \frac{4}{3} = -\frac{1}{3}, b = -1 - 2d = -1 + \frac{2}{3} = -\frac{1}{3}; i_{1111} = 1 - \frac{1}{3}r - \frac{1}{3}p - \frac{1}{3}s.$[16]

hence $\mathsf{A} = \mathbb{F}1 \oplus \mathbb{F}r \oplus \mathbb{F}s \oplus \mathbb{F}p.$

---

[15] $(\frac{1}{3}r + \frac{1}{3}p + \frac{1}{3}s)^2 = \frac{1}{9}r^2 + \frac{1}{9}p^2 + \frac{1}{9}s^2 + \frac{2}{9}rp + \frac{2}{9}ps + \frac{2}{9}sr = \frac{1}{9}r + \frac{1}{9}p + \frac{1}{9}s + \frac{2}{9}p + \frac{2}{9}s + \frac{2}{9}r = \frac{1}{3}r + \frac{1}{3}p + \frac{1}{3}s.$

[16] $(1 - \frac{1}{3}r - \frac{1}{3}p - \frac{1}{3}s)^2 = 1 + \frac{1}{9}r^2 + \frac{1}{9}p^2 + \frac{1}{9}s^2 - \frac{2}{3}r - \frac{2}{3}p - \frac{2}{3}s + \frac{2}{9}rp + \frac{2}{9}ps + \frac{2}{9}sr = 1 + \frac{1}{9}r + \frac{1}{9}p + \frac{1}{9}s - 23r - \frac{2}{93}s - \frac{2}{3}p + \frac{2}{9}p + \frac{2}{9}s + \frac{2}{9}r = 1 - \frac{1}{3}r - \frac{1}{3}p - \frac{1}{3}s$

COROLLARY 7.24. *Denoting* $\mathbf{S} = \{r, p, s\}$, $\mathsf{I} \subset \mathbb{F}\{\overline{\mathbf{S}}'\}$ *the ideal of relations for the rock-paper-scissors operation,* $\mathsf{A} := \mathbb{F}\langle r, p, s\rangle/\mathsf{I}$, $\mathbf{N}(\mathsf{I}) = \{1, r, p, s\}$, $P_0 = (0,0,0) \in \mathbb{F}^3$, $P_i := (\delta_{i1}, \delta_{i2}, \delta_{i3}) \in \mathbb{F}^3, 1 \le i \le 3$, $\mathbf{X} := \{P_i : 0 \le i \le 3\} \subset \mathbb{F}^3$, $I(\mathbf{X}) := \{f \in \mathbb{F}\{\overline{\mathbf{S}}'\} : L_P(f) = 0, P \in \mathbf{X}\}$ *it holds that*

(1) $\mathsf{A} \cong \mathbb{F}[\mathbf{N}(\mathsf{I})] = \mathrm{Span}_{\mathbb{F}}\{1, r, p, s\}$

(2) *The Auzinger-Stetter matrices* [4] *related to such representation are*

$$A_r = {}_rA = \begin{array}{r|cccc} & 1 & r & p & s \\ \hline v_0 = 1 & 0 & 1 & 0 & 0 \\ v_1 = r & 0 & 1 & 0 & 0 \\ v_2 = p & 0 & 0 & 1 & 0 \\ v_3 = s & 0 & 1 & 0 & 0 \end{array} \qquad A_p = {}_pA = \begin{array}{r|cccc} & 1 & r & p & s \\ \hline v_0 = 1 & 0 & 0 & 1 & 0 \\ v_1 = r & 0 & 0 & 1 & 0 \\ v_2 = p & 0 & 0 & 1 & 0 \\ v_3 = s & 0 & 0 & 0 & 1 \end{array}$$

$$A_s = {}_sA = \begin{array}{r|cccc} & 1 & r & p & s \\ \hline v_0 = 1 & 0 & 0 & 0 & 1 \\ v_1 = r & 0 & 1 & 0 & 0 \\ v_2 = p & 0 & 0 & 0 & 1 \\ v_3 = s & 0 & 0 & 0 & 1 \end{array}$$

(3) *The reduced Gröbner basis of* I *is*

$$G = \{r \times p - p, p \times s - s, s \times r - r, p \times r - p, s \times p - s, r \times s - r\} \cup \{x \times x - x, x \in \mathbf{S}\}.$$

(4) *There is the following Cerlienco-Mureddu Correspondence*

$$1 \leftrightarrow P_0, r \leftrightarrow P_1, p \leftrightarrow P_2, s \leftrightarrow P_3$$

between $\mathbf{N}(\mathsf{I})$ *and* $\mathbf{X}$.

(5) $\dim_{\mathbb{F}}(L_i, 0 \le i \le 3) = 4$.

(6) $\mathbf{N}(I(\mathbf{X})) = \mathbf{S} \cup \{1\}$.

(7) $I(\mathbf{X}) = \mathsf{I}$.

Note that G satisfies $G = \pi(G)$ for each permutation π acting on $\{r, p, s\}$ which is sufficient to prove the following

LEMMA 7.25. *For* $\digamma \in \left\{\mathbf{A1}, \mathbf{A2}, \mathbf{F}\right\}$ *we have* $I^\circ_{\text{alt}} = I^\circ_{\digamma} = 0$ *and* $A^\circ_{\text{alt}} = A^\circ_{\digamma} = \mathbb{F}\{\overline{\mathbf{S}}'\}$.

PROOF. It holds

A1: $(r \circ r) \circ s - r \circ (r \circ s) = r \circ s - r \circ r = r - r = 0,$
A2: $(r \circ s) \circ s - r \circ (s \circ s) = r \circ s - r \circ s = r - r = 0,$
F: $(r \circ s) \circ r - r \circ (s \circ r) = r \circ r - r \circ r = r - r = 0.$

□

LEMMA 7.26. *For* $\digamma \in \left\{\mathbf{M1}, \mathbf{M2}, \mathbf{M3}\right\}$ *we have*

$$I^\circ_{\text{alt}} = I^\circ_{\digamma} = \mathbb{I}(p - r, s - r) \text{ and } A^\circ_{\text{alt}} = A^\circ_{\digamma} = \mathbb{F}\{\overline{\{\mathbf{r}\}}\}.$$

PROOF. It holds

M1: $(p \circ (r \circ p)) \circ s - p \circ (r \circ (p \circ s)) = (p \circ p) \circ s - p \circ (r \circ s) = p \circ s - p \circ r = s - p,$
M2: $((s \circ p) \circ r) \circ p - s \circ (p \circ (r \circ p)) = (s \circ p) \circ r - s \circ (p \circ p) = s \circ r - s \circ p = r - s,$
M3: $((p \circ r) \circ (s \circ p)) - p \circ ((r \circ s) \circ p)) = (p \circ s) - p \circ (r \circ p) = s - p \circ p = s - p.$

□

Let us now verify the effect of the cross product $\times$ on $\mathbf{S} = \{r, p, s\}$, on the magma $\mathbb{F}\{\overline{\mathbf{S}}'\} \cong \mathrm{Span}_{\mathbb{F}}\{1, r, p, s\}$; we thus consider

$$\mathbb{F}\{\overline{\mathbf{S}}'\} \cong \mathrm{Span}_{\mathbb{F}}\{1, r, p, s\} = \mathrm{Span}_{\mathbb{F}}\{e, r, p, s\}$$

and define on $\mathbf{S}' = \{e, r, p, s\}$ $\cdot : \overline{\mathbf{S}}' \times \overline{\mathbf{S}}' \mapsto \overline{\mathbf{S}}'$ according the following relation

$$H := \begin{array}{c|cccc} & e & r & p & s \\ \hline v_0 = e & e & r & p & s \\ v_1 = r & r & r & p & r \\ v_2 = p & p & p & p & s \\ v_3 = s & s & r & s & s \end{array}$$

getting for $u = -\sum_{i=0}^{3} a_i v_i$ and $v = -\sum_{i=0}^{3} b_i v_i$ the formula

$$\begin{aligned} u \cdot v &= (a_0, a_1, a_2, a_3) H (b_0, b_1, b_2, b_3)^T \\ &= (a_0, a_1, a_2, a_3)(b_0 e + b_1 r + b_2 p + b_3 s, (b_0 + b_1 + b_3) r + b_2 p, \\ &\qquad (b_0 + b_1 + b_2) p + b_3 s, (b_0 + b_2 + b_3) s + b_1 r) \\ &= a_0 b_0 e + (a_0 b_1 + a_1 b_0 + a_1 b_1 + a_1 b_3 + a_3 b_1) r \\ &\quad + (a_0 b_2 + a_1 b_2 + a_2 b_0 + a_2 b_1 + a_2 b_2) p + (a_0 b_3 + a_2 b_3 + a_3 b_0 + a_3 b_2 + a_3 b_3) s. \end{aligned}$$

COROLLARY 7.27. *Denoting* $\mathbf{S} = \{e, r, p, s\}$, $\mathsf{I} \subset \mathbb{F}\{\overline{\mathbf{S}}\}$ *the ideal of the relations,* $\mathsf{A} := \mathbb{F}\langle e, r, p, s\rangle/\mathsf{I}$, $\mathbf{N}(\mathsf{I}) = \{e, r, p, s\}$, $P_i := (\delta_{i1}, \delta_{i2}, \delta_{i3}, \delta_{i4}) \in \mathbb{F}^4, 1 \leq i \leq 4$, $\mathbf{X} := \{P_i : 0 \leq i \leq 3\} \subset \mathbb{F}^3$, $I(\mathbf{X}) := \{f \in \mathbb{F}\{\overline{\mathbf{S}}\} : L_P(f) = 0, P \in \mathbf{X}\}$ *it holds that*

(1) $\mathsf{A} \cong \mathbb{F}[\mathbf{N}(\mathsf{I})] = \mathrm{Span}_{\mathbb{F}}\{e, r, p, s\}$
(2) *The Auzinger-Stetter matrices* [**4**] *related to such representation are*

$$A_r = {}_rA = \begin{array}{c|cccc} & e & r & p & s \\ \hline v_0 = e & 0 & 1 & 0 & 0 \\ v_1 = r & 0 & 1 & 0 & 0 \\ v_2 = p & 0 & 0 & 1 & 0 \\ v_3 = s & 0 & 1 & 0 & 0 \end{array} \qquad A_p = {}_pA = \begin{array}{c|cccc} & e & r & p & s \\ \hline v_0 = e & 0 & 0 & 1 & 0 \\ v_1 = r & 0 & 0 & 1 & 0 \\ v_2 = p & 0 & 0 & 1 & 0 \\ v_3 = s & 0 & 0 & 0 & 1 \end{array}$$

$$A_s = {}_sA = \begin{array}{c|cccc} & e & r & p & s \\ \hline v_0 = e & 0 & 0 & 0 & 1 \\ v_1 = r & 0 & 1 & 0 & 0 \\ v_2 = p & 0 & 0 & 0 & 1 \\ v_3 = s & 0 & 0 & 0 & 1 \end{array} \quad \text{and } {}_eA = A_e = \begin{array}{c|cccc} & e & r & p & s \\ \hline v_1 = e & 1 & 0 & 0 & 0 \\ v_2 = r & 0 & 1 & 0 & 0 \\ v_3 = p & 0 & 0 & 1 & 0 \\ v_4 = s & 0 & 0 & 0 & 1 \end{array}$$

(3) *The reduced Gröbner basis of* I *is*

$$\begin{aligned} G = \{ &r \times p - p, p \times s - s, s \times r - r, p \times r - p, s \times p - s, r \times s - r, e \times e - e\} \\ &\cup \{x \times x - x, x \times e - x, e \times x - x : x \in \{r, p, s\}\}. \end{aligned}$$

(4) *There is the following Cerlienco-Mureddu Correspondence*

$$e \leftrightarrow P_1, r \leftrightarrow P_1, p \leftrightarrow P_2, s \leftrightarrow P_3$$

between $\mathbf{N}(\mathsf{I})$ *and* $\mathbf{X}$
(5) $\dim_{\mathbb{F}}(L_i, 0 \leq i \leq 3) = 4$.
(6) $\mathbf{N}(I(\mathbf{X})) = \mathbf{S} \cup \{1\}$
(7) $I(\mathbf{X}) = \mathsf{I}$.

LEMMA 7.28. *For $\mathcal{F} \in \left\{\mathbf{A1}, \mathbf{A2}, \mathbf{F}\right\}$ we have $I^{\circ}_{\text{alt}} = I^{\circ}_{\mathcal{F}} = 0$ and $A^{\circ}_{\text{alt}} = A^{\circ}_{\mathcal{F}} = \mathbb{F}\{\overline{\mathbf{S}}'\}$.*

PROOF. It holds

A1: $(r \circ r) \circ s - r \circ (r \circ s) = r \circ s - r \circ r = r - r = 0,$
A2: $(r \circ s) \circ s - r \circ (s \circ s) = r \circ s - r \circ s = r - r = 0,$
F: $(r \circ s) \circ r - r \circ (s \circ r) = r \circ r - r \circ r = r - r = 0.$

□

LEMMA 7.29. *For $\mathcal{F} \in \left\{\mathbf{M1}, \mathbf{M2}, \mathbf{M3}\right\}$ we have*

$$I^{\circ}_{\text{alt}} = I^{\circ}_{\mathcal{F}} = \mathbb{I}(p - r, s - r) \text{ and } A^{\circ}_{\text{alt}} = A^{\circ}_{\mathcal{F}} = \mathbb{F}\{\overline{\{\mathbf{r}\}}\}.$$

PROOF. It holds

M1: $(p \circ (r \circ p)) \circ s - p \circ (r \circ (p \circ s)) = (p \circ p) \circ s - p \circ (r \circ s) = p \circ s - p \circ r = s - p,$
M2: $((s \circ p) \circ r) \circ p - s \circ (p \circ (r \circ p)) = (s \circ r) \circ p - s \circ (p \circ p) = r \circ p - s \circ p = p - s,$
M3: $((p \circ r) \circ (s \circ p)) - p \circ ((r \circ s) \circ p) = (p \circ s) - p \circ (r \circ p) = p \circ s - p \circ p = s - p.$

□

7.8. Octonions. Let us now discuss the construction performed by Cayley[**14**] (see also [**25**]) and Graves[**37**] for producing the Octonions

$$\mathbb{O} \cong \mathbb{H} \times \mathbb{H} \cong \mathbb{R} + \text{Span}_{\mathbb{R}}\{i, j, k, l, il, jl, kl\}$$

starting with the $\mathbb{R}$-subalgebra

$$\mathcal{O} := \left\{ \begin{pmatrix} a_0 + a_1 i + a_2 j + a_3 k & -a_4 - a_5 i - a_6 j - a_7 k \\ -a_4 + a_5 i + a_6 j + a_7 k & a_0 - a_1 i - a_2 j - a_3 k \end{pmatrix} \right\} \subset \text{Mat}(2, \mathbb{H})$$

with unit element $E = \begin{pmatrix} 1 & 0 \\ 0 & 1 \end{pmatrix}$ and with the $\mathbb{R}$-algebra isomorphism $\mathbb{O} \to \mathcal{O}$,

$$(a_0, a_1, a_2, a_3, a_4, a_5, a_6, a_7) \mapsto \begin{pmatrix} a_0 + a_1 i + a_2 j + a_3 k & -a_4 - a_5 i - a_6 j - a_7 k \\ -a_4 + a_5 i + a_6 j + a_7 k & a_0 - a_1 i - a_2 j - a_3 k \end{pmatrix}$$

$=: M_{a_0 + a_1 i + a_2 j + a_3 k + a_4 l + a_5 il + a_6 jl + a_7 kl}$, they imposed on $\mathbb{O}$ a product $\circ : \mathbb{O} \times \mathbb{O} \mapsto \mathbb{O}$ defined for

$$z = a + bl, w = c + dl \in \mathbb{O} \cong \mathbb{H} \oplus \mathbb{H}l, a, b, c, d \in \mathbb{H}$$

as

$$z \circ w = (a + bl) \circ (c + dl) = (ac - bd) + (ad - bc)l,$$

which can be interpreted (as it was remarked by Zorn [**74**]) as the restriction on $\mathcal{O}$ of the matrix product of $\text{Mat}(2, \mathbb{H})$. Such product can be expanded as follows:

$$\begin{aligned} wz &= (a_0 + a_1 i + a_2 j + a_3 k + a_4 l + a_5 il + a_6 jl + a_7 kl)(b_0 + b_1 i + b_2 j + b_3 k \\ &\qquad + b_4 l + b_5 il + b_6 jl + b_7 kl) \\ &= (a_0 b_0 - a_1 b_1 - a_2 b_2 - a_3 b_3 - a_4 b_4 - a_5 b_5 - a_6 b_6 - a_7 b_7) \\ &= (a_0 b_1 + a_1 b_0 + b_2 a_3 - a_3 b_2 + a_4 b_5 - a_5 b_4 - a_6 b_7 + a_7 b_6) i \\ &= (a_0 b_2 - a_1 b_3 + a_2 b_0 + a_3 b_1 + a_4 b_6 + a_5 b_7 - a_6 b_4 - a_7 b_5) j \\ &= (a_0 b_3 + a_1 b_2 - a_2 b_1 + a_3 b_0 + a_4 b_7 - a_5 b_6 + a_6 b_5 - a_7 b_4) k \\ &= (a_0 b_4 - a_1 b_5 - a_2 b_6 - a_3 b_7 + a_4 b_0 + a_5 b_1 + a_6 b_2 + a_7 b_3) l \\ &= (a_0 b_5 + a_1 b_4 - a_2 b_7 + a_3 b_6 - a_4 b_1 + a_5 b_0 - a_6 b_3 + a_7 b_2) il \\ &= (a_0 b_6 - a_1 b_7 + a_2 b_4 - a_3 b_5 - a_4 b_2 + a_5 b_3 + a_6 b_0 + a_7 b_1) jl \end{aligned}$$

$$= (a_0b_7 + a_1b_6 + a_2b_5 + a_3b_4 - a_4b_3 - a_5b_2 + a_6b_1 + a_7b_0)kl$$

and summarized as

	1	v_1	v_2	v_3	v_4	v_5	v_6	v_7
$v_0 = 1$	1	v_1	v_2	v_3	v_4	v_5	v_6	v_7
$v_1 = i$	v_1	-1	v_3	$-v_2$	v_5	$-v_4$	v_7	$-v_6$
$v_2 = j$	v_2	$-v_3$	-1	v_1	v_6	v_7	$-v_4$	$-v_5$
$v_3 = k$	v_3	v_2	$-v_1$	-1	v_7	$-v_6$	v_5	$-v_4$
$v_4 = l$	v_4	$-v_5$	$-v_6$	$-v_7$	-1	v_1	v_2	v_3
$v_5 = il$	v_5	v_4	$-v_7$	v_6	$-v_1$	-1	$-v_3$	v_2
$v_6 = jl$	v_6	v_7	v_4	$-v_5$	$-v_2$	v_3	-1	$-v_1$
$v_7 = kl$	v_7	$-v_6$	v_5	v_4	$-v_3$	$-v_2$	v_1	-1

$\mathcal{O}$ is a 8-dimensional, non commutative and even not-associative,[17] but division algebra.

7.9. Steiner Triple System STS(7). Recall that the Steiner Triple System STS(7) [**69**] is a set $\mathbf{S} = \{a, b, c, d, e, f, g\}$ equipped with a family of subsets (called blocks) of 3 elements such that every pair of elements in $\mathbf{S}$ is in exactly one of the blocks; unique (up to isomorphism) STS(7) has the following block's set:

$$\mathbf{T} := \Big\{\{a, b, c\}, \{a, d, e\}, \{a, f, g\}, \{b, d, f\}, \{b, e, g\}, \{c, d, g\}, \{c, e, f\}\Big\}$$

and it is called the Fano Plane [**26**] and is a natural model for the octonions obtained considering each block as a Hamiltonian triple thus on the magma $\mathbb{F}\{\overline{\mathbf{S}}\} = \mathbb{F}\langle a, b, c, d, e, f, g\rangle$ freely generated by the set $\mathbf{S} = \{a, b, c, d, e, f, g\}$.

The relation between the octonians, STS(7) and the model granted by division on $\{x_1, x_2, x_3\}$ can be submarized as

σ		$\mathcal{O}$	$STS(7)$
1	1	1	1
2	x_1	i	f
3	x_2	j	e
4	x_1x_2	k	g
5	x_3	l	d
6	x_1x_3	il	b
7	x_2x_3	jl	a
8	$x_1x_2x_3$	kl	c

References

[1] J. Apel and W. Lassner, *An Algorithm for calculations in enveloping fields of Lie algebras*, Proc. Int. Conf. on Comp. Algebra and its Appl. n Theoretical Physics JINR **D11-85-792**, Dubna (1985) 231–241

[2] J. Apel and W. Lassner, *Computation and simplification in Lie fields*, EUROCAL '87 (Leipzig, 1987), Lecture Notes in Comput. Sci., vol. 378, Springer, Berlin, 1989, pp. 468–478, DOI 10.1007/3-540-51517-8_156. MR1033324

[3] Maria Emilia Alonso, Maria Grazia Marinari, and Teo Mora, *The big mother of all dualities: Möller algorithm*, Comm. Algebra **31** (2003), no. 2, 783–818, DOI 10.1081/AGB-120017343. MR1968924

[17] $v_5 \circ (v_6 \circ v_7) = v_5 \circ (-v_4) = v_8 \neq -v_8 = v_2 \circ v_7 = (v_5 \circ v_6) \circ v_7$.

[4] W. Auzinger and H. J. Stetter, *An elimination algorithm for the computation of all zeros of a system of multivariate polynomial equations*, Numerical mathematics, Singapore 1988, Internat. Schriftenreihe Numer. Math., vol. 86, Birkhäuser, Basel, 1988, pp. 11–30. MR1022943

[5] P. Bäck, and J. Richter, *Hilbert's basis theorem and simplicity for non-associative skew Laurent polynomial rings and related rings*, Preprint, `arXiv:2207.07994`, 2022.

[6] George M. Bergman, *The diamond lemma for ring theory*, Adv. in Math. **29** (1978), no. 2, 178–218, DOI 10.1016/0001-8708(78)90010-5. MR506890

[7] Cristina Bertone, Francesca Cioffi, Paolo Lella, and Margherita Roggero, *Upgraded methods for the effective computation of marked schemes on a strongly stable ideal*, J. Symbolic Comput. **50** (2013), 263–290, DOI 10.1016/j.jsc.2012.07.006. MR2996879

[8] M. A. Borges-Trenard, M. Borges-Quintana, and T. Mora, *Computing Gröbner bases by FGLM techniques in a non-commutative setting*, J. Symbolic Comput. **30** (2000), no. 4, 429–449, DOI 10.1006/jsco.1999.0415. MR1784751

[9] B. Buchberger, *Ein Algorithmus zum Auffinden der Basiselemente des Restklassenringes nach einem nulldimensionalen Polynomideal*, Ph. D. Thesis, Innsbruck (1965)

[10] B. Buchberger, *A theoretical basis for the reduction of polynomials to canonical forms*, ACM SIGSAM Bull. **10** (1976), no. 3, 19–29, DOI 10.3982/te103bm. MR463136

[11] B. Buchberger, *Some Properties of Gröbner Bases*, SIGSAM Bull. **10** 4 (1976), 19–24

[12] B. Buchberger, *A criterion for detecting unnecessary reductions in the construction of Gröbner-bases*, Symbolic and algebraic computation (EUROSAM '79, Internat. Sympos., Marseille, 1979), Lecture Notes in Comput. Sci., vol. 72, Springer, Berlin-New York, 1979, pp. 3–21. MR575678

[13] B. Buchberger, *Miscellaneous Results on Groebner Bases for Polynomial Ideals II.* Technical Report 83/1, University of Delaware, Department of Computer and Information Sciences, 1983. p. 31

[14] Arthur Cayley, (1845), *On Jacobi's Elliptic functions, in reply to the Rev. Brice Bronwin; and on Quaternions*, Philosophical Magazine, 26: 208–211, DOI 10.1080/14786444508645107

[15] Michela Ceria, *Bar code for monomial ideals*, J. Symbolic Comput. **91** (2019), 30–56, DOI 10.1016/j.jsc.2018.06.012. MR3860883

[16] M. Ceria, *Bar code: a visual representation for finite sets of terms and its applications*, Math. Comput. Sci. **14**, 497–513 (2020). `https://doi.org/10.1007/s11786-019-00425-4`

[17] Michela Ceria and Teo Mora, *Buchberger-Zacharias theory of multivariate Ore extensions*, J. Pure Appl. Algebra **221** (2017), no. 12, 2974–3026, DOI 10.1016/j.jpaa.2017.02.011. MR3666735

[18] Michela Ceria and Teo Mora, *Buchberger-Weispfenning theory for effective associative rings*, J. Symbolic Comput. **83** (2017), 112–146, DOI 10.1016/j.jsc.2016.11.008. MR3645647

[19] Michela Ceria and Ferdinando Mora, *A degroebnerization approach to algebraic statistics*, ISSAC '22—Proceedings of the 2022 International Symposium on Symbolic and Algebraic Computation, ACM, New York, [2022] ©2022, pp. 419–428. MR4488853

[20] Michela Ceria, Samuel Lundqvist, and Teo Mora, *Degröbnerization: a political manifesto*, Appl. Algebra Engrg. Comm. Comput. **33** (2022), no. 6, 675–723, DOI 10.1007/s00200-022-00586-z. MR4511195

[21] M. Ceria and F. Mora, *De Nugis Groebnerialium 4: Rump, Ufnarovski, Zacharias*

[22] L. Cerlienco and M. Mureddu, *Algoritmi combinatori per l'interpolazione polinomiale in dimensione* ≥ 2, Publ. I.R.M.A. Strasbourg, 1993, 461/S-24 Actes 24e Séminaire Lotharingien, p.39-76.

[23] L. Cerlienco and M. Mureddu, *From algebraic sets to monomial linear bases by means of combinatorial algorithms*, Discrete Math. **139** (1995), no. 1-3, 73–87, DOI 10.1016/0012-365X(94)00126-4. Formal power series and algebraic combinatorics (Montreal, PQ, 1992). MR1336833

[24] W. A. de Graaf and J. Wisliceny, *Constructing bases of finitely presented Lie algebras using Gröbner bases in free algebras*, Proceedings of the 1999 International Symposium on Symbolic and Algebraic Computation (Vancouver, BC), ACM, New York, 1999, pp. 37–43, DOI 10.1145/309831.309853. MR1802065

[25] L. E. Dickson, *On quaternions and their generalization and the history of the eight square theorem*, Ann. of Math. (2) **20** (1919), no. 3, 155–171, DOI 10.2307/1967865. MR1502549

[26] G. Fano, (1892), *Sui postulati fondamentali della geometria proiettiva*, Giornale di Matematiche, 30: 106–132

[27] Jeffrey B. Farr and Shuhong Gao, *Computing Gröbner bases for vanishing ideals of finite sets of points*, Applied algebra, algebraic algorithms and error-correcting codes, Lecture Notes in Comput. Sci., vol. 3857, Springer, Berlin, 2006, pp. 118–127, DOI 10.1007/11617983_11. MR2243500

[28] J. C. Faugère, P. Gianni, D. Lazard, and T. Mora, *Efficient computation of zero-dimensional Gröbner bases by change of ordering*, J. Symbolic Comput. **16** (1993), no. 4, 329–344, DOI 10.1006/jsco.1993.1051. MR1263871

[29] Bálint Felszeghy, Balázs Ráth, and Lajos Rónyai, *The lex game and some applications*, J. Symbolic Comput. **41** (2006), no. 6, 663–681, DOI 10.1016/j.jsc.2005.11.003. MR2208506

[30] R. Gebauer and H. M. Möller, *Buchberger's algorithm and staggered linear bases*, Proc. SYMSAC'86 (1986), 218-221, ACM

[31] Rüdiger Gebauer and H. Michael Möller, *On an installation of Buchberger's algorithm*, J. Symbolic Comput. **6** (1988), no. 2-3, 275–286, DOI 10.1016/S0747-7171(88)80048-8. Computational aspects of commutative algebra. MR988418

[32] P. Gordan, *Neuer Beweis des Hilbertschen Satzes über homogene Funktionen*, Gottingen Nachr. (1899), 240–242

[33] P. Gordan, *Les invariants des formes binaries*, Journal de Mathématiques Pure et Appliés (5e séries) **6** (1900), 141–156

[34] H.-D. Ebbinghaus, H. Hermes, F. Hirzebruch, M. Koecher, K. Mainzer, J. Neukirch, A. Prestel, and R. Remmert, *Numbers* **123** (1990), xviii+395, DOI 10.1007/978-1-4612-1005-4. With an introduction by K. Lamotke; Translation edited and with a preface by J. H. Ewing; Readings in Mathematics. MR1066206

[35] L. Gerritzen, *On Non-Associative Gröbner Bases* Symposium in Honor of Bruno Buchberger's 60th Birthday, RISC-Linz , 2002

[36] Lothar Gerritzen, *Tree polynomials and non-associative Gröbner bases*, J. Symbolic Comput. **41** (2006), no. 3-4, 297–316, DOI 10.1016/j.jsc.2003.09.005. MR2202553

[37] John T. Graves, *On a Connection between the General Theory of Normal Couples and the Theory of Complete Quadratic Functions of Two Variables*, Phil. Mag., 26: 315–320, (1845) DOI 10.1080/14786444508645136

[38] Sir W. R. Hamilton, (1853), `https://archive.org/details/lecturesonquater00hami` Lectures on Quaternions Dublin: Hodges and Smith

[39] Sir W. R. Hamilton and W. E. Hamilton, (ed) (1866), "[`https://archive.org/details/elementsquaterni00hamirich` Elements of Quaternions]" London: Longmans, Green, & Co.

[40] Anne Heyworth, *One-sided noncommutative Gröbner bases with applications to computing Green's relations*, J. Algebra **242** (2001), no. 2, 401–416, DOI 10.1006/jabr.2001.8801. MR1848951

[41] E. N. Kuz'min and I. P. Shestakov, *Non-associative structures [MR1060322 (91i:17001)]*, 1995, DOI 10.1007/978-3-662-06292-0_2. MR1360006

[42] A. Kandri-Rody and V. Weispfenning, *Noncommutative Gröbner bases in algebras of solvable type*, J. Symbolic Comput. **9** (1990), no. 1, 1–26, DOI 10.1016/S0747-7171(08)80003-X. MR1044911

[43] Roberto La Scala and Viktor Levandovskyy, *Letterplace ideals and non-commutative Gröbner bases*, J. Symbolic Comput. **44** (2009), no. 10, 1374–1393, DOI 10.1016/j.jsc.2009.03.002. MR2543425

[44] Roberto La Scala and Viktor Levandovskyy, *Skew polynomial rings, Gröbner bases and the letterplace embedding of the free associative algebra*, J. Symbolic Comput. **48** (2013), 110–131, DOI 10.1016/j.jsc.2012.05.003. MR2980469

[45] M. G. Marinari, T. Mora, and H. M. Moeller, *Groebner bases of ideals given by dual bases*, Proc. ISSAC '91, ACM (1991) 55-6

[46] M. G. Marinari, H. M. Möller, and T. Mora, *Gröbner bases of ideals defined by functionals with an application to ideals of projective points*, Appl. Algebra Engrg. Comm. Comput. **4** (1993), no. 2, 103–145, DOI 10.1007/BF01386834. MR1223853

[47] H. Michael Möller, *On the construction of Gröbner bases using syzygies*, J. Symbolic Comput. **6** (1988), no. 2-3, 345–359, DOI 10.1016/S0747-7171(88)80052-X. Computational aspects of commutative algebra. MR988422

[48] H. M. Möller and B. Buchberger, *The construction of multivariate polynomials with preassigned zeros*, Computer algebra (Marseille, 1982), Lecture Notes in Comput. Sci., vol. 144, Springer, Berlin-New York, 1982, pp. 24–31. MR680050

[49] T. Mora, *Solving Polynomial Equation Systems*, 4 Vols., Cambridge University Press, I DOI 10.1017/CBO9780511542831 (2003), II DOI 10.1017/CBO9781107340954 (2005), III DOI 10.1017/CBO9781139015998 (2015), IV DOI 10.1017/CBO9781316271902 (2016).

[50] Teo Mora, *An introduction to commutative and noncommutative Gröbner bases*, Theoret. Comput. Sci. **134** (1994), no. 1, 131–173, DOI 10.1016/0304-3975(94)90283-6. Second International Colloquium on Words, Languages and Combinatorics (Kyoto, 1992). MR1299371

[51] Teo Mora and Lorenzo Robbiano, *Points in affine and projective spaces*, Computational algebraic geometry and commutative algebra (Cortona, 1991), Sympos. Math., XXXIV, Cambridge Univ. Press, Cambridge, 1993, pp. 106–150, DOI 10.1016/S0883-9417(09)90009-7. MR1253990

[52] T. Mora, *The FGLM Problem and Moeller's Algorithm on Zero-dimensional Ideals*, in M.Sala et al., Groebner Bases, Coding, and Cryptography. Springer (2009)27-46

[53] B. Mourrain, *Bezoutian and quotient ring structure*, J. Symbolic Comput. **39** (2005), no. 3-4, 397–415, DOI 10.1016/j.jsc.2004.11.010. MR2168289

[54] Patrik Nordbeck, *Canonical subalgebra bases in non-commutative polynomial rings*, Proceedings of the 1998 International Symposium on Symbolic and Algebraic Computation (Rostock), ACM, New York, 1998, pp. 140–146, DOI 10.1145/281508.281595. MR1805178

[55] P. Nordbeck, *Canonical bases for Subalgebras of Factor Algebras*, Computer Science Journal of Moldavia **7** (1999), 63–79, A.C.M.

[56] Luquan Pan, *On the D-bases of polynomial ideals over principal ideal domains*, J. Symbolic Comput. **7** (1989), no. 1, 55–69, DOI 10.1016/S0747-7171(89)80006-9. MR984271

[57] Bertrand Nguefack and Emmanuel Pola, *Effective Buchberger-Zacharias-Weispfenning theory of skew polynomial extensions of subbilateral coherent rings*, J. Symbolic Comput. **99** (2020), 50–107, DOI 10.1016/j.jsc.2019.03.003. MR4051943

[58] Oystein Ore, *Linear equations in non-commutative fields*, Ann. of Math. (2) **32** (1931), no. 3, 463–477, DOI 10.2307/1968245. MR1503010

[59] Oystein Ore, *Theory of non-commutative polynomials*, Ann. of Math. (2) **34** (1933), no. 3, 480–508, DOI 10.2307/1968173. MR1503119

[60] Michael Pesch, *Two-sided Gröbner bases in iterated Ore extensions*, Symbolic rewriting techniques (Ascona, 1995), Progr. Comput. Sci. Appl. Logic, vol. 15, Birkhäuser, Basel, 1998, pp. 225–243. MR1624592

[61] G. Pistone, E. Riccomagno, and M. P. Rogantin, *Methods in algebraic statistics for the design of experiments*, Optimal design and related areas in optimization and statistics, Springer Optim. Appl., vol. 28, Springer, New York, 2009, pp. 97–132, DOI 10.1007/978-0-387-79936-0_5. MR2513348

[62] F. L. Pritchard, *A syzygies approach to non-commutative Groebner bases*, Preprint (1994)

[63] F. Leon Pritchard, *The ideal membership problem in non-commutative polynomial rings*, J. Symbolic Comput. **22** (1996), no. 1, 27–48, DOI 10.1006/jsco.1996.0040. MR1421258

[64] Saeed Rajaee, *Non-associative Gröbner bases*, J. Symbolic Comput. **41** (2006), no. 8, 887–904, DOI 10.1016/j.jsc.2006.04.005. MR2246715

[65] Lorenzo Robbiano and Moss Sweedler, *Subalgebra bases*, Commutative algebra (Salvador, 1988), Lecture Notes in Math., vol. 1430, Springer, Berlin, 1990, pp. 61–87, DOI 10.1007/BFb0085537. MR1068324

[66] David Shannon and Moss Sweedler, *Using Gröbner bases to determine algebra membership, split surjective algebra homomorphisms determine birational equivalence*, J. Symbolic Comput. **6** (1988), no. 2-3, 267–273, DOI 10.1016/S0747-7171(88)80047-6. Computational aspects of commutative algebra. MR988417

[67] D. A. Spear, *A constructive approach to commutative ring theory*, Proc. of the 1977 MACSYMA Users' Conference, NASA CP-2012 (1977), 369–376

[68] M. Stener, *Moufang Loops General theory and visualization of non-associative Moufang loops of order 16*, Thesis, Uppsala University, 2016.

[69] J. Steiner, *Combinatorische Aufgaben* (German), J. Reine Angew. Math. **45** (1853), 181–182, DOI 10.1515/crll.1853.45.181. MR1578819

[70] Michael Stillman and Harrison Tsai, *Using SAGBI bases to compute invariants*, J. Pure Appl. Algebra **139** (1999), no. 1-3, 285–302, DOI 10.1016/S0022-4049(99)00015-8. Effective methods in algebraic geometry (Saint-Malo, 1998). MR1700547

[71] Weispfenning V., *Finite Groebner bases in non-noetherian Skew Polynomial Rings* Proc. ISSAC'92 (1992), 320–332, A.C.M.

[72] M. Wiesinger-Widi, *Groebner Bases and Generalized Sylvester Matrices.* Ph.D. Thesis, Johannes Kepler University, Institute for Symbolic Computation, submitted 2014

[73] G. Zacharias, *Generalized Gröbner bases in commutative polynomial rings*, Bachelor's thesis, M.I.T. (1978)

[74] Max Zorn, *Alternativkörper und quadratische systeme* (German), Abh. Math. Sem. Univ. Hamburg **9** (1933), no. 1, 395–402, DOI 10.1007/BF02940661. MR3069613

Department of Mechanics, Mathematics and Management, Politecnico di Bari, Italy, 70125

Email address: michela.ceria@gmail.com

Department of Mathematics, University of Genoa, Italy, 16100

Email address: theofmora@gmail.com

Contemporary Mathematics
Volume **826**, 2025
https://doi.org/10.1090/conm/826/16572

Perspectively decomposable modules

Soumitra Das, Yasser Ibrahim, Özgür Taşdemir, and Mohamed Yousif

ABSTRACT. A module M is called perspectively indecomposable if, M does not contain proper perspectively related submodules A and B with $A \cap B = 0$. Where two submodules A and B of M are called perspectively related, and denoted by $A \sim B$, if $M = A \oplus C = B \oplus C$, for a submodule $C \subseteq M$. Every indecomposable module is perspectively indecomposable, but the converse is not true. Moreover, M is called perspectively decomposable (PD-module) if, $M = A \oplus B$ for every pair of submodules A and B of M with $A \sim B$ and $A \cap B = 0$. Examples are provided to show that the class of PD-modules lies strictly between the classes of summand-square-free and $C4$-modules. We will show that every PD-module is a finite direct sum of perspectively indecomposable submodules. As an application, we prove that if M is a PD-module with the finite exchange, then M is clean and has the full exchange. This extends earlier results on the Crawley-Jónsson's open question that asks whether the finite exchange property of a module implies the full exchange property.

Introduction

Two submodules A and B of a right R-module M are said to be perspectively related, and denoted by $A \sim B$, if they have a common complement, i.e. there exists a direct-summand $C \subseteq M$ such that $M = A \oplus C = B \oplus C$. Clearly, if $A \sim B$ then $A \cong B$. The module M is called perspective if every two isomorphic direct-summands of M are perspectively related. A ring R is right perspective if R as a right R-module is perspective. The notion of perspectivity was thoroughly investigated in [**5**], [**9**], [**10**], [**14**], [**15**], and [**16**]. Subsequently, in [**13**], two direct-summands A and B of a right R-module M are called strongly-perspective (s-perspective, for short), and denoted by $A \overset{s}{\sim} B$, if every complement of A is a complement of B and vice-versa, i.e. if for any $X \subseteq M$, $M = A \oplus X$ iff $M = B \oplus X$. A module M is called s-perspective if every two isomorphic direct-summands of M are s-perspective, and a ring R is called s-perspective if R as a right R-module is s-perspective. The notion of s-perspectivity was first introduced by Mohamed and Müller in [**19**, Definition 1.30], using the terminology superspective. Moreover, the notion of s-perspectivity was thoroughly investigated by Khurana and Nielsen in [**15**] under the name "strongly IC rings", i.e. every regular element

2020 *Mathematics Subject Classification.* Primary 16D40, 16D50, 16D60; Secondary 16L30, 16L60, 16P20, 16P40, 16P60.

Key words and phrases. Perspective Rings and Modules; Cancellation, Substitution and Stable Range 1; Exchange and Clean Rings; Regular, Unit-Regular and Strongly Regular Rings; Square-free and Dual-Square-Free modules; Abelian and Quasi-duo Rings.

is strongly regular. The notion of strongly IC was originally introduced and studied by Khurana, Lam and Wang in [**14**]. Several interesting characterization of s-perspective (strongly IC) rings appeared in [**15**]. Clearly, s-perspective modules are perspective, but the converse need not be true. Indeed, if $M = A \oplus B$ with $A \cong B$, then by a result of Handelman in [**10**, Page 2], $A \sim B$, and by the definition of s-perspectivity, A and B can not be s-perspective.

Motivated by Handelman's observation above, we will call a module M perspectively indecomposable (PI-module) if, M does not contain proper submodules A and B of M with $A \sim B$ and $A \cap B = 0$. Moreover, we will say that a module M is perspectively decomposable (PD-module) if, $M = A \oplus B$ for every pair of perspectively related submodules A and B of M (i.e. $A \sim B$) with $A \cap B = 0$. A ring R is called a right PD-ring if R_R is a PD-module. In some sense, the notion of perspectively decomposable may be viewed as a converse to Handelman's observation above.

We will show below that a module M is perspectively indecomposable if and only if M is summand-square-free. We will also show that PD-modules are Dedekind-finite, and provide examples to show that the class of PD-modules lies strictly between the classes of summand-square-free and $C4$-modules. Moreover, we will show that every proper submodule of a PD-module is summand-square-free (perspectively indecomposable). We use the aforementioned result to prove that every PD-module is either a summand-square-free or a direct sum of two isomorphic summand-square-free modules. We provide examples of PD-modules that are infinite direct sums of indecomposable modules. We then show that, if M is a PD-module with the finite exchange, then M is clean and has the full exchange. The aforementioned result provides new classes of exchange modules that are clean, and extends almost all the known results on the Crawley-Jónsson's open question [**3**] that asks whether the finite exchange property of a module implies the full exchange property. Where, a right R-module is said to satisfy the (full) exchange property (in this case M is called an exchange module) if for every right R-module L and any two direct sum decompositions $L = M \oplus N = \oplus_{i \in I} N_i$, there exist submodules $K_i \subseteq N_i$ such that $L = M \oplus (\oplus_{i \in I} K_i)$. If this holds only for $|I| < \infty$, then M is said to satisfy the finite exchange property. A ring R for which R_R has the finite exchange property is called an exchange ring. Exchange rings are left-right symmetric and are closely related to another interesting class of rings called clean rings that was first introduced by Nicholson in [**20**], where a ring R is called clean if every element is the sum of an idempotent and a unit, and a module M is called clean if $End(M)$ is a clean ring. Nicholson [**20**] proved that every clean ring is an exchange ring, and an abelian ring (a ring with central idempotents) is clean iff it is an exchange ring.

Throughout, all rings R are associative with unity and all modules are unitary R-modules. For a module M, we use $rad(M)$, $E(M)$ and $End(M_R)$ to denote the Jacobson radical, the injective hull and the endomorphism ring of M, respectively. If $M = R$, we write $J(R) = rad(R)$. We write $N \subseteq M$ if N is a submodule of M, $N \subseteq^{ess} M$ if N is an essential submodule of M, $N \subseteq^{\oplus} M$ if N is a direct summand of M, and $N \ll M$ if N is a small submodule of M. A submodule N of M is called proper if $0 \neq N \subsetneq M$. Two right R-modules M and N are called orthogonal (factor-orthogonal) if, no nonzero submodule (factor module) of M is isomorphic to a submodule (factor module) of N. We will denote by $idem(R)$ to the set of idempotents of R, and by $U(R)$ to the units of R.

1. Perspectively decomposable modules

If $M = A \oplus B$ and $f : A \longrightarrow B$ is a homomorphism, then the graph submodule of M, denoted by $G(A,B,f)$, is the submodule given by $G(A,B,f) := \{a + f(a) : a \in A\}$.

LEMMA 1.1. *Let $M = A \oplus B$ and $f : A \to B$ be a homomorphism. Then:*

(1) $M = G(A,B,f) \oplus B$; *in particular* $A \sim G(A,B,f)$.
(2) $A \oplus Imf = A + G(A,B,f)$. *In particular, if f is an epimorphism, then* $M = A + G(A,B,f)$.
(3) $\ker f = G(A,B,f) \cap A$. *In particular, if f is a monomorphism, then* $0 = G(A,B,f) \cap A$ *and* $A \oplus Imf = A \oplus G(A,B,f)$.

PROOF. (1) If $x \in M$, then $x = a+b = a+f(a)+b-f(a) \in G(A,B,f)+B$ and so $M = G(A,B,f) + B$. Now, if $a + f(a) = b \in G(A,B,f) \cap B$, then $a = b - f(a) \in A \cap B = 0$ and so $b = a + f(a) = 0 + f(0) = 0$. Thus $G(A,B,f) \cap B = 0$ and $M = G(A,B,f) \oplus B$, as required. Now, since $M = A \oplus B = G(A,B,f) \oplus B$, $A \sim G(A,B,f)$.
(2) Clearly, $A+G(A,B,f) \subseteq A \oplus Imf$. Now, if $x = a+f(a') \in A \oplus Imf$, then $x = a-a'+a'+f(a') \in A+G(A,B,f)$ and so $A \oplus Imf \subseteq A+G(A,B,f)$. Hence $A \oplus Imf = A + G(A,B,f)$. Clearly, if f is an epimorphism, then $M = A + G(A,B,f)$.
(3) If $a = a' + f(a') \in G(A,B,f) \cap A$, then $a - a' = f(a') \in A \cap B = 0$ and so $a = a' \in \ker f$. Thus $G(A,B,f) \cap A \subseteq \ker f$. Now, if $a \in \ker f$, then $a = a + f(a) \in G(A,B,f) \cap A$, and so $\ker f \subseteq G(A,B,f) \cap A$. Hence $\ker f = G(A,B,f) \cap A$. The last statement is now clear.

□

Now, the next result follows directly from Lemma 1.1.

COROLLARY 1.2 ([**10**, Page 2]). *If $M = A \oplus B$ with $A \overset{f}{\cong} B$, then there exists a submodule C of M with $M = A \oplus C = B \oplus C$; i.e, $A \sim B$.*

PROOF. The proof follows directly by setting $C := G(A,B,f)$ in Lemma 1.1 above, where $f : A \longrightarrow B$ is the given isomorphism. □

The following example shows that the converse of Handelman's observation (Corollary 1.2) need not be true in general.

EXAMPLE 1.3. Consider the $\mathbb{Z}$-module $M = \mathbb{Z}_2 \oplus \mathbb{Z}_{2^3}$, where $\mathbb{Z}_n = \mathbb{Z}/n\mathbb{Z}$. Let $A = \langle(\bar{1},\bar{0})\rangle$, $B = \langle(\bar{1},\bar{2^2})\rangle$ and $C = \langle(\bar{0},\bar{1})\rangle$. Clearly, $M = A \oplus C = B \oplus C$ i.e. $A \sim B$ with $A \cap B = 0$. However, $M \neq A \oplus B$.

Motivated by Handelman's observation and the above example, we introduce the following two definitions:

DEFINITION 1.4. A module M is called perspectively indecomposable if, M does not contain proper submodules A and B of M with $A \sim B$ and $A \cap B = 0$.

DEFINITION 1.5. A module M is called perspectively decomposable (PD-module) if, $M = A \oplus B$ for every pair of proper submodules A and B of M with $A \sim B$ and $A \cap B = 0$. A ring R a right PD-ring if R_R is a PD-module.

In some sense, the notion of perspectively decomposable modules may be viewed as a converse of Handelman's observation [**10**, Page 2] above. Clearly every perspectively indecomposable module is a PD-module, but the converse is not true. For, if S is a simple module, then $M = S \oplus S$ is a PD-module that is not perspectively indecomposable. Moreover, we will show below that a module M is perspectively indecomposable if and only if M is summand-square-free. Where, a module M is called square-free (SF-module) if it contains no non-zero isomorphic submodules A and B with $A \cap B = 0$. The module M is called summand-square-free (SSF-module) if the submodules A and B in the aforementioned definition are direct summands of M. Clearly, every indecomposable module is a summand-square-free module, and if $M = \mathbb{Z}_p \oplus \mathbb{Z}_q$, where p and q are distinct primes and $\mathbb{Z}_n = \mathbb{Z}/n\mathbb{Z}$, then M is a summand-square-free module that is not indecomposable. Furthermore, we will also show that every proper direct summand of a PD-module is summand-square-free (perspectively indecomposable), and use this result to prove that every PD-module is either a summand-square-free or a direct sum of two isomorphic summand-square-free modules. In particular, every PD-module is a finite direct sum of summand-square-free (perspectively indecomposable) modules. We conclude from these observations that the class of indecomposable modules is a proper subclass of that of summand-square-free (perspectively indecomposable) modules, and that the class of summand-square-free modules (perspectively indecomposable modules) is strictly contained in that of PD-modules. Moreover, since abelian rings (idempotents are central) are easily seen to be summand-square-free, we infer that abelian rings are PD-rings; in particular strongly regular rings are PD-rings. However, since there are examples of regular rings that are not Dedekind-finite, we infer from Proposition 1.19 below that in general regular rings need not be PD-rings. More examples will be provided after the next proposition.

PROPOSITION 1.6. *The following conditions on a right R-module M are equivalent:*

(1) *M is a PD-module.*
(2) *If $M = A \oplus B$ and $\alpha : A \to B$ is a non-zero monomorphism, where $A, B \subseteq M$, then α is an isomorphism.*
(3) *If $M = A \oplus B$ and $\alpha : A \to B$ is a non-zero homomorphism with* $\ker \alpha \subseteq^{\oplus} A$*, where $A, B \subseteq M$, then α is an epimorphism.*
(4) *If $0 \neq A, B \subseteq M$, $A \cong B$, $A \subseteq^{\oplus} M$ and $A \cap B = 0$, then $M = A \oplus B$.*
(5) *If $0 \neq A \subseteq^{\oplus} M$, $B \subseteq^{\oplus} M$, $A \cong B$ and $A \cap B = 0$, then $M = A \oplus B$.*

PROOF. (1) $\Rightarrow$ (2). Let $M = A \oplus B$ and $\alpha : A \to B$ be a monomorphism. By Lemma 1.1, $A \sim G(A, B, \alpha)$ with $A \cap G(A, B, \alpha) = 0$. Inasmuch as M is a PD-module and A & B are proper (since $\alpha \neq 0$), it follows that $M = A \oplus G(A, B, \alpha)$, and so $\operatorname{Im} \alpha = B$ and α is an isomorphism, as required. (2) $\Rightarrow$ (3). Let $M = A \oplus B$ and $\alpha : A \to B$ a non-zero homomorphism with $0 \neq \ker \alpha \subseteq^{\oplus} A$. Write $A = \ker \alpha \oplus K$ for a submodule $K \subseteq A$. We claim that the restriction map $\alpha|_K : K \to B \hookrightarrow (\ker \alpha \oplus B)$ is a non-zero monomorphism. For, if $\alpha|_K(K) = 0$, then $K \subseteq \ker \alpha$ and so $K = \ker \alpha \cap K = 0$ which in turn gives that $A = \ker \alpha$, a contradiction since α is a non-zero. Now, Inasmuch as

$$M = A \oplus B = \ker \alpha \oplus K \oplus B = K \oplus (\ker \alpha \oplus B),$$

we infer from the hypothesis that, $\operatorname{Im} \alpha = \operatorname{Im} \alpha|_K = \ker \alpha \oplus B$, a contradiction. This shows that $\ker \alpha = 0$, and so α is an isomorphism by the hypothesis.

(2) $\Rightarrow$ (4). Let $0 \neq A, B \subseteq M$ with $A \cap B = 0$ and $B \overset{\alpha}{\cong} A \subseteq^{\oplus} M$. We need to show that $M = A \oplus B$. Write $M = A \oplus A'$ for a submodule $A' \subseteq M$. If $\pi : A \oplus A' \to A'$ is the natural projection, then clearly $A \oplus B = A \oplus \pi(B)$. Inasmuch as $M = A \oplus A'$ and $\pi \circ \alpha^{-1} : A \to A'$ is a non-zero monomorphism, we infer from the hypothesis that $\mathrm{Im}(\pi o \alpha^{-1}) = \pi(B) = A'$. Therefore $M = A \oplus A' = A \oplus \pi(B) = A \oplus B$.

(3) $\Rightarrow$ (2), (4) $\Rightarrow$ (5) and (5) $\Rightarrow$ (1) are clear. $\square$

PROPOSITION 1.7. *The following conditions on a right R-module M are equivalent:*

(1) *M is a PD-module.*
(2) *If $0 \neq A, B \subseteq M$, $A \cong B$, $A \subseteq^{\oplus} M$ and $B \neq A \cap B \subseteq^{\oplus} M$, then $M = A + B$.*
(3) *If $0 \neq A \subseteq^{\oplus} M$, $B \subseteq^{\oplus} M$, $A \cong B$ and $B \neq A \cap B \subseteq^{\oplus} M$, then $M = A + B$.*

PROOF. (1) $\Rightarrow$ (2). Let $0 \neq A, B \subseteq M$, $A \cap B \subseteq^{\oplus} M$ and $B \overset{\sigma}{\cong} A \subseteq^{\oplus} M$. We need to show that $M = A + B$. Write $M = A \oplus T$ for a submodule $T \subseteq M$, and let $\pi : A \oplus T \to T$ be the natural projection. Clearly, $A + B = A + \pi(B)$. Now consider the restriction $\pi|_B : B \to T$. Now, $\pi|_B \circ \sigma^{-1} : A \to T$ is a non-zero homomorphism with $\ker(\pi|_B \circ \sigma^{-1}) = \sigma(A \cap B) \subseteq^{\oplus} A$. Since $M = A \oplus T$, we infer from the hypothesis and Proposition 1.6, that $\mathrm{Im}(\pi|_B \circ \sigma^{-1}) = \pi(B) = T$. This shows that $M = A \oplus T = A \oplus \pi(B) = A + B$, as required. (2) $\Rightarrow$ (3). Obvious.

(3) $\Rightarrow$ (1). Follows from Proposition 1.6. $\square$

EXAMPLE 1.8. Consider the $\mathbb{Z}$-module $M = \mathbb{Z} \oplus \mathbb{Z}$. Since the monomorphism $\alpha : \mathbb{Z} \to \mathbb{Z}$ given by $\alpha(x) = 2x$ is not an isomorphism, we infer from Proposition 1.6, that M is not a PD-module. More generally, if $M = \mathbb{Z} \oplus n\mathbb{Z}$, $n \geq 1$, then M is not a PD-module, as the inclusion map $n\mathbb{Z} \to \mathbb{Z}$ is not an isomorphism.

A module M is called a $C2$-module if every submodule isomorphic to a (direct) summand of M is itself a summand of M. Every $C2$-module M is a $C3$-module, i.e. the sum of any two summands of M with zero intersection is a summand of M. In [**6**], these notions were extended to the more general one of $C4$-module as follows:

DEFINITION 1.9 ([**6**]). A module M is called a $C4$-module if, whenever A_1 and A_2 are submodules of M with $M = A_1 \oplus A_2$ and $f : A_1 \to A_2$ is an R-homomorphism with $Kerf \subseteq^{\oplus} A_1$, we have $Imf \subseteq^{\oplus} A_2$.

The next corollary is an immediate consequence of Proposition 1.6 above.

COROLLARY 1.10. *Every PD-module is a C4-module.*

In Proposition 1.11, we provide several characterizations for perspectively indecomposable modules. In particular, we show that perspectively indecomposable modules are precisely the summand-square-free modules.

PROPOSITION 1.11. *The following statements are equivalent for a right R-module M:*

(1) *M is perspectively indecomposable.*
(2) *M is summand-square-free.*

(3) *M is a $C4$-module and there is no non-zero monomorphism $f : N^2 \longrightarrow M$ with* $\operatorname{Im} f \subseteq^{\oplus} M$.
(4) *If $B \cong A \subseteq^{\oplus} M$, $B \subseteq M$, and $A \cap B = 0$, then $A = B = 0$.*
(5) *If $M = A \oplus A' = B \oplus B'$ and $A \cap B = A \cap B' = 0$, then $A = 0$.*
(6) *There is no non-zero summands A, B of M with $A \cap B = 0$ and $M/A \cong M/B$.*

PROOF. (1) $\Rightarrow$ (2). Assume to the contrary that M is not summand-square-free, and let $0 \neq A, B \subseteq M$, $A \cong B$, $A \subseteq^{\oplus} M$, $B \subseteq^{\oplus} M$ and $A \cap B = 0$. Since M is a PD-module, we infer from Proposition 1.6, that $M = A \oplus B$. Inasmuch $A \cong B$, we infer from Corollary 1.2 that $A \sim B$, a contradiction, since $A \cap B = 0$ and M is perspectively indecomposable. (2) $\Rightarrow$ (1). Let $A, B \subseteq^{\oplus} M$ with $A \sim B$ and $A \cap B = 0$. Then, clearly, $A \cong B$. Since M is summand-square-free, $A = B = 0$, and hence M is perspectively indecomposable, as required.

(2) $\Rightarrow$ (3). Clearly, M is a $C4$-module. Now, let $f : K_1 \oplus K_2 \to M$ be a monomorphism with $K_1 \cong K_2$ and $\operatorname{Im} f \subseteq^{\oplus} M$. Then $\operatorname{Im} f = f(K_1) \oplus f(K_2) = A_1 \oplus A_2$ where $A_i := f(K_i)$. Clearly, $A_1 \cong A_2$, and since M is summand-square-free, $A_1 = A_2 = 0$. Hence $f = 0$, as required.

(3) $\Rightarrow$ (4). Let A and B be two submodules of M with $A \cap B = 0$ and $B \overset{f}{\cong} A \subseteq^{\oplus} M$. Since M is a $C4$-module, $A \oplus B \subseteq^{\oplus} M$. Therefore, clearly, the map $g : B \oplus B \longrightarrow M$, defined by $g(b, b') = f(b) + b'$ for all $b, b' \in B$, is a well-defined monomorphism with $\operatorname{Im} g = A \oplus B \subseteq^{\oplus} M$. Consequently, by the hypothesis, $g = 0$ and hence $A = B = 0$ as required.

(4) $\Rightarrow$ (5). Let $M = A \oplus A' = B \oplus B$ with $A \cap B = A \cap B' = 0$. Consider the natural projection $\pi : B \oplus B' \longrightarrow B'$. Since $A \cap B = 0$, $\pi(A) \cong A \subseteq^{\oplus} M$. Since $\pi(A) \cap A \subseteq B' \cap A = 0$, we infer from the hypothesis that $A = \pi(A) = 0$, as required.

(5) $\Rightarrow$ (1). Let $A, B \subseteq^{\oplus} M$ with $A \sim B$ and $A \cap B = 0$. This means that $M = A \oplus C = B \oplus C$ for some submodules C of M. Now, by the hypothesis, $A = B = 0$ and hence M is perspectively indecomposable.

(1) $\Rightarrow$ (6). Let A and B be two summands of M with $A \cap B = 0$ and $M/A \cong M/B$. We need to show that $A = 0$. To see this, write $M = A \oplus N = B \oplus B'$, where $N, B' \subseteq M$. Let $\pi : B \oplus B' \longrightarrow B'$ be the projection of M onto B'. Since $A \cap B = 0$, $A \oplus B = \pi(A) \oplus B$ and so $A \cong \pi(A)$. Therefore, there is a monomorphism $h : A \to N$ given by the composition of the following maps $A \cong \pi(A) \hookrightarrow B' \cong M/B \cong M/A \cong N$. By Lemma 1.1, clearly, $M = A \oplus N = G(A, N, h) \oplus N$ with $\ker h = G(A, N, h) \cap A = 0$. Thus $A \sim G(A, N, h)$ with $G(A, N, h) \cap A = 0$. Since M is perspectively indecomposable, $A = G(A, N, h) = 0$. Similarly, $B = 0$, as required.

(6) $\Rightarrow$ (2). Let A and B be two summands of M with $A \cap B = 0$ and $A \overset{g}{\cong} B$. Write, $M = A \oplus N$. Let $\pi : A \oplus N \longrightarrow N$ be the projection of M on N. Since $A \cap B = 0$, $A \oplus B = A \oplus \pi(B)$ and consequently, $B \cong \pi(B)$. Therefore, we have a monomorphism $h : A \to N$ given by the composition of the following maps $A \overset{g}{\cong} B \cong \pi(B) \overset{\hookrightarrow}{\longrightarrow} N$. Then, by Lemma 1.1, $M = A \oplus N = G(A, N, h) \oplus N$ with $\ker h = G(A, N, h) \cap A = 0$. Thus A & $G(A, N, h)$ are two summands of M with $A \cap G(A, N, h) = 0$ and $M/A \cong N \cong M/G(A, N, h)$. Therefore, by hypothesis, $A = 0$ and so $B = 0$. Hence M is summand-square-free. $\square$

Now, we have the following hierarchy:

Indecomposable $\to SSF \longleftrightarrow$ Perspectively Indecomposable $\to PD \to C4$

Next, we will show that the class of PD-modules lies strictly inside that of $C4$-modules. Indeed, we will provide an example of a $C3$-module ($C4$-module) that is not a PD-module. Observe first that if $\mathbb{K}$ is a field, then $M = \mathbb{K} \oplus \mathbb{K}$ is a PD-module. However, this is not the case if $M = \mathbb{K} \oplus \mathbb{K} \oplus \mathbb{K}$ as shown below.

EXAMPLE 1.12. Let $\mathbb{K}$ be a field and consider the $\mathbb{K}$-vector space $V = \mathbb{K}^3 = \mathbb{K}\oplus\mathbb{K}\oplus\mathbb{K}$. If $A = \langle(1,0,0)\rangle, B = \langle(0,0,1)\rangle$ and $C = \langle(0,1,0),(1,1,1)\rangle$ are subspaces of V, then clearly $V = A \oplus C = B \oplus C$ with $A \cap B = 0$. However, $V \neq A \oplus B$. This shows that semisimple modules need not be PD-modules. In particular, quasi-injective modules (and hence $C3$-modules, $C4$-modules) need not be PD-modules.

While every PD-module is a $C4$-module, we will show in the next example that PD-modules need not be $C3$-modules.

EXAMPLE 1.13 ([**18**]). Let $R := \mathbb{F}_2[x_1, x_2, \ldots]\langle e, f\rangle$, subject to the following relations $e^2 = e$, $f^2 = f$, $ex_1 = fx_1 = 0$, $efx_2 = fex_2 = 0$, $efex_3 = fefx_3 = 0, \ldots$. For ease of notation set $R_n := \mathbb{F}_2[x_n, x_{n+1}, \ldots]$, $e_{2n} = (ef)^n$, $e_{2n+1} = (ef)^n e$, $f_{2n} = (fe)^n$, $f_{2n+1} = (fe)^n f$ for each $n \geq 1$. (We also set $e_0 = f_0 = 1$). It was shown in [**18**, Example 6.1] that R_R is square-free (and hence a right PD-module), $eR \cap fR = 0$ and $eR + fR$ is not a summand of R_R. This says that R_R is not a $C3$-module.

EXAMPLE 1.14. Observe that if $M = \mathbb{Z}_{p^\infty} \oplus \mathbb{Z}_{p^\infty}$, then M is an example of an injective PD-module that is not summand-square-free.

The following proposition gives a sufficient condition for a $C4$-module to be a PD-module:

PROPOSITION 1.15. *The following are equivalent for a right R-module M:*

(1) *M is a PD-module*
(2) *M is a $C4$-module and every proper direct summand of M is a summand-square-free module (perspectively indecomposable).*

PROOF. (1) $\Rightarrow$ (2). By Corollary 1.10, M is a $C4$-module. Now, let D be a proper direct summand of M that is not summand-square-free. Then there exist two non-zero summands A and B of D with $A \cong B$ and $A \cap B = 0$. Inasmuch as $D \subseteq^{\oplus} M$, we infer that $A \subseteq^{\oplus} M$, $B \subseteq^{\oplus} M$. Since M is a PD-module, it follows from Proposition 1.6 that $M = A \oplus B \subseteq D$, a contradiction. (2) $\Rightarrow$ (1). Let A and B be proper summands of M with $A \sim B$ and $A \cap B = 0$. Since M is a $C4$-module, we infer from [**7**, Theorem 2.2], $A \oplus B \subseteq^{\oplus} M$. If $A \oplus B$ were a proper submodule of M, then by the hypothesis, $A \oplus B$ is a summand-square-free module. Thus $A = B = 0$, a contradiction. Hence $M = A \oplus B$ is a PD-module, as required. □

REMARK 1.16. As a result of Proposition 1.15 above, any direct sum of more than two copies of the same module is not a PD-module.

THEOREM 1.17. *If M is a PD-module, then either M is summand-square-free or $M = A \oplus B$, with $A \cong B$ and both A, B are summand-square-free modules. In particular, every PD-module M is a finite direct sum of summand-square-free (perspectively indecomposable) modules.*

PROOF. We are done if M is a summand-square-free module. Otherwise, assume that M is not summand-square-free, and let $A, B \subseteq M$, $A \cong B$, $A \subseteq^{\oplus} M$, $B \subseteq^{\oplus} M$ and $A \cap B = 0$. Since M is a PD-module, we infer from Proposition 1.6, that $M = A \oplus B$. By Proposition 1.15, $M = A \oplus B$ is a finite direct sum of summand-square-free modules, as required. □

The following is an example of a PD-module that is an infinite direct sum of indecomposables.

EXAMPLE 1.18. Observe that Theorem 1.17 says that every PD-module is a finite direct sum of perspectively indecomposable modules. However, there are examples of PD-modules that are infinite direct sums of indecomposables. Indeed, if $\mathcal{P}$ is an infinite set of distinct primes $p \in \mathbb{Z}$, and M is the $\mathbb{Z}$-module $\oplus_{p\in\mathcal{P}}\mathbb{Z}/p\mathbb{Z}$, then M is a square-free module and hence a PD-module. By Theorem 1.17 above, $M = \oplus_{i=1}^{n} M_i$ a finite direct sum of perspectively indecomposable modules. However, since M is an infinite direct sum of indecomposables, we infer that at least one of the M_i's is not indecomposable.

A module M is called a Dedekind-finite (DF-module) if M is not isomorphic to any proper direct summand of itself, i.e. if $M \cong M \oplus N$ for some module N, then $N = 0$. Clearly, every summand-square-free module is *Dedekind-finite, and we* will show in the next result that every PD-module is Dedekind-finite. But first, observe that injective modules need not be Dedekind-finite. Indeed, there are examples of regular self-injective rings that are not Dedekind-finite. In particular, $C4$-modules need not be Dedekind-finite.

PROPOSITION 1.19. *If M is a PD-module, then M is Dedekind-finite.*

PROOF. Assume to the contrary $M = N \oplus L$ where $N \cong M$ and $L \neq 0$. Then by Proposition 1.15, N is a summand-square-free module. Since $N \cong M$, it follows that M is a summand-square-free module, and hence Dedekind-finite, a contradiction. This shows that M is Dedekind-finite. □

The next result is an immediate consequence of Proposition 1.6. Recall first that a module M is called co-Hopfian if every monomorphism from M into M is an isomorphism.

COROLLARY 1.20. *If $M \oplus M$ is a PD-module, then:*

(1) *M is summand-square-free.*
(2) *M is co-Hopfian.*
(3) *M is a $C2$-module.*

PROOF. 1. Follows from Proposition 1.15. 2. Follows from from Proposition 1.6.

3. Follows from [**7**, Proposition 2.15] and Corollary 1.10. □

If $M = \mathbb{Z}$, the ring of integers, then clearly M is a summand-square-free module. However, by Proposition 1.6, $M \oplus M = \mathbb{Z} \oplus \mathbb{Z}$ is not a PD-module, as the $\mathbb{Z}$-monomorphism $\alpha : \mathbb{Z} \to \mathbb{Z}$, given by $\alpha(x) = 2x$, is not an isomorphism. This shows that the converse to (1) of Corollary 1.20 above need not be true. In the next example we show that the converse to either (2) or (3) of Corollary 1.20 above need not be true.

EXAMPLE 1.21. Let $\mathbb{K}$ be a field and $M = \mathbb{K} \oplus \mathbb{K}$. Then by Proposition 1.15, $M \oplus M$ is not a PD-module. However, clearly M is both a $C2$-module and a co-Hopfian module.

In the next result we address the question of when is the direct sum of PD-modules is again a PD-module. But first we need Lemmas 1.22 and 1.23.

LEMMA 1.22 ([**12**, Lemma 2.4]). *Let $M = \oplus_{i\in I} M_i$ with $Hom(M_i, M_j) = 0$ for every $i \neq j \in I$. If A and B are summands of M, then $A \sim B$ in M if and only if $A \cap M_i \sim B \cap M_i$ in M_i for all $i \in I$.*

LEMMA 1.23 ([**8**, Lemma 2.2]). *If $M = \oplus_{i\in I} M_i$ with $Hom(M_i, M_j) = 0$ for every $i \neq j \in I$, then $K = \oplus_{i\in I}(K \cap M_i)$ for every summand K of M.*

PROPOSITION 1.24. *Let $M = \oplus_{i\in I} M_i$ be a direct sum of submodules M_i with $Hom(M_i, M_j) = 0$ for every $i \neq j$ in I. Then M is a PD-module if and only if each M_i is a PD-module.*

PROOF. The necessity follows from Proposition 1.15. For the sufficiency, assume that each M_i is a PD-module, and let $M = A \oplus C = B \oplus C$ with $A \cap B = 0$, for submodules $A, B, C \subseteq M$. We need to show that $M = A \oplus B$. By Lemma 1.22, for each $i \in I$, $A \cap M_i \sim B \cap M_i$ in M_i with $(A \cap M_i) \cap (B \cap M_i) = 0$. Since each M_i is a PD-module, we infer that $M_i = (A \cap M_i) \oplus (B \cap M_i)$, $i \in I$. Now, by Lemma 1.23, it follows that:

$$M = \oplus_{i\in I} M_i = \oplus_{i\in I}((A \cap M_i) \oplus (B \cap M_i)) = (\oplus_{i\in I}(A \cap M_i)) \oplus (\oplus_{i\in I}(B \cap M_i)) = A \oplus B,$$

as required □

The next result can be proved with an argument similar to the one used in [**2**, Proposition 4.2].

PROPOSITION 1.25. *Let $\{R_\lambda : \lambda \in \Lambda\}$ be any collection of rings, and let $R = \prod_{\lambda\in\Lambda} R_\lambda$. Then R is a right PD-ring if and only if every R_λ is a right PD-ring.*

A module M is said to satisfy the $C1$-condition if every submodule of M is essential in a direct summand. M is called quasi-continuous if it satisfies both the $C1$- and $C3$-conditions. M is called automorphism-invariant (auto-invariant) if it is invariant under any automorphism of its injective hull. In [**11**], the notion of a Utumi-module (U-module) was introduced as a non-trivial and simultaneous generalization of quasi-continuous, square-free and auto-invariant modules, where a right R-module M is called a U-module if, whenever A and B are submodules of M with $A \cong B$ and $A \cap B = 0$, there exist two summands K and T of M such that $A \subseteq^{ess} K$, $B \subseteq^{ess} T$ and $K \oplus T \subseteq^{\oplus} M$. In the next result we show that if M is a U-module, then M is a PD-module if and only if $E(M)$ is a PD-module. But first we need Lemma 1.26.

LEMMA 1.26. *If M is a U-module, then $M = A \oplus B \oplus K$ such that $A \cong B$, $A \oplus B$ is quasi-injective, and K is square-free. In particular, if M is both a U-module and a PD-module, then either M is square-free or quasi-injective.*

PROOF. The first assertion was established in [**11**, Theorem 3.11], and the second assertion follows from Proposition 1.15. □

PROPOSITION 1.27. *If M is a U-module, then M is a PD-module if and only if $E(M)$ is a PD-module.*

PROOF. ($\Rightarrow$). By Lemma 1.26, M is either square-free or quasi-injective. If M is square-free, then clearly $E(M)$ is square-free, and hence a PD-module. Now, assume that M is quasi-injective, and let $M_1, M_2 \subseteq E(M)$ with $M_1 \sim M_2$ and $M_1 \cap M_2 = 0$. We need to show that $E(M) = M_1 \oplus M_2$. Write $E(M) = M_1 \oplus M_3 = M_2 \oplus M_3$, for a submodule $M_3 \subseteq E(M)$. By [**19**, Theorem 2.8(4)],

$$M = (M_1 \cap M) \oplus (M_3 \cap M) = (M_2 \cap M) \oplus (M_3 \cap M).$$

Clearly, $(M_1 \cap M) \sim (M_2 \cap M)$ with $(M_1 \cap M) \cap (M_2 \cap M) = 0$. Inasmuch as M is a PD-module, we infer that:

$$M = (M_1 \cap M) \oplus (M_2 \cap M).$$

Therefore,

$$E(M) = E(M_1 \cap M) \oplus E(M_2 \cap M) = E(M_1) \oplus E(M_2) = M_1 \oplus M_2,$$

as required.

($\Leftarrow$). Let $M_1, M_2 \subseteq M$ with $M_1 \sim M_2$ and $M_1 \cap M_2 = 0$. We need to show that $M = M_1 \oplus M_2$. Write $M = M_1 \oplus M_3 = M_2 \oplus M_3$ for a submodule $M_3 \subseteq M$. Then

$$E(M) = E(M_1) \oplus E(M_3) = E(M_2) \oplus E(M_3).$$

This says that $E(M_1) \sim E(M_2)$ with $E(M_1) \cap E(M_2) = 0$. Since $E(M)$ is a PD-module, $E(M) = E(M_1) \oplus E(M_2)$. Now an application of [**19**, Theorem 2.8(4)], ensures that:

$$M = (M \cap E(M_1)) \oplus (M \cap E(M_2))$$

Inasmuch as $M_i \subseteq^{ess} (M \cap E(M_i))$ and $M_i \subseteq^{\oplus} M$, $1 \leq i \leq 2$, we infer that $M_i = (M \cap E(M_i))$ and so $M = M_1 \oplus M_2$, as required. □

2. PD-modules, exchange property and perspective modules

In his interesting paper [**21**], Pace Nielsen proved that if M is a square-free module, then M satisfies the finite exchange iff M is a clean module, iff M satisfies the full exchange. Nielsen's result was extended to summand-square-free modules in [**13**]. In the next theorem, we will extend Nielsen's result to PD-modules.

THEOREM 2.1. *If M is a PD-module, then M has the finite exchange property iff M is clean, iff M has the full exchange property.*

PROOF. By Proposition 1.17, M is a finite direct sum of summand-square-free modules. Inasmuch as the notions of clean and exchange are inherited by finite direct sums, the result follows immediately from [**13**, Corollary 2.13]. □

EXAMPLE 2.2. If $M = S \oplus S$, where S is a simple module, then clearly M is an exchange PD-module that is not summand-square-free.

EXAMPLE 2.3. If $M = \mathbb{K} \oplus \mathbb{K} \oplus \mathbb{K}$ is a $\mathbb{K}$-vector space, where $\mathbb{K}$ is a field, then M is a clean (and hence full exchange) semisimple module that is not a PD-module as shown in Example 1.12.

EXAMPLE 2.4. If $\mathcal{P}$ is an infinite set of distinct primes $p \in \mathbb{Z}$, and M is the $\mathbb{Z}$-module $\oplus_{p \in \mathcal{P}} \mathbb{Z}/p\mathbb{Z}$, then M is an exchange, clean, and square-free module (and hence a PD-module). However, M is an infinite direct sum of indecomposables.

A module M is said to have the cancellation property if whenever $M \oplus X \cong M \oplus Y$, then $X \cong Y$. The module M is said to have the internal cancellation property (IC for short) if whenever $M = A \oplus X = B \oplus Y$ with $A \cong B$, then $X \cong Y$. Every module with the cancellation property satisfies the internal cancellation property but the converse need not be true in general. A module M is said to have the substitution property if for every module N with decompositions $N = N_1 \oplus H = N_2 \oplus K$ with $N_1 \cong M \cong N_2$, there exists a submodule C of N such that $N = C \oplus H = C \oplus K$. By a result of Warfield in [**22**, Theorem 2.1], a module M has the substitution property iff $End(M)$ has stable range 1, where a ring R is said to have stable range 1 ($sr(R) = 1$ for short) if, for any elements $a, b \in R$ with $Ra+Rb = R$, there is an element $y \in R$ such that $a+yb$ is a unit of R. Rings having stable range one include semilocal rings, unit-regular rings, strongly π-regular rings and exchange rings with artinian primitive factors. Moreover, in [**22**, Theorem 1.9], every module with the substitution property is perspective, where a right R-module M is called perspective if every two isomorphic direct-summands of M are perspectively related, and a ring R is right perspective if R as a right R-module is perspective.

The following notion was introduced in [**13**] as a weaker version of the full (finite) internal exchange property.

DEFINITION 2.5. A right R-module M is said to satisfy the weak internal exchange property if for every internal direct sum decomposition $M = \oplus_{i\in I} M_i$ and every summand $N \subseteq^{\oplus} M$, with $N \cong M_j$ for some $j \in I$, there exist submodules $M_i' \subseteq M_i$, $i \in I$, such that $M = \oplus_{i\in I} M_i' \oplus N$. If this holds only for $|I| < \infty$, then M is said to satisfy the weak finite internal exchange property.

By removing the condition $N \cong M_j$ in the above definition, we obtain the classical and well-known definition of the (finite) internal exchange property of a module M. It was shown in [**7**, Proposition 2.22] that if M is a $C4$-module with the finite internal exchange property, then M is actually a $C3$-module. Using the same arguments provided in the proof of [**7**, Proposition 2.22], we can show the following proposition. The last statement of the proposition is an immediate consequence of Corollary 1.10.

PROPOSITION 2.6. *If M is a $C4$-module with the weak finite internal exchange property, then M is a $C3$-module. In particular, every PD-module with the weak finite internal exchange property is a $C3$-module.*

PROPOSITION 2.7. *Every PD-module with the weak finite internal exchange property is perspective.*

PROOF. Let $M = A \oplus X = B \oplus Y$ with $A \cong B$. Since M has the finite weak exchange property, $M = A \oplus B_1 \oplus Y_1$ with $B = B_1 \oplus B_2$, $Y = Y_1 \oplus Y_2$, $A = A_1 \oplus A_2$ and $A_1 \cong B_1$ & $A_2 \cong B_2$. Consequently, $M = A \oplus B_1 \oplus Y_1 = A_1 \oplus A_2 \oplus B_1 \oplus Y_1$. If $A_1 \oplus B_1$ is a proper submodule of M, then by Proposition 1.15, $A_1 \oplus B_1$ is a summand-square-free module, and so $A_1 = B_1 = 0$. Thus $M = A \oplus Y_1 \cong B \oplus Y_1 \subseteq^{\oplus} M$. Since M is Dedekind-finite, by Proposition 1.19, $M = B \oplus Y_1$. Inasmuch as $M = B \oplus Y$, we infer that $Y = Y_1$. Therefore, $M = A \oplus Y = B \oplus Y$; that is $A \sim B$. Now, suppose that $A_1 \oplus B_1$ is not a proper submodule of M. Then $M = A_1 \oplus B_1$. Consequently, $A_2 = Y_1 = 0$ and so $B_2 = 0$ as $A_2 \cong B_2$. Thus $B = B_1$ and $A = A_1$. Therefore, $M = A \oplus B$. Since $A \cong B$, $A \sim B$ by Corollary 1.2, and so M is a perspective module. □

Clearly the ring of integers $\mathbb{Z}$ is a PD-ring with the (weak) finite internal exchange property that does not have stable range 1 (and so doesn't satisfy the substitution property). However in the case of a PD-module with the finite exchange property the situation is different as shown in Proposition 2.8.

PROPOSITION 2.8. *If M is a PD-module with the finite exchange property, then M has the substitution and cancellation properties, and its endomorphism ring has stable range* 1. *In particular, M is a perspective module.*

PROOF. By Theorem 1.17, either M is summand-square-free or $M = A \oplus B$, with $A \cong B$ and both A, B are summand-square-free modules. If M is a summand-square-free module, the result follows from [**13**, Theorem 4.16]. If $M = A \oplus B$, with $A \cong B$ and both A, B are summand-square-free modules, then both A and B satisfy the finite exchange property. Now, by [**13**, Theorem 4.16], both A and B have the substitution and cancellation properties, and their endomorphism rings have stable range 1. By, [**4**, 15.6], $M = A \oplus B$ satisfy the substitution, and hence the cancellation, property. By [**22**, Theorem 2.1], $End(M)$ has stable range 1. The last statement follows from Proposition 2.7. □

The following is an example of an artinian ring (hence exchange with stable range 1) that is not a PD-ring.

EXAMPLE 2.9. If $R = \begin{bmatrix} F & F \\ 0 & F \end{bmatrix}$, where F is a field, then clearly R is semilocal and hence has stable range 1. Now, if $A = \left\{ \begin{bmatrix} 0 & a \\ 0 & a \end{bmatrix} : a \in F \right\}$, $B = \begin{bmatrix} 0 & 0 \\ 0 & F \end{bmatrix}$ and $C = \begin{bmatrix} F & F \\ 0 & 0 \end{bmatrix}$. Then clearly $R = A \oplus C = B \oplus C$ with $A \cap B = 0$. However, $A \oplus B = \begin{bmatrix} 0 & F \\ 0 & F \end{bmatrix} \neq R$.

In [**13**], two direct-summands A and B of a right R-module M are called strongly-perspective (s-perspective, for short), and denoted by $A \overset{s}{\sim} B$, if every complement of A is a complement of B and vice-versa, i.e. if for any $X \subseteq M$, $M = A \oplus X$ iff $M = B \oplus X$. A module M is called s-perspective if every two isomorphic direct-summands of M are s-perspective, and a ring R is called s-perspective if R as a right R-module is s-perspective. Clearly, s-perspective modules are perspective, and it was shown in [**13**] that the converse need not be true. We provide in the next example an s-perspective ring that is not a PD-ring.

EXAMPLE 2.10 ([**18**, Example 4.8] and [**13**, Example 4.13]). Let $R = F_2\langle x, y, z \rangle$, subject to the relations

$$x^2 = x,\ y^2 = y,\ xy = y,\ yx = x,\ yz = xz,\ z^2 = 0,\ zxz = 0,$$

It was shown in [**18**, Example 4.8] that R is a right summand-square-free ring and hence R is a right PD-ring. Moreover, in [**13**, Example 4.13], it was proved that R is s-perspective. However, in [**7**, Example 2.12], it was shown that R is not a left $C4$-module and hence is not a left PD-ring.

The next result is a consequence of [**13**, Proposition 4.5] and Corollary 1.10.

PROPOSITION 2.11. *The following conditions on a right R-module M are equivalent:*

(1) M *is an s-perspective* $C4$*-module;*
(2) M *is an s-perspective* PD*-module;*
(3) M *is summand-square-free and has the weak internal exchange property.*

3. Corner rings, endomorphism rings and matrix rings

We start this section by showing that the notion of perspectively indecomposable (decomposable) modules is a Morita invariant property of modules. To see this, let us consider the following condition:
$(*)$ "there is no non-zero monomorphism $f : N^2 \longrightarrow M$ with $\operatorname{Im} f \subseteq^{\oplus} M$".

PROPOSITION 3.1. *Being perspectively indecomposable is a Morita invariant property of modules.*

PROOF. By [**7**, Remark 2.4], the notion of a $C4$-module is a Morita-invariant property of modules. Now, by Proposition 1.11, it remains to show that condition $(*)$ is a Morita invariant of modules. To see this, let R and S be two Morita equivalent rings, and let $F : Mod\text{-}R \to Mod\text{-}S$ and $G : Mod\text{-}S \to Mod\text{-}R$ be category equivalences. Let M_R be a module with condition $(*)$ and consider the following exact sequence:

$$0 \to T \oplus T \xrightarrow{f} F(M).$$

with $T \in Mod\text{-}S$ and $\operatorname{Im} f \subseteq^{\oplus} F(M)$. Now, by [**1**, Proposition 21.4],

$$0 \to G(T \oplus T) \cong G(T) \oplus G(T) \xrightarrow{\theta} M$$

is exact in $Mod\text{-}R$ with $G(\operatorname{Im}\theta) \subseteq^{\oplus} M$. Since M satisfies condition $(*)$, $G(T) = 0$, and so $T = 0$. Thus $F(M)$ satisfies condition $(*)$, as required. □

The next result is an immediate consequence of Proposition 1.15, Proposition 3.1 and [**7**, Remark 2.4].

PROPOSITION 3.2. *Being a* PD*-module is a Morita invariant property of modules.*

The next example shows that matrix rings over PD-rings need not be PD-rings, in particular the notion of a PD-ring is not Morita invariant. But first we need three lemmas and a proposition.

LEMMA 3.3 ([**9**, Corollary 5.1]). *Let* M_R *be a right* R*-module and* $S = End(M_R)$. *For* $e, f \in Idem(S)$, *the following are equivalent:*
(1) $eS \oplus fS = S$;
(2) $eM \oplus fM = M$;
(3) $S(1-e) \oplus S(1-f) = S$.

LEMMA 3.4 ([**9**, Corollary 5.2] and [**17**, Page 316]). *Let* M_R *be a right* R*-module and* $S = End(M_R)$. *For* $e, f \in Idem(S)$, *we have* $eS \cong fS$ *if and only if* $eM \cong fM$. *Moreover,* $eS \sim fS$ *if and only if* $eM \sim fM$.

LEMMA 3.5 ([**2**, Lemma 3.2]). *Let* M_R *be a right* R*-module and* $S = End(M_R)$. *For* $e, f \in Idem(S)$,
(1) $eM + fM = eM \oplus (1-e)fM$.
(2) $eM \oplus fM \subseteq^{\oplus} M$ *if and only if* $(1-e)fM \subseteq^{\oplus} M$.
(3) $(1-e)M = (1-f)M$ *and* $eM \cap fM = 0$ *if and only if* $ker e = ker f = ker(e-f)$.

PROPOSITION 3.6. *If $S := End(M)$ is a right PD-ring, then M_R is a PD-module. In particular, if $M_n(R)$ is a right PD-ring, then $R_R^{(n)}$ is a right PD-module.*

PROOF. Let $A, B \subseteq M$ with $A \sim B$ and $A \cap B = 0$. We need to show that $M = A \oplus B$. Inasmuch as $A \subseteq^{\oplus} M$ and $B \subseteq^{\oplus} M$, we can write $A = eM$ and $B = fM$, where e and f are idempotents of S. By Lemma 3.4, $eS \sim fS$. Moreover, $eS \cap fS = 0$. Inasmuch as S is a right PD-ring, we infer that $S = eS \oplus fS$. Now, by Lemma 3.3, $M = eM \oplus fM$. The last assertion follows from the fact that $End(R_R^{(n)}) \cong M_n(R)$. □

In the next two examples we show that the converse to Proposition 3.6 need not be true, and that the notion of a PD-ring is not a Morita invariant property.

EXAMPLE 3.7. If $M = \mathbb{Z}_{p^\infty} \oplus \mathbb{Z}_{p^\infty}$, then $M_{\mathbb{Z}}$ is an injective PD-module, but $S = M_2(\hat{\mathbb{Z}}_p) \cong End_{\mathbb{Z}}(M)$ is not a right PD-ring, where $\hat{\mathbb{Z}}_p$ is the ring of p-adic integers. For, if $e = \begin{pmatrix} 1 & 0 \\ 0 & 0 \end{pmatrix}$ and $f = \begin{pmatrix} 1 & 0 \\ p & 0 \end{pmatrix}$, then $e^2 = e$, $f^2 = f$, $eS \cap fS = 0$, and $eS \overset{\sigma}{\cong} fS$ where $\sigma(e) = f$. However, $eS \oplus fS = \begin{pmatrix} \hat{\mathbb{Z}}_p & \hat{\mathbb{Z}}_p \\ p\hat{\mathbb{Z}}_p & p\hat{\mathbb{Z}}_p \end{pmatrix} \neq S$. This shows that S is not a right PD-ring.

EXAMPLE 3.8. Inasmuch as $\mathbb{Z}$ is indecomposable, it is a PD-module. Let $R = \mathbb{M}_2(\mathbb{Z})$, and consider the following three right ideals:

$$A = \begin{pmatrix} \mathbb{Z} & \mathbb{Z} \\ 0 & 0 \end{pmatrix}, C = \begin{pmatrix} 0 & 0 \\ \mathbb{Z} & \mathbb{Z} \end{pmatrix}, \text{ and } B = \left\{ \begin{pmatrix} x & y \\ 2x & 2y \end{pmatrix} : x, y \in \mathbb{Z} \right\}$$

Clearly, $R = A \oplus C = B \oplus C$ and $A \cap B = 0$. This says that $A \sim B$ with $A \cap B = 0$. However, $A \oplus B = \begin{pmatrix} \mathbb{Z} & \mathbb{Z} \\ 2\mathbb{Z} & 2\mathbb{Z} \end{pmatrix} \neq R$. Hence $R = \mathbb{M}_2(\mathbb{Z})$ is not a right PD-ring. This shows that the notion of a PD-ring is not a Morita invariant property.

PROPOSITION 3.9 ([**2**, Lemma 3.1 and Proposition 3.3] and [**18**, Lemma 4.5]). *Let M be a right R-module and $S = End_R(M)$. Then the following are equivalent:*

(1) *M is a PD-module;*
(2) *If $e, f \in idem(S)$, with $ker\ e = ker\ f = ker(e-f)$, then $eM \oplus (1-e)fM = M$;*
(3) *If $e, f \in idem(S)$ with $eM \sim fM$ and $eM \cap fM = 0$, then there exists $g^2 = g \in S$ such that $gM = eM$ and $(1-g)M = fM$.*

PROOF. (1) ⇒ (2) Let $e, f \in idem(S)$, with $ker\ e = ker\ f = ker(e - f)$. By Lemma 3.5, $(1 - e)M = (1 - f)M$ and $eM \cap fM = 0$. This means that $eM \sim fM$ in M. Since M is a PD-module, $M = eM \oplus fM$. Again, by Lemma 3.5, $M = eM \oplus (1 - e)fM$, as required. (2) ⇒ (1) Let $M = A \oplus X = B \oplus X$ with $A \cap B = 0$ for submodules A, B and X of M. We need to show that $M = A \oplus B$. By [**1**, Proposition 5.7], $A = eM$, $B = fM$, and $X = (1 - e)M = (1 - f)M$, where e and f are idempotents of S. Now, by Lemma 3.5, $kere = kerf = ker(e - f)$, and so by the hypothesis, $M = eM \oplus (1 - e)fM$. Again, by Lemma 3.5, $M = A \oplus B = eM \oplus fM$, as required.

(1) $\Rightarrow$ (3). Let $e, f \in idem(S)$ with $eM \sim fM$ and $eM \cap fM = 0$. Since M is a PD-module, $M = eM \oplus fM$. Let $g : M \to eM$ be the projection map with $ker g = fM$. Then $gM = eM$ and $fM = ker g = (1-g)M$.

(3) $\Rightarrow$ (1). Let $A, B \subseteq M$ with $A \sim B$ and $A \cap B = 0$. We need to show that $M = A \oplus B$. Inasmuch as $A \subseteq^{\oplus} M$ and $B \subseteq^{\oplus} M$, we can write $A = eM$ and $B = fM$, where $e, f \in idem(S)$. By the hypothesis, there exists an idempotent $g \in S$ such that $gM = eM$ and $(1-g)M = fM$. Now, $M = gM \oplus (1-g)M = eM \oplus fM = A \oplus B$, as required. □

PROPOSITION 3.10. *Let M be a right PD-module such that for every pair of idempotents $e, f \in S = End(M)$ with $eS \cap fS = 0$ we have $eM \cap fM = 0$. Then S is a right PD-ring.*

PROOF. Let A and B be direct summands of S with $A \sim B$ and $A \cap B = 0$. We need to show that $S = A \oplus B$. Write $A = eS$ and $B = fS$, for some idempotents e, $f \in S$. Now, by Lemma 3.4, $eM \sim fM$ with $eM \cap fM = 0$. Inasmuch as M is a PD-module, $M = eM \oplus fM$. By Lemma 3.3, $S = eS \oplus fS = A \oplus B$, as required. □

PROPOSITION 3.11 (See [**2**, Theorem 3.4]). *Let M be a right R-module with $S = End_R(M)$. Then S is a right PD-ring, if M is a PD-module and one of the following equivalent conditions is satisfied.*

(1) *$r_M(\phi) = r_S(\phi)M$ for every $\phi \in S$, where $r_M(\phi) := \{m \in M : \phi m = 0\}$ and $r_S(\phi) := \{s \in S : \phi s = 0\}$*
(2) *For any $\alpha \in S$, $ker \alpha$ is generated by M; i.e., $ker\ \alpha = \Sigma\{\theta(M) : \theta \in S,\ \theta(M) \subseteq ker\ \alpha\}$.*

PROOF. Clearly (1) and (2) are equivalent. Now, assume that (1) is satisfied and let $eS \sim fS$ with $eS \cap fS = 0$, where $e, f \in idem(S)$. We need to show that $eS \oplus fS = S$. Write $S = eS \oplus X = fS \oplus X$, for a right ideal $X \subseteq S$. Then $X = (1-e)S = (1-f)S$. Since $(1-e)S = (1-f)S$ and $eS \cap fS = 0$, we have $r_S(e) = r_S(f) = r_S(e-f)$. Now, by the hypothesis, $r_M(e-f) = r_S(e-f)M = r_S(f)M = (1-f)M = (1-e)M$. We claim that $eM \cap fM = 0$. To see this, let $x = em = fm' \in eM \cap fM$. Then $m - em = (1-e)m = m - fm' = (1-f)y$ for some $y \in M$. This implies that $em = fm' = fm$, and so $(e-f)m = 0$. This means that $m \in r_M(e-f) = r_S(e-f)M = (1-e)M$, and so $x = em = 0$, proving the claim. Inasmuch as $eS \sim fS$, we infer from Lemma 3.4, that $eM \sim fM$. Since M is a PD-module, it follows that $eM \oplus fM = M$. Again, by Lemma 3.4, $eS \oplus fS = S$, as required. □

PROPOSITION 3.12. *A module M is a PD-module if for any $e^2 = e \in S := End(M)$ and $v \in U(S)$ the equation $ex + [1-(1-e)v]y = 1$ is solvable for x and y in S.*

PROOF. By Proposition 3.6, we need to show that $S := End(M)$ is a right PD-ring. Let A and B be direct summands of S with $A \sim B$ and $A \cap B = 0$. We need to show that $S = A \oplus B$. Write $A = eS$ and $B = fS$, for some idempotents e, $f \in S$. Inasmuch as $A \sim B$, $S = eS \oplus C = fS \oplus C$, for a right ideal C of S. Now, $C = (1-e)S = (1-f)S$, and by [**17**, Exercise 21.4, p. 295], $(1-f) = (1-e)v$ for some $v \in U(S)$. By the hypothesis, the equation $ex + fy = ex + [1-(1-e)v]y = 1$ is solvable for x and y in S. This means that $eS + fS = S$, as required. □

While matrix rings over PD-rings need not be PD-rings, we will show below that corner rings over PD-rings are again PD-rings.

PROPOSITION 3.13. *If R is a right PD-ring, then so is eRe for any idempotent $e \in R$ such that $ReR = R$.*

PROOF. Since summands of PD-modules are again PD-modules, eR is a PD-module. Write $S := eRe \cong End_R(eR)$. By Proposition 3.10, we need to show that $f(eR) \cap g(eR) = 0$ for any pair of idempotents $f, g \in S$ with $fS \cap gS = 0$. To see this, let $ft = gs \in f(eR) \cap g(eR)$, where $t, s \in eR$. But then, for every $x \in R$, $fetxe = gesxe \in fS \cap gS = 0$. This implies that $fetRe = 0$, and so $fetReR = 0$. Inasmuch as $ReR = R$, we infer that $fetR = 0$, and so $ft = fet = 0$, as required. □

We end the paper by providing an example that shows that the notion of PD-rings is not left-right symmetric.

EXAMPLE 3.14 ([**18**, Example 4.8.]). Let F be a field and consider the associative algebra:

$$R := F\langle x, y, z \mid x^2 = x, y^2 = y, xy = y, yx = x, yz = xz, z^2 = 0, zxz = 0\rangle$$

It was shown in [**18**, Example 4.8.] that R as a right R-module is summand-square-free, and hence R is a right PD-ring. Moreover, it was also shown that the two isomorphic direct summands Rx and Ry satisfy $Rx \cap Ry = 0$. Inasmuch as $Rx \oplus Ry \neq R$, we infer from Proposition 1.6 that R is not a left PD-ring.

Acknowledgment

The first-named author is thankful to Dr. Seshadri Chintapalli for his constant support and suggestions while doing this project, and to IISER Berhampur for providing facilities to carry out this work with a Post-doctoral Fellowship.

References

[1] Frank W. Anderson and Kent R. Fuller, *Rings and categories of modules*, Graduate Texts in Mathematics, Vol. 13, Springer-Verlag, New York-Heidelberg, 1974. MR417223

[2] Meltem Altun-Özarslan, Yasser Ibrahim, A. Çiğdem Özcan, and Mohamed Yousif, *C4- and D4-modules via perspective direct summands*, Comm. Algebra **46** (2018), no. 10, 4480–4497, DOI 10.1080/00927872.2018.1448838. MR3847129

[3] Peter Crawley and Bjarni Jónsson, *Refinements for infinite direct decompositions of algebraic systems*, Pacific J. Math. **14** (1964), 797–855. MR169806

[4] Christian Lomp, Narayanaswami Vanaja, and Robert Wisbauer, *Lifting modules* (2006), xiv+394. Supplements and projectivity in module theory. MR2253001

[5] Alexander J. Diesl, Samuel J. Dittmer, and Pace P. Nielsen, *Idempotent lifting and ring extensions*, J. Algebra Appl. **15** (2016), no. 6, 1650112, 16, DOI 10.1142/S0219498816501127. MR3479816

[6] Nanqing Ding, Yasser Ibrahim, Mohamed Yousif, and Yiqiang Zhou, *D4-modules*, J. Algebra Appl. **16** (2017), no. 9, 1750166, 25, DOI 10.1142/S0219498817501663. MR3661633

[7] Nanqing Ding, Yasser Ibrahim, Mohamed Yousif, and Yiqiang Zhou, *C4-modules*, Comm. Algebra **45** (2017), no. 4, 1727–1740, DOI 10.1080/00927872.2016.1222412. MR3576690

[8] Gabriella D'Este, Derya Keskin Tütüncü, and Rachid Tribak, *D3-modules versus D4-modules—applications to quivers*, Glasg. Math. J. **63** (2021), no. 3, 697–723, DOI 10.1017/S0017089520000452. MR4301050

[9] Shelly Garg, Harpreet K. Grover, and Dinesh Khurana, *Perspective rings*, J. Algebra **415** (2014), 1–12, DOI 10.1016/j.jalgebra.2013.09.055. MR3229504

[10] David Handelman, *Perspectivity and cancellation in regular rings*, J. Algebra **48** (1977), no. 1, 1–16, DOI 10.1016/0021-8693(77)90289-7. MR447329

[11] Yasser Ibrahim and Mohamed Yousif, *Utumi modules*, Comm. Algebra **46** (2018), no. 2, 870–886, DOI 10.1080/00927872.2017.1339064. MR3764904

[12] Yasser Ibrahim and Mohamed Yousif, *U-modules with transitive perspectivity*, Comm. Algebra **49** (2021), no. 10, 4501–4512, DOI 10.1080/00927872.2021.1922699. MR4296853

[13] Yasser Ibrahim and Mohamed Yousif, *Perspectivity, exchange and summand-dual-square-free modules*, Comm. Algebra **50** (2022), no. 6, 2488–2506, DOI 10.1080/00927872.2021.2008414. MR4413847

[14] Dinesh Khurana, T. Y. Lam, and Zhou Wang, *Rings of square stable range one*, J. Algebra **338** (2011), 122–143, DOI 10.1016/j.jalgebra.2011.03.034. MR2805184

[15] Dinesh Khurana and Pace P. Nielsen, *Perspectivity and von Neumann regularity*, Comm. Algebra **49** (2021), no. 12, 5483–5499, DOI 10.1080/00927872.2021.1948555. MR4328552

[16] Dinesh Khurana and Pace P. Nielsen, *Transitivity of perspectivity*, Algebr. Represent. Theory **25** (2022), no. 1, 281–287, DOI 10.1007/s10468-020-10020-y. MR4368587

[17] T. Y. Lam, *A first course in noncommutative rings*, 2nd ed., Graduate Texts in Mathematics, vol. 131, Springer-Verlag, New York, 2001, DOI 10.1007/978-1-4419-8616-0. MR1838439

[18] Ryszard Mazurek, Pace P. Nielsen, and Michał Ziembowski, *Commuting idempotents, square-free modules, and the exchange property*, J. Algebra **444** (2015), 52–80, DOI 10.1016/j.jalgebra.2015.07.015. MR3406169

[19] Saad H. Mohamed and Bruno J. Müller, *Continuous and discrete modules*, London Mathematical Society Lecture Note Series, vol. 147, Cambridge University Press, Cambridge, 1990, DOI 10.1017/CBO9780511600692. MR1084376

[20] W. K. Nicholson, *Lifting idempotents and exchange rings*, Trans. Amer. Math. Soc. **229** (1977), 269–278, DOI 10.2307/1998510. MR439876

[21] Pace P. Nielsen, *Square-free modules with the exchange property*, J. Algebra **323** (2010), no. 7, 1993–2001, DOI 10.1016/j.jalgebra.2009.12.035. MR2594659

[22] R. B. Warfield Jr., *Cancellation of modules and groups and stable range of endomorphism rings*, Pacific J. Math. **91** (1980), no. 2, 457–485. MR615693

Department of Mathematical Sciences, Indian Institute of Science Education and Research (IISER), Berhampur 760010, Odisha, India
Email address: soumitrad330@gmail.com, soumitrad@iiserbpr.ac.in

Department of Mathematics, Faculty of Science, Taibah University, Madinah, Kingdom of Saudi Arabia and Department of Mathematics, Faculty of Science, Cairo University, Giza, Egypt
Email address: yabdelwahab@taibahu.edu.sa, yfibrahim@sci.cu.edu.eg

Department of Business Administration, Trakya University, Faculty of Economics and Administrative Sciences, Balkan Campus, 22030 Edirne, Turkey
Email address: ozgurtasdemir@trakya.edu.tr

Department of Mathematics, The Ohio State University, Lima, Ohio 45804
Email address: yousif.1@osu.edu

Contemporary Mathematics
Volume **826**, 2025
https://doi.org/10.1090/conm/826/16571

Invariant for sets of Pfister forms

Adam Chapman and Ilan Levin

ABSTRACT. We associate an $(n_1 + \cdots + n_t - k(t-1))$-fold Pfister form to any t-tuple of k-linked Pfister forms of dimensions $2^{n_1}, \ldots, 2^{n_t}$, and prove its invariance under the different symbol presentations of the forms with a common k-fold sub-symbol. We then show that it vanishes when the forms are actually $(k+1)$-linked or when the characteristic is 2 and the forms are inseparably k-linked. We study whether the converse statements hold or not.

1. Introduction

Let F be a field. When $char(F) \neq 2$, an n-fold form is a quadratic form of the shape

$$\langle\langle \alpha_1, \ldots, \alpha_n \rangle\rangle = \perp_{c_1,\ldots,c_n \in \{0,1\}} \langle (-1)^{c_1+\cdots+c_n} \alpha_1^{c_1} \cdot \ldots \cdot \alpha_n^{c_n} \rangle,$$

and when $\mathrm{char}(F) = 2$, of the shape

$$\langle\langle \alpha_1, \ldots, \alpha_{n-1}, \alpha_n]] = \perp_{c_1,\ldots,c_{n-1} \in \{0,1\}} \alpha_1^{c_1} \cdot \ldots \cdot \alpha_{n-1}^{c_{n-1}} \cdot [1, \alpha_n],$$

$$\text{where } [1, \alpha] = x^2 + xy + \alpha y^2.$$

Pfister forms correspond to the generators of the Galois cohomology groups $H^n(F, \mu_2)$ in characteristic not 2 or the Kato-Milne cohomology groups $H_2^n(F)$ in characteristic 2 (see [**9**], [**15**] and [**13**]).

When $\mathrm{char}(F) \neq 2$, there is a well-defined map associating to any sum of m symbols in $H^n(F, \mu_2)$, a single symbol in $H^{mn}(F, \mu_2)$ (see [**12**, Appendix A]), which is literally the tensor product of the symbols in the sum (as bilinear forms). There is no characteristic 2 analogue to this. The invariant we propose here can serve as a weak characteristic-free analogue, which associates an invariant to sets of Pfister forms (instead of sums of symbols in the cohomology groups), under the condition they share a certain common quadratic Pfister factor.

Pfister forms $\psi_1, \ldots, \psi_m$ are called **inseparably** k-linked if there exists a k-fold **bilinear** Pfister form B which is a Pfister factor of all the forms $\psi_1, \ldots, \psi_m$, i.e., there exist quadratic Pfister forms $\pi_1, \ldots, \pi_m$ such that $\psi_1 = B \otimes \pi_1, \ldots, \psi_m = B \otimes \pi_m$. Similarly, Pfister forms $\psi_1, \ldots, \psi_m$ are called **separably** k-linked if there exists a k-fold **quadratic** Pfister form π which is a Pfister factor of all the forms

2020 *Mathematics Subject Classification.* Primary 19D45; Secondary 11E04, 11E81, 13A35, 16K20, 20G10.

Key words and phrases. Algebraic K-Theory, Milnor K-Theory, Symmetric Bilinear Forms, Quadratic Forms, Symbol Length, Quaternion Algebras.

$\psi_1, \ldots, \psi_m$, i.e., there exist bilinear Pfister forms $B_1, \ldots, B_m$ such that $\psi_1 = B_1 \otimes \pi, \ldots, \psi_m = B_m \otimes \pi$.

When the forms are over a field of characteristic different from 2, the two notions coincide. The case of characteristic 2 is more subtle. By [**6**, Theorem 4.7], any finite system of inseparably linked quaternion algebras are also separably linked, and thus every finite number of quadratic Pfister forms that are inseparably k-linked are also separably k-linked. The converse is false (see [**14**] for example).

The study of linkage proved to be useful in the study of the u-invariant and other related arithmetic invariants: when every two 2-fold quadratic Pfister forms are 1-linked, the u-invariant is either 0,1,2,4 or 8 ([**10**] and [**4**]), and having $u(F) \leq 4$ is equivalent to 1-linkage of every 3 quadratic 2-fold Pfister forms when $\mathrm{char}(F) \neq 2$ ([**5**]). When $\mathrm{char}(F) = 2$, $u(F) \leq 4$ if and only if every 2 quadratic 2-fold Pfister forms are inseparably 1-linked ([**1**]). For the connection between linkage of pairs of quadratic n-fold Pfister forms for $n \geq 3$ and other similar arithmetic invariants, see [**8**].

The term "k-linkage" without specifying whether it is separable or inseparable can be by default considered separable.

In this paper we associate an $(n_1 + \cdots + n_t - k(t-1))$-fold Pfister form to any t-tuple of k-linked Pfister forms of dimensions $2^{n_1}, \ldots, 2^{n_t}$, and prove its invariance under the different symbol presentations of the forms with a common k-fold sub-symbol. As applications to the study of linkage, we show that it vanishes when the forms are actually $(k+1)$-linked or when the characteristic is 2 and the forms are inseparably k-linked. We study whether the converse statements hold or not.

2. Conditions for k-linkage

PROPOSITION 2.1. *Let F be a field of characteristic 2. Given separably k-linked Pfister forms φ and ψ, they are also inseparably k-linked $\iff$ ω is isotropic, where $\omega = \theta_\varphi \perp \theta_\psi \perp \langle 1 \rangle$, with θ_φ and θ_ψ such that $\varphi = \pi \perp \theta_\varphi$ and $\psi = \pi \perp \theta_\psi$, where π is a k-fold quadratic Pfister form.*

PROOF. Write $\varphi = \langle\langle \beta_1, \ldots, \beta_n, \alpha_1, \ldots, \alpha_k]]$ and $\psi = \langle\langle \gamma_1, \ldots, \gamma_m, \alpha_1, \ldots, \alpha_k]]$. Denote $\pi = \langle\langle \alpha_1, \ldots, \alpha_k]]$. By definition of a pure Pfister subform,

$$\varphi' = \langle\langle \beta_1, \ldots, \beta_n, \alpha_1, \ldots, \alpha_{k-1}\rangle\rangle' \otimes [1, \alpha_k] \perp \langle 1 \rangle = \langle\langle \alpha_1, \ldots, \alpha_{k-1}\rangle\rangle' \otimes [1, \alpha_k] \perp \theta_\varphi \perp \langle 1 \rangle$$

and

$$\psi' = \langle\langle \gamma_1, \ldots, \gamma_m, \alpha_1, \ldots, \alpha_{k-1}\rangle\rangle' \otimes [1, \alpha_k] \perp \langle 1 \rangle = \langle\langle \alpha_1, \ldots, \alpha_{k-1}\rangle\rangle' \otimes [1, \alpha_k] \perp \theta_\psi \perp \langle 1 \rangle.$$

We know that $\langle 1, 1 \rangle \cong \langle 1, 0 \rangle$ (since $x^2 + y^2 = (x+y)^2$), hence $\varphi' \perp \psi' = (2^k - 2) \times \mathbb{H} \perp \theta_\varphi \perp \theta_\psi \perp \langle 1 \rangle = (2^k - 2) \times \mathbb{H} \perp \omega$. Now, ω is isotropic $\iff$ $i_W(\varphi' \perp \psi') \geq 2^k - 1$, and by [**11**, Theorem 2.5.5], $i_W(\varphi' \perp \psi') \geq 2^k - 1 \iff \varphi$ and ψ are inseparably k-linked. This completes the proof. □

PROPOSITION 2.2. *Let F be a field of arbitrary characteristic. Given k-linked Pfister forms φ and ψ, they are $(k+1)$-linked $\iff$ ω is isotropic, where $\varphi \perp -\psi = 2^k \times \mathbb{H} \perp \omega$*

PROOF. First, write $\varphi = \pi \perp \theta_\varphi$ and $\psi = \pi \perp \theta_\psi$. Then, $\varphi \perp -\psi = \pi \perp -\pi \perp \theta_\varphi \perp -\theta_\psi$. So, $\varphi \perp -\psi = 2^k \times \mathbb{H} \perp \omega$, where $\omega = \theta_\varphi \perp -\theta_\psi$.

($\Longrightarrow$): If φ and ψ are $(k+1)$-linked, then $i_W(\varphi \perp -\psi) \geq 2^{k+1}$, and therefore $i_W(\omega) \geq 2^k$, and in particular, ω is isotropic.

($\Longleftarrow$): Suppose $\omega = \theta_\varphi \perp -\theta_\psi$ is isotropic. Then θ_φ and $-\theta_\psi$ represent a common element $a \in F^\times$. Let us claim that $\langle\langle -a\rangle\rangle \otimes \pi$ is a Pfister factor of both φ and ψ. By [**9**, Lemma 23.1], $\langle\langle -a\rangle\rangle \otimes \pi$ is a subform of φ and ψ, and so the forms are $(k+1)$-linked. $\square$

3. The invariant

Fix a field F in the background, and let PF denote the set of quadratic Pfister forms over that field.

LEMMA 3.1. *Given integers $n > k \geq 1$, a quadratic n-fold Pfister form ψ and two quadratic k-fold factors φ_1 and φ_2, there exists a bilinear Pfister form ρ for which $\psi = \rho \otimes \varphi_1$ and $\psi = \rho \otimes \varphi_2$.*

PROOF. It is enough to show that there exists $\gamma \in F^\times$ for which both $\langle\langle\gamma\rangle\rangle \otimes \varphi_1$ and $\langle\langle\gamma\rangle\rangle \otimes \varphi_2$ are subforms of ψ. Then the process can be repeated for $\langle\langle\gamma\rangle\rangle \otimes \varphi_1$ and $\langle\langle\gamma\rangle\rangle \otimes \varphi_2$, and so on, as long as the two forms we obtain are of smaller dimension than $\dim \psi$. The process terminates when the dimension of $\langle\langle\gamma\rangle\rangle \otimes \varphi_1$ and $\langle\langle\gamma\rangle\rangle \otimes \varphi_2$ is the same as $\dim \psi$, and then all three forms are isometeric to each other. We continue with the assumption that ψ is anisotropic, because otherwise we can simply take $\gamma = 1$.

Recall that ψ is endowed with an underlying symmetric bilinear form $B_\psi(v, w) = \psi(v + w) - \psi(v) - \psi(w)$ for any $v, w \in F^{\times 2^n} = \underbrace{F \times \cdots \times F}_{2^n \text{ times}}$, which gives rise to the notation of orthogonality. This symmetric bilinear form is non-degenerate in the sense that there is no vector in $F^{\times 2^n}$ that is perpendicular to all the vectors in $F^{\times 2^n}$. Now, consider the subspaces V_1 and V_2 of $F^{\times 2^n}$ corresponding to the subforms φ_1 and φ_2 of ψ. Note that V_1 and V_2 share at least the vector $(1, 0, \ldots, 0)$ whose image under ψ is 1. Write $V_1^\perp$ and $V_2^\perp$ for the orthogonal complements of V_1 and V_2 in $F^{\times 2^n}$. Their dimension is $2^n - 2^k$. Since they are both subspaces of $W = \mathrm{Span}_F\{(1, 0, \ldots, 0)\}^\perp$ and $\dim V_1^\perp + \dim V_2^\perp > \dim W$, the intersection $V_1^\perp \cap V_2^\perp$ is nontrivial. Take $v \in V_1^\perp \cap V_2^\perp \setminus \{\vec{0}\}$ and set $\gamma = -\psi(v)$. Then, by [**9**, Proposition 6.15 and 15.7], both $\langle\langle\gamma\rangle\rangle \otimes \varphi_1$ and $\langle\langle\gamma\rangle\rangle \otimes \varphi_2$ are subforms of ψ. $\square$

DEFINITION 3.2. Define $PF^2(k) = \{(\psi_1, \psi_2) \in PF \times PF : \psi_1 \text{ and } \psi_2 \text{ are } k\text{-linked}\}$.

DEFINITION 3.3 (The Invariant). Define $\mathcal{F}_k^2 : PF^2(k) \to PF$ by:

$$(\psi_1, \psi_2) \mapsto \varphi_1 \otimes \varphi_2 \otimes \pi$$

where π is a k-fold Pfister form for which $\psi_1 = \varphi_1 \otimes \pi$ and $\psi_2 = \varphi_2 \otimes \pi$ for some bilinear Pfister forms φ_1 and φ_2.

THEOREM 3.4. *$\mathcal{F}_k^2$ is well-defined.*

PROOF. Let us use colors to make it easier to follow the proof. Suppose $\psi_1 = \varphi_1 \otimes \pi = \tau_1 \otimes \mu$ and $\psi_2 = \varphi_2 \otimes \pi = \tau_2 \otimes \mu$, where π and μ are k-Pfister forms. We want to show that $\mathcal{F}_k^2(\psi_1, \psi_2) = \varphi_1 \otimes \varphi_2 \otimes \pi = \tau_1 \otimes \tau_2 \otimes \mu$. Since π and μ are sub-Pfister forms of ψ_1 and ψ_2, and thus by Lemma 3.1 there exist ρ_1 and ρ_2 such that $\varphi_1 \otimes \pi = \psi_1 = \rho_1 \otimes \pi = \rho_1 \otimes \mu = \psi_1 = \tau_1 \otimes \mu$ and $\varphi_2 \otimes \pi = \psi_2 = \rho_2 \otimes \pi = \rho_2 \otimes \mu = \psi_2 = \tau_2 \otimes \mu$.

And now: $\varphi_1 \otimes \varphi_2 \otimes \pi = \rho_1 \otimes \varphi_2 \otimes \pi = \rho_1 \otimes \varphi_2 \otimes \pi = \rho_1 \otimes \tau_2 \otimes \mu = \rho_1 \otimes \tau_2 \otimes \mu = \tau_1 \otimes \tau_2 \otimes \mu$, as desired. $\square$

We can now define the invariant for finite sequences of forms inductively. Denote $PF^m(k) = \{(\psi_1, \dots, \psi_m) \in \underbrace{PF \times \cdots \times PF}_{m} : \text{all } \psi_j \text{ are } k\text{-linked}\}$. Now, let

$$\mathcal{F}_k^m : PF^m(k) \to PF$$

defined by: $\mathcal{F}_k^m(\psi_1, \dots \psi_m) = \mathcal{F}_k^2(\mathcal{F}_k^{m-1}(\psi_1, \dots, \psi_{m-1}), \psi_m)$. It is still invariant, since $\mathcal{F}_k^{m-1}(\psi_1, \dots, \psi_{m-1})$ is a Pfister form that is k-linked to ψ_m. When the number m is obvious from the context, we shall write simply $\mathcal{F}_k$.

REMARK 3.5. In the special case where:

(1) $\sqrt{-1} \in F$,
(2) $\psi_1, \dots, \psi_t$ are n-fold, and
(3) they are separably $(n-1)$-linked,

we have $\Sigma_{\psi_1, \dots, \psi_t} = \mathcal{F}_{n-1}(\psi_1, \dots, \psi_t)$, where $\Sigma_{\psi_1, \dots, \psi_t}$ is the invariant defined in [**7**] for tight sets of quadratic Pfister forms. The following section thus generalizes the results from [**7**, Section 5].

4. From separable to inseparable linkage

In this section, F is a field of characteristic 2.

THEOREM 4.1. *Let $\psi_1 \in PF_{m+k}$ and $\psi_2 \in PF_{n+k}$ be k-linked Pfister forms (that is, separably k-linked).*

(1) *If ψ_1 and ψ_2 are also inseparably k-linked, then their k-invariant is trivial, i.e., $\mathcal{F}_k(\psi_1, \psi_2)$ is hyperbolic.*
(2) *If the k-invariant of ψ_1 and ψ_2 is trivial, and $m = 1$ or $n = 1$ (that is, at least one of them is a $(k+1)$-fold Pfister form), then the forms are inseparably k-linked.*

PROOF.

(1) Let $\psi_1 = \langle\langle a_1, \dots, a_m\rangle\rangle \otimes \pi$ and $\psi_2 = \langle\langle b_1, \dots, b_n\rangle\rangle \otimes \pi$. Consider $\mathcal{F}_k(\langle\langle a_1\rangle\rangle \otimes \pi, \langle\langle b_1\rangle\rangle \otimes \pi) = \langle\langle a_1\rangle\rangle \otimes \langle\langle b_1\rangle\rangle \otimes \pi$. By [CGV, theorem 5.3], since $\langle\langle a_1\rangle\rangle \otimes \pi, \langle\langle b_1\rangle\rangle \otimes \pi$ are inseparably k-linked, $\langle\langle a_1\rangle\rangle \otimes \langle\langle b_1\rangle\rangle \otimes \pi$ is hyperbolic, but it is also a Pfister factor of $\mathcal{F}_k(\psi_1, \psi_2)$, hence it is isotropic, hence hyperbolic, so it's trivial.
(2) Again, let $\psi_1 = \langle\langle a_1, \dots, a_m\rangle\rangle \otimes \pi$ and $\psi_2 = \langle\langle b_1, \dots, b_n\rangle\rangle \otimes \pi$. Now, denote by θ_1 , θ_2 the unique form (up to isometry) such that:
$$\psi_1 = \pi \perp \theta_1$$
$$\psi_2 = \pi \perp \theta_2$$
Note that $\dim(\theta_1) = 2^{m+k} - 2^k$ and $\dim(\theta_2) = 2^{n+k} - 2^k$. Define $\omega = \langle 1\rangle \perp \theta_1 \perp \theta_2$. Note that $\dim(\omega) = 2^{m+k} + 2^{n+k} - 2^{k+1} + 1$, and moreover, ω is a subform of $\mathcal{F}_k(\psi_1, \psi_2)$. Now, without loss of generality, suppose $n = 1$. Then $\dim(\omega) = 2^{m+k} + 1 > 2^{m+k} = \frac{1}{2}2^{m+k+1} = i_W(\mathcal{F}_k(\psi_1, \psi_2))$, where the last equality is true since $\dim(\mathcal{F}_k(\psi_1, \psi_2)) = 2^{m+k+1}$ and it is hyperbolic by assumption. Thus, ω is isotropic, and hence by Proposition 2.1, ψ_1 and ψ_2 are inseparably k-linked, as desired.

□

REMARK 4.2. Part (1) can be stated for a finite system of k-linked Pfister forms, following the exact same argument as in the proof.

EXAMPLE 4.3 (Counter-example for (2) with $n, m \geq 2$). Let ψ_1 $(m+k)$-fold Pfister form and ψ_2 $(n+k)$-fold Pfister form such that they are separably k-linked, but not inseparably k-linked. Then by Proposition 2.1, ω is anisotropic. Now, define $\delta = \mathcal{F}_k(\psi_1, \psi_2)$. Note that $\dim(\omega) = 2^{m+k} + 2^{n+k} - 2^{k+1} + 1$, and $\dim(\delta) = 2^{n+m+k}$. Now, notice that when $m, n \geq 2$, $2^{m+k} + 2^{n+k} - 2^{k+1} + 1 < 2^{n+m+k-1} + 1$, hence $2\dim(\omega) - 2 < \dim(\delta)$, which means that $2\dim(\omega) - 1 \leq \dim(\delta)$, and so by the separation theorem [**9**, Theorem 26.5], $\omega_{F(\delta)}$ is still anisotropic, hence ψ_1 and ψ_2 over $F(\delta)$ are not inseparably k-linked (Proposition 2.1), but their invariant over $F(\delta)$ is trivial.

EXAMPLE 4.4 (Counter-example for (2) with more than 2 forms, with at least one of them being $(k+1)$-fold). Let $F = \mathbb{F}_2^{sep}(x, y)$. Consider the following three forms: $\langle\langle \beta, \alpha]]$, $\langle\langle \alpha, \beta]]$, $\langle\langle \beta, \alpha\beta]]$. They are all separably 1-linked, but not inseparably 1-linked (see [**2**, Section 5]). Their invariant is trivial because they are pair-wise inseparably 1-linked. In order to see why they are separably 1-linked, one can repeat the argument in [**3**, Example 2.4], utilizing the fact that this field is C_2 to show that they are either separably 1-linked or inseparably 1-linked, and then either explain that since they are not inseparably 1-linked, they must be separably 1-linked, or use [**6**] to recall that inseparably linkage implies separable linkage anyway.

5. From k- to $(k+1)$-linkage

THEOREM 5.1. *Let F be a field of any characteristic, with $-1 \in (F^\times)^2$. Let ψ_1 and ψ_2 be $(k+1)$-linked Pfister forms. Then their k-invariant is trivial, i.e., $\mathcal{F}_k(\psi_1, \psi_2)$ is hyperbolic.*

PROOF. Let $\psi_1 = \langle\langle a, b_1, \dots, b_m\rangle\rangle \otimes \pi$ and $\psi_2 = \langle\langle a, c_1, \dots, c_n\rangle\rangle \otimes \pi$, where π is a k-fold Pfister form. Since $-1 \in (F^\times)^2$, $\langle\langle a\rangle\rangle = \langle\langle -a\rangle\rangle$, we get that $\mathcal{F}_k(\psi_1, \psi_2) = \langle\langle a, b_1, \dots, b_m\rangle\rangle \otimes \langle\langle -a, c_1, \dots, c_n\rangle\rangle \otimes \pi$. This is clearly an isotropic Pfister form, and thus it is hyperbolic. □

REMARK 5.2. For any ψ that is k-linked to $\langle\langle -1\rangle\rangle^m$ for some positive integer m bigger than k, the invariant $\mathcal{F}_k(\psi, \langle\langle -1\rangle\rangle^m)$ is trivial.

EXAMPLE 5.3 (Counter-example for Theorem 5.1 without assuming $-1 \in (F^\times)^2$). A trivial example for that would be $F = \mathbb{R}$ and take the n-fold Pfister forms $\psi_1 = \psi_2 = \langle\langle -1, \dots, -1\rangle\rangle$. For each $k \in \{1, \dots, n-1\}$, their k-invariant is $\mathcal{F}_k(\psi_1, \psi_2) = \langle\langle \underbrace{-1, \dots, -1}_{2n-k \text{ times}}\rangle\rangle$. This invariant is nontrivial, despite the fact that the forms are also $(k+1)$-linked.

EXAMPLE 5.4 (Counter-example for the converse of Theorem 5.1). Let ψ_1 m-fold Pfister form and ψ_2 n-fold Pfister form over a field F, k-linked but not $(k+1)$-linked. Write $\psi_1 = \varphi_1 \otimes \pi = \mu_1 \perp \pi$ and $\psi_2 = \varphi_2 \otimes \pi = \mu_2 \perp \pi$. By Proposition 2.2, $\omega = \mu_1 \perp \mu_2$ is anisotropic. Denote $\delta = \mathcal{F}_k(\psi_1, \psi_2)$. Extending scalars to $F(\delta)$, we get that δ is now hyperbolic, so the invariant is trivial. But, $\dim(\omega) = 2^n + 2^m - 2^{k+1} \leq 2^{n+m-k-1} < 2^{n+m-k} = \dim(\delta)$, and by the separation theorem [**9**, Theorem 26.5], ω is anisotropic, and by Proposition 2.2, ψ_1 and ψ_2 are not $(k+1)$-linked (over $F(\delta)$), although their invariant is trivial.

Acknowledgment

The authors wish to thank the referee for the careful reading of the manuscript and the helpful comments.

References

[1] R. Baeza, *Comparing u-invariants of fields of characteristic* 2, Bol. Soc. Brasil. Mat. **13** (1982), no. 1, 105–114, DOI 10.1007/BF02584739. MR692281

[2] Adam Chapman, *Common slots of bilinear and quadratic Pfister forms*, Bull. Aust. Math. Soc. **98** (2018), no. 1, 38–47, DOI 10.1017/S0004972718000229. MR3819970

[3] Adam Chapman, *Linkage of sets of quaternion algebras in characteristic 2*, Comm. Algebra **49** (2021), no. 9, 3680–3684, DOI 10.1080/00927872.2021.1902532. MR4290101

[4] Adam Chapman and Andrew Dolphin, *Differential forms, linked fields, and the u-invariant*, Arch. Math. (Basel) **109** (2017), no. 2, 133–142, DOI 10.1007/s00013-017-1059-7. MR3673632

[5] Adam Chapman, Andrew Dolphin, and David B. Leep, *Triple linkage of quadratic Pfister forms*, Manuscripta Math. **157** (2018), no. 3-4, 435–443, DOI 10.1007/s00229-017-0996-6. MR3858411

[6] Adam Chapman, Mathieu Florence, and Kelly McKinnie, *Common splitting fields of symbol algebras*, Manuscripta Math. **171** (2023), no. 3-4, 649–662, DOI 10.1007/s00229-022-01401-2. MR4597712

[7] Adam Chapman, Shira Gilat, and Uzi Vishne, *Linkage of quadratic Pfister forms*, Comm. Algebra **45** (2017), no. 12, 5212–5226, DOI 10.1080/00927872.2017.1298776. MR3691360

[8] Adam Chapman and Kelly McKinnie, *The u^n-invariant and the symbol length of $H_2^n(F)$*, Proc. Amer. Math. Soc. **147** (2019), no. 2, 513–521, DOI 10.1090/proc/14308. MR3894891

[9] Richard Elman, Nikita Karpenko, and Alexander Merkurjev, *The algebraic and geometric theory of quadratic forms*, American Mathematical Society Colloquium Publications, vol. 56, American Mathematical Society, Providence, RI, 2008, DOI 10.1090/coll/056. MR2427530

[10] Richard Elman and T. Y. Lam, *Quadratic forms and the u-invariant. II*, Invent. Math. **21** (1973), 125–137, DOI 10.1007/BF01389692. MR417053

[11] Frédéric Faivre, *Liaison des formes de Pfister et corps de fonctions de quadriques en caractéristique* 2, 2006, Thesis (Ph.D.)–Université de Franche-Comté.

[12] Bruno Kahn, *Comparison of some field invariants*, J. Algebra **232** (2000), no. 2, 485–492, DOI 10.1006/jabr.2000.8380. MR1792742

[13] Kazuya Kato, *Symmetric bilinear forms, quadratic forms and Milnor K-theory in characteristic two*, Invent. Math. **66** (1982), no. 3, 493–510, DOI 10.1007/BF01389226. MR662605

[14] T. Y. Lam, *On the linkage of quaternion algebras*, Bull. Belg. Math. Soc. Simon Stevin **9** (2002), no. 3, 415–418. MR2016580

[15] Vladimir Voevodsky, *Motivic cohomology with* $\mathbf{Z}/2$*-coefficients*, Publ. Math. Inst. Hautes Études Sci. **98** (2003), 59–104, DOI 10.1007/s10240-003-0010-6. MR2031199

School of Computer Science, Academic College of Tel-Aviv-Yaffo, Rabenu Yeruham St., P.O.B 8401 Yaffo, 6818211, Israel
Email address: `adam1chapman@yahoo.com`

Department of Mathematics, Bar-Ilan University, Ramat Gan, 55200, Israel
Email address: `ilan7362@gmail.com`

Contemporary Mathematics
Volume **826**, 2025
https://doi.org/10.1090/conm/826/16573

Codes constructed from matrices associated with Lie algebras

Joshua D. Carey and Steven T. Dougherty

Abstract. We construct self-orthogonal codes over rings from binary weight matrices (i.e. weight matrices modulo 2) of fundamental Lie modules. We show that the binary weight matrix of any type A_n fundamental Lie module produces a self-orthogonal code of rank n over any Frobenius ring with characteristic 2 given certain divisibility conditions. Additionally, we show that using the ring R_h, self-dual codes of length $2^h\binom{n}{k}$ can be constructed from Lie algebras.

1. Introduction

There is a long history of using combinatorial objects to generate codes. For example, there is an extensive literature on the relationship between codes and finite designs, especially finite planes. For a complete description of this connection see [**1**]. There are numerous other connections between combinatorial objects and codes (see [**2**]). Constructions of interesting codes have also been made from graphs and association schemes, see, for example, [**6**] and [**7**] and the references therein. In this paper, we are going to construct codes from binary weight matrices of Lie algebras and examine their properties.

Self-orthogonal codes, that is codes which satisfy $C \subseteq C^{\perp}$, where $C^{\perp}$ is the orthogonal under the Euclidean inner-product, are a widely studied class of linear codes. The extremal case of self-orthogonal codes are self-dual codes, that is codes which satisfy $C = C^{\perp}$. These codes are one of the most widely studied families of codes in algebraic coding theory. There are numerous reasons for this. One of the reasons is that the codes are interesting in themselves in terms of coding theory, most especially the optimal codes (that is codes with the highest possible minimum weight among all self-dual codes). Specifically, the extremal self-dual codes (those who reach the bound set by invariant theory applied to the weight enumerators) are of great importance. For example, see the open question on the Type II $[72, 36, 16]$ code in [**5**]. Another reason for their importance is that self-dual codes have a strong connection to the theory of designs via the Assmus-Mattson Theorem (see [**3**]), that is, interesting and important designs are constructed from the codewords of a given weight in certain self-dual codes, especially the extremal ones. A third reason is that there is a canonical connection between self-dual codes over various rings and unimodular lattices in certain ambient spaces, which are important structures in the geometry of numbers and in number theory in general. See [**4**] for a complete description of self-dual codes in a very general setting or Chapter 19 in [**10**] for a foundational description of self-dual codes. See also [**9**] for a later description of self-dual codes in a classical setting.

Numerous constructions of self-orthogonal and self-dual codes have been given in various settings. In this work, we expand this greatly by using weight matrices from Lie algebras to construct self-orthogonal and self-dual codes. We build on the work first done in [**12**] generalizing the constructions in this work significantly.

1.1. Codes. Let R be a finite commutative Frobenius ring. Recall that a Frobenius ring R has its character module generated by a single character, that is $\widehat{R} = \langle \chi \rangle$ for some character χ of R. We restrict ourselves to Frobenius rings so that the MacWilliams relations hold and as such the fundamental results about the cardinality of the orthogonal also hold. We assume that all rings have a multiplicative identity. A code of length m over R is a set C with $C \subseteq R^m$. If the code is also a submodule of R^m then the code is said to be linear. If a code over a ring R is isomorphic as a module to R^s, for some non-negative integer s, then the code is said to be a free code of rank s. If R is a field then all linear codes are free and the rank is the dimension. Of course, over a ring that is not a field, any non-trivial ideal is a linear code that is not free.

The ambient space has the standard Euclidean inner-product, that is for $\mathbf{v}, \mathbf{w} \in R^m$, $[\mathbf{v}, \mathbf{w}] = \sum v_i w_i$. The orthogonal of the code C is defined as $C^\perp = \{\mathbf{v} \mid [\mathbf{v}, \mathbf{w}] = 0, \forall \mathbf{w} \in C\}$. The dual code $C^\perp$ is a linear code even if C is not. It is known (see [**4**]) that a linear code over a commutative Frobenius ring R satisfies the MacWilliams relations which in turn implies the important result that $|C||C^\perp| = |R|^m$. Codes over rings that are not Frobenius may not satisfy this result and in fact they may not satisfy the double annihilator condition that $(C^\perp)^\perp = C$.

If $C \subseteq C^\perp$, then C is said to be self-orthogonal. If $C = C^\perp$ then C is said to be self-dual. A self-orthogonal binary code is said to be doubly-even if the weights of all vectors are congruent to $0 \pmod 4$. It is immediate that any $(0,1)$-matrix whose rows are orthogonal with the property that the weight of all rows is congruent to $0 \pmod 4$ generates a doubly-even code. We note that both conditions are necessary to generate a doubly-even code. A binary self-dual code that is doubly-even is said to be a Type II code and a binary self-dual code with at least one singly-even vector is said to be a Type I code. It is well known that Type II codes exist if and only if the length is a multiple of 8 and Type I codes exist for all even lengths.

We know that the MacWilliams relations hold over finite commutative rings and that $C = C^\perp$ for self-dual codes. These results place strong restrictions on the weight enumerators of self-dual codes via invariant theory. See Chapter 19 of [**10**] for a description. This gives a powerful tool in determining when a self-dual code is optimal.

We now define a family of rings. We use this family as the alphabet for codes since they have a canonical connection to binary codes. These rings have been studied extensively and are generally denoted by R_k. We shall change k to h here since k will be used in terms of Lie algebras. Define the following ring for $h \geq 1$. Let

$$R_h = \mathbb{F}_2[u_1, u_2, \ldots, u_h]/\langle u_i^2 = 0, u_i u_j = u_j u_i \rangle. \tag{1.1}$$

The ring R_h has cardinality 2^{2^h} and is a finite local commutative ring. It is easy to see that this ring is Frobenius. Namely, the ring has a unique maximal ideal given by $\langle u_1, u_2, \ldots, u_h \rangle$. For any subset $A \subseteq \{1, 2, \ldots, h\}$ let

$$u_A := \prod_{i \in A} u_i \tag{1.2}$$

with the convention that $u_\emptyset = 1$. Then any element of R_h can be represented as

$$\sum_{A \subseteq \{1, \ldots, h\}} c_A u_A,$$

where $c_A \in \mathbb{F}_2$. Then a character defined by $\chi(\sum_{A \subseteq \{1,\ldots,h\}} c_A u_A) = \prod(-1)^{c_A}$ is non-trivial on the minimal ideal. This gives that χ is a generating character and the ring is Frobenius. See **[4]** for details.

We define the Gray map inductively, extending it naturally from the Gray map on R_1. For $\mathbf{c} \in R_h^n$, we can write $\mathbf{c} = \mathbf{c}_1 + u_h \mathbf{c}_2$ with $\mathbf{c}_1, \mathbf{c}_2 \in R_{h-1}^n$, then we can define

$$\varphi_h(\mathbf{c}) = (\varphi_{h-1}(\mathbf{c}_2), \varphi_{h-1}(\mathbf{c}_1) + \varphi_{h-1}(\mathbf{c}_2)),$$

with φ_0 being the identity map on $\mathbb{F}_2$.

It is shown in **[8]**, that if C is a self-dual code over R_h of length m, then $\varphi_h(C)$ is a self-dual code of length $2^k m$ over $\mathbb{F}_2$, which follows from the fact that $\varphi_h(C^\perp) = \varphi_h(C)^\perp$ for linear codes C over R_h.

2. Codes from $C_{n,k}$

In this section, we shall describe the codes that are formed from the special matrices, $C_{n,k}$. These matrices arise the representation theory of Lie Algebras, although Lie theory will not be needed in this paper. Due to the vast connections between Lie Algebras and other areas of mathematics, these matrices likely arise in other interesting ways which we do not explore here.

2.1. The matrices $C_{n,k}$. The matrices $C_{n,k}$ are, up to a potential reordering of rows and/or columns, the binary weight matrices (i.e. weight matrices modulo 2) of the fundamental Lie module V^{λ_k} over a type A_{n-1} Lie algebra. In **[12]**, Xu examines, among other things, codes generated by $C_{n,k}$ given certain restrictions on n and k. Xu's results are obtained, in part, using the representation theory of Lie Algebras. In this paper, we will expand most of Xu's results for these codes for all values of n and k. Further, we will do so using only combinatorial methods, save Lemma 2.4 below.

For $n \geq 2$, Let $C_{n,k}$ be the binary weight matrix (i.e. the weight matrix mod 2) of $\mathfrak{sl}_n$ acting on the module V^{λ_k}. Since this is a binary matrix, we can replace $h_i = E_{ii} - E_{(i+1)(i+1)}$ with $h_i = E_{ii} + E_{(i+1)(i+1)}$. As a result of the action of E_{ii}, we can construct this matrix in the following purely combinatorial way, which will later aid with our computations. Let $X = \{x_1, x_2, \ldots, x_n\}$ be some ordered set and let $X_k = \{Y \subset X \mid |Y| = k\}$, that is the set of all subsets of X of cardinality k. Then $|X_k| = \binom{n}{k}$. Give X_k some order, say $X_k = \{X_{k_1}, X_{k_2}, \ldots, X_{k_{\binom{n}{k}}}\}$, and let $B_{n,k} = (b_{i,j})$ be the $n \times \binom{n}{k}$ matrix where $b_{i,j} = 1$ if and only if $x_i \in X_{k_j}$ and 0 otherwise. Let ζ_i denote the ith row of $B_{n,k}$. Working in $\mathbb{F}_2^{\binom{n}{k}}$, let $\xi_i = \zeta_i + \zeta_{i+1}$.

DEFINITION 2.1. The matrix $C_{n,k}$, $n \geq 2$, is an $(n-1)$ by $\binom{n}{k}$ matrix whose i^{th} row is ξ_i.

The matrices $C_{n,k}$ are, up to a potential reordering of rows and/or columns, the binary weight matrices (i.e. weight matrices modulo 2) of the fundamental Lie module V^{λ_k} over a type A_n Lie algebra.

We note here that most of the formulas we derive below will only work for $n > k+1$. However, in the special case $n = k+1$, we have that

$$C_{k+1,k} = \begin{pmatrix} 1 & 1 & 0 & 0 & \cdots & 0 & 0 & 0 \\ 0 & 1 & 1 & 0 & \cdots & 0 & 0 & 0 \\ & & & & \vdots & & & \\ 0 & 0 & 0 & 0 & \dots & 0 & 1 & 1 \end{pmatrix}. \tag{2.1}$$

Since $C_{k+1,k}$ has such an easily defined structure, we will be able to obtain counterparts of our formulas for this special case with relative ease. In particular, this matrix will cover the only case where $n = 2$ as this implies the only possible value for k is 1.

Since 0 and 1 are in any ring, we can make the following definition.

DEFINITION 2.2. Let R be a finite commutative Frobenius ring. Then $\mathfrak{C}_{n,k}$ is the row space of $C_{n,k}$ in $R^{\binom{n}{k}}$. If the ring is $\mathbb{F}_2$, then we denote by $\mathcal{C}_{n,k}$, the binary code generated by $C_{n,k}$ in $\mathbb{F}_2^{\binom{n}{k}}$.

In [**12**], only binary and ternary codes were considered but there is no reason to have such a great restriction on the codes constructed. Of course, since the codes $\mathfrak{C}_{n,k}$ and $\mathcal{C}_{n,k}$ are row spaces they are linear codes. We shall endeavor to determine the properties of these codes.

THEOREM 2.3. *Let R be a finite commutative Frobenius ring. The code $\mathfrak{C}_{k+1,k}$ is a free code of rank k and length $k+1$ over R and $\mathfrak{C}_{k+1,k}^{\perp} = \langle (1,-1,1,-1,\dots)\rangle$. The code $\mathcal{C}_{k+1,k}$ is the code E_{k+1} which consists of all even weight vectors.*

PROOF. First we note that $\binom{k+1}{k} = k+1$ which gives the length. It is easy to see that the rows are linearly independent. This gives that the rank is k since there are k rows in the generator matrix. Then the orthogonal must be a free code and have rank 1. The orthogonal then is $\langle (1,-1,1,-1,\dots)\rangle$ by noticing that this vector is orthogonal to every generator of the code. For the binary case, the orthogonal is generated by the all-one vector so the code must consist of every codeword of even weight. □

2.2. Hamming weights in $\mathcal{C}_{n,k}$. Recall that ζ_i denotes the ith row of $B_{n,k}$.

LEMMA 2.4. *Let $u = \zeta_{i_1} + \zeta_{i_2} + \cdots + \zeta_{i_{2t}} \in \mathcal{C}_{n,k}$ with $i_1, \cdots, i_{2t} \in \{1,\dots,n\}$. Then* $\mathrm{wt}(u) = \mathrm{wt}(\zeta_1 + \cdots + \zeta_{2t})$.

PROOF. This result, which was obtained by Xu in [**12**] is due to formulas 1.10 and 1.11 in that work. It is expanded upon in the proof of Theorem 2.1 in [**12**]. □

Let $u \in \mathcal{C}_{n,k}$. Recall that $\xi_i = \zeta_i + \zeta_{i+1}$. Then $u = \xi_{i_1} + \xi_{i_2} + \cdots + \xi_{i_\ell} = \zeta_{j_1} + \zeta_{j_2} + \cdots + \zeta_{j_{2t}}$ for some $1 \le \ell \le n-1$ and $1 \le t \le \lfloor \frac{n}{2} \rfloor$. By Lemma 2.4, for any $j_1, j_2, \dots, j_{2t} \in \{1,\dots,n\}$ we have that

$$\mathrm{wt}(\zeta_{i_1} + \cdots + \zeta_{i_{2t}}) = \mathrm{wt}(\zeta_1 + \cdots + \zeta_{2t}). \tag{2.2}$$

For t as above, let Z_t be the matrix whose rows are $\zeta_1, \zeta_2, \dots, \zeta_{2t}$. Additionally, define $\mathbf{0} := (0,0,\dots,0)$ and $\mathbf{1} := (1,1,\dots,1)$.

THEOREM 2.5. *Let $u \in \mathcal{C}_{n,k}$ such that $u \neq \mathbf{0}, \mathbf{1}$. Then:*

$$\mathrm{wt}(u) = \sum_{i=0}^{(p-L)/2} \binom{2t}{p-2i}\binom{n-2t}{k-(p-2i)}$$

where

$$L = \min\{r \mid 1 \leq r \leq p, r \text{ is odd and } n-2t \geq k-r\},$$
$$p = 2\left\lceil \frac{m}{2} \right\rceil - 1 \text{ and } m = \min(2t, k).$$

PROOF. By our remarks above, there exists $t \in \{1, \dots, \lfloor \frac{n}{2} \rfloor\}$ such that $\mathrm{wt}\, u = \mathrm{wt}(\zeta_1 + \zeta_2 + \cdots + \zeta_{2t})$. There will be a one in the ith entry of u if and only if the ith column of Z_t contains an odd number of ones. Thus, we can find the weight of u by counting how many columns of Z_t contain an odd number of ones.

The maximum number of ones in any column of Z_t will be the minimum of $2t$ and k. Let $Y_1 = \{x_1, \dots, x_{2t}\}$ and $Y_2 = \{x_{2t+1}, \dots, x_n\}$. There will be $r < m$ ones in column i if and only if X_{k_i} contains r elements from Y_1 and $k-r$ elements from Y_2, which is in turn possible if and only if $n - 2t \geq k - r$. Thus, the minimum number of ones in any column of Z_t will be $\min\{r \mid 1 \leq r \leq m \text{ and } n-2t \geq k-r\}$.

The number of ways for X_{k_i} to have r elements from Y_1 and $k - r$ elements from Y_2 is

$$\binom{2t}{r}\binom{n-2t}{k-r}$$

but we only care about columns with an odd number of ones. Hence, letting p be the largest odd number less than or equal to m and $L = \min\{r \mid 1 \leq r \leq p, r \text{ is odd and } n-2t \geq k-r\}$, we have:

$$\mathrm{wt}(u) = \sum_{i=0}^{(p-L)/2} \binom{2t}{p-2i}\binom{n-2t}{k-(p-2i)}.$$

□

COROLLARY 1. *Let ξ_i be as above. Then*

$$\mathrm{wt}(\xi_i) = 2\binom{n-2}{k-1}.$$

PROOF. Using Theorem 2.5, set $t = 1$. Since $n \geq k+1$, it follows that $n - 2t = n - 2 \geq k - 1$ and so $L = 1$. If $k = 1$ then $m = 1$, otherwise if $k > 1$ then $m = 2$. In both cases, this gives that $p = 1$ and the conclusion follows immediately. □

Setting $k = 2$ and $k = 3$ in Corollary 1 yields equations (2.13) and (2.21) from [**12**], respectively.

2.2.1. *Special case:* $k = 2$. Let $u \in \mathcal{C}_{n,2}$ such that $u \neq \mathbf{0}, \mathbf{1}$. If n is even then $\zeta_1 + \zeta_2 + \cdots + \zeta_n = \mathbf{0}$. Thus, $2t < n$ and $n - 2t \geq 1 = k - 1$, giving that $L = 1$. Further, since $t \geq 1$ we have $m = 2$ and, additionally, $p = 1$. Thus, by Theorem 2.5

$$\mathrm{wt}(u) = \binom{2t}{1}\binom{n-2t}{1} = 2t(n-2t). \tag{2.3}$$

Restricting further to the special case where n is even, we recover formula (2.19) in [**12**].

2.2.2. *Special case:* $k = 3$. Let $u \in \mathcal{C}_{n,3}$ such that $u \neq \mathbf{0}, \mathbf{1}$. If $t = 1$ then $u = \xi_i$ and Corollary 1 gives us that $\mathrm{wt}(u) = 2\binom{n-2}{2} = n - 2$. Now assume $t > 1$. Then $m = 3$. We will first assume $n - 2t \geq k - 1 = 2$. Then $L = 1$ and, since $p = 3$, we have:

$$\mathrm{wt}(u) = \binom{2t}{3}\binom{n-2t}{0} + \binom{2t}{1}\binom{n-2t}{2} \tag{2.4}$$

$$= \frac{1}{3}t(2t-1)(2t-2) + t(n-2t)(n-2t-1). \tag{2.5}$$

If $n - 2t < k - 1$ then $n - 2t = 0$ or 1. In the former case, this would mean $u = \mathbf{1}$. In the latter, this gives us that $L = 3$ and so

$$\mathrm{wt}(u) = \binom{2t}{3} = \frac{1}{3}t(2t-1)(2t-2). \tag{2.6}$$

Notice that if $n - 2t = 1$, then Equation (2.5) simplifies to Equation (2.6) and so (2.5) covers both the cases $n - 2t = 1$ and $n - 2t > 1$. We note that (2.5) confirms Equation (2.25) in [**12**].

2.3. Orthogonality. We now consider the orthogonality of rows in the matrix $C_{n,k}$. Since these vectors are restricted to elements from the set $\{0, 1\}$ we shall use $\mathbf{v} \cdot \mathbf{w}$ to be $\mathbf{v} \cdot \mathbf{w} = \sum v_i w_i$ where the sum is done in $\mathbb{Z}$ rather than over a finite ring R.

THEOREM 2.6. *Let $n > k + 1$ and let ξ_i and ξ_j represent the ith and j^{th} rows respectively of $C_{n,k}$ as above with $j > i + 1$. Then*

$$\xi_i \cdot \xi_j = 4\binom{n-4}{k-2}.$$

PROOF. Notice that ξ_i will have a 1 in column r if and only if X_{k_r} contains x_i or x_{i+1}, but not both. So for ξ_i and ξ_j to both have a 1 in column r, it must be that X_{k_r} satisfies one of the following four cases:

(i) $x_i, x_j \in X_{k_r}$ and $x_{i+1}, x_{j+1} \notin X_{k_r}$
(ii) $x_{i+1}, x_j \in X_{k_r}$ and $x_i, x_{j+1} \notin X_{k_r}$
(iii) $x_i, x_{j+1} \in X_{k_r}$ and $x_{i+1}, x_j \notin X_{k_r}$
(iv) $x_{i+1}, x_{j+1} \in X_{k_r}$ and $x_i, x_j \notin X_{k_r}$.

In each case, we have identified two elements of X_{k_r} and two non-elements of X_{k_r}. This leaves $k - 2$ more elements in X_{k_r} and $n - 4$ elements of X from which to choose. Thus, for each of these four cases, there are $\binom{n-4}{k-2}$ ways to chose the remaining elements of X_{k_r}, giving the formula above. $\square$

Next, we consider the case where $j = i + 1$.

THEOREM 2.7. *Let $n > k + 1$ and ξ_i be as in Theorem 2.6. Then*

$$\xi_i \cdot \xi_{i+1} = \binom{n-2}{k-1}.$$

PROOF. We will proceed as in the proof of Theorem 2.6 above. The vectors ξ_i and ξ_{i+1} will both have a 1 in column r when one of the following two cases holds:

(i) $x_i, x_{i+2} \in X_{k_r}$ and $x_{i+1} \notin X_{k_r}$,
(ii) $x_{i+1} \in X_{k_r}$ and $x_i, x_{i+2} \notin X_{k_r}$.

Thus,

$$\xi_i \cdot \xi_{i+1} = \binom{n-3}{k-2} + \binom{n-3}{k-1} = \binom{n-2}{k-1}.$$

□

Because of the structure of (2.1), we immediately have the following:

THEOREM 2.8. *Let $n = k+1$, $j > i+1$ and ξ_i and ξ_j be as in Theorems 2.6 and 2.7. Then, we have $\xi_i \cdot \xi_j = 0$ and $\xi_i \cdot \xi_{i+1} = 1$.*

2.3.1. *Special cases, $k = 2, 3$.* If $k = 2$ or 3, Theorems 2.6 and 2.7 immediately give (2.15), (2.16), (2.22) and (2.23) in [**12**].

2.4. Rank of weight matrices.

LEMMA 2.9. *Let $\tilde{I}_n$ be the n by n matrix whose i,j-entry is 0 for $i+j = n+1$ and 1 otherwise. Then the rank of $\tilde{I}_n$ over a ring of characteristic c is n when $n-1$ is not divisible by c and $n-1$ when $n-1$ is divisible by c.*

PROOF. Using the first row to eliminate the others we have:

$$(2.7) \qquad \begin{pmatrix} 1 & 1 & \cdots & 1 & 1 & 0 \\ 1 & 1 & \cdots & 1 & 0 & 1 \\ 1 & 1 & \cdots & 0 & 1 & 1 \\ & & \vdots & & & \\ 0 & 1 & \cdots & 1 & 1 & 1 \end{pmatrix} \sim \begin{pmatrix} 1 & 1 & \cdots & 1 & 1 & 0 \\ 0 & 0 & \cdots & 0 & -1 & 1 \\ 0 & 0 & \cdots & -1 & 0 & 1 \\ & & \vdots & & & \\ -1 & 0 & \cdots & 0 & 0 & 1 \end{pmatrix}.$$

Next, adding the bottom $n-1$ rows to the top we have:

$$(2.8) \qquad \begin{pmatrix} 0 & 0 & \cdots & 0 & 0 & n-1 \\ 0 & 0 & \cdots & 0 & -1 & 1 \\ 0 & 0 & \cdots & -1 & 0 & 1 \\ & & \vdots & & & \\ -1 & 0 & \cdots & 0 & 0 & 1 \end{pmatrix} \sim \begin{pmatrix} -1 & 0 & \cdots & 0 & 0 & 1 \\ 0 & -1 & \cdots & 0 & 0 & 1 \\ & & \vdots & & & \\ 0 & 0 & \cdots & 0 & -1 & 1 \\ 0 & 0 & \cdots & 0 & 0 & n-1 \end{pmatrix}.$$

This matrix has rank n if c does not divide $n-1$ and rank $n-1$ if c does divide $n-1$. □

Recall that $B_{n,k} = (b_{i,j})$ is the $n \times \binom{n}{k}$ matrix where $b_{i,j} = 1$ if and only if $x_i \in X_{k_j}$ and 0 otherwise.

LEMMA 2.10. *The $n \times \binom{n}{k}$ matrix $B_{n,k}$ has rank n over a ring of characteristic c when $k-1$ is not divisible by c and rank $n-1$ or greater when $k-1$ is divisible by c.*

PROOF. The columns of $B_{n,k}$ are precisely the elements of the set of all n-dimensional, $(0,1)$-column vectors. If $n = k+1$ then using the construction of $B_{n,k}$ above, we can choose an ordering of X_k so that $B_{k+1,k} = \tilde{I}_{k+1}$. In this case our results immediately follow from Lemma 2.9. If $n > k+1$ then we can choose an ordering of X_k so that

$$B_{n,k} = \left(\begin{array}{c|cccc|c}
\tilde{I}_k & \tilde{I}_{k-1} & \tilde{I}_{k-1} & \cdots & \tilde{I}_{k-1} & * \\
 & 0\cdots0 & 0\cdots0 & & 0\cdots0 & \\ \hline
 & 1\cdots1 & 0\cdots0 & & 0\cdots0 & \\
 & 0\cdots0 & 1\cdots1 & & 0\cdots0 & \\
O_{n-k,k} & \vdots & 0\cdots0 & \cdots & 0\cdots0 & * \\
 & \vdots & \vdots & & \vdots & \\
 & \vdots & \vdots & & 0\cdots0 & \\
 & 0\cdots0 & 0\cdots0 & & 1\cdots1 &
\end{array}\right) \tag{2.9}$$

where $O_{a,b}$ is the $a \times b$ matrix whose entries are all 0. Lemma 2.9 gives that (2.9) is row equivalent to:

$$B_{n,k} = \left(\begin{array}{c|cccc|c}
A & \tilde{I}_{k-1} & \tilde{I}_{k-1} & \cdots & \tilde{I}_{k-1} & * \\
 & 0\cdots0 & 0\cdots0 & & 0\cdots0 & \\ \hline
 & 1\cdots1 & 0\cdots0 & & 0\cdots0 & \\
 & 0\cdots0 & 1\cdots1 & & 0\cdots0 & \\
O_{n-k,k} & \vdots & 0\cdots0 & \cdots & 0\cdots0 & * \\
 & \vdots & \vdots & & \vdots & \\
 & \vdots & \vdots & & 0\cdots0 & \\
 & 0\cdots0 & 0\cdots0 & & 1\cdots1 &
\end{array}\right) \tag{2.10}$$

where A is the second matrix in (2.8) above with n replaced by k. If c does not divide $k-1$, then (2.10) is in echelon form and so the rank of $B_{n,k}$ is n. If, however, c divides $k-1$, then we move row k to the bottom, putting (2.10) into echelon form, although it is unknown if this row contains any non-zero entries. Thus $B_{n,k}$ is either of rank $n-1$ or n. □

COROLLARY 2.

(i) *If* $\operatorname{rk} B_{n,k} = n$, *then the set* $\{\zeta_1, \zeta_2, \ldots, \zeta_n\}$ *is linearly independent.*
(ii) *If* $\operatorname{rk} B_{n,k} = n-1$ *then* $\{\zeta_2, \zeta_3, \ldots, \zeta_n\}$ *is linearly independent and* $\zeta_1 = -\zeta_2 - \zeta_3 - \cdots - \zeta_k$.

PROOF. Recall that ζ_i is defined to be the ith row of $B_{n,k}$. Then, if $\operatorname{rk} B_{n,k} = n$, (i) follows immediately from Lemma 2.10. If $\operatorname{rk} B_{n,k} = n-1$, then (ii) follows from equation (2.10) and the proof of Lemma 2.9. □

THEOREM 2.11. *Let R be a ring of characteristic c. Then the code generated by $C_{n,k}$ over R will be a $[\binom{n}{k}, n-1]$ code if one of the following three scenarios occurs:*

(i) *c does not divide $k-1$*
(ii) *c divides $k-1$ and $\operatorname{rk} B_{n,k} = n$*
(iii) *c divides $k-1$, $\operatorname{rk} B_{n,k} = n-1$ and k is odd.*

Otherwise, $C_{n,k}$ will generate a $[\binom{n}{k}, n-2]$ code.

PROOF. First assume that $\operatorname{rk} B_{n,k} = n$. Recall that $\xi_i = \zeta_i + \zeta_{i+1}$ for $i = 1, \ldots, n-1$. If $\operatorname{rk} B_{n,k} = n$, then by Corollary 2, we will be done if we can show that the $(n-1) \times n$ matrix

$$\begin{pmatrix} 1 & 1 & 0 & 0 & \cdots & 0 & 0 & 0 \\ 0 & 1 & 1 & 0 & \cdots & 0 & 0 & 0 \\ & & & & \vdots & & & \\ 0 & 0 & 0 & 0 & \dots & 0 & 1 & 1 \end{pmatrix} \tag{2.11}$$

has rank $n-1$, but this is immediate.

Now assume that $\operatorname{rk} B_{n,k} = n-1$. Then by Corollary 2, $\xi_1 = \zeta_1 + \zeta_2 = (-\zeta_2 - \zeta_3 - \cdots - \zeta_k) + \zeta_2 = -\zeta_3 - \cdots - \zeta_k$ and so we will be done if we can determine the rank of the $(n-1) \times (n-1)$ matrix

$$\begin{pmatrix} 0 & 1 & 1 & 1 & \cdots & 1 & 0 & \cdots & 0 & 0 & 0 \\ 1 & 1 & 0 & 0 & \cdots & 0 & 0 & \cdots & 0 & 0 & 0 \\ 0 & 1 & 1 & 0 & \cdots & 0 & 0 & \cdots & 0 & 0 & 0 \\ & & & & \vdots & & & \vdots & & & \\ 0 & 0 & 0 & 0 & \cdots & 0 & 0 & \cdots & 0 & 1 & 1 \end{pmatrix}. \tag{2.12}$$

If k is even then we can use rows $3, 5, \ldots, k-3$ and $k-1$ to eliminate row 1, yielding

$$\begin{pmatrix} 0 & 0 & 0 & 0 & \cdots & 0 & 0 & \cdots & 0 & 0 & 0 \\ 1 & 1 & 0 & 0 & \cdots & 0 & 0 & \cdots & 0 & 0 & 0 \\ 0 & 1 & 1 & 0 & \cdots & 0 & 0 & \cdots & 0 & 0 & 0 \\ & & & & \vdots & & & \vdots & & & \\ 0 & 0 & 0 & 0 & \cdots & 0 & 0 & \cdots & 0 & 1 & 1 \end{pmatrix} \tag{2.13}$$

which is clearly rank $n-2$. If k is odd, then we can use rows $3, 5, \ldots, k-4$ and $k-2$ to eliminate row 1, giving us:

$$\begin{pmatrix} 0 & 0 & 0 & 0 & \cdots & 0 & 1 & 0 & \cdots & 0 & 0 & 0 \\ 1 & 1 & 0 & 0 & \cdots & 0 & 0 & 0 & \cdots & 0 & 0 & 0 \\ 0 & 1 & 1 & 0 & \cdots & 0 & 0 & 0 & \cdots & 0 & 0 & 0 \\ & & & & \vdots & & & & \vdots & & & \\ 0 & 0 & 0 & 0 & \cdots & 0 & 0 & 0 & \cdots & 0 & 1 & 1 \end{pmatrix}. \tag{2.14}$$

Next, starting with row $n-2$ and working upward, replace each of rows 2 through $n-2$ with its difference with the subsequent row to obtain:

$$\begin{pmatrix} 0 & 0 & 0 & 0 & \cdots & 0 & 1 & 0 & \cdots & 0 & 0 & 0 & 0 \\ 1 & 0 & 0 & 0 & \cdots & 0 & 0 & 0 & \cdots & 0 & 0 & 0 & (-1)^n \\ & & & & \vdots & & & & \vdots & & & & \\ 0 & 0 & 0 & 0 & \cdots & 0 & 1 & 0 & \cdots & 0 & 0 & 0 & (-1)^{n-k+1} \\ & & & & \vdots & & & & \vdots & & & & \\ 0 & 0 & 0 & 0 & \cdots & 0 & 0 & 0 & \cdots & 1 & 0 & 0 & 1 \\ 0 & 0 & 0 & 0 & \cdots & 0 & 0 & 0 & \cdots & 0 & 1 & 0 & -1 \\ 0 & 0 & 0 & 0 & \cdots & 0 & 0 & 0 & \cdots & 0 & 0 & 1 & 1 \end{pmatrix}. \tag{2.15}$$

Now move row $k+1$ to the bottom and move row 1 to row $k+1$:

$$\begin{pmatrix} 1 & 0 & 0 & 0 & \cdots & 0 & 0 & 0 & \cdots & 0 & 0 & 0 & (-1)^n \\ 0 & 1 & 0 & 0 & \cdots & 0 & 0 & 0 & \cdots & 0 & 0 & 0 & (-1)^{n-1} \\ & & & & \vdots & & & & \vdots & & & & \\ 0 & 0 & 0 & 0 & \cdots & 0 & 1 & 0 & \cdots & 0 & 0 & 0 & 0 \\ & & & & \vdots & & & & \vdots & & & & \\ 0 & 0 & 0 & 0 & \cdots & 0 & 0 & 0 & \cdots & 1 & 0 & 0 & 1 \\ 0 & 0 & 0 & 0 & \cdots & 0 & 0 & 0 & \cdots & 0 & 1 & 0 & -1 \\ 0 & 0 & 0 & 0 & \cdots & 0 & 0 & 0 & \cdots & 0 & 0 & 1 & 1 \\ 0 & 0 & 0 & 0 & \cdots & 0 & 1 & 0 & \cdots & 0 & 0 & 0 & (-1)^{n-k+1} \end{pmatrix}, \tag{2.16}$$

and subtract row $k+1$ from row $n-1$ to obtain:

$$\begin{pmatrix} 1 & 0 & 0 & 0 & \cdots & 0 & 0 & 0 & \cdots & 0 & 0 & 0 & (-1)^n \\ 0 & 1 & 0 & 0 & \cdots & 0 & 0 & 0 & \cdots & 0 & 0 & 0 & (-1)^{n-1} \\ & & & & \vdots & & & & \vdots & & & & \\ 0 & 0 & 0 & 0 & \cdots & 0 & 1 & 0 & \cdots & 0 & 0 & 0 & 0 \\ & & & & \vdots & & & & \vdots & & & & \\ 0 & 0 & 0 & 0 & \cdots & 0 & 0 & 0 & \cdots & 1 & 0 & 0 & 1 \\ 0 & 0 & 0 & 0 & \cdots & 0 & 0 & 0 & \cdots & 0 & 1 & 0 & -1 \\ 0 & 0 & 0 & 0 & \cdots & 0 & 0 & 0 & \cdots & 0 & 0 & 1 & 1 \\ 0 & 0 & 0 & 0 & \cdots & 0 & 0 & 0 & \cdots & 0 & 0 & 0 & (-1)^{n-k+1} \end{pmatrix}, \tag{2.17}$$

which is in echelon form and thus rank $n-1$. □

COROLLARY 3. *The code $C_{n,k}$ is a $\left[\binom{n}{k}, n-1\right]$ code.*

PROOF. If $k-1$ is odd then our result is immediate by Theorem 2.11. If $k-1$ is even and $\operatorname{rk} B_{n,k} = n$ then we also get what we want by Theorem 2.11. Finally, if $k-1$ is even and $\operatorname{rk} B_{n,k} = n-1$ then because k is odd, Theorem 2.11 still gives the desired result. □

3. Code construction

We shall now examine the properties of the codes $\mathfrak{C}_{n,k}$, defined over various rings, and of the code $C_{n,k}$, defined over the finite field of order 2.

THEOREM 3.1. *Let $n > k+1$. Let R be a finite commutative Frobenius ring of characteristic c. Assume the following conditions are met.*

- *c divides $2\binom{n-2}{k-1}$,*
- *c divides $4\binom{n-4}{k-2}$,*
- *c divides $\binom{n-2}{k-1}$.*

Then $\mathfrak{C}_{n,k}$ is a self-orthogonal code of rank $n-1$ if either c does not divide $k-1$ or c divides $k-1$ and $\operatorname{rk} B_{n,k} = n$. Otherwise, it is a self-orthogonal code of rank $n-2$.

PROOF. Corollary 1 gives that the inner-product of any row with itself is $2\binom{n-2}{k-1}$. Theorem 2.6 gives the inner-product of distinct rows when $j > i+1$ is $4\binom{n-4}{k-2}$, and Theorem 2.7 gives the inner-product of distinct rows when $j = i+1$ is $\binom{n-2}{k-1}$.

The rank is given by Theorem 2.11. □

THEOREM 3.2. *Let n be an integer such that $n-4$ is prime and $n-4 > k-1$ and $n-4 > n-k-1$, where $n \geq 9$. Then $\mathfrak{C}_{n,k}$ is a self-orthogonal code over any finite Frobenius ring with characteristic $n-4$.*

PROOF. We note that $n-4$ must be at least 5 for this to work so we know the characteristic is not 2. Then the conditions of Theorem 3.1 reduce to the characteristic must divide $\binom{n-4}{k-2}$ and $\binom{n-2}{k-1}$.

If $n-4$ is prime then $n-4$ divides $\binom{n-4}{k-2}$ unless $k-2$ is either $n-4$ or 0. This cases are eliminated by requiring that $n-4 > n-k-1$ since it implies that $k > 2$.

Next, we have that $n-4$ divides the numerator of $\binom{n-2}{k-1}$ but does not divide the denominator of $\binom{n-2}{k-1}$ since $n-4 > k-1$ and $n-4 > n-k-1$. Therefore, the characteristic $n-4$ divides both $\binom{n-4}{k-2}$ and $\binom{n-2}{k-1}$ and Theorem 3.1 gives the result. □

EXAMPLE 3.3. Let $n = 9$, then $n-4 = 5$ is prime. Then for $k = 4$ we have $5 > 4$ and $5 > 9-4-1 = 4$. Therefore, $\mathfrak{C}_{9,4}$ is a self-orthogonal code of rank 7 over $\mathbb{F}_5$.

COROLLARY 4. *If $n \geq k+2$, $\binom{n-2}{k-1}$ is even, and R has characteristic 2 then the code $C_{n,k}$ is self-orthogonal.*

PROOF. It is immediate that $2\binom{n-2}{k-1}$ and $4\binom{n-4}{k-2}$ are even. Then since $\binom{n-2}{k-1}$ is even, we have satisfied the conditions of Theorem 3.1. □

COROLLARY 5. *If $n \geq k+2$, $\binom{n-2}{k-1}$ is even, and $R = \mathbb{F}_2$, then $C_{n,k}$ is doubly-even and self-orthogonal.*

PROOF. Corollary 5 gives that the code is self-orthogonal. Then the weights of every row in $C_{n,k}$ is $2\binom{n-2}{k-1}$ which is divisible by 4 if $\binom{n-2}{k-1}$ is even. If every row of a generator matrix is doubly-even and the rows are mutually orthogonal, then it generates a doubly even code. □

LEMMA 3.4. *Let G be a binary matrix that generates a binary code D with $D \subseteq D^{\perp} \subseteq \mathbb{F}_2^m$. Let E be the code over R_h generated by G. Then the code $C = \langle E, u_i D^{\perp} \rangle$, for $1 \leq i \leq h$, is a self-dual code over R_h.*

PROOF. Since G generates a binary self-orthogonal code E is also self-orthogonal since the characteristic of R_h is 2. Let $\mathbf{v}_j \in E$, $\mathbf{w}_j \in D^{\perp}$. Note that Then

$$\begin{aligned}
[\mathbf{v}_1 + \alpha_1 \mathbf{w}_1, \mathbf{v}_2 + \alpha_2 \mathbf{w}_2] &= [\mathbf{v}_1, \mathbf{v}_2] + 2u_i[\mathbf{v}_1, \mathbf{w}_2] + u_i^2[\mathbf{w}_2, \mathbf{w}_2] \\
&= 0 + 0 + 0[\mathbf{w}_2, \mathbf{w}_2] \\
&= 0.
\end{aligned}$$

Therefore the code is self-orthogonal.

The ideal $\langle u_i \rangle$ has cardinality $2^{2^{h-1}}$. Assume the dimension of D is r. Then $|C| = (2^{2^h})^r \cdot (2^{2^{h-1}})^{m-2r} = (2^{2^{h-1}})^{2r}(2^{2^{h-1}})^{m-2r} = (2^{2^{h-1}})^m$. This gives that $|C|^2 = |R_h|^m$ and so the code C is self-dual. □

THEOREM 3.5. *Assume $n \geq k+2$, $\binom{n-2}{k-1}$ is even, let $C_{n,k}$ be the code generated by $C_{n,k}$ over $\mathbb{F}_2$. Then the code $\mathfrak{D}_{n,k} = \langle C_{n,k}, u_i \mathfrak{C}_{n,k}^{\perp} \rangle$ is a self-dual code of length $m = \binom{n}{k}$ over R_h and $\varphi_h(\mathfrak{D}_{n,k})$ is a binary self-dual code of length $2^h m$.*

PROOF. We know that $\mathcal{C}_{n,k}$ is a binary self-orthogonal code by Corollary 5. The result follows from Lemma 3.4. $\square$

COROLLARY 6. *If n is even then $\mathfrak{D}_{n,\frac{n-2}{2}}$ is a self-dual code over R_h for $h > 0$ and $\varphi_h(\mathfrak{D}_{n,k})$ is a binary self-dual code of length $2^h m$.*

PROOF. We know that $\binom{n}{\frac{n}{2}} = \binom{n-1}{\frac{n}{2}-1} + \binom{n-1}{\frac{n}{2}} = 2\binom{n-1}{\frac{n}{2}}$. Therefore, the conditions of Theorem 3.5 are fulfilled. $\square$

COROLLARY 7. *If $n - 2 = 2^s$ for some s and $1 < k < n-1$ then $\mathfrak{D}_{n,\frac{n-2}{2}}$ is a self-dual code over R_h for $h > 0$ and $\varphi_h(\mathfrak{D}_{n,k})$ is a binary self-dual code of length $2^h m$.*

PROOF. If $n-2$ is a power of 2 and $k-1$ is not 0 or $n-2$ then $\binom{n-2}{k-1}$ is even. Therefore, the conditions of Theorem 3.5 are fulfilled. $\square$

REMARK 3.6. In the cases $k = 2$ and $k = 3$, Xu calculates minimum distance by expressing $\mathrm{wt}(u)$ as a function of t (as described in Lemma 2.4) and using the first derivative test to identify extrema. In these two cases, $\mathrm{wt}(u)$ is a quadratic and cubic polynomial respectively. As k increases, so does the degree of $\mathrm{wt}(u)$ as a polynomial in the variable t, making this technique more difficult.

4. Conclusion

We have studied codes over finite commutative Frobenius rings generated by the weight matrices arising from Lie algebras read modulo 2. We have shown that given certain restrictions these codes generate self-orthogonal codes and can be used to construct self-dual codes. Moreover, we have determined the ranks of the codes generated by these matrices. There are several avenues of research open in this vein. One such avenue is to examine the codes generated when the weight matrices are not read modulo 2 but rather viewed as matrices with entries from $\{0, -1, 1\}$. Another avenue is to explore computationally the codes that are created to determine if optimal codes can be constructed. In the other direction, it is of interest to take the properties of the codes generated to determine if anything can be said about the generating Lie algebra.

Acknowledgment

The authors are very grateful to the referee and the editor André Leroy for their helpful suggestions which have improved this paper.

References

[1] E. F. Assmus Jr. and J. D. Key, *Designs and their codes*, Cambridge Tracts in Mathematics, vol. 103, Cambridge University Press, Cambridge, 1992, DOI 10.1017/CBO9781316529836. MR1192126

[2] E. F. Assmus Jr. and H. F. Mattson Jr., *Coding and combinatorics*, SIAM Rev. **16** (1974), 349–388, DOI 10.1137/1016056. MR359982

[3] E. F. Assmus Jr. and H. F. Mattson Jr., *New 5-designs*, J. Combinatorial Theory **6** (1969), 122–151. MR272647

[4] Steven T. Dougherty, *Algebraic coding theory over finite commutative rings*, SpringerBriefs in Mathematics, Springer, Cham, 2017, DOI 10.1007/978-3-319-59806-2. MR3675844

[5] Steven Dougherty, Jon-Lark Kim, and Patrick Solé, *Open problems in coding theory*, Noncommutative rings and their applications, Contemp. Math., vol. 634, Amer. Math. Soc., Providence, RI, 2015, pp. 79–99, DOI 10.1090/conm/634/12692. MR3307391

[6] Steven T. Dougherty, Jon-Lark Kim, and Patrick Solé, *Double circulant codes from two class association schemes*, Adv. Math. Commun. **1** (2007), no. 1, 45–64, DOI 10.3934/amc.2007.1.45. MR2262767

[7] M. Bilal, J. Borges, S. T. Dougherty, and C. Fernández-Córdoba, *Self-dual codes from 3-class association schemes*, Appl. Algebra Engrg. Comm. Comput. **26** (2015), no. 3, 227–250, DOI 10.1007/s00200-014-0238-z. MR3349061

[8] Steven Dougherty, Bahattin Yıldız, and Suat Karadeniz, *Self-dual codes over R_k and binary self-dual codes*, Eur. J. Pure Appl. Math. **6** (2013), no. 1, 89–106. MR3022323

[9] W. Cary Huffman and Vera Pless, *Fundamentals of error-correcting codes*, Cambridge University Press, Cambridge, 2003, DOI 10.1017/CBO9780511807077. MR1996953

[10] F. J. MacWilliams and N. J. A. Sloane, *The theory of error-correcting codes. I* **Vol. 16** (1977), i–xv and 1–369. MR0465509

[11] Jay A. Wood, *Duality for modules over finite rings and applications to coding theory*, Amer. J. Math. **121** (1999), no. 3, 555–575. MR1738408

[12] Xiaoping Xu, *Representations of the Lie algebras and coding theory*, J. Lie Theory **22** (2012), no. 3, 647–682. MR3012148

Department of Mathematics, Misericordia University, Dallas, Pennsylvania 18612
Email address: `jcarey.phd@gmail.com`

Department of Mathematics, University of Scranton, Scranton, Pennsylvania 18510
Email address: `prof.steven.dougherty@gmail.com`

Contemporary Mathematics
Volume **826**, 2025
https://doi.org/10.1090/conm/826/16574

Symplectically self-orthogonal additive codes and their applications to quantum stabilizer codes

Steven T. Dougherty, Navin Kashyap, Tania Sidana, Serap Şahinkaya, and Deniz Ustun

Abstract. In this paper, for a given duality M of the additive group of a finite commutative Frobenius ring $\mathcal{R}$, we define an M-symplectic inner-product on $\mathcal{R}^{2n}$. For any pair of dualities M and M' of $\mathcal{R}$, we show that an M-symplectically self-orthogonal additive code is symplectically isometric to a M'-symplectically self-orthogonal additive code. As a consequence, we observe that for a given duality M of $\mathcal{R}$, M-symplectically self-orthogonal additive codes over $\mathcal{R}$ give rise to the construction of quantum stabilizer codes. Given this relationship, we then describe constructions for self-orthogonal additive codes over a ring $\mathcal{R}$ with any duality.

1. Introduction

Classically, linear codes were the most important class of codes for correcting errors in electronic communication. More recently, a great deal of attention has been given to the construction of codes in a quantum setting. In this setting, it is additive codes (codes closed under addition but not necessarily closed under scalar multiplication) that are useful. Namely, additive codes are the family of codes that are most useful for the construction of quantum error correcting codes (QECCs). In an early paper on the topic, Calderbank et al. [**3**] established an equivalence between quantum stabilizer codes over qubits and additive codes over the finite field $\mathbb{F}_4$, which are self-orthogonal with respect to the trace inner-product. Later, this equivalence was generalized for quantum stabilizer codes over qudits in terms of non-binary additive codes that are self-orthogonal with respect to a certain trace-alternating form [**14**, Section IV-B], [**16**]. The stabilizer framework to construct quantum stabilizer codes was proposed by Gottesman [**12**]. Quantum stabilizer codes are the most widely studied class of QECCs. In fact, the first QECC constructed by Shor [**17**], as well as QECCs constructed from a method of using self-orthogonal binary classical error-correcting codes, proposed independently by Calderbank and Shor [**4**] and Steane [**18**] were quantum stabilizer codes.

2020 *Mathematics Subject Classification.* Primary 94B99, 94B05; Secondary 94B60.
Key words and phrases. Additive codes, quantum codes, self-orthogonal codes.
The third author was supported by an IoE postdoctoral fellowship awarded by the Indian Institute of Science.
The fourth author was supported by T.U.B.I.T.A.K (Scientific and Technological Research Council of Turkey) (Grant No: 1059B192000947).

Subsequently, the stabilizer framework was extended to higher-dimensional qudit spaces defined via a finite field or ring. The extension of the formalism to construct quantum stabilizer codes over qudits was executed by Ashikhmin and Knill [**2**], and Ketkar et al. [**14**] when qudits are defined via a finite field, and was done by Nadella and Klappenecker [**15**], and Gluesing-Luerssen and Pllaha [**13**] when qudits are defined via a finite commutative Frobenius ring. In particular, QECCs of length n over qudits defined via a finite commutative Frobenius ring $\mathcal{R}$ can be constructed from self-orthogonal additive codes of length $2n$ over $\mathcal{R}$, where the self-orthogonality is with respect to a χ-symplectic inner-product defined from a generating character χ of $\mathcal{R}$.

In this paper, we adapt the methods from [**13, 15**] to construct quantum stabilizer codes from additive codes over $\mathcal{R}$ that are self-orthogonal with respect to certain dualities other than the canonical duality obtained from the character group of the additive group of $\mathcal{R}$. Any finite abelian group G is isomorphic to the group $\widehat{G}$ of its characters (i.e., group homomorphisms from G to $\mathbb{C}^*$). By fixing an isomorphism between G and $\widehat{G}$, we can define a character table, which we term a "duality" of G. Any such duality can be used to define various orthogonals for additive codes over G [**10**]. For a fixed duality M of $\mathcal{R}$, we define an M-symplectic inner-product on $\mathcal{R}^{2n}$. Our main contribution is a result which states that for any pair of dualities M and M' of $\mathcal{R}$, an M-symplectically self-orthogonal additive code is symplectically isometric to an M'-symplectically self-orthogonal additive code. As a consequence of this and using the known equivalence of quantum stabilizer codes with self-orthogonal additive codes with respect to a canonical symplectic inner-product, we observe that given any duality M of $\mathcal{R}$, additive codes that are M-symplectically self-orthogonal give rise to the construction of quantum stabilizer codes. This is useful in the search for quantum stabilizer codes, as we will illustrate by means of an example (Example 4.4).

This paper is organized as follows. In Section 2, we establish the basic definitions and notations needed to describe our main results. In Section 3, we show that for any pair of dualities M and M' of $\mathcal{R}$, an M-symplectically self-orthogonal additive code is symplectically isometric to an M'-symplectically self-orthogonal additive code. Then, in Section 4, we provide applications of this result to the construction of QECCs. Given the use of additive self-orthogonal codes in this construction, in Section 5, we give constructions of additive self-orthogonal codes including an analogue of the neighbor construction and the building up construction for additive codes. These constructions are related to symplectically self-orthogonal additive codes.

2. Preliminaries

Let $\mathcal{R}$ be a finite commutative Frobenius ring with cardinality $|\mathcal{R}| = d$ and characteristic $\text{char}(\mathcal{R}) = e$. A subset $\mathcal{C}$ of $\mathcal{R}^n$ is said to be an *additive code* of length n over $\mathcal{R}$ if $\mathcal{C}$ is an additive subgroup of $\mathcal{R}^n$. A subset $\mathcal{C}$ of $\mathcal{R}^n$ is said to be a *linear code* of length n over $\mathcal{R}$ if $\mathcal{C}$ is an $\mathcal{R}$-module. Clearly, a linear code is additive but an additive code need not be linear.

For each $r = (r_1, r_2, \cdots, r_n) \in \mathcal{R}^n$, the Hamming weight $\text{wt}_H(r)$ of r is defined as $\text{wt}_{\text{H}}(r) := |\{i : r_i \neq 0\}|$. The minimum Hamming weight of a non-empty subset $A \subseteq \mathcal{R}^n$ is defined as $d_{\text{H}}(A) := \min\{\text{wt}_{\text{H}}(a) : a \in A \setminus \{0\}\}$.

Let $\widehat{\mathcal{R}}$ denote the module of all characters of $\mathcal{R}$. In terms of coding theory, the most important property of a Frobenius ring is that it has a generating character, namely an additive character χ such that $\widehat{\mathcal{R}} = \mathcal{R} \cdot \chi$. That is, χ is a generating character if, for any $a \in \mathcal{R}$, there is an associated character χ_a such that $a \cdot \chi = \chi_a(b) = \chi(ab)$.

For additive codes, we are really only interested in the additive group of $\mathcal{R}$. In this setting, we have that $\widehat{\mathcal{R}}$ is a group and isomorphic to $(\mathcal{R}, +)$ but not in a canonical manner. Specifically, we can choose any isomorphism between $(\mathcal{R}, +)$ and $\widehat{\mathcal{R}}$ to give a duality. We can use any generating character χ of $\mathcal{R}$ to give an isomorphism: $r \longleftrightarrow r \cdot \chi$. But there is nothing particularly special about isomorphisms induced by generating characters; other isomorphisms are just as useful. Any isomorphism $\phi : \mathcal{R} \to \widehat{\mathcal{R}}$ gives rise to a character table M as follows. List the elements of $\mathcal{R}$ as $\{r_1, r_2, \ldots, r_d\}$. Denote the isomorphism between $(\mathcal{R}, +)$ and its character group by ϕ, and set $\phi(r_i) = \chi_{r_i}$. Namely, the character χ_{r_i} is the image of r_i under the isomorphism ϕ. Then, the rows of the character table M are indexed by $\phi(r_i)$ and the columns are indexed by r_i for $i = 1, 2, \ldots, d$. That is, the (i, j)-th entry of M is equal to $\chi_{r_i}(r_j) = \phi(r_i)(r_j)$. We call such a character table M a *duality* of $(\mathcal{R}, +)$. This duality gives rise to an orthogonal and a function that acts like an inner-product which we shall describe shortly. The orthogonal will act in the same way that the usual orthogonal acts for linear codes. That is, the MacWilliams relations hold and as a consequence the cardinality of the code times the cardinality of the orthogonal gives the cardinality of the ambient space, which is the analog of the familiar result that the dimension of a linear code plus the dimension of its orthogonal is the dimension of the ambient space.

Note that if $\phi : \mathcal{R} \to \widehat{\mathcal{R}}$ is defined by $\phi(r) = r \cdot \psi$ for some generating character ψ of $\mathcal{R}$, then the corresponding duality M of $\mathcal{R}$ is of the form $M_{i,j} = \psi(r_i r_j)$. A duality of $\mathcal{R}$ of this form is called a *canonical* duality of $\mathcal{R}$. The canonical duality, unlike an arbitrary duality, takes into consideration the ring as a whole and not just the additive group.

EXAMPLE 2.1. Recall the trace function $\mathrm{tr}_{4/2}(a) = a + a^2$ for each $a \in \mathbb{F}_4$. The six possible dualities of $\mathbb{F}_4$ are as follows: for all $a, b \in \mathbb{F}_4$,

- M_1 is defined via $\chi_a(b) = (-1)^{\mathrm{tr}_{4/2}(ab)}$.
- M_2 is defined via $\chi_a(b) = (-1)^{\mathrm{tr}_{4/2}(\omega ab)}$.
- M_3 is defined via $\chi_a(b) = (-1)^{\mathrm{tr}_{4/2}(\omega^2 ab)}$.
- M_4 is defined via $\chi_a(b) = (-1)^{\mathrm{tr}_{4/2}(ab^2)}$.
- M_5 is defined via $\chi_a(b) = (-1)^{\mathrm{tr}_{4/2}(\omega ab^2)}$.
- M_6 is defined via $\chi_a(b) = (-1)^{\mathrm{tr}_{4/2}(\omega^2 ab^2)}$.

Note that for $\ell = 1, 2, 3$, $(M_\ell)_{i,j} = \psi_\ell(r_i r_j)$, for $r_i, r_j \in \mathbb{F}_4$ with $\psi_\ell(b) = (-1)^{\mathrm{tr}_{4/2}(\omega^{\ell-1} b)}$. As any character of $\mathbb{F}_4$ that does not contain a non-trivial ideal in its kernel is a generating character (see [**5**]), the dualities M_1, M_2 and M_3 are canonical. The other dualities are non-canonical.

We can define an inner-product on $\mathcal{R}^n$ with respect to a duality M as follows. For $\mathbf{v} = (v_i), \mathbf{w} = (w_i) \in \mathcal{R}^n$,

$$[\mathbf{v}, \mathbf{w}]_M = \chi_{\mathbf{v}}(\mathbf{w}) = \prod \chi_{v_i}(w_i).$$

We say that two vectors $\mathbf{v}, \mathbf{w} \in \mathcal{R}^n$ are *orthogonal* with respect to M if $[\mathbf{v}, \mathbf{w}]_M = 1$. We use this to define an orthogonality of codes in $\mathcal{R}^n$.

DEFINITION 2.2. Fix a duality M of $\mathcal{R}$. For a subset $\mathcal{C} \subseteq \mathcal{R}^n$, the *orthogonal* of $\mathcal{C}$ with respect to M is defined as $\mathcal{C}^M := \{\mathbf{x} \in \mathcal{R}^n \mid [\mathbf{x}, \mathbf{c}]_M = 1 \text{ for all } \mathbf{c} \in \mathcal{C}\}$.

Note that for any subset $\mathcal{C} \subseteq \mathcal{R}^n$, the orthogonal $\mathcal{C}^M$ is an additive code in $\mathcal{R}^n$. If $\mathcal{C}$ is an additive code in $\mathcal{R}^n$, then $|\mathcal{C}||\mathcal{C}^M| = \mathcal{R}^n$ (see **[5]** for complete details). An additive code $\mathcal{C} \subseteq \mathcal{R}^n$ is called *self-orthogonal* with respect to M if $\mathcal{C} \subseteq \mathcal{C}^M$. An additive code $\mathcal{C} \subseteq \mathcal{R}^n$ is called *self-dual* with respect to M if $\mathcal{C} = \mathcal{C}^M$.

Referring back to Example 2.1, it can be verified that, for a linear code $\mathcal{C}$ over $\mathbb{F}_4$, $\mathcal{C}^{M_1} = \mathcal{C}^{M_2} = \mathcal{C}^{M_3}$, and $\mathcal{C}^{M_4} = \mathcal{C}^{M_5} = \mathcal{C}^{M_6}$.

If a duality satisfies $\chi_a(b) = \chi_b(a)$ for all $a, b \in \mathcal{R}$ then the duality is said to be symmetric. In this case, the matrix M satisfies $M = M^T$. In general, we have $(C^M)^{M^T} = C$. Therefore, it is desirable for the duality to be symmetric so that the code satisfies the double annihilator condition, that is $(C^M)^M = C$ for all additive codes C. See **[6, 9]** for a discussion of the possible dualities on finite abelian groups.

2.1. A symplectic inner-product on $\mathcal{R}^{2n}$. We now introduce a *twisted* inner-product on $\mathcal{R}^{2n}$, defined with respect to an arbitrary duality M of $\mathcal{R}$. From this point on, we are generally concerned with additive codes of even block length. In this context, we write $(\mathbf{a}, \mathbf{b}) \in \mathcal{R}^{2n}$ as a shorthand for $(\mathbf{a}, \mathbf{b}) \in \mathcal{R}^n \times \mathcal{R}^n$, i.e., $\mathbf{a}$ and $\mathbf{b}$ individually belong to $\mathcal{R}^n$. For a fixed duality M of $\mathcal{R}$, we define an M-symplectic inner-product on $\mathcal{R}^{2n}$ as follows: for $(\mathbf{a}, \mathbf{b})$ and $(\mathbf{a}', \mathbf{b}')$ in $\mathcal{R}^{2n}$,

$$[(\mathbf{a}, \mathbf{b}), (\mathbf{a}', \mathbf{b}')]_{M\S} := [\mathbf{b}, \mathbf{a}']_M [\mathbf{b}', -\mathbf{a}]_M = \prod_{i=1}^{n} \chi_{b_i}(a_i') \chi_{b_i'}(-a_i). \tag{2.1}$$

We then define an orthogonality on $\mathcal{R}^{2n}$ using this M-symplectic inner-product.

DEFINITION 2.3. Fix a duality M of $\mathcal{R}$. For a subset C of $\mathcal{R}^n$, we define $C^{M\S} := \{(\mathbf{a}, \mathbf{b}) \in \mathcal{R}^{2n} \mid [(\mathbf{a}, \mathbf{b}), (\mathbf{a}', \mathbf{b}')]_{M\S} = 1 \text{ for all } (\mathbf{a}', \mathbf{b}') \in C\}$ to be the *M-symplectic orthogonal* of C.

Note that for any subset $C \subseteq \mathcal{R}^{2n}$, the M-symplectic orthogonal $C^{M\S}$ is an additive code. An additive code $\mathcal{C}$ is said to be *M-symplectically self-orthogonal* if $\mathcal{C} \subseteq \mathcal{C}^{M\S}$ and $\mathcal{C}$ is said to be *M-symplectically self-dual* if $\mathcal{C} = \mathcal{C}^{M\S}$. The *symplectic weight* $\mathrm{wt_s}(\mathbf{x}, \mathbf{y})$ of a vector $(\mathbf{x}, \mathbf{y}) = (x_1, x_2, \ldots, x_n, y_1, y_2, \ldots, y_n) \in \mathcal{R}^{2n}$ is defined as $\mathrm{wt_s}(\mathbf{x}, \mathbf{y}) := |\{i : (x_i, y_i) \neq 0\}|$. The *minimum symplectic weight* $d_{\mathrm{s}}(A)$ of a non-empty subset A of $\mathcal{R}^{2n}$ is defined as $d_{\mathrm{s}}(A) := \min\{\mathrm{wt_s}(\mathbf{a}) : \mathbf{a} \in A \setminus \{0\}\}$.

REMARK 2.4. Recall that for any duality M of $\mathcal{R}$ of the form $M_{i,j} = \chi(r_i r_j)$ for some generating character χ of $\mathcal{R}$, we have $[(\mathbf{a}, \mathbf{b}), (\mathbf{a}', \mathbf{b}')]_{M\S} = \prod \chi_{b_i}(a_i') \chi_{b_i'}(-a_i)$ $= \prod \chi(b_i \cdot a_i' - b_i' \cdot a_i)$ for $(\mathbf{a}, \mathbf{b}), (\mathbf{a}', \mathbf{b}') \in \mathcal{R}^{2n}$. A duality M of $\mathcal{R}$ of the form $M_{i,j} = \chi(r_i r_j)$ for some generating character χ is called a canonical duality obtained from a generating character. Thus, if $M_{i,j} = \chi(r_i r_j)$, then for any additive code $\mathcal{C}$, the M-symplectic orthogonal $\mathcal{C}^{M\S} = \{(\mathbf{a}, \mathbf{b}) \in \mathcal{R}^{2n} \mid \prod \chi(b_i \cdot a_i' - b_i' \cdot a_i) = 1 \text{ for all } (\mathbf{a}', \mathbf{b}') \in \mathcal{C}\}$. Suppose that M' is another duality of $\mathcal{R}$ of the form $M'_{i,j} = \phi(r_i r_j)$ for some generating character ϕ of $\mathcal{R}$. Since χ and ϕ are two generating characters of $\mathcal{R}$, there exists a unit $u \in \mathcal{R}$ such that $\chi = u \cdot \phi$. Now, by definition of $\mathcal{C}^{M\S}$ and $\mathcal{C}^{M'\S}$, we get $\mathcal{C}^{M'\S} = \{(u\mathbf{a}, u\mathbf{b}) \mid (\mathbf{a}, \mathbf{b}) \in \mathcal{C}^{M\S}\}$. Thus the elements of $\mathcal{C}^{M'\S}$ and $\mathcal{C}^{M\S}$ differ by the unit multiple u, hence have same symplectic weights. In fact, if $\mathcal{C}$ is a linear code over $\mathcal{R}$, then $\mathcal{C}^{M'\S} = \mathcal{C}^{M\S}$.

2.2. QECCs derived from $\mathcal{R}$. In this subsection, we introduce the bare minimum concepts related to QECCs that are needed for the purposes of this paper. For a thorough exposition of QECCs (specifically, quantum stabilizer codes) derived from finite commutative Frobenius rings, we refer the reader to [**13, 15**]. Once again, we take $\mathcal{R}$ to be a finite commutative Frobenius ring $\mathcal{R}$ with cardinality d.

An n-qudit quantum code derived from $\mathcal{R}$ is a subspace of $(\mathbb{C}^d)^{\otimes n} \cong \mathbb{C}^{d^n}$. An $((n, K, D))_d$ QECC is an n-qudit quantum code derived from $\mathcal{R}$, having dimension K (as a subspace of $\mathbb{C}^{d^n}$) and minimum distance D. We do not formally define the notion of minimum distance of a quantum code here, referring the reader instead to [**13**]. Loosely speaking, a quantum code has minimum distance D if it can detect all (Pauli) errors of weight at most $D-1$ and cannot detect some error of weight D. The subscript d records the fact that the individual qudits of the quantum code live in $\mathbb{C}^d$. Thus, d is sometimes referred to as the *local dimension* of the qudits. For simplicity, we will henceforth drop the subscript d from our notation.

3. Symplectic isometries between additive codes

Gluesing-Luerssen and Pllaha [**13**] studied a symplectic isometry between self-orthogonal *linear* codes in $\mathcal{R}^{2n}$ with respect to a symplectic inner-product. This notion can be fruitfully extended to additive codes.

DEFINITION 3.1. Let M and M' be a given pair of dualities of $\mathcal{R}$. Let $\rho : \mathcal{R}^{2n} \to \mathcal{R}^{2n}$ be an additive map. Then ρ is said to be a symplectic isometry if we have $\mathrm{wt_s}((\mathbf{a},\mathbf{b})) = \mathrm{wt_s}(\rho((\mathbf{a}',\mathbf{b}')))$ and $[(\mathbf{a},\mathbf{b}),(\mathbf{a}',\mathbf{b}')]_{M\S} = [\rho((\mathbf{a},\mathbf{b})),\rho((\mathbf{a}',\mathbf{b}'))]_{M'\S}$ for all $(\mathbf{a},\mathbf{b})$ and $(\mathbf{a}',\mathbf{b}')$ in $\mathcal{R}^{2n}$. Two additive codes $\mathcal{C}$ and $\mathcal{C}'$ of $\mathcal{R}^{2n}$ are said to be symplectically isometric if there exists a symplectic isometry $\rho : \mathcal{R}^{2n} \to \mathcal{R}^{2n}$ such that $\rho(\mathcal{C}) = \mathcal{C}'$.

In this section, we show that for any pair of dualities M and M' of $\mathcal{R}$, an M-symplectically self-orthogonal additive code is symplectically isometric to an M'-symplectically self-orthogonal additive code. Thus there exists an M-symplectically self-orthogonal additive code if and only if there exists an M'-symplectically self-orthogonal additive code. To start, we need the following result.

THEOREM 3.2. *For any pair of dualities M and M' of $\mathcal{R}$, there exists an additive bijective map $\rho : \mathcal{R}^{2n} \to \mathcal{R}^{2n}$ that is a symplectic isometry, such that for any additive code $\mathcal{C}$ in $\mathcal{R}^{2n}$, we have $\rho(\mathcal{C})^{M'\S} = \rho(\mathcal{C}^{M\S})$.*

PROOF. Let M and M' correspond to isomorphisms ϕ and ψ between $\mathcal{R}$ and $\widehat{\mathcal{R}}$, respectively. Define a map $\sigma : \mathcal{R} \to \mathcal{R}$ as $\sigma(r_i) := \psi^{-1}(\phi(r_i))$ for each $r_i \in \mathcal{R}$. As ϕ, ψ are isomorphisms between $\mathcal{R}$ and $\widehat{\mathcal{R}}$, σ is bijective and additive. For $\mathbf{b} = (b_1, \dots, b_n) \in \mathcal{R}^n$, set $\sigma(\mathbf{b}) = (\sigma(b_1), \dots, \sigma(b_n))$, $\sigma^{-1}(\mathbf{b}) = (\sigma^{-1}(b_1), \dots, \sigma^{-1}(b_n))$, $\psi(\mathbf{b}) = (\psi(b_1), \dots, \psi(b_n))$ and $\phi(\mathbf{b}) = (\phi(b_1), \dots, \phi(b_n))$. Now using the map σ, define a map $\rho : \mathcal{R}^{2n} \to \mathcal{R}^{2n}$ as

$$\rho((\mathbf{a},\mathbf{b})) := (\mathbf{a}, \sigma(\mathbf{b})) = (a_1, \dots, a_n, \sigma(b_1), \dots \sigma(b_n))$$

for all $(\mathbf{a},\mathbf{b}) = (a_1, \dots, a_n, b_1, \dots, b_n) \in \mathcal{R}^{2n}$. As σ is bijective, additive, and Hamming weight preserving, ρ is also bijective, additive and symplectic weight preserving. Note that for $(\mathbf{a},\mathbf{b}),(\mathbf{a}',\mathbf{b}')\in\mathcal{R}^{2n}$,

$$\begin{aligned}[(\mathbf{a},\mathbf{b}),(\mathbf{a}',\mathbf{b}')]_{M\S} &= \phi(\mathbf{b})(\mathbf{a}')\phi(\mathbf{b}')(-\mathbf{a})\\ &= \psi(\sigma(\mathbf{b}))(\mathbf{a}')\psi(\sigma(\mathbf{b}'))(-\mathbf{a})\\ &= [\rho((\mathbf{a},\mathbf{b})),\rho((\mathbf{a}',\mathbf{b}'))]_{M'\S}.\end{aligned}$$

Thus ρ is a symplectic isometry. If $\mathcal{C}$ is an additive code of $\mathcal{R}^{2n}$, then $\rho(\mathcal{C})$ is also an additive code of $\mathcal{R}^{2n}$.

Next, note that $\mathcal{C}^{M\S}=\{(\mathbf{a},\mathbf{b})\in\mathcal{R}^{2n}\mid\phi(\mathbf{b})(\mathbf{a}')\phi(\mathbf{b}')(-\mathbf{a})=1$ for all $(\mathbf{a}',\mathbf{b}')\in\mathcal{C}\}$ and $\rho(\mathcal{C})^{M'\S}=\{(\mathbf{a},\mathbf{b})\in\mathcal{R}^{2n}\mid\psi(\mathbf{b})(\mathbf{a}')\psi(\mathbf{b}')(-\mathbf{a})=1$ for all $(\mathbf{a}',\mathbf{b}')\in\rho(\mathcal{C})\}$. By definition of the map ρ, we get $\rho(\mathcal{C})^{M'\S}=\{(\mathbf{a},\mathbf{b})\in\mathcal{R}^{2n}\mid\psi(\mathbf{b})(\mathbf{a}')\psi(\sigma(\mathbf{b}'))(-\mathbf{a})$ $=1$ for all $(\mathbf{a}',\mathbf{b}')\in\mathcal{C}\}$. Note that for $(\mathbf{a},\mathbf{b}),(\mathbf{a}',\mathbf{b}')\in\mathcal{R}^{2n}$, $\phi(\mathbf{b})(\mathbf{a}')\phi(\mathbf{b}')(-\mathbf{a})=\psi(\sigma(\mathbf{b}))(\mathbf{a}')\psi(\sigma(\mathbf{b}'))(-\mathbf{a})$ and $\psi(\mathbf{b})(\mathbf{a}')\psi(\sigma(\mathbf{b}'))(-\mathbf{a})=\phi(\sigma^{-1}(\mathbf{b}))(\mathbf{a}')\phi((\mathbf{b}'))(-\mathbf{a})$. From this, we get $\rho(\mathcal{C}^{M\S})\subseteq\rho(\mathcal{C})^{M'\S}$ and $\rho(\mathcal{C})^{M'\S}\subseteq\rho(\mathcal{C}^{M\S})$. Thus $\rho(\mathcal{C}^{M\S})=\rho(\mathcal{C})^{M'\S}$, which completes the proof of the theorem. □

PROPOSITION 3.3. *Fix a duality M of $\mathcal{R}$. For an additive code $\mathcal{C}\subseteq\mathcal{R}^{2n}$, we have $|\mathcal{C}||\mathcal{C}^{M\S}|=|\mathcal{R}^{2n}|$.*

PROOF. Define a duality M' of $\mathcal{R}$ as $M'_{i,j}=\chi(r_ir_j)$. By Theorem 3.2, there exists a bijective map ρ such that $\rho(\mathcal{C})$ is an additive code and $\rho(\mathcal{C})^{M'\S}=\rho(\mathcal{C}^{M\S})$. By Remark 2.4, $\rho(\mathcal{C})^{M'\S}=\{(\mathbf{a},\mathbf{b})\in\mathcal{R}^{2n}\mid\chi(\mathbf{b}\cdot\mathbf{a}'-\mathbf{b}'\cdot\mathbf{a})=1$ for all $(\mathbf{a}',\mathbf{b}')\in\mathcal{C}\}$. From this and using Lemma 6 of Nadella and Klappenecker [**15**], we get $|\rho(\mathcal{C})||\rho(\mathcal{C})^{M'\S}|=|\mathcal{R}^{2n}|$. Thus

$$|\mathcal{C}||\mathcal{C}^{M\S}|=|\rho(\mathcal{C})||\rho(\mathcal{C}^{M\S})|=|\rho(\mathcal{C})||\rho(\mathcal{C})^{M'\S}|=|\mathcal{R}^{2n}|.$$

□

This proposition is very important as it shows that $M\S$ satisfies the same cardinality condition that an arbitrary duality does for additive codes and the standard duality of linear codes in terms of the cardinality of codes.

Theorem 3.4 follows immediately from Theorem 3.2.

THEOREM 3.4. *For any pair of dualities M and M' of $\mathcal{R}$, an M-symplectically self-orthogonal additive code is symplectically isometric to an M'-symplectically self-orthogonal additive code. As a consequence, there exists an M-symplectically self-orthogonal additive code if and only if there exists an M'-symplectically self-orthogonal additive code.*

4. Symplectically self-orthogonal additive codes and their application to QECCs

In this section, we bring together the various threads introduced in the previous sections, of M-symplectically self-orthogonal codes, symplectic isometries, and QECCs. When M is a canonical duality of $\mathcal{R}$ obtained from a generating character (see Remark 2.4), we have the following result of Nadella and Klappenecker [**15**], re-phrased in our language.

THEOREM 4.1 ([**15**], Theorem 7). *Let M be a fixed duality of $\mathcal{R}$ such that $M_{i,j} = \psi(r_i r_j)$ for some generating character ψ of $\mathcal{R}$. Let $\mathcal{C} \subseteq \mathcal{R}^{2n}$ be an M-symplectically self-orthogonal additive code. Then, from $\mathcal{C}$, we can construct an $((n, d^n/|\mathcal{C}|, D))$ QECC derived from $\mathcal{R}$, where the minimum distance is*

$$D = \begin{cases} d_s(\mathcal{C}) & \text{if } \mathcal{C} = \mathcal{C}^{M\S} \\ d_s(\mathcal{C}^{M\S} \setminus \mathcal{C}) & \text{if } \mathcal{C} \subsetneq \mathcal{C}^{M\S}. \end{cases}$$

By virtue of Theorem 3.2, the above theorem of Nadella and Klappenecker immediately extends to a more general result that, for an arbitrary duality M of $\mathcal{R}$, encapsulates the relationship between M-symplectically self-orthogonal additive codes and QECCs derived from $\mathcal{R}$.

THEOREM 4.2. *Fix a duality M of $\mathcal{R}$. Let $\mathcal{C} \subseteq \mathcal{R}^{2n}$ be an M-symplectically self-orthogonal additive code. Then from $\mathcal{C}$, we can construct an $((n, d^n/|\mathcal{C}|, D))$ QECC derived from $\mathcal{R}$, where*

$$D = \begin{cases} d_s(\mathcal{C}) & \text{if } \mathcal{C} = \mathcal{C}^{M\S} \\ d_s(\mathcal{C}^{M\S} \setminus \mathcal{C}) & \text{if } \mathcal{C} \subsetneq \mathcal{C}^{M\S}. \end{cases}$$

In view of Theorems 3.4 and 4.2, we make the following observation.

REMARK 4.3. For any pair of dualities M and M' of $\mathcal{R}$, by Theorem 3.4, there exists an additive code $\mathcal{C}$ that is M-symplectically self-orthogonal iff there exists an additive code $\mathcal{C}'$ that is M'-symplectically self-orthogonal. As a consequence, if $\mathcal{Q}$ is a quantum stabilizer code constructed using Theorem 4.2 from an additive code $\mathcal{C}$ that is M-symplectically self-orthogonal, then the same $\mathcal{Q}$ can just as well be constructed from some additive code $\mathcal{C}'$ that is M'-symplectically self-orthogonal. In particular, the same $\mathcal{Q}$ can be constructed from some additive code $\mathcal{C}''$ that is M''-symplectically self-orthogonal for some duality M'' of $\mathcal{R}$ such that $M''_{i,j} = \psi(r_i r_j)$ for some generating character ψ of $\mathcal{R}$. In other words, M'' is a canonical duality obtained from a generating character ψ. Thus, for the purpose of constructing quantum stabilizer codes from additive codes over $\mathcal{R}$, it suffices to consider additive codes that are symplectically self-orthogonal with respect to a canonical duality obtained from a generating character, and apply the standard construction of Theorem 4.1. This observation was also made by Ezerman et al. [**11**] in the context of their "CSS-like" code constructions.

In the example below, we demonstrate the use of Theorem 4.2 by constructing a QECC from a linear code over $\mathbb{F}_4$ that is M_4-symplectically self-dual.

EXAMPLE 4.4. Let $\mathbb{F}_4 = \{0, 1, \omega, \omega^2\}$. Recall that the duality M_4 is defined via $\chi_a(b) = (-1)^{\mathrm{tr}_{4/2}(ab^2)}$. The duality M_1, given by $\chi_a(b) = (-1)^{\mathrm{tr}_{4/2}(ab)}$ is a canonical duality as $(M_1)_{i,j} = (-1)^{\mathrm{tr}_{4/2}(r_i r_j)}$ for $r_i, r_j \in \mathbb{F}_4$. Consider the linear code $\mathcal{C}$ over $\mathbb{F}_4$ generated by $\begin{bmatrix} \omega^2 & \omega & 1 & \omega^2 & \omega & 1 \\ \omega & \omega^2 & \omega & \omega & \omega^2 & \omega \\ \omega^2 & \omega & \omega^2 & \omega^2 & \omega & \omega^2 \end{bmatrix}$. It is straightforward to verify that $\mathcal{C}$ is an (ω, ω)-multi-twisted code[1], and $\mathcal{C} = \mathcal{C}^{M_4\S}$. By Theorem 4.2, from $\mathcal{C}$, we can construct a $((3, 1, 1))$ QECC $\mathcal{Q}$ derived from $\mathbb{F}_4$.

[1]A linear code $\mathcal{C} \subseteq \mathbb{F}_4^6$ is (ω, ω)-multi-twisted if it has a property that $(a_0, a_1, a_2, a_3, a_4, a_5) \in \mathcal{C}$ implies $(\omega a_2, a_0, a_1, \omega a_5, a_3, a_4) \in \mathcal{C}$.

The same QECC $\mathcal{Q}$ can also be constructed from an additive code $\mathcal{C}'$ that is M_1-symplectically self-dual. By Theorem 3.2, the code $\mathcal{C}'$ can be obtained from $\mathcal{C}$ by applying a mapping $\rho : \mathbb{F}_4^6 \to \mathbb{F}_4^6$ defined by $\rho(\mathbf{a}, \mathbf{b}) := (\mathbf{a}, \mathbf{b}^2)$, where, for $\mathbf{b} = (b_0, b_1, b_2) \in \mathbb{F}_4^3$, we set $\mathbf{b}^2 := (b_0^2, b_1^2, b_2^2)$. Clearly, ρ is an additive and bijective map, and hence, by Theorem 3.2, $\rho(\mathcal{C})$ is additive and $\rho(\mathcal{C})^{M_1 \S} = \rho(\mathcal{C}^{M_4 \S})$. As $\mathcal{C}$ is M_4-symplectically self-dual, $\rho(\mathcal{C})$ is M_1-symplectically self-dual. Thus from $\mathcal{C}' = \rho(\mathcal{C})$, by using Theorem 4.1, we get the same $((3,1,1))$ QECC $\mathcal{Q}$ as above. Note, however, that the code $\mathcal{C}'$ is merely additive, and not linear, over $\mathbb{F}_4$, and it lacks the (ω, ω)-multi-twisted structure.

Example 4.4 shows that despite the fact that it suffices to consider additive codes that are symplectically self-orthogonal with respect to a canonical duality of $\mathcal{R}$, there can be advantages to considering additive codes that are symplectically self-orthogonal with respect to a non-canonical duality of $\mathcal{R}$. As the example shows, the code $\mathcal{C}$ is $\mathbb{F}_4$-linear and has an additional multi-twisted property, while its counterpart $\mathcal{C}' = \rho(\mathcal{C})$ is merely additive with no obvious additional structure. While searching for good codes in practice, it is often easier to look for good linear codes with some structural properties rather than additive codes in the wild.

We end this section with two special instances of our construction of QECCs. The first is a CSS-style construction.

4.1. A CSS-style construction of QECCs.

THEOREM 4.5. *Let M be a fixed symmetric duality of $\mathcal{R}$, i.e., $M_{i,j} = M_{j,i}$. Let $\mathcal{C}_1$ and $\mathcal{C}_2$ be two additive codes in $\mathcal{R}^n$ such that $\mathcal{C}_1 \subseteq \mathcal{C}_2^M$. Then there exists an $((n, d^n/(|\mathcal{C}_1||\mathcal{C}_2|), D))$ QECC derived from $\mathcal{R}$, where*

$$D = \begin{cases} \min\{\mathrm{wt}_H(\mathbf{c}) \mid \mathbf{c} \in (\mathcal{C}_1 \cup \mathcal{C}_2) \setminus \{\mathbf{0}\}\} & \text{if } \mathcal{C}_1 = \mathcal{C}_2^M, \\ \min\{\mathrm{wt}_H(\mathbf{c}) \mid \mathbf{c} \in (\mathcal{C}_2^M \setminus \mathcal{C}_1) \cup (\mathcal{C}_1^M \setminus \mathcal{C}_2)\} & \text{if } \mathcal{C}_1 \subsetneq \mathcal{C}_2^M. \end{cases}$$

PROOF. Since M is a symmetric duality, we have that $\chi_{r_i}(r_j) = \chi_{r_j}(r_i)$. Let $\mathcal{C} = \mathcal{C}_1 \oplus \mathcal{C}_2$. Clearly, $\mathcal{C}$ is an additive code of $\mathcal{R}^{2n}$. We assert that $\mathcal{C}^{M\S} = \mathcal{C}_2^M \oplus \mathcal{C}_1^M$. To prove this assertion, recall that $(\mathbf{a}, \mathbf{b}) \in \mathcal{C}^{M\S}$ if and only if $\prod \chi_{b_i}(a_i') = \prod \chi_{b_i'}(a_i)$ for all $(\mathbf{a}', \mathbf{b}') \in \mathcal{C}$. Thus $(\mathbf{a}, \mathbf{b}) \in \mathcal{C}^{M\sigma}$ if and only if $\prod \chi_{b_i}((c_1)_i) = \prod \chi_0(a_i)$ and $\prod \chi_{b_i}(0) = \prod \chi_{(c_2)_i}(a_i)$ for all $(\mathbf{c}_1, 0), (0, \mathbf{c}_2) \in \mathcal{C}$. From this and using the fact that $\prod \chi_{(c_2)_i}(a) = \prod \chi_{a_i}((c_2)_i)$, we get $(\mathbf{a}, \mathbf{b}) \in \mathcal{C}^{M\S}$ if and only if $\prod \chi_{b_i}((c_1)_i) = 1$ and $\prod \chi_{(c_2)_i}(a_i) = 1$ for all $\mathbf{c}_1 \in \mathcal{C}_1$ and $\mathbf{c}_2 \in \mathcal{C}_2$. Hence, $(\mathbf{a}, \mathbf{b}) \in \mathcal{C}^{M\S}$ if and only if $\mathbf{a} \in \mathcal{C}_2^M$ and $\mathbf{b} \in \mathcal{C}_1^M$. This proves the assertion.

Now, using the fact that M is a symmetric duality and $\mathcal{C}_1 \subseteq \mathcal{C}_2^M$, we have $\mathcal{C}_2 \subseteq \mathcal{C}_1^M$; thus, $\mathcal{C} \subseteq \mathcal{C}^{M\S} = \mathcal{C}_2^M \oplus \mathcal{C}_1^M$. Then, $\mathcal{C}$ is an M-symplectically self-orthogonal code in $\mathcal{R}^{2n}$. Hence, via Theorem 4.2, there exists an $((n, d^n/|\mathcal{C}|, D))$ QECC derived from $\mathcal{R}$, where $D = d_s(\mathcal{C})$ if $\mathcal{C}^{M\S} = \mathcal{C}$, and $D = d_s(\mathcal{C}^{M\S} \setminus \mathcal{C})$ if $\mathcal{C} \subsetneq \mathcal{C}^{M\S}$.

Further using the fact that $\mathcal{C}^{M\S} = \mathcal{C}_2^M \oplus \mathcal{C}_1^M$, we have $\mathcal{C}^{M\S} = \mathcal{C}$ if and only if $\mathcal{C}_2^M = \mathcal{C}_1$. Thus when $\mathcal{C}_2^M = \mathcal{C}_1$, we have $D = \min\{\mathrm{wt}_H(\mathbf{c}) \mid \mathbf{c} \in (\mathcal{C}_1 \cup \mathcal{C}_2) \setminus \{\mathbf{0}\}\}$. Further, when $\mathcal{C}_2^M \subsetneq \mathcal{C}_1$, we have $D = \min\{\mathrm{wt}_H(\mathbf{c}) \mid \mathbf{c} \in (\mathcal{C}_2^M \setminus \mathcal{C}_1) \cup (\mathcal{C}_1^M \setminus \mathcal{C}_2)\}$. $\square$

4.2. QECCs from self-orthogonal additive codes over $\mathrm{GR}(p^b, 2m)$. There is a well-known equivalence between (a) construction using Theorem 4.1 of n-qudit QECCs derived from the field $\mathbb{F}_q$, and (b) additive codes of length n over $\mathbb{F}_{q^2}$ that are self-orthogonal with respect to a certain trace-alternating form [**3**], [**14**, Section IV-B]. Recently, Sidana and Kashyap [**16**] extended this equivalence to additive codes over Galois rings. In this subsection, we will show that the problem of constructing n-qudit QECCs derived from $\mathrm{GR}(p^b, m)$ using Theorem 4.2 is equivalent to that of finding additive codes of length n over $\mathrm{GR}(p^b, 2m)$, which are self-orthogonal with respect to certain dualities. This equivalence paves the way for the use of self-orthogonal additive codes (with respect to certain dualities) over $\mathrm{GR}(p^b, 2m)$ for the purpose of constructing QECCs derived from $\mathrm{GR}(p^b, m)$.

We begin by observing that $\mathrm{GR}(p^b, 2m)$ is a degree-2 extension of $\mathrm{GR}(p^b, m)$. In fact, there exists a monic polynomial $g(x) \in \mathrm{GR}(p^b, m)[x]$ of degree 2 such that its reduction $(g(x) \mod p) \in \mathbb{F}_{p^m}[x]$ is irreducible and primitive over $\mathbb{F}_{p^m}$ and $g(x)$ divides $x^{p^{2m}-1} - 1$ in $\mathrm{GR}(p^b, m)[x]$. Thus $\mathrm{GR}(p^b, 2m) \cong \mathrm{GR}(p^b, m)[x]/\langle g(x)\rangle$. Set $\delta = x + \langle g(x)\rangle$. Clearly, $g(\delta) = 0$, and each element of

$$\mathrm{GR}(p^b, 2m)(\cong \mathrm{GR}(p^b, 2m)[x]/\langle g(x)\rangle)$$

can be a unique written as $r = r_0 + r_1\delta$ with $r_0, r_1 \in \mathrm{GR}(p^b, m)$. Thus $\mathrm{GR}(p^b, 2m)$ is a free module of rank 2 over $\mathrm{GR}(p^b, m)$ with $\{1, \delta\}$ as a basis (see [**19**, Ch. 14]). Define $\pi : \mathrm{GR}(p^b, m)^{2n} \to \mathrm{GR}(p^b, 2m)^n$ as follows: $\pi(a_1, \ldots, a_n, b_1, \ldots, b_n) = (a_1 + \delta b_1, \ldots, a_n + \delta b_n)$. The map π is clearly additive and bijective, thus a $\mathbb{Z}_{p^b}$-module isomorphism. Throughout this section, whenever we consider a tuple $(\mathbf{a}, \mathbf{b}) \in \mathrm{GR}(p^b, m)^{2n}$ and its image $\pi(\mathbf{a}, \mathbf{b}) = \mathbf{a} + \delta\mathbf{b} \in \mathrm{GR}(p^b, 2m)^n$, it is to be implicitly understood that $\mathbf{a} + \delta\mathbf{b} = (a_1 + \delta b_1, \ldots, a_n + \delta b_n) \in \mathrm{GR}(p^b, 2m)^n$. Clearly, $wt_s((\mathbf{a}, \mathbf{b})) = w_H(\pi((a, b))$, thus π is an isometry in this sense.

Let M be a fixed duality of $\mathrm{GR}(p^b, m)$. We use the duality M to define an orthogonality of $\mathrm{GR}(p^b, 2m)^n$ via a lifted duality $\widehat{M}$ of $\mathrm{GR}(p^b, 2m)$. To this end, taking inspiration from (2.1), we define for each $a + \delta b \in \mathrm{GR}(p^b, 2m)$, a character $\rho_{a+\delta b}$ of $\mathrm{GR}(p^b, 2m)$ as follows: for all $a' + \delta b' \in \mathrm{GR}(p^b, 2m)$, $\rho_{a+\delta b}(a' + \delta b') := \chi_b(a')\chi_{b'}(-a)$. This defines a character table (or duality) $\widehat{M}$ of $\mathrm{GR}(p^b, 2m)$, the rows of which are indexed by $\rho(a + \delta b)$, as $a + \delta b$ ranges over the elements of $\mathrm{GR}(p^b, 2m)$, and the columns of which are indexed by the elements of $\mathrm{GR}(p^b, 2m)$. Note that if M is a symmetric duality, then

$$\begin{aligned}\rho_{a'+\delta b'}(a + \delta b) &:= \chi_{b'}(a)\chi_b(-a')\\ &= \chi_b(-a')\chi_{-b'}(-a)\\ &=: \rho_{a+\delta b}(-a' - \delta b').\end{aligned}$$

Thus, when $\mathrm{GR}(p^b, 2m)$ has characteristic 2, i.e., $\mathrm{GR}(p^b, 2m)$ is a finite ring of characteristic 2, M being symmetric implies that $\widehat{M}$ is symmetric; but this implication does not hold in general.

By Definition 2.2, the orthogonal of an additive code $\mathcal{C} \subseteq \mathbb{F}_{q^2}^n$, with respect to the duality $\widehat{M}$, is the additive code $\mathcal{C}^{\widehat{M}} = \{\mathbf{a} + \delta\mathbf{b} \in \mathbb{F}_{q^2}^n \mid \prod \rho_{a_i+\delta b_i}(a_i' + \delta b_i') = 1 \text{ for all } \mathbf{a}' + \delta\mathbf{b}' \in \mathcal{C}\}$. Moreover, $\mathcal{C}$ is said to be self-orthogonal with respect to $\widehat{M}$ if $\mathcal{C} \subseteq \mathcal{C}^{\widehat{M}}$.

As π is a $\mathbb{Z}_{p^b}$-module isomorphism, a subset $\mathcal{C}$ of $\mathrm{GR}(p^b, m)^{2n}$ is an additive code over $\mathrm{GR}(p^b, m)$ of length $2n$ if and only if $\pi(\mathcal{C})$ is an additive code over $\mathrm{GR}(p^b, 2m)$ of length n. The proposition below follows immediately from the relevant definitions.

PROPOSITION 4.6. *Let $\mathcal{C} \subseteq \mathrm{GR}(p^b, m)^{2n}$ be an additive code, and let $\overline{\mathcal{C}} = \pi(\mathcal{C})$ be its image in $\mathrm{GR}(p^b, 2m)^n$ under the mapping π. Then $\mathcal{C} \subseteq \mathrm{GR}(p^b, m)^{2n}$ is M-symplectically self-orthogonal (i.e., $\mathcal{C} \subseteq \mathcal{C}^{M\S}$) if and only if $\overline{\mathcal{C}}$ is self-orthogonal with respect to $\widehat{M}$ (i.e., $\overline{\mathcal{C}} \subseteq \overline{\mathcal{C}}^{\widehat{M}}$).*

The following result is essentially a restatement of Theorem 4.2 re-phrased in terms of self-orthogonality with respect to the duality $\widehat{M}$.

THEOREM 4.7. *[Equivalent to Theorem 4.2] Let $\mathcal{C} \subseteq \mathrm{GR}(p^b, 2m)^n$ be an additive code that is self-orthogonal with respect to $\widehat{M}$. Then, from $\mathcal{C}$, we can construct an $((n, p^{bmn}/|\mathcal{C}|, D))$ QECC derived from $\mathrm{GR}(p^b, m)$, where*

$$D = \begin{cases} d_{\mathrm{H}}(\mathcal{C}) & \text{if } \mathcal{C} = \mathcal{C}^{\widehat{M}}, \\ d_{\mathrm{H}}(\mathcal{C}^{\widehat{M}} \setminus \mathcal{C}) & \text{if } \mathcal{C} \subsetneq \mathcal{C}^{\widehat{M}}. \end{cases}$$

5. Construction of self-orthogonal codes with respect to a given duality

Given the importance of self-orthogonal codes in Theorem 4.2 and Theorem 4.7, in terms of constructions of codes, which is the aim of this present paper, we shall now study some techniques for producing self-orthogonal codes over a Frobenius ring for an arbitrary duality.

For a given code C we define the hull of the code with respect to the duality M to be:

$$Hull_M(C) = C \cap C^M. \tag{5.1}$$

This definition for additive codes was given in [**10**], but first appeared for linear codes with the Euclidean duality in [**1**].

THEOREM 5.1. *Let C be an additive code in G^n with duality M. Then $Hull_M(C)$ is an additive self-orthogonal code with respect to M.*

PROOF. If $\mathbf{v}, \mathbf{w} \in Hull_M(C)$ then $\mathbf{v}, \mathbf{w} \in C$ and $\mathbf{v}, \mathbf{w} \in C^M$. This gives that $[\mathbf{v}, \mathbf{w}]_M = 1$. Therefore $Hull_M(C)$ is self-orthogonal with respect to M. Moreover, $\mathbf{v} + \mathbf{w} \in C$ and $\mathbf{v} + \mathbf{w} \in C^M$ since these codes are additive. Therefore, $\mathbf{v} + \mathbf{w} \in Hull_M(C)$. This gives that $Hull_M(C)$ is an additive self-orthogonal code with respect to M. □

For matrices A, B, define

$$(A \otimes_M B)_{ij} = \prod \chi_{A_{i,k}} B_{k,j}. \tag{5.2}$$

Notice that A, B have entries in the group G but the entries in the matrix $A \otimes_M B$ are in the Complex numbers $\mathbb{C}$. It is immediate that if $\mathcal{G}$ generates a self-orthogonal code then $\mathcal{G} \otimes_M \mathcal{G}^T$ is the all-one matrix.

All of the entries in this matrix are complex numbers that are a k-th root of unity. Fix a primitive k-th root of unity ξ_k. Then every element in the matrix $A \otimes_M B$ is of the for ξ_k^r for some $r \in \mathbb{Z}_k$. Define $\Phi_k(\xi_k^r) = r$ where Φ maps to $\mathbb{Z}_k$.

If the ring R is the finite field $\mathbb{F}_{p^e}$, then we can define for an additive code C, $\mathrm{rank}_p(C)$ as the rank as a module over $\mathbb{F}_p$. The code C then has cardinality $p^{\mathrm{rank}_p(C)}$.

THEOREM 5.2. *Let C be an additive code with generator matrix $\mathcal{G}$ over $\mathbb{F}_{p^e}$ equipped with the duality M. Then we have*

$$\operatorname{rank}_p(C) - \operatorname{rank}_p(Hull_M(C)) = \operatorname{rank}_p(\Phi_p(\mathcal{G} \otimes_M \mathcal{G}^T)). \tag{5.3}$$

PROOF. We have that $\operatorname{rank}_p(\Phi_p(\mathcal{G} \otimes_M \mathcal{G}^T))$ gives the p-rank of the space of vectors generated by $\mathcal{G}$ that are not mutually orthogonal. It follows that

$$\operatorname{rank}_p(C) - \operatorname{rank}_p(Hull_M(C)) = \operatorname{rank}_p(\Phi_p(\mathcal{G} \otimes_M \mathcal{G}^T)).$$

□

THEOREM 5.3. *Let C be an additive code over $\mathbb{F}_{p^2} = \mathbb{F}_p(\omega)$ equipped with the duality M and generated by*

$$\mathcal{G} = \begin{pmatrix} I_{k_1} & A \\ \omega I_{k_2} & B \end{pmatrix},$$

where A, B are defined over $\mathbb{F}_p(\omega)$. Let

$$A \otimes_M A^T = \begin{pmatrix} \alpha & 1 & 1 & 1 & \cdots & 1 \\ 1 & \alpha & 1 & 1 & \cdots & 1 \\ 1 & 1 & \alpha & 1 & \cdots & 1 \\ \vdots & \vdots & \vdots & \vdots & \vdots & \vdots \\ 1 & 1 & 1 & 1 & \cdots & \alpha \end{pmatrix}, \text{ where } \alpha = \frac{1}{\chi_1(1)}, \tag{5.4}$$

$$B \otimes_M B^T = \begin{pmatrix} \beta & 1 & 1 & 1 & \cdots & 1 \\ 1 & \beta & 1 & 1 & \cdots & 1 \\ 1 & 1 & \beta & 1 & \cdots & 1 \\ \vdots & \vdots & \vdots & \vdots & \vdots & \vdots \\ 1 & 1 & 1 & 1 & \cdots & \beta \end{pmatrix}, \text{ where } \beta = \frac{1}{\chi_\omega(\omega)}, \tag{5.5}$$

$$A \otimes_M B^T = \begin{pmatrix} \gamma & 1 & 1 & 1 & \cdots & 1 \\ 1 & \gamma & 1 & 1 & \cdots & 1 \\ 1 & 1 & \gamma & 1 & \cdots & 1 \\ \vdots & \vdots & \vdots & \vdots & \vdots & \vdots \\ 1 & 1 & 1 & 1 & \cdots & \gamma \end{pmatrix}, \text{ where } \gamma = \frac{1}{\chi_1(\omega)}, \tag{5.6}$$

$$B \otimes_M A^T = \begin{pmatrix} \theta & 1 & 1 & 1 & \cdots & 1 \\ 1 & \theta & 1 & 1 & \cdots & 1 \\ 1 & 1 & \theta & 1 & \cdots & 1 \\ \vdots & \vdots & \vdots & \vdots & \vdots & \vdots \\ 1 & 1 & 1 & 1 & \cdots & \theta \end{pmatrix}, \text{ where } \theta = \frac{1}{\chi_\omega(1)}. \tag{5.7}$$

Then C is a self-orthogonal code with respect to duality M over $\mathbb{F}_p(\omega)$ with additive $\mathbb{F}_p$ rank $k_1 + k_2$. In particular, if $|\langle G \rangle| = p^n$ then C is a self-dual code with respect to duality M over $\mathbb{F}_p(\omega)$.

PROOF. We know that if $\mathcal{G}$ generates a self-orthogonal code then $\mathcal{G} \otimes_M \mathcal{G}^T$ is the all-one matrix.

$$\mathcal{G} \otimes_M \mathcal{G}^T = \begin{pmatrix} I_{k_1} & A \\ \omega I_{k_2} & B \end{pmatrix} \otimes_M \begin{pmatrix} I_{k_1} & \omega I_{k_2} \\ A^T & B^T \end{pmatrix} = \left(\begin{array}{c|c} (I_{k_1} \otimes I_{k_1}) \bullet (A \otimes A^T) & (I_{k_1} \otimes \omega I_{k_2}) \bullet (A \otimes B^T) \\ \hline (\omega I_{k_2} \otimes I_{k_1}) \bullet (B \otimes A^T) & (\omega I_{k_2} \otimes \omega I_{k_2}) \bullet (B \otimes B^T) \end{array} \right)$$

$$= \left(\begin{array}{cccc|cccc} \chi_1(1)\langle A_1, A_1\rangle_M & \langle A_1, A_2\rangle_M & \cdots & \langle A_1, A_{k_1}\rangle_M & \chi_1(\omega)\langle A_1, B_1\rangle_M & \langle A_1, B_2\rangle_M & \cdots & \langle A_1, B_{k_2}\rangle_M \\ \langle A_2, A_1\rangle_M & \chi_1(1)\langle A_2, A_2\rangle_M & \cdots & \langle A_2, A_{k_1}\rangle_M & \langle A_2, B_1\rangle & \chi_1(\omega)\langle A_2, B_2\rangle_M & \cdots & \langle A_2, B_{k_2}\rangle_M \\ \vdots & \vdots & \cdots & \vdots & \vdots & \vdots & \cdots & \vdots \\ \langle A_{k_1}, A_1\rangle_M & \langle A_{k_1}, A_2\rangle_M & \cdots & \chi_1(1)\langle A_{k_1} A_{k_1}\rangle_M & \langle A_{k_1}, B_1\rangle_M & \langle A_{k_1}, B_2\rangle_M & \cdots & \chi_1(\omega)\langle A_{k_1}, B_{k_2}\rangle_M \\ \hline \chi_\omega(1)\langle B_1, A_1\rangle_M & \langle B_1, A_2\rangle_M & \cdots & \langle B_1, A_{k_1}\rangle_M & \chi_\omega(\omega)\langle B_1, B_1\rangle_M & \langle B_1, B_2\rangle_M & \cdots & \langle B_1, B_{k_2}\rangle_M \\ \langle B_2, A_1\rangle_M & \chi_\omega(1)\langle B_2, A_2\rangle_M & \cdots & \langle_M B_2, A_{k_1}\rangle_M & \langle B_2, B_1\rangle_M & \chi_\omega(\omega)\langle B_2, B_2\rangle_M & \cdots & \langle B_2, B_{k_2}\rangle_M \\ \vdots & \vdots & \cdots & \vdots & \vdots & \vdots & \cdots & \vdots \\ \langle B_{k_2}, A_1\rangle_M & \langle B_{k_2}, A_2\rangle_M & \cdots & \chi_\omega(1)\langle B_{k_2} A_{k_1}\rangle_M & \langle B_{k_2}, B_1\rangle_M & \langle B_{k_2}, B_2\rangle_M & \cdots & \chi_\omega(\omega)\langle B_{k_2}, B_{k_2}\rangle_M \end{array} \right)$$

where $\bullet$ denotes the componetwise multiplication of matrices and A_i, B_i denote the i-th rows of the corresponding matrices. Then, Equation (5.4) implies that

$$\begin{aligned} &(I_{k_1} \otimes I_{k_1}) \bullet (A \otimes A^T) \\ &= \begin{pmatrix} \chi_1(1) & 1 & \cdots & 1 \\ 1 & \chi_1(1) & \cdots & 1 \\ \vdots & \vdots & \vdots & \vdots \\ 1 & 1 & \cdots & \chi_1(1) \end{pmatrix} \bullet \begin{pmatrix} \langle A_1, A_1\rangle_M & \langle A_1, A_2\rangle_M & \cdots & \langle A_1, A_{k_1}\rangle_M \\ \langle A_2, A_1\rangle_M & \langle A_2, A_2\rangle_M & \cdots & \langle A_2, A_{k_1}\rangle_M \\ \vdots & \vdots & \vdots & \vdots \\ \langle A_{k_1}, A_1\rangle_M & \langle A_{k_1}, A_2\rangle_M & \cdots & \langle A_{k_1}, A_{k_1}\rangle_M \end{pmatrix} \\ &= \begin{pmatrix} \chi_1(1) & 1 & \cdots & 1 \\ 1 & \chi_1(1) & \cdots & 1 \\ \vdots & \vdots & \vdots & \vdots \\ 1 & 1 & \cdots & \chi_1(1) \end{pmatrix} \bullet \begin{pmatrix} \alpha & 1 & \cdots & 1 \\ 1 & \alpha & \cdots & 1 \\ \vdots & \vdots & \vdots & \vdots \\ 1 & 1 & \cdots & \alpha \end{pmatrix} \\ &= \begin{pmatrix} 1 & 1 & \cdots & 1 \\ 1 & 1 & \cdots & 1 \\ \vdots & \vdots & \vdots & \vdots \\ 1 & 1 & \cdots & 1 \end{pmatrix}. \end{aligned}$$

Similarly, Equation (5.5) implies that

$$\begin{aligned} &(\omega I_{k_2} \otimes \omega I_{k_2}) \bullet (B \otimes B^T) \\ &= \begin{pmatrix} \chi_\omega(\omega) & 1 & \cdots & 1 \\ 1 & \chi_\omega(\omega) & \cdots & 1 \\ \vdots & \vdots & \vdots & \vdots \\ 1 & 1 & \cdots & \chi_\omega(\omega) \end{pmatrix} \bullet \begin{pmatrix} \beta & 1 & \cdots & 1 \\ 1 & \beta & \cdots & 1 \\ \vdots & \vdots & \vdots & \vdots \\ 1 & 1 & \cdots & \beta \end{pmatrix} = \begin{pmatrix} 1 & 1 & \cdots & 1 \\ 1 & 1 & \cdots & 1 \\ \vdots & \vdots & \vdots & \vdots \\ 1 & 1 & \cdots & 1 \end{pmatrix}, \end{aligned}$$

Equation (5.6) implies that

$$\begin{aligned} &(I_{k_1} \otimes \omega I_{k_2}) \bullet (A \otimes B^T) \\ &= \begin{pmatrix} \chi_1(\omega) & 1 & \cdots & 1 \\ 1 & \chi_1(\omega) & \cdots & 1 \\ \vdots & \vdots & \vdots & \vdots \\ 1 & 1 & \cdots & \chi_1(\omega) \end{pmatrix} \bullet \begin{pmatrix} \gamma & 1 & \cdots & 1 \\ 1 & \gamma & \cdots & 1 \\ \vdots & \vdots & \vdots & \vdots \\ 1 & 1 & \cdots & \gamma \end{pmatrix} = \begin{pmatrix} 1 & 1 & \cdots & 1 \\ 1 & 1 & \cdots & 1 \\ \vdots & \vdots & \vdots & \vdots \\ 1 & 1 & \cdots & 1 \end{pmatrix}, \end{aligned}$$

and Equation (5.7) implies that

$$\begin{aligned} &(\omega I_{k_2} \otimes I_{k_1}) \bullet (B \otimes A^T) \\ &= \begin{pmatrix} \chi_\omega(1) & 1 & \cdots & 1 \\ 1 & \chi_\omega(1) & \cdots & 1 \\ \vdots & \vdots & \vdots & \vdots \\ 1 & 1 & \cdots & \chi_\omega(1) \end{pmatrix} \bullet \begin{pmatrix} \theta & 1 & \cdots & 1 \\ 1 & \theta & \cdots & 1 \\ \vdots & \vdots & \vdots & \vdots \\ 1 & 1 & \cdots & \theta \end{pmatrix} = \begin{pmatrix} 1 & 1 & \cdots & 1 \\ 1 & 1 & \cdots & 1 \\ \vdots & \vdots & \vdots & \vdots \\ 1 & 1 & \cdots & 1 \end{pmatrix}. \end{aligned}$$

Then $\mathcal{G} \otimes_M \mathcal{G}^T$ is the all-one matrix which completes the proof. $\square$

THEOREM 5.4. *Let R be a finite Frobenius ring. Let C be an additive code over R with $C \subseteq C^M$ for a given duality M. Let $\mathbf{v}$ be a vector in R^n that is not in C^M with $\langle \mathbf{v}, \mathbf{v} \rangle_M = 1$. Let*

$$C_0 = \{\mathbf{w} | \mathbf{w} \in C, \langle \mathbf{w}, \mathbf{v} \rangle_M = 1\}.$$

Let $D = \langle C_0, \mathbf{v} \rangle$, where this is the additive code generated by taking all possible sums of each elements. Then $D \subseteq D^M$, that is D is a self-orthogonal code with respect to M.

PROOF. If $\mathbf{w}_1, \mathbf{w}_2 \in D$ then $\mathbf{w}_i = \mathbf{u}_i + \alpha_i \mathbf{v}$, for $\mathbf{u}_i \in C_0$ and $\alpha_i \in \{0, 1\}$. Then

$$\begin{aligned}\langle \mathbf{w}_1, \mathbf{w}_2 \rangle_M =& \langle \mathbf{u}_1 + \alpha_1 \mathbf{v}, \mathbf{u}_2 + \alpha_2 \mathbf{v} \rangle_M \\ =& \langle \mathbf{u}_1, \mathbf{u}_2 \rangle_M \langle \alpha_1 \mathbf{v}, \alpha_2 \mathbf{v}, \rangle_M \langle \mathbf{u}_1, \alpha_2 \mathbf{v}, \rangle_M \langle \mathbf{u}_2, \alpha_1 \mathbf{v}, \rangle_M \\ =& 1(1)(1)(1) = 1.\end{aligned}$$

Therefore, $D \subseteq D^M$. □

We call this neighbor construction, i.e, $D = N(C, \mathbf{v})$. If C is an additive code over $\mathbb{F}_{p^2}$, and $C_0 = \{\mathbf{w} | \mathbf{w} \in C, \langle \mathbf{w}, \mathbf{v} \rangle_M = 1\}$, then

$$C_0^M = \langle C^M, \mathbf{v} \rangle.$$

Then $|C_0||C_0^M| = |R|^n$. This gives $|C_0| = \frac{|C|}{p}$ and $|D| = |C|$, where $D = \langle C_0, \mathbf{v} \rangle$. Note that if $\mathbf{v}$ is in the code, then the neighbor construction simply gives back the original code, so we are justified using the notation even when $\mathbf{v} \in C$.

THEOREM 5.5. *Let R be a finite Frobenius ring. Given any two-self-orthogonal codes C and D over R with $|C| = |D|$, then there exists $\mathbf{v}_1, \mathbf{v}_2, \ldots, \mathbf{v}_t$ with*

$$C_i = N(C_{i-1}, \mathbf{v}_i), \ C = C_0 \text{ and } D = C_t.$$

That is any two self-orthogonal codes are reachable by the neighbor construction.

PROOF. It is possible to take $\mathbf{v}_1, \mathbf{v}_2, \ldots, \mathbf{v}_t$ to be the additive generators of D. This is not necessarily the shortest path but shows some path exists. □

Let C be a self-orthogonal code with respect to M of length n over R. Let $\mathbf{v}_1 \in R^n$ and let $C_0 = \{\mathbf{w} \mid \mathbf{w} \in C, \langle \mathbf{w}, \mathbf{v}_1 \rangle_M = 1\}$. Then there exists a vector $\mathbf{v}_2$ with $C = C_0 + \mathbf{v}_2$. Let $\mathcal{G}_0$ generates C_0 and

$$\mathcal{G} = \left(\begin{array}{c|c} \mathbf{v}_1 & \mathbf{w}_1 \\ \mathbf{v}_2 & \mathbf{w}_2 \\ \hline & 0 \cdots 0 \\ \mathcal{G}_0 & \vdots \\ & 0 \cdots 0 \end{array} \right). \tag{5.8}$$

Choose $\mathbf{w}_1, \mathbf{w}_2$ so that

$$\langle \mathbf{w}_1, \mathbf{w}_1 \rangle_M = \frac{1}{\langle \mathbf{v}_1, \mathbf{v}_1 \rangle_M}, \ \langle \mathbf{w}_2, \mathbf{w}_2 \rangle_M = 1 \text{ and } \langle \mathbf{w}_1, \mathbf{w}_2 \rangle_M = \frac{1}{\langle \mathbf{v}_1, \mathbf{v}_2 \rangle_M}.$$

Then $\mathcal{G}$ generates a self-orthogonal code of length $n + length(\mathbf{w}_i)$.

We have proven the following.

THEOREM 5.6. *Let R be a finite Frobenius ring. Let $\mathbf{v}_1 \in R^n$ and let $C_0 = \{\mathbf{w} | \mathbf{w} \in C, \langle \mathbf{w}, \mathbf{v}_1 \rangle_M = 1\}$. Then $\mathcal{G}$ generates a self-orthogonal code. This is known as building-up construction.*

5.1. Dualities over $\mathbb{F}_4$. It is known that different dualities can be studied for additive codes, see for example [**6–9**]. It was shown in [**10**] that there are exactly four possible symmetric dualities over the additive group $\mathbb{F}_4$. They are given in the following tables.

M_1	0	ω	1	ω^2
0	1	1	1	1
ω	1	-1	1	-1
1	1	1	-1	-1
ω^2	1	-1	-1	1

M_2	0	ω	1	ω^2
0	1	1	1	1
ω	1	1	-1	-1
1	1	-1	1	-1
ω^2	1	-1	-1	1

M_3	0	ω	1	ω^2
0	1	1	1	1
ω	1	1	-1	-1
1	1	-1	-1	1
ω^2	1	-1	1	-1

M_4	0	ω	1	ω^2
0	1	1	1	1
ω	1	-1	-1	1
1	1	-1	1	-1
ω^2	1	1	-1	-1

Example 5.7 shows that, different dualities can provide better minimum weight.

EXAMPLE 5.7. Let C be an additive code over $\mathbb{F}_4$ with generator matrix

$$\mathcal{G} = \begin{pmatrix} 1 & 0 & \omega & \omega^2 \\ \omega & 0 & \omega^2 & 1 \\ 0 & 1 & \omega^2 & \omega \\ 0 & \omega & 1 & \omega^2 \end{pmatrix}.$$

Then $\mathcal{G}$ generates an additive self-dual code with respect to duality M_1 with weight enumerator $W_C(x, y) = x^4 + 12xy^3 + 3y^4$. The minimum weight of self-dual code C is 3, but it is known that the minimum weight of linear self-dual additive codes of length 4 over $\mathbb{F}_4$ is upper bounded by 2. Therefore, an additive code that is self-orthogonal for a given size can have a larger minimum weight than all linear self-orthogonal codes with the same size.

References

[1] E. F. Assmus Jr. and J. D. Key, *Affine and projective planes*, Discrete Math. **83** (1990), no. 2-3, 161–187, DOI 10.1016/0012-365X(90)90003-Z. MR1065696

[2] Alexei Ashikhmin and Emanuel Knill, *Nonbinary quantum stabilizer codes*, IEEE Trans. Inform. Theory **47** (2001), no. 7, 3065–3072, DOI 10.1109/18.959288. MR1872869

[3] A. Robert Calderbank, Eric M. Rains, P. W. Shor, and Neil J. A. Sloane, *Quantum error correction via codes over* GF(4), IEEE Trans. Inform. Theory **44** (1998), no. 4, 1369–1387, DOI 10.1109/18.681315. MR1665774

[4] A. R. Calderbank and P. Shor, *Good quantum error-correcting codes exist*, Phys. Rev. A, **54**,(1996) pp. 1098–1105.

[5] Steven T. Dougherty, *Algebraic coding theory over finite commutative rings*, SpringerBriefs in Mathematics, Springer, Cham, 2017, DOI 10.1007/978-3-319-59806-2. MR3675844

[6] Steven T. Dougherty, *Dualities for codes over finite abelian groups*, Adv. Math. Commun. **18** (2024), no. 6, 1827–1841, DOI 10.3934/amc.2023023. MR4795442

[7] Steven T. Dougherty, Jon-Lark Kim, and Nari Lee, *Additive self-dual codes over fields of even order*, Bull. Korean Math. Soc. **55** (2018), no. 2, 341–357, DOI 10.4134/BKMS.b160842. MR3789447

[8] Steven T. Dougherty and Sara Myers, *Orthogonality from group characters*, Involve **14** (2021), no. 4, 555–570, DOI 10.2140/involve.2021.14.555. MR4332568

[9] Steven T. Dougherty and Serap Şahinkaya, *Dualities over the cross product of the cyclic groups of order 2*, Adv. Math. Commun. **18** (2024), no. 5, 1531–1546, DOI 10.3934/amc.2023005. MR4770737

[10] Steven T. Dougherty, Serap Şahinkaya, and Deniz Ustun, *Additive complementary dual codes from group characters*, IEEE Trans. Inform. Theory **68** (2022), no. 7, 4444–4452, DOI 10.1109/tit.2022.3162181. MR4449053

[11] Martianus Frederic Ezerman, Somphong Jitman, San Ling, and Dmitrii V. Pasechnik, *CSS-like constructions of asymmetric quantum codes*, IEEE Trans. Inform. Theory **59** (2013), no. 10, 6732–6754, DOI 10.1109/TIT.2013.2272575. MR3106859

[12] D. Gottesman, *Stabilizer codes and quantum error correction*, Ph.D. dissertation, CalTech, 1997.

[13] Heide Gluesing-Luerssen and Tefjol Pllaha, *On quantum stabilizer codes derived from local Frobenius rings*, Finite Fields Appl. **58** (2019), 145–173, DOI 10.1016/j.ffa.2019.04.004. MR3946987

[14] Avanti Ketkar, Andreas Klappenecker, Santosh Kumar, and Pradeep Kiran Sarvepalli, *Nonbinary stabilizer codes over finite fields*, IEEE Trans. Inform. Theory **52** (2006), no. 11, 4892–4914, DOI 10.1109/TIT.2006.883612. MR2300363

[15] S. Nadella and A. Klappenecker, *Stabilizer codes over Frobenius rings*, in Proc. 2012 IEEE International Symposium on Information Theory (ISIT 2012), pp. 165–169.

[16] T. Sidana and N.Kashyap, *Entanglement-assisted quantum error-correcting codes over local Frobenius rings*, `arXiv:2202.00248`, 2023.

[17] P. W. Shor, *Scheme for reducing decoherence in quantum computer memory*, Phys. Rev. A, **52**, (1995),2493–2496.

[18] A. M. Steane, *Error correcting codes in quantum theory*, Phys. Rev. Lett. **77** (1996), no. 5, 793–797, DOI 10.1103/PhysRevLett.77.793. MR1398854

[19] Zhe-Xian Wan, *Lectures on finite fields and Galois rings*, World Scientific Publishing Co., Inc., River Edge, NJ, 2003, DOI 10.1142/5350. MR2008834

Department of Mathematics, University of Scranton,Scranton, Pennsylvania 18518
Email address: prof.steven.dougherty@gmail.com

Department of Electrical Communication Engineering, Indian Institute of Science, Bangalore 560012, India
Email address: nkashyap@iisc.ac.in

Department of Electrical Communication Engineering, Indian Institute of Science, Bangalore 560012, India
Email address: taniasidana@iisc.ac.in

Faculty of Engineering, Department of Natural and Mathematical Sciences, Tarsus University, Mersin, Turkey
Email address: serap@tarsus.edu.tr

Faculty of Engineering, Department of Computer Engineering, Tarsus University, Mersin, Turkey
Email address: denizustun@tarsus.edu.tr

Contemporary Mathematics
Volume **826**, 2025
https://doi.org/10.1090/conm/826/16575

Semidirect products in universal algebra

Alberto Facchini and David Stanovský

ABSTRACT. First of all, we recall the well known notion of semidirect product both for classical algebraic structures (like groups and rings) and for more recent ones (digroups, left skew braces, heaps, trusses). Then we analyse the concept of semidirect product for an arbitrary algebra A in a variety of given type. An inner semidirect decomposition $A = B \ltimes \omega$ of A consists of a subalgebra B of A and a congruence ω on A such that B is a set of representatives of the congruence classes of A modulo ω. An outer semidirect product is the restriction to B of a functor from a suitable category $\mathcal{C}_B$ containing B, called the enveloping category of B, to the category Set_* of pointed sets.

1. Introduction

The standard construction of semidirect product $K \rtimes Y$ of two groups K and Y has been extended in a number of directions in Category Theory. One possibility is that it can be viewed as the lax 2-colimit of the diagram $K\colon Y \to \mathsf{Cat}$ (see [**11,16**]). As a second point of view, it can be seen as a Grothendick construction [**15,17**]. In this paper we present a different point of view, a new one, from the point of view of Universal Algebra. We begin (Section 2) by presenting several examples, both the classical ones (groups, Lie algebras, etc.) and some more recent ones (digroups, skew braces, heaps, trusses). This is in order to allow the reader to understand our subsequent definitions. Then we pass to consider *inner* semidirect products of algebras. Let A be an algebra in a variety $\mathcal{V}$ of type $\mathcal{F}$. Inner semidirect-product decompositions of A consist of a subalgebra B of A and a congruence ω on A. One has a semidirect-product decomposition $A = B \ltimes \omega$ when the subalgebra B is a set of representatives of the congruence classes of A modulo ω. There is a one-to-one correspondence between semidirect-product decompositions of A and idempotent endomorphisms of A (Corollary 3.2). In the study of semidirect-product decompositions of A, one immediately sees that an important role is played by the category Set_* of pointed sets (Remark 3.3). Another important role is played by totally idempotent elements of B, that is, the elements b of B for which the singletons $\{b\}$ are subalgebras of B. For instance, in the variety of lattices all elements are totally idempotent, in the variety of semigroups the totally idempotent elements

2020 *Mathematics Subject Classification.* Primary 08B25, 08A05.

The first author was partially supported by Ministero dell'Università e della Ricerca (Progetto di ricerca di rilevante interesse nazionale "Categories, Algebras: Ring-Theoretical and Homological Approaches (CARTHA)"), and the Department of Mathematics "Tullio Levi-Civita" of the University of Padua (Research programme DOR1828909 "Anelli e categorie di moduli").

are the idempotent ones, in the varieties of monoids and groups the unique totally idempotent element is the identity. We then pass to the study of *outer* semidirect products. Given an algebra B in a variety $\mathcal{V}$ of type $\mathcal{F}$, an outer semidirect product of B assigns a pointed set (A_b, a_b) to each element $b \in B$, and assigns to each n-ary function symbol $f \in \mathcal{F}$ and each n-tuple $(b_1, \dots, b_n) \in B^n$ a morphism $f_{(b_1,\dots,b_n)}\colon (A_{b_1}, a_{b_1}) \times \cdots \times (A_{b_n}, a_{b_n}) \to (A_{f(b_1,\dots,b_n)}, a_{f(b_1,\dots,b_n)})$ in Set_*. On the disjoint union $A = \bigcup_{b\in B}(A_b \times \{b\})$, there are natural n-ary operations defined by

$$f((a_{b_1}, b_1), \dots, (a_{b_n}, b_n)) = (f_{(b_1,\dots,b_n)}(a_{b_1}, \dots, a_{b_n}), f(b_1, \dots, b_n))$$

for each n-ary function symbol $f \in \mathcal{F}$, so that A becomes an algebra of type $\mathcal{F}$, and we require this algebra A to be in the variety $\mathcal{V}$. Every such semidirect product $B \to \mathsf{Set}_*$ can be extended to a functor $\mathcal{C}_B \to \mathsf{Set}_*$ from a suitable *enveloping category* $\mathcal{C}_B$ of B to the category Set_* of pointed sets (see 4.1).

2. Examples and basic notions

2.1. Groups. The following is well known to all of us, but we repeat it here to fix the basic notions and notations. For a group G, a normal subgroup K of G and a subgroup $Y \leq G$, the fact that G is the *(inner) semidirect product* of K and Y can be equivalently stated saying that:

(a) $G = KY$ and $K \cap Y = \{1_G\}$.
(b) For every $g \in G$, there is a unique pair $(k, y) \in K \times Y$ such that $g = ky$.
(c) For every $g \in G$, there is a unique pair $(k, y) \in K \times Y$ such that $g = yk$.
(d) There is an idempotent endomorphism of G with kernel K and image Y.
(e) There is a homomorphism of G onto Y that is the identity on Y and has kernel K.
(f) The mapping $Y \to G/K$, $y \mapsto yK$, is an isomorphism.

Given any such semidirect-product decomposition $G = KY$ with $K \cap Y = \{1_G\}$, there is an *action* of Y on K, that is, a group homomorphism $\phi\colon Y \to \mathrm{Aut}(K)$, defined by $\phi(y)(k) = yky^{-1}$ for every $y \in Y$, $k \in K$.

Conversely, given any two groups K and Y and any group homomorphism $\phi\colon Y \to \mathrm{Aut}(K)$, it is possible to define a group structure on the cartesian product $K \times Y := \{(k, y) \mid k \in K,\ y \in Y\}$ defining the operation by

$$(k, y)(k', y') = (k\phi_y(k'), yy').$$

This group is the *outer semidirect product* of K and Y, denoted by $K \rtimes_\phi Y$.

Any inner semidirect product $G = KY$ is isomorphic to the outer semidirect product $K \rtimes_\phi Y$ with ϕ the action of Y on K defined by conjugation:

$$\phi(y)(k) = yky^{-1}$$

for every $y \in Y$, $k \in K$.

For a fixed group G, there is a bijection between the set E of all idempotent endomorphisms of G

$$E := \{ e \mid e\colon G \to G \text{ an endomorphism of } G, \text{ and } e^2 = e \}$$

and the set A of all pairs (K, Y), where K is a normal subgroup of G, Y is a subgroup of G, and G is the inner semidirect product $K \rtimes Y$:

$$A := \{ (K,Y) \mid K \trianglelefteq G,\ Y \leq G,\ G = KY,\ K \cap Y = \{1_G\} \}.$$

In this bijection, an idempotent endomorphism e of G corresponds to the pair $(\ker(e), \mathrm{im}(e))$.

For an additive *abelian* group G, idempotent endomorphisms $e\colon G \to G$ of G correspond bijectively to the pairs (K, Y), where $K = \ker(e)$, $Y = \mathrm{im}(e)$ and $G = K \oplus Y$ (direct-sum decomposition). This also holds for vector spaces and right modules G over a ring: there is a one-to-one correspondence between the set of all idempotent endomorphisms of G and the set of all pairs (K, Y) of substructures of G for which $G = K \oplus Y$.

2.2. Lie algebras. For Lie algebras, semidirect products are sometimes called "inessential extensions" [**5**, Chapitre Premier, §1, N. 7, Définition 6 and N. 8]. Inner semidirect-product decompositions of a Lie algebra L over a commutative ring k with identity are exactly the direct-sum decomposition $L = K \oplus Y$ of L as a k-module, where K is an ideal of L and Y is a Lie subalgebra of L. Such decompositions of L bijectively correspond to idempotent endomorphisms of L.

Given any such semidirect-product decomposition $L = K \oplus Y$ of L, the adjunction $\mathrm{ad}_L\colon L \to \mathrm{Der}(L)$, defined by $\mathrm{ad}_L(x)(y) = [x, y]$, induces by restriction an *action* of Y on K, that is, a Lie algebra homomorphism $\phi\colon Y \to \mathrm{Der}(K)$, defined by $\phi(y)(k) = [y, k]$ for every $y \in Y$, $k \in K$, into the Lie algebra $\mathrm{Der}(K)$ of derivations of K.

Conversely, given any two Lie k-algebras K and Y and any Lie algebra morphism $\phi\colon Y \to \mathrm{Der}(K)$, it is possible to define a Lie algebra structure on the k-module $K \oplus Y$ (direct sum of the two k-modules K and L), defining the Lie bracket as $[(k, y), (k', y')] = ([k, k'] + \phi_y(k') - \phi_{y'}(k), [y, y'])$ [**5**, Chapitre premier, §1, N. 8]. This Lie algebra is the outer semidirect product of K and Y.

2.3. Rings. For an associative ring R (not-necessarily with an identity), idempotent endomorphisms of R are in a one-to-one correspondence with the pairs (K, S), where K is an ideal of R, S is a subring of R, and $R = K \oplus S$ as an abelian additive group. Now the natural R-R-bimodule structure ${}_RR_R$ on the ring R induces an S-K-bimodule structure ${}_SK_K$ and a K-S-bimodule structure ${}_KK_S$ on K. Correspondingly, we have a ring homomorphism $\lambda\colon S \to \mathrm{End}_{\mathsf{Mod}\text{-}K}(K_K)$ and a ring antihomomorphism $\rho\colon S \to \mathrm{End}_{K-\mathsf{Mod}}({}_KK)$ such that $\lambda(s) \circ \rho(t) = \rho(t) \circ \lambda(s)$ and $\rho(s)(x)y = x\lambda(s)(y)$ for all $s, t \in S$ and $x, y \in K$ (these conditions immediately follow from the associativity of multiplication in R).

Conversely, given two associative rings K and S, a ring homomorphism

$$\lambda\colon S \to \mathrm{End}_{\mathsf{Mod}\text{-}K}(K_K)$$

and a ring antihomomorphism

$$\rho\colon S \to \mathrm{End}_{K-\mathsf{Mod}}({}_KK)$$

such that $\lambda(s) \circ \rho(t) = \rho(t) \circ \lambda(s)$ and $\rho(s)(x)y = x\lambda(s)(y)$ for every $s, t \in S$ and every $x, y \in K$, it is possible to define a multiplication on the abelian group $K \oplus S$ setting $(k, s)(k', s') = (kk' + \lambda(s)(k') + \rho(s')(k), ss')$.

Let us check associativity of multiplication in $K \oplus S$. We have

$$\begin{aligned}((k, s)(k', s'))(k'', s'') &= (kk' + \lambda(s)(k') + \rho(s')(k), ss')(k'', s'') \\ &= ((kk')k'' + \lambda(s)(k')k'' + \rho(s')(k)k'' + \lambda(ss')(k'') \\ &\quad +\rho(s'')(kk' + \lambda(s)(k') + \rho(s')(k)), (ss')s'')\end{aligned}$$

and

$$\begin{aligned}(k,s)((k',s')(k'',s'')) &= (k,s)(k'k''+\lambda(s')(k'')+\rho(s'')(k'),s's'')\\ &= (k(k'k''+\lambda(s')(k'')+\rho(s'')(k'))+\lambda(s)(k'k''+\lambda(s')(k'')\\ &\quad +\rho(s'')(k'))+\rho(s's'')(k),s(s's'').\end{aligned}$$

Now

$$(kk')k'' = k(k'k'')$$

because K is an associative ring,

$$\lambda(s)(k')k'' = \lambda(s)(k'k'')$$

because $\lambda\colon S\to \mathrm{End}_{\mathrm{Mod}\text{-}K}(K_K)$,

$$\rho(s')(k)k'' = k\lambda(s')(k''),$$

$$\lambda(ss')(k'') = \lambda(s)\lambda(s')(k'')$$

because λ is a ring morphism,

$$\rho(s'')(kk') = k\rho(s'')(k')$$

because $\rho\colon S\to \mathrm{End}_{K-\mathsf{Mod}}({}_KK)$,

$$\rho(s'')\lambda(s)(k') = \lambda(s)\rho(s'')(k')$$

because $\lambda(s)\circ\rho(s'') = \rho(s'')\circ\lambda(s)$,

$$\rho(s'')\rho(s')(k) = \rho(s's'')(k)$$

because ρ is a ring antihomomorphism, and

$$(ss')s'' = s(s's'')$$

because S is an associative ring.

Notice that now we need *two* "actions" λ and ρ, a left one and a right one, while for groups and Lie algebras one action was sufficient. This is due to the fact that groups and Lie algebras are isomorphic to their opposite (the antiisomorphisms are $g\mapsto g^{-1}$ for groups and $x\mapsto -x$ for Lie algebras). This does not occur for associative algebras.

What we have seen in this subsection about associative rings can be easily extended to the case of associative C-algebras, where C is any commutative ring with identity.

2.4. Not-necessarily associative algebras. Let C be a commutative ring with identity 1. With the term *C-algebra* we now mean a C-module A with a further C-bilinear operation

$$\cdot\colon A\times A\to A,\qquad (r,r')\mapsto r\cdot r' = rr'$$

(see [**5**] and [**1**, Section 2]). Thus C-algebras generalize at the same time the Lie algebras we have considered in 2.2 and the associative rings considered in 2.3. For a C-algebra A, idempotent endomorphisms of A are in a one-to-one correspondence with the pairs (K,B), where K is an ideal of A, B is a sub-C-algebra of A, and $A=K\oplus B$ as a C-module. Then we have two C-module morphisms $\lambda,\rho\colon B\to \mathrm{End}_{\mathrm{Mod}\text{-}C}(K)$.

Conversely, given two C-algebras K and S, and two C-module morphisms

$$\lambda,\rho\colon B\to \mathrm{End}_{\mathrm{Mod}\text{-}C}(K),$$

it is possible to define a multiplication on the C-module $K \oplus B$ setting $(k,b)(k',b') = (kk' + \lambda(b)(k') + \rho(b')(k), bb')$ for every $k, k' \in K$ and every $b, b' \in B$.

For applications of idempotent endomorphisms in the study of algebras with a bilinear form, see [**9**].

2.5. Digroups. The content of this subsection is taken from [**10**]. A *digroup* is a triple $(A, *, \circ)$, where $(A, *)$ and $(A, \circ)$ are groups (not-necessarily abelian) for which $1_{(A,*)} = 1_{(A,\circ)}$. An *ideal* in a digroup $(A, *, \circ)$ is a subset I of A such that I is a normal subgroup of $(A, *)$, I is a normal subgroup of $(A, \circ)$, and $a * I = a \circ I$ for every $a \in A$ (the cosets modulo I with respect to the two operations coincide). For a digroup $(A, *, \circ)$, there is a one-to-one correspondence between the set of all ideals of A and the set of all the equivalence relations on A compatible with both the operations $*$ and $\circ$. It associates with any equivalence relation $\sim$ on the set A compatible with both the operations $*$ and $\circ$ of A the equivalence class $[1_A]_\sim$ of the identity of A, which is an ideal of the digroup A. Conversely, it associates with any ideal I of A, the relation $\sim_I$ on A defined, for every $a, b \in A$, by $a \sim_I b$ if $a * b^{-*} \in I$. Here a^{-*} denotes the inverse of a in the group $(D, *)$, and $a^{-\circ}$ will denote the inverse of a in $(D, \circ)$.

Since $1_{(A,*)} = 1_{(A,\circ)}$, the kernel of a morphism is defined as the inverse image of $1_{(A,*)} = 1_{(A,\circ)}$. The kernel of any morphism is an ideal, as is easily seen.

There is a forgetful functor U of the category DiGp of digroups into the category Set_* of pointed sets, which assigns to each digroup D the pointed set $(D, 1)$. Notice that if we have any two group structures $(D, *, 1_*)$ and $(D, \circ, 1_\circ)$ on the same set D, and we take any isomorphism $\varphi\colon (D, 1_*) \to (D, 1_\circ)$ in Set_*, we can lift the group structure of $(D, \circ, 1_\circ)$ to the pointed set $(D, 1_*)$, obtaining a digroup structure $(D, *, \circ, 1_*)$ on the set D.

Given a digroup $(D, *, \circ)$ and an element a of D, let $\lambda_a^D\colon D \to D$ denote the mapping defined by

$$\lambda_a^D(b) = a^{-*} * (a \circ b)$$

for every $b \in D$. Then every λ_a^D is an automorphism of $U(D)$ in the category Set_*. Its inverse is the isomorphism $(D, 1) \to (D, 1)$, $c \mapsto a^{-\circ} \circ (a * c)$. (This is the mapping λ_a^D relative to the digroup $(D, \circ, *)$ with the operation $*$ and $\circ$ swapped. Notice that the digroups $(D, *, \circ)$ and $(D, \circ, *)$ are not isomorphic in the category DiGp.)

The mapping $\lambda^D\colon (D, 1) \to (\mathrm{Aut}_{\mathsf{Set}_*}(D, 1), \mathrm{id}_D)$, $\lambda^D\colon a \in D \mapsto \lambda_a^D$, is a morphism in the category Set_* of the pointed set $(D, 1)$ into the automorphism group $\mathrm{Aut}_{\mathsf{Set}_*}(D, 1)$ of all automorphisms of the pointed set $(D, 1)$. The property "$a * I = a \circ I$ for every $a \in D$" in the definition of ideal I of a digroup D is equivalent to "$\lambda_a^D(I) \subseteq I$ for every $a \in D$".

Any intersection of ideals is an ideal, so that we get a complete lattice $\mathcal{I}(A)$ of ideals for any digroup A. In particular, every subset X of A generates an ideal, the intersection of the ideals that contain X.

Fix a digroup $(D, *, \circ)$ and an idempotent endomorphism e of D. Let B be the image of e, so that B is a subdigroup of D, and let I be the kernel of e, which is an ideal of D. Since e is an idempotent group endomorphism with respect to both group structures on D, we get that D is a semidirect product of B and I with respect to both groups structures on D, so $D = I * B = \{ i * b \mid i \in I, b \in B \}$, $D = I \circ B = \{ i \circ b \mid i \in I, b \in B \}$ and $I \cap B = \{1_D\}$.

PROPOSITION 2.1 ([**10**, Proposition 2.2]). *Let $(D, *, \circ)$ be a digroup, B a subdigroup of D and I an ideal of D. The following conditions are equivalent:*

(1) *$D = B \circ I$ and $B \cap I = \{1_D\}$.*
(2) *For every $a \in D$, there is a unique pair $(b, i_1) \in B \times I$ such that $a = b \circ i_1$.*
(3) *For every $a \in D$, there is a unique pair $(b, i_2) \in B \times I$ such that $a = i_2 \circ b$.*
(4) *$D = B * I$ and $B \cap I = \{1_D\}$.*
(5) *For every $a \in D$, there is a unique pair $(b, i_3) \in B \times I$ such that $a = b * i_3$.*
(6) *For every $a \in D$, there is a unique pair $(b, i_4) \in B \times I$ such that $a = i_4 * b$.*
(7) *There is an idempotent digroup endomorphism e of D whose image is B and whose kernel is I.*

The digroup D is the *inner semidirect product* of its subdigroup B and its ideal I if any of the equivalent conditions of the previous proposition holds.

By Proposition 2.1, every element $a \in D$ can be written in a unique way in the four forms $a = b \circ i_1 = i_2 \circ b = b * i_3 = i_4 * b$, where $b = e(a)$ and $i_1, i_2, i_3, i_4 \in I$. Now it is easy to prove [**10**] that the elements i_2, i_3, i_4 depend only on i_1, via the formulas

$$i_2 = (\phi_{\circ b})^{-1}(i_1), \qquad i_3 = \lambda_b(i_1), \qquad \text{and} \qquad i_4 = (\phi_{*b})^{-1}(\lambda_b(i_1)). \tag{2.1}$$

Here and in the rest of this subsection we will write conjugation on the right, as one often does in Group Theory, so that, for any two groups $(Y, \otimes)$ and $(K, \otimes)$ and any group antihomomorphism $\phi_\otimes \colon (Y, \otimes) \to \mathrm{Aut}_{\mathsf{Gp}}(K, \otimes)$, we denote by $(Y \ltimes_{\phi_\otimes} K, \otimes)$ the semidirect product of Y and K via $\phi_\otimes$, where $\otimes$ is defined by

$$(y, k) \otimes (y', k') := (y \otimes y', \phi_{\otimes y'}(k) \otimes k').$$

Suppose we have a digroup $(D, *, \circ)$. We want to describe the semidirect-product decompositions of D. Equivalently, we want to study the idempotent digroup endomorphisms e of D. Such idempotent endomorphisms e have a kernel K and an image Y, and D turns out to be a semidirect product $D = Y \ltimes K$. By the formulas (2.1), three mappings immediately appear in a natural way:

(1) The group antihomomorphism $\phi_* \colon (Y, *) \to \mathrm{Aut}_{\mathsf{Gp}}(K, *)$ related to the semidirect-product decomposition $D = Y \ltimes K$ of the group $(D, *)$. It is defined by $\phi_* \colon y \mapsto \phi_{*y}$, where $\phi_{*y}(k) = y^{-*} * k * y$ for every $y \in Y$ and every $k \in K$, and is induced by the conjugation in the group $(D, *)$.

(2) The group antihomomorphism $\phi_\circ \colon (Y, \circ) \to \mathrm{Aut}_{\mathsf{Gp}}(K, \circ)$ related to the semidirect-product decomposition $D = Y \ltimes K$ of the group $(D, \circ)$. It is defined by $\phi_\circ \colon y \mapsto \phi_{\circ y}$, where $\phi_{\circ y}(k) = y^{-\circ} \circ k \circ y$ for every $y \in Y$ and $k \in K$, and is induced by the conjugation in the group $(D, \circ)$.

(3) The morphism $\Lambda \colon (Y, 1) \to (\mathrm{Aut}_{\mathsf{Set}_*}(K, 1), \mathrm{id}_K)$ in the category Set_*, defined by $\Lambda \colon y \mapsto \Lambda_y$, where $\Lambda_y(k) = y^{-*} * (y \circ k)$ for every $y \in Y$, $k \in K$. It is induced by the morphism $\lambda \colon (D, 1) \to (\mathrm{Aut}_{\mathsf{Set}_*}(D, 1), \mathrm{id}_D)$ (λ is a morphism in Set_*), relative to the digroup $(D, *, \circ)$. Here $\mathrm{Aut}_{\mathsf{Set}_*}(X, x_0)$ denotes the group of all automorphisms of the pointed set (X, x_0). Each Λ_y is a permutation of K because its two-sided inverse is the mapping $K \to K$, $k \mapsto y^{-\circ} \circ (y * k)$. Notice that $\Lambda_1 \colon K \to K$ is the identity mapping of K and $\Lambda_y(1) = 1$ for every $y \in Y$.

THEOREM 2.2 ([**10**, Theorem 3.1]). *Let $(D, *, \circ)$ be any digroup, and suppose that D is the semidirect product of its subdigroup Y and its ideal K (i.e., that the equivalent conditions of Proposition 2.1 are satisfied). Let ϕ_*, $\phi_\circ$ and Λ be the three mappings defined above. Then there is a digroup isomorphism*

$$\alpha\colon (Y \times K, +, \circ) \to (D, *, \circ)$$

defined by $\alpha(y, k) = y \circ k$ for every $y \in Y$ and every $k \in K$, where the operations on the digroup $(Y \times K, +, \circ)$ are defined by

$$(y, k) + (y', k') = (y * y', (\Lambda_{y*y'})^{-1}(\phi_{*y'}(\Lambda_y(k)) * \Lambda_{y'}(k'))) \tag{2.2}$$

and

$$(y, k) \circ (y', k') = (y \circ y', \phi_{\circ y'}(k) \circ k') \tag{2.3}$$

for every $(y, k), (y', k') \in Y \times K$.

The operations $+$ and $\circ$ in the statement of Theorem 2.2 may seem quite complicate, but their naturality appears as soon as we look at how they define the products between an element of the form $(y, 1)$ and an element of the form $(k, 1)$. In fact, it is very easy to see, from the two formulas in Theorem 2.2, that for every $y \in Y$ and every $k \in K$:

$$\begin{array}{l} (y, 1) \circ (1, k) = (y, k) \\ (1, k) \circ (y, 1) = (y, \phi_{\circ y}(k)) \\ (y, 1) + (1, k) = (y, (\Lambda_y)^{-1}(k)) \\ (1, k) + (y, 1) = (y, (\Lambda_y)^{-1}\phi_{*y}(k)). \end{array} \tag{2.4}$$

Of these four identities, the first is a trivial identity, the second allows to define the antihomomorphism $\phi_\circ$, the third allows to define the morphism Λ, and the fourth, in view of the third, allows to define the antihomomorphism ϕ_*. Clearly, the last three identities in (2.4) correspond to the identities in (2.1).

Finally, here is the outer semidirect product of two digroups:

THEOREM 2.3 ([**10**, Theorem 4.1]). *Let $(Y, *, \circ)$ and $(K, *, \circ)$ be digroups, $\phi_*\colon (Y, *) \to \mathrm{Aut}_{\mathsf{Gp}}(K, *)$ and $\phi_\circ\colon (Y, \circ) \to \mathrm{Aut}_{\mathsf{Gp}}(K, \circ)$ be two group antihomomorphisms, and $\Lambda\colon (Y, 1) \to (S_K, \mathrm{id})$ be a morphism in Set_*. On the cartesian product $Y \times K$ define two operations via*

$$(y, k) + (y', k') = (y * y', (\Lambda_{y*y'})^{-1}(\phi_{*y'}(\Lambda_y(k)) * \Lambda_{y'}(k'))) \tag{2.5}$$

and

$$(y, k) \circ (y', k') = (y \circ y', \phi_{\circ y'}(k) \circ k') \tag{2.6}$$

for every $y, y' \in Y$ and $k, k' \in K$. Then $(Y \times K, +, \circ)$ is a digroup.

COROLLARY 2.4. *Let $(Y, *, \circ)$ and $(K, *, \circ)$ be digroups, $\phi_*\colon (Y, *) \to \mathrm{Aut}_{\mathsf{Gp}}(K, *)$ and $\phi_\circ\colon (Y, \circ) \to \mathrm{Aut}_{\mathsf{Gp}}(K, \circ)$ be two group antihomomorphisms, $\Lambda\colon (Y, 1) \to (S_K, \mathrm{id})$ be a morphism in Set_*, and construct the correponding semidirect product $(Y \times K, +, \circ)$. Then $(Y \times K, +, \circ)$ is canonically the direct product of the digroups Y and K if and only if the three mappings ϕ_*, $\phi_\circ$ and Λ are all the constant mapping that sends every element of Y to the identity id_K of K.*

PROOF. If $(Y \times K, +, \circ)$ is the direct product, then

$$k \circ k' = \phi_{\circ y'}(k) \circ k' \quad \text{and} \quad k * k' = (\Lambda_{y*y'})^{-1}(\phi_{*y'}(\Lambda_y(k)) * \Lambda_{y'}(k')) \tag{2.7}$$

for every $y, y' \in Y$ and $k, k' \in K$. Now the first equation in (2.7) is clearly equivalent to $k = \phi_{\circ y'}(k)$ for every k, i.e. that $\phi_\circ$ is the constant mapping that sends every element of Y to id_K. The second equation in (2.7) can be rewritten as

$$\phi_{*y'}(\Lambda_y(k)) * \Lambda_{y'}(k') = \Lambda_{y*y'}(k * k'),$$

and from this equation we get the equalities

$$(2.8) \quad \phi_{*y'}(\Lambda_y(k)) = \Lambda_{y*y'}(k) \text{ (for } k' = 1) \quad \text{and} \quad \Lambda_{y'}(k') = \Lambda_{y*y'}(k') \text{ (for } k = 1).$$

From the last equation in (2.8), we get for $y = (y')^{-*}$ that

$$\Lambda_{y'}(k') = \Lambda_{(y')^{-*}*y'}(k') = k',$$

i.e., that Λ is the constant mapping that sends every element of Y to id_K. From this and the first equation in (2.8) we get that $\phi_{*y'}$ is also the identity automorphism of K.

The inverse implication is trivial. □

2.6. Left skew braces. A *(left) skew brace* is a triple $(A, *, \circ)$, where $(A, *)$ and $(A, \circ)$ are groups (not necessarily abelian) and

$$(2.9) \qquad a \circ (b * c) = (a \circ b) * a^{-*} * (a \circ c)$$

for every $a, b, c \in A$. It is not difficult to show that left skew braces are digroups, so that all what we have seen for digroups applies to left skew braces.

Clearly, left skew braces form an algebraic variety, whose homomorphisms $A \to A'$ are the mappings $f \colon A \to A'$ such that $f(a * b) = f(a) * f(b)$ and $f(a \circ b) = f(a) \circ f(b)$ for every $a, b \in A$. In particular, we have the category SKB of all left skew braces.

LEMMA 2.5 ([**3**]). *A digroup $(A, *, \circ)$ is a left skew brace if and only if the mapping $\lambda \colon (A, \circ) \to \mathrm{Aut}(A, *)$, given by $\lambda \colon a \mapsto \lambda_a$, where $\lambda_a(b) = a^{-*} * (a \circ b)$, is a group morphism.*

We say that the left skew brace D is the *inner semidirect product* of its subbrace B and its ideal I if any of the equivalent conditions of Proposition 2.1 holds.

There is a bijection between the set of all idempotent left skew brace endomorphisms of a left skew brace D and the set of all pairs (B, I), where B is a subleft skew brace of D, I is an ideal of D and D is the semidirect product of B and I. This bijection associates with each idempotent endomorphism e the pair $(e(D), \ker(e))$.

If $(D, *, \circ)$ is a left skew brace, e is an idempotent group endomorphism of either $(D, *)$ or $(D, \circ)$, the kernel of e is an ideal of the left skew brace $(D, *, \circ)$ and the image of e is a subleft skew brace, then e is an idempotent left skew brace endomorphism of $(D, *, \circ)$.

PROPOSITION 2.6 ([**10**, Proposition 4.2]). *Let $(Y, *, \circ)$ and $(K, *, \circ)$ be left skew braces and suppose that there are two group antihomomorphisms $\phi_* \colon (Y, *) \to \mathrm{Aut}_{\mathsf{Gp}}(K, *)$ and $\phi_\circ \colon (Y, \circ) \to \mathrm{Aut}_{\mathsf{Gp}}(K, \circ)$, and a group homomorphism $\Lambda \colon (Y, \circ) \to \mathrm{Aut}_{\mathsf{Gp}}(K, *)$. Then their semidirect product (as digroups) is a left skew brace if and only if*

$$\begin{aligned}\phi_{\circ(y'*y'')}(k) \circ (\Lambda_{y'*y''})^{-1}(\phi_{*y''}(\Lambda_{y'}(k')) * \Lambda_{y''}(k''))\\ = (\Lambda_{y\circ(y'*y'')})^{-1}(\phi_{*\lambda^Y_y(y'')}(\Lambda_{y\circ y'}(\phi_{\circ y'}(k) \circ k')\\ *\Lambda_y(k)^{-*})) * \Lambda_{y\circ y''}(\phi_{\circ y''}(k) \circ k'').\end{aligned}$$

Consider the inclusion functor $S\colon \mathsf{SKB} \to \mathsf{DiGp}$ from the category of left skew braces to the category of digroups. This functor S has a left adjoint, which assigns with every digroup D the quotient digroup D/I, where I is the ideal of the digroup D generated by the subset $\{\,(a \circ b) * a^{-1} * (a \circ c) * (a \circ (b * c))^{-*} \mid a, b, c \in D\,\}$. Clearly, D/I is a left skew brace, and the category SKB of left skew braces is a reflective subcategory of the category DiGp of digroups.

The commutator of two ideals I and J of a left skew brace A is the ideal of A generated by the commutator $[I, J]_{(A,*)}$ in the group $(A, *)$, the commutator $[I, J]_{(A,\circ)}$ in the group $(A, \circ)$, and all the elements $(i \circ j)^{-*} * i * j$ with $i \in I$ and $j \in J$ [**6**]. Notice the clear relation between these three sets whose union generates the commutator and the three conditions necessary and sufficient for a semidirect product of digroups (left skew braces) to be a direct product (Corollary 2.4).

LEMMA 2.7. *For every pair of ideals I and J of a left skew brace A, one has $[I, J] = [J, I]$.*

PROOF. For groups, one has that $[I, J]_{(A,*)} = [J, I]_{(A,*)}$ and $[I, J]_{(A,\circ)} = [J, I]_{(A,\circ)}$. Hence it remains to show that if $(A, *, \circ)$ is a left skew brace with the groups $(A, *)$ and $(A, \circ)$ abelian and $(i \circ j)^{-*} * i * j = 1$ for every $i \in I$ and $j \in J$, then $(j \circ i)^{-*} * j * i = 1$ for every $i \in I$ and $j \in J$. Now $(i \circ j)^{-*} * i * j = 1$ implies that $i \circ j = i * j$, so that $j \circ i = i \circ j = i * j = j * i$. Hence $(j \circ i)^{-*} * j * i = 1$. □

Recall that in the lattice of all the ideals of a left skew brace A, one has that $I \vee J = I * J = I \circ J$ and $I \wedge J = I \cap J$. This is a multiplicative lattice [**8**] in which multiplication is commutative in view of Lemma 2.7.

LEMMA 2.8. *If I, J, K are ideals of a left skew brace A, then $[I, J * K] = [I, J] * [I, K]$.*

PROOF. From $J * K \supseteq J, K$, it follows that $[I, J * K] \supseteq [I, J], [I, K]$. Therefore $[I, J * K] \supseteq [I, J] * [I, K] = [I, J] \circ [I, K]$. Now it is known that the equality $[I, JK] = [I, J][I, K]$ holds for normal subgroups I, J, K of a group G. Hence, to conclude the proof, it suffices to prove that if $(A, *, \circ)$ is a brace in which both the groups $(A, *)$ and $(A, \circ)$ are abelian, and $i * j = i \circ j$, $i * k = i \circ k$ for every $i \in I$, $j \in J$ and $k \in K$, then $i * (j \circ k) = i \circ (j \circ k)$ for every i, j, k. Now $i * j = i \circ j$ for every i and j can be restated saying that J is contained in the kernel of the group morphism $\lambda|^I\colon (A, \circ) \to \mathrm{Aut}(I, *)$. Similarly, K is contained in the kernel of the group morphism $\lambda|^I$. Therefore $J \circ K$ is contained in that kernel, that is $i * x = i \circ x$ for every $x \in J \circ K = J * K$, as desired. □

The center of a left skew brace A is the greatest ideal Z of A such that $[Z, A] = 1$. Hence $Z = \{\, z \in A \mid a * z = z * a,\ a \circ z = z \circ a \text{ and } a * z = a \circ z \text{ for every } a \in A\,\}$. Again, notice the relation between these three conditions for elements in the center of a left skew brace and the three conditions necessary and sufficient for a semidirect product of digroups (in particular, of left skew braces) to be a direct product (Corollary 2.4).

2.7. Heaps. The content of this subsection is taken from [**2**], to which we refer for further details. A *heap* is a set X endowed with a ternary operation $[-, -, -]\colon X \times X \times X \to X$ that is a *Mal'tsev operation*, that is, $[x, x, y] = y$ and $[x, y, y] = x$ for every $x, y \in X$, and associative (i.e., $[[x, y, z], w, u] = [x, y, [z, w, u]]$ for every $x, y, z, w, u \in X$).

As usual in Universal Algebra, a mapping $f\colon (X,[-,-,-]) \to (X',[-,-,-])$ between two heaps is a *heap homomorphism* if

$$f([x,x',x'']) = [f(x), f(x'), f(x'')]$$

for every $x, x', x'' \in X$, and a subset S of a heap is a *subheap* if $[x,y,z] \in S$ for every $x, y, z \in S$.

For example, fix any group G and define a ternary operation $[-,-,-]$ on G setting $[x,y,z] = xy^{-1}z$ for every $x,y,z \in G$. Then $(G,[-,-,-])$ is a heap. Moreover, every non-empty heap is of this form, and there is a natural functor $(G,\cdot) \mapsto (G,[-,-,-])$ of the category of groups into the categories of heaps. Nevertheless these two categories are not equivalent.

A subheap S of a heap X is a *normal subheap* if it is non-empty and $[[x,e,s],x,e] \in S$ for every $x \in X$ and every $e, s \in S$.

There is an onto mapping $\{\text{normal subheaps}\} \to \{\text{congruences}\}$, $S \mapsto \sim_S$, which is not a bijection in general.

More precisely, let X be a heap. On the set $\mathcal{N}(X)$ of all normal subheaps of X define a pre-order $\preceq$ setting, for all $M, N \in \mathcal{N}(X)$, $M \preceq N$ if for every $x, y \in X$ and $s \in M$ such that $[x,y,s] \in M$ there exists $t \in N$ such that $[x,y,t] \in N$. Let $\simeq$ be the equivalence relation on $\mathcal{N}(X)$ associated to the pre-order $\preceq$. Then the partially ordered set $\mathcal{N}(X)/\simeq$ is order isomorphic to the partially ordered set $\mathcal{C}(X)$ of all congruences of the heap X.

PROPOSITION 2.9. *Let $X \neq \emptyset$ be a heap, Y be a subheap of X, and ω a congruence on X. The following conditions are equivalent:*

(a) *Y is a set of representatives of the equivalence classes of X modulo ω, that is, $Y \cap [x]_\omega$ is a singleton for every $x \in X$.*
(b) *There exists an idempotent heap endomorphism of X whose image is Y and whose kernel is ω.*
(c) *For every $a \in X$ and every $c \in Y$ there exist a unique element $b \in X$ and a unique element $d \in Y$ such that $a = [b,c,d]$ and $b\,\omega\,c$.*
(d) *For every $a \in X$ and every $c \in Y$ there exist a unique element $b \in X$ and a unique element $d \in Y$ such that $a = [d,c,b]$ and $b\,\omega\,c$.*
(e) *The mapping $g\colon Y \to X/\omega$, defined by $g(y) = [y]_\omega$ for every $y \in Y$, is a heap isomorphism.*

If the equivalent conditions of Proposition 2.9 are satisfied, the heap X is called the semidirect product of ω and Y, denoted $X = \omega \rtimes Y$.

PROPOSITION 2.10. *Let $(X,[-,-,-])$ be a heap, f be an idempotent heap endomorphism on X. Let ω be the kernel of f, so that ω is a congruence of the heap X, and let e be a fixed element of $Y := f(X)$. Then $[e]_\omega = f^{-1}(e)$ and there is an action, i.e., a heap homomorphism,*

$$\alpha\colon (Y,[-,-,-]) \to \mathrm{Aut}_{\mathsf{Heap}}([e]_\omega),$$

defined by $\alpha_y(k) = [y,e,[k,y,e]]$ for every $y \in Y$ and $k \in [e]_\omega$. Moreover, α_e is the identity automorphism of $[e]_\omega$.

The operation $[-,-,-]$ on $\mathrm{Aut}_{\mathsf{Heap}}([e]_\omega)$ is defined by $[f,g,h] = f \circ g^{-1} \circ h$ for every $f, g, h \in \mathrm{Aut}_{\mathsf{Heap}}([e]_\omega)$.

We are now ready to present the outer semidirect product of heaps:

PROPOSITION 2.11. *Let K and Y be heaps. Let $\alpha\colon Y \to \mathrm{Aut}_{\mathsf{Heap}}(K)$, $\alpha\colon b \in Y \mapsto \alpha_b$, be a heap morphism, and let y be an element of Y that is mapped by α to the identity automorphism α_y of K. On the cartesian product $K \times Y$ define a ternary operation $[-,-,-]$ setting*

$$[(k_1, y_1), (k_2, y_2), (k_3, y_3)] := ([k_1, \alpha_{[y_1,y_2,y]}(k_2), \alpha_{[y_1,y_2,y]}(k_3)], [y_1, y_2, y_3])$$

for every $(k_1, y_1), (k_2, y_2), (k_3, y_3) \in K \times Y$. Then:

(a) *$K \times Y$ is a heap and $K \times \{y\}$ is a normal subheap of $K \times Y$ isomorphic to K.*
(b) *For every element $k \in K$, $\{k\} \times Y$ is a subheap of $K \times Y$ isomorphic to Y and the mapping $f\colon K \times Y \to K \times Y$ defined by $f(a, b) = (k, b)$ for every $(a, b) \in K \times Y$ is an idempotent heap endomorphism of $K \times Y$.*

It is clear how to define the (direct) product of two heaps.

PROPOSITION 2.12. *Let X be a heap, ω be a congruence on X, Y be a subheap of X, and suppose $X = \omega \rtimes Y$. Let f be the corresponding idempotent heap endomorphism of X. Fix an element $e \in Y$ and set $K := [e]_\omega$. Then the following conditions are equivalent:*

(a) *The subheap Y of X is normal.*
(b) *$X = K \times Y$ (direct product of heaps).*
(c) *$[y, e, k] = [k, e, y]$ for every $y \in Y$ and every $k \in K$.*
(d) *There is an idempotent heap endomorphism g of X whose image is K and such that $g^{-1}(e) = Y$.*
(e) *The heap morphism $\alpha\colon Y \to \mathrm{Aut}_{\mathsf{Heap}}(K)$ of Proposition 2.10 is constantly equal to the identity mapping of K.*

In particular, if the equivalent conditions of the previous Proposition 2.12 hold, then the commutator $[\sim_Y, \omega]$ is the conguence $=$ on X.

2.8. Left near-trusses. A *left near-truss* $(X, [-,-,-], \cdot)$ is a set X endowed with a ternary operation $[-,-,-]$ and a binary operation $\cdot$, such that $(X, [-,-,-])$ is a heap, $(X, \cdot)$ is a semigroup, and *left distributivity* holds, that is,

$$x \cdot [y, z, w] = [x \cdot y, x \cdot z, x \cdot w]$$

for every $x, y, z, w \in X$. Similarly for *right near-trusses*, where left distributivity is replaced by *right distributivity*: $[y, z, w] \cdot x = [y \cdot x, z \cdot x, w \cdot x]$ for every $x, y, z, w \in X$. Clearly, the category of left near-trusses is isomorphic to the category of right near-trusses, it suffices to associate to any left near-truss $(X, [-,-,-], \cdot)$ its opposite right near-truss $(X, [-,-,-], \cdot^{\mathrm{op}})$.

PROPOSITION 2.13. *Let $X \neq \emptyset$ be a left near-truss, Y be a subnear-truss of X, and ω a congruence on the left near-truss X. The following conditions are equivalent:*

(a) *Y is a set of representatives of the equivalence classes of X modulo ω (i.e., $Y \cap [x]_\omega$ is a singleton for every $x \in X$).*
(b) *There exists an idempotent left near-truss endomorphism of X whose image is Y and whose kernel is ω.*
(c) *For every $a \in X$ and every $c \in Y$ there exist a unique element $b \in X$ and a unique element $d \in Y$ such that $a = [d, c, b]$ and $b\,\omega\, c$.*
(d) *The mapping $g\colon Y \to X/\omega$, defined by $g(y) = [y]_\omega$ for every $y \in Y$, is a left near-truss isomorphism.*

3. Inner semidirect product in Universal Algebra

With all the previous examples in mind, it is now clear what must be defined as inner semidirect product in Universal Algebra. Fix a variety $\mathcal{V}$ of type $\mathcal{F}$.

THEOREM 3.1. *Let A be an algebra, B be a subalgebra of A, and ω be a congruence on A. The following conditions are equivalent:*

(a) *B is a set of representatives for the elements of A modulo ω, that is, for every $a \in A$ the intersection $[a]_\omega \cap B$ is a singleton.*

(b) *There is an idempotent endomorphism of A whose image is B and whose kernel is ω.*

(c) *There is a surjective homomorphism $A \to B$ that is the identity on B and whose kernel is ω.*

(d) *The canonical mapping $B \to A/\omega$, $b \mapsto [b]_\omega$, is an isomorphism.*

PROOF. (a) $\Rightarrow$ (b) Suppose that (a) holds. Let $e\colon A \to A$ be the mapping that associates with any $a \in A$ the unique element in $[a]_\omega \cap B$. It is clear that the image of this mapping is contained in B and, conversely, if $b \in B$, then $b = e(b)$. Hence $B = e(A)$.

Let us show that the mapping e is a homomorphism. Let $p\colon A^n \to A$ be any n-ary operation of A and $a_1, \dots, a_n$ be elements of A. In order to show that $e(p(a_1, \dots, a_n)) = p(e(a_1), \dots, e(a_n))$, notice that both $e(p(a_1, \dots, a_n))$ and $p(e(a_1), \dots, e(a_n))$ are elements of B because $B = e(A)$ and B is a subalgebra of A. Hence, in order to show that they are equal, it suffices to show that $e(p(a_1, \dots, a_n)) \,\omega\, p(e(a_1), \dots, e(a_n))$. Clearly $e(a) \,\omega\, a$ for every $a \in A$, so that $e(p(a_1, \dots, a_n)) \,\omega\, p(a_1, \dots, a_n) \,\omega\, p(e(a_1), \dots, e(a_n))$ because ω is a congruence of A. This concludes the proof that $e\colon A \to A$ is a homomorphism. Finally, $e(a) \,\omega\, a$ implies that $[e(a)]_\omega = [a]_\omega$, so $[e(a)]_\omega \cap B = [a]_\omega \cap B$, whence $e(e(a)) = e(a)$. Therefore e is an idempotent endomorphism.

Finally, it is clear that two elements a, a' of A are mapped to the same element of B if and only if $[a]_\omega = [a']_\omega$, that is, if and only if $a \,\omega\, a'$. This proves that the kernel of e is exactly ω.

(b) $\Rightarrow$ (c) If $e\colon A \to A$ is an idempotent endomorphism of A whose image is B and whose kernel is ω, its corestriction $e|^B\colon A \to B$ is a surjective homomorphism whose kernel is still ω. The mapping $e|^B$ is the identity on B, because if $b \in B$, there exists $a \in A$ such that $e(a) = b$. Then $b = e(a) = e(e(a)) = e(b) = e|^B(b)$.

(c) $\Rightarrow$ (d) If $g\colon A \to B$ is a surjective homomorphism with kernel ω that is the identity on B, then there is an isomorphism $\overline{g}\colon A/\omega \to B$ such that $\overline{g}([b]_\omega) = b$ for every $b \in B$. Its inverse $\overline{g}^{-1}\colon B \to A/\omega$ is the canonical mapping $B \to A/\omega$, $b \mapsto [b]_\omega$, and is an isomorphism.

(d) $\Rightarrow$ (a) because (a) is just a restatement of (d). $\square$

If the equivalent conditions in the statement of Theorem 3.1 hold, we will say that A is the *(inner) semidirect-product* of its subalgebra B and its congruence ω, and write $A = B \ltimes \omega$.

COROLLARY 3.2. *Given an algebra A there is a one-to-one correspondence between:*

(a) *The set of all idempotent endomorphisms of A.*
(b) *The set of all pairs (B, ω), where B is a subalgebra of A, ω is a congruence on A, and $A = B \ltimes \omega$.*

REMARK 3.3. Assume $A = B \ltimes \omega$. By Theorem 3.1(a), the subalgebra B is a set of representatives for the congruence classes of A modulo ω. Each congruence class of A modulo ω is of the form $A_b := [b]_\omega$ for a unique $b \in B$, and A is the disjoint union of all its subsets A_b. Moreover, each A_b has a privileged point b (because $A_b \cap B$ is the singleton $\{b\}$), so that (A_b, b) is a pointed set. For every n-ary operation $p\colon A^n \to A$ of A and for every $b_1, \dots, b_n \in B$, one has that $p(b_1, \dots, b_n) \in B$ (this exactly means that B is a subalgebra of A) and $p(A_{b_1} \times \cdots \times A_{b_n}) \subseteq A_{p(b_1,\dots,b_n)}$ (this is equivalent to "ω is a conguence for A"). Notice that the restriction $A_{b_1} \times \cdots \times A_{b_n} \to A_{p(b_1,\dots,b_n)}$ of p is a morphism of pointed sets, that is, a morphism in the category Set_*. Hence the equivalent conditions of Theorem 3.1 say that there is a partition of A into pointed sets, this partition is compatible with the operations on A, and defines a sort of "grading" on A.

In the pairs (B, ω) relative to semidirect-product decompositions of an algebra A, B can be any subalgebra of A that is also a homomorphic image of A. Relatively to such a subalgebra B, ω can be any "complement" of B in A, that is, the kernel of any homomorphism $A \to B$ that is the identity on B.

By Remark 3.3, if we have a semidirect-product decomposition $A = B \ltimes \omega$, then *the algebra B acts on the corresponding pointed partition of A*, in the sense that for every n-ary function symbol $f \in \mathcal{F}$ and every n-tuple $(b_1, \dots, b_n) \in B^n$, there is a morphism $f_{(b_1,\dots,b_n)}\colon (A_{b_1}, a_{b_1}) \times \cdots \times (A_{b_n}, a_{b_n}) \to (A_{f(b_1,\dots,b_n)}, a_{f(b_1,\dots,b_n)})$ in the category Set_*. It is defined, for every $(x_1, \dots, x_n) \in A_{b_1} \times \cdots \times A_{b_n}$, by $f_{(b_1,\dots,b_n)}(x_1, \dots, x_n) = f(x_1, \dots, x_n)$ (as we have said above, each $f_{(b_1,\dots,b_n)}$ is just the operation f, but restricted to $A_{b_1} \times \cdots \times A_{b_n} \to A_{f(b_1,\dots,b_n)}$).

3.1. The special role of constant endomorphisms in semidirect products. Until now we have stressed the role of idempotent endomorphisms in the study of semidirect-product decompositions of an algebra A. Among idempotent endomorphisms of A, the easiest but probably most important ones are the *constant endomorphisms* of A, that is the endomorphisms $f\colon A \to A$ such that $f(a) = f(a')$ for every $a, a' \in A$. Equivalently, the endomorphisms $f\colon A \to A$ for which there exists an element $\overline{a} \in A$ such that $f(a) = \overline{a}$ for every $a \in A$. Clearly, every constant endomorphism is an idempotent endomorphism.

We say that an element a of an algebra A is *totally idempotent* if $p(a, a, \dots, a) = a$ for every operation p of A.

PROPOSITION 3.4. *Let A be an algebra. There are canonical one-to-one correspondences between the following three sets:*

(a) *The set of all constant endomorphisms of A.*
(b) *The set of all subalgebras of A of cardinality one.*
(c) *The set of all totally idempotent elements of A.*

The set E of all idempotent endomorphisms of A can be partially ordered setting, for all $e, f \in E$, $e \leq f$ if $e(A) \subseteq f(A)$ and, for all $a, a' \in A$, $f(a) = f(a') \Rightarrow e(a) = e(a')$. The partially ordered set $(E, \leq)$ has the identity automorphism id_A of A as its greatest element. If A has constant endomorphisms, they are all minimal elements of $(E, \leq)$.

THEOREM 3.5. *Let A be an algebra, $A = B \ltimes \omega$ be a semidirect-product decomposition of A, and e the corresponding idempotent endomorphism of A. The following conditions are equivalent for a congruence class K of A modulo ω:*

(a) *K is a subalgebra of A.*
(b) *The element of $K \cap B$ is totally idempotent.*
(c) *In the partially ordered set E of all idempotent endomorphisms of E, there exists a constant endomorphism $e_0 \in E$ such that $e_0 \leq e$.*

PROOF. If (a) holds, then the idempotent endomorphism $e\colon A \to A$, which associates to each congruence class modulo ω the unique element of the congruence class which also belongs to B, restricts to a constant endomorphism $e|_K^K\colon K \to K$, which associates to every element of K the unique element of $K \cap B$. Since images of homomorphisms are subalgebras, it follows that $K \cap B$ is a subalgebra of B, so that its unique element is totally idempotent. This proves (b).

If (b) holds, then the constant mapping $e_0\colon A \to A$ that sends all elements of A to the totally idempotent element k_0 is a constant endomorphism of A. Moreover $e_0 \leq e$ because $e_0(A) = \{k_0\} \subseteq B = e(A)$ and, for all $a, a' \in A$, the implication $e(a) = e(a') \Rightarrow e_0(a) = e_0(a') = k_0$ holds trivially. Thus (c) holds.

Assume that (c) holds. Let e_0 be a constant endomorphism of A with $e_0 \leq e$. If k_0 is the unique element in the image of e_0, then $k_0 \in e(A) = B$. Moreover k_0 is a totally idempotent element because the image $\{k_0\}$ of the homomorphism e_0 is a subalgebra of A. Now K is a subalgebra of A, because if f is an n-ary operation of A and $k_1, \ldots, k_n \in K$, then $k_1 \,\omega\, k_0, \ldots, k_n \,\omega\, k_0$, so that $f(k_1, \ldots, k_n) \,\omega\, f(k_0, \ldots, k_0) = k_0$ because ω is a congruence. Therefore $f(k_0, \ldots, k_0) \in K$, and (a) holds. □

COROLLARY 3.6. *Let A be an algebra and $A = B \ltimes \omega$ be a semidirect-product decomposition of A. Then the canonical bijection $B \to A/\omega$ of Theorem 3.1(d) induces by restriction a bijection between the set of all the elements of B that are totally idempotent and the set of all congruence classes of A modulo ω that are subalgebras of A.*

For instance, in the variety of groups, the unique totally idempotent element of any group is its identity. And of all the cosets gN modulo a normal subgroup N, the unique coset that is a subgroup is the subgroup N.

REMARK 3.7. The setting described until now in this Section 3 can be described in an evocative way by the diagram

$$\omega \longrightarrow A \underset{\iota}{\overset{\varphi}{\rightleftarrows}} B, \tag{3.1}$$

where φ is the surjective homomorphism described in Theorem 3.1(c), ω is the kernel of φ, $\iota\colon B \hookrightarrow A$ is the inclusion of the subalgebra B into the algebra A, ι is a right inverse of φ, and $\iota\varphi$ is the corresponding idempotent endomorphism of A.

In several cases the congruence ω on A can be described (can be determined) by one of its congruence classes K that turns out to be a subalgebra of A. This is the case of normal subgroups, or of two-sided ideals of a ring (not-necessarily with an identity). On the one hand this allows to replace diagram (3.1) with the diagram

$$K \hookrightarrow A \underset{\iota}{\overset{\varphi}{\rightleftarrows}} B. \tag{3.2}$$

On the other hand this also allows us to talk of "action of the algebra B on the algebra K". This is possible whenever K is the congruence class modulo ω of a totally idempotent element.

3.2. The example of groups. Let $\mathcal{V}$ be the varieties of groups (three operations, one binary, one unary, and one 0-ary). Let A be a group. Every idempotent endomorphism e of A corresponds to a semidirect-product decomposition $A = \omega \rtimes B$, where B is a subgroup of A and ω a congruence on A. In A there is a unique totally idempotent element: the identity 1_A of A. Since the variety $\mathcal{V}$ of groups is ideal determined, the congruence ω corresponds to a normal subgroup N of A, the congruence class of 1_A modulo ω. The corresponding partition of A into pointed subsets is the partition of A in cosets Na. More precisely, it is the partition into pointed sets $\mathcal{N} = \{ (Nb, b) \mid b \in B \}$. For each of the three n-ary operations $p\colon A^n \to A$ on A and each $b_1, \dots, b_n \in B$, we have the restriction $A_{b_1} \times \cdots \times A_{b_n} \to A_{p(b_1,\dots,b_n)}$, which is a morphism in the category Set_*. In this case of groups, these restrictions are morphisms of pointed sets $(Nb_1, b_1) \times \cdots \times (Nb_n, b_n) \to (Nb_1 \dots b_n, b_1 \dots b_n)$. But due to the canonical bijections $N \to Nb$, $n \mapsto nb$, these restrictions can be equivalently described as:

(1) Mappings $g_{b_1,b_2}\colon N \times N \to N$. They are defined by the multiplications on the corresponding cosets, so that $g_{b_1,b_2}(n_1, n_2) = (n_1 b_1)(n_2 b_2)(b_1 b_2)^{-1} = n_1 b_1 n_2 b_1^{-1}$. Notice that, therefore, all these mappings g_{b_1,b_2} do not depend on b_2, but only on b_1. Moreover, $g_{b_1,1_B}(1_N, -)\colon N \to N$ is the conjugation by b_1, written on the left like in Subsection 2.1 (it is a group homomorphism). Finally notice that $g_{b_1,b_2}(n_1, n_2) = n_1 g_{b_1,1_B}(1_N, n_2)$, so that all the mappings g_{b_1,b_2} can be expressed using the only mappings $g_{b_1,1_B}(1_N, -)$ and multiplication in N, that is, the mappings $g_{b_1,1_B}(1_N, -)$ are sufficient to parametrize all the mappings g_{b_1,b_2}.

(2) The 0-ary operation corresponds to the element 1_N in the pointed set $(N, 1_B)$.

(3) The mappings $h_b\colon N \to N$ are defined by the inverse on the corresponding cosets, so that $h_b(n) = (nb)^{-1}b = b^{-1}n^{-1}b = g_{(b^{-1},1_B)}(1_N, n^{-1})$. Thus all the mappings h_b are also completely determined by the only mappings $g_{b_1,1_B}(1_N, -)$.

Thus, for the special case of groups, all the data $\{g_{(b_1,b_2)}, h_b, 1_N\}$ are completely determined by the group homomorphisms $B \to \mathrm{Aut}_{\mathsf{Gp}}(N)$, $b \mapsto g_{b,1_B}(1_N, -)$. But notice that the correct notion of "action" of an algebra B on an algebra N in the case of one binary operation as is the case of semigroups for instance, is as a collection of morphisms $(A_{b_1}, b_1) \times (A_{b_2}, b_2) \to (A_{b_1 b_2}, b_1 b_2)$ indexed in B^2.

4. Outer semidirect product

Let B be an algebra in a variety $\mathcal{V}$ of type $\mathcal{F}$. We want to construct an algebra A in $\mathcal{V}$ that is a semidirect product of B and a congruence ω on A. To this end, we take as our initial data the algebra B in the variety and an indexed family $\mathcal{N} = \{ (A_b, a_b) \mid b \in B \}$ of pointed sets, indexed in B. Set

$$A := \bigcup_{b \in B} (A_b \times \{b\}) \subseteq \left(\bigcup_{b \in B} A_b \right) \times B.$$

For every n-ary function symbol $f \in \mathcal{F}$ and every $b_1, \dots, b_n \in B$, let

$$f_{(b_1,\dots,b_n)}\colon (A_{b_1}, a_{b_1}) \times \cdots \times (A_{b_n}, a_{b_n}) \to (A_{f(b_1,\dots,b_n)}, a_{f(b_1,\dots,b_n)})$$

be a morphism in Set_*, that is, a mapping $f_{(b_1,\dots,b_n)}\colon A_{b_1} \times \cdots \times A_{b_n} \to A_{f(b_1,\dots,b_n)}$ such that $f_{(b_1,\dots,b_n)}(a_{b_1}, \dots, a_{b_n}) = a_{f(b_1,\dots,b_n)}$. Then A becomes an algebra of type $\mathcal{F}$ with the operations defined by

$$f((a_{b_1}, b_1), \dots, (a_{b_n}, b_n)) = (f_{(b_1,\dots,b_n)}(a_{b_1}, \dots, a_{b_n}), f(b_1, \dots, b_n)).$$

The outer semidirect products of B and $\mathcal{N}$ are indexed only by those mappings $f_{(b_1,\dots,b_n)}$ for which the corresponding operations f on A satisfy the identities of $\mathcal{V}$. For any such choice of $\mathcal{N}$ and mappings $f_{(b_1,\dots,b_n)}$, we will call the resulting algebra A the *(outer) semidirect product* parametrized by $\mathcal{N}$ and the family of mappings $F = \{ f_{(b_1,\dots,b_{n_f})} \mid f \in \mathcal{F},\ n_f$ the arity of $f,\ b_1,\dots,b_{n_f} \in B \}$. It will be denoted by $A = \mathcal{N} \rtimes_F B$. Also, we will denote this assignment of the pointed set (A_b, a_b) to each element $b \in B$ and of the mapping $f_{(b_1,\dots,b_n)}$ to each n-ary function symbol $f \in \mathcal{F}$ and each $b_1,\dots,b_n \in B$ in the form $F\colon B \to \mathsf{Set}_*$ because of the similarity of the outer semidirect product we have just defined with a functor into the category Set_* (see Subsection 4.1 below).

LEMMA 4.1. *Let A be an algebra admitting an inner semidirect-product decomposition $A = \omega \rtimes B$. Then A is isomorphic to the outer semidirect product $\mathcal{N} \rtimes_F B$ where $\mathcal{N} = \{ ([b]_\omega, b) \mid b \in B \}$ and F consists of the restrictions of the operations p of A to $[b_1]_\omega \times \dots \times [b_{n_p}]_\omega \to [p(b_1,\dots,b_{n_p})]_\omega$.*

PROOF. It is straighforward to check that this is a well defined outer semidirect product and that the mapping $a \mapsto (a, b)$, where b is the unique element of B such that $a \in [b]_\omega$, is an isomorphism $A \to \mathcal{N} \rtimes_F B$. □

An *action* of B on $\mathcal{N}$ is therefore an element of

$$\prod_{f\in\mathcal{F}} \left(\prod_{(b_1,\dots,b_n)\in B^n} \hom_{\mathsf{Set}_*}\big((A_{b_1}, a_{b_1}) \times \dots \times (A_{b_n}, a_{b_n}), (A_{f(b_1,\dots,b_n)}, a_{f(b_1,\dots,b_n)})\big) \right) \tag{4.1}$$

for which the corresponding operations f on A satisfy the identities of $\mathcal{V}$. Thus the outer semidirect product is constructed from an algebra B, an indexed family $\mathcal{N} = \{ (A_b, a_b) \mid b \in B \}$ of pointed sets indexed in B, and a family F of morphisms $f_{(b_1,\dots,b_n)}\colon (A_{b_1}, a_{b_1}) \times \dots \times (A_{b_n}, a_{b_n}) \to (A_{f(b_1,\dots,b_n)}, a_{f(b_1,\dots,b_n)})$ in Set_*, one for each n-ary function symbol $f \in \mathcal{F}$ and each n-tuple $(b_1,\dots,b_n) \in B^n$.

4.1. Outer semidirect products can be extended to functors. In Category Theory, semidirect product of groups has been generalized to the so-called *Grothendieck construction*. The Grothendieck construction can be applied to any functor from any small category to the category Cat of small categories.

As a matter of fact, there is a similarity between the definition of functor and our definition of outer semidirect product of arbitrary algebras. In order to define a functor F from a category $\mathcal{C}$ to a category $\mathcal{D}$, we must:

(1) associate with each object X in $\mathcal{C}$ an object $F(X)$ in $\mathcal{D}$;
(2) associate with each morphism $f\colon X \to Y$ in $\mathcal{C}$ a morphism $F(f)\colon F(X) \to F(Y)$ in $\mathcal{D}$;
(3) and two very natural conditions must hold.

To define an outer semidirect product, we need an algebra B in a variety $\mathcal{V}$ of type $\mathcal{F}$, and then we must:

(1) associate with each element $b \in B$ a pointed set (A_b, a_b);
(2) associate with each operation $f \in \mathcal{F}$, n_f-ary operation say, and each n_f-tuple $(b_1,\dots,b_{n_f}) \in B^{n_f}$, a morphism $f_{(b_1,\dots,b_{n_f})}\colon (A_{b_1}, a_{b_1}) \times \dots \times (A_{b_{n_f}}, a_{b_{n_f}}) \to (A_{f(b_1,\dots,b_{n_f})}, a_{f(b_1,\dots,b_{n_f})})$ in Set_*,

in such a way that if we define $A := \dot{\bigcup}_{b\in B} A_b = \{ (x,b) \mid b \in B,\ x \in A_b \}$ and the operations $f\colon A^{n_f} \to A$ on A via

$$f((a_{b_1}, b_1), \ldots, (a_{b_n}, b_n)) = (f_{(b_1,\ldots,b_n)}(a_{b_1}, \ldots, a_{b_n}), f(b_1, \ldots, b_n)),$$

then

(3) the identities defining the variety $\mathcal{V}$ must hold for the algebra A.

We all know that the first definition in MacLane's "Categories for the Working Mathematician" [**14**, p. 7] is that of metagraph: A *metagraph* consists of *objects* $a, b, c, \ldots$, *arrows* $f, g, h, \ldots$, and two operations: (1) Domain, which assigns to each arrow f an object $a = \operatorname{dom} f$; (2) Codomain, which assigns to each arrow f an object $b = \operatorname{cod} f$. In order to describe here now algebras and their semidirect product, we need the more general notion of metahypergraph: A *metahypergraph* consists of *objects* $a, b, c, \ldots$, *arrows* $f, g, h, \ldots$, and two operations: (1) Domain, which assigns to each arrow f an n_f-tuple $(a_1, \ldots, a_{n_f}) = \operatorname{dom} f$ of objects, where n_f is a non-negative integer that depends on f; (2) Codomain, which assigns to each arrow f an object $b = \operatorname{cod} f$. Then an outer semidirect product is a "functor" from a metaphypergraph to the category Set_*. More precisely, every algebra A with operations f_γ of arity n_γ can be viewed as a metaphypergraph with objects the elements of A, and one arrow $f_{\gamma,(a_1,\ldots,a_{n_\gamma})}$ for each operation f_γ of A and each n_γ-tuple $(a_1, \ldots, a_{n_\gamma}) \in A^{n_\gamma}$. The domain of $f_{\gamma,(a_1,\ldots,a_{n_\gamma})}$ is the n_γ-tuple $(a_1, \ldots, a_{n_\gamma})$, and its codomain is the element $f_\gamma(a_1, \ldots, a_{n_\gamma})$ of A.

Now the metagraph constructed in the previous paragraph (we will denote it by A, the same symbol used for the algebra) can be extended to a category $\mathcal{C}_A$ which we will call the *enveloping category* of the metahypergraph A. The objects of $\mathcal{C}_A$ are all n-tuples $(a_1, \ldots, a_n)$ of elements of the algebra A for an arbitrary integer $n \geq 0$. A morphism $(a_1, \ldots, a_n) \to (b_1, \ldots, b_m)$ is any m-tuple $(p_1, \ldots, p_m)$ of n-ary terms [**7**, Definitions 10.1 and 10.2] such that $p_j(a_1, \ldots, a_n) = b_j$ for every $j = 1, 2, \ldots, m$. Composition of morphisms is defined by

$$(q_1, \ldots, q_r) \circ (p_1, \ldots, p_m) = (q_1(p_1, \ldots, p_m), q_2(p_1, \ldots, p_m), \ldots, q_r(p_1, \ldots, p_m)).$$

THEOREM 4.2. *Any outer semidirect product $F\colon A \to \mathsf{Set}_*$ can be extended to a functor $G\colon \mathcal{C}_A \to \mathsf{Set}_*$ by defining*

$$G(a_1, \ldots, a_n) = F(a_1) \times \cdots \times F(a_n)$$

and

$$\begin{aligned} G(p_1, \ldots, p_m) &= (F(p_1), \ldots, F(p_m)) \\ &= \textstyle\prod_{j=1}^{m} \left(F(p_j)\colon F(a_1) \times \cdots \times F(a_n) \to F(p_j(a_1, \ldots, a_n))\right). \end{aligned}$$

4.2. The category $\mathcal{S}_B$ of all outer semidirect products of an algebra B. Let B be an algebra in a variety $\mathcal{V}$ of type $\mathcal{F}$. It is possible to construct a category $\mathcal{S}_B$ as follows.

(a) The objects of $\mathcal{S}_B$ are all semidirect products $F\colon B \to \mathsf{Set}_*$, where

$$F(b) = (A_b, a_b)$$

for every $b \in B$,

$$f^F_{(b_1,\ldots,b_n)}\colon (A_{b_1}, a_{b_1}) \times \cdots \times (A_{b_n}, a_{b_n}) \to (A_{f(b_1,\ldots,b_n)}, a_{f(b_1,\ldots,b_n)})$$

for every n_f-ary function symbol $f \in \mathcal{F}$ and every $(b_1, \ldots, b_n) \in B^n$, and $A = \dot{\bigcup}_{b\in B}(A_b \times \{b\})$ is an algebra in $\mathcal{V}$.

(b) A morphism $\varphi\colon F \to F'$, where $F, F'\colon B \to \mathsf{Set}_*$ are semidirect products of B, consists of a morphism $\varphi_b\colon F(b) \to F'(b)$ in Set_* for every $b \in B$, in such a way that for every $f \in \mathcal{F}$ and every n_f-uple $(b_1, \dots, b_n) \in B^n$ the diagram

$$
\begin{array}{ccc}
(A_{b_1}, a_{b_1}) \times \cdots \times (A_{b_n}, a_{b_n}) & \xrightarrow{f^F_{(b_1,\dots,b_n)}} & (A_{f(b_1,\dots,b_n)}, a_{f(b_1,\dots,b_n)}) \\
\Big\downarrow{\scriptstyle \varphi_{b_1}\times\cdots\times\varphi_{b_n}} & & \Big\downarrow{\scriptstyle \varphi_{f(b_1,\dots,b_n)}} \\
(A'_{b_1}, a'_{b_1}) \times \cdots \times (A'_{b_n}, a'_{b_n}) & \xrightarrow[f^{F'}_{(b_1,\dots,b_n)}]{} & (A'_{f(b_1,\dots,b_n)}, a'_{f(b_1,\dots,b_n)})
\end{array}
\tag{4.2}
$$

commutes.

Also recall the definition of *category* $\mathcal{P}_B$ *of pointed objects over* B [**4**, p. 28]:

(a) Its objects are all triples (A, α, β), where $A \in \mathcal{V}$, $\alpha\colon B \to A$, $\beta\colon A \to B$ are homomorphisms in $\mathcal{V}$ and $\beta\alpha = 1_B$.

(b) Morphisms $\gamma\colon (A, \alpha, \beta) \to (A', \alpha', \beta')$ are all morphisms $\gamma\colon A \to A'$ in $\mathcal{V}$ for which the diagram

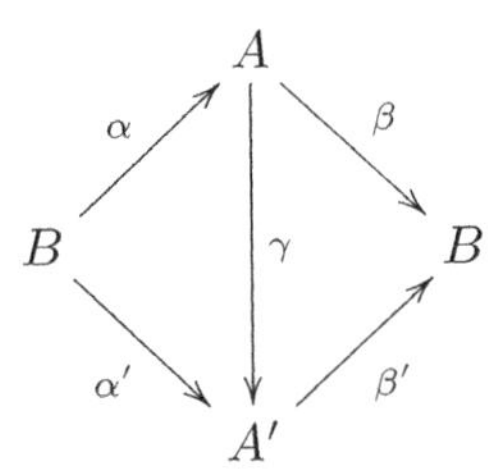

commutes (i.e., such that $\gamma\alpha = \alpha'$ and $\beta'\gamma = \beta$).

(c) Composition $\gamma' \circ \gamma$ of morphisms is composition of mappings.

It is straightforward to check that:

THEOREM 4.3. *The two categories* $\mathcal{S}_B$ *and* $\mathcal{P}_B$ *are equivalent. If* (A, α, β) *is an object of* $\mathcal{P}_B$*, the corresponding object* $F\colon B \to \mathsf{Set}_*$ *is defined by* $F(b) = (\beta^{-1}(b), \alpha(b))$ *for every* $b \in B$ *and each*

$$
f^F_{(b_1,\dots,b_n)}\colon (\beta^{-1}(b_1), \alpha(b_1)) \times \cdots \times (\beta^{-1}(b_{n_f}), \alpha(b_{n_f})) \to \\
\to (\beta^{-1}(f(b_1, \dots, b_n)), \alpha(f(b_1, \dots, b_n)))
$$

is the restriction of the operation $f^A\colon A^n \to A$ *in* A.

Notice that the commutativity of all the squares (4.2) exactly means that the corresponding mapping $A = \dot{\bigcup}_{b\in B} F(b) \to A' = \dot{\bigcup}_{b\in B} F'(b)$ is a homomorphism in $\mathcal{V}$.

4.3. Action of an algebra B on another algebra K. At the beginning of this Section 4, we have defined A as $\bigcup_{b\in B}(A_b \times \{b\})$, so that there is an embedding $A \hookrightarrow \left(\bigcup_{b\in B} A_b\right) \times B$. A particular case of this construction of semidirect product is that in which the indexed family $\mathcal{N} = \{\, (A_b, a_b) \mid b \in B \,\}$ is such that there exist a set C and an element $c \in C$ such that $A_b = C \times \{b\}$ and $a_b = (c, b)$ for every $b \in B$ (i.e., the semidirect product $F\colon B \to \mathsf{Set}_*$ constantly assigns the pointed set (C, c) with every element $b \in B$). In this case one has a bijection $A \to C \times B$. For example, suppose that B has a totally idempotent element $\overline{b}$, and let K be another algebra in $\mathcal{V}$ with a totally idempotent element $\overline{k} \in K$. Then any corresponding outer semidirect product consists in determining all structures of algebra in $\mathcal{V}$ over the cartesian product $K\times B$ for which the mapping $K\times B \to K\times B$, $(k,b)\mapsto(\overline{k},b)$, is an idempotent endomorphism, and the two mappings $K\to K\times B$, $k\mapsto(k,\overline{b})$ and $B\to K\times B$, $b\mapsto(\overline{k},b)$ are two algebra embeddings.

The best case is the following. Assume that we have a pointed variety $\mathcal{V}$ (i.e., a variety with exactly one constant) and the subset consisting of only that constant is a subalgebra. Equivalently, in the variety there is a constant 0 such that, for every n-ary operation $f\in\mathcal{F}$, $f(0,0,\dots,0)=0$. In such a variety the singletons are the zero objects. For these algebras there is a Galois correspondence between congruence classes of zero (ideals, normal subalgebras) and congruences.

Let B be an algebra in such a variety $\mathcal{V}$ of type $\mathcal{F}$. Take as our initial data the algebra B in the variety and an indexed family $\{\,A_b \mid b\in B\,\}$ of nonempty sets. With the axiom of choice, fix an element $a_b\in A_b$ for each $b\in B$, and construct the disjoint union $A:=\dot{\bigcup}_{b\in B}A_b$.

Given any two algebras B and K in $\mathcal{V}$, an *action* of B on K is any indexed family m_f $(f\in\mathcal{F})$ of mappings $m_f\colon B^{n_f}\to \hom_{\mathsf{Set}_*}(K^{n_f},K)$ such that the operations

$$f((k_1,b_1),\dots,(k_{n_f},b_{n_f}))=(m_f(b_1,\dots,b_{n_f})(k_1,\dots,k_{n_f}),f(b_1,\dots,b_{n_f}))$$

give an algebra structure on the cartesian product $B\times K$ in such a way that its subalgebras $B\times\{0_K\}$ and $\{0_B\}\times K$ turn out to be canonically isomorphic to the algebras B and K, respectively.

This is the case for Ω-groups, which were introduced by Higgins in [**12**]. A variety of Ω-groups is a variety which is pointed and has amongst its operations and identities those of the variety of groups. Examples of Ω-groups are abelian groups, non-unital rings, commutative algebras, modules, and Lie algebras. Also see [**13**, Example (2)].

4.4. When is a semidirect product a direct product. Let now $\mathcal{V}$ be a variety of type $\mathcal{F}$, and fix two algebras B and K in $\mathcal{V}$ with two totally idempotent elements $\overline{b}\in B$ and $\overline{k}\in K$. Construct a semidirect product $F_K\colon B\to\mathsf{Set}_*$ with $F_K(b)=(K,\overline{k})$ for every $b\in B$, so that $A:=K\times B$, and F_K associates with any n-ary function symbol $f\in\mathcal{F}$ and every n-tuple $(b_1,\dots,b_n)\in B^n$, a morphism $f_{(b_1,\dots,b_n)}\colon K^n\to K$ in Set_*, i.e. a mapping with $f_{(b_1,\dots,b_n)}(\overline{k},\dots,\overline{k})=\overline{k}$. On $A=K\times B$ the semidirect-product operations are defined setting

$$f((k_1,b_1),\dots,(k_n,b_n))=(f_{(b_1,\dots,b_n)}(k_1,\dots,k_n),f^B(b_1,\dots,b_n)).$$

We leave to the reader to verify that:

PROPOSITION 4.4. *The semidirect-product algebra $A=K\times B$ is canonically the direct product of K and B if and only if $f_{(b_1,\dots,b_n)}=f^K$, f^K the n-ary operation on K, for every n-ary function symbol $f\in\mathcal{F}$ and every $(b_1,\dots,b_n)\in B^n$.*

5. Examples

5.1. The example of groups. The elements of (4.1) that satisfy the identities of the variety $\mathcal{V}$ parametrize the outer semidirect products of B and $\mathcal{N}$ exactly in the same way as for groups the semidirect products of two groups B and N are parametrized by the group homomorphisms $B\to\mathrm{Aut}_{\mathsf{Gp}}(N)$.

Let us see this case more in detail. In the case of semidirect product of groups, B is a group (an algebra with one binary operation, one unary operation, and one 0-ary operation), and the unique totally idempotent element of B is the identity

1_B of B. Also, the variety of groups is ideal determined, so that the congruence ω will correspond to a normal subgroup N of A, and the corresponding partition of A is the partition into cosets. Thus if, for the indexed family $\mathcal{N} = \{ (A_b, a_b) \mid b \in B \}$ of pointed sets, the sets A_b are not all equipotent, then no family of functions $f_{(b_1,\dots,b_n)} \colon (A_{b_1}, a_{b_1}) \times \cdots \times (A_{b_n}, a_{b_n}) \to (A_{f(b_1,\dots,b_n)}, a_{f(b_1,\dots,b_n)})$ will satisfy the axioms of groups. Hence we can exclude that case immediately and suppose, without loss of generality, that $A_b = N \times \{b\}$ and $a_b = (1_N, b)$ for some fixed group N. Then $A = \bigcup_{b\in B}(A_b \times \{b\}) = \bigcup_{b\in B}(N \times \{(b,b)\})$, so that A is the cartesian product of N and the diagonal $\Delta(B \times B)$ of $B \times B$. The diagonal $\Delta(B \times B)$ is clearly in one-to-one correspondence with B itself, so that we can suppose that $A = N \times B$, the cartesian product of N and B.

For the binary function symbol $p \in \mathcal{F}$ denoting the product and every pair $(b_1, b_2) \in B^2$, we must now fix a morphism

$$p_{(b_1,b_2)} \colon (A_{b_1}, a_{b_1}) \times (A_{b_2}, a_{b_2}) \to (A_{b_1b_2}, a_{b_1b_2})$$

in Set_*. In our case, this means that we have a mapping

$$p_{(b_1,b_2)} \colon (N \times \{b_1\}) \times (N \times \{b_2\}) \to (N \times \{b_1b_2\})$$

such that $p_{(b_1,b_2)}((1_N, b_1), (1_N, b_2)) = (1_N, b_1b_2)$. Due to the canonical bijections $N \to N \times \{b\}$, we can equivalently fix mappings $g_{(b_1,b_2)} \colon N \times N \to N$ such that $g_{(b_1,b_2)}(1_N, 1_N) = 1_N$. Similarly, for the unary operation, we must fix mappings $h_b \colon N \to N$ such that $h_b(1_N) = 1_N$. For the 0-ary operation we take the constant 1_N in $N \cong N \times \{1_B\}$. Now we must impose that these data $\{g_{(b_1,b_2)}, h_b, 1_N\}$ define on the disjoint union $A = N \times B$ a group structure, i.e., that the three axioms of groups are satisfied. As it is easy to see, they correspond to the three conditions

(1) $g_{(b_1b_2,b_3)}(g_{(b_1,b_2)}(n_1,n_2),n_3) = g_{(b_1,b_2b_3)}(n_1, g_{(b_2,b_3)}(n_2,n_3))$ for every $b_1, b_2, b_3 \in B$, $n_1, n_2, n_3 \in N$;
(2) $g_{(1_B,b)}(1_N, n) = n$ for every $b \in B$, $n \in N$;
(3) $g_{(b^{-1},b)}(h(n), n) = 1_N$ for every $b \in B$, $n \in N$.

Then the disjoint union $A = N \times B$ becomes a group with respect to the multiplication defined by

$$(n_1, b_1)(n_2, b_2) = (g_{(b_1,b_2)}(n_1, n_2), b_1b_2).$$

There is a bijection between the set of all group morphisms $B \to \mathrm{Aut}_{\mathsf{Gp}}(N)$ and the set of all data $\{g_{(b_1,b_2)}, h_b, 1_N\}$. It associates to $\{g_{(b_1,b_2)}, h_b, 1_N\}$ the group morphism $\gamma \colon B \to \mathrm{Aut}_{\mathsf{Gp}}(N)$, $\gamma \colon b \mapsto \gamma_b := g_{(b,1_B)}(1_N, -)$, and to any group morphism $\gamma \colon B \to \mathrm{Aut}_{\mathsf{Gp}}(N)$ the mappings $g_{(b_1,b_2)} \colon N \times N \to N$ and $h_b \colon N \to N$ defined by $g_{(b_1,b_2)}(n_1, n_2) = n_1\gamma_{b_1}(n_2)$ and $h_b(n) = \gamma_{b^{-1}}(n^{-1})$.

In order to see that this is a bijection, suppose that $\gamma \colon B \to \mathrm{Aut}_{\mathsf{Gp}}(N)$ is a group homomorphism. Construct the semidirect product $G := B \ltimes_\gamma N$. Then G is a group, so that we can apply to it what we saw in Example 3.2, i.e., that the restrictions of the operations on G satisfy the conditions (1), (2), (3) above.

5.2. The example of digroups. The case of digroups is pretty similar to that of groups. Let $(B, *, \circ)$ and $(N, *, \circ)$ be two digroups (it would be more careful to denote digroups as $(B, *, \circ, {}^{-*}, {}^{-\circ}, 1_B)$, where the operations $*$ and $\circ$ are binary, the operations ${}^{-*}$ and ${}^{-\circ}$ are unary, and 1_B is 0-ary). Similarly to our construction of semidirect product in this paper (but now writing conjugation on the right like in Subsection 2.5), we construct a digroup structure on the cartesian product $B \times N$, making use of mappings $g^+_{(b_1,b_2)}\colon N\times N\to N$, $g^\circ_{(b_1,b_2)}\colon N\times N\to N$, $h^+_b\colon N\to N$, $h^\circ_b\colon N\to N$, and the constant $(1_B,1_N)\in B\times N$, imposing that they define a digroup structure on $B\times N$.

Notice that on $B\times N$ we now have that $(b_1,n_1)\circ(b_2,n_2)=b_1\circ n_1\circ b_2\circ n_2=b_1\circ b_2\circ b_2^{-1}\circ n_1\circ b_2\circ n_2=(b_1\circ b_2)\circ(\phi_{\circ\, b_2}(n_1)\circ n_2)=(b_1\circ b_2,\phi_{\circ\, b_2}(n_1)\circ n_2)$, so that

$$g^\circ_{(b_1,b_2)}(n_1,n_2)=\phi_{\circ\, b_2}(n_1)\circ n_2. \tag{5.1}$$

(This is exactly the equality (2.6).) Again we find that $g^\circ_{(b_1,b_2)}(n_1,n_2)$ does not depend on b_1. Moreover, $g^\circ_{(1_B,b)}(n,1_N)=\phi_{\circ\, b}(n)$, so that

$$\phi_{\circ\, b}=g^\circ_{(1_B,b)}(-,1_N), \tag{5.2}$$

which yields a group antihomomorphism $(B,\circ)\to\mathrm{Aut}_{\mathsf{Gp}}(B,\circ)$.

For the operation $+$ on $B\times N$ we have from (2.5) that

$$g^+_{(b_1,b_2)}(n_1,n_2)=(\Lambda_{b_1*b_2})^{-1}(\phi_{*b_2}(\Lambda_{b_1}(n_1))*\Lambda_{b_2}(n_2)). \tag{5.3}$$

In particular, $g^+_{(b,1_B)}(1_N,n)=(\Lambda_b)^{-1}(\Lambda_b(1_N)*\Lambda_{1_B}(n))=(\Lambda_b)^{-1}(n)$, so that $(\Lambda_b)^{-1}=g^+_{(b,1_B)}(1_N,-)$, or, equivalently,

$$\Lambda_b=(g^+_{(b,1_B)}(1_N,-))^{-1}. \tag{5.4}$$

Finally, $g^+_{(1_B,b)}(n,1_N)=(\Lambda_b)^{-1}(\phi_{*b}(\Lambda_{1_B}(n))*\Lambda_b(1_N))=(\Lambda_b)^{-1}(\phi_{*b}(n))$, so that

$$\phi_{*b}(n)=\Lambda_b(g^+_{(1_B,b)}(n,1_N)). \tag{5.5}$$

The three formulas (5.2), (5.4) and (5.5) allow determine $\phi_\circ$, Λ and ϕ_* as functions of the mappings g^+ and g°.

Conversely, assume that $\phi_\circ$, ϕ_* and Λ are given. From (5.1), it is possible to determine the mappings g^+, and from (5.3) it is possible to determine the mappings g°. We leave to the reader to check that $h^\circ_b(n)=\phi_{\circ\, b^{-\circ}}(n^{-\circ})$ and $h^+_b(n)=(\Lambda_{b^{-*}})^{-1}(\phi_{*\, b^{-*}}((\Lambda_b(n))^{-*}))$.

References

[1] Meltem Altun-Özarslan and Alberto Facchini, *The Krull-Schmidt-Remak-Azumaya Theorem for G-groups*, Rings, modules and codes, Contemp. Math., vol. 727, Amer. Math. Soc., [Providence], RI, [2019] ©2019, pp. 25–38, DOI 10.1090/conm/727/14622. MR3938137

[2] M. J. Arroyo Paniagua and A. Facchini, *Heaps and trusses*, in "Recent Progress in Ring and Factorization Theory", M. Brešar, A. Geroldinger, B. Olberding and D. Smertnig Eds., Springer Proc. in Math. & Statistics **477**, Springer Cham; arXiv:2308.00527, 2023.

[3] David Bachiller, *Extensions, matched products, and simple braces*, J. Pure Appl. Algebra **222** (2018), no. 7, 1670–1691, DOI 10.1016/j.jpaa.2017.07.017. MR3763276

[4] Francis Borceux, *A survey of semi-abelian categories*, Galois theory, Hopf algebras, and semi-abelian categories, Fields Inst. Commun., vol. 43, Amer. Math. Soc., Providence, RI, 2004, pp. 27–60, DOI 10.1090/fic/043/02. MR2075580

[5] N. Bourbaki, *Éléments de mathématique. Fasc. XXVI. Groupes et algèbres de Lie. Chapitre I: Algèbres de Lie* (French), Actualités Scientifiques et Industrielles [Current Scientific and Industrial Topics], No. 1285, Hermann, Paris, 1971. Seconde édition. MR271276

[6] Dominique Bourn, Alberto Facchini, and Mara Pompili, *Aspects of the category SKB of skew braces*, Comm. Algebra **51** (2023), no. 5, 2129–2143, DOI 10.1080/00927872.2022.2151609. MR4561474

[7] Stanley Burris and H. P. Sankappanavar, *A course in universal algebra*, Graduate Texts in Mathematics, vol. 78, Springer-Verlag, New York-Berlin, 1981. MR648287

[8] Alberto Facchini, Carmelo Antonio Finocchiaro, and George Janelidze, *Abstractly constructed prime spectra*, Algebra Universalis **83** (2022), no. 1, Paper No. 8, 38, DOI 10.1007/s00012-021-00764-z. MR4367895

[9] Alberto Facchini and Leila Heidari Zadeh, *Algebras with a bilinear form, and idempotent endomorphisms*, J. Algebra **628** (2023), 434–451, DOI 10.1016/j.jalgebra.2023.03.028. MR4575074

[10] Alberto Facchini and Mara Pompili, *Semidirect products of digroups and skew braces*, Bull. Belg. Math. Soc. Simon Stevin **31** (2024), no. 1, 40–53, DOI 10.36045/j.bbms.230825. MR4745565

[11] John W. Gray, *The existence and construction of lax limits*, Cahiers Topologie Géom. Différentielle **21** (1980), no. 3, 277–304. MR591387

[12] P. J. Higgins, *Groups with multiple operators*, Proc. London Math. Soc. (3) **6** (1956), 366–416, DOI 10.1112/plms/s3-6.3.366. MR82492

[13] George Janelidze, László Márki, and Walter Tholen, *Semi-abelian categories*, J. Pure Appl. Algebra **168** (2002), no. 2-3, 367–386, DOI 10.1016/S0022-4049(01)00103-7. Category theory 1999 (Coimbra). MR1887164

[14] Saunders Mac Lane, *Categories for the working mathematician*, 2nd ed., Graduate Texts in Mathematics, vol. 5, Springer-Verlag, New York, 1998. MR1712872

[15] Saunders Mac Lane and Ieke Moerdijk, *Sheaves in geometry and logic*, Universitext, Springer-Verlag, New York, 1994. A first introduction to topos theory; Corrected reprint of the 1992 edition. MR1300636

[16] *Are semi-direct products categorical (co)limits?*, in MathOverflow, `https://mathoverflow.net/questions/96078/are-semi-direct-products-categorical-colimits`

[17] R. W. Thomason, *Homotopy colimits in the category of small categories*, Math. Proc. Cambridge Philos. Soc. **85** (1979), no. 1, 91–109, DOI 10.1017/S0305004100055535. MR510404

Dipartimento di Matematica "Tullio Levi-Civita", Università di Padova, Via Trieste 63, 35121 Padova, Italy
Email address: `facchini@math.unipd.it`

Department of Algebra, Faculty of Mathematics and Physics, Charles University, Sokolovská 83, 18675 Praha 8, Czech Republic
Email address: `stanovsk@karlin.mff.cuni.cz`

Contemporary Mathematics
Volume **826**, 2025
https://doi.org/10.1090/conm/826/16576

On dual relative CS-Baer modules

Septimiu Crivei, Derya Keskin Tütüncü, and Rachid Tribak

ABSTRACT. We prove that every supplement submodule of a quasi-projective dual CS-Rickart module is a direct summand. We also prove that for any $\mathbb{Z}$-module M, $\mathrm{Hom}_{\mathbb{Z}}(M,\mathbb{Z})=0$ if and only if $\mathbb{Z}_{\mathbb{Z}}$ is dual M-CS-Baer (Rickart) if and only if M has no nonzero cyclic torsion-free direct summands. Finally, we describe some classes all of whose modules are dual self-CS-Baer.

1. Introduction and preliminaries

Throughout this paper, all rings will be unitary rings with identity and all modules will be unitary right modules. Let M be a module and N a submodule of M. Then $N \leq M$, $N \leq_d M$ and $N \ll M$ mean that N is a submodule of M, N is a direct summand of M and N is a small submodule of M, respectively. Also, $E(M)$ will be the injective hull of M and $\mathrm{Rad}(M)$ will be the Jacobson radical of M.

Let M and N be modules over a unitary ring. Then M is called *dual N-CS-Rickart* if the image of every homomorphism $f : N \to M$ lies above a direct summand of M in the sense that $\mathrm{Im}(f)/L$ is small in M/L for some direct summand L of M [**1**]. And M is called *dual N-CS-Baer* if for every family $(f_i)_{i\in I}$ of homomorphisms $f_i : N \to M$, $\sum_{i\in I} \mathrm{Im}(f_i)$ lies above a direct summand of M (see [**3**] for definitions and more information). Let M be a module and N a submodule of M. Then N is called a *supplement* submodule of M if there exists a submodule K of M such that $M = N + K$ and $N \cap K$ is small in N, equivalently, N is minimal with the property that $M = N + K$. In this case we also say that N is a supplement of K in M. A module M is called *lifting* if every supplement submodule of M is a direct summand of M and for every two submodules K and L of M with $M = K + L$, K has a supplement N which is contained in L. Note that M is lifting if and only if every submodule K of M lies above a direct summand of M (see [**7**] and [**11**] for definitions and more information). Recall that a module M is said to be *semiregular* if every finitely generated submodule of M lies above a direct summand of M (see, e.g., [**10**, p. 153]).

We can give some examples of dual CS-Rickart modules and dual CS-Baer modules as follows:

(i) Let R be a right hereditary right noetherian ring. Then every injective right R-module is dual self-CS-Baer by [**6**, Theorem 2.3].

2020 *Mathematics Subject Classification.* Primary 16D10; Secondary 16L30.
Key words and phrases. Dual CS-Baer module, lifting module, perfect ring, semiperfect ring.

(ii) Let R be a right hereditary ring. Then every injective right R-module is dual self-CS-Rickart by [**9**, Example 2.2(1)].
(iii) Clearly, any module which has a von Neumann regular endomorphism ring is dual self-CS-Rickart.
(iv) Let M be a nonzero module such that $\mathrm{Rad}(M) = M$ and M has no proper nonzero submodules N with $\mathrm{Rad}(N) = N$ (for example, the $\mathbb{Z}$-modules $\mathbb{Q}$ and the Prüfer p-group $\mathbb{Z}(p^\infty)$, where p is a prime integer). Let $f : M \to M$ be an endomorphism of M. Then $\mathrm{Im} f \cong M/\mathrm{Ker} f$ and since $\mathrm{Rad}(M/\mathrm{Ker} f) = M/\mathrm{Ker} f$, $\mathrm{Rad}(\mathrm{Im} f) = \mathrm{Im} f$. By hypothesis, $\mathrm{Im} f = 0$ or $\mathrm{Im} f = M$. Therefore, M is dual self-CS-Baer.
(v) Let M be a module. By definitions, it is not hard to see that M is lifting if and only if M is dual K-CS-Baer for all submodules K of M if and only if M is dual K-CS-Rickart for all submodules K of M. Therefore, $\mathbb{Z}/q^n\mathbb{Z}$ and $\mathbb{Z}(p^\infty)$ are relatively dual CS-Baer, for any prime integers p, q and any positive number n. In this direction, we can prove that for a module M, M is dual R_R-CS-Baer if and only if M is a lifting module. For, assume that M is dual R_R-CS-Baer. Let $m \in M$. Consider the homomorphism $f : R \to M$ defined by $f(r) = mr$ for all $r \in R$. Then $\mathrm{Im} f = mR$. Now, let $N \leq M$. Clearly, $N = \sum_{x \in N} xR$. Therefore, there exists a direct summand T of M such that $T \leq N$ and $N/T \ll M/T$. This means that M is lifting.
(vi) Let R be a commutative noetherian ring and let $M = mR$ be a cyclic R-module. Then the following are equivalent by [**9**, Corollary 2.10]:
    (a) M is dual self-CS-Baer.
    (b) M is dual self-CS-Rickart.
    (c) M is semiregular.
    (d) M is a lifting module.
(vii) If M and N are two modules with $\mathrm{Hom}_R(N, M) = 0$, then clearly M is dual N-CS-Baer. Therefore, for any different two maximal ideals M_1 and M_2 of a commutative noetherian ring R, $E(R/M_1)$ is dual $E(R/M_2)$-CS-Baer. Also, $\mathbb{Z}_\mathbb{Z}$ is dual $\mathbb{Q}_\mathbb{Z}$-CS-Baer, $\mathbb{Z}_\mathbb{Z}$ is dual $\mathbb{Q}/\mathbb{Z}$-CS-Baer and $\mathbb{Z}_\mathbb{Z}$ is dual $\mathbb{Z}(p^\infty)$-CS-Baer.
(viii) Let I be a finitely generated ideal of a commutative ring R. Let M be a dual self-CS-Baer R-module. By adapting the proof of [**9**, Corollary 2.30] and using [**3**, Corollary 8.2] we can deduce that there exist dual self-CS-Baer submodules K and N of M such that $M = N \oplus K$, $N = NI$ and $KI = K \cap MI \ll K$.
(ix) By adapting the proof of [**9**, Corollary 2.31], and using [**3**, Corollary 8.2] and (viii), we can deduce that there exist dual self-CS-Baer submodules N and K of M such that $M = N \oplus K$, $\mathrm{Rad}(N) = N$ and $\mathrm{Rad}(K) \ll K$, if R is a commutative semilocal ring with the finitely generated Jacobson radical $J(R)$ and if M is a dual self-CS-Baer R-module.

In this paper, we prove that every supplement submodule of a quasi-projective dual CS-Rickart module is a direct summand (Proposition 2.3). Then we give an example showing that a quasi-projective dual CS-Rickart module may not be lifting (Example 2.4). Later we investigate the relative dual CS-Baer property of the $\mathbb{Z}$-module $\mathbb{Z}$. In this direction, we prove that for any $\mathbb{Z}$-module M, $\mathrm{Hom}_\mathbb{Z}(M, \mathbb{Z}) = 0$

if and only if $\mathbb{Z}_{\mathbb{Z}}$ is dual M-CS-Baer (Rickart) if and only if M has no nonzero cyclic torsion-free direct summands (Proposition 2.7).

Following [**8**], a module M is called a *dual essentially Baer module* if for any right ideal I of $\mathrm{End}_R(M)$, $\sum_{\varphi \in I} \mathrm{Im}\,\varphi$ lies above a direct summand of M. We prove that a dual essentially Baer module is precisely the same as a dual self-CS-Baer module (Proposition 2.11). Finally, we discuss some classes whose modules are dual self-CS-Baer (Propositions 2.12 and 2.16).

2. Results

Let M and N be two R-modules. As we noticed in (vii) from the introduction, if $\mathrm{Hom}_R(N, M) = 0$, then clearly M is dual N-CS-Baer. Now, we give more examples illustrating that M and N are relatively dual CS-Baer, where $\mathrm{Hom}_R(N, M) = 0$ or $\mathrm{Hom}_R(M, N) = 0$.

EXAMPLE 2.1.

(i) Let R be a prime right Goldie ring such that R is not right primitive. Let X be a torsion-free divisible right R-module and Y a torsion semisimple right R-module. Then X and Y are relatively dual CS-Baer modules. For, $\mathrm{Hom}_R(X, Y) = \mathrm{Hom}_R(Y, X) = 0$ by [**5**, Corollary 2.16]. Therefore X is dual Y-CS-Baer and Y is dual X-CS-Baer.

(ii) Let R be prime PI-ring which is not artinian. Let X be a torsion-free divisible right R-module and Y a torsion semisimple right R-module. Then X and Y are relatively dual CS-Baer modules. For, $\mathrm{Hom}_R(X, Y) = \mathrm{Hom}_R(Y, X) = 0$ by [**5**, Corollary 2.17]. Therefore X is dual Y-CS-Baer and Y is dual X-CS-Baer.

(iii) Let R be a right Ore domain with right quotient division ring $Q \neq R$. Then R is dual Q-CS-Baer. For, $\mathrm{Hom}_R(Q, R) = 0$ by [**5**, Lemma 3.4]. Therefore R is dual Q-CS-Baer.

REMARK 2.2. Since $\mathbb{Z}/p\mathbb{Z}$ is lifting (p is any prime integer), $\mathbb{Z}/p\mathbb{Z}$ is dual $\mathbb{Z}$-CS-Baer. But $\mathrm{Hom}_{\mathbb{Z}}(\mathbb{Z}, \mathbb{Z}/p\mathbb{Z})$ is nonzero.

Now, we give the following first main result of this paper.

PROPOSITION 2.3. *Let M be a quasi-projective module. If M is dual self-CS-Rickart, then every supplement submodule of M is a direct summand of M.*

PROOF. Let N be a supplement submodule of M. Then there exists a submodule K of M such that $M = N + K$ and $N \cap K \ll N$. Namely, N is minimal with the property that $M = N + K$. Clearly, we have the natural epimorphism $\pi : M \to M/K$ and the epimorphism $\alpha : N \to M/K$ with $\mathrm{Ker}(\alpha) = N \cap K$. Since M is N-projective, we have the homomorphism $\beta : M \to N$ such that $\alpha\beta = \pi$. Therefore $\alpha\beta(M) = \pi(M) = \alpha(N)$ and so $\beta(M) + \mathrm{Ker}(\alpha) = N$. Since $\mathrm{Ker}(\alpha) = N \cap K \ll N$, $\beta(M) = N$. Therefore, we have a homomorphism $\beta : M \to M$ with $\beta(M) = N$. Since M is dual self-CS-Rickart, there exists a direct summand K' of M such that $K' \leq N$ and $N/K' \ll M/K'$. Now, $M/K' = N/K' + (K + K')/K'$. Since $N/K' \ll M/K'$, $M/K' = (K + K')/K'$ and hence $M = K + K'$. By the minimality of N, $K' = N$. Consequently, N is a direct summand of M. □

EXAMPLE 2.4. In Proposition 2.3, M does not have to be a lifting module. For, let R be a von Neumann regular ring which is not semisimple (e.g., we can take R to be the ring $\prod_{n=1}^{\infty} \mathbb{Z}/2\mathbb{Z}$). Then R_R is dual self-CS-Rickart, but it is not lifting by [**9**, Example 2.14(ii)].

In the following we deal with the indecomposable dual CS-Baer modules.

PROPOSITION 2.5. *Let M and N be two nonzero modules. Consider the following conditions:*

(i) *M is indecomposable and dual N-CS-Baer.*
(ii) *For every family of homomorphisms $\{f_i \mid f_i : N \to M\}_{i\in I}$, either $\sum_{i\in I} f_i(N) = M$ or $\sum_{i\in I} f_i(N) \ll M$.*
(iii) *M is indecomposable.*
(iv) *M is dual N-CS-Baer.*

Then (i) $\Rightarrow$ (ii) *and* (ii) $\Rightarrow$ (iv). *If there is an epimorphism $\alpha : N \to M$, then* (ii) $\Rightarrow$ (iii). *In particular, if there is an epimorphism $\alpha : N \to M$, then* (i) $\Leftrightarrow$ (ii).

PROOF. (i) $\Rightarrow$ (ii): Suppose that M is indecomposable and dual N-CS-Baer. Let $\{f_i \mid f_i : N \to M\}_{i\in I}$ be a family of homomorphisms. Since M is dual N-CS-Baer, there exists a direct summand T of M such that $T \le \sum_{i\in I} f_i(N)$ and $\sum_{i\in I} f_i(N)/T \ll M/T$. Since M is indecomposable, either $T = 0$ or $T = M$. Therefore, either $\sum_{i\in I} f_i(N) \ll M$ or $M = \sum_{i\in I} f_i(N)$.

(ii) $\Rightarrow$ (iv): Assume that for every family of homomorphisms $\{f_i \mid f_i : N \to M\}_{i\in I}$, either $\sum_{i\in I} f_i(N) = M$ or $\sum_{i\in I} f_i(N) \ll M$. Let $\{f_i \mid f_i : N \to M\}_{i\in I}$ be any family of homomorphisms. By assumption either $\sum_{i\in I} f_i(N) = M$ or $\sum_{i\in I} f_i(N) \ll M$. Therefore, M is dual N-CS-Baer.

(ii) $\Rightarrow$ (iii): Assume that $M = M_1 \oplus M_2$ for some submodules M_1 and M_2 of M. Considering the epimorphism $\alpha : N \to M$, the projection map $\pi : M \to M_1$ and the inclusion map $i : M_1 \to M$, we have the epimorphism $i\pi\alpha : N \to M_1$. By hypothesis, $(i\pi\alpha)(N) = M_1$ is either small in M or $M_1 = M$. Therefore, either $M_1 = 0$ or $M_1 = M$. This means that M is indecomposable.

The last statement is clear. □

The following is an immediate corollary of the above proposition.

COROLLARY 2.6. *Let M be a nonzero module. The following are equivalent:*

(i) *M is indecomposable and dual self-CS-Baer.*
(ii) *For every family of homomorphisms $\{f_i \mid f_i : M \to M\}_{i\in I}$, either $\sum_{i\in I} f_i(M) = M$ or $\sum_{i\in I} f_i(M) \ll M$.*

Now, we give the second main result of this paper.

PROPOSITION 2.7. *The following are equivalent for a $\mathbb{Z}$-module M:*

(i) *$\mathbb{Z}_{\mathbb{Z}}$ is dual M-CS-Baer.*
(ii) *$\mathbb{Z}_{\mathbb{Z}}$ is dual M-CS-Rickart.*
(iii) *M has no nonzero cyclic torsion-free direct summands.*
(iv) $\mathrm{Hom}_{\mathbb{Z}}(M, \mathbb{Z}) = 0$.

PROOF. (i) $\Rightarrow$ (ii): It is clear.

(ii) $\Rightarrow$ (iii): Assume that M has an element x such that $x\mathbb{Z} \le_d M$ and $\mathbb{Z}_{\mathbb{Z}} \cong x\mathbb{Z}$. Let $\pi : M \to x\mathbb{Z}$ be the projection map and let $f : x\mathbb{Z} \to \mathbb{Z}_{\mathbb{Z}}$ be an isomorphism. Then $f\pi : M \to \mathbb{Z}_{\mathbb{Z}}$ is an epimorphism. Let $\alpha \in \mathbb{Z}$ be a nonzero element which is different from 1. Consider the $\mathbb{Z}$-homomorphism $g : \mathbb{Z}_{\mathbb{Z}} \to \mathbb{Z}_{\mathbb{Z}}$ defined by $g(a) = \alpha a$ for all $a \in \mathbb{Z}$. Then $gf\pi : M \to \mathbb{Z}_{\mathbb{Z}}$ and $\mathrm{Im}(gf\pi) = \alpha\mathbb{Z}$, which is nonzero and proper. By (ii), there is a direct summand I of $\mathbb{Z}_{\mathbb{Z}}$ such that $I \leq \alpha\mathbb{Z}$ and $\alpha\mathbb{Z}/I \ll \mathbb{Z}/I$. Then either $I = 0$ or $I = \mathbb{Z}$. If $I = 0$, then $\alpha\mathbb{Z} \ll \mathbb{Z}$ and so $\alpha = 0$, a contradiciton. If $I = \mathbb{Z}_{\mathbb{Z}}$, then $\alpha\mathbb{Z} = \mathbb{Z}$, a contradiction. Therefore, M has no nonzero cyclic torsion-free direct summands.

(iii) $\Rightarrow$ (iv): Assume that there exists a nonzero homomorphism $f : M \to \mathbb{Z}_{\mathbb{Z}}$. Thus $\mathrm{Im}\, f = a\mathbb{Z}$ for some nonzero $a \in \mathbb{Z}$. Now, $M/\mathrm{Ker}\, f \cong a\mathbb{Z} \cong \mathbb{Z}$. Hence $\mathrm{Ker}\, f$ is a direct summand of M, since $M/\mathrm{Ker}\, f$ is projective. Let $M = \mathrm{Ker}\, f \oplus A$ for some submodule A of M. Then $A \cong \mathbb{Z}$, which contradicts our assumption (iii). Thus $\mathrm{Hom}_{\mathbb{Z}}(M, \mathbb{Z}) = 0$.

(iv) $\Rightarrow$ (i): It is clear. □

Now, we can give the following example.

EXAMPLE 2.8. Consider a $\mathbb{Z}$-module $M = \mathbb{Q}^{(I)} \oplus T$, where T is a torsion $\mathbb{Z}$-module and I is an index set. Suppose that $\mathbb{Z}$ is not dual M-CS-Baer. By Proposition 2.7, there is a cyclic submodule L of M such that $L \cong \mathbb{Z}$ and $L \leq_d M$. Let $M = L \oplus K$ for some submodule K of M. Since T is torsion, $T \subseteq K$. Hence, $T \leq_d K$. Let $K = T \oplus N$ for some submodule N of K. Thus, $M = L \oplus T \oplus N$. Therefore, $L \oplus N \cong \mathbb{Q}^{(I)}$. This means that $\mathbb{Z} \cong L$ is injective, a contradiction. It follows that $\mathbb{Z}$ is dual M-CS-Baer.

PROPOSITION 2.9. *Let R be a right hereditary right noetherian right max ring (e.g., R is a right hereditary right artinian ring). Let M be an R-module and let E be an injective R-module. Then the following are equivalent:*

(i) *E is dual M-CS-Rickart.*
(ii) *E is dual M-CS-Baer.*

PROOF. (i) $\Rightarrow$ (ii): Let A be a nonempty subset of $\mathrm{Hom}_R(M, E)$. By assumption, for any $f \in A$, there exist a direct summand K_f of E and a small submodule L_f of E such that $\mathrm{Im}\, f = K_f \oplus L_f$. Then $\sum_{f\in A} \mathrm{Im}\, f = \sum_{f\in A} K_f + \sum_{f\in A} L_f$. Since R is a right hereditary right noetherian ring, it follows that $K = \sum_{f\in A} K_f$ is an injective module. Moreover, $L = \sum_{f\in A} L_f \subseteq \mathrm{Rad}(E)$ and hence $L \ll E$ as R is a right max ring. Therefore, $\sum_{f\in A} \mathrm{Im}\, f$ lies above a direct summand of E (see [**11**, 41.11]). Consequently, E is dual M-CS-Baer.

(ii) $\Rightarrow$ (i): This is clear by definitions. □

EXAMPLE 2.10. For any field F and any positive integer n, the ring of all upper triangular $n \times n$ matrices with entries in F satisfies the hypotheses of the preceding proposition.

Following [**8**], a module M is called a *dual essentially Baer module* if for any right ideal I of $\mathrm{End}_R(M)$, $\sum_{\varphi\in I} \mathrm{Im}\, \varphi$ lies above a direct summand of M. Clearly, every dual self-CS-Baer is dual essentially Baer. Next, we prove that a dual essentially Baer module is precisely the same as a dual self-CS-Baer module.

PROPOSITION 2.11. *An R-module M is dual essentially Baer if and only if M is dual self-CS-Baer.*

PROOF. The sufficiency is clear. Conversely, suppose that M is dual essentially Baer, and let $S = \mathrm{End}_R(M)$. Let A be a nonempty subset of S and consider the right ideal $I = \sum_{f \in A} fS$ of S. It is easily checked that $\sum_{\varphi \in A} \operatorname{Im} \varphi = \sum_{\varphi \in I} \operatorname{Im} \varphi$. Hence $\sum_{\varphi \in A} \operatorname{Im} \varphi$ lies above a direct summand of M. Therefore M is dual self-CS-Baer. □

Finally, we give some further results on classes all of whose modules are dual self-CS-Baer. A useful result in this direction is the following one.

PROPOSITION 2.12. *Let $\mathcal{C}$ be a class of modules which is closed under binary direct sums and contains all projective modules. Then the following are equivalent:*

(i) *Every module in $\mathcal{C}$ is dual self-CS-Baer.*
(ii) *Every module in $\mathcal{C}$ is dual self-CS-Rickart.*
(iii) *Every module in $\mathcal{C}$ is lifting.*

PROOF. This follows from [**4**, Theorem 5.1] together with the fact that every lifting module is dual self-CS-Baer. □

When applied to the classes of all modules and all projective modules, it yields the following corollaries respectively.

COROLLARY 2.13. *The following are equivalent for a ring R with Jacobson radical $J(R)$:*

(i) *Every R-module is dual self-CS-Baer.*
(ii) *Every R-module is dual self-CS-Rickart.*
(iii) *Every R-module is lifting.*
(iv) *R is a left and right artinian serial ring with $J(R)^2 = 0$.*

PROOF. Use Proposition 2.12 and [**2**, 29.10]. □

COROLLARY 2.14. *The following are equivalent for a ring R:*

(i) *Every projective R-module is dual self-CS-Baer.*
(ii) *Every projective R-module is dual self-CS-Rickart.*
(iii) *Every projective R-module is lifting.*
(iv) *R is right perfect.*

PROOF. This follows by Proposition 2.12 and [**4**, Corollary 5.4]. □

In this direction we can give the following easy observation, as well.

PROPOSITION 2.15. *The following are equivalent for a ring R:*

(i) *Every finitely generated quasi-projective R-module is lifting.*
(ii) *Every finitely generated quasi-projective (free) R-module is dual self-CS-Baer.*
(iii) *R is semiperfect.*

PROOF. By definitions, [**3**, Proposition 5.9] and [**7**, Theorem 4.41]. □

Recall that a module M is called *weakly cohereditary* if every homomorphic image of M is the direct sum of a small submodule and an injective submodule (e.g., see [**4**]).

PROPOSITION 2.16. *Let R be a right noetherian ring. Then the following are equivalent:*

(i) *Every injective R-module is dual self-CS-Baer.*
(ii) *Every injective R-module is dual self-CS-Rickart.*

(iii) *Every injective R-module is weakly cohereditary.*

PROOF. (i)$\Rightarrow$(ii): This is clear.

(ii)$\Rightarrow$(iii): This follows by [**4**, Theorem 5.9].

(iii)$\Rightarrow$(i): Assume that every injective R-module is weakly cohereditary, and let M be an injective R-module. Let $(f_i)_{i\in I}$ be a family of endomorphisms of M. Since R is right noetherian, $M^{(I)}$ is injective. By hypothesis, the homomorphic image $\sum_{i\in I}\mathrm{Im}(f_i)$ of $M^{(I)}$ has a direct sum decomposition $\sum_{i\in I}\mathrm{Im}(f_i) = S\oplus E$ for some small submodule S and injective submodule E. By taking the pushout of the inclusion $\sum_{i\in I}\mathrm{Im}(f_i)\to M$ and the projection $\sum_{i\in I}\mathrm{Im}(f_i)\to S$, we obtain an induced commutative diagram:

$$
\begin{array}{ccccccccc}
 & & 0 & & 0 & & & & \\
 & & \downarrow & & \downarrow & & & & \\
 & & E & = & E & & & & \\
 & & \downarrow & & \downarrow & & & & \\
0 & \longrightarrow & \sum_{i\in I}\mathrm{Im}(f_i) & \longrightarrow & M & \longrightarrow & C & \longrightarrow & 0\\
 & & \downarrow & & \downarrow & & \| & & \\
0 & \longrightarrow & S & \longrightarrow & N & \longrightarrow & C & \longrightarrow & 0\\
 & & \downarrow & & \downarrow & & & & \\
 & & 0 & & 0 & & & &
\end{array}
$$

with exact rows and columns. Since E is injective, we have $M = E\oplus N$. It follows that $S \ll N$ by [**2**, 2.2 (5),(6)], which implies that $\sum_{i\in I}\mathrm{Im}(f_i)$ lies above the direct summand E of M. Hence M is self-CS-Baer. □

References

[1] A. N. Abyzov and T. H. N. Nhan, *CS-Rickart modules*, Lobachevskii J. Math. **35** (2014), no. 4, 317–326, DOI 10.1134/S199508021404009X. MR3284620

[2] John Clark, Christian Lomp, Narayanaswami Vanaja, and Robert Wisbauer, *Lifting modules*, Frontiers in Mathematics, Birkhäuser Verlag, Basel, 2006. Supplements and projectivity in module theory. MR2253001

[3] Septimiu Crivei, Derya Keskin Tütüncü, Simona Maria Radu, and Rachid Tribak, *CS-Baer and dual CS-Baer objects in abelian categories*, J. Algebra Appl. **22** (2023), no. 10, Paper No. 2350220, 30, DOI 10.1142/S0219498823502201. MR4641868

[4] Septimiu Crivei and Simona Maria Radu, *CS-Rickart and dual CS-Rickart objects in abelian categories*, Bull. Belg. Math. Soc. Simon Stevin **29** (2022), no. 1, 97–120, DOI 10.36045/j.bbms.210902. MR4546031

[5] Derya Keskin Tütüncü, Berke Kaleboğaz, and Patrick F. Smith, *Direct sums of semi-projective modules*, Colloq. Math. **127** (2012), no. 1, 67–81, DOI 10.4064/cm127-1-5. MR2945777

[6] Derya Keskin Tütüncü, Patrick F. Smith, and Sultan Eylem Toksoy, *On dual Baer modules*, Ring theory and its applications, Contemp. Math., vol. 609, Amer. Math. Soc., Providence, RI, 2014, pp. 173–184, DOI 10.1090/conm/609/12081. MR3204359

[7] Saad H. Mohamed and Bruno J. Müller, *Continuous and discrete modules*, London Mathematical Society Lecture Note Series, vol. 147, Cambridge University Press, Cambridge, 1990, DOI 10.1017/CBO9780511600692. MR1084376

[8] Ch. Kh. N. Nyan, *Essentially Baer modules* (Russian, with English and Russian summaries), Chebyshevskiĭ Sb. **16** (2015), no. 3, 355–375. MR3656999

[9] Rachid Tribak, *Dual CS-Rickart modules over Dedekind domains*, Algebr. Represent. Theory **23** (2020), no. 2, 229–250, DOI 10.1007/s10468-018-09845-5. MR4097313

[10] Askar Tuganbaev, *Rings close to regular*, Mathematics and its Applications, vol. 545, Kluwer Academic Publishers, Dordrecht, 2002, DOI 10.1007/978-94-015-9878-1. MR1958361

[11] Robert Wisbauer, *Foundations of module and ring theory*, Revised and translated from the 1988 German edition, Algebra, Logic and Applications, vol. 3, Gordon and Breach Science Publishers, Philadelphia, PA, 1991. A handbook for study and research. MR1144522

Department of Mathematics, Babeş-Bolyai University, Str. M. Kogălniceanu 1, 400084 Cluj-Napoca, Romania
Email address: septimiu.crivei@ubbcluj.ro

Department of Mathematics, Hacettepe University, 06800 Beytepe, Ankara, Türkiye
Email address: keskin@hacettepe.edu.tr

Centre Régional des Métiers de L'Education et de la Formation (CRMEF)-Tanger, Avenue My Abdelaziz, Souani, B.P.:3117, Tangier, Morocco
Email address: tribak12@yahoo.com

Contemporary Mathematics
Volume **826**, 2025
https://doi.org/10.1090/conm/826/16577

On the excellence for inseparable quartic extensions

Ahmed Laghribi and Diksha Mukhija

Abstract. Let F be a field of characteristic 2. When K/F is a field extension of degree two or three, it is known to be excellent for quadratic as well as bilinear forms. In this paper, we discuss the excellence property when K/F is an inseparable quartic extension.

1. Introduction

Throughout this paper F denotes a field of characteristic 2. A field extension K/F is called *excellent* for quadratic forms if for any quadratic form φ over F, the anisotropic part $(\varphi_K)_{an}$ of the K-form φ_K is defined over F, i.e., there exists a quadratic form ψ over F such that $(\varphi_K)_{an}$ is isometric to ψ_K. The same definition can be formulated for bilinear forms. Moreover, since we are in characteristic 2, we need to consider a refinement of the above definition in the case of quadratic forms (introduced in [**19**]). More precisely, the extension K/F is called (r,s)-*excellent* if for any F-quadratic form φ of type (r,s), the K-form $(\varphi_K)_{\mathrm{an}}$ is defined over F (recall that a quadratic form is of type (r,s) if it is of dimension $2r+s$ and its quasilinear part has dimension s). In particular, an extension which is excellent is necessary (r,s)-excellent for any pair of positive integers r and s. Moreover, the excellence property holds for quadratic forms of type $(0,s)$ for any positive integer s, i.e., totally singular quadratic forms and any field extension of F [**12**, Proposition 8.1(iii)]. Other known examples of excellent extensions of F are: A purely transcendental extension, an algebraic extension of odd degree [**9**, Corollary 18.5], and multiquadratic purely inseparable extensions. This last result is due to Hoffmann for quadratic and bilinear forms [**10**], [**11**].

For extensions given by function fields of quadrics, the case of 2-dimensional quadratic forms is covered by quadratic extensions, which are excellent. For conics, one knows that any extension given by the function field of a nonsingular conic is excellent for both bilinear and quadratic forms [**13**, Cor. 5.7]. Recently in [**19**] the two authors proved that the extension given by the function field a singular conic is excellent for bilinear forms, and it is neither $(2,s+1)$-excellent nor $(3+m,s)$-excellent for $m,s\geq 0$.

Our aim in this paper is to study the excellence of inseparable quartic extensions for quadratic forms. These extensions split in four types as given in Section 3. We prove in Corollary 7.2 that quartic simple purely inseparable extensions are

2020 *Mathematics Subject Classification.* Primary 11E04, 11E81.

generally not (r,s)-excellent when $r \geq 2$ and $s \geq 0$. To this end, we study the isotropy over such extensions using the work [**5**] of Aravire, Laghribi and O'Ryan. For quartic inseparable extensions which are simple but not purely inseparable, we use a transfer map that they introduce in the setting of Kato-Milne cohomology (Corollary 5.6) to prove that these extensions are generally not (r,s)-excellent when $r \geq 2$ and $s \geq 1$ (Corollary 7.2). The other two types concern biquadratic extensions. The excellence of purely inseparable biquadratic extensions is due to Hoffmann [**11**] as we said before. In [**8**] Bingöl, Chapman and the first author proved that mixed biquadratic extensions are generally not excellent. For completeness we also give the proof that any separable biquadratic extension is not (r,s)-excellent when $r \geq 2$ and $s \geq 0$ (Example 6.1). This proof is an extension to characteristic 2 of that given by Sivatski in characteristic not 2, and is based on a result of Rowen about indecomposable central simple algebras of degree 8 and exponent 2.

2. Background and preliminaries

2.1. On quadratic forms. Any quadratic form φ over F can be written up to isometry as follows:

$$\varphi \simeq [a_1,b_1] \perp [a_2,b_2] \perp \ldots \perp [a_r,b_r] \perp \langle c_1\rangle \perp \ldots \perp \langle c_s\rangle, \tag{2.1}$$

where $\simeq$ and $\perp$ denote the isometry and orthogonal sum of quadratic forms, and $[a,b]$ (*resp.* $\langle a\rangle$) denotes the quadratic form $ax^2+xy+by^2$ (*resp.* ax^2). The quadratic form $\langle c_1\rangle \perp \ldots \perp \langle c_s\rangle$ is unique up to isometry, we call it the quasilinear part of φ, and denote it by $\mathrm{ql}(\varphi)$. As in equation (2.1), the form φ is called of type (r,s). We say that φ is:

- nonsingular (*resp.* totally singular) if $s=0$ (*resp.* $r=0$),
- singular if $s>0$.

For $a_1,\ldots,a_n \in F$, let $\langle a_1,\ldots,a_n\rangle$ denote the totally singular quadratic form $\langle a_1\rangle \perp \ldots \perp \langle a_n\rangle$.

The Arf invariant $\Delta(\varphi)$ of a nonsingular form $\varphi \simeq [a_1,b_1] \perp \ldots \perp [a_r,b_r]$ is defined to be $\sum_{i=1}^r a_ib_i + \wp(F)$ in $F/\wp(F)$, where $\wp(F)=\{\alpha^2+\alpha \mid \alpha\in F\}$.

An Albert form is a nonsingular quadratic form of dimension 6 and trivial Arf invariant.

A quadratic form φ of underlying F-vector space V is called isotropic if there exists $v \in V\setminus\{0\}$ such that $\varphi(v)=0$, otherwise φ is called anisotropic.

For an integer $n\geq 0$ and φ a quadratic form, let $n\times\varphi$ denote the quadratic form $\underbrace{\varphi \perp \ldots \perp \varphi}_{n \text{ times}}$.

Recall that any quadratic form φ over F uniquely decomposes as follows:

$$\varphi \simeq \varphi_{an} \perp i\times[0,0] \perp j\times\langle 0\rangle,$$

where φ_{an} is an anisotropic quadratic form. We call φ_{an} the anisotropic part of φ, and the integer i (*resp.* j) is called the Witt index (*resp.* the defect index) of φ. We denote i and j by $i_W(\varphi)$ and $i_d(\varphi)$, respectively. The form $\varphi_{an} \perp i\times[0,0]$ is called the nondefective part of φ.

Two quadratic forms φ_1 and φ_2 are called Witt-equivalent, denoted by $\varphi_1 \sim \varphi_2$, if there exist $m, n \in \mathbb{N}$ such that $\varphi_1 \perp m \times [0,0] \simeq \varphi_2 \perp n \times [0,0]$.

A quadratic form φ of underlying vector space V represents a scalar $\alpha \in F$ if there exists $v \in V$ such that $\varphi(v) = \alpha$. We denote by $D_F(\varphi)$ the set of scalars in $F^* := F \setminus \{0\}$ represented by φ.

For $a, b \in F$ such that $b \neq 0$, let $[a, b)$ be the F-quaternion algebra generated by two elements i and j subject to the relations: $i^2 + i = a$, $j^2 = b$ and $jij^{-1} = i + 1$.

For a quadratic form φ, we denote by $C(\varphi)$ its Clifford algebra. Explicitly, if φ is nonsingular given as follows $\varphi \simeq a_1[1, b_1] \perp \cdots \perp a_r[1, b_r]$ for $a_i, b_i \in F$ such that $a_i \neq 0$ for any $1 \leq i \leq r$, then $C(\varphi)$ is isomorphic to the tensor product $[b_1, a_1) \otimes_F \cdots \otimes_F [b_r, a_r)$.

We will need the following cancellation result:

PROPOSITION 2.1 ([**16**, Proposition 1.2] for (1), [**12**, Lemma 2.6] for (2)). *Let φ_1, φ_2 be two quadratic forms (possibly singular). Suppose that one of the following conditions holds:*

(1) *$\varphi_1 \perp \psi \simeq \varphi_2 \perp \psi$ for some nonsingular form ψ.*
(2) *$\varphi_1 \perp s \times \langle 0 \rangle \simeq \varphi_2 \perp s \times \langle 0 \rangle$ for some integer $s \geq 0$ such that $i_d(\varphi_1) = i_d(\varphi_2) = 0$.*

Then, $\varphi_1 \simeq \varphi_2$.

We recall the definition of domination of quadratic forms that will often appear in the upcoming sections [**12**, Definition 3.4].

DEFINITION 2.2. Let φ and ψ be F-quadratic forms of underlying vector spaces V and W, respectively. We say that φ is dominated by ψ if there exists an injective isometry $(V, \varphi) \longrightarrow (W, \psi)$. In this case, we write $\varphi < \psi$.

For $p \in F[x_1, \ldots, x_n]$ an irreducible polynomial, let $F(p)$ be the field of fractions of the quotient ring $F[x_1, \ldots, x_n]/(p)$. We call it the function field of p. We take $F(p) = F$ when p is reducible.

Recall that a polynomial given by an F-quadratic form φ is reducible iff the nondefective part of φ is either of type $(0, 1)$, or of type $(1, 0)$ and isometric to $[0, 0]$.

For a field extension K/F and a quadratic form φ over F, let φ_K denote the quadratic form φ viewed as a quadratic form over K by scalar extension.

Let $W(F)$ (*resp.* $W_q(F)$) be the Witt ring of regular symmetric F-bilinear forms (*resp.* the Witt group of nonsingular F-quadratic forms). The group $W_q(F)$ is endowed with a $W(F)$-module structure as follows. For B (*resp.* φ) a regular F-bilinear form of underlying vector space V (*resp.* a nonsingular F-quadratic form of underlying vector space W), we associate a nonsingular quadratic form $B \otimes \varphi$ defined on $V \otimes_F W$ by:

$$B \otimes \varphi(v \otimes w) = B(v, v)\varphi(w) \text{ for any } (v, w) \in V \times W \tag{2.2}$$

and whose polar form is $B \otimes B_\varphi$, where B_φ is the polar form of φ.

For $a_1, \ldots, a_n \in F^*$, let $\langle a_1, \ldots, a_n \rangle_b$ be the diagonal bilinear form defined by:

$$((x_1, \ldots, x_n), (y_1, \ldots, y_n)) \mapsto \sum_{i=1}^{n} a_i x_i y_i.$$

An n-fold bilinear Pfister form is a bilinear form of type $\langle 1, a_1 \rangle_b \otimes \cdots \otimes \langle 1, a_n \rangle_b$ for some $a_i \in F^*$; we write $\langle\langle a_1, \cdots, a_n \rangle\rangle_b$ for short. An $(n+1)$-fold quadratic

Pfister form is a nonsingular quadratic form of type $\langle\langle a_1, \cdots, a_n\rangle\rangle_b \otimes [1, b]$ for some $a_i \in F^*$, $b \in F$, we write $\langle\langle a_1, \cdots, a_n; b]]$ for short.

Any quadratic Pfister form π is round, i.e., $x \in F^*$ is represented by π if and only if $\pi \simeq x\pi$. The same fact is true for bilinear Pfister forms. The set of quadratic forms isometric (*resp.* similar) to n-fold quadratic Pfister forms will be denoted by P_nF (*resp.* GP_nF).

DEFINITION 2.3. A quadratic form φ is said to be a Pfister neighbor if there exist a quadratic Pfister form π and $a \in F^*$ such that $a\varphi \prec \pi$ and $2\dim\varphi > \dim\pi$.

We summarize in the following proposition some properties of Pfister neighbors.

PROPOSITION 2.4 ([**17**, Proposition 3.1]).

(1) *If $\pi \in P_nF$ and φ are anisotropic. Then, φ is a Pfister neighbor of π if and only if $\dim\varphi > 2^{n-1}$ and $\pi_{F(\varphi)}$ is isotropic.*

(2) *If φ is a Pfister neighbor of π, then we have the following:*
   (a) *φ is isotropic if and only if π is also isotropic.*
   (b) *π is uniquely determined by φ up to isometry.*

For any integer $n \geq 1$, let I^nF be the n-th power of the fundamental ideal IF of $W(F)$. We take $I^0F = WF$. For any integer $n \geq 0$, let $I_q^{n+1}F$ denote the $W(F)$-submodule of $W_q(F)$ given by $I^nF \otimes W_q(F)$. This submodule is additively generated by forms similar to $(n+1)$-fold quadratic Pfister forms.

For any field extension L/F, we denote by $W_q(L/F)$ the kernel of the homomorphism $W_q(F) \to W_q(L)$ induced by scalar extension. Similarly, let $I_q^n(L/F)$ (*resp.* $\overline{I}_q^n(L/F)$) be the kernel of the homomorphism $I_q^nF \longrightarrow I_q^nL$ (*resp.* $\overline{I}_q^nF \longrightarrow \overline{I}_q^nL$) induced by scalar extension.

We recall the following result known as the Hauptsatz of Arason and Pfister:

THEOREM 2.5 ([**6**]). *Let φ be a nonzero anisotropic quadratic form that belongs to I_q^nF. Then, $\dim\varphi \geq 2^n$.*

2.2. On differential forms. For any integer $n \geq 1$, let $\Omega_F^n = \overset{n}{\wedge}\,\Omega_F^1$ where Ω_F^1 is the F-vector space of absolute 1-differential forms, i.e., the F-vector space generated by the symbols $\mathrm{d}\,a$ with $a \in F$ subject to the relations: $\mathrm{d}(a+b) = \mathrm{d}\,a + \mathrm{d}\,b$ and $\mathrm{d}(ab) = a\,\mathrm{d}\,b + b\,\mathrm{d}\,a$ for any $a, b \in F$. We set $\Omega_F^0 = F$. The differential operator $\mathrm{d} : \Omega_F^n \longrightarrow \Omega_F^{n+1}$ is given by: $a\,\mathrm{d}\,a_1 \wedge \cdots \wedge \mathrm{d}\,a_n \mapsto \mathrm{d}\,a \wedge \mathrm{d}\,a_1 \wedge \cdots \wedge \mathrm{d}\,a_n$. The Artin-Schreier operator $\wp : \Omega_F^n \longrightarrow \Omega_F^n/\mathrm{d}\,\Omega_F^{n-1}$ is defined by: $a\frac{\mathrm{d}\,a_1}{a_1} \wedge \cdots \wedge \frac{\mathrm{d}\,a_n}{a_n} \mapsto (a^2 - a)\frac{\mathrm{d}\,a_1}{a_1} \wedge \cdots \wedge \frac{\mathrm{d}\,a_n}{a_n} + \mathrm{d}\,\Omega_F^{n-1}$. Let $H_p^{n+1}(F)$ denote the cokernel of $\wp$. Then, we have

$$H_p^{n+1}(F) = \Omega_F^n/(\wp(\Omega_F^n) + \mathrm{d}\,\Omega_F^{n-1}).$$

To translate some results from differential forms to quadratic forms and vice-versa, we will use the Kato's isomorphism [**15**]:

THEOREM 2.6. *Let F be a field of characteristic 2. For every integer $n \geq 1$ there exists an isomorphism:*

$$f_n : \overline{I}_q^nF \to H_2^{n+1}(F)$$

$$\overline{\langle\langle a_1, \ldots, a_n; b]]} \mapsto \overline{b\frac{\mathrm{d}\,a_1}{a_1} \wedge \ldots \wedge \frac{\mathrm{d}\,a_n}{a_n}}. \tag{2.3}$$

2.3. Preliminaries. In this section, we recall some results to be useful in the upcoming sections.

PROPOSITION 2.7 ([**1**, Cor. 2.8]). *Let φ be an anisotropic nonsingular F-quadratic form and $K = F(\sqrt{b})$ for $b \in F \setminus F^2$. Then φ_K is hyperbolic if and only if φ is divisible by $\langle 1, b\rangle_b$.*

PROPOSITION 2.8. *Let L be a field extension of F, and $z \in F$. Let $\langle\langle x, y]]$ be an anisotropic 2-fold Pfister form over F. We have:*

(1) The following two statements are equivalent:

(a) *$x[1, y] \perp \langle 1, z\rangle$ is isotropic over L.*

(b) *either $z \in L^2$, or $z \notin L^2$ and $\langle\langle x, y]]_{L(\sqrt{z})}$ is isotropic.*

(2) If L/F is a purely inseparable extension, then statements (a) and (b) are equivalent to the following:

(c) *either $z \in L^2$, or $z \notin L^2$ and there exists $t \in F$ such that the Albert form $x[1, y] \perp z[1, t] \perp [1, y + t]$ is hyperbolic over L.*

PROOF. (1) Let $\varphi = x[1, y] \perp \langle 1, z\rangle$.

(b) $\Longrightarrow$ (a) If $z \in L^2$, then $\langle 1, z\rangle_L$ is isotropic, and thus φ_L is isotropic. So suppose $z \notin L^2$ and $\langle\langle x, y]]_{L(\sqrt{z})}$ is isotropic. We have $\langle\langle x, y]]_{L(\sqrt{z})} \sim 0$.

(i) If $\langle\langle x, y]]_L$ is isotropic, then φ_L is isotropic because the Pfister neighbor $(x[1, y] \perp \langle 1\rangle)_L$ is isotropic and dominated by φ_L.

(ii) If $\langle\langle x, y]]_L$ is anisotropic, then we get by Proposition 2.7 and the roundness of Pfister forms that $\langle\langle x, y]]_L \simeq \langle\langle z, t]]$ for some $t \in L$. Adding the form $\langle 1, z\rangle_L$ on both sides of the previous isometry, we get $\varphi_L \perp [0, 0] \simeq 2 \times [0, 0] \perp \langle 1, z\rangle_L$. Canceling a hyperbolic plane (Proposition 2.1) implies that φ_L is isotropic.

(a) $\Longrightarrow$ (b) Suppose that φ_L is isotropic. If $i_d(\varphi_L) = 1$, then $\langle 1, z\rangle_L$ is isotropic and $z \in L^2$. If $i_d(\varphi_L) = 0$, then $\varphi_L \simeq [0, 0] \perp \langle 1, z\rangle_L$. Extending to $L(\sqrt{z})$ yields

$$(x[1, y] \perp \langle 1\rangle)_L \perp \langle 0\rangle \simeq [0, 0] \perp \langle 1\rangle_L \perp \langle 0\rangle .$$

Canceling the form $\langle 0\rangle$ (Proposition 2.1), we get $(x[1, y] \perp \langle 1\rangle)_L \simeq [0, 0] \perp \langle 1\rangle_L$. Hence, $\langle\langle x, y]]_{L(\sqrt{z})}$ is isotropic.

(2) Now we assume that L/F is purely inseparable.

(b) $\Longrightarrow$ (c) Suppose that $z \notin L^2$ and $\langle\langle x, y]]_{L(\sqrt{z})}$ is isotropic. Without loss of generality, we may suppose that L/F is of finite degree. As we did in the proof of the implication (b) $\Longrightarrow$ (a), there exists $t \in L$ such that $\langle\langle x, y]]_L \simeq \langle\langle z, t]]$ (we take $t = 0$ when $\langle\langle x, y]]_L$ is isotropic). Since L/F is purely inseparable of finite degree, there exists an integer $m \geq 1$ such that $L^{2^m} \subset F$. Hence, we may suppose $t \in F$ because $[1, t] \simeq [1, t^{2^m}]$. Consequently, $(\langle\langle x, y]] \perp \langle\langle z, t]])_L$ is hyperbolic, in particular $(x[1, y] \perp z[1, t] \perp [1, y + t])_L$ is hyperbolic.

(c) $\Longrightarrow$ (b) Suppose that the Albert form $(x[1, y] \perp z[1, t] \perp [1, y + t])_L$ is hyperbolic for some $t \in F$. Passing to $L(\sqrt{z})$ and using the fact that

$$(z[1, t] \perp [1, y + t])_{L(\sqrt{z})} \simeq ([1, t] \perp [1, y + t])_{L(\sqrt{z})} \simeq ([0, 0] \perp [1, y])_{L(\sqrt{z})},$$

we deduce that $(x[1, y] \perp [0, 0] \perp [1, y])_{L(\sqrt{z})} \simeq 3 \times [0, 0]$. It follows from Proposition 2.1(1) that $(x[1, y] \perp [1, y])_{L(\sqrt{z})} \simeq 2 \times [0, 0]$, and thus $\langle\langle x, y]]_{L(\sqrt{z})}$ is isotropic. $\square$

We recall the following result which is well known:

LEMMA 2.9. *Let L/F be a purely inseparable field extension and $x \in F$. If $x \in \wp(L)$, then $x \in \wp(F)$. In particular, if φ is an F-quadratic form such that $\Delta(\varphi_L) = 0$, then $\Delta(\varphi) = 0$.*

PROOF. We may reduce to the case where L/F is finite. Hence, $L^{2^n} \subset F$ for some integer $n \geq 1$. Since $x \in \wp(L)$, we have $x^{2^n} \in \wp(F)$ and thus there exists $y \in F$ such that $x^{2^n} = y^2 + y$. Since $x^{2^n}, y^2 \in F^2$, we get $y \in F^2$. Hence, $y = z^2$ for some $z \in F$. This implies that $x^{2^{n-1}} = z^2 + z \in \wp(F)$. If $n = 1$, then $x \in \wp(F)$ and we are done. If not, we repeat the same argument until we get $x \in \wp(F)$. □

3. Isotropy of quadratic forms of dimension ≤ 4 over quartic extensions

Let us recall that an inseparable quartic field extension K of F is one of the following mutually exclusive types (we refer to [**14**, Section 3] for more details):

(Type 1) *Simple purely inseparable:* $K = F(\sqrt[4]{b})$ for some $b \in F \setminus F^2$.
(Type 2) *Simple but not purely inseparable:* $K = F(\alpha)$ such that α is a root of an irreducible polynomial $x^4 + ax^2 + b \in F[x]$ such that $\langle 1, a, b\rangle$ is anisotropic over F.
(Type 3) *Purely inseparable biquadratic:* $K = F(\sqrt{d_1}, \sqrt{d_2})$ for $d_1, d_2 \in F \setminus F^2$.
(Type 4) *Mixed biquadratic:* $K = F(\sqrt{b}, \alpha)$ such that $b \in F \setminus F^2$ and α is a root of an irreducible polynomial $x^2 + x + a \in F[x]$.

In this section we discuss the isotropy of quadratic forms of small dimension over quartic extensions of type 1. This will aid us in inspecting whether this field extension is excellent or not for quadratic forms. As we will see, we are mostly interested in the isotropy of nonsingular quadratic forms of dimension 4, but for completeness we consider the isotropy for all quadratic forms of dimension at most 4. Recall that quartic extensions of type 3 are excellent by a result of Hoffmann [**11**].

For the rest of this section, we fix $K = F(\sqrt[4]{b})$ such that $b \in F \setminus F^2$. We recall some results that will help us to classify F-quadratic forms of dimension ≤ 4 that become isotropic over K.

LEMMA 3.1 ([**17**, Lemma 4.3]). *Let φ be a nonsingular quadratic form of dimension 4 and $a \in F$ such that $\Delta(\varphi) = a + \wp(F)$. Let $L = F(\wp^{-1}(a))$. Then, φ_L is isotropic if and only if φ is isotropic.*

We deduce from the general result [**5**, Theorem 1.4] the following theorem:

THEOREM 3.2. *Let K be the quartic extension as above, and $c \in F \setminus \wp(F)$. Then, we have:*

$$H_2^2(K(\wp^{-1}(c))/F) = H_2^2(K/F) + \overline{cG_K^1(1)}, \tag{3.1}$$

where $G_K^1(1)$ is the subgroup of Ω_F^1 additively generated by the differentials $d_{x,y} := \dfrac{x^2}{x^2+by^2}\dfrac{\mathrm{d}x}{x} + \dfrac{by^2}{x^2+by^2}\dfrac{\mathrm{d}y}{y}$ for $x, y \in F^*$.

This theorem can be rewritten as follows:

COROLLARY 3.3. *Let K be the quartic extension as above, and $c \in F \setminus \wp(F)$. Then, we have:*

$$H_2^2(K(\wp^{-1}(c))/F) = H_2^2(K/F) + \overline{c\nu_F(1)} + \overline{c\langle d_{x,1} \mid x \in F^*\rangle}. \tag{3.2}$$

PROOF. It suffices to prove that $\nu_F(1) + \langle d_{x,1} \mid x \in F^* \rangle = G_K^1(1)$. In fact, the group $\nu_F(1)$ is contained in $G_K^1(1)$ because $d_{x,x} = \frac{\mathrm{d}\,x}{x}$ for all $x \in F^*$. Conversely, let $x, y \in F^*$. We have

$$\begin{aligned} d_{x,y} &= \frac{x^2}{x^2+by^2}\frac{\mathrm{d}\,x}{x} + \frac{by^2}{x^2+by^2}\frac{\mathrm{d}\,y}{y} \\ &= \frac{(xy^{-1})^2}{(xy^{-1})^2+b}\frac{\mathrm{d}(xy^{-1})}{xy^{-1}} + \frac{(xy^{-1})^2}{(xy^{-1})^2+b}\frac{\mathrm{d}(y^{-1})}{y^{-1}} + \frac{b}{(xy^{-1})^2+b}\frac{\mathrm{d}\,y}{y} \\ &= \frac{(xy^{-1})^2}{(xy^{-1})^2+b}\frac{\mathrm{d}(xy^{-1})}{xy^{-1}} + \frac{\mathrm{d}\,y}{y} \in \nu_F(1) + \langle d_{x,1} \mid x \in F^* \rangle . \qquad \square \end{aligned}$$

Now applying the Kato's isomorphism (2.3) to equation (3.2), we obtain:

COROLLARY 3.4. *Let K be the quartic extension as above, $c \in F \setminus \wp(F)$. Then:*

$$\overline{I}_q^2(K(\wp^{-1}(c))/F) = \overline{I}_q^2(K/F) + \overline{IF \otimes [1,c]} + \overline{\sum_i \langle 1, x_i\rangle_b \otimes \left[1, \frac{cx_i^2}{x_i^2+b}\right]}, \tag{3.3}$$

where $x_i \in F^*$.

To complete the kernels given in Corollary 3.4, we recall the kernels $\overline{I}_q^2(K/F)$ and $I_q^2(K/F)$ studied by Sobiech:

THEOREM 3.5. *Let K be the quartic extension as above. We have:*

(1) *([**22**, Corollary 5.2]) The kernel $\overline{I}_q^2(K/F)$ is additively generated by the classes $\overline{\langle\langle u, bu^2]]}$ and $\overline{\langle\langle b, v]]}$ for $u, v \in F^*$.*
(2) *([**22**, Theorem 5.3, Remark 5.5]) The kernel $I_q^2(K/F)$ coincides with the $W(F)$-module generated by the forms $\langle\langle u, bu^2]]$ and $\langle\langle b, v]]$ for $u, v \in F^*$.*

Now we treat the isotropy of F-quadratic forms of dimension ≤ 4 over K.

THEOREM 3.6. *Let φ be an anisotropic F-quadratic form of dimension ≤ 4, and let K be as mentioned above. Then, φ_K is isotropic if and only if one of the following conditions holds:*

(1) *φ is totally singular and $\alpha\langle 1, b\rangle \prec \varphi$ for some $\alpha \in F^*$.*
(2) *φ is a Pfister neighbor of a 2-fold Pfister form which is isotropic over K.*
(3) *$\varphi \simeq x[1,y] \perp \langle 1, z\rangle$ for some $x \neq 0, y, z \in F$ such that either $z \in F^2(b)$, or $z \notin F^2(b)$ and there exists $t \in F$ such that the Albert form $x[1,y] \perp z[1,t] \perp [1, y+t]$ is hyperbolic over K.*
(4) *φ is of type $(2,0)$ and there exist $\psi \in I_q^2(K/F)$ and $x_1, \ldots, x_s, y \in F$ for some integer $s \geq 1$ such that*

$$\varphi \perp y[1,c] \perp \psi \perp \sum_{i=1}^{s} \langle 1, x_i\rangle_b \otimes \left[1, \frac{cx_i^2}{x_i^2+b}\right] \in I_q^3F, \tag{3.4}$$

where $c \in F$ satisfies $\Delta(\varphi) = c + \wp(F)$.

PROOF. When φ is of type $(2,0)$, we take $c \in F$ such that $\Delta(\varphi) = c + \wp(F)$, and consider the separable quadratic extension $F(\alpha)$ given by $\alpha^2 + \alpha = c$.

– If φ is totally singular, then φ_K is isotropic iff $\alpha\langle 1, b\rangle \prec \varphi$ for some $\alpha \in F^*$ [**18**, Lemma 2.10]. For the rest of the proof we suppose that φ is not totally singular.

– If φ is of type $(1,0)$ and φ_K is isotropic, then $\Delta(\varphi_K) = 0$. Lemma 2.9 implies $\Delta(\varphi) = 0$, i.e., φ is isotropic, which is excluded.

– If φ is of type $(1,1)$. Modulo a scalar we may suppose $\varphi \simeq x[1,y] \perp \langle 1\rangle$ for some $x, y \in F^*$. Since φ is a Pfister neighbor of $\pi := \langle\langle x, y]]$, it follows that φ_K is isotropic iff π_K is isotropic (Proposition 2.4).

– If φ is of type $(1,2)$, then the statement is a consequence of Proposition 2.8.

– Now suppose that φ is of type $(2,0)$ and φ_K isotropic. It follows that $\varphi_{K(\alpha)}$ is hyperbolic since $\Delta(\varphi_{F(\alpha)}) = 0$. In particular, we have $(\varphi \perp [1,c])_{K(\alpha)} \sim 0$, and thus $\varphi \perp [1,c] + I_q^3F \in \overline{I}_q^2(K(\alpha)/F)$. We now use Corollary 3.4 with Theorem (2) to get:

$$\varphi \perp [1,c] \perp \psi \perp \sum_{i=1}^{s} \langle 1, x_i\rangle_b \otimes \left[1, \frac{cx_i^2}{x_i^2 + b}\right] \perp \rho \otimes [1,c] \in I_q^3F,$$

where $\psi \in I_q^2(K/F)$ and $\rho \in IF$. Let $y \in F^*$ be such that $\det\rho = yF^{*2}$ and use $\rho \perp \langle 1, y\rangle_b \in I^2F$ to get:

$$\varphi \perp y[1,c] \perp \psi \perp \sum_{i=1}^{s} \langle 1, x_i\rangle_b \otimes \left[1, \frac{cx_i^2}{x_i^2 + b}\right] \in I_q^3F.$$

Conversely, if φ satisfies statement (4), then extending the equation (3.4) to $K(\alpha)$ yields $\varphi_{K(\alpha)} \sim 0$ by the Hauptsatz. Hence, Lemma 3.1 implies that φ_K is isotropic. □

REMARK 3.7. In the formulation of the statements (2), (3) and (4) of Theorem 3.6, we use some forms satisfying the hyperbolicity hypothesis over K. These forms are classified by Theorem 3.5.

4. Nonexcellence of quartic extensions of type 1

Our inspiration to build an example of a non-excellent quartic simple purely inseparable extension K/F is motivated by statement (4) of Theorem 3.6 where some forms in $P_2(F)$ that become hyperbolic over K appear in equation (3.4). Starting with such a 2-fold Pfister form and modifying the Arf invariant, we get our example of nonexcellence as follows:

EXAMPLE 4.1. Let F_0 be a field of characteristic 2 and x, y, z be variables over F_0. Over the field $F := F_0(x, y, z)$, let us consider the F-quadratic form:

$$\varphi = \left[1, z^2 + \frac{z^2x^2}{x^2 + y}\right] \perp x\left[1, \frac{z^2x^2}{x^2 + y}\right].$$

Then, for $K = F(\sqrt[4]{y})$ we have the following:

(1) φ is anisotropic over F.
(2) φ is isotropic over K.
(3) $(\varphi_K)_{\mathrm{an}}$ is not defined over F.

In particular, the extension K/F is not $(2,0)$-excellent.

PROOF. (1) Let us take $u = \frac{z^2x^2}{x^2+y}$. Then, $\varphi = [1, z^2 + u] \perp x[1, u]$ viewed as a quadratic form over $F_0(x, y, z) = F_0(x, u, z)$ (obviously, x, u, z remain variables over F_0). Since the forms $[1, z^2 + u]$ and $[1, u]$ are anisotropic over $F_0(u, z)$, we deduce that φ is anisotropic over $F = F_0(x, u, z)$.

(2) Extending φ to K, we get

$$\begin{aligned}\varphi_K &= \left[1, z^2 + \frac{z^2x^2}{x^2+y}\right]_K \perp x\left[1, \frac{z^2x^2}{x^2+y}\right]_K \\ &\simeq \left[1, z + \frac{zx}{x+\sqrt{y}}\right] \perp x\left[1, \frac{zx}{x+\sqrt{y}}\right] \\ &\sim [1,z]_K \perp \langle 1, x\rangle_b \otimes \left[1, \frac{zx}{x+\sqrt{y}}\right].\end{aligned}$$

Using the fact that $\sqrt{y} \in K^2$ and [**2**, Lemma 2.1], we get

$$\begin{aligned}\varphi_K &\sim [1,z]_K \perp \langle 1, x+\sqrt{y}\rangle_b \otimes [1,z] \\ &\sim ((x+\sqrt{y})[1,z])_K.\end{aligned}$$

Moreover, the anisotropy of $[1,z]$ over F implies that $[1,z]_K$ is anisotropic since K/F is purely inseparable. Hence, φ_K is isotropic and $(\varphi_K)_{\text{an}} \simeq ((x+\sqrt{y})[1,z])_K$.

(3) Suppose that $(\varphi_K)_{\text{an}}$ is defined over F. Then, there exists $\alpha \in F^*$ such that

$$(\langle 1, x+\sqrt{y}\rangle_b \otimes [1,z])_K \sim (\alpha[1,z])_K. \tag{4.1}$$

Consider the F-linear map $s : K \to F$ given by: $1 \mapsto 0$, $\sqrt[4]{y} \mapsto 0$, $(\sqrt[4]{y})^3 \mapsto 0$ and $\sqrt{y} \mapsto 1$. We compute $C_1 := s_*(\langle \alpha\rangle_b)$ and $C_2 := s_*(\langle x+\sqrt{y}\rangle_b)$ with respect to the ordered F-basis $\{1, \sqrt[4]{y}, \sqrt{y}, (\sqrt[4]{y})^3\}$ of K, we obtain

$$C_1 = \begin{pmatrix} 0 & \alpha & 0 & 0 \\ \alpha & 0 & 0 & 0 \\ 0 & 0 & \alpha & 0 \\ 0 & 0 & 0 & \alpha y\end{pmatrix}, \quad C_2 = \begin{pmatrix} 1 & x & 0 & 0 \\ x & y & 0 & 0 \\ 0 & 0 & x & y \\ 0 & 0 & y & xy\end{pmatrix}.$$

Applying the transfer map s_* to (4.1) and using the Frobenius reciprocity, we get $C_1 \otimes [1,z] \simeq C_2 \otimes [1,z]$. Since $C_1 \sim \begin{pmatrix} \alpha & 0 \\ 0 & \alpha y\end{pmatrix}$, it follows that $C_2 \otimes [1,z]$ is isotropic over F. But C_2 is anisotropic over F since its corresponding totally singular form $\widetilde{C_2} \simeq \langle 1, y, x, xy\rangle$ is anisotropic over F. Since C_2 is defined over $F_0(x,y)$, the form $C_2 \otimes [1,z]$ remains anisotropic over $F = F_0(x,y)(z)$, which is a contradiction. This finishes the proof of the example. □

5. Nonexcellence of quartic extensions of type 2

In this section, we build an example of a quartic extension K/F of type 2 which is not excellent. Our example will not be inspired from the isotropy problem as for the quartic extensions of type 1, because we have no answer to the isotropy of nonsingular F-quadratic forms of dimension 4 over inseparable quartic extensions of type 2. For this reason our example will be given in dimension 5 using a transfer argument. More precisely, starting from the usual trace map of the separable quadratic extension of F contained in K, we first construct in the setting of Kato-Milne cohomology a transfer map from K to a quadratic inseparable extension of F. This construction is stated in a general setting in the subsection 5.1 below.

5.1. A generalized transfer map. Let $K = F(\alpha)$ be an extension of F which is inseparable but not purely inseparable. One knows that $\min(\alpha, F) = g(x^{2^l})$, where $l \geq 1$ and $g(x) \in F[x]$ is a separable polynomial of degree $n \geq 2$. Let us write $g(x) = x^n + a_{n-1}x^{n-1} + \cdots + a_1 x + a_0$ and let $L = F^2(a_0, \ldots, a_{n-1})$ with $[L : F^2] = 2^d$. Then, there exist $0 \leq i_1 < i_2 < \ldots < i_d \leq n-1$ such that $L = F^2(a_{i_1}, \ldots, a_{i_d})$, and thus $a_{i_1}, \ldots, a_{i_d}$ are 2-independent. Moreover, we suppose that n is even. Note that there exists an odd integer $i \in \{1, \ldots, n-1\}$ such that $a_i \neq 0$ (because $g(x)$ is separable). Let $\beta = \alpha^{2^l}$.

We thus have a sequence of fields $F \subset M := F(\beta) \subset K = F(\alpha) = M(\sqrt[2^l]{\beta})$, where M/F is separable and K/M is purely inseparable. Moreover, we choose a 2-basis $\mathcal{B}$ of F such that $\mathcal{B} = \{e_1 = a_{i_1}, \ldots, e_d = a_{i_d}, \cdots\} = \{e_i \mid i \in I\}$ for a suitable set I. Since M/F is separable, $\mathcal{B}$ remains a 2-basis of M. Since $a_{i_1}, \ldots, a_{i_d}$ are 2-dependent over K, we may choose $\mathcal{C} := (\mathcal{B} \setminus \{a_{i_1}\}) \cup \{\alpha\}$ as a 2-basis of K. We have $\Omega_M^m = \Omega_F^m \oplus \bigoplus_{1 \leq i \leq n-1} \beta^i \Omega_F^m$. In fact, since M/F is separable we can replace β by β^2, that is, $\Omega_M^m = \Omega_F^m \oplus \bigoplus_{1 \leq i \leq n-1} \beta^{2i} \Omega_F^m$.

NOTATION 5.1. (1) Let $\mathcal{L}_F^m$ be the subgroup of Ω_F^m whose differentials are free from $\mathrm{d}\, a_{i_1}$.

(2) Clearly, $(\Omega_M^m)_K = \{(w_0 + \sum_{i=1}^{n-1} \beta^{2i} w_i)_K \mid w_i \in \mathcal{L}_F^m\}$. We denote the group $(\Omega_M^m)_K$ by $\widehat{\Omega}_K^m$.

(3) Let S denote the field $F(\sqrt{a_{i_2}}, \cdots, \sqrt{a_{i_d}})$.

We then have the following result:

LEMMA 5.2. *Let S and K be as mentioned above. Then, there exists a well-defined F-linear map $s : \widehat{\Omega}_K^m \to (\Omega_F^m)_S$ given by:*

$$\left(w_0 + \sum_{i=1}^{n-1} \beta^{2i} w_i\right)_K \mapsto \left(\sum_{i=1}^{n-1} \mathrm{Tr}(\beta^{2i}) w_i\right)_S$$

where $w_i \in \mathcal{L}_F^m$ for all $0 \leq i \leq n-1$.

PROOF. Suppose $(w_0 + \sum_{i=1}^{n-1} \beta^{2i} w_i)_K = (w_0' + \sum_{i=1}^{n-1} \beta^{2i} w_i')_K$, where $w_i, w_i' \in \mathcal{L}_F^m$. Hence, $\sum_{i=0}^{n-1} \beta^{2i} w_i + \sum_{i=0}^{n-1} \beta^{2i} w_i' \in \mathrm{Ker}(\Omega_M^m \to \Omega_K^m)$. Using [4, Proposition 5.3], we get in Ω_M^m:

$$\sum_{i=0}^{n-1} \beta^{2i}(w_i + w_i') = \mathrm{d}\,\beta \wedge (\eta_0 + \mathrm{d}\, a_{i_1} \wedge \eta_1), \tag{5.1}$$

such that $\eta_0 \in M.\mathcal{L}_F^{m-1}$, $\eta_1 \in M.\mathcal{L}_F^{m-2}$. Since β is a root of $g(x)$ and there exists an odd integer $i \in \{1, \ldots, n-1\}$ such that $a_i \neq 0$, we pass to the derivative of $g(\beta) = 0$, we get $\mathrm{d}\,\beta \in \sum_{j=1}^{d} M.\mathrm{d}\, a_{i_j}$. Note that the left hand side of equation (5.1) is free from $\mathrm{d}\, a_{i_1}$, we thus get $\sum_{i=0}^{n-1} \beta^{2i}(w_i + w_i') \in \sum_{j=2}^{d} M.\mathrm{d}\, a_{i_j} \wedge \eta_0$. Hence, for all $0 \leq i \leq n-1$, we have $w_i + w_i' \in \sum_{j=2}^{d} \mathrm{d}\, a_{i_j} \wedge \Omega_F^{m-1}$, which implies that $(w_i + w_i')_S = 0$. Consequently, $(\sum_{i=1}^{n-1} \mathrm{Tr}(\beta^{2i}) w_i)_S = (\sum_{i=1}^{n-1} \mathrm{Tr}(\beta^{2i}) w_i')_S$. Hence, s is a well-defined map. □

LEMMA 5.3. *We keep the same notations as mentioned above. Then, $\mathrm{d}\,\widehat{\Omega}_K^m \subset \widehat{\Omega}_K^{m+1}$ and $s(\mathrm{d}\,\widehat{\Omega}_K^m) \subset \mathrm{d}(\Omega_F^m)_S$.*

PROOF. Let $w = (w_0 + \sum_{i=1}^{n-1} \beta^{2i} w_i)_K \in \widehat{\Omega}_K^m$, where $w_i \in \mathcal{L}_F^m$. Since we are over K, we may suppose that $w_i = \sum_\sigma c_\sigma \frac{\mathrm{d}\, e_\sigma}{e_\sigma}$ is free from $\mathrm{d}\, a_{i_1}$ and each $c_\sigma \in F$ does not contain an odd power of a_{i_1} in its expansion in $\mathcal{B}$. Hence, $\mathrm{d}\, w_i \in \mathcal{L}_F^{m+1}$ for all $0 \leq i \leq n-1$. Consequently, $\mathrm{d}\, w = (\mathrm{d}\, w_0 + \sum_{i=1}^{n-1} \beta^{2i} \mathrm{d}\, w_i)_K \in \widehat{\Omega}_K^{m+1}$. We have

$$
\begin{aligned}
s(\mathrm{d}\, w) &= \left(\sum_{i=1}^{n-1} \mathrm{Tr}(\beta^{2i}) \mathrm{d}\, w_i \right)_S \\
&= \left(\sum_{i=1}^{n-1} (\mathrm{Tr}(\beta^i))^2 \mathrm{d}\, w_i \right)_S \\
&= \mathrm{d} \left(\left(\sum_{i=1}^{n-1} (\mathrm{Tr}(\beta^i))^2 w_i \right)_S \right) \in \mathrm{d}(\Omega_F^m)_S. \qquad \square
\end{aligned}
$$

LEMMA 5.4. *We keep the same notations as mentioned above. Then,* $s(\wp(\widehat{\Omega}_K^m)) \subset \wp((\Omega_F^m)_S) + \mathrm{d}(\Omega_F^m)_S$.

PROOF. Let $w = (w_0 + \sum_{i=1}^{n-1} \beta^{2i} w_i)_K \in \widehat{\Omega}_K^m$, where $w_i \in \mathcal{L}_F^m$. We have $\wp(w) = \wp(w_0)_K + \sum_{i=1}^{n-1} \wp(\beta^{2i} w_i)_K$, that is,

$$
\wp(w) = \wp(w_0)_K + \sum_{i=1}^{n-1} ((\beta^{2i})^2 w_i^{[2]} + \beta^{2i} w_i)_K + \mathrm{d}\, \lambda \tag{5.2}
$$

for some $\lambda \in \Omega_K^{m-1}$.

We claim that $\mathrm{d}\, \lambda \in \mathrm{d}\, \widehat{\Omega}_K^{m-1}$. In fact, we have $\lambda \in \Omega_K^{m-1} = K.\Omega_M^{m-1} \oplus K.\Omega_M^{m-2} \wedge \mathrm{d}\, \alpha$. Recall that $K.\Omega_M^{m-1} = \Omega_M^{m-1} \oplus \bigoplus_{i=1}^{2^l-1} \alpha^i \Omega_M^{m-1}$. Thus

$$
\lambda \in \Omega_M^{m-1} \oplus \bigoplus_{i=1}^{2^l-1} \alpha^i \Omega_M^{m-1} \oplus K.\Omega_M^{m-2} \wedge \mathrm{d}\, \alpha.
$$

Note that the w_i are free from α and $\mathrm{d}\, \alpha$, so we get $\mathrm{d}\, \lambda = (\mathrm{d}\, \mu)_K$, where $\mu \in \Omega_M^{m-1}$. Now since equation (5.2) is considered over K and each w_i belongs to $\mathcal{L}_F^m$, we may take $\mu = \delta_0 + \sum_{i=1}^{n-1} \beta^{2i} \delta_i$ such that $\delta_i = \sum_\sigma c_{i,\sigma} \frac{\mathrm{d}\, e_\sigma}{e_\sigma} \in \mathcal{L}_F^{m-1}$ and each $c_{i,\sigma}$ has no odd power of a_{i_1} in its expansion in $\mathcal{B}$. Hence, $\mathrm{d}\, \lambda \in \mathrm{d}\, \widehat{\Omega}_K^{m-1}$. Lemma 5.3 implies that $s(\mathrm{d}\, \lambda) \in \mathrm{d}(\Omega_F^{m-1})_S$.

Now we apply the map s to equation (5.2), we get:

$$
\begin{aligned}
s(\wp(w)) &= \sum_{i=1}^{n-1} (\mathrm{Tr}((\beta^{2i})^2) w_i^{[2]} + \mathrm{Tr}(\beta^{2i}) w_i) + s(\mathrm{d}\, \lambda) \\
&= \sum_{i=1}^{n-1} ((\mathrm{Tr}(\beta^{2i}))^2 w_i^{[2]} + \mathrm{Tr}(\beta^{2i}) w_i) + s(\mathrm{d}\, \lambda) \\
&= \wp(\sum_{i=1}^{n-1} \mathrm{Tr}(\beta^{2i}) w_i) + s(\mathrm{d}\, \lambda) \\
&= \wp(s(w)) + s(\mathrm{d}\, \lambda) \in \wp((\Omega_F^m)_S) + \mathrm{d}(\Omega_F^m)_S. \qquad \square
\end{aligned}
$$

NOTATION 5.5. For any field extension E_2/E_1, let

$$
H_{2,E_1}^{m+1}(E_2) := (\Omega_{E_1}^m)_{E_2} / (\wp((\Omega_{E_1}^m)_{E_2}) + \mathrm{d}(\Omega_{E_1}^{m-1})_{E_2}).
$$

A consequence of the last three lemmas is the following corollary:

COROLLARY 5.6. *Let K, M and S be as mentioned above. Then, the map s defined above extends to a group homomorphism $\widetilde{\mathrm{Tr}}_*: H^{m+1}_{2,M}(K) \to H^{m+1}_{2,F}(S)$ given by: $\overline{w} \mapsto \overline{s(w)}$.*

5.2. An application. Our aim in this subsection is to apply the above construction to get a counterexample to the excellence of quartic inseparable extensions of type 2. So let $K = F(\alpha)$ be a quartic extension of F, where $\min(\alpha, F) = x^4 + bx^2 + d \in F[x]$ such that $\langle 1, b, d\rangle$ is anisotropic.

As in Section 5.1, we let: $L = F^2(b,d)$ and $M = F(\beta)$, where $\beta = \alpha^2$. Our 2-basis of F is $\mathcal{B} = \{b, d\} \cup \{e_i \mid i \in I\}$, and $\mathcal{C} = \{b, \alpha\} \cup \{e_i \mid i \in I\}$ is a 2-basis of K. We consider the sequence of fields: $F \subset M := F(\beta) \subset K = M(\sqrt{\beta})$. Finally, we take S the field $F(\sqrt{b})$.

EXAMPLE 5.7. We keep the same notations as above. Let $\varphi = [1, t_1^{-1}] \perp [b, t_2^{-1}] \perp \langle d\rangle$ be a quadratic form over $E := F((t_1))((t_2))$ the field of iterated Laurent series in the variables t_1, t_2. Then, we have:

(1) φ is anisotropic over E.
(2) $\varphi_{E(\alpha)}$ is isotropic but $(\varphi_{E(\alpha)})_{an}$ is not defined over E.

PROOF. (1) It is easy to verify that φ is anisotropic over E since the form $\langle 1, b, d\rangle$ is anisotropic over F.

(2) Extending φ to $E(\alpha)$, we get:

$$\begin{aligned}\varphi_{E(\alpha)} &\simeq [\alpha^4, \alpha^{-4}t_1^{-1}] \perp [b\alpha^2, \alpha^{-2}t_2^{-1}] \perp \langle d\rangle \\ &\simeq [\alpha^4 + b\alpha^2, \alpha^{-4}t_1^{-1}] \perp [b\alpha^2, \alpha^{-2}t_2^{-1} + \alpha^{-4}t_1^{-1}] \perp \langle d\rangle \\ &\simeq [\alpha^4 + b\alpha^2 + d, \alpha^{-4}t_1^{-1}] \perp [b\alpha^2, \alpha^{-2}t_2^{-1} + \alpha^{-4}t_1^{-1}] \perp \langle d\rangle \\ &\simeq \mathbb{H} \perp b[1, bt_2^{-1} + b\alpha^{-2}t_1^{-1}] \perp \langle d\rangle\,.\end{aligned}$$

CLAIM. The 2-fold Pfister form $\pi := \langle\langle bd, bt_2^{-1} + b\alpha^{-2}t_1^{-1}]]$ is not defined over E.

This claim implies that the form $\psi := b[1, bt_2^{-1} + b\alpha^{-2}t_1^{-1}] \perp \langle d\rangle$ is anisotropic and not defined over E. This proves that $(\varphi_{E(\alpha)})_{\mathrm{an}} \simeq \psi$ is not defined over E. So it remains to prove the claim.

Suppose that $\langle\langle bd, bt_2^{-1} + b\alpha^{-2}t_1^{-1}]]$ is defined over E. Then, there exists $\theta := x[1, y] \perp x'[1, z]$ an E-quadratic form such that $\pi \simeq \theta_{E(\alpha)}$. Comparing Arf invariants yields $\Delta(\theta_{E(\alpha)}) = 0$. Lemma 2.9 implies that $\Delta(\theta_{E(\beta)}) = 0$. Hence, $\Delta(\theta) \in \{\wp(E), b^{-2}d + \wp(E)\}$ because $M = F(\beta) = F(\wp^{-1}(b^{-2}d))$. If necessary, we may change z by $z + b^{-2}d$ and suppose that $\theta \in GP_2(E)$. By roundness, we take $\theta \simeq \langle\langle x, y]]$. Now we have this isometry of 2-fold Pfister forms:

$$\langle\langle x, y]]_{E(\alpha)} \simeq \langle\langle bd, bt_2^{-1} + b\beta^{-1}t_1^{-1}]]_{E(\alpha)}$$

(recall that $\beta = \alpha^2$). Consequently, we get in $H_2^2(E(\alpha))$

$$\overline{\left(y\frac{\mathrm{d}x}{x}\right)}_{E(\alpha)} = \overline{\left((bt_2^{-1} + bt_1^{-1}\beta^{-1})\frac{\mathrm{d}(bd)}{bd}\right)}_{E(\alpha)} \overset{(*)}{=} \overline{\left((b^2t_2^{-2} + b^2t_1^{-2}\beta^{-2})\frac{\mathrm{d}(bd)}{bd}\right)}_{E(\alpha)}.$$

$(*)$: We apply the Frobenius operator.

Hence, there exist $u \in \Omega^2_{E(\alpha)}$ and $v \in \Omega^1_{E(\alpha)}$ such that

$$\overline{\left(y\frac{\mathrm{d}\,x}{x}\right)}_{E(\alpha)} + \overline{\left((b^2t_2^{-2} + b^2t_1^{-2}\beta^{-2})\frac{\mathrm{d}(bd)}{bd}\right)}_{E(\alpha)} = \wp(u) + \mathrm{d}\,v. \tag{5.3}$$

The extension $E(\alpha)/E(\beta)$ is quadratic inseparable. Hence, using the same arguments as in the proof of [**3**, Proposition 2], we may suppose that $u \in \Omega^2_{E(\beta)}$ and $v \in \Omega^1_{E(\beta)}$. This means that we have

$$\overline{\left(y\frac{\mathrm{d}\,x}{x}\right)}_{E(\alpha)} = \overline{\left((b^2t_2^{-2} + b^2t_1^{-2}\beta^{-2})\frac{\mathrm{d}(bd)}{bd}\right)}_{E(\alpha)} \text{ in } H^2_{2,E(\beta)}(E(\alpha)). \tag{5.4}$$

Notice that $\mathrm{Tr}(\beta^{-2}) = b^2d^{-2}$. Now applying $\widetilde{\mathrm{Tr}}_*$ to equation (5.4), we get:

$$0 = \overline{\left(b^4t_1^{-2}d^{-2}\frac{\mathrm{d}(bd)}{bd}\right)}_{E(\sqrt{b})} = \overline{\left(b^2t_1^{-1}d^{-1}\frac{\mathrm{d}(bd)}{bd}\right)}_{E(\sqrt{b})} \text{ in } H_{2,E}(E(\sqrt{b})).$$

Using Kato's isomorphism (2.3) with the Hauptsatz, we get $\langle\langle bd, b^2t_1^{-1}d^{-1}]]_{E(\sqrt{b})} \sim 0$, that is, $[1, b^2t_1^{-1}d^{-1}]_{E(\sqrt{b})} \simeq bd[1, b^2t_1^{-1}d^{-1}]_{E(\sqrt{b})}$. Now taking the first residue form with respect to the t_1-adic valuation of $E(\sqrt{b})$, we get:

$$\langle 1\rangle_{F((t_2))(\sqrt{b})} \simeq \langle bd\rangle_{F((t_2))(\sqrt{b})} \simeq \langle d\rangle_{F((t_2))(\sqrt{b})}\,.$$

In particular, the form $\langle 1, d\rangle_{F(\sqrt{b})}$ is isotropic, which implies that $\langle 1, b, d\rangle$ is isotropic, a contradiction. □

6. Nonexcellence of separable biquadratic extensions

For completeness we mention the case of biquadratic separable extensions. In general, such extensions are not excellent for quadratic forms. The counterexample concerns dimension 4 and is essentially due to Sivatski [**21**]. He assumed the characteristic to be different from 2 but the construction goes through basically without any change to characteristic 2 with the help of a result of Rowen.

By [**20**, Section 3], there exists a field F of characteristic 2 with an indecomposable F-algebra D of degree 8 and exponent 2. Moreover, by [**20**, Theorem 2] we have $M_2(D) \simeq \otimes_{i=1}^4 [a_i, b_i)$ with $a_i, b_i \in F^*$ such that D is split by the triquadratic separable extension $M := F(\wp^{-1}(a_1), \wp^{-1}(a_2), \wp^{-1}(a_3))$. Then, we have the following example showing the nonexcellence of biquadratic separable extensions:

EXAMPLE 6.1. Let $K = F(\wp^{-1}(a_1), \wp^{-1}(a_2))$ and consider the F-quadratic form $\varphi = b_4[1, a_4] \perp [1, a_3 + a_4]$. Then, we have:

(1) φ is anisotropic over F.
(2) φ_K is isotropic.
(3) $(\varphi_K)_{\mathrm{an}}$ is not defined over F.

In particular, K/F is not $(2, 0)$-excellent.

PROOF. Let D, M be as mentioned above.

(1) Suppose that φ is isotropic over F. Then, there exist $x \in F^*$ such that $b_4[1, a_4] \simeq x[1, a_4]$ and $[1, a_3 + a_4] \simeq x[1, a_3 + a_4]$. Hence, $[a_4, b_4) \simeq [a_4, x)$ and

$[a_3 + a_4, x) \sim 0$, that is, $[a_3, x) \simeq [a_4, x)$. Thus,

$$\begin{aligned} M_2(D) &\simeq [a_1, b_1) \otimes [a_2, b_2) \otimes [a_3, b_3) \otimes [a_4, b_4) \\ &\simeq [a_1, b_1) \otimes [a_2, b_2) \otimes [a_3, b_3) \otimes [a_4, x) \\ &\simeq [a_1, b_1) \otimes [a_2, b_2) \otimes [a_3, b_3) \otimes [a_3, x) \\ &\simeq [a_1, b_1) \otimes [a_2, b_2) \otimes M_2([a_3, b_3 x)) \\ &\simeq M_2([a_1, b_1) \otimes [a_2, b_2) \otimes [a_3, b_3 x)). \end{aligned}$$

This implies that $D \simeq [a_1, b_1) \otimes [a_2, b_2) \otimes [a_3, b_3 x)$, a contradiction to the hypothesis that D is indecomposable. Thus, φ is anisotropic over F.

(2) Recall that D is split by M, that is, $D_M \sim 0$. Thus $[a_4, b_4)_M = 0$, consequently $\varphi_M \sim ([1, a_4] \perp b_4[1, a_4])_M \sim 0$. It follows from Lemma 3.1 that φ_K is isotropic. Since $\triangle(\varphi_K) \neq 0$, we get $\dim(\varphi_K)_{\mathrm{an}} = 2$.

(3) Suppose that $(\varphi_K)_{\mathrm{an}}$ is defined over F. Then, there exist $\alpha \in F^*$ such that $(\varphi_K)_{\mathrm{an}} \simeq (\alpha[1, a_3])_K$. Hence, $(\varphi \perp \alpha[1, a_3])_K \sim 0$. It follows, from a result of Baeza [**7**, Corollary 4.16], that

$$\varphi \perp \alpha[1, a_3] \sim \rho_1 \otimes [1, a_1] \perp \rho_2 \otimes [1, a_2], \tag{6.1}$$

for suitable bilinear forms ρ_1 and ρ_2. Taking the Clifford algebra in equation (6.1), we get

$$[a_4, b_4) \sim [a_3, \alpha) \otimes [a_1, r) \otimes [a_2, s),$$

where $r, s \in F^*$ satisfy $\det \rho_1 = rF^{*2}$ and $\det \rho_2 = sF^{*2}$. Substituting $[a_4, b_4)$ in $M_2(D)$ yields:

$$D \simeq [a_1, rb_1) \otimes [a_2, sb_2) \otimes [a_3, \alpha b_3),$$

and thus D is decomposable, a contradiction. □

7. Extending nonexcellence results to higher dimensional forms

The nonexcellence results that we established for quartic extensions concern quadratic forms of type $(2, 0)$ or $(2, 1)$. Based on a generic argument, we have that these extensions are also not (r, s)-excellent for larger r and s. In particular, combining [**19**, Lemmas 4.5, 4.6], we have the following result:

LEMMA 7.1. *Let L/F be a field extension which is not (r, s)-excellent. Let $x_1, \dots, x_{2m+n}$ be variables over L, and let M (resp. N) be the field of iterated Laurent series $F((x_1)) \cdots ((x_{2m+n}))$ (resp. $L((x_1)) \cdots ((x_{2m+n}))$). Then, the extension N/M is not $(r + m, s + n)$-excellent.*

We summarize below what we know up to now for quartic extensions:

COROLLARY 7.2.

(1) *Quartic extensions of type 1 are generally not (r, s)-excellent when $r \geq 2$ and $s \geq 0$.*
(2) *Quartic extensions of type 2 are generally not (r, s)-excellent when $r \geq 2$ and $s \geq 1$.*
(3) *Quartic extensions of type 3 are (r, s)-excellent for any $r, s \geq 0$.*
(4) *Biquadratic separable extensions are generally not (r, s)-excellent when $r \geq 2$ and $s \geq 0$.*

PROOF. Statement (3) is given in [**11**]. For statements (1), (2) and (4) we combine Lemma 7.1 with Examples 4.1, 5.7 and 6.1, respectively. □

Now, we are left to address the following question to completely classify inseparable quartic extensions:

QUESTION 7.3. Is it true that any quartic extension of type 2 is $(2,0)$-excellent?

In contrast of what we did for quartic inseparable extensions of type 1, and in order to answer this question, it seems natural to study the isotropy in dimension 4 over quartic inseparable extensions of type 2.

The generic argument proposed above extends nonexcellence for same type of field extensions to higher dimensional quadratic forms. Naturally, one can ask the question if such an extension is also possible for higher degree field extensions. To be precise, we ask the following:

QUESTION 7.4. Let $F \subseteq L \subseteq K$ be a tower of fields such that the extension L/F is not excellent. What can one say about the excellence or nonexcellence of K/F?

While K/F is not excellent when K/L is purely transcendental or odd-degree algebraic extension, and excellent when K is algebraic closure of L for trivial reasons. It is not as straightforward to answer for non-trivial cases as in Example 7.5.

EXAMPLE 7.5. Let t_1, t_2, t_3, t_4 be independent variables over $\mathbb{F}_2$ (the finite field with two elements). Let $\gamma = [t_1, t_2] \perp [t_3, t_4] \perp [1, t_1t_2 + t_3t_4]$ be an Albert quadratic form over the rational function field $F := \mathbb{F}_2(t_1, t_2, t_3, t_4)$. Let Q be the totally singular quadratic form $\langle 1, t_1, t_3\rangle$. Then, $F(Q)/F$ is not $(3,0)$-excellent but $F(Q)(\sqrt{t_1})/F$ is excellent.

PROOF. Consider the tower of field extensions $F \subseteq F(Q) \subseteq F(Q)(\sqrt{t_1})$. The fact that the extension $F(Q)/F$ is not $(3,0)$-excellent was proved in [**19**, Proposition 4.3]. Since $Q_{F(\sqrt{t_1})} \simeq \langle 0, 1, t_3\rangle$, we get the following equalities:

$$F(Q)(\sqrt{t_1}) = F(\sqrt{t_1})(Q) = F(\sqrt{t_1})(\langle 0, 1, t_3\rangle) = F(x, y)(\langle \sqrt{t_1}, \sqrt{t_3}\rangle),$$

for some variables x, y over F. By the result of Hoffmann [**11**], we get

$$F(x, y)(\langle \sqrt{t_1}, \sqrt{t_3}\rangle)/F$$

is excellent. Hence, $F(Q)(\sqrt{t_1})/F$ is excellent. □

This example reiterates why the question of excellence is obstinate to study in general.

References

[1] Hamza Ahmad, *On quadratic forms over inseparable quadratic extensions*, Arch. Math. (Basel) **63** (1994), no. 1, 23–29, DOI 10.1007/BF01196294. MR1277906

[2] R. Aravire and R. Baeza, *The behavior of the ν-invariant of a field of characteristic 2 under finite extensions*, Rocky Mountain J. Math. **19** (1989), no. 3, 589–600, DOI 10.1216/RMJ-1989-19-3-589. Quadratic forms and real algebraic geometry (Corvallis, OR, 1986). MR1043232

[3] Roberto Aravire and Ahmed Laghribi, *Results on Witt kernels of quadratic forms for multi-quadratic extensions*, Proc. Amer. Math. Soc. **141** (2013), no. 12, 4191–4197, DOI 10.1090/S0002-9939-2013-11730-2. MR3105862

[4] Roberto Aravire, Ahmed Laghribi, and Manuel O'Ryan, *The behavior of differential forms under purely inseparable extensions*, J. Pure Appl. Algebra **223** (2019), no. 1, 439–457, DOI 10.1016/j.jpaa.2018.03.019. MR3833469

[5] Roberto Aravire, Ahmed Laghribi, and Manuel O'Ryan, *Cohomological kernels of mixed extensions in characteristic 2*, J. Algebra **542** (2020), 249–276, DOI 10.1016/j.jalgebra.2019.09.012. MR4019082

[6] Ricardo Baeza, *Ein Teilformensatz für quadratische Formen in Charakteristik* 2 (German), Math. Z. **135** (1973/74), 175–184, DOI 10.1007/BF01189354. MR337765

[7] Ricardo Baeza, *Quadratic forms over semilocal rings*, Lecture Notes in Mathematics, Vol. 655, Springer-Verlag, Berlin-New York, 1978. MR491773

[8] K. Bingöl, A. Chapman and A. Laghribi, *Mixed multiquadratic splitting fields*, arXiv:2406.14339.

[9] Richard Elman, Nikita Karpenko, and Alexander Merkurjev, *The algebraic and geometric theory of quadratic forms*, American Mathematical Society Colloquium Publications, vol. 56, American Mathematical Society, Providence, RI, 2008, DOI 10.1090/coll/056. MR2427530

[10] Detlev W. Hoffmann, *Witt kernels of bilinear forms for algebraic extensions in characteristic 2*, Proc. Amer. Math. Soc. **134** (2006), no. 3, 645–652, DOI 10.1090/S0002-9939-05-08175-X. MR2180880

[11] Detlev W. Hoffmann, *Witt kernels of quadratic forms for multiquadratic extensions in characteristic 2*, Proc. Amer. Math. Soc. **143** (2015), no. 12, 5073–5082, DOI 10.1090/proc/12651. MR3411127

[12] Detlev W. Hoffmann and Ahmed Laghribi, *Quadratic forms and Pfister neighbors in characteristic 2*, Trans. Amer. Math. Soc. **356** (2004), no. 10, 4019–4053, DOI 10.1090/S0002-9947-04-03461-0. MR2058517

[13] Detlev W. Hoffmann and Ahmed Laghribi, *Isotropy of quadratic forms over the function field of a quadric in characteristic 2*, J. Algebra **295** (2006), no. 2, 362–386, DOI 10.1016/j.jalgebra.2004.02.038. MR2194958

[14] Detlev W. Hoffmann and Marco Sobiech, *Witt kernels and Brauer kernels for quartic extensions in characteristic two*, J. Pure Appl. Algebra **219** (2015), no. 10, 4619–4634, DOI 10.1016/j.jpaa.2015.02.034. MR3346508

[15] Kazuya Kato, *Symmetric bilinear forms, quadratic forms and Milnor K-theory in characteristic two*, Invent. Math. **66** (1982), no. 3, 493–510, DOI 10.1007/BF01389226. MR662605

[16] Manfred Knebusch, *Specialization of quadratic and symmetric bilinear forms, and a norm theorem*, Acta Arith. **24** (1973), 279–299, DOI 10.4064/aa-24-3-279-299. MR349582

[17] Ahmed Laghribi, *Certaines formes quadratiques de dimension au plus 6 et corps des fonctions en caractéristique 2* (French, with English summary), Israel J. Math. **129** (2002), 317–361, DOI 10.1007/BF02773169. MR1910948

[18] Ahmed Laghribi, *The norm theorem for totally singular quadratic forms*, Rocky Mountain J. Math. **36** (2006), no. 2, 575–592, DOI 10.1216/rmjm/1181069468. MR2234821

[19] Ahmed Laghribi and Diksha Mukhija, *The excellence of function fields of conics*, Proc. Amer. Math. Soc. **152** (2024), no. 5, 1915–1923, DOI 10.1090/proc/16690. MR4728462

[20] Louis H. Rowen, *Division algebras of exponent* 2 *and characteristic* 2, J. Algebra **90** (1984), no. 1, 71–83, DOI 10.1016/0021-8693(84)90199-6. MR757082

[21] A. S. Sivatski, *Nonexcellence of multiquadratic field extensions*, J. Algebra **275** (2004), no. 2, 859–866, DOI 10.1016/j.jalgebra.2003.07.014. MR2052642

[22] Marco Sobiech, *The behavior of differential, quadratic and bilinear forms under purely inseparable field extensions*, J. Algebra **499** (2018), 151–182, DOI 10.1016/j.jalgebra.2017.11.033. MR3758498

Univ. Artois, UR 2462, Laboratoire de Mathématiques de Lens (LML), F-62300 Lens, France

Email address: ahmed.laghribi@univ-artois.fr

Department of Mathematical and Statistical Sciences, University of Alberta, 632 Central Academic Building Edmonton, Alberta T6G 2G1 Canada

Email address: dmukhija@ualberta.ca

Contemporary Mathematics
Volume **826**, 2025
https://doi.org/10.1090/conm/826/16578

Is A_1 of type B_2?

S Launois and I Oppong

ABSTRACT. By a theorem of Dixmier, primitive quotients of enveloping algebras of finite-dimensional complex nilpotent Lie algebras are isomorphic to Weyl algebras. In view of this result, it is natural to consider simple quotients of positive parts of quantized enveloping algebras (and more generally of uniparameter Quantum Nilpotent Algebras) as quantum analogues of Weyl algebras. In this note, we study the Lie algebra of derivations of the simple quotients of $U_q^+(\mathfrak{so}_5)$ of Gelfand-Kirillov dimension 2. For a specific family of such simple quotients, we prove that all derivations are inner (as in the case of Weyl algebras) whereas all other such algebras are quantum Generalized Weyl Algebras over a commutative Laurent polynomial algebra in one variable and have a first Hochschild cohomology group of dimension 1.

1. Introduction

Let $\mathbb{K}$ be a field of characteristic 0 and let q be a non-zero scalar which is not a root of unity. Weyl algebras are central objects in noncommutative algebra due to their appearences in a range of areas e.g. representation theory of Lie algebras; combinatorics; quantum mechanics; $\mathcal{D}$-modules to name a few. Despite many efforts over the last 60 years, the structure of Weyl algebras is still proving mysterious. For instance, only the automorphism group of the first Weyl algebra $A_1(\mathbb{K})$ has been computed (by Dixmier in [**10**]) and only the irreducible representations of $A_1(\mathbb{K})$ have been classified in a highly technical tour de force by Block [**6**]. These two questions remain open for higher Weyl algebras.

Since Weyl algebras are simple, their endomorphisms are injective, and it is an open conjecture of Dixmier [**10**, Problème 1] that all endomorphisms are actually automorphisms. This conjecture is (stably) equivalent to the celebrated Jacobian Conjecture (see [**5**, **24**]), and it is still open even in the case of the first Weyl algebra $A_1(\mathbb{K})$, to the best of the authors' knowledge.

Given its central role in noncommutative algebra, it is not surprising that mathematicians have tried to generalise and/or deform Weyl algebras. This led in particular to the notion of Generalized Weyl Algebras (GWAs for short), a class of noncommutative algebras introduced by Bavula in the 1990s (see [**3**]). This class of algebras includes in particular two families of (quantum) algebras that are deforming Weyl algebras: quantum Weyl algebras and Weyl-Hayashi algebras. The

2020 *Mathematics Subject Classification.* Primary 16T20, 17B37; Secondary 16W25.
Key words and phrases. Quantized enveloping algebras, quantum Weyl algebras, primitive ideals, derivations.

later appear as simple quotients of the quantum Heisenberg algebra $U_q^+(\mathfrak{sl}_3)$ [**13**], and were studied for instance in [**1, 14**]. A well-known theorem of Dixmier [**11**, Théorème 4.7.9] shows that primitive quotients of enveloping algebras of nilpotent Lie algebras are isomorphic to Weyl algebras. In many respects, (uniparameter) Quantum Nilpotent Algebras (QNA for short), also known as CGL extensions [**12, 20**], are behaving very much as enveloping algebras of nilpotent Lie algebras, and indeed include all quantum Schubert cells $U_q[w]$ which are deformations of the enveloping algebras of $\mathfrak{n}^+ \cap w(\mathfrak{n}^-)$, where $\mathfrak{n}^{\pm}$ are the nilradicals of the standard opposite Borel subalgebras of a symmetrizable Kac-Moody algebra $\mathfrak{g}$ and where w is an element of the Weyl group of $\mathfrak{g}$.

As such, it is natural to think of primitive/simple quotients of QNAs as "quantum" Weyl algebras. This point of view is reinforced by the fact that primitive quotients of uniparameter QNAs have even Gelfand-Kirillov dimension [**4**, Theorem 1.6]. With this idea in mind, Lopes [**23**] studied the primitive ideals of $U_q^+(\mathfrak{sl}_n)$ coming from the 0-stratum, that is the primitive ideals that do not contain any H-eigenvectors for the natural action by automorphisms of the torus $H := (\mathbb{K}^*)^{n-1}$ on $U_q^+(\mathfrak{sl}_n)$. These primitive ideals are all maximal of even Gelfand-Kirillov dimension, and so feature some common properties with Weyl algebras. However, they also feature notable differences: they contain non-trivial units and not all derivations are inner in general. It is worth noting that, in the case of $\mathfrak{sl}_3$, these primitive quotients are the Weyl-Hayashi algebras and can be presented as GWAs over a (commutative) Laurent polynomial ring $\mathbb{K}[h^{\pm 1}]$. The presence of non-trivial units was exploited in [**16**] to prove that all endomorphisms of Weyl-Hayashi algebras (and more generally of quantum GWAs over a commutative Laurent polynomial algebra) are automorphisms. In this note, we study the case of the positive part $U_q^+(\mathfrak{so}_5)$ of the quantized enveloping algebra of $\mathfrak{so}_5$, and this will lead us to a "quantum deformation of $A_1(\mathbb{K})$". Recently, the authors did study the case of $U_q^+(G_2)$ [**22**] to obtain quantum/algebraic deformations of the second Weyl algebra.

In [**19**], primitive ideals of $U_q^+(\mathfrak{so}_5)$ were classified. This led in particular to three non-isomorphic families of simple quotients of $U_q^+(\mathfrak{so}_5)$ of Gelfand-Kirillov dimension 2. Out of these three families, we retrieve the Weyl-Hayashi algebras as a first family, another family of quantum GWAs over a commutative Laurent polynomial ring, and a third family $B_{\alpha,\beta}$, with $\alpha, \beta \in \mathbb{K}^*$, which cannot be presented as a GWA over a commutative Laurent polynomial ring. This last family can be presented as the $\mathbb{K}$-algebra generated by e_1, e_2, e_3 and e_4 subject to the following relations:

$$e_2e_1 = q^{-2}e_1e_2 \qquad e_4e_1 = q^2e_1e_4 - q^2e_2 \qquad e_4e_2 = q^{-2}e_2e_4 + (1-q^{-2})\beta$$

and

$$\frac{q^6}{q^4-1}e_2^2 = \frac{-(q+q^{-1})}{1-q^{-2}}\alpha - \beta e_1 + e_1e_2e_4,$$

where $\alpha, \beta \in \mathbb{K}^*$.

Before we go any further, it is worth noting that these relations are just expressing the fact that $B_{\alpha,\beta}$ is the quotient of $U_q^+(\mathfrak{so}_5)$ by the maximal ideal $\langle \chi_1 - \alpha, \chi_2 - \beta \rangle$ of $U_q^+(\mathfrak{so}_5)$, where χ_1 and χ_2 are the generators of the center of $U_q^+(\mathfrak{so}_5)$ (see [**19**, Proposition 2.7]).

For the first two families of simple quotients of $U_q^+(\mathfrak{so}_5)$, since they can be presented as quantum GWAs over a Laurent polynomial ring, their endomorphisms were proved to be automorphisms in [**16**], and derivations were computed in [**15**], proving in particular that not all derivations are inner (whereas all derivations of a Weyl algebra are inner).

Our main aim in this note is to show that all derivations of $B_{\alpha,\beta}$ when $\alpha\beta \neq 0$ are inner. This makes the algebra $B_{\alpha,\beta}$ with $\alpha\beta \neq 0$ into a quantum deformation of the first Weyl algebra which is simple, has Gelfand-Kirillov dimension 2, has no non-trivial units, and has all its derivations inner. This somehow justifies the claim that "$A_1(\mathbb{K})$ is of type B_2" since none of the primitive quotients of the quantum Heisenberg algebra $U_q^+(\mathfrak{sl}_3)$ have all these properties.

This paper is organised as follows. We recall some basic properties of $U_q^+(\mathfrak{so}_5)$ and $B_{\alpha,\beta}$ in Sections 2 and 3 respectively. More importantly, we study Cauchon's deleting derivation algorithm [**8**] for the QNA $U_q^+(\mathfrak{so}_5)$ in Section 2. This theory allows us to construct an embedding:

$$B_{\alpha,\beta} \subset R = B_{\alpha,\beta}[e_4^{-1}]$$

and to present R as a GWA. This presentation of R is obtained through the elements f_1 and f_2 constructed at the first step of the deleting derivation algorithm (DDA for short) for $U_q^+(\mathfrak{so}_5)$. We then extend any derivation D of $B_{\alpha,\beta}$ uniquely to a derivation of the GWA R. Thanks to a result of Kitchin [**15**, Proposition 1.3], derivations of R are known to be the sum of an inner derivation and a scalar derivation (i.e. a derivation that acts on the generators f_1, f_2, e_3 and e_4 of R by multiplication by scalars). Having a perfect description of D as a derivation R, we then describe it as a derivation of $B_{\alpha,\beta}$ by controlling the effect of the DDA on our derivation D. Similarly to Weyl algebras, we conclude that every derivation of $B_{\alpha,\beta}$ is inner provided $\alpha\beta \neq 0$.

2. Deleting derivation algorithm for the quantum nilpotent algebra $U_q^+(\mathfrak{so}_5)$

2.1. $U_q^+(\mathfrak{so}_5)$ **as a quantum nilpotent algebra.** The quantized enveloping algebra $U_q^+(\mathfrak{so}_5)$ is the $\mathbb{K}$-algebra generated by two indeterminates E_1 and E_4 subject to the quantum Serre relations:

$$E_1^2E_4 - (q^2 + q^{-2})E_1E_4E_1 + E_4E_1^2 = 0$$
$$E_4^3E_1 - (q^2 + 1 + q^{-2})E_4^2E_1E_4 + (q^2 + 1 + q^{-2})E_4E_1E_4^2 - E_1E_4^3 = 0.$$

To each reduced decomposition of the longest element of the Weyl group of $\mathfrak{so}_5$, we can associate a PBW basis for $U_q^+(\mathfrak{so}_5)$ by using Lusztig automorphisms. In the present case, this leads us to introduce the following elements (see [**9**, Appendix A] with $w_0 = s_\beta s_\alpha s_\beta s_\alpha$, where the root vectors E_β, $E_{\alpha+\beta}$, $E_{2\alpha+\beta}$ and E_α from [**9**] corresponds to our E_1, E_2, E_3 and E_4 respectively):

$$E_2 := E_1E_4 - q^{-2}E_4E_1$$

and

$$E_3 := \frac{1}{q+q^{-1}}\left(E_2E_4 - E_4E_2\right).$$

By results of Levendorskii-Soibelman and Lusztig (see [**7**, Theorem I.6.8]), the monomials $E_1^{i_1}E_2^{i_2}E_3^{i_3}E_4^{i_4}$ with $i_1, i_2, i_3, i_4 \in \mathbb{Z}_{\geq 0}$ form a PBW basis of $U_q^+(\mathfrak{so}_5)$, so that $U_q^+(\mathfrak{so}_5)$ can be seen as the $\mathbb{K}$-algebra generated by E_1, E_2, E_3 and E_4 subject to the relations below:

$$E_2E_1 = q^{-2}E_1E_2 \qquad E_4E_1 = q^2E_1E_4 - q^2E_2 \quad E_3E_1 = E_1E_3 + \frac{q-q^3}{1+q^2}E_2^2$$

$$E_4E_2 = E_2E_4 - (q+q^{-1})E_3 \quad E_3E_2 = q^{-2}E_2E_3 \qquad E_4E_3 = q^{-2}E_3E_4.$$

This presentation leads to the following presentation of $U_q^+(\mathfrak{so}_5)$ as an iterated Ore extension:

$$B := U_q^+(\mathfrak{so}_5) = \mathbb{K}[E_1][E_2;\sigma_2][E_3;\sigma_3,\delta_3][E_4;\sigma_4,\delta_4],$$

where σ_2 is an automorphism of $\mathbb{K}[E_1]$ defined by $\sigma_2(E_1) = q^{-2}E_1$; σ_3 is an automorphism of $\mathbb{K}[E_1][E_2;\sigma_2]$ defined by $\sigma_3(E_1) = E_1$, $\sigma_3(E_2) = q^{-2}E_2$; δ_3 is a σ_3-derivation of $\mathbb{K}[E_1][E_2;\sigma_2]$ defined by $\delta_3(E_1) = (q-q^3)/(1+q^2)E_2^2$, $\delta_3(E_2) = 0$; σ_4 is an automorphism of $\mathbb{K}[E_1][E_2;\sigma_2][E_3;\sigma_3,\delta_3]$ defined by $\sigma_4(E_1) = q^2E_1$, $\sigma_4(E_2) = E_2$, $\sigma_4(E_3) = q^{-2}E_3$ and δ_4 is a σ_4-derivation of $\mathbb{K}[E_1][E_2;\sigma_2][E_3;\sigma_3,\delta_3]$ defined by $\delta_4(E_1) = -q^2E_2$, $\delta_4(E_2) = -(q+q^{-1})E_3$, $\delta_4(E_3) = 0$. Note in particular that B is a noetherian domain.

This presentation of $U_q^+(\mathfrak{so}_5)$ as an iterated Ore extension makes $U_q^+(\mathfrak{so}_5)$ into a so-called (uniparameter) QNA, (see for instance [**8**, Proposition 6.1.2 and Lemme 6.2.1]). At this stage, we do not need to recall the definition of a QNA. However, we point out that, first, this implies that all primes of B are completely prime, and then that this will allow us to apply the deleting derivation theory of Cauchon [**8**] to $U_q^+(\mathfrak{so}_5)$ with this iterated Ore extension structure. It is worth noting that the iterated Ore extension presentation of $U_q^+(\mathfrak{so}_5)$ used in [**2**, **19**] does not allow to apply Cauchon's deleting derivation theory.

2.2. Deleting derivation algorithm of $U_q^+(\mathfrak{so}_5)$. Since the iterated Ore extension B is a QNA, similarly to [**22**, Sec 2.3], Cauchon's theory of deleting derivation algorithm (DDA for short) (see [**8**]) can be used to construct the following elements of the skew field of fractions $\mathrm{Frac}(B)$ of B:

$$\begin{aligned}
&E_{i,5} := E_i \quad (i \in [1,4]) && E_{1,4} = E_1 + p_1E_2E_4^{-1} + p_2E_3E_4^{-2}\\
&E_{2,4} = E_2 + p_3E_3E_4^{-1} && E_{3,4} = E_4\\
&E_{4,4} = E_4 && E_{1,3} = E_{1,4} + p_4E_{2,4}^2E_{3,4}^{-1}\\
&T_1 := E_{1,2} = E_{1,3} && T_2 := E_{2,2} = E_{2,3} = E_{2,4}\\
&T_3 := E_{3,2} = E_{3,3} = E_{3,4} = E_3 && T_4 := E_{4,2} = E_{4,3} = E_{4,4} = E_4,
\end{aligned}$$

where

$$p_1 = \frac{1}{q^{-2}-1} = \frac{q^2}{1-q^2}, \qquad p_2 = \frac{q^{-1}+q^{-3}}{(1+q^{-2})(1-q^{-2})^2} = \frac{q^3}{(q^2-1)^2},$$

$$p_3 = \frac{-(q+q^{-1})}{1-q^{-2}} = \frac{-(q^3+q)}{q^2-1}, \qquad p_4 = \frac{q-q^3}{(1-q^{-4})(1+q^2)} = \frac{-q^5}{(1+q^2)^2}.$$

Recall from [**2**, Lemma 3.1] that the center of B is a polynomial algebra $Z(B) = \mathbb{K}[\chi_1,\chi_2]$, where

$$\chi_1 := T_1T_3 = E_{1,4}E_{3,4} + p_4E_{2,4}^2 = E_1E_3 + p_4E_2^2$$

and

$$\chi_2 := T_2T_4 = E_2E_4 + p_3E_3.$$

Note that these two central generators take a simpler form when expressed in the new elements T_i.

3. Simple quotients of $U_q^+(B_2)$

Primitive/maximal ideals of $U_q^+(\mathfrak{so}_5)$ were classified in [**19**], where it is proved (see [**19**, Propositions 2.6 and 2.7]) that the only maximal ideals of height 2 are the ideals $\langle \chi_1 - \alpha, \chi_2 - \beta \rangle$ with $(\alpha, \beta) \in \mathbb{K}^2 \setminus \{(0,0)\}$.

Our aim in this note is to compute the derivations of the corresponding simple quotients of $U_q^+(\mathfrak{so}_5)$. So we set:

$$B_{\alpha,\beta} := \frac{B}{\langle \chi_1 - \alpha, \chi_2 - \beta \rangle},$$

where $(\alpha, \beta) \in \mathbb{K}^2 \setminus \{(0,0)\}$.

The algebra $B_{\alpha,\beta}$ is a simple noetherian domain with Gelfand-Kirillov dimension 2 ([**19**, Proposition 2.8]) and its center is reduced to $\mathbb{K}$ by [**19**, Proposition 3.7]. From [**19**, Observation 3.1], we also have that $B_{\alpha,\beta}$ is a quantum deformation of the first Weyl algebra $A_1(\mathbb{K})$, hence $B_{\alpha,\beta}$ can be thought of as a quantum first Weyl algebra.

In the case where α or β is zero (but not both), then $B_{0,\beta}$ and $B_{\alpha,0}$ are Generalized Weyl Algebras (GWA) over a commutative Laurent polynomial ring in one variable by [**19**, Propositions 3.9 and 3.10]. As such, their derivations were computed in [**15**]. Thus, from now on, we focus our attention on the case where $\alpha\beta \neq 0$.

Denote the canonical image of each E_i by e_i in $B_{\alpha,\beta}$. We have that $B_{\alpha,\beta}$ satisfies the following commutation relations:

$$e_2e_1 = q^{-2}e_1e_2 \qquad e_4e_1 = q^2e_1e_4 - q^2e_2 \qquad e_3e_1 = e_1e_3 + \frac{q-q^3}{1+q^2}e_2^2$$

$$e_3e_2 = q^{-2}e_2e_3 \qquad e_4e_2 = e_2e_4 - (q+q^{-1})e_3 \qquad e_4e_3 = q^{-2}e_3e_4,$$

and

$$\begin{aligned} e_3 &= k_1\beta - k_1e_2e_4 \\ e_2^2 &= \alpha k_2 - \beta k_1k_2e_1 + k_1k_2e_1e_2e_4, \end{aligned} \tag{3.1}$$

where $k_1 = 1/p_3 = \dfrac{1-q^2}{q^3+q}$ and $k_2 = 1/p_4 = -q^{-5}(1+q^2)^2$.

The following relations can easily be checked by induction.

LEMMA 3.1. *For each $i \in \mathbb{N}$:*

(1) $e_3e_1^i = e_1^ie_3 + d[i]e_1^{i-1}e_2^2$, *where* $d[i] = \dfrac{(q-q^3)(1-q^{-4i})}{(1-q^{-4})(1+q^2)}$. *Note in particular that* $d[0] = 0$.

(2) $e_4e_1^i = q^{2i}e_1^ie_4 - f[i]e_1^{i-1}e_2$, *where* $f[i] = \dfrac{q^{2i}(1-q^{-4i})}{1-q^{-4}}$. *Note in particular that* $f[0] = 0$.

A linear basis of $B_{\alpha,\beta}$ was obtained in [**19**]. Noting that our generators E_1, E_2 and E_4 of $U_q^+(\mathfrak{so}_5)$ correspond to the generators e_1, e_3 and e_2 of $U_{q^{-1}}^+(B_2)$ in [**19**], the linear basis of $B_{\alpha,\beta}$ from [**19**, Proposition 3.2] can be expressed as follows with the notation of this note.

PROPOSITION 3.2. *The set $\mathcal{E} = \{e_1^i e_4^j, e_1^i e_2 e_4^j \mid i, j \in \mathbb{N}\}$ is a $\mathbb{K}$-basis of $B_{\alpha,\beta}$.*

4. Derivations of $B_{\alpha,\beta}$

In this section, we aim to compute the set of $\mathbb{K}$-derivations of the algebra $B_{\alpha,\beta}$. Our strategy is to prove that a localization of $B_{\alpha,\beta}$ is a quantum GWA. This isomorphism was obtained through the first step of the DDA since this GWA presentation is naturally expressed in terms of the elements produced by the first step of the DDA, that is $E_{1,4}$, $E_{2,4}$, E_3 and E_4.

Once this isomorphism is established, we can extend any derivation of $B_{\alpha,\beta}$ to a derivation of this GWA. It then follows from work of Kitchin [**15**] that derivations of a (simple) quantum GWA are the sum of an inner derivation and a central/scalar derivation. To conclude, we will need to control the effect of the deleting derivation algorithm on our derivation.

Unless otherwise stated, we assume that $\alpha\beta \neq 0$.

4.1. Embedding $B_{\alpha,\beta}$ into a quantum GWA.

At this stage, we recall that the (quantum) generalized Weyl algebra $\mathbb{K}[h^{\pm 1}](\sigma_q, a)$ associated to the automorphism σ_q of $\mathbb{K}[h^{\pm 1}]$ defined by $\sigma_q(h) = q^2 h$ and to a nonzero element $a \in \mathbb{K}[h^{\pm 1}]$ is the $\mathbb{K}[h^{\pm 1}]$-algebra generated by two indeterminates x and y subject to the relations

$$yx = a, xy = \sigma_q(a), xh = q^2 hx, \text{ and } yh = q^{-2} hy.$$

Observe that $E_4 \notin \langle \chi_1 - \alpha, \chi_2 - \beta \rangle$. Indeed, otherwise, E_4 and $[E_4, E_2] = -(q + q^{-1})E_3$ would belong to $\langle \chi_1 - \alpha, \chi_2 - \beta \rangle$ and then $\chi_2 = E_2 E_4 + p_3 E_3$ would also belong to $\langle \chi_1 - \alpha, \chi_2 - \beta \rangle$, a contradiction since $\beta \neq 0$.

Set $e_4 := E_4 + \langle \chi_1 - \alpha, \chi_2 - \beta \rangle \in B_{\alpha,\beta}$. Since the multiplicative system generated by E_4 in B satisfies the Ore condition on both side by [**8**, Lemme 2.1], it follows that $\{e_4^i \mid i \in \mathbb{N}\}$ is a multiplicative system of regular elements of $B_{\alpha,\beta}$ satisfying the Ore condition on both sides.

We set $R := B_{\alpha,\beta}[e_4^{-1}] = \frac{B[E_4^{-1}]}{\langle \chi_1 - \alpha, \chi_2 - \beta \rangle [E_4^{-1}]}$. Our aim in this section is to prove that R is a GWA of the form $\mathbb{K}[h^{\pm 1}](\sigma_q, a)$. To prove this, we will make use of the elements created by the first step of the DDA. Namely, recall that the first step of the algorithm leads to the following elements of $\mathrm{Frac}(B)$:

$$E_{1,4} = E_1 + p_1 E_2 E_4^{-1} + p_2 E_3 E_4^{-2} \text{ and } E_{2,4} = E_2 + p_3 E_3 E_4^{-1}.$$

Recall also that

$$\chi_1 = E_{1,4} E_3 + p_4 E_{2,4}^2 \text{ and } \chi_2 = E_{2,4} E_4.$$

This leads us to introduce the following elements of R:

$$f_1 := e_1 + p_1 e_2 e_4^{-1} + p_2 e_3 e_4^{-2}$$

and

$$f_2 := e_2 + p_3 e_3 e_4^{-1}.$$

Clearly, $f_1, f_2, e_3, e_4^{\pm 1}$ can be expressed in terms of the generators $e_1, e_2, e_3, e_4^{\pm 1}$ of R. Note that the converse is also true since $e_2 = f_2 - p_3 e_3 e_4^{-1}$ and $e_1 = f_1 - p_1 e_2 e_4^{-1} - p_2 e_3 e_4^{-2} = f_1 - p_1 (f_2 - p_3 e_3 e_4^{-1}) e_4^{-1} - p_2 e_3 e_4^{-2}$.

At this point, it is worth noting that the fact that $\chi_2 = \beta$ in $B_{\alpha,\beta}$ forces f_2 to be invertible in R and $e_4 := \beta f_2^{-1}$. As a consequence, R is generated by $f_1, f_2^{\pm 1}$

and e_3, and one can easily show using various universal properties that

$$R \simeq \frac{\mathbb{K}\langle f_1, f_2^{\pm 1}, e_3\rangle}{\langle f_1 f_2 = q^2 f_2 f_1, e_3 f_2 = q^{-2} f_2 e_3, e_3 f_1 = f_1 e_3 + \frac{q-q^3}{1+q^2} f_2^2, f_1 e_3 - \frac{q^5}{(1+q^2)^2} f_2^2 = \alpha\rangle}.$$

We are ready to prove that R is indeed a quantum GWA.

PROPOSITION 4.1. *R is isomorphic to the GWA $\mathbb{K}[h^{\pm 1}](\sigma_q, \alpha + \frac{q}{(q^2+1)^2}h^2)$ by an isomorphism that sends f_2 to h, f_1 to x and e_3 to y.*

PROOF. By universal property, it is easy to check that there exists an homomorphism

$$\phi : \mathbb{K}[h^{\pm 1}](\sigma_q, \alpha + \frac{q}{(q^2+1)^2}h^2) \to R$$

defined by $\phi(h) = f_2$, $\phi(x) = f_1$ and $\phi(y) = e_3$. By the above discussion, this homomorphism is surjective. Moreover, the Gelfand-Kirillov dimension of the domains R and $\mathbb{K}[h^{\pm 1}](\sigma_q, \alpha + \frac{q}{(q^2+1)^2}h^2)$ is equal to 2. This follows from [**19**, Proposition 2.8] for R. For the GWA $\mathbb{K}[h^{\pm 1}](\sigma_q, \alpha + \frac{q}{(q^2+1)^2}h^2)$, this follows from the fact that it contains the quantum plane $\mathbb{K}_{q^2}[h, x]$ and is contained in the quantum torus $\mathbb{K}_{q^2}[h^{\pm 1}, x^{\pm 1}]$ (and both have Gelfand-Kirillov dimension 2). Hence ϕ is an isomorphism by [**17**, Proposition 3.15]. □

As a GWA, the algebra R has a linear basis formed by the monomials $\{f_2^i f_1^j \mid i, j \in \mathbb{N}\} \cup \{f_2^i e_3^j \mid i, j \in \mathbb{N} \text{ with } j \geq 1\}$. However, it also possesses a linear basis inherited from $B_{\alpha,\beta}$, see Proposition 4.2.

PROPOSITION 4.2. *The set $\mathcal{E}' = \{e_1^i e_4^j, e_1^i e_2 e_4^j \mid (i, j) \in \mathbb{N} \times \mathbb{Z}\}$ is a $\mathbb{K}$-basis of R.*

4.2. Derivations of $B_{\alpha,\beta}$ when $\alpha\beta \neq 0$. Let $\mathrm{Der}(B_{\alpha,\beta})$ denote the set of $\mathbb{K}$-derivations of $B_{\alpha,\beta}$ and $D \in \mathrm{Der}(B_{\alpha,\beta})$. One can extend D to a derivation of the GWA R (still denoted by D). Derivations of quantum GWA have been described in [**15**] and we deduce from the isomorphism ϕ from Proposition 4.1 the following result:

LEMMA 4.3. *There exist $x \in R$ and $\lambda \in \mathbb{K}$ such that $D = \mathrm{ad}_x + \delta_\lambda$, where $\delta_\lambda \in \mathrm{Der}(R)$ is defined by $\delta_\lambda(f_2) = 0$, $\delta_\lambda(f_1) = \lambda f_1$ and $\delta_\lambda(e_3) = -\lambda e_3$.*
Moreover $\delta_\lambda(e_4) = 0$.

PROOF. This follows from [**15**, Proposition 2.3] and the fact that $e_4 = \beta f_2^{-2}$. □

We are now ready to describe D as a derivation of $B_{\alpha,\beta}$.

LEMMA 4.4.

(1) $x \in B_{\alpha,\beta}$;
(2) $\lambda = 0$;
(3) $D = \mathrm{ad}_x$.

PROOF. (1) Since $x \in R = B_{\alpha,\beta}[e_4^{-1}]$, it follows from Proposition 4.2 that one can write

$$x = \sum_{(i,j)\in I_1} a_{i,j} e_1^i e_4^j + \sum_{(i,j)\in I_2} b_{i,j} e_1^i e_2 e_4^j,$$

where $a_{i,j}, b_{i,j} \in \mathbb{K}$ are families of scalars and, I_1 and I_2 are finite subsets of $\mathbb{N} \times \mathbb{Z}$. Write $x = x_+ + x_-$, where

$$x_+ = \sum_{\substack{(i,j)\in I_1 \\ j\geq 0}} a_{i,j} e_1^i e_4^j + \sum_{\substack{(i,j)\in I_2 \\ j\geq 0}} b_{i,j} e_1^i e_2 e_4^j$$

and

$$x_- = \sum_{\substack{(i,j)\in I_1 \\ j<0}} a_{i,j} e_1^i e_4^j + \sum_{\substack{(i,j)\in I_2 \\ j<0}} b_{i,j} e_1^i e_2 e_4^j.$$

Clearly, $x_+ \in B_{\alpha,\beta}$. We want to show that $x_- = 0$. Suppose that $x_- \neq 0$. Then, there exists a minimum $j_0 < 0$ such that $a_{i,j_0} \neq 0$ or $b_{i,j_0} \neq 0$ for some $i \geq 0$, and $a_{i,j} = b_{i,j} = 0$ for all $j < j_0$. Write

$$x_- = \sum_{\substack{(i,j)\in I_1 \\ j_0\leq j<0}} a_{i,j} e_1^i e_4^j + \sum_{\substack{(i,j)\in I_2 \\ j_0\leq j<0}} b_{i,j} e_1^i e_2 e_4^j.$$

Now, $D(e_3) = \mathrm{ad}_{x_+}(e_3) + \mathrm{ad}_{x_-}(e_3) + \delta_\lambda(e_3) \in B_{\alpha,\beta}$. Since $\delta_\lambda(e_3) = -\lambda e_3 \in B_{\alpha,\beta}$, we obtain that $w := \mathrm{ad}_{x_-}(e_3) = x_- e_3 - e_3 x_- \in B_{\alpha,\beta}$. This implies that $w e_4^{-j_0-1} \in B_{\alpha,\beta}$, that is:

$$\sum_{\substack{(i,j)\in I_1 \\ j_0\leq j<0}} a_{i,j} e_1^i e_4^j e_3 e_4^{-j_0-1} + \sum_{\substack{(i,j)\in I_2 \\ j_0\leq j<0}} b_{i,j} e_1^i e_2 e_4^j e_3 e_4^{-j_0-1}$$
$$- \sum_{\substack{(i,j)\in I_1 \\ j_0\leq j<0}} a_{i,j} e_3 e_1^i e_4^{j-j_0-1} - \sum_{\substack{(i,j)\in I_2 \\ j_0\leq j<0}} b_{i,j} e_3 e_1^i e_2 e_4^{j-j_0-1} \in B_{\alpha,\beta}.$$

Using the relation $e_3 e_4 = q^2 e_4 e_3$, this leads to:

$$\sum_{\substack{(i,j)\in I_1 \\ j_0\leq j<0}} a_{i,j} q^{-2j} e_1^i e_3 e_4^{j-j_0-1} + \sum_{\substack{(i,j)\in I_2 \\ j_0\leq j<0}} b_{i,j} q^{-2j} e_1^i e_2 e_3 e_4^{j-j_0-1}$$
$$- \sum_{\substack{(i,j)\in I_1 \\ j_0\leq j<0}} a_{i,j} e_3 e_1^i e_4^{j-j_0-1} - \sum_{\substack{(i,j)\in I_2 \\ j_0\leq j<0}} b_{i,j} e_3 e_1^i e_2 e_4^{j-j_0-1} \in B_{\alpha,\beta}.$$

The only terms that could not belong to $B_{\alpha,\beta}$ in the above sums are those for which the exponents of e_4 is negative. Since $j - j_0 - 1 \geq 0$ for all $j > j_0$, we obtain that:

$$\sum_{(i,j_0)\in I_1} a_{i,j_0} q^{-2j_0} e_1^i e_3 e_4^{-1} + \sum_{(i,j_0)\in I_2} b_{i,j_0} q^{-2j_0} e_1^i e_2 e_3 e_4^{-1}$$
$$- \sum_{(i,j_0)\in I_1} a_{i,j_0} e_3 e_1^i e_4^{-1} - \sum_{(i,j_0)\in I_2} b_{i,j_0} e_3 e_1^i e_2 e_4^{-1} \in B_{\alpha,\beta}.$$

This is equivalent to

$$\sum_{(i,j_0)\in I_1} a_{i,j_0} q^{-2j_0} e_1^i e_3 + \sum_{(i,j_0)\in I_2} b_{i,j_0} q^{-2j_0} e_1^i e_2 e_3$$
$$- \sum_{(i,j_0)\in I_1} a_{i,j_0} e_3 e_1^i - \sum_{(i,j_0)\in I_2} b_{i,j_0} e_3 e_1^i e_2 \in B_{\alpha,\beta} e_4.$$

Since $e_3 - k_1\beta \in B_{\alpha,\beta}e_4$ by (3.1), this leads to

$$\sum_{(i,j_0)\in I_1} a_{i,j_0}q^{-2j_0}k_1\beta e_1^i + \sum_{(i,j_0)\in I_2} b_{i,j_0}q^{-2j_0}k_1\beta e_1^i e_2 - \sum_{(i,j_0)\in I_1} a_{i,j_0}e_3e_1^i - \sum_{(i,j_0)\in I_2} b_{i,j_0}e_3e_1^i e_2 \in B_{\alpha,\beta}e_4.$$

Using the straightening relation between e_3 and e_1^i from Lemma 3.1 and the relation $e_3e_2 = q^{-2}e_2e_3$, we get:

$$\sum_{(i,j_0)\in I_1} a_{i,j_0}q^{-2j_0}k_1\beta e_1^i + \sum_{(i,j_0)\in I_2} b_{i,j_0}q^{-2j_0}k_1\beta e_1^i e_2 - \sum_{(i,j_0)\in I_1} a_{i,j_0}e_1^i e_3 - \sum_{(i,j_0)\in I_1} a_{i,j_0}d[i]e_1^{i-1}e_2^2 - \sum_{(i,j_0)\in I_2} b_{i,j_0}q^{-2}e_1^i e_2e_3 - \sum_{(i,j_0)\in I_2} b_{i,j_0}d[i]e_1^{i-1}e_2^3 \in B_{\alpha,\beta}e_4.$$

Since $e_3 - k_1\beta \in B_{\alpha,\beta}e_4$ and $e_2^2 - \alpha k_2 + \beta k_1k_2e_1 \in B_{\alpha,\beta}e_4$ by (3.1), we obtain:

$$\begin{aligned} w' := &\sum_{(i,j_0)\in I_1} a_{i,j_0}q^{-2j_0}k_1\beta e_1^i + \sum_{(i,j_0)\in I_2} b_{i,j_0}q^{-2j_0}k_1\beta e_1^i e_2 - \sum_{(i,j_0)\in I_1} a_{i,j_0}k_1\beta e_1^i \\ &- \sum_{(i,j_0)\in I_1} a_{i,j_0}d[i]\alpha k_2e_1^{i-1} + \sum_{(i,j_0)\in I_1} a_{i,j_0}d[i]\beta k_1k_2e_1^i - \sum_{(i,j_0)\in I_2} b_{i,j_0}q^{-2}k_1\beta e_1^i e_2 \\ &- \sum_{(i,j_0)\in I_2} b_{i,j_0}d[i]\alpha k_2e_1^{i-1}e_2 + \sum_{(i,j_0)\in I_2} b_{i,j_0}d[i]q^{-2}\beta k_1k_2e_1^i e_2 \in B_{\alpha,\beta}e_4. \end{aligned}$$

(Notice that a coefficient q^{-2} has appeared in the last sum since we had to use the relation $e_2e_1 = q^{-2}e_1e_2$.)

Note that all the monomials involved in the above sums are elements of the basis $\mathcal{E}$ of $B_{\alpha,\beta}$ not involving e_4. However, for an element to belong to $B_{\alpha,\beta}e_4$, it needs to have all the coefficients of the basis elements e_1^i and $e_1^i e_2$ to be equal to 0. So, if i is maximum such that $(i,j_0) \in I_1$ and $a_{i,j_0} \neq 0$, we obtain by inspection of the coefficient of e_i^i in w':

$$a_{i,j_0}q^{-2j_0}k_1\beta - a_{i,j_0}k_1\beta + a_{i,j_0}d[i]\beta k_1k_2 = 0.$$

Since $k_1\beta \neq 0$ and $d[i]k_2 = 1 - q^{-4i}$, this leads to

$$a_{i,j_0}(q^{-2j_0} - q^{-4i}) = 0.$$

Since q is not a root of unity and $j_0 < 0$ whereas $i \geq 0$, we conclude that $a_{i,j_0} = 0$. This is a contradiction, and so all coefficients a_{i,j_0} are equal to 0.

By definition of j_0, this means that there exists $i \in \mathbb{N}$ such that $b_{i,j_0} \neq 0$. Let choose such i maximal. Then the coefficients of $e_1^i e_2$ in w' must be equal to 0 for w' to belong to $B_{\alpha,\beta}e_4$. This leads to

$$b_{i,j_0}q^{-2j_0}k_1\beta - b_{i,j_0}q^{-2}k_1\beta + b_{i,j_0}d[i]q^{-2}\beta k_1k_2 = 0.$$

Since $k_1\beta \neq 0$ and $d[i]k_2 = 1 - q^{-4i}$, this leads to

$$b_{i,j_0}\left(q^{-2j_0} - q^{-2} + q^{-2}(1 - q^{-4i})\right) = 0,$$

that is

$$b_{i,j_0}(q^{-2j_0} - q^{-4i-2}) = 0.$$

Again, $q^{-2j_0} - q^{-4i-2} \neq 0$ since q is not a root of unity, and $-2j_0 > 0$ while $-4i - 2 < 0$. Hence, $b_{i,j_0} = 0$, a contradiction with our choice of j_0.

Hence $x_- = 0$, as desired.

(2) Recall that $\delta_\lambda(e_3) = -\lambda e_3$. Also, from Lemma 4.3, $\delta_\lambda(e_4) = \delta(e_2) = 0$ and $\delta_\lambda(f_1) = \lambda f_1$. Recall that $e_2 = f_2 - p_3 e_3 e_4^{-1}$ in R. Therefore, $\delta_\lambda(e_2) = -p_3\lambda e_3 e_4^{-1}$. Now, $D(e_2) = \mathrm{ad}_x(e_2) + \delta_\lambda(e_2) \in B_{\alpha,\beta}$. Since $\mathrm{ad}_x(e_2) \in B_{\alpha,\beta}$, we have that $\delta_\lambda(e_2) = -p_3\lambda e_3 e_4^{-1} \in B_{\alpha,\beta}$. This implies that $p_3\lambda e_3 \in B_{\alpha,\beta}e_4$. Hence, $p_3\lambda k_1\beta - p_3\lambda k_1 e_2 e_4 \in B_{\alpha,\beta}e_4$ since $e_3 = k_1\beta - k_1 e_2 e_4$ by (3.1). Clearly, $e_2e_4 \in B_{\alpha,\beta}e_4$, it follows that $p_3\lambda k_1\beta \in B_{\alpha,\beta}e_4$. Since $k_1, p_3, \beta \neq 0$, this forces $\lambda = 0$. Otherwise, we would have a contradiction given the basis $\mathcal{E}$ of $B_{\alpha,\beta}$ from Proposition 3.2.

(3) Since $\lambda = 0$, then δ_λ coincides with the 0 derivation on a generating set for R. Hence $\delta_\lambda = 0$ and $D = \mathrm{ad}_x$ with $x \in B_{\alpha,\beta}$ as desired. □

4.3. Summary. We are now ready to state the main result of this note which shows that the simple quotients $B_{\alpha,\beta}$ present similarities with the first Weyl algebra when $\alpha\beta \neq 0$. Before we proceed, let us recall that the first Hochschild cohomology group of $B_{\alpha,\beta}$ (denoted by $\mathrm{HH}^1(B_{\alpha,\beta})$) is defined by

$$\mathrm{HH}^1(B_{\alpha,\beta}) := \frac{\mathrm{Der}(B_{\alpha,\beta})}{\mathrm{InnDer}(B_{\alpha,\beta})},$$

where the subset $\mathrm{InnDer}(B_{\alpha,\beta}) := \{\mathrm{ad}_x \mid x \in B_{\alpha,\beta}\}$ of $\mathrm{Der}(B_{\alpha,\beta})$ is the set of inner derivations of $B_{\alpha,\beta}$.

Recall that $\mathrm{HH}^1(B_{\alpha,\beta})$ is a free module over $Z(B_{\alpha,\beta}) = \mathbb{K}$.

THEOREM 4.5. *Assume that* $\alpha, \beta \neq 0$.

(1) *Every derivation of* $B_{\alpha,\beta}$ *is inner and* $\mathrm{HH}^1(B_{\alpha,\beta}) = 0$.
(2) $\dim \mathrm{HH}^1(B_{\alpha,0}) = \dim \mathrm{HH}^1(B_{0,\beta}) = 1$.

PROOF. The first point was proved in Lemma 4.4. For the second point, observe that $B_{\alpha,0}$ and $B_{0,\beta}$ are both quantum GWAs by [**18**, Propositions 3.9 and 3.10] and so the result follows from [**15**, Proposition 1.3]. □

5. Open questions

As explained in the introduction, simple quotients of (uniparameter) QNAs can be thought of as "algebraic deformation" of Weyl algebras. However, some of them may behave differently to Weyl algebras. For instance, they may contain nontrivial units, and this fact was for instance used to prove that endomorphisms of GWAs over a Laurent polynomial ring in one variable are automorphisms (see [**16**]). Closer to the main topic of this paper, some simple quotients have only inner derivations whereas some have larger first Hochschild cohomology group. This leads us to the following natural questions:

QUESTION 5.1. Can we classify simple quotients of uniparameter QNAs for which all derivations are inner?

QUESTION 5.2. If A is a simple quotient of a uniparameter QNA, are all endomorphisms of A automorphisms?

General results regarding Question 5.1 have been obtained while preparing this note and will be published in a forthcoming publication [**21**].

References

[1] J. Alev and F. Dumas, *Rigidité des plongements des quotients primitifs minimaux de $U_q(\mathrm{sl}(2))$ dans l'algèbre quantique de Weyl-Hayashi* (French), Nagoya Math. J. **143** (1996), 119–146, DOI 10.1017/S002776300000595X. MR1413010

[2] Nicolás Andruskiewitsch and François Dumas, *On the automorphisms of $U_q^+(\mathfrak{g})$*, Quantum groups, IRMA Lect. Math. Theor. Phys., vol. 12, Eur. Math. Soc., Zürich, 2008, pp. 107–133, DOI 10.4171/047-1/4. MR2432991

[3] V. V. Bavula, *Generalized Weyl algebras and their representations* (Russian), Algebra i Analiz **4** (1992), no. 1, 75–97; English transl., St. Petersburg Math. J. **4** (1993), no. 1, 71–92. MR1171955

[4] J. Bell, S. Launois, and N. Nguyen, *Dimension and enumeration of primitive ideals in quantum algebras*, J. Algebraic Combin. **29** (2009), no. 3, 269–294, DOI 10.1007/s10801-008-0132-5. MR2496308

[5] Alexei Belov-Kanel and Maxim Kontsevich, *The Jacobian conjecture is stably equivalent to the Dixmier conjecture* (English, with English and Russian summaries), Mosc. Math. J. **7** (2007), no. 2, 209–218, 349, DOI 10.17323/1609-4514-2007-7-2-209-218. MR2337879

[6] Richard E. Block, *The irreducible representations of the Lie algebra $\mathfrak{sl}(2)$ and of the Weyl algebra*, Adv. in Math. **39** (1981), no. 1, 69–110, DOI 10.1016/0001-8708(81)90058-X. MR605353

[7] Ken A. Brown and Ken R. Goodearl, *Lectures on algebraic quantum groups*, Advanced Courses in Mathematics. CRM Barcelona, Birkhäuser Verlag, Basel, 2002, DOI 10.1007/978-3-0348-8205-7. MR1898492

[8] Gérard Cauchon, *Effacement des dérivations et spectres premiers des algèbres quantiques* (French, with English summary), J. Algebra **260** (2003), no. 2, 476–518, DOI 10.1016/S0021-8693(02)00542-2. MR1967309

[9] W. A. de Graaf, *Computing with quantized enveloping algebras: PBW-type bases, highest-weight modules and R-matrices*, J. Symbolic Comput. **32** (2001), no. 5, 475–490, DOI 10.1006/jsco.2001.0479. MR1858005

[10] Jacques Dixmier, *Sur les algèbres de Weyl* (French), Bull. Soc. Math. France **96** (1968), 209–242. MR242897

[11] Jacques Dixmier, *Algèbres enveloppantes* (French), Cahiers Scientifiques [Scientific Reports], Fasc. XXXVII, Gauthier-Villars Éditeur, Paris-Brussels-Montreal, Que., 1974. MR498737

[12] Kenneth R. Goodearl and Milen T. Yakimov, *Quantum cluster algebras and quantum nilpotent algebras*, Proc. Natl. Acad. Sci. USA **111** (2014), no. 27, 9696–9703, DOI 10.1073/pnas.1313071111. MR3263301

[13] Takahiro Hayashi, *q-analogues of Clifford and Weyl algebras—spinor and oscillator representations of quantum enveloping algebras*, Comm. Math. Phys. **127** (1990), no. 1, 129–144. MR1036118

[14] Ellen E. Kirkman and Lance W. Small, *q-analogs of harmonic oscillators and related rings*, Israel J. Math. **81** (1993), no. 1-2, 111–127, DOI 10.1007/BF02761300. MR1231181

[15] Andrew P. Kitchin, *Derivations of quantum and involution generalized Weyl algebras*, J. Algebra Appl. **23** (2024), no. 4, Paper No. 2450076, 21, DOI 10.1142/S0219498824500762. MR4688843

[16] Andrew P. Kitchin and Stéphane Launois, *Endomorphisms of quantum generalized Weyl algebras*, Lett. Math. Phys. **104** (2014), no. 7, 837–848, DOI 10.1007/s11005-014-0691-4. MR3210235

[17] Günter R. Krause and Thomas H. Lenagan, *Growth of algebras and Gelfand-Kirillov dimension*, Revised edition, Graduate Studies in Mathematics, vol. 22, American Mathematical Society, Providence, RI, 2000, DOI 10.1090/gsm/022. MR1721834

[18] S. Launois, *On the automorphism groups of q-enveloping algebras of nilpotent Lie algebras*, From Lie Algebras to Quantum Groups, 125–143, 2006.

[19] Stéphane Launois, *Primitive ideals and automorphism group of $U_q^+(B_2)$*, J. Algebra Appl. **6** (2007), no. 1, 21–47, DOI 10.1142/S0219498807002053. MR2302694

[20] S. Launois, T. H. Lenagan, and L. Rigal, *Quantum unique factorisation domains*, J. London Math. Soc. (2) **74** (2006), no. 2, 321–340, DOI 10.1112/S0024610706022927. MR2269632

[21] S. Launois, S. A. Lopes and I. Oppong, *On the first Hochschild cohomology group of quantum nilpotent algebras*, In preparation.

[22] S. Launois and I. Oppong, *Derivations of a family of quantum second Weyl algebras*, Bull. Sci. Math. **184** (2023), Paper No. 103257, 43, DOI 10.1016/j.bulsci.2023.103257. MR4571198
[23] Samuel A. Lopes, *Primitive ideals of* $U_q(\mathfrak{sl}_n^+)$, Comm. Algebra **34** (2006), no. 12, 4523–4550, DOI 10.1080/00927870600936682. MR2273722
[24] Yoshifumi Tsuchimoto, *Preliminaries on Dixmier conjecture*, Mem. Fac. Sci. Kochi Univ. Ser. A Math. **24** (2003), 43–59. MR1967342

Université de Caen Normandie, UMR 6139 LMNO, 14032 Caen, France
Email address: stephane.launois@unicaen.fr

School of Computing and Mathematical Sciences, University of Greenwich, Old Naval College, Park Row, Greenwich, London, SE10 9LS, United Kingdom
Email address: I.Oppong@greenwich.ac.uk

Contemporary Mathematics
Volume **826**, 2025
https://doi.org/10.1090/conm/826/16589

Essentially quasi-duo rings

Christian Lomp, Mohamed Yousif, and Yiqiang Zhou

Abstract. A ring with identity is called essentially right quasi-duo if every essential maximal right ideal of it is a two-sided ideal. Essentially right quasi-duo rings generalize essentially right duo rings, a notion that arose in the study of hypercyclic rings, and right quasi-duo rings, as introduced by S.H. Brown. We prove that a ring R is essentially right quasi-duo if and only if R is semisimple or $R/\operatorname{Soc}(R_R)$ is right quasi-duo. Although it is still unknown, whether a right quasi-duo ring is left quasi-duo, we provide an example of an essentially right quasi-duo ring that is not essentially left quasi-duo. Furthermore, while exchange right quasi-duo rings are known to be clean, there exist exchange essentially right quasi-duo rings that are not clean. A thorough study of essentially right quasi-duo rings is carried out and their relationship to skew power series rings, trivial extensions and formal triangular matrix rings is explored.

1. Introduction

Throughout, rings are associative with identity and modules are unitary. It is well-known that the following three conditions on a ring R are equivalent to R being semisimple Artinian: Every right ideal of R is a direct summand; every essential right ideal of R is a direct summand; every maximal right ideal of R is a direct summand. Moreover, because an essential maximal right ideal of a non-trivial ring can't be a direct summand, every essential maximal right ideal of R being a direct summand implies that R is a trivial ring. Here is another set of conditions on a ring R:

(1) Every right ideal of R is an ideal;
(2) Every essential right ideal of R is an ideal;
(3) Every maximal right ideal of R is an ideal;
(4) Every essential maximal right ideal of R is an ideal.

2020 *Mathematics Subject Classification.* Primary 16D10, 16D50, 16L30, 16S50.

Key words and phrases. Essentially quasi-duo ring, quasi-duo ring, semilocal ring, ring with cyclic injective hull.

The first author was partially supported by CMUP, member of LASI, which is financed by national funds through FCT - Fundação para a Ciência e a Tecnologia, I.P., under the projects with reference UIDB/00144/2020 and UIDP/00144/2020.

The third author was supported in part by a Discovery Grant (RGPIN-2022-03783) from NSERC of Canada.

In the literature, rings with (1) and rings with (3) are called right duo rings and, respectively, right quasi-duo rings. The definitions are due to Feller [**1**] and, respectively, Brown [**2**, Remark 3.21]. Right duo rings and right quasi-duo rings have been extensively studied (see, e.g. [**7**, **12**, **14**, **15**, **19**, **23**]). Rings with (2) are termed essentially right duo rings and were motivated by the study of hypercyclic rings in [**17**]. In this paper, we initiate a study of rings satisfying (4), and these rings are called essentially right quasi-duo rings. They include right quasi-duo rings, essentially right duo rings, semisimple Artinian rings and some other classes of rings. The essentially right quasi-duo property of various ring extensions is considered in Section 2 and, as consequences, many examples of rings with this property are presented. Essentially right quasi-duo rings are characterized as semisimple Artinian rings or those rings R such that $R/\operatorname{Soc}(R_R)$ is right quasi-duo (Theorem 2.11). Moreover, essentially right quasi-duo rings that are regular and right CS are completely determined (Corollary 2.17). Section 3 presents a characterization of a formal triangular matrix ring that is essentially right quasi-duo (Theorem 3.1). This result is used to construct an essentially right quasi-duo ring that is not essentially left quasi-duo (Example 3.3), and to establish the structure of an essentially right quasi-duo semilocal ring (Theorem 3.7). In Section 4, we revisit the open question whether, for a ring R, the injective hull $E(R_R)$ being cyclic implies the (right) injectivity of R. It is known that the answer is "yes" if R is right quasi-duo. Here this is extended to R being essentially right quasi-duo in Theorem 4.3, as a consequence of the result that, for a finitely generated projective module M, every essential maximal submodule of M is fully invariant if and only if M does not contain proper submodules X, Y such that $M = X + Y$ and $M/X \cong M/Y$ is singular (Corollary 4.2).

For a ring R, the Jacobson radical, the right socle and the left socle of R are denoted by $J(R)$, $\operatorname{Soc}(R_R)$ and $\operatorname{Soc}({}_RR)$, respectively. The $n \times n$ matrix ring (resp. upper triangular matrix ring, or lower triangular matrix ring) over R is denoted by $\mathbb{M}_n(R)$ (resp. $\mathbb{UT}_n(R)$ or $\mathbb{LT}_n(R)$). The injective hull of the regular module R_R is denoted by $E(R_R)$. We write homomorphisms of right modules on the left of their arguments.

2. Essentially quasi-duo rings

DEFINITION 2.1. A ring R is called essentially right quasi-duo if every essential maximal right ideal of R is an ideal.

Essentially right quasi-duo rings include right quasi-duo rings and essentially right duo rings.

EXAMPLE 2.2. There exist essentially right quasi-duo rings that are neither essentially right duo nor right quasi-duo.

PROOF. Let D be a division ring and R be an integral domain that is not a field. Then $\mathbb{M}_2(D)$ is an essentially quasi-duo ring that is not right quasi-duo and $\mathbb{LT}_2(R)$ is a quasi-duo ring that is not essentially right duo (see [**17**]). Hence $\mathbb{M}_2(D) \oplus \mathbb{LT}_2(R)$ is an essentially quasi-duo ring that is neither essentially right duo nor right quasi-duo. □

Indeed, there exist indecomposable essentially right quasi-duo rings that are neither essentially right duo nor right quasi-duo (see Example 3.3).

Next we study the essentially quasi-duo property of some standard ring extensions.

PROPOSITION 2.3.

(1) *If R is essentially right quasi-duo (resp. essentially right duo), then so is R/I for any proper ideal I of R.*
(2) *If $R = \prod_{i=1}^{n} R_i$ is a direct product of rings, then R is essentially right quasi-duo (resp. essentially right duo) if and only if so is each R_i.*
(3) *If $n \geq 2$, then $\mathbb{M}_n(R)$ is essentially right quasi-duo (resp. essentially right duo) if and only if R is a semisimple Artinian ring.*

PROOF. (1) If T is an essential (maximal) right ideal of R/I, then $T = K/I$ where K is an essential (maximal) right ideal of R. So K is an ideal of R and hence T is an ideal of R/I.

(2) The necessity is by (1). For the sufficiency, let T be an essential (maximal) right ideal of R. Then $T = \prod_{i=1}^{n} TR_i$ where, for each i, TR_i is an essential (maximal) right ideal of R_i or $TR_i = R_i$. So each TR_i is an ideal of R_i. Hence T is an ideal of R.

(3) It is known that a ring is semisimple Artinian if and only if each of its maximal right ideals is a direct summand. Assume that R is not semisimple Artinian. Then R has a maximal right ideal I that is not a direct summand of R_R. Thus I is an essential maximal right ideal. So $\begin{pmatrix} I & I & \cdots & I \\ R & R & \cdots & R \\ \vdots & \vdots & \cdots & \vdots \\ R & R & \cdots & R \end{pmatrix}$ is an essential maximal right ideal, but not an ideal, of $\mathbb{M}_n(R)$. Hence $\mathbb{M}_n(R)$ is not essentially right quasi-duo.

If R is semisimple Artinian, then $\mathbb{M}_n(R)$ is semisimple Artinian, and hence has itself as the only essential right ideal. So $\mathbb{M}_n(R)$ is essentially right duo. □

For an endomorphism σ of a ring R, let $R[[t;\sigma]]$ (resp. $R[t;\sigma]$) be the ring of left skew power series (resp. left polynomials) over R with multiplication subject to the relation $ta = \sigma(a)t$ for all $a \in R$.

PROPOSITION 2.4. *Let R be a ring, $\sigma : R \to R$ be an endomorphism of rings and $n \geq 1$. The following are equivalent:*

(1) *$R[[t;\sigma]]$ is essentially right quasi-duo (resp. right quasi-duo).*
(2) *$R[t;\sigma]/(t^{n+1})$ is essentially right quasi-duo (resp. right quasi-duo).*
(3) *R is right quasi-duo.*

PROOF. (1) $\Rightarrow$ (2). Since $R[t;\sigma]/(t^{n+1})$ is a homomorphic image of $R[[t;\sigma]]$, the implication follows from Proposition 2.3(1).

(2) $\Rightarrow$ (3). Let I be a maximal right ideal of R. Then $T := I + Rt + \ldots + Rt^n$ is an essential maximal right ideal of $R[t;\sigma]/(t^{n+1})$, so it is an ideal. It follows that I is an ideal of R. So R is right quasi-duo.

(3) $\Rightarrow$ (1). Let T be an (essential) maximal right ideal of $R[[t;\sigma]]$. Then $T = I + R[[t;\sigma]]t$ where I is a maximal right ideal of R. So I is an ideal of R, and hence T is an ideal of $R[[t;\sigma]]$. □

It is known that a left duo ring need not be right duo (see [**3**]). Below we show that, for an endomorphism $\sigma : R \to R$, if $R[t;\sigma]/(t^{n+1})$ is essentially right duo, then R is right duo; but the converse is not true as seen below. Below it is also seen that an essentially left duo ring need not be essentially right duo.

EXAMPLES 2.5.

(1) Let V_D be a vector space over a division ring D and $R = \text{End}(V_D)$. Then R is essentially right quasi-duo (resp. essentially right duo) if and only if $\dim(V_D) < \infty$.
(2) If R is a simple essentially right quasi-duo ring, then R is a simple Artinian ring.
(3) If R is an essentially right quasi-duo ring with $R/J(R)$ simple and idempotents lift modulo $J(R)$, then R is local or simple Artinian.
(4) An infinite direct product of essentially right quasi-duo (resp. essentially right duo) rings need not be essentially right quasi-duo (resp. essentially right duo).
(5) An essentially left duo ring need not be essentially right duo.
(6) Let F be a field with an endomorphism $\sigma : F \to F$ such that $0 \neq \sigma(F) \subsetneqq F$. Then $F[t;\sigma]/(t^{n+1})$ is essentially right duo if and only if $n = 1$.

PROOF. (1) The sufficiency is clear. If $\dim(V_D) = \infty$, then $V_D \cong V_D^2$, so $R = \text{End}(V_D) \cong \text{End}(V_D^2) \cong \mathbb{M}_2(R)$. Since R is not semisimple Artinian, $\mathbb{M}_2(R)$ is not essentially right quasi-duo by Proposition 2.3(3), so R is not essentially right quasi-duo.

(2) It suffices to show that R is semisimple Artinian. Assume that R is not semisimple Artinian. Then R has an essential maximal right ideal I, so I is a proper ideal of R. Hence $I = 0$ because R is simple, contradicting that I is essential in R_R. So R is semisimple Artinian.

(3) Since $R/J(R)$ is a simple essentially right quasi-duo ring, $R/J(R)$ is simple Artinian by (2), so $R/J(R) \cong \mathbb{M}_n(D)$ where $n \geq 1$ and D is a division ring. Since idempotents lift modulo $J(R)$, $R \cong \mathbb{M}_n(S)$ where $S/J(S) \cong D$. If $n = 1$, then $R \cong S$ is local. If $n \geq 2$, then $J(S) = 0$ by Proposition 2.3(3), so R is simple Artinian.

(4) Let $R = \prod_{i=1}^{\infty} R_i$ where $R_i = \mathbb{M}_2(F_i)$ and F_i is a field. Then each R_i is essentially right duo, while $R \cong \mathbb{M}_2(\prod_{i=1}^{\infty} F_i)$ is not essentially right quasi-duo, since $\prod_{i=1}^{\infty} F_i$ is not semisimple Artinian.

(5) By [**11**, Theorem 1], the skew power series ring $R := D[[x;\sigma]]$, where D is a commutative principal ideal domain and σ is a monomorphism from the quotient field of D to D, is left duo. If σ is not onto, then R is not right duo by [**3**, Examples, p.157]. Indeed, R is not essentially right duo, because $I := \sigma(D)x + Dx^2 + Dx^3 + \dots$ is an essential right ideal of R that is not an ideal.

(6) Let $R = F[t;\sigma]/(t^{n+1})$. If $n \geq 2$, then $\sigma(F)t + Ft^2 + \dots + Ft^n$ is an essential right ideal of R, which is not a left ideal. So R is not essentially right duo.

If $n = 1$, let I be an essential right ideal of R. We show that I is an ideal. We can assume that $I \neq R$. Then $I \subseteq Ft$. For any $0 \neq a \in F$, $0 \neq at \in R$, so there exists $u \in F$ such that $a\sigma(u)t \in I$. Then $at = a\sigma(u)\sigma(u^{-1})t = (a\sigma(u)t)u^{-1} \in I$. This shows that $I = Ft$, which is an ideal of R. □

PROPOSITION 2.6. *Let R be a ring and $0 \neq e^2 = e \in R$.*

(1) *If R is essentially right duo, then so is eRe.*
(2) *If R is essentially right quasi-duo, then so is eRe.*

PROOF. (1) Let I be an essential right ideal of eRe. Then IR is a submodule of $(eR)_R$. There exists a submodule Y of $(eR)_R$ such that $IR \oplus Y$ is essential in $(eR)_R$. Then $K := IR \oplus Y \oplus (1-e)R$ is an essential right ideal of R. Since R is essentially right duo, K is an ideal of R, so eKe is an ideal of eRe. Since $I \cap Ye \subseteq IR \cap Y = 0$, $I \cap Ye = 0$. So $Ye = 0$, because Ye is a right ideal of eRe and I is an essential right ideal of eRe. Hence $eKe = IRe + Ye = IRe = IeRe = I$, so I is an ideal of eRe.

(2) Let I be an essential maximal right ideal of eRe. Noting that $IR = eR$ would imply $e \in eRe = eIRe = I(eRe) = I$, we have $IR \subsetneq eR$. Let X be a maximal submodule of $(eR)_R$ such that $IR \subseteq X$. Then $K := X + (1-e)R$ is a maximal right ideal of R. We next verify that $I = Xe$ and X is an essential submodule of $(eR)_R$. Firstly, $I \subseteq Xe \subseteq eRe$. Secondly, $Xe \neq eRe$: If $Xe = eRe$, then $e \in Xe \subseteq X$, showing $X = eR$, a contradiction. Since I is a maximal right ideal of eRe, it must be that $I = Xe$. Assume that X is not essential in $(eR)_R$. Then $X \oplus Y = eR$ where Y is a simple submodule of $(eR)_R$, and so $Xe \oplus Ye = eRe$. It follows that $Ye = 0$, so $Xe = eRe$, a contradiction. So X is essential in $(eR)_R$. Hence, K is an essential right ideal of R. Since R is essentially quasi-duo, K is an ideal of R. Hence $I = Xe = eXe = eKe$ is an ideal of eRe. □

Let C be a (unital) subring of a ring D with $1_C = 1_D$. The set

$$\mathcal{R}[D, C] := \{(d_1, \ldots, d_n, c, c, \ldots) : c \in C, n \geq 1, d_i \in D \text{ for } i = 1, \ldots, n\}$$

is a ring where addition and multiplication are defined component-wise.

PROPOSITION 2.7. *If the ring $\mathcal{R}[D, C]$ is essentially right quasi-duo (resp. essentially right duo), then C and D are essentially right quasi-duo (resp. essentially right duo). The converse is false.*

PROOF. Since D and C are homomorphic images of $\mathcal{R}[D, C]$, the necessity follows from Proposition 2.3(1).

If $D = C = \mathbb{M}_2(F)$ where F is a field, then D and C are essentially right duo. But, since $\mathcal{R}[F, F]$ is not semisimple Artinian, $\mathcal{R}[D, C] \cong \mathbb{M}_2(\mathcal{R}[F, F])$ is not essentially right quasi-duo. □

If M is a left module over a ring R, let $\mathbf{l}_R(M) = \{a \in R : aM = 0\}$.

PROPOSITION 2.8. *Let R be a ring and M be an (R, R)-bimodule. Then the trivial extension $R \propto M$ is essentially right quasi-duo if and only if every maximal right ideal I of R such that $I \cap aR \neq 0$ for any $0 \neq a \in \mathbf{l}_R(M)$ is an ideal.*

PROOF. ($\Rightarrow$). If I is a maximal right ideal of R such that $I \cap aR \neq 0$ for any $0 \neq a \in \mathbf{l}_R(M)$, then $I \propto M$ is an essential maximal right ideal of $R \propto M$, so it is an ideal. Hence I is an ideal of R.

($\Leftarrow$). Let T be an essential maximal right ideal of $R \propto M$. Then $T = I \propto M$ with I a maximal right ideal of R. Moreover, T essential in $R \propto M$ implies that $I \cap aR \neq 0$ for any $0 \neq a \in \mathbf{l}_R(M)$. So I is an ideal of R by our assumption, and hence T is an ideal of $R \propto M$. □

It is well-known that a ring R is right quasi-duo if and only if $R/J(R)$ is right quasi-duo. If R is essentially right quasi-duo, then so is $R/J(R)$ by Proposition 2.3(1); but the converse is not true.

EXAMPLE 2.9. Let R be a semilocal ring with $J(R) \neq 0$. Then, for $T = \mathbb{M}_n(R)$ where $n \geq 2$, $T/J(T) \cong \mathbb{M}_n(R/J(R))$ is essentially right duo, but T is not essentially right quasi-duo.

In [**24**], a submodule N of a module M is called δ-small in M if $M \neq X + N$ for any proper submodule X with M/X singular. The sum of all δ-small submodules of a module M is denoted by $\delta(M)$. As shown in [**24**], $\delta(R_R)$ is an ideal of R which is the intersection of all essential maximal right ideals of R, so $\operatorname{Soc}(R_R) \subseteq \delta(R_R)$ and $J(R) \subseteq \delta(R_R)$; indeed, $J(R/\operatorname{Soc}(R_R)) = \delta(R_R)/\operatorname{Soc}(R_R)$.

Below is an example of a ring R such that $R/\operatorname{Soc}(R_R) = R/J(R) = R/\delta(R_R)$ is essentially right quasi-duo but R is not essentially right quasi-duo.

EXAMPLE 2.10. Let $R = \mathbb{M}_2(F) \propto \mathbb{M}_2(F)$ be the trivial extension where F is a field. Then $\operatorname{Soc}(R_R) = 0 \propto \mathbb{M}_2(F) = J(R)$, which is an essential right ideal of R. Since $R/\operatorname{Soc}(R_R) \cong \mathbb{M}_2(F)$, $J(R/\operatorname{Soc}(R_R)) = 0$, so $\delta(R_R) = \operatorname{Soc}(R_R)$. Hence $R/\delta(R_R) = R/\operatorname{Soc}(R_R) = R/J(R)$ is an essentially right quasi-duo ring. Let $I = \{(a,x) \in R : a \in \begin{pmatrix} F & F \\ 0 & 0 \end{pmatrix}, x \in \mathbb{M}_2(F)\}$. Then I is an essential maximal right ideal of R, but not an ideal. So R is not essentially right quasi-duo.

Essentially right quasi-duo rings are characterized below.

THEOREM 2.11. *The following are equivalent for a ring R:*

(1) *R is essentially right quasi-duo.*
(2) *R is semisimple Artinian or $R/\delta(R_R)$ is right quasi-duo.*
(3) *R is semisimple Artinian or $R/\operatorname{Soc}(R_R)$ is right quasi-duo.*

PROOF. (1) $\Rightarrow$ (3). Assume that R is not semisimple Artinian. Then $\operatorname{Soc}(R_R) \neq R$. Let $K/\operatorname{Soc}(R_R)$ be a maximal right ideal of $R/\operatorname{Soc}(R_R)$. Then K is an essential maximal right ideal of R. So K is an ideal of R, and hence $K/\operatorname{Soc}(R_R)$ is an ideal of $R/\operatorname{Soc}(R_R)$.

(3) $\Rightarrow$ (2). By [**24**], $\delta(R_R)/\operatorname{Soc}(R_R) = J(R/\operatorname{Soc}(R_R))$, so the implication follows from this and the fact that factor rings of a right quasi-duo ring are right quasi-duo.

(2) $\Rightarrow$ (1). We can assume that R is not semisimple Artinian. So $R/\delta(R_R)$ is right quasi-duo. If K is an essential maximal right ideal of R, then $\delta(R_R) \subseteq K$, so $K/\delta(R_R)$ is a maximal right ideal of $R/\delta(R_R)$. Hence $K/\delta(R_R)$ is an ideal of $R/\delta(R_R)$, and it follows that K is an ideal of R. □

Note that Theorem 2.11 provides an alternative proof of Proposition 2.6(2), because if $e \in R$ is an idempotent of an essentially right quasi-duo ring R, then in case R is semisimple, also eRe is semisimple. In case $R/\operatorname{Soc}(R_R)$ is right quasi-duo, then by Proposition 2.6(1) also $\overline{e}R/\operatorname{Soc}(R_R)\overline{e}$ is right quasi-duo, where $\overline{e} = e + \operatorname{Soc}(R_R)$. Since $\overline{e}\,(R/\operatorname{Soc}(R_R))\,\overline{e} \simeq eRe/\operatorname{Soc}(eRe_{eRe})$, we obtain that eRe is essentially right quasi-duo.

It follows from Theorem 2.11 that, for a ring R with $\operatorname{Soc}(R_R) = 0$, R is essentially right quasi-duo if and only if it is right quasi-duo. A ring R is reduced if it contains no nonzero nilpotent elements.

COROLLARY 2.12. *If R is an essentially right quasi-duo ring that is not semisimple Artinian, then $R/\delta(R_R)$ is a reduced ring.*

PROOF. By Theorem 2.11, $R/\delta(R_R)$ a right quasi-duo ring. Since $J(R/\delta(R_R)) = 0$, it follows from [**23**, Corollary 2.4] that $R/\delta(R_R)$ is reduced. □

COROLLARY 2.13. *A semilocal ring R is essentially right quasi-duo if and only if R is semisimple Artinian or $R/\delta(R_R)$ is a finite direct product of division rings.*

PROOF. The sufficiency follows from Theorem 2.11. For the necessity, assume that R is not semisimple Artinian. Then $J(R) \subseteq \delta(R_R) \subsetneq R$ by [**24**]. So $R/\delta(R_R)$ is semisimple Artinian, and right quasi-duo (by Theorem 2.11). By Corollary 2.12, $R/\delta(R_R)$ is reduced, so it must be a finite direct product of division rings. □

A (von Neumann) regular ring is right quasi-duo if and only if it is strongly regular (see [**23**]).

COROLLARY 2.14. *A regular ring R is essentially right quasi-duo if and only if either R is semisimple Artinian, or $R/\operatorname{Soc}(R_R)$ is strongly regular.*

PROOF. The sufficiency follows from Theorem 2.11. For the necessity, assume that R is not semisimple Artinian. Then $R/\operatorname{Soc}(R_R)$ is right quasi-duo by Theorem 2.11. Since R is regular, $R/\operatorname{Soc}(R_R)$ is regular. So $R/\operatorname{Soc}(R_R)$ is strongly regular by [**23**, Theorem 2.7]. □

The quasi-duo property of skew polynomial rings is extensively discussed in [**14**].

COROLLARY 2.15. *Let $\sigma : R \to R$ be an endomorphism of ring R. Then $R[t;\sigma]$ is essentially right quasi-duo if and only if $R[t;\sigma]$ is right quasi-duo.*

PROOF. We only need to verify the necessity. For any nonzero right ideal T of $R[t;\sigma]$, Tt is a nonzero right ideal properly contained in T, so the right socle of $R[t;\sigma]$ is zero. Hence the necessity follows from Theorem 2.11. □

The ring in Example 2.5(1) is right self-injective regular. We next characterize right self-injective regular rings that are essentially right quasi-duo. Some related notions need be recalled. A ring R is directly finite if $ab = 1$ in R always implies $ba = 1$. An idempotent e in a regular ring R is called an Abelian idempotent if eRe is an Abelian ring (i.e. all idempotents of eRe are central), and is called a directly finite idempotent if eRe is a directly finite ring. An idempotent e in a right self-injective regular ring is called a faithful idempotent if 0 is the only central idempotent orthogonal to e.

THEOREM 2.16. *Let R be a right self-injective regular ring. Then R is essentially right quasi-duo (resp. essentially right duo) if and only if R is a direct product of a semisimple Artinian ring and a strongly regular ring.*

PROOF. Since semisimple Artinian rings and strongly regular rings are all essentially right duo, the sufficiency follows by Proposition 2.3(2).

To show the necessity, let R be essentially right quasi-duo. By [**5**, Theorem 10.22], there is a decomposition of rings

$$R = A \times B \times C,$$

where A is of type I_f (i.e. A is a directly finite ring containing a faithful Abelian idempotent), B is of type II_f (i.e. B is a directly finite ring that contains a faithful directly finite idempotent but contains no nonzero Abelian idempotents), and C

is purely infinite (i.e. C contains no nonzero directly finite central idempotents). By [**5**, Theorem 10.24], $A \cong \prod_{i=1}^{\infty} \mathbb{M}_i(S_i)$ where each S_i is a strongly regular ring. Let $S = \prod_{i=2}^{\infty} \mathbb{M}_i(S_i)$. Then S has a corner ring isomorphic to $\prod_{i=2}^{\infty} \mathbb{M}_2(S_i) \cong \mathbb{M}_2(\prod_{i=2}^{\infty} S_i)$. By Proposition 2.6, S is essentially right quasi-duo, so $\prod_{i=2}^{\infty} S_i$ is semisimple Artinian by Proposition 2.3(3). Hence there exists $n_0 > 2$ such that $S_i = 0$ for all $i > n_0$. It follows that $S = \prod_{i=2}^{n_0} \mathbb{M}_i(S_i)$ is semisimple Artinian. So $A \cong S_1 \times S$ is a direct product of a strongly regular ring and a semisimple Artinian ring.

Since B is of type II_f, by [**5**, Proposition 10.28] there exists $e^2 = e \in B$ such that $B_B \cong (eB \oplus eB)_B$, so $B \cong \text{End}((eB)^2_B) \cong \mathbb{M}_2(eBe)$, which is essentially right quasi-duo. So eBe is semisimple Artinian, and hence B is semisimple Artinian. Since B is of type II_f, it must be that $B = 0$.

Since C is purely infinite, by [**5**, Theorem 10.16] $C_C \cong (C \oplus C)_C$, so $C \cong \mathbb{M}_2(C)$, which is essentially right quasi-duo. So C is a semisimple Artinian ring. It must be that $C = 0$.

Therefore, $R \cong A \cong S_1 \times S$ is a direct product of a strongly regular ring and a semisimple Artinian ring. □

A ring R is called right CS if every submodule of R_R is essential in a direct summand of R_R, and a right CS ring R is called right continuous if the module R_R satisfies $C2$, i.e. every submodule of R_R isomorphic to a direct summand is itself a direct summand.

COROLLARY 2.17. *A regular right CS ring R is essentially right quasi-duo (resp. essentially right duo) if and only if it is the direct product of a semisimple Artinian ring and a strongly regular ring.*

PROOF. The sufficiency is clear. For the necessity, since R is regular, the module R_R satisfies $C2$, so R is right continuous (and regular). By [**18**, Theorem 3.1, Corollary 3.13], $R \cong A \times B$ where A is right self-injective regular and B is strongly regular. Since A is essentially right quasi-duo, by Theorem 2.16 A is a direct product of a semisimple Artinian ring and a strongly regular ring. Hence R is a direct product of a semisimple Artinian ring and a strongly regular ring. □

3. Formal triangular matrix rings and applications

While Corollary 2.13 gives a characterization of essentially right quasi-duo semilocal rings, here we will attain the structure of essentially right quasi-duo semilocal rings. To do so, we need to characterize when a formal triangular matrix ring is essentially right quasi-duo.

THEOREM 3.1. *Let A, B be rings, M be an (A, B)-bimodule. The following are equivalent:*

(1) *The formal triangular matrix ring $\begin{pmatrix} A & M \\ 0 & B \end{pmatrix}$ is essentially right quasi-duo.*
(2) *Every maximal right ideal I of A such that $I \cap aA \neq 0$ for any $0 \neq a \in \mathbf{l}_A(M)$ is an ideal, and B is essentially right quasi-duo.*

PROOF. (1) ⇒ (2). Let $R = \begin{pmatrix} A & M \\ 0 & B \end{pmatrix}$. Being an image of R, B is essentially right quasi-duo. If I is a maximal right ideal of A such that $I \cap aA \neq 0$ for any

$0 \neq a \in \mathbf{l}_A(M)$, then $\begin{pmatrix} I & M \\ 0 & B \end{pmatrix}$ is an essential maximal right ideal of R, so it is an ideal. Hence I is an ideal of A.

(2) $\Rightarrow$ (1). Let T be an essential maximal right ideal of R. Then, by [**6**, Proposition 2.1], $T = \begin{pmatrix} I & M \\ 0 & B \end{pmatrix}$ with I a maximal right ideal of A or $T = \begin{pmatrix} A & M \\ 0 & K \end{pmatrix}$ with K a maximal right ideal of B. If $T = \begin{pmatrix} A & M \\ 0 & K \end{pmatrix}$, then T essential in R_R implies that K is essential in B_B. So K is an ideal of B, and hence T is an ideal of R. If $T = \begin{pmatrix} I & M \\ 0 & B \end{pmatrix}$, then T essential in R_R implies that $I \cap aA \neq 0$ for any $0 \neq a \in \mathbf{l}_A(M)$. So I is an ideal of A by (2), and hence T is an ideal of R. $\square$

COROLLARY 3.2. *Let R be a ring and $n \geq 2$. Then $\mathbb{UT}_n(R)$ is essentially right quasi-duo if and only if R is right quasi-duo*

PROOF. Write $\mathbb{UT}_n(R) = \begin{pmatrix} A & M \\ 0 & B \end{pmatrix}$, where $A = R, B = \mathbb{UT}_{n-1}(R)$ and $M = R^{n-1}$. Then $\mathbf{l}_R(M) = 0$. So the necessity follows from Theorem 3.1. For the sufficiency, B is essentially right quasi-duo by induction. Hence $\mathbb{UT}_n(R)$ is essentially right quasi-duo by Theorem 3.1. $\square$

As mentioned earlier, an essentially right duo (resp. right duo) ring need not be essentially left duo (resp. left duo). It is still an open question whether a right quasi-duo ring is left quasi-duo. Here we show that an essentially right quasi-duo ring need not be essentially left quasi-duo.

EXAMPLE 3.3. Let $R = \begin{pmatrix} A & M \\ 0 & B \end{pmatrix}$ be the subring $\begin{pmatrix} F & F & F \\ 0 & F & F \\ 0 & F & F \end{pmatrix}$ of $\mathbb{M}_3(F)$, where F is a field with $A = F$ and $B = \mathbb{M}_2(F)$. Since A is right quasi-duo and B is essentially right quasi-duo, R is essentially right quasi-duo by Theorem 3.1. Let $I = \begin{pmatrix} F & F & F \\ 0 & 0 & F \\ 0 & 0 & F \end{pmatrix}$. One can check that I is an essential maximal left ideal of R that is not an ideal. So R is not essentially left quasi-duo. Indeed, $\operatorname{Soc}(R_R) = \begin{pmatrix} 0 & F & F \\ 0 & F & F \\ 0 & F & F \end{pmatrix}$ and $\operatorname{Soc}({}_RR) = \begin{pmatrix} F & F & F \\ 0 & 0 & 0 \\ 0 & 0 & 0 \end{pmatrix}$. So $R/\operatorname{Soc}(R_R) \cong F$ is right quasi-duo, but $R/\operatorname{Soc}({}_RR) \cong \mathbb{M}_2(F)$ is not left quasi-duo.

Note that the ring R in Example 3.3 is Artinian and can even be finite (when F is a finite field). We know that $\operatorname{Soc}(R_R) = \operatorname{Soc}({}_RR)$ for any regular ring (indeed for any semiprime ring) R. Thus, by Corollary 2.14, a regular ring is essentially right quasi-duo if and only if it is essentially left quasi-duo. This result has an extension. Recall that a ring R is an exchange ring if for any $a \in R$, there exists $e^2 = e$ such that $e \in aR$ and $1 - e \in (1 - a)R$.

PROPOSITION 3.4. *Let R be a ring with $\delta(R_R) = \delta({}_RR)$ such that $R/\delta(R_R)$ is an exchange ring. Then R is essentially right quasi-duo if and only if it is essentially left quasi-duo.*

PROOF. Let R be essentially right quasi-duo. It suffices to show that R is essentially left quasi-duo. We can assume that R is not semisimple Artinian. So, by Theorem 2.11, $R/\operatorname{Soc}(R_R)$ is right quasi-duo. Moreover, $\frac{R/\operatorname{Soc}(R_R)}{J(R/\operatorname{Soc}(R_R))} \cong R/\delta(R_R)$ is an exchange ring. By [**12**, Theorem 4.6], we infer that $R/\operatorname{Soc}(R_R)$ is left quasi-duo. So, being an image of $R/\operatorname{Soc}(R_R)$, $R/\delta({}_RR) = R/\delta(R_R)$ is left quasi-duo. Hence R is essentially left quasi-duo by Theorem 2.11. □

We remark that, for a ring R with $S := \operatorname{Soc}(R_R) = \operatorname{Soc}({}_RR)$, $\delta(R_R)/\operatorname{Soc}(R_R) = J(R/S) = \delta({}_RR)/\operatorname{Soc}({}_RR)$, so $\delta(R_R) = \delta({}_RR)$.

COROLLARY 3.5. *Let R be a ring with $\operatorname{Soc}(R_R) = \operatorname{Soc}({}_RR)$ such that $R/J(R)$ is an exchange ring. Then R is essentially right quasi-duo if and only if it is essentially left quasi-duo.*

For an essentially left and right quasi-duo ring R, the ideal $\delta(R_R)$ need not coincide with $\delta({}_RR)$ as the following example shows: Let $R = \mathbb{UT}_2(F)$ where F is a field. Then R is essentially left and right quasi-duo ring, but $\delta(R_R) = \operatorname{Soc}(R_R) = \begin{pmatrix} 0 & F \\ 0 & F \end{pmatrix}$ and $\delta({}_RR) = \operatorname{Soc}({}_RR) = \begin{pmatrix} F & F \\ 0 & 0 \end{pmatrix}$.

Next we prove a structure theorem of essentially right quasi-duo semilocal rings. The proof of the following lemma is routine.

LEMMA 3.6. *Let M be a semisimple module. If every maximal submodule of M is fully-invariant, then M is a direct sum of non-isomorphic simple modules.*

Because homomorphisms of right modules are written on the left of their arguments, for two right modules M and N, $\operatorname{Hom}(M,N)$ is an $(\operatorname{End}(N), \operatorname{End}(M))$-bimodule, $\operatorname{Hom}(N,M)$ is an $(\operatorname{End}(M), \operatorname{End}(N))$-bimodule, and

$$\operatorname{End}(M \oplus N) \cong \begin{pmatrix} \operatorname{End}(M) & \operatorname{Hom}(N,M) \\ \operatorname{Hom}(M,N) & \operatorname{End}(N) \end{pmatrix}.$$

THEOREM 3.7. *Let R be a semilocal ring. Then R is essentially right quasi-duo if and only if $R \cong \begin{pmatrix} A & M \\ 0 & B \end{pmatrix}$, where $A/J(A)$ is a finite direct product of division rings, B is a semisimple Artinian ring, and M is an (A,B)-bimodule.*

PROOF. Let R be a semilocal ring with Jacobson radical J. By [**16**, Theorem 3.5(c)], $R_R = P \oplus Q$, where Q is semisimple projective and P contains J as an essential submodule. In particular, the radical of P is $\operatorname{Rad}(P) = J$ and $\operatorname{Hom}(P,Q) = \operatorname{Hom}(P/J,Q) = 0$, as Q is semisimple projective and P/J is semisimple singular. Hence we have

$$R \cong \operatorname{End}(R_R) \cong \begin{pmatrix} \operatorname{End}(P) & \operatorname{Hom}(Q,P) \\ 0 & \operatorname{End}(Q) \end{pmatrix}.$$

Then $B := \operatorname{End}(Q)$ is semisimple Artinian and $A := \operatorname{End}(P)$ is semilocal.

Assume R is essentially right quasi-duo, then A is essentially right quasi-duo by Proposition 2.6. Let M/J be a maximal submodule of P/J and let $f \in \operatorname{End}(P/J)$. Because M contains J and J is essential in P, $M \oplus Q$ is an essential maximal right ideal of R. By hypothesis $M \oplus Q$ is fully-invariant in R, which shows that M is fully-invariant in P. Let $\pi : P \to P/J$ be the canonical epimorphism. Since P is projective, there exists $g \in \operatorname{End}(P)$ such that $\pi h = f\pi$. Hence $h(M) \subseteq M$, and so $f(M/J) = f\pi(M) = \pi h(M) \subseteq \pi(M) = M/J$. So M/J is fully-invariant in P/J.

By Lemma 3.6, P/J is a direct sum of non-isomorphic (singular) simple modules and therefore, by [**22**, 22.2], $A/J(A) \cong \mathrm{End}(P/J)$ is a (finite) direct product of division rings.

($\Leftarrow$). Let $R = \begin{pmatrix} A & M \\ 0 & B \end{pmatrix}$, where $A/J(A)$ is a finite direct product of division rings, B is a semisimple Artinian ring and M is an (A, B)-bimodule. Then A is right quasi-duo, and B is essentially right quasi-duo. So R is essentially right quasi-duo by Theorem 3.1. □

The opposite of a ring R is denoted by R^o. For a bimoddule ${}_AM_B$, let $M^o = \{x^o : x \in M\}$. Then M^o is a (B^o, A^o)-bimodule where, for $a^o \in A^o, b^o \in B^o$ and $x^o, y^o \in M^o$, $x^o + y^o = (x+y)^o$, $x^o a^o = (ax)^o$ and $b^o x^o = (xb)^o$. Moreover, if $R = \begin{pmatrix} A & M \\ 0 & B \end{pmatrix}$, then $R^o \cong \begin{pmatrix} A^o & 0 \\ M^o & B^o \end{pmatrix}$ via $\begin{pmatrix} a & x \\ 0 & b \end{pmatrix}^o \mapsto \begin{pmatrix} a^o & 0 \\ x^o & b^o \end{pmatrix}$.

A ring R is called left pseudo-Frobenius if R is left self-injective, semilocal and the socle $\mathrm{Soc}({}_RR)$ is an essential submodule of ${}_RR$.

COROLLARY 3.8. *A left pseudo-Frobenius ring R is essentially right quasi-duo if and only if $R \cong A \times B$ where B is semisimple Artinian and $A/J(A)$ is a finite direct product of division rings.*

PROOF. The sufficiency is by Proposition 2.3(2). For the necessity, by Theorem 3.7 we have $R \cong \begin{pmatrix} A & M \\ 0 & B \end{pmatrix}$, where $A/J(A)$ is a finite direct product of division rings, B is semisimple Artinian and M is an (A, B)-bimodule. Then $R^o \cong \begin{pmatrix} A^o & 0 \\ M^o & B^o \end{pmatrix}$ is right pseudo-Frobenius, and so $M^o = 0$ by [**6**, Theorem 5.6]. So $M = 0$ and hence $R \cong A \times B$. □

4. A question on rings with cyclic injective hull

It is an open question whether the injective hull $E(R_R)$ of R_R cyclic implies that R is right self-injective (see [**10**]). Some partial answers are presented in [**10**], one of which states that a right quasi-duo ring R with $E(R_R)$ cyclic is right self-injective. Here we extend this result to an essentially right quasi-duo ring.

A submodule N of a module M is said to be fully-invariant if $f(N) \subseteq N$ for all $f \in \mathrm{End}(M)$. Following [**20**], a module P is called pseudo-projective if for any module X and any epimorphisms $f, g \in \mathrm{Hom}(P, X)$, there exists $h \in \mathrm{End}(P)$ such that $f = gh$. Note that a module P is projective if, for any modules M, X and any epimorphism $f \in \mathrm{Hom}(P, X)$ and any homomorphism $g \in \mathrm{Hom}(M, X)$, there exists $h \in \mathrm{Hom}(P, M)$ such that $f = gh$ (see [**21**]). Quasi-projective modules are pseudo-projective, but the converse is not true (see [**4**]). The next result is motivated by the work in [**9**], and condition (1) can be compared with the notion of dual-square-free modules: A module M is called dual-square-free if M has no proper submodules A and B with $M = A + B$ and $M/A \cong M/B$ (see [**8**]).

THEOREM 4.1. *The following are equivalent for a finitely generated pseudo-projective module M:*

(1) *M does not contain proper submodules X and Y such that $M = X + Y$, and $M/X \cong M/Y$ is singular.*
(2) *Every maximal submodule X of M with M/X singular is fully-invariant.*

PROOF. (1) $\Rightarrow$ (2). Assume that a maximal submodule X of M with M/X singular is not fully-invariant. Then $f(X) \nsubseteq X$ for some $f \in \operatorname{End}(M)$. Let $\pi : M \to M/X$ be the canonical map. Then $\pi f : M \to M/X$ is an epimorphism. So $\frac{M}{X} \cong \frac{M}{\ker(\pi f)} = \frac{M}{f^{-1}(X)}$, which is singular and simple. So $f^{-1}(X)$ is also a maximal submodule of M. Since $f(X) \nsubseteq X$, $X \nsubseteq f^{-1}(X)$. It follows that $M = X + f^{-1}(X)$, so (1) fails.

(2) $\Rightarrow$ (1). Assume that (1) fails. Then $M = A + B$ where A, B are proper submodules with $M/A \overset{\sigma}{\cong} M/B$ singular. Let Y be a maximal submodule of M with $B \subseteq Y$ (since M is finitely generated, such a Y exists), and let $\phi : M/B \to M/Y$ be the canonical map. Then $\phi\sigma : M/A \to M/Y$ is an epimorphism. Write $\ker(\phi\sigma) = X/A$. Then $M = X + Y$, and $M/Y \cong M/X$ which is singular and simple. Hence we can assume that both A and B are maximal submodules. Let $\pi_A : M \to M/A$ and $\pi_B : M \to M/B$ be the canonical maps. The pseudo-projectivity of M ensures that there exists a homomorphism $\lambda : M \to M$ such that $\sigma\pi_A\lambda = \pi_B$. Thus, $\frac{M}{B} = \frac{A+B}{B} = \pi_B(A) = \sigma\pi_A\lambda(A) = \sigma(\frac{\lambda(A)+A}{A})$. Since σ is an isomorphism, it follows that $\frac{M}{A} = \frac{\lambda(A)+A}{A}$, i.e. $M = \lambda(A) + A$. So $\lambda(A) \nsubseteq A$. Hence A is not fully-invariant, a contradiction to (2). □

The implication "(1) $\Rightarrow$ (2)" in Theorem 4.1 does not need any hypothesis. Note that, for a maximal submodule X of a projective module P, P/X is singular if and only if X is essential in M. Theorem 4.1 has a quick consequence.

COROLLARY 4.2. *Let M be a finitely generated projective module. The following are equivalent:*

(1) *M does not contain proper submodules X and Y such that $M = X + Y$, and $M/X \cong M/Y$ is singular.*
(2) *Every essential maximal submodule of M is fully-invariant.*

Theorem 4.3 extends a result in [**10**] that a right quasi-duo ring R with $E(R_R)$ cyclic is right self-injective.

THEOREM 4.3. *Let $E(R_R)$ be cyclic. If R is essentially right quasi-duo, then R_R is injective.*

PROOF. Assume on the contrary that R_R is not injective. Write $E := E(R_R) = xR$, and let $i : R \to E$ be the inclusion, $\beta : R \to E$, $r \mapsto xr$, be the natural epimorphism. Since E is injective, there exists a homomorphism $f : E \to E$ such that $\beta = fi$. Thus, $f(E) = \beta(R) = (fi)(R) = f(R)$, so $E = R + \ker(f)$. Let $\pi : E \to E/R$ be the natural homomorphism. Since $\ker(f) \subseteq \ker(\pi f)$, we have $E = R + \ker(\pi f)$, with $E/R \cong E/\ker(\pi f)$ being singular. Then $R = \beta^{-1}(R) + \beta^{-1}(\ker(\pi f))$ is a sum of two proper submodules. Moreover, $R/\beta^{-1}(R) \cong E/R$ and $R/\beta^{-1}(\ker(\pi f)) \cong E/\ker(\pi f)$ are all singular. So R is not essentially right quasi-duo by Corollary 4.2. □

There exists a regular ring R such that $R/\operatorname{Soc}(R_R)$ is a field, but R is not clean (see [**13**, Example 1]). This shows that an exchange essentially right quasi-duo ring need not be clean, in contrast to a known result that an exchange right quasi-duo ring is always clean (see [**9**]).

Acknowledgments

Part of this work was carried out when Yiqiang Zhou was visiting the University of Porto, and he thanks the host for the warm hospitality received there.

References

[1] Edmund H. Feller, *Properties of primary noncommutative rings*, Trans. Amer. Math. Soc. **89** (1958), 79–91, DOI 10.2307/1993133. MR98763

[2] S. H. Brown, *Rings over which every simple module is rationally complete*, Canadian J. Math. **25** (1973), 693–701, DOI 10.4153/CJM-1973-070-0. MR379587

[3] R. C. Courter, *Finite-dimensional right duo algebras are duo*, Proc. Amer. Math. Soc. **84** (1982), no. 2, 157–161, DOI 10.2307/2043655. MR637159

[4] Gabriella D'Este and Derya Keskin Tütüncü, *Pseudo projective modules which are not quasi projective and quivers*, Taiwanese J. Math. **22** (2018), no. 5, 1083–1090, DOI 10.11650/tjm/180401. MR3859367

[5] K. R. Goodearl, *von Neumann regular rings*, 2nd ed., Robert E. Krieger Publishing Co., Inc., Malabar, FL, 1991. MR1150975

[6] A. Haghany and K. Varadarajan, *Study of formal triangular matrix rings*, Comm. Algebra **27** (1999), no. 11, 5507–5525, DOI 10.1080/00927879908826770. MR1713049

[7] Chan Huh, Sung Hee Jang, Chol On Kim, and Yang Lee, *Rings whose maximal one-sided ideals are two-sided*, Bull. Korean Math. Soc. **39** (2002), no. 3, 411–422, DOI 10.4134/BKMS.2002.39.3.411. MR1920699

[8] Nanqing Ding, Yasser Ibrahim, Mohamed Yousif, and Yiqiang Zhou, *D4-modules*, J. Algebra Appl. **16** (2017), no. 9, 1750166, 25, DOI 10.1142/S0219498817501663. MR3661633

[9] Yasser Ibrahim and Mohamed Yousif, *Dual-square-free modules*, Comm. Algebra **47** (2019), no. 7, 2954–2966, DOI 10.1080/00927872.2018.1543429. MR3960376

[10] Y. Ibrahim and M. F. Yousif, *Rings with cyclic injective hulls*, Comm. Algebra **48**(12) (2020), 5110–5122.

[11] Arun Vinayak Jategaonkar, *Left principal ideal domains*, J. Algebra **8** (1968), 148–155, DOI 10.1016/0021-8693(68)90040-9. MR218387

[12] T. Y. Lam and Alex S. Dugas, *Quasi-duo rings and stable range descent*, J. Pure Appl. Algebra **195** (2005), no. 3, 243–259, DOI 10.1016/j.jpaa.2004.08.011. MR2114274

[13] Tsiu-Kwen Lee, Zhong Yi, and Yiqiang Zhou, *An example of Bergman's and the extension problem for clean rings*, Comm. Algebra **36** (2008), no. 4, 1413–1418, DOI 10.1080/00927870701866374. MR2406593

[14] André Leroy, Jerzy Matczuk, and Edmund R. Puczyłowski, *Quasi-duo skew polynomial rings*, J. Pure Appl. Algebra **212** (2008), no. 8, 1951–1959, DOI 10.1016/j.jpaa.2008.01.002. MR2414695

[15] André Leroy, Jerzy Matczuk, and Edmund R. Puczyłowski, *A description of quasi-duo $\mathbb{Z}$-graded rings*, Comm. Algebra **38** (2010), no. 4, 1319–1324, DOI 10.1080/00927870902897954. MR2656578

[16] Christian Lomp, *On semilocal modules and rings*, Comm. Algebra **27** (1999), no. 4, 1921–1935, DOI 10.1080/00927879908826539. MR1679679

[17] Christian Lomp, Mohamed Yousif, and Yiqiang Zhou, *On hypercyclic rings*, J. Algebra Appl. **23** (2024), no. 8, Paper No. 2550051, 14, DOI 10.1142/S0219498825500513. MR4763148

[18] Saad H. Mohamed and Bruno J. Müller, *Continuous and discrete modules*, London Mathematical Society Lecture Note Series, vol. 147, Cambridge University Press, Cambridge, 1990, DOI 10.1017/CBO9780511600692. MR1084376

[19] E. R. Puczyłowski, *On quasi-duo rings*, Noncommutative rings and their applications, Contemp. Math., vol. 634, Amer. Math. Soc., Providence, RI, 2015, pp. 243–252, DOI 10.1090/conm/634/12703. MR3307402

[20] A. K. Tiwary and B. M. Pandeya, *Pseudo projective and pseudo injective modules*, Indian J. Pure Appl. Math. **9** (1978), no. 9, 941–949. MR491817

[21] Wenting Tong, *On pseudo-projective modules* (English, with English and Chinese summaries), J. Math. Res. Exposition **16** (1996), no. 3, 355–358. MR1408573

[22] Robert Wisbauer, *Foundations of module and ring theory*, Revised and translated from the 1988 German edition, Algebra, Logic and Applications, vol. 3, Gordon and Breach Science Publishers, Philadelphia, PA, 1991. A handbook for study and research. MR1144522

[23] Hua-Ping Yu, *On quasi-duo rings*, Glasgow Math. J. **37** (1995), no. 1, 21–31, DOI 10.1017/S0017089500030342. MR1316960

[24] Yiqiang Zhou, *Generalizations of perfect, semiperfect, and semiregular rings*, Algebra Colloq. **7** (2000), no. 3, 305–318, DOI 10.1007/s10011-000-0305-9. MR1810586

Department of Mathematics, University of Porto, 4000 Porto, Portugal
Email address: clomp@fc.up.pt

Department of Mathematics, The Ohio State University, Lima, Ohio 45804
Email address: yousif.1@osu.edu

Department of Mathematics and Statistics, Memorial University of Newfoundland, St.John's, Newfoundland and Labrador A1C 5S7, Canada
Email address: zhou@mun.ca

Contemporary Mathematics
Volume **826**, 2025
https://doi.org/10.1090/conm/826/16579

Amenable bases for algebras of entangled polynomials

Sergio R. López-Permouth and Ashley H. Pallone

Abstract. The study of amenability and simplicity of bases is extended, as is also the exploration of the properties of the recently introduced algebras of entangled polynomials. We consider modular bases for m-nomials, which can be built up based on some component polynomial bases, and see that when the components are amenable and pairwise mutually congenial the resulting modular basis is left amenable. We explore the possibility that the converse of this result may also hold.

1. Preliminaries

1.1. Introduction. The notion of an amenable basis for an algebra of infinite dimension was introduced in [**1**] with the intention to provide a potential mechanism to extend the module structure of the algebra to a larger vector space. The expression amenable bases is used to refer to those cases when the extension is feasible. Many questions arise naturally in this context for algebras in general but, perhaps more excitingly, also for familiar algebras like the algebra $K[x]$ of polynomials in a single variable.

As introduced in [**1**], a simple basis is an amenable basis that is smallest in a certain way; for convenience, we include the precise description in Definition 1.6. Subsection 1.2 presents a brief summary of definitions and results pertaining the notions of amenability and simplicity, as well as of the related concept of congeniality of bases.

It was shown in [**1**] that the *monomial basis* $\{ x^k \mid k \geq 0 \}$, consisting of all powers of x, is both amenable and simple. One may expect a similar behavior from the monomial basis $\{ x^n \mid n \in \mathbb{Z} \}$ of the algebra $K[x, x^{-1}]$ of Laurent polynomials. However, that basis is not simple, as it was shown in Proposition 1 of [**2**]. An alternative interesting and natural direction in which to look for analog results is by considering monomial bases for algebras of polynomials in several variables. In that alternative setting, the subject of [**5**], one faces additional options by looking at those algebras not only as algebras but also as modules over algebras of polynomials with fewer variables, expanding the scope of the inquiry in the direction of [**7**], [**8**], and [**9**], where amenable bases of infinite dimensional modules over arbitrary algebras were introduced.

2020 *Mathematics Subject Classification.* Primary 16S99, 15A30; Secondary 16D99, 15B99.
Key words and phrases. Entangled polynomials, m-nomials, amenable bases, congeniality of bases, proper congeniality, simple bases, infinite-dimensional algebras.

In this paper, we continue the investigation of amenability of bases by looking at those notions in the context of a family of algebras, the m-nomials, which were recently introduced in our paper [**6**]. While these algebras are similar in several ways to algebras of polynomials in a single variable and share some properties with them, they also exhibit striking differences. In particular, algebras of m-nomials are not commutative. For that reason, we conclude this introduction with Subsection 1.3, in order to enumerate not only some of the most remarkable common features of polynomials and m-nomials, but also some of their differences, as presented originally in [**6**].

The remainder of the paper deals, in this order, with the following topics.

(1) In Section 2, we explore natural choices of bases for the various algebras of entangled polynomials that are analog to the the monomial bases for polynomials. Our candidates fit within a mold of bases for m-nomials that may be built from polynomial bases; we call such bases *modular* and the bases from which they are built their *components.*
(2) In Section 3, we analyze the amenability of modular bases. It is shown, in Theorem 3.3 and Theorem 3.4, that amenable bases for the algebra of polynomials may be combined to yield left amenable bases of m-nomials. Theorem 3.4, in particular, shows that modular bases are left amenable bases when their component bases are mutually congenial.
(3) In Section 4, we offer a thorough analysis of several related concepts but, for the sake of convenience, we do so only in the more limited scope of the algebra of 2-nomials. This is done with the intention of exploring possibilities and setting the groundwork for future work. The two subsections of Section 4 consider examples and questions of amenability, congeniality, and simplicity.

To close this introductory subsection, we introduce some unifying standard notation. Let V be a vector space with basis $\mathcal{B}$. For $v \in V$, the notation $(v)_b$ refers to the coefficient of $b \in \mathcal{B}$ in the representation $v = \sum_{b\in\mathcal{B}}(v)_b$ of v with respect to the basis $\mathcal{B}$. Similarly, the notation $[v]_\mathcal{B}$ refers to the coordinates of v with respect to the basis $\mathcal{B}$. For $v \in V$, the notation $[v]_\mathcal{B}$ refers to the coordinates of v with respect to the basis $\mathcal{B}$. Given another basis $\mathcal{C}$ of V and linear transformation $T : V \to V$, the matrix $[T]_\mathcal{B}^\mathcal{C}$ refers to the representation of T with respect to the bases $\mathcal{B}$ and $\mathcal{C}$. That is, $[T]_\mathcal{B}^\mathcal{C}$ has rows indexed by the elements of $\mathcal{C}$ and columns indexed by the elements of $\mathcal{B}$ so that the bth column of $[T]_\mathcal{B}^\mathcal{C}$ is $[T(b)]_\mathcal{C}$. Notice that for any linear transformation $T : V \to V$, the matrix $[T]_\mathcal{B}^\mathcal{C}$ is column-finite and for any $v \in V$, the identity $[T]_\mathcal{B}^\mathcal{C}[v]_\mathcal{B} = [T(v)]_\mathcal{C}$ holds. Given any two bases $\mathcal{B}$ and $\mathcal{C}$ of V, the matrix $Q = [I]_\mathcal{B}^\mathcal{C}$ representing the identity map $I : V \to V$ with respect to the bases $\mathcal{B}$ and $\mathcal{C}$ serves as a change of basis matrix, in the sense that, as $Q^{-1} = [I]_\mathcal{C}^\mathcal{B}$, the identity $[T]_\mathcal{C}^\mathcal{C} = Q[T]_\mathcal{B}^\mathcal{B}Q^{-1}$ holds for any endomorphism $T : V \to V$. To simplify notation, we write $[T]_\mathcal{B}$ in place of $[T]_\mathcal{B}^\mathcal{B}$.

1.2. Amenable, congenial, and simple bases. For convenience and ease of reference, we list here notions, examples, and results from [**1**] which are needed in the sequel. Throughout this paper, K is a field and A is an infinite dimensional K-algebra with basis $\mathcal{B}$. Associated with A is another K-vector space ${}_\mathcal{B}P = \prod_{b\in\mathcal{B}} Kb$, which is the direct product of copies of K indexed by $\mathcal{B}$. A logical question that arises in this context is for what bases does ${}_\mathcal{B}P$ have a (left) module structure over A that naturally extends that of the (left) A-module ${}_AA$. The concept of (left) amenability was first introduced in [**1**] to answer this question. Because the algebras considered in this paper are not commutative, we emphasize the distinction between *left* and *right* amenability.

The notion of amenable bases and their corresponding *basic extensions* was extended to infinite dimensional modules over arbitrary algebras in [**7**], [**8**], and [**9**]. However, for the purposes of this paper, it suffices to focus on bases for infinite dimensional algebras (that is, the specific case when the module is the algebra itself).

DEFINITION 1.1. A basis $\mathcal{B}$ for an infinite dimensional K-algebra A is **left amenable** if for every $r \in A$ and $c \in \mathcal{B}$, there exist only finitely many $d \in \mathcal{B}$ such that $(rd)_c \neq 0$. Equivalently, $\mathcal{B}$ is left amenable if every left multiplication linear transformation $l_r(v) = rv$ has a row-finite representation $[l_r]_{\mathcal{B}}$ with respect to $\mathcal{B}$. A **right amenable** basis is defined similarly.

In the context of commutative algebras, every left amenable basis is also right amenable and so we simply refer to them as amenable bases. In the specific case where $A = K[x]$, it suffices to check the single matrix $[l_x]_{\mathcal{B}}$ is row-finite to determine if $\mathcal{B}$ is amenable.

EXAMPLE 1.2. The standard basis $\mathcal{B} = \{\, x^k \mid k \geq 0 \,\}$ and $\mathcal{C} = \{\, 1 \,\} \cup \{\, x^k + x^{k+1} \mid k \geq 0 \,\}$ are amenable bases of $K[x]$ because the matrices

$$[l_x]_{\mathcal{B}} = \begin{bmatrix} 0 & 0 & 0 & 0 & \cdots \\ 1 & 0 & 0 & 0 & \cdots \\ 0 & 1 & 0 & 0 & \cdots \\ 0 & 0 & 1 & 0 & \cdots \\ \vdots & \vdots & \ddots & \ddots & \ddots \end{bmatrix} \qquad [l_x]_{\mathcal{C}} = \begin{bmatrix} -1 & 0 & 0 & 0 & \cdots \\ 1 & 0 & 0 & 0 & \cdots \\ 0 & 1 & 0 & 0 & \cdots \\ 0 & 0 & 1 & 0 & \cdots \\ \vdots & \vdots & \ddots & \ddots & \ddots \end{bmatrix}$$

are row-finite, respectively.

EXAMPLE 1.3. For any $\alpha, \beta \in K$, where $\beta \neq 0$, the set $\mathcal{P}_{(\alpha,\beta)} = \{\, (\alpha + \beta x)^k \mid k \geq 0 \,\}$ is a basis for $K[x]$, referred to as a *Pascal basis*. Amenability, in this case, can also most conveniently be checked by observing the multiplication by $\alpha + \beta x$ rather than by x. In doing so, it is straightforward to see that the Pascal bases are all amenable.

By design, a (left) amenable basis $\mathcal{B}$ is one for which ${}_{\mathcal{B}}P$ is a (left) A-module extension of ${}_AA$. In this case, we call ${}_{\mathcal{B}}P$ a *basic module*. In order to attain a sufficient condition for isomorphic basic modules, the related notion of congeniality was introduced also in [**1**].

DEFINITION 1.4. Let $\mathcal{B}$ and $\mathcal{C}$ be two bases of the same infinite dimensional K-algebra.

(1) The basis $\mathcal{B}$ is **congenial** to the basis $\mathcal{C}$ if for every $c \in \mathcal{C}$ there exists only finitely many $b \in \mathcal{B}$, when represented in terms of $\mathcal{C}$, that use a nonzero coefficient for c. Equivalently, $\mathcal{B}$ is congenial to $\mathcal{C}$ if the change of basis matrix $[I]_{\mathcal{B}}^{\mathcal{C}}$ is row-finite.
(2) If $\mathcal{B}$ is congenial to $\mathcal{C}$ and $\mathcal{C}$ is not congenial to $\mathcal{B}$, then $\mathcal{B}$ is **properly congenial** $\mathcal{C}$.
(3) If $\mathcal{B}$ is congenial to $\mathcal{C}$ and $\mathcal{C}$ is congenial to $\mathcal{B}$, then $\mathcal{B}$ and $\mathcal{C}$ are **mutually congenial**.

(4) If neither $\mathcal{B}$ is congenial to $\mathcal{C}$ nor $\mathcal{C}$ is congenial to $\mathcal{B}$, then $\mathcal{B}$ and $\mathcal{C}$ are **discordant**.

EXAMPLE 1.5. Consider the bases $\mathcal{B} = \{ x^k \mid k \geq 0 \}$ and $\mathcal{C} = \{ 1 \} \cup \{ x^k + x^{k+1} \mid k \geq 0 \}$ of $K[x]$ from Example 1.2. We see that $\mathcal{C}$ is properly congenial to $\mathcal{B}$ because the matrices

$$[I]_{\mathcal{C}}^{\mathcal{B}} = \begin{bmatrix} 1 & 1 & 0 & 0 & \cdots \\ 0 & 1 & 1 & 0 & \cdots \\ 0 & 0 & 1 & 1 & \ddots \\ 0 & 0 & 0 & 1 & \ddots \\ \vdots & \vdots & \vdots & \ddots & \ddots \end{bmatrix} \qquad [I]_{\mathcal{B}}^{\mathcal{C}} = \begin{bmatrix} 1 & -1 & 1 & -1 & \cdots \\ 0 & 1 & -1 & 1 & \ddots \\ 0 & 0 & 1 & -1 & \ddots \\ 0 & 0 & 0 & 1 & \ddots \\ \vdots & \vdots & \vdots & \ddots & \ddots \end{bmatrix}$$

are row-finite and not row-finite, respectively.

While mutual congeniality is sufficient for the basic modules induced by two (left) amenable bases to be isomorphic, the condition is not necessary. Examples of isomorphic basic modules induced by non-congenial (even discordant) bases are given in [**3**]. Isomorphisms between basic modules that leave the algebra invariant are characterized in Theorem 2.8 of [**4**].

When $\mathcal{B}$ is (properly) congenial to $\mathcal{C}$, the matrix $[I]_{\mathcal{B}}^{\mathcal{C}}$ is called a *(proper) congeniality matrix*. It was shown in [**1**] that if $\mathcal{B}$ and $\mathcal{C}$ are both (left) amenable and $\mathcal{B}$ is properly congenial to $\mathcal{C}$, then the linear transformation represented by the proper congeniality matrix is a module epimorphism. This result gives the appearance of maximal and minimal basic modules, and motivates the following notion of simple bases.

DEFINITION 1.6. A left amenable basis $\mathcal{B}$ is called **left simple** if it is not properly congenial to any other left amenable basis. A **right simple** basis is defined similarly.

The veracity of the following example is proven in [**1**].

EXAMPLE 1.7. All Pascal bases of Example 1.3 are simple. In particular, the standard basis $\mathcal{B} = \mathcal{P}_{(0,1)} = \{ x^k \mid k \geq 0 \}$ of $K[x]$ is a simple basis.

1.3. Algebras of m-nomials. Many questions remain open about amenable and simple bases over the algebra $K[x]$ and each questions motivates us to ask similar questions for other algebras that may be similar to $K[x]$ in some sense. The algebra $K[x, x^{-1}]$ of Laurent polynomials is one such an algebra. It is not known, at this point, whether $K[x, x^{-1}]$ has simple bases. We do know that, in contrast with the situation that we just mentioned for the polynomial algebra, the *monomial basis*, consisting of all powers (positive, zero, and negative) of x, is not a simple basis of this algebra; this is the subject of Proposition 1 in [**2**]. Similar questions regarding polynomials with several variables are considered in [**5**].

In this subsection, we recall (from [**6**]) the basic properties of m-nomials (also called *entangled polynomials*). These are algebras that resemble polynomials in many ways, while, at the same time, enjoy some very distinct characteristics. Throughout this paper, for the field K, the rings RCFM(K) and CFM(K) are, respectively, the ring of all row and column finite matrices and the ring of all column finite matrices over K. In both cases, we assume that the rows and columns of these matrices are indexed by $\mathbb{Z}^+ = \{1, 2, 3, \dots\}$. It is a well known result that the polynomial ring $K[x]$ can be embedded into $\mathrm{RCFM}(K)$ via the map φ given by

$$\varphi(f(x)) = \begin{bmatrix} f_0 & f_1 & f_2 & \cdots & f_n & 0 & 0 & 0 & \cdots \\ 0 & f_0 & f_1 & f_2 & \cdots & f_n & 0 & 0 & \cdots \\ 0 & 0 & f_0 & f_1 & f_2 & \cdots & f_n & 0 & \cdots \\ \vdots & \vdots & \ddots & \ddots & \ddots & \ddots & \cdots & \ddots & \ddots \end{bmatrix}$$

where $f(x) = \sum_{k=0}^{n} f_k x^k$. As φ is an injective ring homomorphism, $K[x]$ is isomorphic to its image under φ.

We can similarly associate to every m-tuple of polynomials with coefficients in K a single matrix by assigning certain rows to each entry of the m-tuple. Any such matrix in $\mathrm{RCFM}(K)$ is called an m-nomial. It is important to note that this new map, from $(K[x])^m$ to $\mathrm{RCFM}(K)$, is not a ring homomorphism, because multiplication is not entry-wise for the rows of matrices in $\mathrm{RCFM}(K)$. These ideas are made precise in Definition 1.8.

A word of warning ahead of the definition seems appropriate. Our m-tuples are numbered in an unorthodox way, beginning with entries 1 through $m-1$, and concluding with entry 0. We made the decision to do this when we realized that such a small accommodation yielded substantial simplifications in latter discussions. Notice that this alternative numbering allows a better alignment (modulo m) of the *parity* of the number of rows with that of the number of the entries that they represent.

DEFINITION 1.8. An upper triangular matrix $A \in \mathrm{RCFM}(K)$ is an m**-nomial** if there exists m polynomials $f_1(x)$, $f_2(x)$, $\dots$, $f_{m-1}(x)$, and $f_0(x)$, where $f_k(x) = \sum_{l=0}^{n_k} f_{kl} x^l$, such that for $i = qm + r$ with $0 \le r < m$ and $q \in \mathbb{N}$, $a_{ij} = f_{r,j-i}$ and for $j > i + n_r$, $a_{ij} = 0$.

If $A \in \mathrm{RCFM}(K)$ is the m-nomial for the m polynomials $f_1(x)$, $\dots$, $f_{m-1}(x)$, and $f_0(x)$, then we write $A = \big(f_1(x), \dots, f_{m-1}(x), f_0(x)\big)$. The following matrix should help visualize the m-nomial $A = \big(f_1(x), \dots, f_{m-1}(x), f_0(x)\big)$.

$$\begin{bmatrix} f_{10} & f_{11} & f_{12} & \cdots & f_{1n_1} & 0 & 0 & \cdots & 0 & 0 & 0 & \cdots & 0 & \cdots \\ 0 & f_{20} & f_{21} & f_{22} & \cdots & f_{2n_2} & 0 & \cdots & 0 & 0 & 0 & \cdots & 0 & \cdots \\ \vdots & \ddots & \ddots & \ddots & \ddots & \cdots & \ddots & \ddots & \vdots & \vdots & \vdots & \cdots & \vdots & \cdots \\ 0 & \cdots & 0 & f_{00} & f_{01} & f_{02} & \cdots & f_{0n_0} & 0 & 0 & 0 & \cdots & 0 & \cdots \\ 0 & \cdots & 0 & 0 & f_{10} & f_{11} & f_{12} & \cdots & f_{1n_1} & 0 & 0 & \cdots & 0 & \cdots \\ 0 & \cdots & 0 & 0 & 0 & f_{20} & f_{21} & f_{22} & \cdots & f_{2n_2} & 0 & \cdots & 0 & \cdots \\ \vdots & \vdots & \vdots & \vdots & \vdots & \ddots & \ddots & \ddots & \ddots & \cdots & \ddots & \ddots & \vdots & \cdots \\ 0 & \cdots & 0 & 0 & 0 & \cdots & 0 & f_{00} & f_{01} & f_{02} & \cdots & f_{0,n_0} & 0 & \cdots \\ \vdots & \cdots & \vdots & \vdots & \vdots & \cdots & \vdots & \ddots & \ddots & \ddots & \ddots & \cdots & \ddots & \ddots \end{bmatrix}$$

Notice, however, the illusion in this representation that $n_1 \le n_2 \le \cdots \le n_{m-1} \le n_0$ is merely an unintended bi-product of an effort to avoid typographical complications; no such condition on the degrees is purposely trying to be implied.

Through matrix multiplication, the product of two m-nomials entangles their respective m polynomials in a unique way. In fact, the resulting product is itself an m-nomial (see [**6**]).

THEOREM 1.9. *The set of all m-nomials, $K^m[x]$, is a ring with identity for all $m \in \mathbb{Z}^+$.*

2. Building bases for algebras of m-nomials from polynomial bases

Each m-nomial in the algebra of m-nomials is induced by m polynomials from $K[x]$. Because the standard basis of $K[x]$, $\mathcal{B} = \{ x^k \mid k \geq 0 \}$, consists of consecutive powers of the indeterminate x, it is a natural first step in the creation of a standard basis for $K^m[x]$ to consider the m-nomials analogous to x and its powers.

For each $0 \leq \alpha \leq m-1$, let $(x^k)_\alpha$ denote the m-nomial corresponding to the m polynomials $f_i(x) = 0$ $(i \neq \alpha)$ and $f_\alpha(x) = x^k$. Write $(\mathcal{B})_\alpha = \{ (x^k)_\alpha \mid k \geq 0 \}$. Using a capital X instead of the lower case x, we will write X^k_α instead of $(x^k)_\alpha$, when convenient. Our next theorem, which is straightforward to prove, states that the union

$$\mathcal{B}^m = (\mathcal{B})_1 \cup (\mathcal{B})_2 \cup \cdots \cup (\mathcal{B})_{m-1} \cup (\mathcal{B})_0 = \bigcup_{\alpha=1}^{m-1} (\mathcal{B})_\alpha \cup (\mathcal{B})_0$$

is a basis for the algebra of m-nomials. While the union of sets is commutative, we choose to use the notation $\bigcup_{\alpha=1}^{m-1}(\mathcal{B})_\alpha \cup (\mathcal{B})_0$ rather than $\bigcup_{\alpha=0}^{m-1}(\mathcal{B})_\alpha$ to make clear the unorthodox ordering of $\mathcal{B}^m$ which matches that of Definition 1.8.

THEOREM 2.1. *The set $\mathcal{B}^m$ is a basis for $K^m[x]$. We refer to $\mathcal{B}^m$ as the standard basis.*

Similar to the way in which we produced the elements of the standard basis above, for any polynomial $f(x) \in K[x]$ and $0 \leq \alpha \leq m-1$, the notation $\big(f(x)\big)_\alpha$, serves to denote the m-nomial corresponding to the m polynomials $f_i(x) = 0$ $(i \neq \alpha)$ and $f_\alpha(x) = f(x)$. Just as we did with the standard basis, it is possible to build a basis for $K^m[x]$ from any set of m bases of $K[x]$. Let $\mathcal{C}$ be any basis of $K[x]$, then we denote $(\mathcal{C})_\alpha$ the set $\{ (c)_\alpha \mid c \in \mathcal{C} \}$; such sets are clearly linearly independent and will be the building blocks of a particular type of basis for $K^m[x]$.

LEMMA 2.2. *Let $\mathcal{C}$ be a basis of $K[x]$ and $0 \leq \alpha \leq m-1$. Then $(\mathcal{C})_\alpha$ is a linearly independent set of m-nomials.*

To construct a basis for $K^m[x]$, we simply take the union of these building blocks from $\alpha = 0$ to $\alpha = m-1$ using any m bases of $K[x]$. This concept is made more precise in the following theorem, which is an extension of Theorem 2.1.

THEOREM 2.3. *If $\mathcal{C}_1, \ldots, \mathcal{C}_{m-1}, \mathcal{C}_0$ are m bases of $K[x]$, then the set*

$$(\mathcal{C}_1)_1 \cup (\mathcal{C}_2)_2 \cup \cdots \cup (\mathcal{C}_{m-1})_{m-1} \cup (\mathcal{C}_0)_0 = \bigcup_{\alpha=1}^{m-1} (\mathcal{C}_\alpha)_\alpha \cup (\mathcal{C}_0)_0$$

is a basis for $K^m[x]$. We refer to bases of $K^m[x]$ that are built from bases of $K[x]$ in this fashion as modular bases.

PROOF. Each set $(\mathcal{C}_i)_i$ is linearly independent due to Lemma 2.2. By their construction, if $i \neq j$ then the set $(\mathcal{C}_i)_i$ has rows of zeros where the set $(\mathcal{C}_j)_j$ has nonzero rows and the set $(\mathcal{C}_j)_j$ has rows of zeros where the set $(\mathcal{C}_i)_i$ has nonzero rows. Therefore, $\bigcup_{\alpha=0}^{m=1}(\mathcal{C}_\alpha)_\alpha$ is linearly independent. Let $A = \big(f_1(x), \ldots, f_{m-1}(x), f_0(x)\big) \in$

$K^m[x]$ be an m-nomial. Since each $\mathcal{C}_i$ is a basis of $K[x]$, there exist $c_{ik} \in \mathcal{C}_i$, $r_{ik} \in K$, and $n_i \in \mathbb{N}$ such that

$$f_i(x) = r_{i0}c_{i0} + r_{i1}c_{i1} + r_{i2}c_{i2} + \cdots + r_{in_i}c_{in_i} = \sum_{k=0}^{n_i} r_{ik}c_{ik}.$$

Using these representations of $f_i(x)$, A can be written as

$$A = \sum_{i=0}^{m-1}\sum_{k=0}^{n_i} r_{ik}(c_{ik})_i$$

where $(c_{ik})_i \in (\mathcal{C}_i)_i \subset \bigcup_{\alpha=0}^{m=1}(\mathcal{C}_\alpha)_\alpha$ and $r_{ik} \in K$. Hence, $\bigcup_{\alpha=0}^{m=1}(\mathcal{C}_\alpha)_\alpha$ spans $K^m[x]$. □

We call the m bases $\mathcal{C}_1, \mathcal{C}_2, \ldots, \mathcal{C}_{m-1}, \mathcal{C}_0$ from Theorem 2.3 the *components* or the *component bases* of the corresponding modular basis. These components may all be the same, all distinct, or some combination of distinct and repeated bases. We state the result for the case when they are all the same in the following remark, for the record. We also take advantage of that same remark in order to establish such notation that was used in Theorem 2.1 to refer to the standard basis $\mathcal{B}^m$.

REMARK 2.4. Let $\mathcal{C}$ be a basis of $K[x]$. When the components $\mathcal{C}_\alpha = \mathcal{C}$ for $0 \leq \alpha \leq m-1$, we denote the modular basis from Theorem 2.3 by

$$\mathcal{C}^m = \bigcup_{\alpha=1}^{m-1}(\mathcal{C})_\alpha \cup (\mathcal{C})_0.$$

A natural question to ask is in what way is the amenability of a basis in $K[x]$ preserved when that basis is used as a component to create a modular basis for $K^m[x]$. We begin to explore this question in Section 3.

3. Amenable bases for the algebra of m-nomials

Understanding multiplication of the elements of the standard basis is the most natural place to begin our discussion of amenability in the algebra of m-nomials. We establish this in Proposition 3.1.

PROPOSITION 3.1. *Multiplication of elements $X_\alpha^k, X_\beta^l \in \mathcal{B}^m$ is given by*

$$X_\alpha^k X_\beta^l = \begin{cases} X_\alpha^{k+l} & \text{if } \alpha + k \equiv \beta \pmod{m} \\ 0 & \text{otherwise.} \end{cases}$$

PROOF. Denote the entries of X_α^k, X_β^l, and the product $X_\alpha^k X_\beta^l$ by a_{ij}, b_{ij}, and c_{ij}, respectively. Then $a_{r,r+k} = 1$ if $r \equiv \alpha \pmod{m}$, $b_{s,s+l} = 1$ if $s \equiv \beta \pmod{m}$, and $a_{ij} = b_{ij} = 0$ elsewhere. There are two cases to consider.

CASE 1. Suppose $\alpha + k \not\equiv \beta \pmod{m}$. If $i \not\equiv \alpha \pmod{m}$, then $a_{in} = 0$ for $n \geq 1$ and so $c_{ij} = \sum_{n=1}^\infty a_{in}b_{nj} = 0$. If $i \equiv \alpha \pmod{m}$, then $a_{i,i+k} = 1$, $b_{i+k,j} = 0$, and $a_{in} = 0$ for $n \neq i+k$. Therefore,

$$c_{ij} = \sum_{n=1}^{\infty} a_{in}b_{nj} = \sum_{n=1}^{i+k-1} \underbrace{a_{in}}_{0} b_{nj} + \underbrace{a_{i,i+k}}_{1}\underbrace{b_{i+k,j}}_{0} + \sum_{n=i+k+1}^{\infty} \underbrace{a_{in}}_{0} b_{nj} = 0.$$

It follows that $X_\alpha^k X_\beta^l = 0$.

CASE 2. Suppose $\alpha + k \equiv \beta \pmod{m}$. If $i \not\equiv \alpha \pmod{m}$, then $a_{in} = 0$ for $n \geq 1$ and so $c_{ij} = \sum_{n=1}^{\infty} a_{in} b_{nj} = 0$. If $i \equiv \alpha \pmod{m}$ and $j \neq i + k + l$, then $a_{i,i+k} = 1$, $b_{i+k,j} = 0$, and $a_{in} = 0$ for $n \neq i + k$. Therefore,

$$c_{ij} = \sum_{n=1}^{\infty} a_{in} b_{nj} = \sum_{n=1}^{i+k-1} \underbrace{a_{in}}_{0} b_{nj} + \underbrace{a_{i,i+k}}_{1} \underbrace{b_{i+k,j}}_{0} + \sum_{n=i+k+1}^{\infty} \underbrace{a_{in}}_{0} b_{nj} = 0.$$

If $i \equiv \alpha \pmod{m}$ and $j = i + k + l$, then $a_{i,i+k} = b_{i+k,j} = 1$ and $a_{in} = 0$ for $n \neq i + k$. Therefore,

$$c_{ij} = \sum_{n=1}^{\infty} a_{in} b_{nj} = \sum_{n=1}^{i+k-1} \underbrace{a_{in}}_{0} b_{nj} + \underbrace{a_{i,i+k}}_{1} \underbrace{b_{i+k,j}}_{1} + \sum_{n=i+k+1}^{\infty} \underbrace{a_{in}}_{0} b_{nj} = 1.$$

It follows that $X_\alpha^k X_\beta^l = X_\alpha^{k+l}$.

Combining the above two cases gives the desired result. □

Because $K^m[x]$ is not commutative, the distinction between left and right amenability is not superfluous. In the literature, the default emphasis has been on the left version of this property and so we follow that precedent. As with $K[x]$, a simplification of the left amenability criteria is possible for $K^m[x]$. Notice that for $n \geq 0$,

$$X_1^n = \underbrace{X_1^1 X_2^1 \dots X_{m-1}^1 X_0^1 X_1^1 \dots X_{m-1}^1 \cdots X_{m-1}^1 X_0^1 X_1^1 \dots X_{m-1}^1 \cdots X_{n \bmod m}^1}_{n \text{ terms}}$$
$$X_2^n = \underbrace{X_2^1 X_3^1 \dots X_{m-1}^1 X_0^1 X_1^1 \dots X_{m-1}^1 \cdots X_{m-1}^1 X_0^1 X_1^1 \dots X_{m-1}^1 \cdots X_{(n+1) \bmod m}^1}_{n \text{ terms}}$$
$$\vdots$$
$$X_{m-1}^n = \underbrace{X_{m-1}^1 X_0^1 X_1^1 \dots X_{m-1}^1 \cdots X_{m-1}^1 X_0^1 X_1^1 \dots X_{m-1}^1 \cdots X_{(n+m-2) \bmod m}^1}_{n \text{ terms}}$$
$$X_0^n = \underbrace{X_0^1 X_1^1 \dots X_{m-1}^1 X_0^1 X_1^1 \dots X_{m-1}^1 \cdots X_{(n-1) \bmod m}^1}_{n \text{ terms}}$$

and so the products of elements $\{ X_\alpha^1 \mid 0 \leq \alpha \leq m - 1 \}$ generate the standard basis $\mathcal{B}^m$. Therefore, we need only check that the m column-finite matrices $\{ [l_{X_\alpha^1}]_\mathcal{C} \mid 0 \leq \alpha \leq m - 1 \}$ are also row-finite to determine whether a basis $\mathcal{C}$ of $K^m[x]$ is left amenable.

In this paper, we focus exclusively on the modular bases introduced in Theorem 2.3 which are composed of elements of the form $(f(x))_\beta$ for some $f(x) \in K[x]$ and $0 \leq \beta \leq m - 1$. The following lemma describes the left multiplication rule for elements of the standard basis with elements of any modular basis.

LEMMA 3.2. *Let $f(x) \in K[x]$ be a polynomial and $0 \leq \beta \leq m-1$. If $X_\alpha^k \in \mathcal{B}^m$, then*

$$X_\alpha^k (f(x))_\beta = \begin{cases} (x^k f(x))_\alpha & \text{if } \alpha + k \equiv \beta \pmod{m} \\ 0 & \text{otherwise.} \end{cases}$$

PROOF. Since $f(x) \in K[x]$ is a polynomial, there exist $k_i \in K$ and $n \in \mathbb{N}$ such that $f(x) = \sum_{i=0}^{n} k_i x^i$. We use Proposition 3.1 to simplify the product in each case below.

CASE 1. Suppose $\alpha + k \equiv \beta \pmod{m}$. Then

$$X_\alpha^k\big(f(x)\big)_\beta = X_\alpha^k\left(\sum_{i=0}^{n} k_i x^i\right)_\beta = X_\alpha^k\left(\sum_{i=0}^{n} k_i X_\beta^i\right) = \sum_{i=0}^{n} k_i X_\alpha^k X_\beta^i = \sum_{i=0}^{n} k_i X_\alpha^{k+i}$$
$$= \left(\sum_{i=0}^{n} k_i x^{k+i}\right)_\alpha = \left(x^k\left[\sum_{i=0}^{n} k_i x^i\right]\right)_\alpha = \big(x^k f(x)\big)_\alpha.$$

CASE 2. Suppose $\alpha + k \not\equiv \beta \pmod{m}$. Then

$$X_\alpha^k\big(f(x)\big)_\beta = X_\alpha^k\left(\sum_{i=0}^{n} k_i x^i\right)_\beta = X_\alpha^k\left(\sum_{i=0}^{n} k_i X_\beta^i\right) = \sum_{i=0}^{n} k_i X_\alpha^k X_\beta^i = 0.$$

Combining the above two cases gives the desired result. □

As the next result will show, amenability in $K[x]$ is preserved when only that basis is used as the components to build a modular basis of $K^m[x]$.

THEOREM 3.3. *Given a basis $\mathcal{C}$ of $K[x]$, the basis $\mathcal{C}^m$ is left amenable if and only if $\mathcal{C}$ is amenable.*

PROOF. We index the rows and columns of the m matrices $\{\,[l_{X_\alpha^1}]_{\mathcal{C}^m} \mid 0 \le \alpha \le m-1\,\}$ according to the order

$$(\mathcal{C})_1 \cup (\mathcal{C})_2 \cup \cdots \cup (\mathcal{C})_{m-1} \cup (\mathcal{C})_0$$

so that each is a block matrix of the form

$$[l_{X_\alpha^1}]_{\mathcal{C}^m} = \left[\begin{array}{c|c|c|c} a_{11} & a_{12} & \cdots & a_{1m} \\ \hline a_{21} & a_{22} & \cdots & a_{2m} \\ \hline \vdots & \vdots & \ddots & \vdots \\ \hline a_{m1} & a_{m2} & \cdots & a_{mm} \end{array}\right]$$

where a_{ij} has rows and columns indexed by $\mathcal{C}_{i \bmod m}$ and $\mathcal{C}_{j \bmod m}$, respectively. Since $\mathcal{C}$ is a basis of $K[x]$, for each $c \in \mathcal{C}$ there exist $k_i \in K$, $c_i \in \mathcal{C}$, and $m \in \mathbb{N}$ such that $l_x(c) = x \cdot c = \sum_{i=0}^{m} k_i c_i$. These k_i correspond to nonzero entries of the cth column of $[l_x]_{\mathcal{C}}$. Fix α and let $0 \le \beta \le m-1$. Using Lemma 3.2, we see that if $\alpha+1 \not\equiv \beta \pmod{m}$, then $l_{X_\alpha^1}\big((c)_\beta\big) = X_\alpha^1(c)_\beta = 0$ and if $\alpha+1 \equiv \beta \pmod{m}$, then

$$l_{X_\alpha^1}\big((c)_\beta\big) = X_\alpha^1(c)_\beta = (x \cdot c)_\alpha = \left(\sum_{i=0}^{m} k_i c_i\right)_\alpha = \sum_{i=0}^{m} k_i (c_i)_\alpha.$$

It follows that

$$[l_{X^1_\alpha}]_{\mathcal{C}^m} = \left[\begin{array}{c|c|c|c|c|c|c} 0 & 0 & \cdots & 0 & 0 & \cdots & 0 \\ \hline 0 & 0 & \cdots & 0 & 0 & \cdots & 0 \\ \hline \vdots & \vdots & \cdots & \vdots & \vdots & \cdots & \vdots \\ \hline 0 & 0 & \cdots & [l_x]_{\mathcal{C}} & 0 & \cdots & 0 \\ \hline \vdots & \vdots & \cdots & \vdots & \vdots & \cdots & \vdots \\ \hline 0 & 0 & \cdots & 0 & 0 & \cdots & 0 \end{array}\right]$$

where $a_{ij} = [l_x]_{\mathcal{C}}$ if $i \equiv \alpha \pmod{m}$ and $j = \alpha + 1$, and $a_{ij} = 0$ otherwise. Because the left amenability of $\mathcal{C}$ and $\mathcal{C}^m$ are dependent upon whether $[l_x]_{\mathcal{C}}$ and the m matrices $\{\, [l_{X^1_\alpha}]_{\mathcal{C}^m} \mid 0 \le \alpha \le m-1 \,\}$ are row-finite, respectively, the result follows. $\square$

In particular, the previous result proves that the standard basis of $K^m[x]$ is left amenable. Although tedious, it is straightforward to compute the m matrices $\{\, [r_{X^1_\alpha}]_{\mathcal{B}^m} \mid 0 \le \alpha \le m-1 \,\}$ and see that each is row-finite. Therefore, the standard basis is also right amenable. Upon completing these computations, the reader will notice that computing $[r_{X^1_\alpha}]_{\mathcal{C}}$ in general is more difficult than $[l_{X^1_\alpha}]_{\mathcal{C}}$ for a basis $\mathcal{C}$ of $K^m[x]$. Further analysis of Proposition 3.1 indicates why this is the case. If the left element X^k_α is held constant, then there is only one position β in which multiplication on the right by standard basis elements X^l_β results in a nonzero product. On the other hand, if the right element X^l_β is held constant, then every position α has certain standard basis elements X^k_α in which multiplication on the left by X^k_α results in a nonzero product. This complication is another reason for our concentration on left amenability for the purposes of this paper.

Recall from Theorem 2.3 that the components of a modular basis need not all be the same basis. In the case where some or all of the component bases are distinct amenable bases, it is somewhat surprising that the resulting modular basis is not guaranteed to be left amenable in general. Example 4.3 in Section 4.1 is an example of one such basis in $K^2[x]$ which is not left amenable. As the next result will show, we can impose the additional condition of pairwise mutual congeniality between the amenable component bases to ensure that the resulting modular basis is left amenable.

THEOREM 3.4. *Let $\mathcal{C}_1, \mathcal{C}_2, \ldots, \mathcal{C}_{m-1}, \mathcal{C}_0$ be m amenable bases of $K[x]$. If $\mathcal{C}_1, \mathcal{C}_2, \ldots, \mathcal{C}_{m-1}, \mathcal{C}_0$ are pairwise mutually congenial, then*

$$\bigcup_{\alpha=1}^{m=1} (\mathcal{C}_\alpha)_\alpha \cup (\mathcal{C}_0)_0$$

is a left amenable basis for $K^m[x]$.

PROOF. For convenience, let $\mathcal{C} = \bigcup_{\alpha=1}^{m=1} (\mathcal{C}_\alpha)_\alpha \cup (\mathcal{C}_0)_0$. We index the rows and columns of the m matrices $\{\, [l_{X^1_\alpha}]_{\mathcal{C}} \mid 0 \le \alpha \le m-1 \,\}$ according to the order

$$(\mathcal{C}_1)_1 \cup (\mathcal{C}_2)_2 \cup \cdots \cup (\mathcal{C}_{m-1})_{m-1} \cup (\mathcal{C}_0)_0$$

so that each is a block matrix of the form

$$[l_{X_\alpha^1}]_\mathcal{C} = \left[\begin{array}{c|c|c|c} a_{11} & a_{12} & \cdots & a_{1m} \\ \hline a_{21} & a_{22} & \cdots & a_{2m} \\ \hline \vdots & \vdots & \ddots & \vdots \\ \hline a_{m1} & a_{m2} & \cdots & a_{mm} \end{array}\right]$$

where a_{ij} has rows and columns indexed by $(\mathcal{C}_{i \bmod m})_{i \bmod m}$ and $(\mathcal{C}_{j \bmod m})_{j \bmod m}$, respectively. Fix $0 \leq \alpha \leq m-1$ and let $0 \leq \beta \leq m-1$. Since $\mathcal{C}_\alpha$ is a basis of $K[x]$, for each $c_\beta \in \mathcal{C}_\beta$ there exist $k_i \in K$, $c_{\alpha i} \in \mathcal{C}_\alpha$, and $m \in \mathbb{N}$ such that $l_x(c_\beta) = x \cdot c_\beta = \sum_{i=0}^m k_i c_{\alpha i}$. These k_i correspond to nonzero entries of the c_βth column of $[l_x]_{\mathcal{C}_\beta}^{\mathcal{C}_\alpha}$. Using Lemma 3.2, we see that if $\alpha + 1 \not\equiv \beta \pmod{m}$, then $l_{X_\alpha^1}\big((c_\beta)_\beta\big) = X_\alpha^1 (c_\beta)_\beta = 0$ and if $\alpha + 1 \equiv \beta \pmod{m}$, then

$$l_{X_\alpha^1}\big((c_\beta)_\beta\big) = X_\alpha^1 (c_\beta)_\beta = (x \cdot c_\beta)_\alpha = \left(\sum_{i=0}^m k_i c_{\alpha i}\right)_\alpha = \sum_{i=0}^m k_i (c_{\alpha i})_\alpha.$$

It follows that

$$[l_{X_\alpha^1}]_\mathcal{C} = \left[\begin{array}{c|c|c|c|c|c|c} 0 & 0 & \cdots & 0 & 0 & \cdots & 0 \\ \hline 0 & 0 & \cdots & 0 & 0 & \cdots & 0 \\ \hline \vdots & \vdots & \cdots & \vdots & \vdots & \cdots & \vdots \\ \hline 0 & 0 & \cdots & [l_x]_{\mathcal{C}_\beta}^{\mathcal{C}_\alpha} & 0 & \cdots & 0 \\ \hline \vdots & \vdots & \cdots & \vdots & \vdots & \cdots & \vdots \\ \hline 0 & 0 & \cdots & 0 & 0 & \cdots & 0 \end{array}\right]$$

where $a_{ij} = [l_x]_{\mathcal{C}_\beta}^{\mathcal{C}_\alpha}$ if $i \equiv \alpha \pmod{m}$ and $j = \alpha + 1$, and $a_{ij} = 0$ otherwise. Because $\mathcal{C}_\beta$ and $\mathcal{C}_\alpha$ are mutually congenial amenable bases, the product $[I]_{\mathcal{C}_\beta}^{\mathcal{C}_\alpha}[l_x]_{\mathcal{C}_\beta} = [l_x]_{\mathcal{C}_\beta}^{\mathcal{C}_\alpha}$ is row-finite. Thus, $[l_{X_\alpha^1}]_\mathcal{C}$ is row-finite. $\square$

In Section 4, we investigate whether the converse of Theorem 3.4 holds in the more limited scope of the algebra of 2-nomials. Because computations in $K^2[x]$ are more manageable, we also provide examples of left amenable bases and explore the related notions of congeniality and simplicity in that context.

4. Modular bases of 2-nomials

4.1. Congeniality among components of modular bases for 2-nomials.

We open this section by restating Theorem 3.4 in the context of 2-nomials as it will be referenced frequently.

THEOREM 4.1 (For 2-nomials). *Let $\mathcal{C}_1$ and $\mathcal{C}_0$ be two amenable bases of $K[x]$. If $\mathcal{C}_1$ and $\mathcal{C}_0$ are mutually congenial, then $(\mathcal{C}_1)_1 \cup (\mathcal{C}_0)_0$ is a left amenable basis for $K^2[x]$.*

REMARK 4.2. If $\mathcal{C}_1$ and $\mathcal{C}_0$ are bases of $K[x]$, the proof of Theorem 3.4 shows us that the matrices $[l_{X_1^1}]_{(\mathcal{C}_1)_1 \cup (\mathcal{C}_0)_0}$ and $[l_{X_0^1}]_{(\mathcal{C}_1)_1 \cup (\mathcal{C}_0)_0}$ are block matrices of the form

$$[l_{X_1^1}]_{(\mathcal{C}_1)_1\cup(\mathcal{C}_0)_0} = \left[\begin{array}{c|c} 0 & [l_x]_{\mathcal{C}_0}^{\mathcal{C}_1} \\ \hline 0 & 0 \end{array}\right] \qquad [l_{X_0^1}]_{(\mathcal{C}_1)_1\cup(\mathcal{C}_0)_0} = \left[\begin{array}{c|c} 0 & 0 \\ \hline [l_x]_{\mathcal{C}_1}^{\mathcal{C}_0} & 0 \end{array}\right].$$

Notice that transposing each matrix yields

$$([l_{X_1^1}]_{(\mathcal{C}_1)_1\cup(\mathcal{C}_0)_0})^T = \left[\begin{array}{c|c} 0 & [l_x]_{\mathcal{C}_0}^{\mathcal{C}_1} \\ \hline 0 & 0 \end{array}\right]^T = \left[\begin{array}{c|c} 0 & 0 \\ \hline [l_x]_{\mathcal{C}_0}^{\mathcal{C}_1} & 0 \end{array}\right] = [l_{X_0^1}]_{(\mathcal{C}_0)_1\cup(\mathcal{C}_1)_0}$$

$$([l_{X_0^1}]_{(\mathcal{C}_1)_1\cup(\mathcal{C}_0)_0})^T = \left[\begin{array}{c|c} 0 & 0 \\ \hline [l_x]_{\mathcal{C}_1}^{\mathcal{C}_0} & 0 \end{array}\right]^T = \left[\begin{array}{c|c} 0 & [l_x]_{\mathcal{C}_1}^{\mathcal{C}_0} \\ \hline 0 & 0 \end{array}\right] = [l_{X_1^1}]_{(\mathcal{C}_0)_1\cup(\mathcal{C}_1)_0}$$

and so the amenability of $(\mathcal{C}_1)_1 \cup (\mathcal{C}_0)_0$ is equivalent to the amenability of $(\mathcal{C}_0)_1 \cup (\mathcal{C}_1)_0$. Therefore, the results and examples in this section could be duplicated with the positions of the component bases swapped.

Example 4.3 illustrates why the mutually congeniality condition between two amenable bases of $K[x]$ is necessary to guarantee the left amenability of the resulting modular basis.

EXAMPLE 4.3. Recall from Examples 1.2 and 1.5 that the standard basis $\mathcal{B} = \{ x^k \mid k \geq 0 \}$ and $\mathcal{C} = \{ 1 \} \cup \{ x^k + x^{k+1} \mid k \geq 0 \}$ are both amenable bases of $K[x]$ and that $\mathcal{C}$ is properly congenial to $\mathcal{B}$. According to Theorem 2.3, the set

$$(\mathcal{B})_1 \cup (\mathcal{C})_0 = \{ X_1^k \mid k \geq 0 \} \cup \{ (1)_0 \} \cup \{ \left(x^k + x^{k+1}\right)_0 \mid k \geq 0 \}$$

is a basis for $K^2[x]$. Using Proposition 3.1 we see that for $k \geq 0$,

$$\begin{aligned}
&l_{X_1^1}\left(X_1^k\right) = X_1^1 X_1^k = 0 \\
&l_{X_1^1}((1)_0) = X_1^1(1)_0 = X_1^1 X_0^0 = X_1^1 \\
&l_{X_1^1}\left[\left(x^k + x^{k+1}\right)_0\right] = X_1^1\left(x^k + x^{k+1}\right)_0 = X_1^1\left(X_0^k + X_0^{k+1}\right) = X_1^{k+1} + X_1^{k+2}
\end{aligned}$$

and so

$$[l_{X_1^1}]_{(\mathcal{B})_1\cup(\mathcal{C})_0} = \left[\begin{array}{c|cccc} & 0 & 0 & 0 & \cdots \\ & 1 & 1 & 0 & \ddots \\ 0 & 0 & 1 & 1 & \ddots \\ & 0 & 0 & 1 & \ddots \\ & \vdots & \vdots & \ddots & \ddots \\ \hline 0 & & 0 & & \end{array}\right] = \left[\begin{array}{c|c} 0 & [l_x]_{\mathcal{C}}^{\mathcal{B}} \\ \hline 0 & 0 \end{array}\right] = \left[\begin{array}{c|c} 0 & [I]_{\mathcal{C}}^{\mathcal{B}}[l_x]_{\mathcal{C}} \\ \hline 0 & 0 \end{array}\right]$$

is row-finite. Similarly, for $k \geq 0$,

$$\begin{aligned}
&l_{X_0^1}\left(X_1^k\right) = X_0^1 X_1^k = X_0^{k+1} = (-1)^{k+1}(1)_0 + \sum_{i=0}^{k}(-1)^{k+i}\left(x^i + x^{i+1}\right)_0 \\
&l_{X_0^1}((1)_0) = X_0^1(1)_0 = X_0^1 X_0^0 = 0 \\
&l_{X_0^1}\left[\left(x^k + x^{k+1}\right)_0\right] = X_0^1\left(x^k + x^{k+1}\right)_0 = X_0^1\left(X_0^k + X_0^{k+1}\right) = 0
\end{aligned}$$

and so

$$[l_{X_0^1}]_{(\mathcal{B})_1\cup(\mathcal{C})_0} = \left[\begin{array}{ccccc|c} & & 0 & & & 0 \\ \hline -1 & 1 & -1 & \cdots & & \\ 1 & -1 & 1 & \ddots & & \\ 0 & 1 & -1 & \ddots & & 0 \\ 0 & 0 & 1 & \ddots & & \\ \vdots & \vdots & \ddots & \ddots & & \end{array}\right] = \left[\begin{array}{c|c} 0 & 0 \\ \hline [l_x]_{\mathcal{B}}^{\mathcal{C}} & 0 \end{array}\right] = \left[\begin{array}{c|c} 0 & 0 \\ \hline [I]_{\mathcal{B}}^{\mathcal{C}}[l_x]_{\mathcal{B}} & 0 \end{array}\right]$$

is not row-finite. Since at least one of $[l_{X_1^1}]_{(\mathcal{B})_1\cup(\mathcal{C})_0}$ and $[l_{X_0^1}]_{(\mathcal{B})_1\cup(\mathcal{C})_0}$ is not row-finite, it follows that $(\mathcal{B})_1 \cup (\mathcal{C})_0$ is not left amenable.

Given two amenable bases of $K[x]$, a natural follow-up question that arises is whether the converse of Theorem 4.1 is true for the resulting modular basis. That is, if $\mathcal{C}_1$ and $\mathcal{C}_0$ are amenable bases such that $(\mathcal{C}_1)_1 \cup (\mathcal{C}_0)_0$ is left amenable, are $\mathcal{C}_1$ and $\mathcal{C}_0$ mutually congenial? Further inspection of the proof of Theorem 3.4 leads us to conclude that the matrices $[l_x]_{\mathcal{C}_1}^{\mathcal{C}_0}$ and $[l_x]_{\mathcal{C}_0}^{\mathcal{C}_1}$ must both be row-finite if $(\mathcal{C}_1)_1 \cup (\mathcal{C}_0)_0$ is left amenable. However, because it is possible for the product of a row and column finite matrix with a column finite but not row finite matrix to be row-finite, we cannot be certain that the factors $[I]_{\mathcal{C}_1}^{\mathcal{C}_0}$ and $[I]_{\mathcal{C}_0}^{\mathcal{C}_1}$ are row-finite. For this reason, we suspect that the converse does not hold in general but have so far been unable to produce a counterexample.

The next two results show that the converse of Theorem 4.1 does hold if one of the component bases is the standard basis $\mathcal{B}$ due to the known structure of $[l_x]_{\mathcal{C}}^{\mathcal{B}}$ and $[l_x]_{\mathcal{B}}^{\mathcal{C}}$.

LEMMA 4.4. *Let $\mathcal{B}$ be the standard basis and $\mathcal{C}$ be any basis of $K[x]$. Then*

(1) *$[l_x]_{\mathcal{C}}^{\mathcal{B}}$ is the matrix with a row of zeros followed by the matrix $[I]_{\mathcal{C}}^{\mathcal{B}}$. That is,*

$$[l_x]_{\mathcal{C}}^{\mathcal{B}} = \begin{bmatrix} 0 & 0 & 0 & 0 & \cdots \\ & & [I]_{\mathcal{C}}^{\mathcal{B}} & & \end{bmatrix}.$$

(2) *$[l_x]_{\mathcal{B}}^{\mathcal{C}} = \left([I]_{\mathcal{B}}^{\mathcal{C}}\right)^{>1}$ is the matrix obtained from $[I]_{\mathcal{B}}^{\mathcal{C}}$ by removing its first column.*

PROOF. For the standard basis $\mathcal{B}$, the linear transformation matrix $[l_x]_{\mathcal{B}}$ is a row of zeros followed by the infinite identity matrix $I_{\mathbb{Z}_+}$. Therefore, it is easy to see that

$$[l_x]_{\mathcal{C}}^{\mathcal{B}} = [l_x]_{\mathcal{B}}[I]_{\mathcal{C}}^{\mathcal{B}} = \begin{bmatrix} 0 & 0 & 0 & 0 & \cdots \\ & & I_{\mathbb{Z}_+} & & \end{bmatrix}[I]_{\mathcal{C}}^{\mathcal{B}} = \begin{bmatrix} 0 & 0 & 0 & 0 & \cdots \\ & & [I]_{\mathcal{C}}^{\mathcal{B}} & & \end{bmatrix}$$

$$[l_x]_{\mathcal{B}}^{\mathcal{C}} = [I]_{\mathcal{B}}^{\mathcal{C}}[l_x]_{\mathcal{B}} = [I]_{\mathcal{B}}^{\mathcal{C}}\begin{bmatrix} 0 & 0 & 0 & 0 & \cdots \\ & & I_{\mathbb{Z}_+} & & \end{bmatrix} = \left([I]_{\mathcal{B}}^{\mathcal{C}}\right)^{>1}$$

where $\left([I]_{\mathcal{B}}^{\mathcal{C}}\right)^{>1}$ is the matrix obtained from $[I]_{\mathcal{B}}^{\mathcal{C}}$ by removing its first column. □

PROPOSITION 4.5. *Let $\mathcal{B}$ be the standard basis and $\mathcal{C}$ be an amenable basis of $K[x]$. Then $(\mathcal{B})_1 \cup (\mathcal{C})_0$ is a left amenable basis for $K^2[x]$ if and only if $\mathcal{B}$ and $\mathcal{C}$ are mutually congenial.*

PROOF. ($\Longrightarrow$) Since $(\mathcal{B})_1 \cup (\mathcal{C})_0$ is left amenable, the matrices $[l_{X_1^1}]_{(\mathcal{B})_1\cup(\mathcal{C})_0}$ and $[l_{X_0^1}]_{(\mathcal{B})_1\cup(\mathcal{C})_0}$ are both row-finite. According to the proof of Theorem 3.4, we have

$$[l_{X_1^1}]_{(\mathcal{B})_1\cup(\mathcal{C})_0} = \left[\begin{array}{c|c} 0 & [l_x]_{\mathcal{C}}^{\mathcal{B}} \\ \hline 0 & 0 \end{array}\right] \qquad [l_{X_0^1}]_{(\mathcal{B})_1\cup(\mathcal{C})_0} = \left[\begin{array}{c|c} 0 & 0 \\ \hline [l_x]_{\mathcal{B}}^{\mathcal{C}} & 0 \end{array}\right]$$

and so it follows that $[l_x]_{\mathcal{C}}^{\mathcal{B}}$ and $[l_x]_{\mathcal{B}}^{\mathcal{C}}$ must both be row-finite. But then $[I]_{\mathcal{C}}^{\mathcal{B}}$ must be row-finite by Lemma 4.4 since $[l_x]_{\mathcal{C}}^{\mathcal{B}}$ is the matrix with a row of zeros followed by $[I]_{\mathcal{C}}^{\mathcal{B}}$. Looking again to Lemma 4.4, we see that $[l_x]_{\mathcal{B}}^{\mathcal{C}} = \left([I]_{\mathcal{B}}^{\mathcal{C}}\right)^{>1}$ is row-finite. But then the resulting matrix $[I]_{\mathcal{B}}^{\mathcal{C}}$ of adjoining the original first column of $[I]_{\mathcal{B}}^{\mathcal{C}}$ back onto $\left([I]_{\mathcal{B}}^{\mathcal{C}}\right)^{>1}$ would still only have finitely many nonzero elements in each row and so $[I]_{\mathcal{B}}^{\mathcal{C}}$ is row-finite. ($\Longleftarrow$) Theorem 3.4. □

Proposition 4.6 is an interesting result along the same lines as Theorem 4.1 and Proposition 4.5. The subtle difference is that we begin with mutually congenial component bases, rather than amenable component bases, to create a modular basis of $K^2[x]$.

PROPOSITION 4.6. *Let $\mathcal{C}_1$ and $\mathcal{C}_0$ be mutually congenial bases of $K[x]$. Then $(\mathcal{C}_1)_1 \cup (\mathcal{C}_0)_0$ is left amenable if and only if $\mathcal{C}_1$ and $\mathcal{C}_0$ are amenable.*

PROOF. ($\Longrightarrow$) Since $(\mathcal{C}_1)_1 \cup (\mathcal{C}_0)_0$ is left amenable, the matrices $[l_{X_1^1}]_{(\mathcal{C}_1)_1\cup(\mathcal{C}_0)_0}$ and $[l_{X_0^1}]_{(\mathcal{C}_1)_1\cup(\mathcal{C}_0)_0}$ are both row-finite. Using the proof of Theorem 3.4 again, we have

$$[l_{X_1^1}]_{(\mathcal{C}_1)_1\cup(\mathcal{C}_0)_0} = \left[\begin{array}{c|c} 0 & [l_x]_{\mathcal{C}_0}^{\mathcal{C}_1} \\ \hline 0 & 0 \end{array}\right] \qquad [l_{X_0^1}]_{(\mathcal{C}_1)_1\cup(\mathcal{C}_0)_0} = \left[\begin{array}{c|c} 0 & 0 \\ \hline [l_x]_{\mathcal{C}_1}^{\mathcal{C}_0} & 0 \end{array}\right]$$

and so it follows that $[l_x]_{\mathcal{C}_0}^{\mathcal{C}_1}$ and $[l_x]_{\mathcal{C}_1}^{\mathcal{C}_0}$ must both be row-finite. We also know that $[I]_{\mathcal{C}_1}^{\mathcal{C}_0}$ and $[I]_{\mathcal{C}_0}^{\mathcal{C}_1} = \left([I]_{\mathcal{C}_1}^{\mathcal{C}_0}\right)^{-1}$ are row-finite because $\mathcal{C}_1$ and $\mathcal{C}_0$ are mutually congenial. Therefore,

$$[I]_{\mathcal{C}_1}^{\mathcal{C}_0}[l_x]_{\mathcal{C}_0}^{\mathcal{C}_1} = [I]_{\mathcal{C}_1}^{\mathcal{C}_0}\left([I]_{\mathcal{C}_0}^{\mathcal{C}_1}[l_x]_{\mathcal{C}_0}\right) = \left([I]_{\mathcal{C}_1}^{\mathcal{C}_0}[I]_{\mathcal{C}_0}^{\mathcal{C}_1}\right)[l_x]_{\mathcal{C}_0} = [l_x]_{\mathcal{C}_0}$$

and

$$[I]_{\mathcal{C}_0}^{\mathcal{C}_1}[l_x]_{\mathcal{C}_1}^{\mathcal{C}_0} = [I]_{\mathcal{C}_0}^{\mathcal{C}_1}\left([I]_{\mathcal{C}_1}^{\mathcal{C}_0}[l_x]_{\mathcal{C}_1}\right) = \left([I]_{\mathcal{C}_0}^{\mathcal{C}_1}[I]_{\mathcal{C}_1}^{\mathcal{C}_0}\right)[l_x]_{\mathcal{C}_1}) = [l_x]_{\mathcal{C}_1}$$

are both row-finite (as each is the product of row-finite matrices). ($\Longleftarrow$) Theorem 3.4. □

4.2. Congeniality of modular bases for 2-nomials. This last section of the paper is devoted to exploring the notion of congeniality for the family of modular bases introduced in Section 2. In particular, we examine the ways in which the congeniality of component bases in $K[x]$ is preserved when building modular bases for the algebra of 2-nomials.

Consider the subfamily of modular bases of $K^2[x]$ in which the same basis is used for both components (rather than two distinct bases). There is a one-to-one correspondence between the bases of $K[x]$ and the modular bases of $K^2[x]$ in this subfamily given by $\mathcal{C} \to \mathcal{C}^2$. In the next proposition, we show that the congeniality relations for bases in $K[x]$ are preserved for the corresponding modular bases in this subfamily.

PROPOSITION 4.7. *Let $\mathcal{C}$ and $\mathcal{D}$ be bases of $K[x]$. Then $\mathcal{C}$ is properly congenial [resp. mutually congenial or discordant] to $\mathcal{D}$ if and only if $\mathcal{C}^2$ is properly congenial [resp. mutually congenial or discordant] to $\mathcal{D}^2$.*

PROOF. Since $\mathcal{C}$ and $\mathcal{D}$ are bases of $K[x]$, for each $c \in \mathcal{C}$ and $d \in \mathcal{D}$ there exist $r_i, s_i \in K$, $c_i \in \mathcal{C}$, $d_i \in \mathcal{D}$, and $n, m \in \mathbb{N}$ such that $c = \sum_{i=1}^{n} r_i d_i$ and $d = \sum_{i=1}^{m} s_i c_i$. These r_i and s_i correspond to the nonzero entries of the cth column of $[I]_{\mathcal{C}}^{\mathcal{D}}$ and the dth column of $[I]_{\mathcal{D}}^{\mathcal{C}}$, respectively. Since

$$(c)_1 = \left(\sum_{i=1}^{n} r_i d_i\right)_1 = \sum_{i=1}^{n} r_i (d_i)_1 \qquad (d)_1 = \left(\sum_{i=1}^{m} s_i c_i\right)_1 = \sum_{i=1}^{m} s_i (c_i)_1$$

$$(c)_0 = \left(\sum_{i=1}^{n} r_i d_i\right)_0 = \sum_{i=1}^{n} r_i (d_i)_0 \qquad (d)_0 = \left(\sum_{i=1}^{m} s_i c_i\right)_0 = \sum_{i=1}^{m} s_i (c_i)_0$$

it follows that

$$[I]_{\mathcal{C}^2}^{\mathcal{D}^2} = \left[\begin{array}{c|c} [I]_{\mathcal{C}}^{\mathcal{D}} & 0 \\ \hline 0 & [I]_{\mathcal{C}}^{\mathcal{D}} \end{array}\right] \qquad [I]_{\mathcal{D}^2}^{\mathcal{C}^2} = \left[\begin{array}{c|c} [I]_{\mathcal{D}}^{\mathcal{C}} & 0 \\ \hline 0 & [I]_{\mathcal{D}}^{\mathcal{C}} \end{array}\right]$$

and so $[I]_{\mathcal{C}^2}^{\mathcal{D}^2}$ is row-finite if and only if $[I]_{\mathcal{C}}^{\mathcal{D}}$ is row-finite and $[I]_{\mathcal{D}^2}^{\mathcal{C}^2}$ is row-finite if and only if $[I]_{\mathcal{D}}^{\mathcal{C}}$ is row-finite. The result follows. □

It turns out that the congeniality relationship between any two modular bases $(\mathcal{C}_1)_1 \cup (\mathcal{C}_0)_0$ and $(\mathcal{D}_1)_1 \cup (\mathcal{D}_0)_0$ of $K^2[x]$ is determined completely by the combination of congeniality relationships between $\mathcal{C}_1$ and $\mathcal{D}_1$ and between $\mathcal{C}_0$ and $\mathcal{D}_0$. Our next theorem summarizes the congeniality outcome for each possible combination.

THEOREM 4.8. *Let $\mathcal{C}_1$, $\mathcal{C}_0$, $\mathcal{D}_1$, and $\mathcal{D}_0$ be bases of $K[x]$.*

(1) *$\mathcal{C}_1$ is mutually congenial to $\mathcal{D}_1$ and $\mathcal{C}_0$ is mutually congenial to $\mathcal{D}_0$ if and only if $(\mathcal{C}_1)_1 \cup (\mathcal{C}_0)_0$ is mutually congenial to $(\mathcal{D}_1)_1 \cup (\mathcal{D}_0)_0$.*
(2) *If $\mathcal{C}_1$ [resp. $\mathcal{C}_0$] is properly congenial to $\mathcal{D}_1$ [resp. $\mathcal{D}_0$] and $\mathcal{C}_0$ [resp. $\mathcal{C}_1$] is (not necessarily mutually) congenial to $\mathcal{D}_0$ [resp. $\mathcal{D}_1$], then $(\mathcal{C}_1)_1 \cup (\mathcal{C}_0)_0$ properly congenial to $(\mathcal{D}_1)_1 \cup (\mathcal{D}_0)_0$.*
(3) *If $\mathcal{C}_1$ is properly congenial to $\mathcal{D}_1$ and $\mathcal{D}_0$ is properly congenial to $\mathcal{C}_0$, then $(\mathcal{C}_1)_1 \cup (\mathcal{C}_0)_0$ and $(\mathcal{D}_1)_1 \cup (\mathcal{D}_0)_0$ are discordant.*
(4) *If $\mathcal{C}_1$ and $\mathcal{D}_1$ are discordant or $\mathcal{C}_0$ and $\mathcal{D}_0$ are discordant or both, then $(\mathcal{C}_1)_1 \cup (\mathcal{C}_0)_0$ and $(\mathcal{D}_1)_1 \cup (\mathcal{D}_0)_0$ are discordant.*

PROOF. Since $\mathcal{D}_1$ and $\mathcal{D}_0$ are bases of $K[x]$, for each $c_1 \in \mathcal{C}_1$ and $c_0 \in \mathcal{C}_0$ there exist $r_i, s_i \in K$, $d_{1i} \in \mathcal{D}_1$, $d_{0i} \in \mathcal{D}_0$, and $n, m \in \mathbb{N}$ such that $c_1 = \sum_{i=1}^{n} r_i d_{1i}$ and $c_0 = \sum_{i=1}^{m} s_i d_{0i}$. These r_i and s_i correspond to the nonzero entries of the c_1th column of $[I]_{\mathcal{C}_1}^{\mathcal{D}_1}$ and the c_0th column of $[I]_{\mathcal{C}_0}^{\mathcal{D}_0}$, respectively. Since

$$(c_1)_1 = \left(\sum_{i=1}^{n} r_i d_{1i}\right)_1 = \sum_{i=1}^{n} r_i (d_{1i})_1 \qquad (c_0)_0 = \left(\sum_{i=1}^{m} s_i d_{0i}\right)_0 = \sum_{i=1}^{m} s_i (d_{0i})_0$$

it follows that

$$[I]_{(\mathcal{C}_1)_1\cup(\mathcal{C}_0)_0}^{(\mathcal{D}_1)_1\cup(\mathcal{D}_0)_0} = \left[\begin{array}{c|c} [I]_{\mathcal{C}_1}^{\mathcal{D}_1} & 0 \\ \hline 0 & [I]_{\mathcal{C}_0}^{\mathcal{D}_0} \end{array}\right]$$

and so $[I]_{(\mathcal{C}_1)_1\cup(\mathcal{C}_0)_0}^{(\mathcal{D}_1)_1\cup(\mathcal{D}_0)_0}$ is row-finite if and only if $[I]_{\mathcal{C}_1}^{\mathcal{D}_1}$ and $[I]_{\mathcal{C}_0}^{\mathcal{D}_0}$ are both row-finite. Switching the roles of $(\mathcal{C}_1)_1 \cup (\mathcal{C}_0)_0$ and $(\mathcal{D}_1)_1 \cup (\mathcal{D}_0)_0$ in the argument above, we see that

$$[I]_{(\mathcal{D}_1)_1\cup(\mathcal{D}_0)_0}^{(\mathcal{C}_1)_1\cup(\mathcal{C}_0)_0} = \left[\begin{array}{c|c} [I]_{\mathcal{D}_1}^{\mathcal{C}_1} & 0 \\ \hline 0 & [I]_{\mathcal{D}_0}^{\mathcal{C}_0} \end{array}\right]$$

and so $[I]_{(\mathcal{D}_1)_1\cup(\mathcal{D}_0)_0}^{(\mathcal{C}_1)_1\cup(\mathcal{C}_0)_0}$ is row-finite if and only if $[I]_{\mathcal{D}_1}^{\mathcal{C}_1}$ and $[I]_{\mathcal{D}_0}^{\mathcal{C}_0}$ are both row-finite. Each result follows through careful inspection of how combinations of $[I]_{\mathcal{C}_1}^{\mathcal{D}_1}$, $[I]_{\mathcal{C}_0}^{\mathcal{D}_0}$, $[I]_{\mathcal{D}_1}^{\mathcal{C}_1}$, and $[I]_{\mathcal{D}_0}^{\mathcal{C}_0}$ being not row-finite affects the ability of $[I]_{(\mathcal{C}_1)_1\cup(\mathcal{C}_0)_0}^{(\mathcal{D}_1)_1\cup(\mathcal{D}_0)_0}$ and $[I]_{(\mathcal{D}_1)_1\cup(\mathcal{D}_0)_0}^{(\mathcal{C}_1)_1\cup(\mathcal{C}_0)_0}$ to be row-finite. □

For the final result of this paper, Theorem 1.7, we show that the standard basis $\mathcal{B}^2$ of $K^2[x]$ is simple. The proof uses many of the same tools and arguments that are used in the proof that the standard basis of $K[x]$ is simple, which is provided in [**1**]. However, the proof also relies on the facts provided in Remark 4.9 and Lemma 4.10.

REMARK 4.9. Consulting Proposition 3.1, we see that form of the block the matrices $\{ [l_{X_\alpha^n}]_{\mathcal{B}^2} \mid 0 \le \alpha \le 1, n \ge 0 \}$ depends upon the parity of n and whether $\alpha = 0$ or 1.

(1) If n is even, then $[l_{X_1^n}]_{\mathcal{B}^2}$ and $[l_{X_0^n}]_{\mathcal{B}^2}$ are block matrices of the form

$$[l_{X_1^n}]_{\mathcal{B}^2} = \left[\begin{array}{c|c} a_{11} & a_{12} \\ \hline a_{21} & a_{22} \end{array}\right] = \left[\begin{array}{ccccc|c} 0 & 0 & 0 & 0 & \cdots & \\ \vdots & \vdots & \vdots & \vdots & \vdots & 0 \\ 0 & 0 & 0 & 0 & \cdots & \\ & & I_{\mathbb{Z}_+} & & & \\ \hline & & 0 & & & 0 \end{array}\right]$$

$$[l_{X_0^n}]_{\mathcal{B}^2} = \left[\begin{array}{c|c} b_{11} & b_{12} \\ \hline b_{21} & b_{22} \end{array}\right] = \left[\begin{array}{c|ccccc} 0 & & & 0 & & \\ \hline & 0 & 0 & 0 & 0 & \cdots \\ 0 & \vdots & \vdots & \vdots & \vdots & \vdots \\ & 0 & 0 & 0 & 0 & \cdots \\ & & & I_{\mathbb{Z}_+} & & \end{array}\right]$$

where $a_{11} = b_{22}$ has n rows of zeros followed by the infinite identity matrix $I_{\mathbb{Z}_+}$.

(2) If n is odd, then $[l_{X_1^n}]_{\mathcal{B}^2}$ and $[l_{X_0^n}]_{\mathcal{B}^2}$ are block matrices of the form

$$[l_{X_1^n}]_{\mathcal{B}^2} = \left[\begin{array}{c|c} a_{11} & a_{12} \\ \hline a_{21} & a_{22} \end{array}\right] = \left[\begin{array}{c|c} 0 & \begin{array}{ccccc} 0 & 0 & 0 & 0 & \cdots \\ \vdots & \vdots & \vdots & \vdots & \vdots \\ 0 & 0 & 0 & 0 & \cdots \\ & & I_{\mathbb{Z}_+} & & \end{array} \\ \hline 0 & 0 \end{array}\right]$$

$$[l_{X_0^n}]_{\mathcal{B}^2} = \left[\begin{array}{c|c} b_{11} & b_{12} \\ \hline b_{21} & b_{22} \end{array}\right] = \left[\begin{array}{c|c} 0 & 0 \\ \hline \begin{array}{ccccc} 0 & 0 & 0 & 0 & \cdots \\ \vdots & \vdots & \vdots & \vdots & \vdots \\ 0 & 0 & 0 & 0 & \cdots \\ & & I_{\mathbb{Z}_+} & & \end{array} & 0 \end{array}\right]$$

where $a_{12} = b_{21}$ has n rows of zeros followed by the infinite identity matrix $I_{\mathbb{Z}_+}$.

Because $\mathrm{RCFM}(K)$ is a subring of $\mathrm{CFM}(K)$, the product of $A \in \mathrm{RCFM}(K)$ and $B \in \mathrm{CFM}(K) \setminus \mathrm{RCFM}(K)$ need not be row-finite. However, if we impose the additional condition that $A^{-1} \in \mathrm{RCFM}(K)$, we show in Lemma 4.10 that we can be certain $AB \in \mathrm{RCFM}(K)$. We use this subtle but important detail to conclude that a matrix transformed by elementary operations into a form known to be not row-finite is, in fact, not row-finite to begin with.

LEMMA 4.10. *Suppose A and A^{-1} are row and column-finite and B is column-finite, but not row-finite. Then AB is column-finite, but not row-finite.*

PROOF. For the sake of contradiction, suppose AB is row and column-finite. Because the product of two row and column finite matrices is also row and column-finite, it follows that $A^{-1}(AB) = (A^{-1}A)B = B$ is row and column-finite, a contradiction. □

THEOREM 4.11. *The standard basis $\mathcal{B}^2$ for $K^2[x]$ is left simple.*

PROOF. Suppose $\mathcal{C}$ is a basis of $K^2[x]$ and that $\mathcal{B}^2$ is properly congenial to $\mathcal{C}$. Then $A = [I]_{\mathcal{B}^2}^{\mathcal{C}}$ is row-finite and $A^{-1} = [I]_{\mathcal{C}}^{\mathcal{B}^2}$ is not row-finite. Since $\mathcal{B}^2$ and $\mathcal{C}$ are bases of $K^2[x]$, there exist $\mathcal{C}_1, \mathcal{C}_0 \subset \mathcal{C}$ such that $\mathcal{C}_1 \cap \mathcal{C}_0 = \emptyset$, $|\mathcal{C}_1| = |(\mathcal{B})_1| = |\mathbb{N}|$, $|\mathcal{C}_0| = |(\mathcal{B})_0| = |\mathbb{N}|$, and $\mathcal{C} = \mathcal{C}_1 \cup \mathcal{C}_0$. For $0 \le \alpha \le 1$ and $n \ge 0$, it follows that A, A^{-1}, and $[l_{X_\alpha^n}]_{\mathcal{C}}$ are block matrices of the form

$$A = \left[\begin{array}{c|c} a_{11} & a_{12} \\ \hline a_{21} & a_{22} \end{array}\right] \qquad A^{-1} = \left[\begin{array}{c|c} \bar{a}_{11} & \bar{a}_{12} \\ \hline \bar{a}_{21} & \bar{a}_{22} \end{array}\right] \qquad [l_{X_\alpha^n}]_{\mathcal{C}} = \left[\begin{array}{c|c} c_{11} & c_{12} \\ \hline c_{21} & c_{22} \end{array}\right]$$

where a_{ij}, $\bar{a}_{ij}$, and c_{ij} are infinite matrices with rows and columns indexed by $(\mathcal{B})_1$, $(\mathcal{B})_0$, $\mathcal{C}_1$, or $\mathcal{C}_0$. For convenience, enumerate rows and columns of A, A^{-1}, $[l_{X_\alpha^n}]_{\mathcal{C}}$, and $[l_{X_\alpha^n}]_{\mathcal{B}^2}$ by

$$\{1, 2, 3, 4, \ldots, \omega, \omega+1, \omega+2, \omega+3, \omega+4, \ldots\} = \mathbb{Z}_+ \cup \{\omega + z \mid z \in \mathbb{Z}_+\}.$$

Note that since A^{-1} is not row-finite, at least one $\bar{a}_{ij}$ is not row-finite.

Suppose $\bar{a}_{11}$ is not row-finite. Let $k \in \mathbb{Z}_+$ be the first value such that the kth row of $\bar{a}_{11}$ is infinite and let $L = \{l_i\}$ be the set of all values such that there is a nonzero entry in the l_ith row of A beyond the kth position. Since A is invertible, L

is nonempty. Let $t_i > k$ denote value that is the position of the last nonzero entry of the row l_i before the ω position. Because both a_{11} and a_{21} are row-finite, such a $t_i \in \mathbb{Z}_+$ exists for each l_i.

CASE 1. Suppose there exists an $l_e \in L$ such that $t_e - k$ is even. Let $n = t_e - k$. Due to the form of $[l_{X_1^n}]_{\mathcal{B}^2}$ provided in Remark 4.9, we see that

$$[l_{X_1^n}]_{\mathcal{C}} = A\,[l_{X_1^n}]_{\mathcal{B}^2}\,A^{-1} = \left[\begin{array}{c|c} a_{11}^{>n} & 0 \\ \hline a_{21}^{>n} & 0 \end{array}\right]\left[\begin{array}{c|c} \bar{a}_{11} & \bar{a}_{12} \\ \hline \bar{a}_{21} & \bar{a}_{22} \end{array}\right]$$

where $a_{ij}^{>n}$ represents the matrix obtained from a_{ij} by removing the first n columns. It follows that the l_eth row of $[l_{X_1^n}]_{\mathcal{C}}$ (which occurs in either c_{11} or c_{21}) is not row-finite.

CASE 2. Suppose that there is no $l_i \in L$ such that $t_i - k$ is even. Let $t > k$ be the first value before ω such that $t - k$ is even and l a row in L. Notice that A can be transformed through finitely many elementary row operations into the matrix A' with rows identical to A except in row l, where there is a nonzero entry in the tth position followed by zeros in every position after t. Therefore, $E_m \dots E_1 A = A'$ for suitable elementary matrices $E_1, \dots, E_m$. Let $E = E_m \dots E_1$ and $n = t - k$. Due to the form of $[l_{X_1^n}]_{\mathcal{B}^2}$ provided in Remark 4.9, we see that

$$E[l_{X_1^n}]_{\mathcal{C}} = EA[l_{X_1^n}]_{\mathcal{B}^2}A^{-1} = A'[l_{X_1^n}]_{\mathcal{B}^2}A^{-1} = \left[\begin{array}{c|c} a_{11}^{>n} & 0 \\ \hline a_{21}^{>n} & 0 \end{array}\right]\left[\begin{array}{c|c} \bar{a}_{11} & \bar{a}_{12} \\ \hline \bar{a}_{21} & \bar{a}_{22} \end{array}\right]$$

and so the lth row of $E[l_{X_1^n}]_{\mathcal{C}}$ is not row-finite. Because m is finite, E and E^{-1} are both row and column-finite. It follows from Lemma 4.10 that $E^{-1}\left(E[l_{X_1^n}]_{\mathcal{C}}\right) = (E^{-1}E)[l_{X_1^n}]_{\mathcal{C}} = [l_{X_1^n}]_{\mathcal{C}}$ is not row-finite.

Similar arguments show that if $\bar{a}_{12}$, $\bar{a}_{21}$, or $\bar{a}_{22}$ are not row-finite, then there exists an n such that $[l_{X_1^n}]_{\mathcal{C}}$, $[l_{X_0^n}]_{\mathcal{C}}$, or $[l_{X_0^n}]_{\mathcal{C}}$, respectively, are not row-finite. Therefore, it must be that $\mathcal{C}$ is not left amenable and so $\mathcal{B}^2$ is left simple. □

References

[1] L. M. Al-Essa, S. R. López-Permouth, and N. M. Muthana, *Modules over infinite-dimensional algebras*, Linear Multilinear Algebra **66** (2018), no. 3, 488–496, DOI 10.1080/03081087.2017.1301365. MR3760382

[2] Pınar Aydoğdu, Joaquín Díaz Boils, Sergio R. López-Permouth, and Rebin A. Muhammad, *Two value graph magma algebras and amenability*, Algebra and related topics with applications, Springer Proc. Math. Stat., vol. 392, Springer, Singapore, [2022] ©2022, pp. 383–400, DOI 10.1007/978-981-19-3898-6_30. MR4540736

[3] P. Aydoğdu, C. A. Arellano, S. R. López-Permouth, R. Muhammad, and M. Zailaee, *The isomorphism problem for basic modules and the divisibility profile of the algebra of polynomials*, Rev. R. Acad. Cienc. Exactas Fís. Nat. Ser. A Mat. RACSAM **118** (2024), no. 2, Paper No. 49, 15, DOI 10.1007/s13398-023-01547-y. MR4689398

[4] F. A. Ebrahim, S. R. López-Permouth, and M. Zailaee, *Symmetry Questions about Amenability and simplicity of Bases*, Contemp. Math., Amer. Math. Soc., to appear.

[5] F. A. Ebrahim, S. R. López-Permouth, and A. Mozaffarikhah, *Amenable and simple bases in algebras and modules of polynomials with several variables*, Preprint.

[6] Sergio R. López-Permouth and Ashley H. Pallone, *Rings of entangled polynomials*, Rend. Circ. Mat. Palermo (2) **72** (2023), no. 3, 1823–1843, DOI 10.1007/s12215-022-00772-5. MR4576733

[7] Sergio R. López-Permouth and Benjamin Stanley, *Graph magma algebras have no projective bases*, Linear Multilinear Algebra **69** (2021), no. 11, 1997–2005, DOI 10.1080/03081087.2019.1652552. MR4283117

[8] Sergio R. López-Permouth and Benjamin Stanley, *On the amenability profile of infinite dimensional algebras*, Comm. Algebra **49** (2021), no. 5, 2253–2264, DOI 10.1080/00927872.2020.1869245. MR4232497

[9] Sergio R. López-Permouth and Benjamin Stanley, *Topological characterizations of amenability and congeniality of bases*, Appl. Gen. Topol. **21** (2020), no. 1, 1–15, DOI 10.4995/agt.2020.11488. MR4087782

Department of Mathematics, Ohio University, Athens, Ohio 43210
Email address: lopez@ohio.edu

Department of Mathematics, Pennsylvania State University, State College, Pennsylvania 16801
Email address: axp5993@psu.edu

Contemporary Mathematics
Volume **826**, 2025
https://doi.org/10.1090/conm/826/16580

Weakly right $\mathfrak{cP}$-Baer ring

Nasibeh Aramideh and Ahmad Moussavi

ABSTRACT. We define a ring R to be weakly right (left) $\mathfrak{cP}$-Baer if, for any cyclic projective right (left) R-module P, $r_R(P)$ ($\ell_R(P)$) is left (right) s-unital by left (right) semicentral idempotents. This class of rings includes both $\mathfrak{cP}$-Baer and weakly p.q.-Baer rings. The upper triangular matrix ring $T_n(R)$ is weakly $\mathfrak{cP}$-Baer if and only if R is weakly $\mathfrak{cP}$-Baer. The weakly right $\mathfrak{cP}$-Baer condition is studied with respect to polynomial, skew polynomial, and formal power series rings. A characterization of when a 2-by-2 generalized upper triangular matrix ring is weakly right $\mathfrak{cP}$-Baer is provided. The last major result extends a theorem of Kist for commutative Baer rings to weakly $\mathfrak{cP}$-Baer rings, where every prime ideal contains a unique minimal prime ideal without employing topological arguments. We demonstrate that there is an important subclass of weakly $\mathfrak{cP}$-Baer rings, which have a nontrivial subdirect product representation. Examples are provided to illustrate the results.

Introduction

Throughout this paper, all rings are assumed to be associative with unity (denoted by R) and all modules are assumed to be unital right R-modules unless explicitly stated otherwise. A ring R is said to be reduced if it does not have a nonzero nilpotent element. A ring R is considered abelian if every idempotent in R is central (e.g., commutative rings, rings with no nontrivial idempotents, and reduced rings). Let S be a nonempty subset of R. We denote $r_R(S) = \{a \in R \mid sa = 0, \forall s \in S\}$, and it is referred to as the right annihilator of S in R, and $\ell_R(S) = \{a \in R \mid as = 0, \forall s \in S\}$, is the left annihilator of S in R.

We recall from **[25]** (**[31]**) that a ring R is called a Baer ring (right Rickart or right p.p. ring) if the right annihilator of any nonempty (singleton) subset of R is generated by an idempotent. If R is a $*$-ring, then R is called a Baer $*$-ring if the right annihilator of every nonempty subset is generated, as a right ideal, by a projection. Kaplansky proves in **[25]** that the definitions of a Baer ring and a Baer $*$-ring are left-right symmetric. The subject of Baer $*$-rings is essentially pure algebra, with historic roots in operator algebras and lattice theory. Baer $*$-rings are a common generalization of AW^*-algebras and complete $*$-regular rings. The AW^*-algebras are precisely the Baer $*$-rings that happen to be C^*-algebras, and the complete $*$-regular rings are the Baer $*$-rings that happen to be regular in the

2020 *Mathematics Subject Classification.* Primary 16D25, 16D40, 16D70, 16N60, 16S36, 16S50, 16W60.

Key words and phrases. weakly $\mathfrak{cP}$-Baer ring, right APP ring, p.q.-Baer ring, p.p.-ring, Annihilator, minimal prime ideal, left s-unital ideal.

sense of von Neumann. The theory of Baer rings, Baer $*$-rings, and AW^*-algebras (C^*-algebras that are Baer $*$-rings) were studied in [**6**, **25**].

Clark generalized Baer rings in [**20**] and referred to a quasi-Baer ring as one in which the right annihilator of any ideal is generated by an idempotent. This is a nontrivial generalization of the class of Baer rings. For example, prime rings with nonzero right singular ideal are quasi-Baer and not Baer, since Baer rings are nonsingular. The quasi-Baer ring property is left-right symmetric. In [**9**], Birkenmeier and Park introduced a quasi-Baer $*$-ring as a $*$-ring R in which the right annihilator of every ideal is generated by a projection. As in the case of Baer $*$-rings, the involution can be used to show that this notion is left-right symmetric. If R is a commutative nonPrüfer domain then $M_n(R)$, with the transpose involution, is a quasi-Baer $*$-ring, which is not a Baer $*$-ring.

In [**14**], Birkenmeier, Kim, and Park defined a right principally quasi-Baer ring (or simply right p.q.-Baer ring) as a ring in which the right annihilator of every principal right ideal is generated by an idempotent. Equivalently, R is right p.q.-Baer if R modulo, the right annihilator of any principal right ideal is projective. The authors provided several examples in [**14**] to demonstrate that the class of left p.q.-Baer rings is not a subset of the class of right p.p.-rings, and vice versa. Birkenmeier, Park, and Rizvi [**18**] introduced the concept of principally quasi-Baer (p.q.-Baer) $*$-rings as a generalization of quasi-Baer $*$-rings. A p.q.-Baer $*$-ring R is defined as a $*$-ring where every principal right ideal aR satisfies $r_R(aR) = eR$, where e is a projection in R. It follows that $\ell_R(aR) = Rc$ for a projection c. Several authors have explored the spectral theory of such algebraic structures [**16**, **24**, **26**].

These rings have a rich structure theory because of the abundance of their idempotents. Additionally, these rings are naturally found not only in Ring Theory (e.g., right Noetherian right semihereditary rings are Baer rings) but also in other areas of Mathematics, such as Operator Theory (e.g., von Neumann algebras are Baer rings and local multiplier algebras are quasi-Baer rings). In [**8**], as a generalization of right p.q.-Baer rings, Birkenmeier and Heider introduce right $\mathfrak{cP}$-Baer rings, in which every right annihilator of a cyclic projective R-module is generated by an idempotent, and demonstrate that every semiprime ring has a $\mathfrak{cP}$-Baer hull. Among other results, they also investigate the behavior of right $\mathfrak{cP}$-Baer rings with respect to polynomial and formal power series, as well as generalized upper triangular matrix ring extensions.

In this paper, we introduce and examine the concept of weakly $\mathfrak{cP}$-Baer rings. A ring R is weakly right $\mathfrak{cP}$-Baer if for each cyclic projective right R-module P, there exists a nonempty subset E of left semicentral idempotents of R such that $r_R(P) = \bigcup_{c\in E} cR$. It is known that, if P is a cyclic projective right R-module, then $P \cong eR$, where $e = e^2 \in R$. Hence, $r_R(P) = r_R(eR) = r_R(ReR) = r_R(tr(P))$, where $tr(-)$ denote the trace of $(-)$ in R. Therefore, R is a weakly right $\mathfrak{cP}$-Baer ring if for each $e = e^2 \in R$ there exists a nonempty subset E of left semicentral idempotents of R such that $r_R(eR) = \bigcup_{c\in E} cR$. Thus, the class of weakly $\mathfrak{cP}$-Baer rings includes $\mathfrak{cP}$-Baer rings and weakly p.q.-Baer rings, and hence abelian rings, semisimple Artinian rings, local rings, biregular rings, prime rings. Notice that, unlike the right $\mathfrak{cP}$-Baer rings, the class of weakly right $\mathfrak{cP}$-Baer rings also contains both right p.q.-Baer rings and left p.q.-Baer rings. The following table displays the implications, which are irreversible, and demonstrates that the class of weakly right $\mathfrak{cP}$-Baer rings properly contains all other aforementioned classes of Baer-type rings:

$$
\begin{array}{ccccccc}
 & & & & \text{biregular} & \Rightarrow & \text{left p.q.}-\text{Baer} \\
 & & & & \Downarrow & & \Downarrow \\
 & & \text{prime} & & \text{right p.q.}-\text{Baer} & \Rightarrow & \text{weakly p.q.}-\text{Baer} \\
 & & \Downarrow & & \Downarrow & & \Downarrow \\
\text{Baer} & \Rightarrow & \text{quasi}-\text{Baer} & \Rightarrow & \text{right } \mathfrak{cP}-\text{Baer} & \Rightarrow & \text{weakly right } \mathfrak{cP}-\text{Baer} \\
 & & & & \Uparrow & & \\
 & & \text{local} & \Rightarrow & \text{abelian} & & \\
 & & & & \Uparrow & & \\
 & & & & \text{reversible} & & \\
 & & & & \Uparrow & & \\
 & & \text{commutative} & \Rightarrow & \text{symmetric} & & \\
 & & & & \Uparrow & & \\
 & & & & \text{reduced} & &
\end{array}
$$

In Section 1, we investigate weakly right $\mathfrak{cP}$-Baer rings and present several fundamental results for them.

In Section 2, we delve into matrix extensions. We provide a characterization for when a 2-by-2 generalized upper triangular matrix ring is weakly right $\mathfrak{cP}$-Baer. Furthermore, we utilize this result to show that the n-by-n upper triangular matrix ring over R is weakly right $\mathfrak{cP}$-Baer if and only if R is weakly right $\mathfrak{cP}$-Baer.

In Section 3, we examine the relationship between the weakly right $\mathfrak{cP}$-Baer property of a ring R and that of the polynomial ring $R[x]$ and power series ring $R[[x]]$. We show that an α-compatible ring R is weakly right $\mathfrak{cP}$-Baer ring if and only if $R[x;\alpha]$ is weakly right $\mathfrak{cP}$-Baer ring. Additionally, in Theorem 3.9, we establish that when R is a ring and X is a nonempty set of not necessarily commuting indeterminates, R is weakly right $\mathfrak{cP}$-Baer if and only if $R[X]$ is a weakly right $\mathfrak{cP}$-Baer ring.

In the final section, Section 4, we extend a theorem of Kist (see [**27**]) for commutative p.p.-rings to weakly right $\mathfrak{cP}$-Baer rings. We achieve this without employing topological arguments, and instead focus on the condition that every prime ideal contains a unique minimal prime ideal. when R is a weakly right $\mathfrak{cP}$-Baer ring such that $\bar{E}(P) = P$ for all minimal prime ideals P of R, then we show that every prime ideal contains a unique minimal prime ideal, where

$$E(P) = \{a \in R \mid eRa = 0,\ for\ some\ e = e^2 \in R \setminus P\}$$

and

$$\bar{E}(P) = \{a \in R \mid a^n \in E(P),\ \textit{for some positive integer } n\}.$$

Additionally, we provide a complete characterization of a specific class of weakly right $\mathfrak{cP}$-Baer rings that possess a nontrivial subdirect product representation. For more details and examples of Baer and quasi-Baer rings see, [**4**], [**12**], [**13**], [**15**], [**17**], [**23**], [**34**], [**36**], [**35**], [**37**] and [**38**].

1. Basic results and examples

An idempotent $c \in R$ is called left (resp. right) semicentral if $rc = crc$ (resp. $cr = crc$), for all $r \in R$. The set of all left (right) semicentral idempotents of R is denoted by $S_l(R)$ (resp. $S_r(R)$). We define $B(R) = S_l(R) \cap S_r(R)$ as the set of all central idempotents. If R is a semiprime ring, then $S_l(R) = B(R) = S_r(R)$.

Following Tominaga [**42**], a right ideal I of a ring R is said to be left s-unital, if for each $a \in I$ there exists an element $x \in I$ such that $ax = a$. According to Liu and Zhao [**30**], a ring R is called right APP if the right annihilator $r_R(aR)$ is left s-unital as an ideal of R for any element $a \in R$.

According to [**32**, Definition 2.1], an ideal I of R is left s-unital by left semicentral idempotents if for every $a \in I$, $ca = a$ for some $c \in I \cap S_l(R)$, or equivalently, $I = \cup_{c \in E} cR$ for some nonempty subset E of $S_l(R)$. Similarly, we define the right mode. As a generalization of p.q.-Baer rings, the authors in [**32**], introduced the concept of weakly p.q.-Baer rings. A ring R is weakly p.q.-Baer if for each $a \in R$ there exists a nonempty subset E of left semicentral idempotents of R such that $r_R(aR) = \bigcup_{e \in E} eR$. This implies that R modulo the right annihilator of any principal right ideal is flat. The class of weakly p.q.-Baer rings is a natural subclass of the class of APP rings and includes both left p.q.-Baer rings and right p.q.-Baer rings.

An important principal right ideal in a ring R is one generated by an idempotent. Note that every cyclic projective module is isomorphic to such a right ideal. In [**8**], Birkenmeier and Heider introduced the concept of a ring in which every right annihilator of a cyclic projective R-module is generated by an idempotent and provide some basic examples and results. By [**8**], a ring R is called right (left) $\mathfrak{c}\mathfrak{P}$-Baer if for any cyclic projective right (left) R-module P, there exists $c = c^2 \in R$ such that $r_R(P) = cR$, (respectively, $\ell_R(P) = Rc$). The ring R is called $\mathfrak{c}\mathfrak{P}$-Baer if R is both right and left $\mathfrak{c}\mathfrak{P}$-Baer.

Now we introduce a new class of Bear type rings as follows.

Definition 1.1. We say a ring R is weakly right (left) $\mathfrak{c}\mathfrak{P}$-Baer if for any cyclic projective right (left) R-module P, $r_R(P)$ ($\ell_R(P)$) is left (right) s-unital by left (right) semicentral idempotents. A ring R is called weakly $\mathfrak{c}\mathfrak{P}$-Baer if R is both weakly right and left $\mathfrak{c}\mathfrak{P}$-Baer.

Remark 1.2. Note that if $c_1, \ldots, c_n \in S_l(R)(S_r(R))$ for an arbitrary ring R, then for any permutation σ of the set $\{1, \ldots, n\}$ we have $c_{\sigma(1)} \cdots c_{\sigma(n)} \in S_l(R)(S_r(R))$.

Taking this into account, we will have Lemma 1.3.

Lemma 1.3. *If $I_1, \ldots, I_n$ are left s-unital by left semicentral idempotents ideal for some positive integer n, then so is $I = \cap_{i=1}^n I_n$.*

Proof. Let $a \in I = \bigcap_{i=1}^n I_n$. Thus, $a \in I_i$ for each $1 \leq i \leq n$. By hypothesis, there exists $c_i \in I_i \cap S_l(R)$ such that $c_i a = a$. Set $c = c_1 \cdots c_n$, then $ca = a$ and c is left semicentral idempotent such that $c \in I$. □

Proposition 1.4. *The following conditions are equivalent:*

(i) *R is a weakly right $\mathfrak{c}\mathfrak{P}$-Baer ring;*
(ii) *$r_R(eR)$ is left s-unital by left semicentral idempotents for each $e = e^2 \in R$;*
(iii) *For any finitely many idempotent elements $e_1, \ldots, e_n \in R$, $r_R(\sum_{i=1}^n e_i R)$ is left s-unital by left semicentral idempotents.*

Proof. (i) $\Rightarrow$ (ii) Assume that R is a weakly right $\mathfrak{c}\mathfrak{P}$-Baer ring and $e = e^2 \in R$. As eR is a cyclic projective right R-module, then $r_R(eR)$ is left s-unital by left semicentral idempotents.

(ii) ⇒ (iii) As e_i is an idempotent of R for $1 \leq i \leq n$, so by part (*ii*), $r_R(e_iR)$ is left s-unital by left semicentral idempotents. We know that $r_R(\sum_{i=1}^{n} e_iR) = \cap_{i=1}^{n} r_R(e_iR)$, so by Lemma 1.3, $r_R(\sum_{i=1}^{n} e_iR)$ is left s-unital by left semicentral idempotents.

(iii) ⇒ (i) Let P be a cyclic projective right R-module. Then there exists $e = e^2 \in R$ such that $P \cong eR$. Therefore, by (iii), $r_R(P)$ is left s-unital by left semicentral idempotents. □

Examples of rings that are weakly right c𝔓-Baer include right c𝔓-Baer rings, and hence abelian rings, semisimple Artinian rings, local rings, biregular rings, and prime rings. It is evident that every weakly p.q.-Baer ring is also a weakly right (left) c𝔓-Baer ring. Consequently, the class of weakly right (left) c𝔓-Baer rings encompasses both left p.q.-Baer and right p.q.-Baer rings, as depicted in Figure 1.

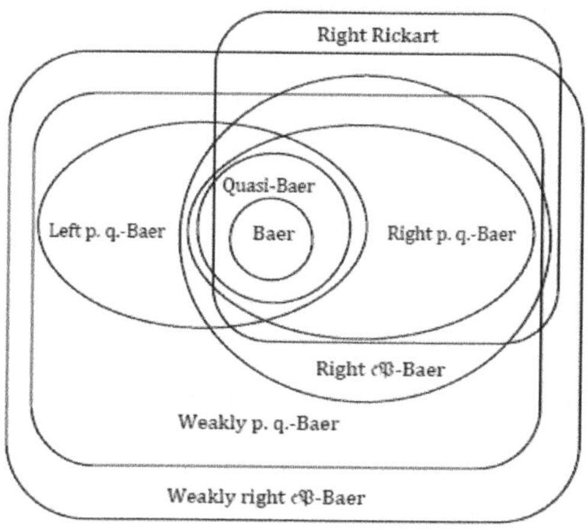

FIGURE 1. Venn diagram of subclasses of weakly right c𝔓-Baer rings

The subsequent examples demonstrate the existence of classes of weakly right (left) c𝔓-Baer rings that are not weakly p.q.-Baer. Thus, the class of weakly right (left) c𝔓-Baer rings properly contains the class of weakly p.q.-Baer rings.

EXAMPLE 1.5. (1) [**12**, Example 2.3]. For a given field F, let

$$R = \Big\{ (a_n)_{n=1}^{\infty} \in \prod_{n=1}^{\infty} F_n \mid a_n \text{ is eventually constant } \Big\},$$

which is a subring of $\prod_{n=1}^{\infty} F_n$, where $F_n = F$ for each natural number $n \in \mathbb{N}$. Thus, R is a commutative von Neumann regular ring (hence reduced) and a c𝔓-Baer ring. By [**8**, Theorem 4.6], $R[[x]]$ is a c𝔓-Baer ring, which shows that it is weakly c𝔓-Baer. However, according to [**30**, Example 2.4], $R[[x]]$ is not a weakly p.q.-Baer ring.

(2) [**14**, Example 3.13]. For an integer $k > 1$, let W be the kth Weyl algebra over a field of characteristic zero. Then W is simple, but neither right nor left hereditary. From [**40**], there exists a positive integer m such that $M_m(W)$ is neither right nor

left p.p.-ring. Assume that

$$S = \Big\{ (a_n)_{n=1}^{\infty} \in \prod_{m=1}^{\infty} M_m(W) \mid \ a_n \text{ is eventually constant } \Big\}.$$

Then S is biregular (hence p.q.-Baer) and weakly p.q.-Baer. Now, if we consider R as in part (1), then $R \bigoplus S$ is a weakly $\mathfrak{c}\mathfrak{P}$-Baer ring, according to Proposition 1.12, but it is not weakly p.q.-Baer according to [**32**, Theorem 2.22].

If a given ring is regular, the conditions weakly p.q.-Baer and weakly $\mathfrak{c}\mathfrak{P}$-Baer are equivalent.

LEMMA 1.6. *Assume that R is a regular ring. Then R is weakly p.q.-Baer if and only if R is weakly right $\mathfrak{c}\mathfrak{P}$-Baer.*

PROOF. This equivalence follows immediately from [**21**, Theorem 1.1]. □

Example 1.7 shows that the class of weakly right $\mathfrak{c}\mathfrak{P}$-Baer rings properly contains the class of right $\mathfrak{c}\mathfrak{P}$-Baer rings.

EXAMPLE 1.7. Consider R as in part (1) of Example 1.5, so R is a left p.q.-Baer ring. Let

$$P = \Big\{ (b_n)_{n=1}^{\infty} \in R \mid \ b_n = 0 \ eventually \Big\},$$

and $\bar{R} = R/P$. If $b = (b_n)_{n=1}^{\infty} \in P$ then $\ell_R(Rb)$ is not subset of P. We consider the following ring

$$S = \begin{pmatrix} \bar{R} & \bar{R} \\ 0 & R \end{pmatrix}.$$

According to [**14**, Corollary 2.4], S is a left p.q.-Baer ring. Therefore, S is weakly p.q.-Baer and thus weakly right $\mathfrak{c}\mathfrak{P}$-Baer. Furthermore, P is not generated as a right ideal by an idempotent. Let $e = 1$ in $\bar{R}$, then $r_R(e\bar{R}) = P$. As a result, it does not satisfy condition (iii) of [**8**, Theorem 5.1], and therefore, S is not a right $\mathfrak{c}\mathfrak{P}$-Baer ring.

Examples 1.8 and 1.9 show that right APP rings need not be weakly right $\mathfrak{c}\mathfrak{P}$-Baer, and vice versa.

EXAMPLE 1.8. We use the Dorroh extension [**29**, Example 1(1)]. Let $S_0 = \mathbb{Z}_2$, the ring of integers modulo 2, $S_1 = \mathbb{Z}_2 * \mathbb{Z}_2, S_2 = S_1 * \mathbb{Z}_2, \ldots, S_n = S_{n-1} * \mathbb{Z}_2, \ldots$, where the operations on S_n are as following: for $(a, \bar{b}), (c, \bar{d}) \in S_n$ with $a, c \in S_{n-1}$,

$$(a, \bar{b}) + (c, \bar{d}) = (a + c, b \bar{+} d) \ and \ (a, \bar{b})(c, \bar{d}) = (ac + bc + da, \bar{bd}),$$

where $n = 1, 2, \ldots$. Then S_n is a Boolean ring of characteristic 2 for each n.

By the ring-monomorphism $f : S_{n-1} \longrightarrow S_n$ defined by $f(x) = (x, \bar{0})$, we can consider S_{n-1} as a subring of S_n. We obtain a direct product $S = \prod_{n=1}^{\infty} S_n$ with $S_1 \subset S_2 \subset S_3 \subset \ldots$. We also consider

$$R = \left\langle \bigoplus_{n=1}^{\infty} S_n, 1_S \right\rangle,$$

which is a $\mathbb{Z}_2$-subalgebra of S generated by $\bigoplus_{n=1}^{\infty} S_n$ and 1_S, where

$$1_S = ((0, \bar{1}), ((0, \bar{0}), \bar{1}), (((0, \bar{0}), \bar{0}), \bar{1}), \ldots).$$

Notice that for $(a, \bar{b}) \in S_1, (a, \bar{b}) = ((a, \bar{b}), \bar{0}) \in S_2, (a, \bar{b}) = (((a, \bar{b}), \bar{0}), \bar{0}) \in S_3$, etc, and R is a Boolean ring (hence it is commutative, reduced, abelian and a

$c\mathfrak{P}$-Baer ring). As $T = R[[x]]$ is a commutative ring, it is $c\mathfrak{P}$-Baer and therefore a weakly $c\mathfrak{P}$-Baer ring. Suppose that T is an APP-ring. Assume that $f(x) = f_0 + f_1x + f_2x^2 + \cdots \in T$, with $f_0 = (1_{S_1}, 0, 0, \ldots), f_1 = (1_{S_1}, 0, 1_{S_3}, 0, \ldots), f_2 = (1_{S_1}, 0, 1_{S_3}, 0, 1_{S_5}, 0, \ldots), \ldots$. Additionally, let $g(x) = g_0 + g_1x + g_2x^2 + \cdots \in T$, with $g_0 = (0, 1_{S_2}, 0, 0, \ldots), g_1 = (0, 1_{S_2}, 0, 1_{S_4}, 0, \ldots), g_2 = (0, 1_{S_2}, 0, 1_{S_4}, 0, 1_{S_6}, 0, \ldots), \ldots$.

Thus, $g(x) \in r_T(f(x)T)$ and so there exists $h(x) = h_0 + h_1x + h_2x^2 + \cdots \in r_T(f(x)T)$ such that $g(x) = h(x)g(x)$. By $f(x)h(x) = 0$ and [**1**, p. 2269], $f_ih_l = 0$ for all i and l. So there exists $m_l \in \mathbb{N}$ such that h_l has the form

$$(a_{l1}, 0, a_{l3}, 0, \ldots, a_{l(2m_l+1)}, 0, 0, \ldots),$$

where $a_{lk} \in S_k$ and $l = 1, 2, \ldots$. On the other hand, $(1_T - h(x))g(x) = 0$ it follows that $(1_T - h_0)g_j = 0$ and $h_lg_j = 0$ for all j and $l \geq 1$. So there exists $n_l \in \mathbb{N}$ such that h_l has the form $(0, a_{l2}, 0, a_{l4}, 0, \ldots, a_{l(2n_l)}, 0, 0, \ldots)$, where $a_{lk} \in S_k$ and $l = 1, 2, \ldots$. Thus, $h_1 = h_2 = \ldots$, and so $h(x) = h_0$. This contradicts with $g_j = h_0g_j$, $j = 0, 1, \ldots$. Therefore, T is not APP.

EXAMPLE 1.9 ([**14**, Example 1.6]). For a field F, let

$$R = \begin{pmatrix} \prod_{n=1}^{\infty} F_n & \oplus_{n=1}^{\infty} F_n \\ \oplus_{n=1}^{\infty} F_n & \langle \oplus_{n=1}^{\infty} F_n, 1 \rangle \end{pmatrix},$$

where $F_n = F$ for $n = 1, 2, \ldots$ and $\langle \oplus_{n=1}^{\infty} F_n, 1 \rangle$ is the subalgebra of $\prod_{n=1}^{\infty} F_n$ generated by $\oplus_{n=1}^{\infty} F_n$ and $1 \in \prod_{n=1}^{\infty} F_n$. Thus, from [**14**] R is a regular ring (hence p.p.), so by [**30**, Proposition 2.3] R is APP but not weakly $c\mathfrak{P}$-Baer (since otherwise R is weakly p.q.-Baer, this contradicts [**32**, Example 2.6]).

PROPOSITION 1.10. *If R is a weakly right $c\mathfrak{P}$-Baer ring, then eRe is a weakly right $c\mathfrak{P}$-Baer ring for every idempotent $e \in R$.*

PROOF. Let $e = e^2 \in R$ be a nonzero element. Also, let $x \in eRe$ be an arbitrary idempotent. We will show that $r_{eRe}(x(eRe))$ is left s-unital by left semicentral idempotents. As R is a weakly right $c\mathfrak{P}$-Baer ring, $r_R(xR)$ is left s-unital by left semicentral idempotents. Assume that $a \in r_{eRe}(x(eRe))$ is an arbitrary element. Since $a = eae$, $x = exe$ and $x(eRe)a = 0$, $xRa = 0$. So, by hypothesis, $a = ca$ for some $c \in r_R(xR) \cap S_l(R)$. Set $c' = ece \in eRe$. Clearly, $a = c'a$. So, it is enough to show that $c' \in r_{eRe}(x(eRe)) \cap S_l(eRe)$. Let $ere \in eRe$ be an arbitrary element. Then we have

$$c'(ere)c' = (ece)(ere)(ece) = e(c(ere)c)e = e((ere)c)e = (ere)c'.$$

On the other hand, $x(eRe)c' = x(eRe)ce = xRce = 0$, which shows that eRe is weakly right $c\mathfrak{P}$-Baer. □

PROPOSITION 1.11. *Let I be an index set and R_i a ring for each $i \in I$. Then, the ring $R = \prod_{i \in I} R_i$ is weakly right $c\mathfrak{P}$-Baer if and only if R_i is weakly right $c\mathfrak{P}$-Baer for every $i \in I$.*

PROOF. ($\Rightarrow$) Assume that $R = \prod_{i \in I} R_i$ is weakly right $c\mathfrak{P}$-Baer and $e_i = e_i^2 \in R_i$ for some $i \in I$. We will prove that $r_{R_i}(e_iR_i)$ is left s-unital by left semicentral idempotents. Let $a_i \in r_{R_i}(e_iR_i)$. Set $e = \iota_i(e_i)$. By our hypothesis, $r_R(eR)$ is left s-unital by left semicentral idempotents. Put $a = \iota_i(a_i)$, so for an arbitrary element r we have $era = \iota_i(e_i)r\iota_i(a_i) = 0$. Thus, there exists $c \in r_R(eR) \cap S_l(R)$ such that $ca = a$. Clearly, c_i is an idempotent of R_i and $c_ia_i = a_i$. Let $r_i \in R_i$ be an arbitrary element. If we set $r = \iota_i(r_i)$, then $0 = erc = \iota_i(e_i)\iota_i(r_i)c$ and $\iota_i(r_i)c = rc = crc = c\iota_i(r_i)c$. Therefore, $e_i r_i c_i = 0$, $r_i c_i = r_i c_i r_i$ and so $c_i \in r_{R_i}(e_i R_i) \cap S_l(R_i)$.

($\Leftarrow$) Let R_i be weakly right $c\mathfrak{P}$-Baer for every $i \in I$ and $e = e^2 \in R$. Thus, $e_i = \pi_i(e)$ is an idempotent for every $i \in I$. Since R_i is weakly right $c\mathfrak{P}$-Baer, $r_{R_i}(e_i R_i)$ is left s-unital by left semicentral idempotents for every $i \in I$. Let $a \in r_R(eR)$ be an arbitrary element. Hence, $a_i = \pi_i(a) \in R_i$ and $a_i \in r_{R_i}(e_i R_i)$ for every $i \in I$. By definition, there exists $c_i \in r_{R_i}(e_i R_i) \cap S_l(R_i)$ such that $c_i a_i = a_i$ for every $i \in I$. Set $c = (c_i)_{i \in I}$. It is easy to see that $ca = a$ and $c \in r_R(eR) \cap S_l(R)$. □

PROPOSITION 1.12. *Let $R = \oplus_{i \in I} R_i$ (when I is an infinite index set, R is a ring without identity), where R_i is a ring for each $i \in I$. Then R is weakly right $c\mathfrak{P}$-Baer if and only if R_i is weakly right $c\mathfrak{P}$-Baer for each $i \in I$.*

PROOF. This can be accomplished with an argument similar to the proof of Proposition 1.12. □

EXAMPLE 1.13. According to [**8**, Example 5.3], there exists a right $c\mathfrak{P}$-Baer ring R_1 that is not right p.q.-Baer. Additionally, let R_2 be the ring S defined in Example 1.7, which is a left p.q.-Baer ring but not a right $c\mathfrak{P}$-Baer ring. Consider the direct sum $R = R_1 \oplus R_2$. By Proposition 1.12, R is a weakly right $c\mathfrak{P}$-Baer ring, but it is not right $c\mathfrak{P}$-Baer according to [**8**, Proposition 1.8].

Let A be a ring, $B \subseteq A$ be a unitary subring (containing the same unity 1_A as in A), $\{A_i\}_{i=1}^{\infty}$ be a countable set of copies of A, D be the direct product of all rings A_i, and let $R := R(A, B)$ be the subring of D generated by the ideal $\bigoplus_{i=1}^{\infty} A_i$ and by the subring $\{(b, b, \ldots) \mid b \in B\}$ (see [**43**]). We have the following result.

PROPOSITION 1.14. *If A is a ring such that $S_l(A) = S_l(B)$, then the ring $R(A, B)$ is weakly right $c\mathfrak{P}$-Baer if and only if A and B are weakly right $c\mathfrak{P}$-Baer.*

PROOF. If $R := R(A, B)$ is a weakly right $c\mathfrak{P}$-Baer ring, then it is easy to see that A and B are weakly right $c\mathfrak{P}$-Baer rings. For the converse, assume that $(e_i)_{i=1}^{\infty} \in R$ is an idempotent and $(a_i)_{i=1}^{\infty} \in r_R((e_i)_{i=1}^{\infty} R)$. So e_i is an idempotent in A for each i. Also, by definition, there exists n such that $e_n = e_{n+1} = \ldots \in B$ and $a_n = a_{n+1} = \ldots \in B$. Thus, we have $e_i A a_i = 0$ for $1 \leq i \leq n$. Since A is weakly right $c\mathfrak{P}$-Baer, there exists $c_i \in r_A(e_i A) \cap S_l(A)$ such that $a_i = c_i a_i$, for $1 \leq i \leq n-1$. Also, since B is weakly right $c\mathfrak{P}$-Baer, there exists $c_n \in r_B(e_n B) \cap S_l(B)$ such that $a_n = c_n a_n$. By assumption, we have $c_n \in r_A(e_n A) \cap S_l(A)$. Hence, $(c_1, c_2, \ldots, c_{n-1}, c_n, c_n, \ldots) \in r_R((e_i)_{i=1}^{\infty} R) \cap S_l(R)$ and $(e_i)_{i=1}^{\infty} = (c_1, c_2, \ldots, c_{n-1}, c_n, c_n, \ldots)(e_i)_{i=1}^{\infty}$. Therefore, $R(A, B)$ is a weakly right $c\mathfrak{P}$-Baer ring and we are done. □

COROLLARY 1.15. *If A is an abelian ring, then the ring $R(A, B)$ is weakly right $c\mathfrak{P}$-Baer.*

COROLLARY 1.16. *Let A be a weakly right $c\mathfrak{P}$-Baer ring. If $r_B(eA) = 0$ for every $0 \neq e = e^2 \in B$, then the ring $R(A, B)$ is a weakly right $c\mathfrak{P}$-Baer ring.*

PROOF. In the proof of Proposition 1.14, the following two cases occur:

(i) If $0 = e_n = e_{n+1} = \ldots \in B$, then $r_B(e_n B) = B$. Take $c_n = 1 \in B$.
(ii) If $0 \neq e_n = e_{n+1} = \ldots \in B$, then $r_B(e_n A) = 0$, by hypothesis. Thus, $a_n = 0$. Take $c_n = 0 \in B$.

In both cases, we have $a_n = c_n a_n$, $c_n \in S_l(A)$ and $e_n A c_n = 0$. Thus, $(a_i)_{i=1}^{\infty} = (c_1, c_2, \ldots, c_{n-1}, c_n, c_n, \ldots)(a_i)_{i=1}^{\infty}$ and $(e_i)_{i=1}^{\infty} R(c_1, c_2, \ldots, c_{n-1}, c_n, c_n, \ldots) = 0$.

Hence R is a weakly right c𝔓-Baer ring, as desired. □

2. Characterizations of weakly right c𝔓-Baer rings of generalized triangular matrix rings

In this section, we will examine the weakly right c𝔓-Baer property of the 2-by-2 generalized upper triangular matrix rings.

LEMMA 2.1. *Let R, S be rings and M be a (S, R)-bimodule. Then the idempotent $e = \begin{pmatrix} c_1 & k \\ 0 & c_2 \end{pmatrix}$ is the left semicentral idempotent in the generalized triangular matrix ring $T = \begin{pmatrix} S & M \\ 0 & R \end{pmatrix}$ if and only if*

(i) $c_1 \in S_l(S)$;
(ii) $c_2 \in S_l(R)$;
(iii) $c_1 k = k$;
(iv) $c_1 m c_2 = m c_2$ *for all* $m \in M$.

LEMMA 2.2 ([**19**, Lemma 3.1]). *Let $\begin{pmatrix} I & N \\ 0 & L \end{pmatrix}$ be an ideal of $T = \begin{pmatrix} S & M \\ 0 & R \end{pmatrix}$. Then*

$$r_T(\begin{pmatrix} I & N \\ 0 & L \end{pmatrix}) = \begin{pmatrix} r_S(I) & r_M(I) \\ 0 & r_R(L) \cap r_R(N) \end{pmatrix}.$$

THEOREM 2.3. *Let $T = \begin{pmatrix} S & M \\ 0 & R \end{pmatrix}$ where S and R are rings, and M is an (S, R)-bimodule. Then T is weakly right c𝔓-Baer if and only if all the following conditions hold:*

(i) *S and R are weakly right c𝔓-Baer;*
(ii) *for each $e = e^2 \in S$, we have*
   (a) *for every $s \in r_S(eS)$ and $m \in r_M(eS)$ there exists a left semicentral idempotent $c \in r_S(eS)$ such that $cs = s$ and $cm = m$;*
   (b) *$r_R(eM)$ is left s-unital by left semicentral idempotents.*

PROOF. (⇒) Assume T is a weakly right c𝔓-Baer ring.

(i) $S \cong \begin{pmatrix} 1_s & 0 \\ 0 & 0 \end{pmatrix} T \begin{pmatrix} 1_s & 0 \\ 0 & 0 \end{pmatrix}$ and $R \cong \begin{pmatrix} 0 & 0 \\ 0 & 1_R \end{pmatrix} T \begin{pmatrix} 0 & 0 \\ 0 & 1_R \end{pmatrix}$ where $\begin{pmatrix} 1_s & 0 \\ 0 & 0 \end{pmatrix}$ and $\begin{pmatrix} 0 & 0 \\ 0 & 1_R \end{pmatrix}$ are idempotents of T. So by Proposition 1.10, R and S are weakly right c𝔓-Baer rings.

(a) Let $e = e^2 \in S$, $s \in r_S(eS)$ and $m \in r_M(eS)$. Thus $e' = \begin{pmatrix} e & 0 \\ 0 & 0 \end{pmatrix} \in T$ is an idempotent. As $e'T = \begin{pmatrix} eS & eM \\ 0 & 0 \end{pmatrix}$, $r_T(e'T) = \begin{pmatrix} r_S(eS) & r_M(eS) \\ 0 & 0 \end{pmatrix}$, by Lemma 2.2. Since $s' = \begin{pmatrix} s & m \\ 0 & 0 \end{pmatrix} \in r_T(e'T)$, $c's' = s'$ for some $c' = \begin{pmatrix} c & n \\ 0 & 0 \end{pmatrix} \in r_T(e'T) \cap S_l(T)$. Therefore, $cs = s$, $cm = m$ and by Lemma 2.1, $c \in r_S(eS) \cap S_l(S)$.

(b) Let $e = e^2 \in S$. Thus $e' = \begin{pmatrix} e & 0 \\ 0 & 0 \end{pmatrix} \in T$ is an idempotent and so $r_T(e'T)$ is left s-unital by left semicentral idempotents. We can easily see that $r_R(eM)$ is left s-unital by left semicentral idempotents.

($\Leftarrow$) Assuming conditions (i) and (ii) are satisfied, consider an arbitrary idempotent $e = \begin{pmatrix} e_1 & k \\ 0 & e_2 \end{pmatrix} \in T$. Therefore, $e_1 = e_1^2 \in S$, $e_2 = e_2^2 \in R$ and $k = e_1 k + k e_2$. We will show that $r_T(eT)$ is left s-unital by left semicentral idempotents. Now since $eT = \begin{pmatrix} e_1 S & e_1 M + kR \\ 0 & e_2 R \end{pmatrix}$ by Lemma 2.2 we have,

$$r_T(eT) = \begin{pmatrix} r_S(e_1 S) & r_M(e_1 S) \\ 0 & r_R(e_2 R) \cap r_R(e_1 M + kR) \end{pmatrix}.$$

Let $f = \begin{pmatrix} f_1 & n \\ 0 & f_2 \end{pmatrix} \in r_T(eT)$. Thus, $f_2 \in r_R(e_2 R) \cap r_R(e_1 M + kR)$ and so $f_2 \in r_R(e_2 R) \cap r_R(e_1 M)$. From parts (i) and (b), $r_R(e_2 R)$ and $r_R(e_1 M)$ are both left s-unital by left semicentral idempotents. Therefore, $r_R(e_2 R) \cap r_R(e_1 M)$ is also left s-unital by left semicentral idempotents, as stated in Lemma 1.3. Thus $c_2 f_2 = f_2$, for some $c_2 \in S_l(R) \cap (r_R(e_2 R) \cap r_R(e_1 M))$. Since $r_T(eT)$ is an ideal of T, $M(r_R(e_2 R) \cap r_R(e_1 M + kR)) \subseteq r_M(e_1 S)$, by [**28**, Proposition 1.17]. As $f_1 \in r_S(e_1 S)$, $n \in r_M(e_1 S)$ and $mc_2 \in r_M(e_1 S)$ for all $m \in M$, so by property (a), there exists $c_1 \in (S_l(S) \cap r_S(e_1 S))$ such that $c_1 f_1 = f_1$, $c_1 n = n$ and $c_1 m c_2 = m c_2$. Put $c = \begin{pmatrix} c_1 & 0 \\ 0 & c_2 \end{pmatrix}$. By Lemma 2.1, it is clear that $c \in S_l(T)$. Since, $e_1 M c_2 = e_1 c_1 M c_2 = 0$ and $kRc_2 = e_1 k c_2 + k e_2 c_2 = e_1 c_1 k c_2 = 0$ we have

$$eTc = \begin{pmatrix} e_1 S c_1 & (e_1 M + kR) c_2 \\ 0 & e_2 R c_2 \end{pmatrix} = 0.$$

Therefore, $c \in r_T(eT)$. Now since $c_1 n = n$, so $cf = f$ and we are done. $\square$

For a positive integer n and $1 \leq i, j \leq n$, E_{ij} denotes the (i,j)-matrix unit in the upper triangular matrix ring $T_n(R)$ over R.

LEMMA 2.4. *Let R be a ring and $c_1, \ldots, c_n \in S_l(R)$, such that $c_i c_j = c_i$, for all $i \leq j$. Then $E = c_1 E_{11} + \cdots + c_n E_{nn} \in S_l(T_n(R))$.*

PROOF. The proof follows from [**32**, Lemma 2.18], using the fact that $c \in S_r(Tn(R))$ if and only if $1 - c \in S_l(Tn(R))$. $\square$

LEMMA 2.5 ([**32**, Lemma 2.14]). *Let J be an ideal of a ring R, $a_1, \ldots, a_n \in J$ and there exist $c_1, \ldots, c_n \in S_l(R) \cap J$, such that $c_i a_i = a_i$ for each $i = 1, \ldots, n$. Then there exists $c \in S_l(R) \cap J$, such that $ca_i = a_i$, for each $i = 1, \ldots, n$. In particular, an ideal J is left s-unital by left semicentral idempotents if and only if for every finitely many elements $a_1, \ldots, a_n \in J$, there exists an idempotent $c \in S_l(R) \cap J$ such that $ca_i = a_i$, for each i.*

COROLLARY 2.6. *Let $T_n(R)$ is the $n \times n$ upper triangular matrix ring over a ring R, where $n \geq 1$ is a positive integer. Then the following are equivalent:*

(i) *R is a weakly right $\mathfrak{c}\mathfrak{P}$-Baer ring,*
(ii) *$T_n(R)$ is weakly right $\mathfrak{c}\mathfrak{P}$-Baer for every positive integer n,*
(iii) *$T_k(R)$ is weakly right $\mathfrak{c}\mathfrak{P}$-Baer for some $k > 1$,*

(iv) *$T_2(R)$ is weakly right c𝔓-Baer.*

PROOF. (i) $\Rightarrow$ (ii) Let R be a weakly right c𝔓-Baer ring. We will prove the required condition using induction on n. It is obvious that $T_1(R)$ is weakly right c𝔓-Baer. Since $T_2(R)$ satisfies the conditions of Theorem 2.3, it is a weakly right c𝔓-Baer ring. Assume that the result holds for all $1 \leq l \leq n$. We will prove that $T_{n+1}(R)$ is a weakly right c𝔓-Baer ring. We know that $T_{n+1}(R) \cong \begin{pmatrix} R & M \\ 0 & T_n(R) \end{pmatrix}$, where $M = (R, \ldots, R)$ (n-tuple). To apply Theorem 2.3, let $e = e^2 \in R$, $a \in r_R(eR)$ and $\begin{pmatrix} a_1 & a_2 & \ldots & a_n \end{pmatrix} \in r_M(eR)$. Thus, we have

$$0 = eR\begin{pmatrix} a_1 & a_2 & \ldots & a_n \end{pmatrix} = \begin{pmatrix} eRa_1 & eRa_2 & \ldots & eRa_n \end{pmatrix},$$

which implies that $a_i \in r_R(eR)$ for all $1 \leq i \leq n$. By Lemma 2.5, since $r_R(eR)$ is left s-unital by left semicentral idempotents, there exists $c \in r_R(eR) \cap S_l(R)$ such that $ca = a$ and $ca_i = a_i$ for all $1 \leq i \leq n$.

Next, we will show that $r_{T_n(R)}(eM)$ is left s-unital by left semicentral idempotents. Suppose $A \in r_{T_n(R)}(eM)$ such that $A = (a_{ij})$ with $1 \leq i, j \leq n$. Let $\begin{pmatrix} r_1 & r_2 & \ldots & r_n \end{pmatrix} \in M$. Thus, we have

$$e(r_1, r_2, \ldots, r_n)A = (er_1a_{11}, er_1a_{12} + er_2a_{22}, \ldots, \sum_{i=1}^{n} er_ia_{in}) = (0, 0, \ldots, 0).$$

Take $r_1 \neq 0$ and $r_i = 0$ for all $i \neq 1$. Then $a_{1i} \in eR$ for all $1 \leq i \leq n$. Similarly, for $r_2 \neq 0$ and $r_i = 0$ for all $i \neq 2$, then $a_{2i} \in eR$ for $2 \leq i \leq n$. By an induction argument, we have $a_{ij} \in eR$ for all $1 \leq i, j \leq n$. As $r_R(eR)$ is left s-unital by left semicentral idempotents, so by Lemma 2.5 there exists $c' \in r_R(eR) \cap S_l(R)$ such that $c'a_{ij} = a_{ij}$ for all $1 \leq i, j \leq n$. Put $C = c'I_n \in T_n(R)$. By Lemma 2.4 we have, $C \in S_l(T_n(R))$. Also, it is clear that $CA = A$ and $C \in r_{T_n(R)}(eM)$.

(ii) $\Rightarrow$ (iii) This implication is clear.

(iii) $\Rightarrow$ (iv) Let $T_k(R)$ be a weakly right c𝔓-Baer ring for some $k > 1$ and $E_{ij} \in T_k(R)$. Put $F = E_{11} + E_{22}$. Then $F = F^2 \in T_k(R)$ and $T_2(R) \cong FT_k(R)F$. Therefore by Proposition 1.10, $T_2(R)$ is a weakly right c𝔓-Baer ring.

(iv) $\Rightarrow$ (i) This implication is similar to (iii) $\Rightarrow$ (iv). □

Corollary 2.7 is an immediate consequence of Theorem 2.3 and provides a rich source of rings which are weakly right c𝔓-Baer rings.

COROLLARY 2.7. *Let S be a weakly right c𝔓-Baer ring and R a unitary subring of S such that $r_S(eR) = 0$ for any $0 \neq e = e^2 \in R$. Then the ring $T = \begin{pmatrix} S & S \\ 0 & R \end{pmatrix}$ is a weakly right c𝔓-Baer ring.*

COROLLARY 2.8. *Let $T = \begin{pmatrix} \bar{S} & \bar{S} \\ 0 & S \end{pmatrix}$, where S is a weakly right c𝔓-Baer ring and $\bar{S} = S/P$ for a prime ideal P such that if $e = e^2 \in P$ then $r_S(eS)$ is not subset of P. Then T is weakly right c𝔓-Baer.*

PROOF. Since $\bar{S}$ is a prime ring, [**14**, Lemma 1.2] yields that $\bar{S}$ is a p.q.-Baer ring and so is weakly right c𝔓-Baer. Thus part (i) of Theorem 2.3 holds. Let $\bar{e} = \bar{e}^2 \in \bar{S}$ such that $\bar{e} = e + P$. Consider the following two cases:

(i) If $e \in P$, then $\bar{e} = P$ and so $r_{\bar{S}}(\bar{e}\bar{S}) = \bar{S}$, which is left s-unital by left semicentral idempotents. Thus part (a) of Theorem 2.3 holds. Moreover, $r_S(\bar{e}\bar{S}) = r_S(P\bar{S}) = S$ is left s-unital by left semicentral idempotents.

(ii) If $e \notin P$, then $r_{\bar{S}}(\bar{e}\bar{S}) = P$, that is left s-unital by left semicentral idempotents. Thus part (a) of Theorem 2.3 holds. Furthermore, $r_S(\bar{e}\bar{S}) = r_S(P\bar{S}) = P$ is left s-unital by left semicentral idempotents.

□

Using the fact that a ring R is weakly right $\mathfrak{cP}$-Baer if and only if $T_n(R)$ is weakly right $\mathfrak{cP}$-Baer for every positive integer n, one can provide many classes of weakly right $\mathfrak{cP}$-Baer rings which are neither right p.q.-Baer nor left p.q.-Baer and also not right $\mathfrak{cP}$-Baer and not left $\mathfrak{cP}$-Baer.

3. Polynomial extensions of weakly right $\mathfrak{cP}$-Baer rings

In this section, we will examine the skew polynomial rings in relation to the weakly right $\mathfrak{cP}$-Baer condition. The skew polynomial ring $R[x;\alpha]$ is defined as a ring where the elements are polynomials over R, with addition defined in the usual way, and multiplication given by $xa = \alpha(a)x$, for any $a \in R$.

LEMMA 3.1 ([**3**, Lemma 2.4]). *Let R be a ring with an endomorphism α such that $\alpha(e) = e$ for any $e = e^2 \in R$. If $e(x) = \sum_{i=0}^{m} e_i x^i \in R[x;\alpha]$ is an idempotent, then $e_i \in Re_0R$ for any $i \geq 0$.*

PROPOSITION 3.2 ([**3**, Proposition 2.5]). *Let R be a ring with an endomorphism α such that $\alpha(c) = c$ for any left semicentral idempotent $c \in R$. If $e(x) \in S_l(R[x;\alpha])$, where $e(x) = e_0 + e_1x + \cdots + e_mx^m$, then:*

(i) *$e_0 \in S_l(R)$,*
(ii) *$e_0e_i = e_i$ for all $0 \leq i \leq m$,*
(iii) *$e_ie_0 = 0$ for all $1 \leq i \leq m$,*
(iv) *$e(x)R[x;\alpha] = e_0R[x;\alpha]$.*

LEMMA 3.3. *Let R be a weakly right $\mathfrak{cP}$-Baer ring with a monomorphism α such that $\alpha(c) = c$ for any left semicentral idempotent $c \in R$. If $eR\alpha^k(a) = 0$ for some integer $k \geq 0$, where $e = e^2 \in R$ and $a \in R$, then $eR\alpha^m(a) = 0$ for any integer $m \geq 0$.*

PROOF. Let $e = e^2 \in R$, $a \in R$ and $eR\alpha^k(a) = 0$ for some integer $k \geq 0$. Since by assumption $r_R(eR)$ is left s-unital by left semicentral idempotents, there exists $c \in S_l(R) \cap r_R(eR)$ such that $c\alpha^k(a) = \alpha^k(a)$. Hence, $\alpha^k(ca) = \alpha^k(a)$ and so $a = ca$. Therefore, $eR\alpha^m(a) = eR\alpha^m(ca) = eRc\alpha^m(a) = 0$, for any integer $m \geq 0$. □

LEMMA 3.4. *Let R be a weakly right $\mathfrak{cP}$-Baer ring with a monomorphism α such that $\alpha(c) = c$ for any left semicentral idempotent $c \in R$. For any idempotent $e(x) = \sum_{i=0}^{m} e_i x^i \in R[x;\alpha]$ and any $p(x) = \sum_{j=0}^{n} a_j x^j \in R[x;\alpha]$, if $e(x)R[x;\alpha]p(x) = 0$ then $e_0R\alpha^l(a_j) = 0$ for any integer $l \geq 0$ and $0 \leq j \leq n$.*

PROOF. Let $e(x) = \sum_{i=0}^{m} e_i x^i \in R[x;\alpha]$ be any idempotent, and $p(x) = \sum_{j=0}^{n} a_j x^j \in R[x;\alpha]$ such that $e(x)R[x;\alpha]p(x) = 0$. Then, we can conclude that $0 = e(x)rp(x) = \sum_{i=0}^{m}\sum_{j=0}^{n}(e_i\alpha^i(ra_j))x^{i+j}$ for all $r \in R$. First, we will prove that $e_0Ra_j = 0$ for each $0 \leq j \leq n$. We proceed by induction on j. For the base case of

$j = 0$, we have $0 = e_0 r a_0$, hence $e_0 R a_0 = 0$. Now assume the result is true for all $1 \leq t < j$. So $e_0 R a_t = 0$ for each $1 \leq t < j$. Using Lemmas 3.1 and 3.3, we have $e_i R \alpha^i(a_t) = 0$ for each $1 \leq t < j$ and any $0 \leq i \leq m$. On the other hand,

$$0 = e_0 r a_j + e_1 \alpha(r a_{j-1}) + \cdots + e_j \alpha^j(r a_0).$$

By the induction hypothesis, we must have $e_0 R a_j = 0$. Therefore, by Lemma 3.3, $e_0 R \alpha^l(a_j) = 0$ for any integer $l \geq 0$ and $0 \leq j \leq n$. □

THEOREM 3.5. *Let R be a ring with a monomorphism α such that $\alpha(e) = e$ for any $e = e^2 \in R$. Then R is weakly right c$\mathfrak{P}$-Baer if and only if $R[x;\alpha]$ is a weakly right c$\mathfrak{P}$-Baer ring.*

PROOF. ($\Rightarrow$) Let R be weakly right c$\mathfrak{P}$-Baer. Let $e(x) = (e(x))^2 \in R[x;\alpha]$ such that $e(x) = \sum_{i=0}^m e_i x^i$. So, $e_0 = e_0^2 \in R$. We will show that $r_{R[x;\alpha]}(e(x)R[x;\alpha])$ is left s-unital by left semicentral idempotents. Assume that $f(x) = \sum_{j=0}^n a_j x^j \in r_{R[x;\alpha]}(e(x)R[x;\alpha])$ is an arbitrary element. Then $e(x)R[x;\alpha]f(x) = 0$. So by Lemma 3.4, $e_0 R a_j = 0$ for all $0 \leq j \leq n$. By hypothesis, $r_R(e_0 R)$ is left s-unital by left semicentral idempotents so by Lemma 2.5 there exists $c \in r_R(e_0 R) \cap S_l(R)$ such that $c a_j = a_j$ for all j. Hence, $c f(x) = \sum_{j=0}^n c a_j x^j = f(x)$. On the other hand, by Lemma 3.1, $e_i \in R e_0 R$ for all $0 \leq i \leq m$ so we have $e_i R c = 0$. Assune that $g(x) = \sum_{l=0}^r b_l x^l \in R[x;\alpha]$ is an arbitrary element. So,

$$e(x)g(x)c = (\sum_{i=0}^m e_i x^i)(\sum_{l=0}^r b_l x^l)c = \sum_{k=0}^{m+r}(\sum_{i+l=k} e_i \alpha^i(b_l)c)x^k = 0.$$

Thus, $c \in r_{R[x;\alpha]}(e(x)R[x;\alpha])$. Also, $cg(x)c = c(\sum_{l=0}^r b_l x^l)c = \sum_{l=0}^r c b_l c x^l = cg(x)$. Therefore, $c \in S_l(R[x;\alpha])$ and we are done.

($\Leftarrow$) Assume that $R[x;\alpha]$ is a weakly right c$\mathfrak{P}$-Baer ring and $c = c^2 \in R$. Let $a \in r_R(cR)$. By Lemma 3.3, $cR\alpha^l(a) = 0$ for any integer $l \geq 0$. Thus, $a \in r_{R[x;\alpha]}(cR[x;\alpha])$, so there exists $e(x) = \sum_{i=0}^m e_i x^i \in S_l(R[x;\alpha]) \cap r_{R[x;\alpha]}(cR[x;\alpha])$ such that $e(x)a = a$. Then, $e_0 a = a$. By Proposition 3.2, we have $e(x)R[x;\alpha] = e_0 R[x;\alpha]$ where $e_0 \in S_l(R)$. Thus, $e_0 \in r_R(cR) \cap S_l(R)$. Therefore, $r_R(cR)$ is left s-unital by left semicentral idempotents and so R is a weakly right c$\mathfrak{P}$-Baer ring. □

In [**23**], the authors introduced α-compatible rings and studied their properties. A ring R is α-compatible if for each $a, b \in R$, $ab = 0$ if and only if $a\alpha(b) = 0$. This may only happen when the endomorphism α is injective. By [**23**], whenever R is an α-compatible ring then $\alpha(e) = e$, for any idempotent $e \in R$. Also, α is compatible if and only if for $a, b \in R$, $\alpha(a)b = 0 \Leftrightarrow ab = 0$.

COROLLARY 3.6. *Let R be an α-compatible ring with an endomorphism α. Then R is weakly right c$\mathfrak{P}$-Baer if and only if $R[x;\alpha]$ is a weakly right c$\mathfrak{P}$-Baer ring.*

COROLLARY 3.7. *A ring R is weakly right c$\mathfrak{P}$-Baer if and only if $R[x]$ is weakly right c$\mathfrak{P}$-Baer.*

THEOREM 3.8 ([**17**, Theorem 2.3]). *Let R be a ring and X a nonempty set of not necessarily commuting indeterminates. Let $\Delta = R[X]$ or $R[[X]]$. Then $e \in S_l(\Delta)$ with $e_0 \in R$ the constant term of e if and only if $e_0 \in S_l(R)$, $e_0 e = e$, and $e_0 = e e_0$.*

THEOREM 3.9. *Let R be a ring and X a nonempty set of not necessarily commuting indeterminates. Then R is weakly right $\mathfrak{cP}$-Baer if and only if $R[X]$ is a weakly right $\mathfrak{cP}$-Baer ring.*

PROOF. ($\Rightarrow$) The result is true when $X = \{x\}$, as shown in the Corollary 3.7. We will prove the result for $X = \{x_1, x_2\}$. The proof for an arbitrary nonempty set of indeterminates X is a straightforward generalization. Let $e = \sum_{i=0}^{m}\sum_{j=0}^{n} e_{(i,j)}x_1^i x_2^j$ is an idempotent of $R[X]$ and that $e_0 = e_{(0,0)}$ is the constant term of e. Thus, $e_0 = e_0^2 \in R$. We will prove that $r_{R[X]}(eR[X])$ is left s-unital by left semicentral idempotents. Let $f = \sum_{k=0}^{p}\sum_{l=0}^{q} a_{(k,l)}x_1^k x_2^l$ be an arbitrary element in $r_{R[X]}(eR[X])$. As $eR[X]f = 0$, for each $r \in R$ we have

$$0 = erf = \sum_{i=0}^{m}\sum_{j=0}^{n}\sum_{k=0}^{p}\sum_{l=0}^{q}(e_{(i,j)}ra_{(k,l)})x_1^i x_2^j x_1^k x_2^l.$$

First we will prove that $e_0 ra_{(k,l)} = 0$ for each $k, l \geq 0$. We proceed by induction on k, l. Note that $e_0 ra_{(0,0)} = 0$. Now assume the result is true for all $1 \leq s < k$ and $1 \leq t < l$. So $e_0 ra_{(s,t)} = 0$ for each $1 \leq s < k$ and $1 \leq t < l$. We have

$$e_0 ra_{(k,l)} + \cdots + e_{(k,0)}ra_{(0,l)} + e_{(1,1)}ra_{(k-1,l-1)} + \cdots + e_{(k,l)}ra_{(0,0)} = 0. \tag{3.1}$$

By multiplying (3.1) on the left side by e_0, we get

$$e_0 ra_{(k,l)} + \cdots + e_0 e_{(k,0)}ra_{(0,l)} + e_0 e_{(1,1)}ra_{(k-1,l-1)} + \cdots + e_0 e_{(k,l)}ra_{(0,0)} = 0.$$

But since $e_0 ra_{(s,t)} = 0$ for $1 \leq s < k$ and $1 \leq t < l$ by the induction hypothesis, it follows that $e_0 ra_{(k,l)} = 0$. Therefore, $e_0 ra_{(k,l)} = 0$ for each $k, l \geq 0$. By Lemma 2.5, there exists a left semicentral idempotent $c \in r_R(e_0 R)$ such that $ca_{(k,l)} = a_{(k,l)}$ for all k, l. Thus, $cf = \sum_{k,l\geq 0}^{m} ca_{(k,l)}x_1^k x_2^l = \sum_{k,l\geq 0}^{m} a_{(k,l)}x_1^k x_2^l = f$. Now, by Theorem 3.8, $c \in S_l(R[X])$. On the other hand, $cR[X]c = R[X]c$. Therefore, $R[X]$ is a weakly right $\mathfrak{cP}$-Baer ring.

($\Leftarrow$) Let $R[X]$ be a weakly right $\mathfrak{cP}$-Baer ring and $c = c^2 \in R$. Thus, $r_{R[X]}(cR[X])$ is left s-unital by left semicentral idempotents. Assume that $a \in r_R(cR)$. Thus, $a \in r_{R[X]}(cR[X])$ so $ea = a$ for some $e \in r_{R[X]}(cR[X])$ where e is a left semicentral idempotent with $e_0 \in R$ the constant term. Thus, $e_0 a = a$. By Theorem 3.8, $e_0 \in S_l(R)$ so it is enough to show that $e_0 \in r_R(cR)$. We have $e \in r_{R[X]}(cR[X])$. Thus $cRe = 0$ and so $cRe_0 = 0$. Then, $e_0 \in r_R(cR)$. Therefore, R is a weakly right $\mathfrak{cP}$-Baer ring and we get the desired result. □

THEOREM 3.10. *Let R be a ring satisfying the descending chain condition on left annihilators. Then R is weakly right $\mathfrak{cP}$-Baer if and only if $R[[x]]$ is weakly right $\mathfrak{cP}$-Baer.*

PROOF. ($\Rightarrow$) Let R be a weakly right $\mathfrak{cP}$-Baer ring. Assume that $e(x) = \sum_{i=0}^{\infty} e_i x^i \in R[[x]]$ is an idempotent and $f(x) = \sum_{j=0}^{\infty} a_j x^j r_{R[[x]]}(e(x)R[[x]])$. Then $e_0 = e_0^2 \in R$. Moreover, $e(x)R[[x]]f(x) = 0$. So $e(x)rf(x) = 0$ for every $r \in R$.

Hence, $\sum_{i=0}^{\infty}(\sum_{j=0}^{\infty} e_i r a_j)x^{i+j} = 0$, and we get

$$e_0 r a_0 = 0 \qquad (1)$$

$$e_0 r a_1 + e_1 r a_0 = 0 \qquad (2)$$

$$\vdots$$

$$e_0 r a_m + e_1 r a_{m-1} + \cdots + e_m r a_0 = 0 \qquad (3)$$

$$\vdots$$

Since $e_0 r a_0 = 0$, one has $a_0 \in r_R(e_0 R)$. By multiplying equation (2) on the left side with e_0, we obtain $e_0 r a_1 + e_0 e_1 r a_0 = 0$. Using (1), we get $e_0 r a_1 = 0$, which implies $a_1 \in r_R(e_0 R)$. Similarly, multiplying equation (3) on the left side with e_0, we can conclude that $a_m \in r_R(e_0 R)$. This process can be continued to give $a_j \in r_R(e_0 R)$ for all $j \geq 0$. On the other hand, consider the descending chain given below:

$$\ell_R(a_0) \supseteq \ell_R(a_0, a_1) \supseteq \ell_R(a_0, a_1, a_2) \supseteq \cdots .$$

According to the hypothesis, there exists a positive integer n such that

$$\ell_R(a_0, a_1, \ldots, a_n) = \ell_R(a_0, a_1, \ldots, a_n, a_{n+1}) = \cdots .$$

Since $a_0, a_1, \ldots, a_n \in r_R(e_0 R)$, by Lemma 2.5, there exists $c \in r_R(e_0 R) \cap S_l(R)$ such that $a_j = c a_j$ for $0 \leq j \leq n$. Thus, $1 - c \in \ell_R(a_0, a_1, \ldots, a_n)$ and so $1 - c \in \ell_R(a_0, a_1, \ldots, a_n, \ldots, a_l)$ for any $l \geq n$. By Theorem 3.8, $c \in S_l(R[[x]])$. Hence, $a_l = c a_l$, for any $l \geq n$. Checking that $c \in r_{R[[x]]}(e(x)R[[x]])$ and $f(x) = cf(x)$ is routine. Therefore, $R[[X]]$ is a weakly right $\mathfrak{cP}$-Baer ring.

($\Leftarrow$) The proof is similar to that of ($\Leftarrow$) of Theorem 3.9. □

4. Minimal prime ideals of weakly right $\mathfrak{cP}$-Baer rings

A proper ideal P of R is called completely prime (or completely semiprime) if $xy \in P$ ($x^2 \in P$) implies that either $x \in P$ or $y \in P$ ($x \in P$). Andrunakievic and Rjabuhin [**2**], as well as Stewart [**41**] independently, have demonstrated that a reduced ring R is a subdirect product of integral domains. Therefore, a proper ideal is completely semiprime if and only if it is an intersection of completely prime ideals.

Assuming P is a prime ideal, the following definitions are crucial for the remainder of this section:

$$E(P) = \{a \in R \mid eRa = 0,\ for\ some\ e = e^2 \in R \setminus P\},$$

$$\bar{E}(P) = \{a \in R \mid a^n \in E(P),\ for\ some\ positive\ integer\ n\},$$

$$E_P = \{a \in R \mid ea = 0,\ for\ some\ e = e^2 \in R \setminus P\},$$

$$\bar{E}_P = \{a \in R \mid a^n \in E_P,\ for\ some\ positive\ integer\ n\}.$$

Lemma 4.1 can be immediately deduced from these definitions.

LEMMA 4.1. *Let P be a prime ideal of R.*

1. *$E(P) \subseteq P$,*
2. *$\mathcal{N}(R) \subseteq \bar{E}(P)$, where $\mathcal{N}(R)$ represents the set of nilpotent elements of the ring R.*
3. *If P is a completely prime ideal, then $E_P \subseteq P$.*

EXAMPLE 4.2. Let F be a field and

$$R = \begin{pmatrix} F & F \\ 0 & F \end{pmatrix}, \ P_1 = \begin{pmatrix} 0 & F \\ 0 & F \end{pmatrix}, \ P_2 = \begin{pmatrix} F & F \\ 0 & 0 \end{pmatrix}.$$

Thus, P_1 and P_2 are all prime ideals of R and satisfy the following properties. We know that

$$I(R) = \left\{ \begin{pmatrix} 0 & 0 \\ 0 & 0 \end{pmatrix}, \begin{pmatrix} 1 & 0 \\ 0 & 1 \end{pmatrix}, \begin{pmatrix} 1 & F \\ 0 & 0 \end{pmatrix}, \begin{pmatrix} 0 & F \\ 0 & 1 \end{pmatrix} \right\}.$$

(i) $E(P_1) = 0$ and $E(P_2) = P_2$,
(ii) $\bar{E}(P_1) = \{A \in R \mid A^n \in E(P_1)$, for some positive integer $n\} = \mathcal{N}(R) = \mathcal{P}(R) \neq P_1$ and $\bar{E}(P_2) = P_2$,
(iii) $E_{P_1} = \left\{ \begin{pmatrix} 0 & yx \\ 0 & x \end{pmatrix} \mid x, y \in F \right\}$, which is not an ideal of R and $E_{P_2} = P_2$,
(iv) $\bar{E}_{P_1} = P_1$ and $\bar{E}_{P_2} = P_2$.

Let $\mathcal{P}(R)$ denote the prime radical of a ring R. A ring is called *2*-primal if $\mathcal{P}(R) = \mathcal{N}(R)$. It is important to note that every commutative ring is *2*-primal. Therefore, *2*-primal rings act as a bridge between commutative and noncommutative ring theory. According to [**7**, Proposition 2.2], subrings of *2*-primal rings are also *2*-primal.

PROPOSITION 4.3. *For a weakly right $c\mathfrak{P}$-Baer ring R with a prime ideal P, the following holds:*
(i) $E(P) = \sum_{c \in P \cap S_l(R)} cR$,
(ii) *If $\bar{E}(P)$ or $\bar{E}_P$ is an ideal, then it is a completely semiprime ideal,*
(iii) *If $\bar{E}(P)$ or $\bar{E}_P$ is an ideal, then every minimal prime ideal belonging to $\bar{E}(P)$ or $\bar{E}_P$, is a completely prime ideal, respectively,*
(iv) $\mathcal{N}(R/E(P)) = \bar{E}(P)/E(P)$.

PROOF. (i) It is enough to prove that $E(P) = \bigcup_{f \in P \cap S_l(R)} fR$. Let $a \in E(P)$, so there exists an idempotent $e \in R \setminus P$ such that $eRa = 0$. Thus, $a \in r_R(eR)$. The hypothesis implies $a = ca$ for some $c \in r_R(eR) \cap S_l(R)$. So, $a \in cR$. On the other hand, $eRc = 0 \in \mathcal{P}(R)$. Thus, $eRc \subseteq P$ and so $c \in P \cap S_l(R)$. Hence, $E(P) \subseteq \bigcup_{c \in P \cap S_l(R)} cR$. Conversely, assume $a \in \bigcup_{c \in P \cap S_l(R)} cR$. Hence, $a \in cR$ for some $c \in P \cap S_l(R)$. Thus, $a = ca$ and so $(1 - c)a = 0$. As $1 - c \in S_r(R)$, we have $(1 - c)Ra = (1 - c)R(1 - c)a = 0$. Then since $1 - c \in R \setminus P$, it follows that $a \in E(P)$. Therefore, $E(P) = \bigcup_{c \in P \cap S_l(R)} cR$, and so $E(P) = \sum_{c \in P \cap S_l(R)} cR$.
(ii) By definition, it is clear.
(iii) This follows immediately from [**7**, Proposition 1.2, Proposition 1.6].
(iv) It follows from definitions. □

THEOREM 4.4. *If R is a 2-primal ring with a minimal prime ideal P, then $P = \bar{E}(P)$.*

PROOF. Since R is *2*-primal, P is completely prime, [**39**, Proposition 1.11]. Thus, $R \setminus P$ is multiplicatively closed. Assume that $B = S_r(R \setminus P)$. So B is a multiplicatively closed set, by Remark 1.2. Let $a \in P$. If $a = 0$, then $a \in \bar{E}(P)$. Now suppose that $a \neq 0$. Let S be the multiplicative system generated by $B \cup \{a\}$. We claim that $0 \in S$. Assume to the contrary that $0 \notin S$. A partial order, the collection of ideals disjoint, with S by inclusion. By Zorn's lemma, we get an ideal M which is maximal disjoint with S. Then M is a prime ideal and M is properly contained in P, a contradiction. Hence, $0 \in S$. Thus, we have, $0 = f_1a^{t_1}\dots f_ka^{t_k}$ for some $f_1, \dots, f_k \in B$ and positive integers $t_1, \dots, t_k$. Since $\mathcal{N}(R)$ is completely semiprime and $f_1a^{t_1}\dots f_ka^{t_k} \in \mathcal{N}(R)$, by [**10**, Lemma 7], $f_1\dots f_ka^t \in \mathcal{N}(R)$, where $t = t_1+\dots+t_k$. If $f = f_1\dots f_k$, then $fa^t \in \mathcal{N}(R)$. Again, since $\mathcal{N}(R)$ is completely semiprime, it follows that $fa \in \mathcal{N}(R)$ and so there is m such that $(fa)^m = 0$. Thus, $fa^m = 0$ and so $fRa^m = fRfa^m = 0$. Therefore, $a \in \bar{E}(P)$ and so $P \subseteq \bar{E}(P)$.

Conversely, Since P is a prime ideal, there exists a minimal prime ideal Q of R such that $Q \subseteq P$. By [**39**, Proposition 1.11], Q is a completely prime ideal. Now, assume that $a \in \bar{E}(P)$. Thus, there exists $e = e^2 \in R \setminus P$ such that $eRa^n = 0$ for some positive integer n. We have $eRa^n = 0 \in \mathcal{P}(R) \subseteq Q$ and so $eRa^n \subseteq Q$. As Q is a prime ideal and $e \notin Q$, we have $a^n \in Q$. Since Q is completely prime, $a \in Q \subseteq P$. Therefore, $\bar{E}(P) \subseteq P$ and we are done. □

LEMMA 4.5. *Let P be a prime ideal of a weakly right c$\mathfrak{P}$-Baer ring R.*

(i) *If $I \subseteq P$ is left s-unital by left semicentral idempotent ideal, then $I \subseteq E(P)$.*
(ii) *P is left s-unital by left semicentral idempotent ideal if and only if $P = E(P)$.*

PROOF. (i) By definition, $I = \cup_{c\in E} cR$ such that $\emptyset \neq E \subseteq I \cap S_l(R) \subseteq P \cap S_l(R)$. Thus, $I \subseteq \cup_{c\in P\cap S_l(R)} cR$ and so $I \subseteq E(P)$.

(ii) This part is a consequence of the definition and part (i). □

LEMMA 4.6. *Let P and Q be prime ideals of a weakly right c$\mathfrak{P}$-Baer ring R such that $Q \subseteq P$. Then $E(P) = E(Q)$.*

PROOF. If $a \in E(P)$, then $eRa = 0$ for some $e = e^2 \in R \setminus P \subseteq R \setminus Q$. Hence, $a \in E(Q)$. For the converse we have, $E(Q) = \cup_{c\in Q\cap S_l(R)} cR \subseteq \cup_{c\in P\cap S_l(R)} cR = E(P)$, by Proposition 4.3. □

LEMMA 4.7. *For a prime ideal P of a weakly right c$\mathfrak{P}$-Baer ring R, the following conditions hold:*

(i) *If $P = \bar{E}(P)$, then $P = \bar{E}_P$,*
(ii) *If $P = \bar{E}_P$, then P is a completely prime ideal and is minimal among completely prime ideals of R. In particular, if $P = \bar{E}_P$ for every minimal prime ideal P of R, then R is a 2-primal ring,*
(iii) *If $P = \bar{E}_P$ and R is a 2-primal ring, then P is a minimal prime ideal of R which is completely prime ideal,*
(iv) *$P = \bar{E}_P$ if and only if $\mathcal{N}(R/E(P)) = P/E(P)$,*
(v) *If $R/E(P)$ is a 2-primal ring, then $P = \bar{E}_P$ if and only if $P/E(P)$ is the unique minimal prime ideal of $R/E(P)$.*

PROOF. (i) Let $P = \bar{E}(P)$. Thus, $\bar{E}(P)$ is a prime ideal. By Proposition 4.3(iii), every minimal prime ideal belonging to $\bar{E}(P)$ is completely prime. Thus, $\bar{E}(P)$ is a 2-primal ideal of R, by [**7**, Proposition 1.6]. Consequently, $\bar{E}(P)$ is a completely prime ideal, so P is also completely prime. Let $a \in \bar{E}_P$. Hence, there exists $e = e^2 \in R \setminus P$ such that $ea^n = 0$ for some positive integer n. So we get $ea^n \in \mathcal{P}(R) \subseteq P$. Therefore, $a^n \in P$ and so $a \in P$. By Proposition 4.3 (v), the converse is clear, and the proof is complete.

(ii) By a similar argument to that used in the proof of part (i), P is completely prime. Let Q be a completely prime ideal of R such that $Q \subseteq P$. Thus, $Q \subseteq P =$

$\bar{E}(P) \subseteq \bar{E}(Q) \subseteq Q$ and so $Q = P$. If $P = \bar{E}_P$ for every minimal prime ideal P of R, then P is completely prime. Therefore, R is a 2-primal ring, by [**39**, Proposition 1.11].

(iii)Let Q be a prime ideal of R such that $Q \subseteq P$. Thus, $Q \subseteq P = \bar{E}(P) \subseteq \bar{E}(Q)$. On the other hand, $\bar{E}(Q) \subseteq Q$, by Theorem 4.4. Hence, P is a minimal prime ideal of R. Since R is 2-primal, P is completely prime, by [**39**, Proposition 1.11].

(iv) This part follows from definition and Proposition 4.3.

(v) The proof follows from Proposition 4.3, using the definition of a 2-primal ring. □

THEOREM 4.8. *Let R be a weakly right $c\mathfrak{P}$-Baer ring such that $\bar{E}(Q) = Q$, for every minimal prime ideal Q of R. Then we have the following:*

(i) *R is 2-primal,*
(ii) *Every prime ideal of R contains a unique minimal prime ideal,*
(iii) *For any prime ideal P of R, the ring $R/E(P)$ is 2-primal and $\mathcal{P}(R/E(P))$ is completely prime.*

PROOF. (i) By Lemma 4.7, the result follows.

(ii) Let P be a prime ideal of R. Assume that $Q_1, Q_2 \subseteq P$ are minimal prime ideals. We have $E(P) = E(Q_1)$ and $E(P) = E(Q_2)$, by Lemma 4.6. Thus, $E(Q_1) = E(Q_2)$ and so $\bar{E}(Q_1) = \bar{E}(Q_2)$. Based on the hypothesis, we have, $Q_1 = Q_2$.

(iii) First we claim that $\bar{E}(P)/E(P)$ is the unique minimal prime ideal of the ring $R/E(P)$. Let $Q/E(P)$ be a prime ideal of $R/E(P)$. By Proposition 4.3 we have, $E(P) \subseteq E(Q)$. Now assume that H is the unique minimal ideal of R contained in P. By Lemma 4.7 and assumption, we have $H = \bar{E}(H) = \bar{E}(P) \subseteq \bar{E}(Q) = \bar{E}(K) = K \subseteq Q$ where K is the unique minimal ideal of R contained in Q. So, $H = K = \bar{E}(P)$. Thus, $\bar{E}(P)/E(P)$ is the unique minimal ideal of $R/E(P)$. Therefore, $\mathcal{N}(R/E(P)) = \bar{E}(P)/E(P) = \mathcal{P}(R/E(P))$ and so, $R/E(P)$ is 2-primal. Now, by assumption and Lemma 4.7, Q is a completely prime ideal. Thus $\bar{E}(P)$ is completely prime, so $\bar{E}(P)/E(P)$ is also completely prime in $R/E(P)$. □

In [**5**], it is mentioned that a ring R satisfies the insertion of factors property IFP if and only if $r_R(x)$ is an ideal of R for all $x \in R$. It is important to note that any ring with IFP is abelian and 2-primal. According to [**8**], abelian rings are right $c\mathfrak{P}$-Baer, which implies that any ring with IFP is a weakly right $c\mathfrak{P}$-Baer ring. Therefore, Corollary 4.9 holds.

COROLLARY 4.9. *Let R be a ring with IFP and P be a prime ideal of R. Then P is a minimal prime ideal of R if and only if $P = \bar{E}(P)$.*

PROOF. This result follows from Theorem 4.4 and Lemma 4.7. □

COROLLARY 4.10. *If R is a 2-primal weakly right $c\mathfrak{P}$-Baer ring and P a prime ideal of R, then $\bar{E}(P)$ is the unique minimal prime ideal of R contained in P and $\bar{E}(P)$ is completely Prime.*

PROOF. By Theorem 4.4, $\bar{E}(B) = B$ for every minimal prime ideal B. So there exists a unique minimal prime ideal Q of R, such that $Q \subseteq P$, by Theorem 4.8. Thus, $Q = \bar{E}(Q) = \bar{E}(P)$, by Lemma 4.6. So $\bar{E}(P)$ is a minimal prime ideal, and hence by Proposition 4.3(iii), $\bar{E}(P)$ is a completely prime ideal. □

PROPOSITION 4.11. *If R is a ring with IFP, then $\bar{E}(P) = \bar{E}_P$ and $\bar{E}(P)$ is an ideal for all prime ideals P of R.*

PROOF. It follows from Proposition 4.3 and [**39**]. □

COROLLARY 4.12. *If R is a ring with IFP and P a prime ideal of R, then $\bar{E}_P$ is the unique minimal prime ideal of R contained in P and $\bar{E}_P$ is completely prime.*

PROOF. This corollary follows from Corollary 4.9 and Proposition 4.11. □

We present a comprehensive characterization of a class of weakly right $c\mathfrak{P}$-Baer rings. For a prime ideal P of a ring R, let $S_P = \{e \mid e \in S_r(R) \cap (R \setminus P)\}$. In a similar way as in the proof of [**11**, Proposition 2.4], S_P is a left denominator set, for every prime ideal P.

LEMMA 4.13. *Let P be a prime ideal of a weakly right $c\mathfrak{P}$-Baer ring R. Then $E(P) = \{a \in R \mid ea = 0 \ for \ some \ e \in S_P\}$.*

PROOF. It can be easily checked that $\{a \in R \mid eRa = 0 \ for \ some \ e \in S_P\} = \{a \in R \mid ea = 0 \ for \ some \ e \in S_P\}$. Assume that $a \in E(P)$ is an arbitrary element. By Proposition 4.3, $a \in cR$ for some $c \in P \cap S_l(R)$. Thus, $a = ca$ and so $(1 - c)a = 0$. Moreover, $1 - c \in R \setminus P$ and $1 - c \in S_r(R)$. Therefore, $a \in \{a \in R \mid ea = 0 \ for \ some \ e \in S_P\}$. The converse inclusion is clear by definition. □

Note that Lemma 4.5 implies that for any prime ideal P of a weakly right $c\mathfrak{P}$-Baer ring R, $E(P)$ is left s-unital by left semicentral idempotents. Consequently, $E(P)$ is right pure, and thus $R/E(P)$ is flat as a right R-module. According to [**22**, Theorem 10.3], for a prime ideal P of R, $R[S_P^{-1}]$ is a left ring of fractions of R.

THEOREM 4.14. *Let R be a weakly right $c\mathfrak{P}$-Baer ring. For a prime ideal P of R, $R[S_P^{-1}] \cong R/E(P)$.*

PROOF. As $R[S_P^{-1}]$ is the left ring of fractions with the left denominator set S_P, there exists a ring homomorphism $\phi : R \longrightarrow R[S_P^{-1}]$ and

$$R[S_P^{-1}] = \{(e)\phi^{-1}(a)\phi \mid a \in R \ and \ e \in S_P\}.$$

Thus the kernel of ϕ is $E(P)$, by Lemma 4.13. Furthermore, for each $e \in S_P$, note that $(e)\phi$ is an idempotent in $R[S_P^{-1}]$ which is invertible. Therefore, $(e)\phi = 1$ for every $e \in S_P$. So $R[S_P^{-1}] = (R)\phi$ which is a homomorphic image of R with the kernel $E(P)$. Therefore, $R[S_P^{-1}] \cong R/E(P)$ and we are done. □

LEMMA 4.15. *If R is a weakly $c\mathfrak{P}$-Baer ring, then $R/E(P)$ is a weakly left $c\mathfrak{P}$-Baer ring.*

PROOF. Let $e + E(P)$ be an idempotent of $R/E(P)$. So e is an idempotent of R. We will show that $\ell_{R/E(P)}((R/E(P))(e + E(P)))$ is right s-unital by right semicentral idempotents. Assume that $a + E(P) \in \ell_{R/E(P)}(R/E(P)(e + E(P)))$ is an arbitrary element. Then, $(a + E(P))(b + E(P))(e + E(P)) = E(P)$ for each $b + E(P) \in R/E(P)$. So $abe \in E(P)$. Hence, there exists an element $f \in S_P$ such that $fabe = 0$, by Lemma 4.13. Since $b \in R$ is arbitrary, $fa \in \ell_R(Re)$. By assumption, $fac = fa$ for some $c \in \ell_R(Re) \cap S_r(R)$. It is easy to see that $(c + E(P)) \in \ell_{R/E(P)}(R/E(P)(e + E(P))) \cap S_l(R/E(P))$. By Theorem 4.14, $(f)\phi(a)\phi(c)\phi = (f)\phi(a)\phi$ and so $(c)\phi(a)\phi = (a)\phi$. Hence, $(c + E(P))(b + E(P)) =$

$(b+E(P))$. Therefore, $\ell_{R/E(P)}(R/E(P)(e+E(P)))$ is right s-unital by right semicentral idempotents, and we are done. □

Recall that a ring R has a representation as a subdirect product of rings S_i (called components), $i \in I$, if and only if $S_i \cong R/K_i$, where K_i is an ideal of R and $\bigcap_{i\in I} K_i = 0$. The representation is called trivial if $K_i = 0$ for at least one $i \in I$ and nontrivial otherwise [**33**, p. 53]. The set of all prime ideals (respectively, minimal prime ideals) of a ring R will be denoted by $Spec(R)$ (respectively, $MinSpec(R)$).

LEMMA 4.16. *If R is a weakly right $\mathfrak{c}\mathfrak{P}$-Baer ring, then R is a subdirect product of $R/E(P)$, where P ranges through all minimal prime ideals.*

PROOF. For a minimal prime ideal P of R, we have that $E(P) = E(Q)$ for every prime ideal Q of R containing P, by Lemma 4.6. Therefore, we have that $\bigcap_{P\in Spec(R)} E(P) = \bigcap_{P\in MinSpec(R)} E(P)$. Now [**11**, Lemma 3.1] yields the result. □

THEOREM 4.17. *Let R be a weakly right $\mathfrak{c}\mathfrak{P}$-Baer ring such that $E(P) \neq 0$ for every minimal prime ideal P of R. Then R has a nontrivial representation as a subdirect product of the right rings of fractions $R[S_P^{-1}]$, where P ranges through all minimal prime ideals.*

PROOF. This result is an immediate consequence of Theorem 4.14 and Lemma 4.16. □

LEMMA 4.18. *Let R be a ring with a nontrivial central idempotent e. Then for every prime ideal P of R, $E(P) \neq 0$.*

PROOF. Note that $eR(1-e) = (1-e)Re = 0$ and for every prime ideal P either $e \in P$ or $(1-e) \in P$. So $e \in E(P)$ or $1-e \in E(P)$ by the definition of $E(P)$. Therefore $E(P) \neq 0$. □

DEFINITION 4.19 ([**8**, Definition 2.1]). We call a ring R right (left) I-prime if for every idempotent $e \in R$ and every right (left) ideal A of R, $eRA = 0 (ARe = 0)$ implies either $A = 0$ or $e = 0$. A ring R is I-prime if it is both left and right I-prime.

Examples of rings which are I-prime are: prime rings, rings in which the only idempotents are 0 and 1 (e.g., local rings) and semiprime rings satisfying a one-sided I-prime condition.

COROLLARY 4.20. *Let R be a a semiprime weakly right $\mathfrak{c}\mathfrak{P}$-Baer ring which is not right I-prime. Then for every prime ideal P of R, $E(P) \neq 0$.*

PROOF. Since R is not right I-prime so by [**8**, Proposition 2.2], there exists a nonzero idempotent $e \in R$ such that $0 \neq r_R(eR)$, and hence there is $0 \neq a \in r_R(eR)$. As R is a weakly right $\mathfrak{c}\mathfrak{P}$-Baer ring, $r_R(eR)$ is left s-unital by left semicentral idempotents. So there exists $c \in r_R(eR) \cap S_l(R)$ such that $ca = a$. Since R is semiprime c is central. Therefore $E(P) \neq 0$ by Lemma 4.18, and the proof is complete. □

COROLLARY 4.21. *Let R be a a semiprime weakly right $\mathfrak{c}\mathfrak{P}$-Baer ring which is not right I-prime. Then R has a nontrivial representation as a subdirect product of the right rings of fractions $R[S_P^{-1}]$, where P ranges through all minimal prime ideals.*

PROOF. This result is a consequence of Theorem 4.17 and Corollary 4.20. □

Conflicts of interests

The authors have no conflicts of interest to declare.

References

[1] D. D. Anderson and Victor Camillo, *Armendariz rings and Gaussian rings*, Comm. Algebra **26** (1998), no. 7, 2265–2272, DOI 10.1080/00927879808826274. MR1626606

[2] V. A. Andrunakievič and Ju. M. Rjabuhin, *Rings without nilpotent elements, and completely prime ideals* (Russian), Dokl. Akad. Nauk SSSR **180** (1968), 9–11. MR230760

[3] Nasibeh Aramideh and Ahmad Moussavi, *$c\mathfrak{P}$-Baer polynomial extensions*, Bull. Iranian Math. Soc. **50** (2024), no. 4, Paper No. 52, 21, DOI 10.1007/s41980-024-00898-5. MR4767571

[4] Efraim P. Armendariz, *A note on extensions of Baer and P.P.-rings*, J. Austral. Math. Soc. **18** (1974), 470–473. MR366979

[5] Howard E. Bell, *Near-rings in which each element is a power of itself*, Bull. Austral. Math. Soc. **2** (1970), 363–368, DOI 10.1017/S0004972700042052. MR263877

[6] Sterling K. Berberian, *Baer *-rings*, Die Grundlehren der mathematischen Wissenschaften, Band 195, Springer-Verlag, New York-Berlin, 1972. MR429975

[7] Gary F. Birkenmeier, Henry E. Heatherly, and Enoch K. Lee, *Completely prime ideals and associated radicals*, Ring theory (Granville, OH, 1992), World Sci. Publ., River Edge, NJ, 1993, pp. 102–129. MR1344225

[8] Gary F. Birkenmeier and Blaise J. Heider, *Annihilators and extensions of idempotent-generated ideals*, Comm. Algebra **47** (2019), no. 3, 1348–1375, DOI 10.1080/00927872.2018.1506462. MR3938560

[9] Gary F. Birkenmeier and Jae Keol Park, *Self-adjoint ideals in Baer $*$-rings*, Comm. Algebra **28** (2000), no. 9, 4259–4268, DOI 10.1080/00927870008827088. MR1772505

[10] Gary F. Birkenmeier, Jin-Yong Kim, and Jae Keol Park, *A connection between weak regularity and the simplicity of prime factor rings*, Proc. Amer. Math. Soc. **122** (1994), no. 1, 53–58, DOI 10.2307/2160840. MR1231028

[11] Gary F. Birkenmeier, Jin Yong Kim, and Jae Keol Park, *A sheaf representation of quasi-Baer rings*, J. Pure Appl. Algebra **146** (2000), no. 3, 209–223, DOI 10.1016/S0022-4049(99)00164-4. MR1742340

[12] Gary F. Birkenmeier, Jin Yong Kim, and Jae Keol Park, *On quasi-Baer rings*, Algebra and its applications (Athens, OH, 1999), Contemp. Math., vol. 259, Amer. Math. Soc., Providence, RI, 2000, pp. 67–92, DOI 10.1090/conm/259/04088. MR1778495

[13] Gary F. Birkenmeier, Jin Yong Kim, and Jae Keol Park, *Polynomial extensions of Baer and quasi-Baer rings*, J. Pure Appl. Algebra **159** (2001), no. 1, 25–42, DOI 10.1016/S0022-4049(00)00055-4. MR1823504

[14] Gary F. Birkenmeier, Jin Yong Kim, and Jae Keol Park, *Principally quasi-Baer rings*, Comm. Algebra **29** (2001), no. 2, 639–660, DOI 10.1081/AGB-100001530. MR1841988

[15] Gary F. Birkenmeier, Jin Yong Kim, and Jae Keol Park, *Quasi-Baer ring extensions and biregular rings*, Bull. Austral. Math. Soc. **61** (2000), no. 1, 39–52, DOI 10.1017/S0004972700022000. MR1819314

[16] Gary F. Birkenmeier, Jae Keol Park, and S. Tariq Rizvi, *Principally quasi-Baer ring hulls*, Advances in ring theory, Trends Math., Birkhäuser/Springer Basel AG, Basel, 2010, pp. 47–61, DOI 10.1007/978-3-0346-0286-0_4. MR2664663

[17] Gary F. Birkenmeier and Jae Keol Park, *Triangular matrix representations of ring extensions*, J. Algebra **265** (2003), no. 2, 457–477, DOI 10.1016/S0021-8693(03)00155-8. MR1987012

[18] Gary F. Birkenmeier, Jae Keol Park, and S. Tariq Rizvi, *Extensions of rings and modules*, Birkhäuser/Springer, New York, 2013, DOI 10.1007/978-0-387-92716-9. MR3099829

[19] Gary F. Birkenmeier, Jae Keol Park, and S. Tariq Rizvi, *Generalized triangular matrix rings and the fully invariant extending property*, Proceedings of the Second Honolulu Conference on Abelian Groups and Modules (Honolulu, HI, 2001), Rocky Mountain J. Math. **32** (2002), no. 4, 1299–1319, DOI 10.1216/rmjm/1181070024. MR1987609

[20] W. Edwin Clark, *Twisted matrix units semigroup algebras*, Duke Math. J. **34** (1967), 417–423. MR214626

[21] K. R. Goodearl, *von Neumann regular rings*, Monographs and Studies in Mathematics, vol. 4, Pitman (Advanced Publishing Program), Boston, Mass.-London, 1979. MR533669

[22] K. R. Goodearl and R. B. Warfield Jr., *An introduction to noncommutative Noetherian rings*, 2nd ed., London Mathematical Society Student Texts, vol. 61, Cambridge University Press, Cambridge, 2004, DOI 10.1017/CBO9780511841699. MR2080008

[23] E. Hashemi and A. Moussavi, *Polynomial extensions of quasi-Baer rings*, Acta Math. Hungar. **107** (2005), no. 3, 207–224, DOI 10.1007/s10474-005-0191-1. MR2148584

[24] M. Henriksen and M. Jerison, *The space of minimal prime ideals of a commutative ring*, Trans. Amer. Math. Soc. **115** (1965), 110–130, DOI 10.2307/1994260. MR194880

[25] Irving Kaplansky, *Rings of operators*, W. A. Benjamin, Inc., New York-Amsterdam, 1968. MR244778

[26] Anil Khairnar and B. N. Waphare, *A sheaf representation of principally quasi-Baer $*$-rings*, Algebr. Represent. Theory **22** (2019), no. 1, 79–97, DOI 10.1007/s10468-017-9758-0. MR3908896

[27] Joseph Kist, *Minimal prime ideals in commutative semigroups*, Proc. London Math. Soc. (3) **13** (1963), 31–50, DOI 10.1112/plms/s3-13.1.31. MR143837

[28] T. Y. Lam, *A first course in noncommutative rings*, Graduate Texts in Mathematics, vol. 131, Springer-Verlag, New York, 1991, DOI 10.1007/978-1-4684-0406-7. MR1125071

[29] Yang Lee, Nam Kyun Kim, and Chan Yong Hong, *Counterexamples on Baer rings*, Comm. Algebra **25** (1997), no. 2, 497–507, DOI 10.1080/00927879708825869. MR1428793

[30] Zhongkui Liu and Renyu Zhao, *A generalization of PP-rings and p.q.-Baer rings*, Glasg. Math. J. **48** (2006), no. 2, 217–229, DOI 10.1017/S0017089506003016. MR2256973

[31] Shûichirô Maeda, *On a ring whose principal right ideals generated by idempotents form a lattice*, J. Sci. Hiroshima Univ. Ser. A **24** (1960), 509–525. MR133348

[32] A. Majidinya and A. Moussavi, *Weakly principally quasi-Baer rings*, J. Algebra Appl. **15** (2016), no. 1, 1650002, 20, DOI 10.1142/S021949881650002X. MR3393931

[33] Neal H. McCoy, *The theory of rings*, Chelsea Publishing Co., Bronx, NY, 1973. Reprint of the 1964 edition. MR393090

[34] Ancel C. Mewborn, *Regular rings and Baer rings*, Math. Z. **121** (1971), 211–219, DOI 10.1007/BF01111594. MR292894

[35] A. R. Nasr-Isfahani and A. Moussavi, *Baer and quasi-Baer differential polynomial rings*, Comm. Algebra **36** (2008), no. 9, 3533–3542, DOI 10.1080/00927870802104337. MR2441131

[36] A. R. Nasr-Isfahani and A. Moussavi, *On Ore extensions of quasi-Baer rings*, J. Algebra Appl. **7** (2008), no. 2, 211–224, DOI 10.1142/S0219498808002771. MR2417042

[37] Adolf Pollingher and Abraham Zaks, *On Baer and quasi-Baer rings*, Duke Math. J. **37** (1970), 127–138. MR252445

[38] S. Tariq Rizvi and Cosmin S. Roman, *Baer and quasi-Baer modules*, Comm. Algebra **32** (2004), no. 1, 103–123, DOI 10.1081/AGB-120027854. MR2036224

[39] Gooyong Shin, *Prime ideals and sheaf representation of a pseudo symmetric ring*, Trans. Amer. Math. Soc. **184** (1973), 43–60 (1974), DOI 10.2307/1996398. MR338058

[40] Lance W. Small, *Semihereditary rings*, Bull. Amer. Math. Soc. **73** (1967), 656–658, DOI 10.1090/S0002-9904-1967-11812-3. MR212051

[41] Patrick N. Stewart, *Semi-simple radical classes*, Pacific J. Math. **32** (1970), 249–254. MR255593

[42] Hisao Tominaga, *On s-unital rings*, Math. J. Okayama Univ. **18** (1975/76), no. 2, 117–134. MR419511

[43] Askar Tuganbaev, *Rings close to regular*, Mathematics and its Applications, vol. 545, Kluwer Academic Publishers, Dordrecht, 2002, DOI 10.1007/978-94-015-9878-1. MR1958361

Department of Pure Mathematics, University of Tarbiat Modarres, Tehran, Iran

Current address: Department of Pure Mathematics, University of Tarbiat Modarres, Tehran, Iran

Email address: `n.aramideh@modares.ac.ir`

Department of Pure Mathematics, University of Tarbiat Modarres, Tehran, Iran

Email address: `moussavi.a@modares.ac.ir`

Contemporary Mathematics
Volume **826**, 2025
https://doi.org/10.1090/conm/826/16581

On the structure of a finitely generated module

Khaerudin Saleh, Intan Muchtadi-Alamsyah, and Pudji Astuti

ABSTRACT. Given a finitely generated module over a principal ideal domain, we present the structure of its endomorphism ring and identification of a fully invariant submodule in terms of the cyclic decomposition of the module. Furthermore, we will use this identification to characterize S-prime submodules.

1. Introduction

The concept of a module over a principal ideal domain cannot be separated from the concept of module decomposition. A finitely generated module over a principal ideal domain can be decomposed into a direct sum of cyclic modules. Cyclic modules have an important role in developing the concept of a linear operator on a finite dimensional vector space, especially in the development of the concept of invariant subspace and its applications.

Meanwhile, many researchers have shown interest in adapting concepts from rings to modules, one of which is the prime ideal concept in rings adapted into a prime submodule. Dauns [**2**] identified prime submodules based on the multiplication of elements in the module with the scalar elements. Furthermore, this concept was generalized to the prime submodule of a left module by Smith and McCasland [**3**].

Sanh et al. [**4**] found prime submodules in line with the concept of prime ideals for commutative rings and noncommutative rings. This prime submodule concept stands in contrast with the prime submodule defined by Dauns [**2**]. They proved that their definition agrees with the properties of the prime ideal [**4**]. They redefined prime submodules by requiring that they are fully invariant and their multiplication is carried out on the ideal of the module endomorphism [**5**]. The prime submodules defined by Sanh et al. are called S-prime submodules.

Several researchers have studied the identification of the endomorphism ring of modules, e.g. Alolaiyan and Alghamdi [**1**]. In contrast, the identification of fully invariant submodules and the ideal of the required endomorphism ring is not widely known. The characteristics of S-prime submodules are related to both, and form a gap for further research.

Based on the above, in this paper, we will identify the ideal of an endomorphism ring and the fully invariant submodules of a finitely generated module over a principal ideal domain.

2020 *Mathematics Subject Classification.* Primary 16S50; Secondary 16D70.

We will first identify the endomorphism ring and its generators based on the decomposition of a module over a principal ideal domain, specifically the endomorphism of cyclic modules. Based on this identification, we will identify the ideal of the endomorphism ring and its fully invariant submodule. Then, we characterize the S-prime submodules and explore the relationship between the S-prime submodules and prime submodules.

2. Finitely generated modules over a principal ideal domain

The concept of S-prime submodules was introduced by Sanh et al. as a generalization of the prime ideals in an associative ring [**4**], as described in Definition 2.1.

DEFINITION 2.1. A proper fully invariant submodule $X \subseteq M$ is called an S-prime submodule of M if every ideal of $End_R(M)$, say I, and every fully invariant submodule U, which satisfies $I(U) \subset X$, implying $U \subseteq X$ or $I(M) \subseteq X$.

Consider $\mathbb{Z}$ as $\mathbb{Z}$-module. Every fully invariant submodule of $\mathbb{Z}$ is always in the form $a\mathbb{Z}$ for $a \in \mathbb{Z}$. Furthermore, let p be a prime element in $\mathbb{Z}$, then every submodule $p\mathbb{Z}$ is a prime submodule and an S-prime submodule.

Note that the endomorphism ring of $\mathbb{Z}_m$ as $\mathbb{Z}$-module is not isomorphic to $\mathbb{Z}$, but every prime submodule and S-prime submodule is exactly the same, i.e., $p\mathbb{Z}_m$ with p is a prime number in $\mathbb{Z}$. Moreover, every submodule of $\mathbb{Z}$ and $\mathbb{Z}_m$ is a fully invariant submodule.

Let R be a principal ideal domain, M an R–module, and $E = End_R(M)$, then we have that $f(m) \in M$ for every $f \in E$ and $m \in M$. Hence, M is an E–module. Furthermore, an R–submodule $X \subseteq M$ is a fully invariant submodule of M if and only if X is an E–submodule of M. Hence, the class of fully invariant submodules and the endomorphism rings of that class have an interesting structure, which we will discuss here.

For any finitely generated module M over a principal ideal domain R, M can be decomposed into

$$M = M_T \oplus M_F,$$

where M_T is the torsion part of M and M_F is a free submodule of M. The endomorphism ring E is isomorphic to

$$\begin{pmatrix} End_R(M_T) & Hom_R(M_F, M_T) \\ 0 & End_R(M_F) \end{pmatrix}.$$

Let $\mu = p_1^{e_1} p_2^{e_2} \dots p_k^{e_k}$ be the order of M_T with $p_1, p_2, \dots, p_k$ as different prime elements. Then M_T can be decomposed into

$$M_T = M_{p_1} \oplus \cdots \oplus M_{p_k}$$

with $M_{p_i} = \{u \in M_T \ : \ p_i^{e_i} u = 0\}$ is a primary submodule with order $p_i^{e_i}$.

Based on this decomposition, the module M_T can be seen as the column vector

$$M_T = \begin{pmatrix} M_{p_1} \\ M_{p_2} \\ \cdots \\ M_{p_k} \end{pmatrix}. \tag{2.1}$$

Therefore, $End_R(M_T)$ is a ring of $k \times k$ matrices whose ij-th entry is an element in $Hom_R(M_{p_i}, M_{p_j})$. Since $Hom_R(M_{p_i}, M_{p_j}) = 0$ for $i, j \in \{1, 2, 3, \ldots k\}$ and $i \neq j$, then $End_R(M_T)$ is:

$$End_R(M_T) = \begin{pmatrix} End_R(M_{p_1}) & 0 & \ldots & 0 \\ 0 & End_R(M_{p_2}) & 0 & \ldots \\ \ldots & 0 & \ldots & 0 \\ 0 & \ldots & 0 & End_R(M_{p_k}) \end{pmatrix}. \tag{2.2}$$

Let M_p be a finitely generated primary module over R with $o(M_p) = p^e$ for a prime p in R and a positive integer e. We know that M_p can be decomposed into cyclic modules as follows:

$$M_p = \langle u_1 \rangle \oplus \cdots \oplus \langle u_m \rangle \quad o(u_i) = p^{e_i}, \qquad e = e_1 \geq \cdots \geq e_m \geq 1 \tag{2.3}$$

Let $\delta_{ij} \in End_R(M_p)$ be defined as:

$$\delta_{ij}(u_\mu) = \begin{cases} p^{\max\{0, e_i - e_j\}} u_i & \text{if } \mu = j \\ 0 & \text{if } \mu \neq j \end{cases} \tag{2.4}$$

for all $\mu \in \{1, 2, 3, \ldots, m\}$.

Now consider $End_R(M_p)$ as a linear combination of $\{\delta_{ij} | i, j \in \{1, 2, 3, \ldots m\}\}$. The endomorphism ring $End_R(M_p)$ as an R-module can be described as:

$$End_R(M_p) = \oplus_{i=1}^m \oplus_{j=1}^m \langle \delta_{ij} \rangle. \tag{2.5}$$

Meanwhile, a free module M_F over R can be decomposed as a direct sum of cyclic submodules as follows:

$$M_F = \langle v_1 \rangle \oplus \langle v_2 \rangle \oplus \langle v_3 \rangle \cdots \oplus \langle v_n \rangle, \tag{2.6}$$

where $\{v_1, v_2, v_3, \ldots, v_n\}$ is a basis of M_F.

Let $\pi_{ij} \in End_R(M_F)$ be defined as:

$$\pi_{ij}(v_\mu) = \begin{cases} v_i & \text{if } \mu = j \\ 0 & \text{if } \mu \neq j \end{cases} \tag{2.7}$$

for all $\mu \in \{1, 2, 3, \ldots, n\}$. We can show that the endomorphism ring $End_R(M_F)$ can be described as:

$$End_R(M_F) = \oplus_{i=1}^n \oplus_{j=1}^n \langle \pi_{ij} \rangle \tag{2.8}$$

Next, by considering M_T as a column vector in Equation (2.1), a module homomorphism from M_F to M_T can be seen as:

$$Hom_R(M_F, M_T) = \begin{pmatrix} Hom_R(M_F, M_{p_1}) \\ Hom_R(M_F, M_{p_2}) \\ \ldots \\ Hom_R(M_F, M_{p_k}) \end{pmatrix}. \tag{2.9}$$

3. Ideals of an endomorphism ring

In this section, we present a characterization of the ideals of the endomorphism ring of a module M according to its decomposition. First, we review the structure of the endomorphism ring of the module under consideration, denoted by $E = End_R(M)$.

THEOREM 3.1. *Let M_T be a finitely generated torsion module over R. If*

$$End_R(M_T) = diag(End_R(M_{p_1}), End_R(M_{p_2}), \ldots, End_R(M_{p_k})),$$

then $I \subseteq End_R(M_T)$ is an ideal in $End_R(M_T)$ if and only if

$$I = diag(I_{p_1}, I_{p_2}, \ldots, I_{p_k})$$

where $I_{p_i} = \{f_i | diag(f_1, f_2, \ldots, f_k) \in I\}$, for all $i \in \{1, 2, \ldots k\}$, is an ideal in $End_R(M_{p_i})$.

PROOF. Let $I_{p_i}, i \in \{1, 2, \ldots k\}$ be an ideal in $End_R(M_{p_i})$ and

$$I = diag(I_{p_1}, I_{p_2}, \ldots, I_{p_k}).$$

We will prove that I is an ideal in $End_R(M_T)$.

Since I_{p_i} is an ideal in $End_R(M_{p_i})$, then $I = diag(I_{p_1}, I_{p_2}, \ldots, I_{p_k})$ is not empty. Suppose $f = diag(f_1, f_2, \ldots, f_k), h = diag(h_1, h_2, \ldots, h_k) \in I$ with $f_i, h_i \in I_{p_i}$. Since I_{p_i} is an ideal $\forall i \in \{1, 2, \ldots k\}$, then $f_i - h_i \in I_{p_i}$ so

$$f - h = diag(f_1 - h_1, f_2 - h_2, \ldots, f_k - h_3) \in I$$

Suppose $f = diag(f_1, f_2, \ldots, f_k) \in I$ and $g = diag(g_1, g_2, \ldots, g_k) \in End_R(M_T)$ with $f_i \in I_{p_i}, g_i \in End_R(M_{p_i})$, then

$$\begin{aligned} fg &= diag(f_1, f_2, \ldots, f_k) diag(g_1, g_2, \ldots, g_k) \\ &= diag(f_1.g_1, f_2.g_2, \ldots, f_k.g_k) \end{aligned}$$

Since $f_i.g_i \in I_{p_i}$ for $i \in \{1, 2, 3, \ldots, k\}$, then $f.g \in I$. In the same way, we can prove that $g.f \in I$. So I is an ideal in $End_R(M_T)$.

On the other hand, suppose I is an ideal in $End_R(M_T)$ and

$$I_{p_i} = \{f_i | diag(f_1, f_2, \ldots, f_k) \in I\}, i \in \{1, 2, \ldots k\},$$

then we get:

$$I \subseteq diag(I_{p_1}, I_{p_2}, \ldots, I_{p_k})$$

We will show that $diag(0 \ldots, 0, I_{p_i}, 0, \ldots, 0) \subseteq I$ for every $i \in \{1, 2, \ldots, k\}$.

Consider $e_i = diag(0, \ldots, 0, 1_i, 0, \ldots, 0) \in End_R(M_T)$ so that for every $g \in End_R(M_T)$ $e_i.g = g.e_i = diag(0, \ldots, 0, g_i, 0, \ldots, 0)$.

Suppose $u_i = diag(0, \ldots, 0, f_i, 0, \ldots, 0) \in diag(0 \ldots, 0, I_{p_i}, 0, \ldots, 0)$ with $f_i \in I_{p_i}$, then there is $f = diag(f_1, f_2, \ldots, f_k) \in I$ so

$$u_i = f.e_i \in I.$$

So $diag(0 \ldots, 0, I_{p_i}, 0, \ldots, 0) \subseteq I$ for every $i \in \{1, 2, \ldots, k\}$ and hence

$$diag(I_{p_1}, I_{p_2}, \ldots, I_{p_k}) \subseteq I.$$

Since $diag(I_{p_1}, I_{p_2}, \ldots, I_{p_k}) = I$, then I_{p_i} is a subgroup of $End_R(M_{p_i})$.

Suppose $f \in I$ and $g \in End_R(M_T)$, then

$$\begin{aligned} f.g &= diag(f_1, f_2, \ldots, f_k) diag(g_1, g_2, \ldots, g_k) \\ &= diag(f_1.g_1, f_2.g_2, \ldots, f_k.g_k) \\ &\in diag(I_{p_1}, I_{p_2}, \ldots, I_{p_k}) \end{aligned}$$

so $f_i.g_i \in I_{p_i}$. In the same way, we get $g_i.f_i \in I_{p_i}$. We conclude that I_{p_i} is an ideal of $End_R(M_{p_i})$. □

By Theorem 3.1, we can see the relation between an ideal of the endomorphism ring of a torsion module and the ideals of the endomorphism rings of its primary modules. By taking $I_{p_i} = 0$ for $i \neq \mu$ we obtain Corollary 3.2.

COROLLARY 3.2. *Let M_T be a finitely generated torsion module over R. If*

$$End_R(M_T) = diag(End_R(M_{p_1}), End_R(M_{p_2}), \ldots, End_R(M_{p_k})),$$

then I_{p_μ} is an ideal in $End_R(M_{p_\mu})$, $\mu \in \{1, 2, \ldots k\}$, if and only if

$$diag(0, \ldots, 0, I_{p_\mu}, 0, \ldots, 0)$$

is an ideal in $End_R(M_T)$.

PROOF. Suppose I_{p_μ} is an ideal in $End_R(M_{p_\mu})$. Then, consider

$$diag(0, \ldots, 0, I_{p_\mu}, 0, \ldots, 0) \subseteq End_R(M_T).$$

Since for $i \neq \mu, I_{p_i} = 0$ is an ideal in $End_R(M_{p_i})$, then, based on Theorem 3.1, $diag(0, \ldots, 0, I_{p_\mu}, 0, \ldots, 0)$ is an ideal in $End_R(M_T)$.

Conversely, according to Theorem 3.1, if $diag(0, \ldots, 0, I_{p_\mu}, 0, \ldots, 0)$ is an ideal in $End_R(M_T)$, then I_{p_μ} is an ideal in $End_R(M_{p_\mu})$. □

Now by Theorem 3.1, we can identify the prime ideals of $End_R(M_T)$.

COROLLARY 3.3. *Let M_T be a finitely generated torsion module over R. If*

$$End_R(M_T) = diag(End_R(M_{p_1}), End_R(M_{p_2}), \ldots, End_R(M_{p_k})),$$

then $I \subseteq End_R(M_T)$ is a prime ideal if and only if

$$I = diag(I_{p_1}, I_{p_2}, \ldots, I_{p_k})$$

where there exist a $\mu \in \{1, 2, \ldots k\}$, such that $I_{p_\mu} \neq End_R(M_{p_\mu})$ is a prime ideal of $End_R(M_{p_\mu})$, and $I_{p_i} = End_R(M_{p_i})$ for all $i \in \{1, 2, \ldots k\}, i \neq \mu$.

PROOF. Suppose

$$I = diag(I_{p_1}, I_{p_2}, \ldots, I_{p_k})$$

where there is exactly one $\mu \in \{1, 2, \ldots k\}$, so $I_{p_\mu} \neq End_R(M_{p_\mu})$ is a prime ideal in $End_R(M_{p_\mu})$, and $I_{p_i} = End_R(M_{p_i})$ for every $i \in \{1, 2, \ldots k\}, i \neq \mu$. We will prove that I is a prime ideal in $End_R(M_T)$.

Because I_{p_μ} is a prime ideal in $End_R(M_{p_\mu})$ and $I_{p_i} = End_R(M_{p_i})$ for every $i \in \{1, 2, \ldots k\}, i \neq \mu$, then based on Theorem 3.1, we get:

$$I = diag(End_R(M_{p_1}), \ldots, End_R(M_{p_{\mu-1}}), I_{p_\mu}, End_R(M_{p_{\mu+1}}), \ldots, End_R(M_{p_k}))$$

is an ideal in $End_R(M_T)$. Next, take any ideals $J = diag(J_{p_1}, J_{p_2}, \ldots, J_{p_k})$ and $H = diag(H_{p_1}, H_{p_2}, \ldots, H_{p_k})$ in $End_R(M_T)$ with $JH \subseteq I$. Since J and H are ideals in $End_R(M_T)$, then J_{p_i} and H_{p_i} are ideals in $End_R(M_{p_i})$ for each $i \in \{1, 2, \ldots k\}$. Next, note that

$$\begin{aligned} J.H &= diag(J_{p_1}, J_{p_2}, \ldots, J_{p_k}).diag(H_{p_1}, H_{p_2}, \ldots, H_{p_k}) \\ &= diag(J_{p_1}.H_{p_1}, J_{p_2}.H_{p_2}, \ldots, J_{p_k}.H_{p_k}) \\ &\subseteq diag(End_R(M_{p_1}), \ldots, End_R(M_{p_{\mu-1}}), I_{p_\mu}, End_R(M_{p_{\mu+1}}), \ldots, End_R(M_{p_k})) \end{aligned}$$

with $J_{p_\mu}H_{p_\mu} \subseteq I_{p_\mu}$. Since I_{p_μ} is a prime ideal in $End_R(M_{p_\mu})$, then $J_{p_\mu} \subseteq I_{p_\mu}$ or $K_{p_\mu} \subseteq I_{p_\mu}$. Therefore,

$$\begin{aligned} J &= diag(J_{p_1}, J_{p_2}, \ldots, J_{p_k}) \\ &\subseteq diag(End_R(M_{p_1}), \ldots, End_R(M_{p_{\mu-1}}), I_{p_\mu}, End_R(M_{p_{\mu+1}}), \ldots, End_R(M_{p_k})) \\ &= I \end{aligned}$$

or

$$\begin{aligned} H &= diag(H_{p_1}, H_{p_2}, \ldots, H_{p_k}) \\ &\subseteq diag(End_R(M_{p_1}), \ldots, End_R(M_{p_{\mu-1}}), I_{p_\mu}, End_R(M_{p_{\mu+1}}), \ldots, End_R(M_{p_k})) \\ &= I. \end{aligned}$$

Otherwise, suppose $I \subseteq End_R(M_T)$ is a prime ideal. We will prove that

$$I = diag(I_{p_1}, I_{p_2}, \ldots, I_{p_k})$$

where there is exactly one $\mu \in \{1, 2, \ldots k\}$ such that $I_{p_\mu} \neq End_R(M_{p_\mu})$ is a prime ideal in $End_R(M_{p_\mu})$, and $I_{p_i} = End_R(M_{p_i})$ for every $i \in \{1, 2, \ldots k\}, i \neq \mu$.

Since $I \subseteq End_R(M_T)$ is an ideal, then $I = diag(I_{p_1}, I_{p_2}, \ldots, I_{p_k})$ where I_{p_i} is an ideal in $End_R(M_{p_i})$ for every $i \in \{1, 2, 3, \ldots, k\}$. Because I is prime, then $I \neq End_R(M_T)$, meaning that there is a $\mu \in \{1, 2, \ldots k\}$ such that $I_{p_\mu} \neq End_R(M_{p_\mu})$.

Suppose there are $\theta, \mu \in \{1, 2, 3, \ldots, k\}, \theta \neq \mu$ such that ideal $I_{p_\theta} \neq End_R(M_{p_\theta})$ and $I_{p_\mu} \neq End_R(M_{p_\mu})$. Then, consider J, H ideals in $End_R(M_T)$ with $J_{p_\theta} = I_{p_\theta}$ and $H_{p_\mu} = I_{p_\mu}$, but $J_{p_\mu} = End_R(M_{p_\mu}), H_{p_\theta} = End_R(M_{p_\theta})$ and $J_{p_i} = H_{p_i} = I_{p_i}$ for $i \neq \mu$ and $i \neq \theta$. Thus, we get $J.H \subseteq I$ but $J \nsubseteq I$ and $H \nsubseteq I$. This is a contradiction with the assumption that I is prime, so there must be exactly one $\mu \in \{1, 2, \ldots k\}$ such that $I_{p_\mu} \neq End_R(M_{p_\mu})$ and $I_{p_i} = End_R(M_{p_i})$ for every $i \in \{1, 2, \ldots k\}, i \neq \mu$.

We will prove that I_{p_μ} is a prime ideal in $End_R(M_{p_\mu})$. Take any H_{p_μ}, J_{p_μ} ideals in $End_R(M_{p_\mu})$ such that $H_{p_\mu}, J_{p_\mu} \subseteq I_{p_\mu}$.

Then, consider $diag(0, \ldots, 0, J_{p_\mu}, 0, \ldots, 0)$ and $diag(0, \ldots, 0, H_{p_\mu}, 0, \ldots, 0)$ as ideals in $End_R(M_T)$ such that

$$diag(0, \ldots, 0, J_{p_\mu}, 0, \ldots, 0).diag(0, \ldots, 0, H_{p_\mu}, 0, \ldots, 0) \subseteq I.$$

Since I is a prime ideal in $End_R(M_T)$, then

$$diag(0, \ldots, 0, J_{p_\mu}, 0, \ldots, 0) \subseteq I$$

or

$$diag(0, \ldots, 0, H_{p_\mu}, 0, \ldots, 0) \subseteq I.$$

Thus, $J_{p_\mu} \subseteq I_{p_\mu}$ or $H_{p_\mu} \subseteq I_{p_\mu}$. So there is exactly one $\mu \in \{1, 2, \ldots k\}$ such that $I_{p_\mu} \neq End_R(M_{p_\mu})$ is a prime ideal in $End_R(M_{p_\mu})$ and $I_{p_i} = End_R(M_{p_i})$ for each $i \in \{1, 2, \ldots k\}, i \neq \mu$. □

4. Ideals in an endomorphism ring of a finitely generated module

In this section, we will investigate the ideals in $End_R(M)$. First, we will discuss the ideals in $End_R(M_F)$, which is an endomorphism ring of a free module over a principal ideal domain R.

By Equation (2.8), $End_R(M_F) = \oplus_{i=1}^n \oplus_{j=1}^n \langle \pi_{ij} \rangle$ with $\pi_{ij} \in End_R(M_F)$ is defined in (2.7). Let I be an ideal in $End_R(M_F)$. Since R is a principal ideal domain, there exists an ideal J in R such that

$$I = \oplus_{i=1}^n \oplus_{j=1}^n \langle J\pi_{ij} \rangle. \tag{4.1}$$

Every ideal J in R can be written as $\langle a\rangle$ for an $a \in R$. Therefore, an ideal in $End_R(M_F)$ is

$$I = a \oplus_{i=1}^{n} \oplus_{j=1}^{n} \langle \pi_{ij} \rangle = aEnd_R(M_F). \tag{4.2}$$

Hence, a prime ideal in $End_R(M_F)$ can be written as $p.End_R(M_F)$ for p, some prime element in R.

Let $M = \begin{pmatrix} M_T \\ M_F \end{pmatrix}$ be a finitely generated module over R. We denote $End_R(M) = E$, $End_R(M_T) = E_T$, $End_R(M_F) = E_F$, and $Hom_R(M_F, M_T) = H_{FT}$. We consider H_{FT} as an E_T-left module and an E_F-right module.

THEOREM 4.1. *For any $I \subseteq E$, I is an ideal of E if and only if*

$$I = \begin{pmatrix} I_{11} & I_{12} \\ 0 & I_{22} \end{pmatrix}.$$

where I_{11} is an ideal of E_T, I_{22} is an ideal of E_F, I_{12} is an E_T-left submodule and an E_F-right submodule of H_{FT} that satisfies

$$I_{11}H_{FT}, H_{FT}I_{22} \subseteq I_{12}.$$

PROOF. Suppose I is an ideal in E and $I_{ij} = \left\{ \lambda_{ij} | \begin{pmatrix} \lambda_{11} & \lambda_{12} \\ 0 & \lambda_{22} \end{pmatrix} \in I \right\}$ with $i, j \in \{1, 2\}$ and $I_{21} = 0$, then we get:

$$I \subseteq \begin{pmatrix} I_{11} & I_{12} \\ 0 & I_{22} \end{pmatrix}.$$

We will prove that $\begin{pmatrix} I_{11} & I_{12} \\ 0 & I_{22} \end{pmatrix} \subseteq I$. Suppose $\lambda \in I$, then $\lambda = \begin{pmatrix} \lambda_{11} & \lambda_{12} \\ 0 & \lambda_{22} \end{pmatrix}$ with $\lambda_{11} \in I_{11}, \lambda_{12} \in I_{12}$ and $\lambda_{22} \in I_{22}$.

Next,

$$\begin{pmatrix} \lambda_{11} & 0 \\ 0 & 0 \end{pmatrix} = \begin{pmatrix} \lambda_{11} & \lambda_{12} \\ 0 & \lambda_{22} \end{pmatrix} \begin{pmatrix} 1_{11} & 0 \\ 0 & 0 \end{pmatrix} \in I$$

$$\begin{pmatrix} 0 & 0 \\ 0 & \lambda_{22} \end{pmatrix} = \begin{pmatrix} 0 & 0 \\ 0 & 1_{22} \end{pmatrix} \begin{pmatrix} \lambda_{11} & \lambda_{12} \\ 0 & \lambda_{22} \end{pmatrix} \in I$$

$$\begin{pmatrix} 0 & \lambda_{12} \\ 0 & 0 \end{pmatrix} = \begin{pmatrix} 1_{11} & 0 \\ 0 & 0 \end{pmatrix} \begin{pmatrix} \lambda_{11} & \lambda_{12} \\ 0 & \lambda_{22} \end{pmatrix} \begin{pmatrix} 0 & 0 \\ 0 & 1_{22} \end{pmatrix} \in I$$

thus $I = \begin{pmatrix} I_{11} & I_{12} \\ 0 & I_{22} \end{pmatrix}$.

Next, because $I = \begin{pmatrix} I_{11} & I_{12} \\ 0 & I_{22} \end{pmatrix}$ is an ideal, then I_{11}, I_{12} and I_{22} are groups, respectively. Then, notice that

$$\begin{pmatrix} I_{11} & I_{12} \\ 0 & I_{22} \end{pmatrix} \begin{pmatrix} E_T & H_{FT} \\ 0 & E_F \end{pmatrix} = \begin{pmatrix} I_{11}E_T & I_{11}H_{FT} + I_{12}E_F \\ 0 & I_{22}E_F \end{pmatrix} \subseteq \begin{pmatrix} I_{11} & I_{12} \\ 0 & I_{22} \end{pmatrix}$$

and

$$\begin{pmatrix} E_T & H_{FT} \\ 0 & E_F \end{pmatrix} \begin{pmatrix} I_{11} & I_{12} \\ 0 & I_{22} \end{pmatrix} = \begin{pmatrix} E_T I_{11} & E_T I_{12} + H_{FT} I_{22} \\ 0 & E_F I_{22} \end{pmatrix} \subseteq \begin{pmatrix} I_{11} & I_{12} \\ 0 & I_{22} \end{pmatrix}.$$

So I_{11} is an ideal in E_T, I_{22} is an ideal in E_F. Then, consider $\begin{pmatrix} 0 & H_{FT} \\ 0 & 0 \end{pmatrix} \subset E$. Then

$$\begin{pmatrix} 0 & H_{FT} \\ 0 & 0 \end{pmatrix}\begin{pmatrix} I_{11} & I_{12} \\ 0 & I_{22} \end{pmatrix} = \begin{pmatrix} 0 & H_{FT}I_{22} \\ 0 & 0 \end{pmatrix} \subseteq \begin{pmatrix} I_{11} & I_{12} \\ 0 & I_{22} \end{pmatrix}$$

and

$$\begin{pmatrix} I_{11} & I_{12} \\ 0 & I_{22} \end{pmatrix}\begin{pmatrix} 0 & H_{FT} \\ 0 & 0 \end{pmatrix} = \begin{pmatrix} 0 & I_{11}H_{FT} \\ 0 & 0 \end{pmatrix} \subseteq \begin{pmatrix} I_{11} & I_{12} \\ 0 & I_{22} \end{pmatrix}.$$

So I_{12} is an E_T-left submodule and an E_F-right submodule of H_{FT} that satisfies the following conditions:

$$I_{11}H_{FT}, H_{FT}I_{22} \subseteq I_{12}.$$

Conversely, if

$$I = \begin{pmatrix} I_{11} & I_{12} \\ 0 & I_{22} \end{pmatrix}$$

where I_{11} is an ideal in E_T, I_{22} is an ideal in E_F, I_{12} is an E_T-left submodule and an E_F-right submodule of H_{FT}, then I is a commutative group. Furthermore, because

$$I_{11}H_{FT}, H_{FT}I_{22} \subseteq I_{12},$$

then

$$\begin{pmatrix} I_{11} & I_{12} \\ 0 & I_{22} \end{pmatrix}\begin{pmatrix} E_T & H_{FT} \\ 0 & E_F \end{pmatrix} = \begin{pmatrix} I_{11}E_T & I_{11}H_{FT} + I_{12}E_F \\ 0 & I_{22}E_F \end{pmatrix} \subseteq \begin{pmatrix} I_{11} & I_{12} \\ 0 & I_{22} \end{pmatrix}$$

and

$$\begin{pmatrix} E_T & H_{FT} \\ 0 & E_F \end{pmatrix}\begin{pmatrix} I_{11} & I_{12} \\ 0 & I_{22} \end{pmatrix} = \begin{pmatrix} E_TI_{11} & E_TI_{12} + H_{FT}I_{22} \\ 0 & E_FI_{22} \end{pmatrix} \subseteq \begin{pmatrix} I_{11} & I_{12} \\ 0 & I_{22} \end{pmatrix}.$$

Therefore, I is an ideal. □

COROLLARY 4.2. *Let $S \subseteq H_{FT}$, then S is a left E_T-submodule and a right E_F-submodule of H_{FT} if and only if $\begin{pmatrix} 0 & S \\ 0 & 0 \end{pmatrix} \subset E$ is an ideal in E.*

PROOF. Suppose

$$\begin{pmatrix} 0 & S \\ 0 & 0 \end{pmatrix}$$

is an ideal in E. Based on Theorem 4.1, S is a left E_T−module and also a right E_F−right module of H_{FT}.

Otherwise, suppose S is a left E_T-submodule and also a right E_F-submodule of H_{FT}. Note that $0 \subseteq E_F$ is an ideal in E_F and $0 \subseteq E_T$ is an ideal in E_T that satisfies

$$0H_{FT}, H_{FT}0 \subseteq S.$$

Based on Theorem 4.1, $\begin{pmatrix} 0 & S \\ 0 & 0 \end{pmatrix}$ is an ideal in E. □

COROLLARY 4.3. $\begin{pmatrix} I_{11} & I_{11}H_{FT} \\ 0 & 0 \end{pmatrix}$ *$\subseteq E$ is an ideal in E if only if I_{11} is an ideal of E_T.*

PROOF. If $\begin{pmatrix} I_{11} & I_{11}H_{FT} \\ 0 & 0 \end{pmatrix} \subseteq E$ is an ideal in E, then according to Theorem 4.1, I_{11} is an ideal in E_T.

Otherwise, suppose I_{11} is an ideal in E_T. Notice that for $I_{11}H_{FT} \subseteq H_{FT}$

$$(I_{11}H_{FT})E_F \subseteq I_{11}H_{FT}$$

and

$$E_T(I_{11}H_{FT}) \subseteq I_{11}H_{FT}.$$

So $I_{11}H_{FT}$ is a left E_T-submodule and a right E_F-submodule of H_{FT}. Next,

$$I_{11}(I_{11}H_{FT}) \subseteq I_{11}H_{FT}$$

and

$$(I_{11}H_{FT})0 \subseteq I_{11}H_{FT}.$$

Therefore, based on Theorem 4.1, $\begin{pmatrix} I_{11} & I_{11}H_{FT} \\ 0 & 0 \end{pmatrix} \subseteq E$ is an ideal in E.

COROLLARY 4.4. *$\begin{pmatrix} 0 & H_{FT}I_{22} \\ 0 & I_{22} \end{pmatrix} \subseteq E$ is an ideal in E if and only if I_{22} is an ideal in E_T.*

PROOF. Suppose $\begin{pmatrix} 0 & H_{FT}I_{22} \\ 0 & I_{22} \end{pmatrix} \subseteq E$ is an ideal in E, hence, based on Theorem 4.1, I_{22} is an ideal in E_T.

Otherwise, suppose I_{22} is an ideal in E_F. Notice that for $H_{FT}I_{22} \subseteq H_{FT}$

$$E_T(H_{FT}I_{22}) \subseteq H_{FT}I_{22}$$

and

$$(H_{FT}I_{22})E_F \subseteq H_{FT}I_{22}.$$

So $H_{FT}I_{22}$ is a left E_T-submodule and a right E_F-submodule of H_{FT}. Next,

$$(H_{FT}I_{22})I_{22} \subseteq H_{FT}I_{22}$$

and

$$0(H_{FT}I_{22}) \subseteq H_{FT}I_{22}.$$

Therefore, based on Theorem 4.1, $\begin{pmatrix} 0 & H_{FT}I_{22} \\ 0 & I_{22} \end{pmatrix} \subseteq E$ is an ideal in E. □

Next, we will discuss the relationship between the prime ideals of $End_R(M)$ and the prime ideals of E_T and E_F.

THEOREM 4.5. *Let E be an endomorphism ring of a finitely generated module M, then $I \subseteq E$ is a prime ideal of E if and only if*

$$I = \begin{pmatrix} I_{11} & H_{FT} \\ 0 & E_F \end{pmatrix}$$

where I_{11} is a prime ideal of E_T or

$$I = \begin{pmatrix} E_T & H_{FT} \\ 0 & I_{22} \end{pmatrix}$$

where I_{22} is a prime ideal of E_F.

PROOF. Suppose I is a prime ideal in E. Since I is an ideal, then

$$I = \begin{pmatrix} I_{11} & I_{12} \\ 0 & I_{22} \end{pmatrix},$$

where I_{11} is an ideal in E_T, I_{22} is an ideal in E_F, I_{12} is a left E_T-submodule and a right E_F-submodule of H_{FT}, which satisfy the conditions $I_{11}H_{FT}, H_{FT}I_{22} \subseteq I_{12}$.

We will prove $I = \begin{pmatrix} I_{11} & H_{FT} \\ 0 & E_F \end{pmatrix}$, where I_{11} is a prime ideal in E_T or $I = \begin{pmatrix} I_{11} & H_{FT} \\ 0 & E_F \end{pmatrix}$, where I_{22} is a prime ideal in E_F.

First, we will prove $I_{12} = H_{FT}$.

Suppose $I_{12} \neq H_{FT}$. Consider $\begin{pmatrix} I_{11} & H_{FT} \\ 0 & 0 \end{pmatrix}$ and $\begin{pmatrix} 0 & H_{FT} \\ 0 & 0 \end{pmatrix}$ as ideals in E. Note that $I_{11}H_{FT} \subseteq I_{12}$ and

$$\begin{pmatrix} I_{11} & H_{FT} \\ 0 & 0 \end{pmatrix} . \begin{pmatrix} 0 & H_{FT} \\ 0 & 0 \end{pmatrix} \subseteq \begin{pmatrix} I_{11} & I_{12} \\ 0 & I_{22} \end{pmatrix} = I.$$

However, $\begin{pmatrix} I_{11} & H_{FT} \\ 0 & 0 \end{pmatrix} \not\subseteq I$ and $\begin{pmatrix} 0 & H_{FT} \\ 0 & 0 \end{pmatrix} \not\subseteq I$. This is a contradiction. Therefore, $I_{12} = H_{FT}$.

Suppose $I_{11} \neq E_T$. We will prove that I_{11} is a prime ideal in E_T. Take any J_{11}, K_{11} ideal in E_T with $J_{11}.K_{11} \subseteq I_{11}$, we will prove $J_{11} \subseteq I_{11}$ or $K_{11} \subseteq I_{11}$.

Since J_{11}, K_{11} are ideals in E_T, then based on Corollary 4.3,

$$\begin{pmatrix} J_{11} & J_{11}H_{FT} \\ 0 & 0 \end{pmatrix}, \begin{pmatrix} K_{11} & K_{11}H_{FT} \\ 0 & 0 \end{pmatrix} \subseteq E$$

each is an ideal in E. Next, notice that

$$J_{11}.K_{11} \subseteq I_{11}$$

and

$$J_{11}.K_{11}H_{FT} \subseteq I_{11}.H_{FT} \subseteq I_{12}$$

so we get:

$$\begin{pmatrix} J_{11} & J_{11}H_{FT} \\ 0 & 0 \end{pmatrix} . \begin{pmatrix} K_{11} & K_{11}H_{FT} \\ 0 & 0 \end{pmatrix} \subseteq \begin{pmatrix} I_{11} & I_{12} \\ 0 & I_{22} \end{pmatrix} = I.$$

Since I is a prime ideal, then $\begin{pmatrix} K_{11} & K_{11}H_{FT} \\ 0 & 0 \end{pmatrix} \subseteq I$ or $\begin{pmatrix} J_{11} & J_{11}H_{FT} \\ 0 & 0 \end{pmatrix} \subseteq I$. Thus, $J_{11} \subseteq I_{11}$ or $K_{11} \subseteq I_{11}$. So I_{11} is a prime ideal in E_T.

Next, suppose $I_{22} \neq E_F$. We will prove that I_{22} is a prime ideal in E_F.

Take any J_{22}, K_{22} ideal in E_F with $J_{22}.K_{22} \subseteq I_{22}$. We will prove $J_{22} \subseteq I_{22}$ or $K_{22} \subseteq I_{22}$.

Since J_{22}, K_{22} are ideals in E_F, then based on Corollary 4.4,

$$\begin{pmatrix} 0 & H_{FT}.J_{22} \\ 0 & A_{22} \end{pmatrix}, \begin{pmatrix} 0 & H_{FT}K_{22} \\ 0 & K_{22} \end{pmatrix} \subseteq E$$

each is an ideal in E. Next, notice that

$$J_{22}.K_{22} \subseteq I_{22}$$

and

$$H_{FT}J_{22}.K_{22} \subseteq H_{FT}I_{22} \subseteq I_{12}$$

so we get:

$$\begin{pmatrix} 0 & H_{FT}.J_{22} \\ 0 & A_{22} \end{pmatrix} . \begin{pmatrix} 0 & H_{FT}K_{22} \\ 0 & K_{22} \end{pmatrix} \subseteq \begin{pmatrix} I_{11} & I_{12} \\ 0 & I_{22} \end{pmatrix} = I.$$

Since I is a prime ideal, then $\begin{pmatrix} 0 & H_{FT}.J_{22} \\ 0 & A_{22} \end{pmatrix} \subseteq \begin{pmatrix} I_{11} & I_{12} \\ 0 & I_{22} \end{pmatrix} = I$ or $\begin{pmatrix} 0 & H_{FT}K_{22} \\ 0 & K_{22} \end{pmatrix} \subseteq \begin{pmatrix} I_{11} & I_{12} \\ 0 & I_{22} \end{pmatrix} = I$. Thus, $J_{22} \subseteq I_{22}$ or $K_{22} \subseteq I_{22}$. So I_{22} is a prime ideal in E_F.

We will prove that $I_{11} = E_T$ or $I_{22} = E_F$.

Suppose $I_{11} \neq E_T$ and $I_{22} \neq E_F$. Then, consider

$$J = \begin{pmatrix} I_{11} & H_{FT} \\ 0 & E_F \end{pmatrix}, K = \begin{pmatrix} E_F & H_{FT} \\ 0 & I_{22} \end{pmatrix} \subseteq E$$

Each is an ideal in E. Next, notice that $I_{11}.E_T = I_{11}$, $I_{11}.H_{FT} + H_{FT}I_{22} \subseteq H_{FT}$ and $E_F.I_{22} = I_{22}$ so that

$$JK = \begin{pmatrix} I_{11} & H_{FT} \\ 0 & E_F \end{pmatrix} . \begin{pmatrix} E_F & H_{FT} \\ 0 & I_{22} \end{pmatrix} \subseteq \begin{pmatrix} I_{11} & H_{FT} \\ 0 & I_{22} \end{pmatrix}.$$

However, $\begin{pmatrix} I_{11} & H_{FT} \\ 0 & E_F \end{pmatrix} \not\subseteq \begin{pmatrix} I_{11} & H_{FT} \\ 0 & I_{22} \end{pmatrix}$ and $\begin{pmatrix} E_F & H_{FT} \\ 0 & I_{22} \end{pmatrix} \not\subseteq \begin{pmatrix} I_{11} & H_{FT} \\ 0 & I_{22} \end{pmatrix}$. This is a contradiction. Thus, $I_{11} = E_T$ or $I_{22} = E_F$.

Therefore, $I = \begin{pmatrix} E_T & H_{FT} \\ 0 & I_{22} \end{pmatrix}$ where I_{22} is a prime ideal in E_F or $I = \begin{pmatrix} I_{11} & H_{FT} \\ 0 & E_F \end{pmatrix}$ where I_{11} is a prime ideal in E_T.

Otherwise, suppose $I = \begin{pmatrix} E_T & H_{FT} \\ 0 & I_{22} \end{pmatrix}$ where I_{22} is a prime ideal in E_F. We will prove that I is a prime ideal in E.

First we will prove that if $I = \begin{pmatrix} E_T & H_{FT} \\ 0 & I_{22} \end{pmatrix}$ where I_{22} is a prime ideal in E_F, then I is a prime ideal in E.

Take any $J = \begin{pmatrix} J_{11} & J_{12} \\ 0 & A_{22} \end{pmatrix}$ and $K = \begin{pmatrix} K_{11} & K_{12} \\ 0 & K_{22} \end{pmatrix}$ ideals in E with $J.K \subseteq I$.

Furthermore, note that

$$\begin{pmatrix} J_{11} & J_{12} \\ 0 & A_{22} \end{pmatrix} . \begin{pmatrix} K_{11} & K_{12} \\ 0 & K_{22} \end{pmatrix} = \begin{pmatrix} J_{11}K_{11} & J_{11}K_{12} + J_{12}K_{22} \\ 0 & A_{22}K_{22} \end{pmatrix} \subseteq \begin{pmatrix} E_T & H_{FT} \\ 0 & I_{22} \end{pmatrix}$$

with $J_{22}K_{22} \subseteq I_{22}$. Since I_{22} is a prime ideal in E_T, then $J_{22} \subseteq I_{22}$ or $K_{22} \subseteq I_{22}$. Thus, $\begin{pmatrix} K_{11} & K_{12} \\ 0 & K_{22} \end{pmatrix} \subseteq \begin{pmatrix} E_T & H_{FT} \\ 0 & I_{22} \end{pmatrix}$ or $\begin{pmatrix} J_{11} & J_{12} \\ 0 & A_{22} \end{pmatrix} \subseteq \begin{pmatrix} E_T & H_{FT} \\ 0 & I_{22} \end{pmatrix}$.

So $I = \begin{pmatrix} E_T & H_{FT} \\ 0 & I_{22} \end{pmatrix}$ is a prime ideal in E.

In the same way, one can prove that $I = \begin{pmatrix} I_{11} & H_{FT} \\ 0 & E_F \end{pmatrix}$ is a prime ideal in E, where I_{11} a prime ideal in E_T. □

5. Fully invariant submodules

In this section we discuss the structure of fully invariant submodules in terms of cyclic decomposition (2.3).

DEFINITION 5.1. Let M be an R-module. A submodule X is called a fully invariant submodule of M if for every $f \in End_R(M)$, the condition $f(X) \subset X$ holds.

According to Definition 5.1, this yields that aM is a fully invariant submodule of M for an $a \in R$. After that we will present the connection between fully invariant submodules and the ideals of $End_R(M)$.

Consider $E = End_R(M)$. Note that for every $m \in M$ and for every endomorphism $f \in E$ there is an $f(m) \in M$. Thus, M can be considered a left module over E.

THEOREM 5.2. *Let M be a finitely generated module over R and $E = End_R(M)$. An $R-$submodule $X \subseteq M$ is fully invariant if and only if X is an E-left submodule of M.*

PROOF. By defining a mapping from $(f, m) \in (E \times M)$ to $f(m) \in M$, M is a left module over E. Let $X \subseteq M$ be a fully invariant submodule of M. We will prove that X is a left E-submodule of M.

Since X is a fully invariant submodule of M, then X is a subgroup of M and $f(X) \subseteq X, \forall f \in E$. So X is a left E-submodule of M.

Otherwise, suppose X is a E-left submodule of M, then X is an $R-$submodule of M and for every $m \in X$ and $f \in E$ we have $fm = f(m) \in X$. Thus, $f(X) \subseteq X$, therefore X is a fully invariant submodule. □

Next, we will show the relationship between a fully invariant submodule of M and an ideal in E.

LEMMA 5.3. *Let M be a finitely generated module over R, $E = End_R(M)$, I an ideal of E, and X a nonempty subset of M. If $IX = \{\sum_{i=1}^{n} g_i x_i | g_i \in I, x_i \in X\}$, then IX is a fully invariant submodule of M.*

PROOF. Suppose $I \subseteq E$ is an ideal, X is a non-empty subset of M, and $IX = \{\sum_{i=1}^{n} g_i x_i | g_i \in I, x_i \in X\}$. We will prove that IX is a fully invariant submodule of M.

Take any $u = \sum_{i=1}^{l} g_i x_i$ and $v = \sum_{i=1}^{k} h_i y_i$ elements of IX with $g_i, h_i \in I, x_i, y_i \in X$. Note that for every $a \in R$ we have $a.1 \in E$, where 1 is the identity element in E, so that $ag_i, ah_i \in I$ and

$$a(u + v) = \sum_{i=1}^{l} ag_i(x_i) + \sum_{i=1}^{k} ah_i(y_i) \in IX.$$

Thus, IX is an $R-$submodule of M.

Next, take any $f \in E$ and $v = \sum_{i=1}^{k} g_i x_i \in IX$ for some $g_i \in I, x_i \in X$. Then

$$f(v) = f(\sum_{i=1}^{k} g_i x_i) = \sum_{i=1}^{k} (fg_i)x_i \in IX.$$

Thus, $f(IX) \subset IX$. Therefore, IX is a fully invariant submodule. □

We can conclude that for every ideal I of E, IM is a fully invariant submodule of M.

LEMMA 5.4. *Let M be a finitely generated module over R, $E = End_R(M)$, I an ideal of E, X a submodule of M, and $I_X = \{f \in E | f(M) \subseteq X\}$. The submodule X is fully invariant if and only if I_X is an ideal of E.*

PROOF. Let X be a submodule of M and consider

$$I_X = \{f \in E | f(M) \subseteq X\}$$

as an ideal in E. We will prove that X is a fully invariant submodule of M.

Suppose X is not a fully invariant submodule of M, then there is an $h \in E$ such that $h(X) \nsubseteq X$. Take any $f \in I_X$, then $f(M) \subseteq X$. Next, notice that

$$h(f(M)) \subseteq h(X) \nsubseteq X$$

This is a contradiction with $hf \in I_X$. Hence, X is fully invariant.

Otherwise, suppose X is a fully invariant submodule of M and consider $I_X = \{f \in E | f(M) \subseteq X\}$. We will prove that I_X is an ideal in E.

Take $0 \in E$ then $0(M) \subseteq X$. So $0 \in I_X$, which means that I_X is not empty.

Take any $f, g \in I_X$, then $\forall v \in M$

$$(f - g)(v) = f(v) - g(v) \in X.$$

So $f - g \in I_X$.

Suppose $h \in E$ and $f \in I_X$. Since X is fully invariant, then $h(X) \subseteq X$. Next, notice that

$$f(M) \subseteq X.$$

So we get:

$$h(f(M)) \subseteq h(X) \subseteq X,$$

and

$$f(h(M)) \subseteq f(M) \subseteq X.$$

This means that $hf \in I_X$ and $fh \in I_X$ for every $h \in E$ and $f \in I_X$. Therefore, I_X is an ideal in E. □

Based on Lemma 5.4, if there is an ideal $J \subseteq E$ in E so that $J(M) = X$, then $J \subseteq I_X$.

Next, we will discuss the fully invariant submodule of a module that decomposes as a finite direct sum of submodules.

THEOREM 5.5. *Let M be a finitely generated module over R and $M = \bigoplus_{i=1}^n M_i$. If $S \subset M$ is a fully invariant submodule of M, then $S = \bigoplus_{i=1}^n S_i$ with $S_i = S \cap M_i$ is a fully invariant submodule of M_i for all i.*

PROOF. Suppose $S \subset M$ is a fully invariant submodule of M, then $\bigoplus_{i=1}^n S_i \subseteq S$, where $S_i = S \cap M_i$. We will prove that $S \subseteq \bigoplus_{i=1}^n S_i$.

Suppose $i \in \{1, 2, \ldots, n\}$. Define $\bar{f}_i \in E$ with $\bar{f}_i(x) = \bar{f}_i(\sum_{j=1}^n x_j) = x_i \in M_i, \forall x = \sum_{j=1}^n x_j \in M$ with $x_j \in M_j$. Suppose $s \in S$, then $s = \sum_{j=1}^n m_j$ with $m_j \in M_j$. Since S is a fully invariant submodule of M, then $\bar{f}_i(s) = m_i \in S$ such that $m_i \in (S \cap M_i) = S_i$. Therefore,

$$s = \sum_{i=1}^n m_i \in \sum_{i=1}^n S_i = \bigoplus_{i=1}^n S_i.$$

Thus, $S \subseteq \bigoplus_{i=1}^n S_i$, so $S = \bigoplus_{i=1}^n S_i$.

Suppose $i \in \{1, 2, 3, \ldots, n\}$. We will prove that S_i is fully invariant in M_i.

Take any $g_i \in End_R(M_i)$ and $s_i \in S_i$, then define $\bar{g}_i \in E$ with $\bar{g}_i(\sum_{j=1}^n x_j) = g_i(x_i)$ for every $x = \sum_{j=1}^n x_j \in M$ with $x_j \in M_j$.

Since S is fully invariant, for every $s_i \in S_i \subset S$ we have $\bar{g}_i(S) \subseteq S$. Hence, $g_i(s_i) = \bar{g}_i(s_i) \in (S \cap M_i) = S_i$. Thus, $g_i(S_i) \subseteq S_i$. Therefore, S_i is fully invariant in M_i. □

Notice that $M = M_T \bigoplus M_F$, and as column vector $\begin{pmatrix} M_T \\ M_F \end{pmatrix}$. If $S \subseteq M$ is a fully invariant submodule of M, then by Theorem 5.5, the submodule $S = S_T \bigoplus S_F$ or $\begin{pmatrix} S_T \\ S_F \end{pmatrix}$ with S_F is fully invariant in M_F and S_T is fully invariant in M_T.

Notice that the converse of Theorem 5.5 is not true. For example, let $M = \begin{pmatrix} M_T \\ M_F \end{pmatrix}$, with M_T is a torsion module and $M_F = \bigoplus_{i=1}^{n} \langle v_i \rangle$ is a free module. And let S be a submodule of M with $S = \begin{pmatrix} 0 \\ M_F \end{pmatrix}$. Clearly, M_F is fully invariant in M_F and 0 is fully invariant in M_T.

We will show that S is not fully invariant. Let $x \in M_T, x \neq 0$, and for some $\mu \in \{1, 2, 3, \ldots, n\}$ define $\lambda_{x\mu} \in H_{FT} = Hom_R(M_F, M_T)$ as follows: for every $j \in \{1, 2, 3, \ldots, n\}$

$$\lambda_{x\mu}(v_j) = \begin{cases} x & \text{if } \mu = j \\ 0 & \text{if } \mu \neq j. \end{cases}$$

We get, $\lambda_{x\mu}(M_F) = \langle x \rangle \neq 0$. Then choose

$$h = \begin{pmatrix} 1 & \lambda_{x\mu} \\ 0 & 0 \end{pmatrix} \in E,$$

with $1 \in E_T$ is the identity map in M_T. Hence,

$$h(S) = \begin{pmatrix} 1 & \lambda_{x\mu} \\ 0 & 0 \end{pmatrix} \begin{pmatrix} 0 \\ M_F \end{pmatrix} = \begin{pmatrix} \langle x \rangle \\ 0 \end{pmatrix} \not\subseteq \begin{pmatrix} 0 \\ M_F \end{pmatrix}.$$

Therefore, S is not fully invariant.

LEMMA 5.6. *Let M_p be a primary module over a principal ideal domain R, X a submodule of M_p, and $I_X = \{f \in End_R(M_p) | f(M) \subseteq X\}$. If X is fully invariant, then $X = I_X M_p$.*

PROOF. Suppose X is a fully invariant submodule of M_p. We will prove that $X = I_X M_p$.

Note that

$$I_X M_p = \{\sum_{i=1}^{n} g_i x_i | g_i \in I_X, x_i \in M_p\}.$$

Since for every $f \in I_X$, we have $f(M) \subseteq X$, then $I_X M_p \subseteq X$.

Furthermore, since X is a fully invariant submodule of M_p, then I_X is an ideal in $End_R(M_p)$ and

$$X = \langle p^{s_1} u_1 \rangle \oplus \cdots \oplus \langle p^{s_m} u_m \rangle \tag{5.1}$$

with

$$0 \leq s_i - s_j \leq e_i - e_j \quad \text{for every} \quad i \leq j. \tag{5.2}$$

Next, notice that for every $i \in \{1, 2, 3, \ldots, m\}$

$$p^{s_i} \delta_{ii}(M_p) = p^{s_i} \langle u_i \rangle \subseteq X.$$

This means that there is a $p^{s_i} \delta_{ii} \in I_X$ so $\langle p^{s_i} u_i \rangle \subseteq I_X M_p$. Thus, $X \subseteq I_X M_p$. Therefore, $X = I_X M_p$. □

THEOREM 5.7. *Let M_p be a primary module over a principal ideal domain R. A submodule X of M_p is fully invariant if and only if there exists an ideal I of $End_R(M_p)$ such that $X = IM_p$.*

PROOF. Suppose X is a fully invariant submodule of M_p. We will prove that there is an ideal I in $End_R(M_p)$ so that $X = IM_p$.

Consider $I = I_X = \{f \in End_R(M_p) | f(M) \subseteq X\}$. Since X is a *fully invariant* submodule of M_p, then based on Lemma 5.4, I is ideal in $End_R(M_p)$. Thus, we get $X = I_X M_p = IM_p$.

On the other hand, suppose there is an ideal I in $End_R(M_p)$ such that $X = IM_p$. Then based on Lemma 5.3, X is a fully invariant submodule of M_p. □

COROLLARY 5.8. *If M_p is a primary module over R, then M_p as a left module over $End_R(M_p)$ is a multiplication module.*

PROOF. Suppose M_p is a primary module over a principal ideal domain R, then M_p is a left module over $End_R(M_p)$. Then, based on Theorem 5.2, any left submodule X over $End_R(M_p)$ is a fully invariant submodule of M_p. Based on Theorem 5.7, $X = IM_p$ with $I = \{f \in End_R(M_p) | f(M_p) \subseteq X\}$ is an ideal in $End_R(M_p)$. Therefore, M_p as a left module over $End_R(M_p)$ is a multiplication module. □

We will further investigate the effect of X as a fully invariant submodule on the structure of I_X.

LEMMA 5.9. *Let M_p be a primary module over a principal ideal domain R and*

$$X = \oplus_{i=1}^{m} \langle p^{s_i} u_i \rangle \tag{5.3}$$

a fully invariant submodule of M_p with

$$0 \le s_i - s_j \le e_i - e_j \quad \textit{for all} \quad i \le j. \tag{5.4}$$

If $I_X = \{f \in End_R(M_p) | f(M) \subseteq X\}$, then

$$I_X = \oplus_{i=1}^{m} \oplus_{j=1}^{m} \langle p^{k_{ij}} \delta_{ij} \rangle \tag{5.5}$$

with

$$k_{ij} = \begin{cases} s_i & \textit{untuk } 1 \le j \le i \\ max\{0, s_i - (e_i - e_j)\} & \textit{for } j > i. \end{cases}$$

PROOF. Suppose X is a fully invariant submodule of M_p, then $I_X M_p = X$. We will show that $I_X = \oplus_{i=1}^{m} \oplus_{j=1}^{m} \langle p^{k_{ij}} \delta_{ij} \rangle$ with

$$k_{ij} = \begin{cases} s_i & \text{for } 1 \le j \le i \\ max\{0, s_i - (e_i - e_j)\} & \text{for } j > i. \end{cases}$$

First, we will show that $\oplus_{i=1}^{m} \oplus_{j=1}^{m} \langle p^{k_{ij}} \delta_{ij} \rangle \subseteq I_X$, i.e., $p^{s_i} \delta_{ij} \in I_X$ for $1 \le j \le i$.

Suppose $i, j \in \{1, 2, 3, \dots m\}$ with $1 \le j \le i$. Take any $\mu \in \{1, 2, 3, \dots m\}$.
If $j \ne \mu$, then $p^{s_i} \delta_{ij}(u_\mu) = 0 \in X$.
If $j = \mu$, then $p^{s_i} \delta_{ij}(u_\mu) = p^{s_i} u_i \in X$.
Hence, $p^{s_i} \delta_{ij} \in I_X$ for $1 \le j \le i$.
Next, we will show that $p^{\max\{0, s_i - (e_i - e_j)\}} \delta_{ij} \in I_X$ for $j > i$.
Suppose $i, j \in \{1, 2, 3, \dots m\}$ with $j > i$.
If $j \ne \mu$, then $p^{\max\{0, s_i - (e_i - e_j)\}} \delta_{ij}(u_\mu) = 0 \in X$.

If $j = \mu$, then

$$p^{\max\{0,s_i-(e_i-e_j)\}}\delta_{ij}(u_\mu) = p^{\max\{0,s_i-(e_i-e_j)\}}p^{e_i-e_j}u_i.$$

Next, notice that if $s_i - (e_i - e_j) \geq 0$

$$p^{\max\{0,s_i-(e_i-e_j)\}}p^{e_i-e_j}u_i = p^{s_i}u_i \in X.$$

Meanwhile, if $s_i - (e_i - e_j) < 0$, then $s_i < (e_i - e_j)$ and

$$p^{\max\{0,s_i-(e_i-e_j)\}}p^{e_i-e_j}u_i = p^{e_i-e_j}u_i \in X.$$

Thus, $p^{\max\{0,s_i-(e_i-e_j)\}}\delta_{ij} \in I_X$ for $j > i$.

Hence, $\oplus_{i=1}^m \oplus_{j=1}^m \langle p^{k_{ij}}\delta_{ij}\rangle \subseteq I_X$.

We will show that $I_X \subseteq \oplus_{i=1}^m \oplus_{j=1}^m \langle p^{k_{ij}}\delta_{ij}\rangle$. Take any $f \in I_X$.

Suppose $f = \sum_{i=1}^m \sum_{j=1}^m \alpha_{ij}\delta_{ij}$ with $\alpha_{ij} \in R$.

We will show $p^{k_{ij}}|\alpha_{ij}, \forall i, j \in \{1, 2, 3, \ldots m\}$

Note that I_X is an ideal such that $\delta_{ii}f\delta_{hh} = \alpha_{ij}\delta_{ij} \in I_X$.

Next, if $1 \leq j \leq i$

$$\alpha_{ij}\delta_{ij}(u_j) = \alpha_{ij}(u_i) \in (X \cap \langle u_i\rangle) = \langle p^{s_i}u_i\rangle.$$

This means $p^{s_i}|\alpha_{ij}$. Since $s_i = k_{ij}$, then $p^{k_{ij}}|\alpha_{ij}$.

If $i < j$, then

$$\alpha_{ij}\delta_{ij}(u_j) = \alpha_{ij}p^{e_i-e_j}(u_i) \in (X \cap \langle u_i\rangle) = \langle p^{s_i}u_i\rangle.$$

Thus, $p^{s_i}|p^{e_i-e_j}\alpha_{ij}$. Note that $\max\{0, s_i - (e_i - e_j)\} = k_{ij}$.

If $s_i - (e_i - e_j) \geq 0$, then $s_i - (e_i - e_j) = k_{ij}$. This means $p^{k_{ij}}|\alpha_{ij}$.

If $s_i - (e_i - e_j) < 0$, then $k_{ij} = 0$. This means $s_i = e_i - e_j$, thus $p^{k_{ij}}|\alpha_{ij}$.

Therefore, $I_X \subseteq \oplus_{i=1}^m \oplus_{j=1}^m \langle p^{k_{ij}}\delta_{ij}\rangle$. □

Next, we will discuss the relationship between a fully invariant submodule in a torsion module M_T, which can be decomposed as primary modules and a fully invariant submodule of a primary module.

THEOREM 5.10. *Let M_T be a finitely generated torsion module over R. By (2.1)*

$$M_T = \begin{pmatrix} M_{p_1} \\ M_{p_2} \\ \cdots \\ M_{p_k} \end{pmatrix}$$

with M_{p_i} is a primary submodule. If X is a submodule of M_T, then X is a fully invariant submodule of M_T if and only if

$$X = \begin{pmatrix} X_1 \\ X_2 \\ \cdots \\ X_{p_k} \end{pmatrix}$$

with X_i is a fully invariant submodule of M_{p_i} for all $i \in \{1.2.3\ldots, k\}$.

PROOF. According to Theorem 5.5, if X is a fully invariant submodule of M_T, then $X = \begin{pmatrix} X_1 \\ X_2 \\ \cdots \\ X_k \end{pmatrix} = \bigoplus_{i=1}^k X_i$, where X_i is a fully invariant submodule of M_{p_i} for each i.

Now suppose for every $i \in \{1.2.3\ldots,k\}$, X_i is a fully invariant submodule of M_{p_i} and $f = diag(f_1, f_2, \ldots f_{p_k}) \in End_R(M_p)$ with $f_i \in End_R(M_{p_i})$. Then $f_i(X_i) \subseteq X_i$ and

$$f(X) = diag(f_1, f_2, \ldots f_{p_k})\begin{pmatrix} X_1 \\ X_2 \\ \ldots \\ X_{p_k} \end{pmatrix} = \begin{pmatrix} f_1(X_1) \\ f_2(X_2) \\ \ldots \\ f_{p_k}(X_{p_k}) \end{pmatrix} \subseteq \begin{pmatrix} X_1 \\ X_2 \\ \ldots \\ X_k \end{pmatrix}.$$

Therefore, X is a *fully invariant* submodule of M_T. □

COROLLARY 5.11. *Let M_T be a finitely generated torsion module over R that satisfies (2.1) and $E_T = End_R(M_T)$. If X is a submodule of M_T, then X is a fully invariant submodule of M_T if and only if there exists an ideal $I = diag(I_{X_1}, I_{X_2}, \ldots I_{X_k})$ of E_T with I_{X_i} is an ideal of $End_R(M_{p_i}), \forall i \in \{1, 2, 3, \ldots k\}$ such that*

$$X = IM_T = \begin{pmatrix} I_{X_1}M_{p_1} \\ I_{X_2}M_{p_2} \\ \ldots \\ I_{X_k}M_{p_k} \end{pmatrix}. \tag{5.6}$$

PROOF. According to Theorem 5.10, X is a fully invariant submodule of M_T if and only if

$$X = \begin{pmatrix} X_1 \\ X_2 \\ \ldots \\ X_k \end{pmatrix}$$

where X_i is a fully invariant submodule of M_{p_i} for each i.

Further, according to Theorem 5.7, X_i is fully invariant in M_{p_i} if and only if there is an ideal I_{X_i} in $End_R M_{p_i}$ such that $X_i = I_{X_i}M_{p_i}$. Then based on Theorem 3.1

$$I = diag(I_{X_1}, I_{X_2}, \ldots, I_{X_k})$$

where I_{X_i} is an ideal in $End_R(M_{p_i})$ if and only if $I \subseteq E_T$ is an ideal in E_T.

Therefore, X is a fully invariant submodule of M_T if and only if there is an $I = diag(I_{X_1}, I_{X_2}, \ldots I_{X_k})$ ideal in E_T such that

$$X = IM_T = \begin{pmatrix} I_{X_1}M_{p_1} \\ I_{X_2}M_{p_2} \\ \ldots \\ I_{X_k}M_{p_k} \end{pmatrix}.$$

□

Next, we will discuss the relationship between the ideal $I_X = \{f \in E_T | f(M_T) \subseteq X\}$ and the ideal I in the fully invariant submodule $X = IM_T$.

COROLLARY 5.12. *Let M_T be a finitely generated torsion module over R that satisfies (2.1) and $E_T = End_R(M_T)$. If*

$$X = \begin{pmatrix} X_1 \\ X_2 \\ \ldots \\ X_k \end{pmatrix}$$

is a fully invariant submodule of M_T *and*

$$I_X = \{f \in E_T | f(M_T) \subseteq X\}$$

then $I_X = diag(I_{X_1}, I_{X_2}, \ldots I_{X_k})$ *with* I_{X_i} *ideal of* $End_R(M_{p_i})$ *such that* $I_{X_i} M_{p_i} = X_i$.

PROOF. Let $X = \begin{pmatrix} X_1 \\ X_2 \\ \ldots \\ X_k \end{pmatrix}$ be fully invariant, then based on Corollary 5.11 there is an

$$I = diag(I_{X_1}, I_{X_2}, \ldots I_{X_k})$$

ideal in E_T where I_{X_i} is an ideal in $End_R(M_{p_i}), \forall i \in \{1, 2, 3, \ldots k\}$ so that $X = IM_T$. We will prove that $I \subseteq I_X$.

Notice that

$$IM_T = \{\sum_{i=1}^{n} f_i m_i | f_i \in I, m_i \in M_T\}.$$

So $I \subseteq I_X$.

We will prove $I_X \subseteq I$.

Take any $f = diag(f_1, f_2, \ldots f_k) \in I_X$ and $m = \begin{pmatrix} m_1 \\ m_2 \\ \ldots \\ m_k \end{pmatrix} \in M_T$ then

$$f(m) = \begin{pmatrix} f_1(m_1) \\ f_2(m_2) \\ \ldots \\ f_k(m_k) \end{pmatrix} \in X = \begin{pmatrix} X_1 \\ X_2 \\ \ldots \\ X_k \end{pmatrix}.$$

Thus, $f_i(m_i) \in X_i, \forall i \in \{1, 2, 3, \ldots k\}$. This means $f_i \in I_{X_i}, \forall i \in \{1, 2, 3, \ldots k\}$, so

$$f = diag(f_1, f_2, \ldots f_k) \in diag(I_{X_1}, I_{X_2}, \ldots I_{X_k}) = I$$

Therefore, $I_X \subseteq I$. □

Next, we will discuss the relationship between a fully invariant submodule of the torsion module M_T and a left submodule of M_T as a left module over E_T.

COROLLARY 5.13. *Let* M_T *be a finitely generated torsion module over* R*, then* M_T *as a left module over* E_T *is a multiplication module.*

PROOF. Suppose M_T is a torsion module over a principal ideal domain R, then M_T is a left module over E_T. Then, based on Theorem 5.2, the left submodule X over E_T is a fully invariant submodule of M_T. Based on Corollary 5.11, fully invariant submodule $X = IM_T$ with $I = \{f \in E_T | f(M_T) \subseteq X\}$ is an ideal in E_T. Therefore, M_T as a left module over E_T is a multiplication module. □

COROLLARY 5.14. *Let* M *be a finitely generated module over* R*,* $E = End_R(M)$*,* $M = \begin{pmatrix} M_T \\ M_F \end{pmatrix}$*,* $S \subseteq M$ *be a submodule of* M*, and*

$$I_S = \{f \in E | f(M) \subseteq S\}.$$

The submodule S *is a fully invariant submodule of* M *if and only if* $S = I_S M$ *with* I_S *is an ideal of* E*.*

PROOF. Consider a module $M = \begin{pmatrix} M_T \\ M_F \end{pmatrix}$ over a principal ideal domain R, $E = End_R(M)$, $I_S = \{f \in E | f(M) \subseteq S\}$ and $S = I_S M$ with I_S is an ideal in E. Since I_S is an ideal in E, then based on Lemma 5.3, S is a fully invariant submodule of M.

Otherwise, suppose $S \subseteq M$ is a fully invariant submodule of M. We will prove $S = I_S M$ with I_S is an ideal in E.

Since S is a submodule of M, then based on Lemma 5.4, I_S is an ideal in E.

Note that

$$I_S = \{f \in E | f(M) \subseteq S\},$$

which means $I_S M \subseteq S$.

Since $S \subseteq M$ is a fully invariant submodule, then based on Theorem 5.5, $S = \begin{pmatrix} S_T \\ S_F \end{pmatrix}$ with S_T fully invariant in M_T and S_F fully invariant in M_F. Thus, $S_T = I_{S_T}(M_T)$ and $S_F = I_{S_F}(M_F)$.

Further, notice that

$$\begin{pmatrix} I_{S_T} & 0 \\ 0 & 0 \end{pmatrix} \begin{pmatrix} M_T \\ M_F \end{pmatrix} = \begin{pmatrix} S_T \\ 0 \end{pmatrix} \subseteq S$$

and

$$\begin{pmatrix} 0 & 0 \\ 0 & I_{S_F} \end{pmatrix} \begin{pmatrix} M_T \\ M_F \end{pmatrix} = \begin{pmatrix} 0 \\ S_F \end{pmatrix} \subseteq S.$$

Meaning $\begin{pmatrix} 0 & 0 \\ 0 & I_{S_F} \end{pmatrix}, \begin{pmatrix} I_{S_T} & 0 \\ 0 & 0 \end{pmatrix} \subseteq I_S$.

While

$$S = \begin{pmatrix} S_T \\ S_F \end{pmatrix} = \begin{pmatrix} 0 \\ S_F \end{pmatrix} + \begin{pmatrix} S_T \\ 0 \end{pmatrix}.$$

Thus, $S \subseteq I_S M$. Therefore, $S = I_S M$. □

By Corollary 5.14, the fully invariant submodule $S = IM$ with $I = \{f \in E | f(M) \subseteq S\}$ is an ideal in E. Thus, if $I_S = \{f \in E | f(M) \subseteq S\} = \{0\}$, then $S = \{0\}$, and if $I = \{f \in E | f(M) \subseteq S\} = E$, then $S = M$.

COROLLARY 5.15. *Let M be a finitely generated module over R, then M as a left module over E is a multiplication module.*

PROOF. Suppose M is an R-module, then M is a left module over E. Furthermore, based on the Theorem 5.2, the left submodule S over E is a fully invariant submodule of M. Based on Corollary 5.14, the fully invariant submodule $S = IM$ with $I = \{f \in E | f(M) \subseteq S\}$ is an ideal in E. Therefore, M as a left module over E is a multiplication module. □

THEOREM 5.16. *Let M be a finitely generated module over R, and $M = \begin{pmatrix} M_T \\ M_F \end{pmatrix}$. A submodule $S \subseteq M$ is fully invariant if and only if $S = \begin{pmatrix} S_T \\ aM_F \end{pmatrix}$ for some $a \in R$, with S_T is fully invariant in M_T and $a(M_T) \subseteq S_T$.*

PROOF. Consider $M = \begin{pmatrix} M_T \\ M_F \end{pmatrix}$ and $S \subseteq M$, a fully invariant submodule of M. We will prove that $S = \begin{pmatrix} S_T \\ aM_F \end{pmatrix}$ for an $a \in R$ with S_T fully invariant in M_T and $a(M_T) \subseteq S_T$.

Based on Theorem 5.5, $S = \begin{pmatrix} S_T \\ S_F \end{pmatrix}$ with S_T fully invariant in M_T and S_F fully invariant in M_F.

Since S_F is fully invariant in M_F, then

$$S_F = aM_F$$

for some $a \in R$.

Furthermore, since S is fully invariant in M, then for every $f = \begin{pmatrix} f_1 & f_2 \\ 0 & f_3 \end{pmatrix} \in E$

$$f(S) = \begin{pmatrix} f_1 & f_2 \\ 0 & f_3 \end{pmatrix} \begin{pmatrix} S_T \\ aM_F \end{pmatrix} \subseteq S$$

with $f_1 \in E_T, f_3 \in E_F$, and $f_2 \in Hom_R(M_F, M_T)$.

Suppose $M_F = \bigoplus_{i=1}^{n} \langle v_i \rangle$. Take any $x \in M_T$, then for a $\mu \in \{1, 2, 3, \ldots, n\}$, define $\lambda_{x\mu} \in H_{FT}$ such that for every $i \in \{1, 2, 3, \ldots, n\}$

$$\lambda_{x\mu}(v_i) = \begin{cases} x & \text{if } \mu = i \\ 0 & \text{if } \mu \neq i. \end{cases}$$

Then, consider

$$h = \begin{pmatrix} 0 & \lambda_{x\mu} \\ 0 & 0 \end{pmatrix} \in E,$$

so that

$$h(S) = \begin{pmatrix} 0 & \lambda_{x\mu} \\ 0 & 0 \end{pmatrix} \begin{pmatrix} S_T \\ aM_F \end{pmatrix} = \begin{pmatrix} \langle ax \rangle \\ 0 \end{pmatrix} \subseteq \begin{pmatrix} S_T \\ aM_F \end{pmatrix}.$$

Since $\langle ax \rangle \subseteq S_T, \forall x \in M_T$, then $aM_T \subseteq S_T$.

On the other hand, if $S = \begin{pmatrix} S_T \\ aM_F \end{pmatrix}$ for an $a \in R$ with S_T fully invariant in M_T and $aM_T \subseteq S_T$, then for every $f \in E$,

$$f(S) = \begin{pmatrix} f_1 & f_2 \\ 0 & f_3 \end{pmatrix} \begin{pmatrix} S_T \\ aM_F \end{pmatrix} \subseteq S.$$

□

6. S-prime submodules

In this section we will explain the characterization of S-prime submodules in primary modules. Our characterization of S-prime submodules is developed in relation with the characterization of fully invariant submodules in terms of the cyclic decomposition of the module.

We will first discuss the relationship between an S-prime submodule M and a prime submodule of M as a module over E.

THEOREM 6.1. *Let M be a finitely generated module over R and $E = End_R(M)$. A submodule X is S-prime if and only if X is a prime submodule of M as a left module over E.*

PROOF. Let X be an S-prime submodule of M. We will prove that X is a prime submodule of M. Since X is an S-prime submodule of M, then $X \neq M$ and is a fully invariant submodule of M. Thus, $X \neq M$ and is a left E-submodule of M.

Take any $f \in E$ and $u \in M$ with

$$fE(u) = fEu \subseteq X.$$

We will prove $f(M) \subseteq X$ or $u \in X$.

Consider $I = E.f.E$ is an ideal in E and $U = E.u$ as a left E-submodule of M. Next, notice that

$$\begin{aligned} IU &= (EfE)(Eu) \\ &\subseteq (EfEu) \\ &\subseteq EX \subseteq X. \end{aligned}$$

Because X is an S-prime submodule of M, $I(U) \subseteq X$ with ideal I in E and U a fully invariant left E-submodule of M, then $IM \subseteq X$ or $U \subseteq X$. Because E contains the identity element, then $f(M) \subseteq X$ or $u \in X$. Hence, X is a prime submodule of M as an E-module.

Otherwise, suppose that X is a prime submodule of M as an E-module. We will prove that X is an S-prime submodule of R-module M.

Since X is a prime submodule of E-module M, then $X \neq M$ and X is an E-left submodule of M. Thus, $X \neq M$ and is a fully invariant submodule of M.

Take any ideal I of E and a fully invariant submodule U of M with $IU \subseteq X$ and $U \not\subseteq X$. We will prove $IM \subseteq X$, i.e., $f(m) \in X, \forall f \in I, m \in M$.

Because $U \not\subseteq X$, there is an $u \in U$ and $u \notin X$.

Take any $f \in I$, then notice that

$$f.E \subseteq I.E \subseteq I$$

so that

$$(f.E)u \subseteq Iu \subseteq IU \subseteq X.$$

Since X is a prime submodule of M as a left module of E, then $f(M) \subseteq X$. Thus, $IM \subseteq X$. Therefore, X is an S-prime submodule. □

Note that if R is considered as an R-module, then R is isomorphic to $End_R(R)$. Thus, based on Theorem 6.1, on R as an R-submodule, X is an S-prime submodule if and only if X is a prime submodule. Next, we will discuss the relationship between an S-prime submodule of module M and a prime ideal in E.

LEMMA 6.2. *Let M be a finitely generated module over R, $E = End_R(M)$, X a fully invariant submodule of M, and*

$$I_X = \{f \in E | f(M) \subseteq X\}.$$

The submodule X is S-prime if and only if I_X is a prime ideal of E with $I_X M = X$ and $X \neq M$.

PROOF. Suppose X is S-prime, we will prove that I_X is a prime ideal of E.

Since X is an S-prime submodule, then $X \neq M$ and X is a fully invariant submodule of M. Based on Corollary 5.14, $X = I_X M$ with $I_X = \{f \in E | f(M) \subseteq X\}$, which is an ideal in E. Next, take any ideals I and J in E so that $IJ \subseteq I_X$.

We will prove $I \subseteq I_X$ or $J \subseteq I_X$.

Since $IJ \subseteq I_X$, then

$$IJM \subset I_X M = X.$$

Note that J is an ideal in E, so JM is a *fully invariant* submodule of M.

Whereas X is an S-prime submodule, if $IJM \subset X$, then $IM \subset X$ or $JM \subset X$. This means $I \subseteq I_X$ or $J \subseteq I_X$. So I_X is a prime ideal in E.

Otherwise, suppose I_X is a prime ideal of E where $X = I_X M$ and $X \neq M$. We will prove that X is an S-prime submodule of M.

Since $X = I_X M \neq M$, then X is fully invariant in M.

Take any fully invariant submodule U of M and an ideal I in E with $IU \subseteq X$. We will prove $IM \subseteq X$ or $U \subseteq X$.

Since U is fully invariant, then $I_U = \{f \in E | f(M) \subseteq U\}$ is an ideal in E such that $I_U M = U$. Thus, we get that

$$(I.I_U)M = IU \subseteq X.$$

This means $I.I_U \subseteq I_X = \{f \in E | f(M) \subseteq X\}$.

Since I_X is a prime ideal, then $I \subseteq I_X$ or $I_U \subseteq I_X$. This means $IM \subseteq X$ or $U = I_U M \subseteq X$.

Therefore, X is an S-prime submodule of M. □

In a similar way, Lemma 6.2 applies to free modules and torsion modules generated over R.

THEOREM 6.3. *Let $M = \begin{pmatrix} M_T \\ M_F \end{pmatrix}$, S be a fully invariant submodule of M. The submodule S is S-prime if and only if $S = \begin{pmatrix} M_T \\ S_F \end{pmatrix}$ with S_F is an S-prime submodule of M_F and*

$$I_S = \begin{pmatrix} E_T & H_{FT} \\ 0 & I_{22} \end{pmatrix}$$

with I_{22} as a prime ideal in E_F.

PROOF. Consider $M = \begin{pmatrix} M_T \\ M_F \end{pmatrix}$ and S is an S-prime submodule of M. We will prove that $S = \begin{pmatrix} M_T \\ S_F \end{pmatrix}$ with S_F is an S-prime submodule in M_F and $I_S = \begin{pmatrix} E_T & H_{FT} \\ 0 & I_{22} \end{pmatrix}$ with I_{22} is a prime ideal in E_F.

Since S is an S-prime submodule of M, then $S \neq M$ and S is a fully invariant submodule of M.

Then, based on Lemma 6.2, $I_S = \{f \in E | f(M) \subseteq S\}$ is a prime ideal of E. Furthermore, based on Theorem 4.5,

$$I_S = \begin{pmatrix} I_{11} & H_{FT} \\ 0 & E_F \end{pmatrix}$$

with I_{11} is a prime ideal in E_T or

$$I_S = \begin{pmatrix} E_T & H_{FT} \\ 0 & I_{22} \end{pmatrix}$$

with I_{22} is a prime ideal in E_F.

Since S is a fully invariant submodule of M, then according to Theorem 5.5, $S = \begin{pmatrix} S_T \\ S_F \end{pmatrix}$ with S_T is fully invariant in M_T and S_F is fully invariant in M_F. Then, based on Corollary 5.14, we get

$$I_S M = S.$$

Thus,

$$\begin{pmatrix} I_{11} & H_{FT} \\ 0 & E_F \end{pmatrix} \cdot \begin{pmatrix} M_T \\ M_F \end{pmatrix} = \begin{pmatrix} S_T \\ S_F \end{pmatrix} \tag{6.1}$$

or

$$\begin{pmatrix} E_T & H_{FT} \\ 0 & I_{22} \end{pmatrix} \cdot \begin{pmatrix} M_T \\ M_F \end{pmatrix} = \begin{pmatrix} S_T \\ S_F \end{pmatrix}. \tag{6.2}$$

In Equation (6.1), $S_T = I_{11}(M_T) + H_{FT}(M_F) = M_T$ and $E_F(M_F) = M_F$. Thus, $S = I_S M = M$. Since S is a fully invariant submodule, then $S \neq M$. Thus, the prime ideal I_S is

$$I_S = \begin{pmatrix} E_T & H_{FT} \\ 0 & I_{22} \end{pmatrix}$$

with I_{22} is a prime ideal in E_F.

In Equation (6.2), $S_T = E_T(M_T) + H_{FT}(M_F) = M_T$. Because I_{22} is a prime ideal in E_F, then, based on Lemma 6.2, $I_{22}(M_F) = S_F$ is an S-prime submodule of M_F.

Hence, $S = \begin{pmatrix} M_T \\ S_F \end{pmatrix}$ with S_F as an S-prime submodule in M_F.

Otherwise, suppose $S = \begin{pmatrix} M_T \\ S_F \end{pmatrix}$ is a fully invariant submodule of M with S_F as an S-prime submodule in M_F and

$$I_S = \begin{pmatrix} E_T & H_{FT} \\ 0 & I_{22} \end{pmatrix}$$

with I_{22} is a prime ideal in E_F.

We will prove that S is an S-prime submodule of M.

Based on Theorem 4.5,

$$I_S = \begin{pmatrix} E_T & H_{FT} \\ 0 & I_{22} \end{pmatrix}$$

is a prime ideal in E, where I_{22} is a prime ideal in E_F.

Next, notice that $I_S M = S$ and $S \neq M$. Thus, based on Lemma 6.2, S is an S-prime submodule of M. Therefore, S is an S-prime submodule of M. □

THEOREM 6.4. *Let M_T be a finitely generated torsion module over R,*

$$M_T = \begin{pmatrix} M_{p_1} \\ M_{p_2} \\ \dots \\ M_{p_k} \end{pmatrix}$$

where $\forall i \in \{1, 2, 3 \dots, k\}$, M_{p_i} is a primary submodule of M_T. If

$$X = \begin{pmatrix} X_1 \\ X_2 \\ \dots \\ X_k \end{pmatrix}$$

is fully invariant submodule of M_T, then X is an S-prime submodule of M_T if and only if there exists a unique $\mu \in \{1, 2, \dots k\}$ such that X_μ is an S-prime submodule of M_{p_μ}, and $X_i = M_{p_i}$ for all $i \in \{1, 2, \dots k\}, i \neq \mu$.

PROOF. Suppose M_T is a finitely generated torsion module over a principal ideal domain R and X is an S-prime submodule of M_T. We will prove that X is an S-prime submodule of M_T if and only if there is exactly one $\mu \in \{1, 2, \ldots k\}$ such that X_μ is an S-prime submodule of M_{p_μ}, and $X_i = M_{p_i}$ for every $i \in \{1, 2, \ldots k\}, i \neq \mu$.

Based on Lemma 6.2, X is an S-prime submodule of M_T if and only if I_X is a prime ideal of E_T.

Then, by Corollary 3.3, $I_X \subseteq E_T$ is a prime ideal if and only if

$$I_X = diag(I_{X_1}, I_{X_2}, \ldots, I_{X_k})$$

where there is exactly one $\mu \in \{1, 2, \ldots k\}$ such that $I_{X_\mu} \neq End_R(M_{p_\mu})$ is a prime ideal in $End_R(M_{p_\mu})$ and $I_{X_i} = End_R(M_{p_i})$ for every $i \in \{1, 2, \ldots k\}, i \neq \mu$. Thus,

$$X = \begin{pmatrix} X_1 \\ X_2 \\ \ldots \\ X_\mu \\ \ldots \\ X_k \end{pmatrix} = \begin{pmatrix} End_R(M_{p_1})M_{p_1} \\ End_R(M_{p_2})M_{p_2} \\ \ldots \\ I_{X_\mu} M_{p_\mu} \\ \ldots \\ End_R(M_{p_k})M_{p_k} \end{pmatrix}.$$

Since I_{X_μ} is a prime ideal in $End_R(M_{p_\mu})$ and $I_{X_\mu} M_{p_\mu} = X_\mu$, then, based on Lemma 6.2, X_μ is an S-prime submodule of M_{p_μ}. Then, for every $i \neq \mu$ $X_i = End_R(M_{p_i})M_{p_i}$, hence $X_i = M_{p_i}$.

Hence, X is an S-prime submodule of M_T if and only if there is exactly one $\mu \in \{1, 2, \ldots k\}$, such that X_μ is an S-prime submodule of M_{p_μ} and $X_i = M_{p_i}$ for every $i \in \{1, 2, \ldots k\}, i \neq \mu$. □

THEOREM 6.5. *Let M be a finitely generated module over R. If a submodule X is S-prime, then X is a prime submodule of M.*

PROOF. Let X be an S-prime submodule of M. We will prove that X is a prime submodule of M.

Since X is an S-prime submodule of M, then according to Theorem 6.3, the submodule $X = \begin{pmatrix} M_T \\ S_F \end{pmatrix}$ with S_F as an S-prime submodule in M_F. Since S_F is an S-prime submodule in M_F, then the submodule S_F is prime in M_F.

Suppose $r \in R$ and $m \in M$ with $rm \in X$, we will prove $rM \subseteq X$ or $m \in X$.

Note that

$$m = \begin{pmatrix} m_1 \\ m_2 \end{pmatrix} \in \begin{pmatrix} M_T \\ M_F \end{pmatrix} = M$$

and

$$rm = \begin{pmatrix} rm_1 \\ rm_2 \end{pmatrix} \in X = \begin{pmatrix} M_T \\ S_F \end{pmatrix}$$

with $rm_1 \in M_T$ and $rm_2 \in S_F$. Because submodule S_F is prime, then $rM_F \subseteq S_F$ or $m_2 \in S_F$.

Suppose $rM_F \subseteq S_F$ and $m' \in M$. Next, note that

$$rm' = \begin{pmatrix} rm'_1 \\ rm'_2 \end{pmatrix} \in \begin{pmatrix} rM_T \\ rM_F \end{pmatrix} \subseteq \begin{pmatrix} M_T \\ S_F \end{pmatrix}.$$

This means $rM \subseteq X$.

Next, suppose $m_2 \in S_F$, then

$$m = \begin{pmatrix} m_1 \\ m_2 \end{pmatrix} \in \begin{pmatrix} M_T \\ S_F \end{pmatrix} = X.$$

This means $m \in X$. Thus, $rM \subseteq X$ or $m \in X$. So X is a prime submodule of M. The opposite does not apply because not all prime submodules are fully invariant submodules. □

References

[1] H. A. Alolaiyan and A. M. Alghamdi, *On the Endomorphism Ring of a Noetherian Chain Module*, British Journal of Mathematics and Computer Science, **14(1)**(BJMCS.23383), 1–7, 2016.

[2] J. Dauns, *Prime modules*, J. Reine Angew. Math. **298** (1978), 156–181, DOI 10.1515/crll.1978.298.156. MR498715

[3] R. L. McCasland and P. F. Smith, *Prime submodules of Noetherian modules*, Rocky Mountain J. Math. **23** (1993), no. 3, 1041–1062, DOI 10.1216/rmjm/1181072540. MR1245463

[4] N. V. Sanh, N. A. Vu, K. F. U. Ahmed, S. Asawasamrit, and L. P. Thao, *Primeness in module category*, Asian-Eur. J. Math. **3** (2010), no. 1, 145–154, DOI 10.1142/S1793557110000106. MR2654814

[5] N. V.Sanh, O. Arunphalungsanti, S. Chairat, and N. T. Bach, *On Fully Bounded Noetherian Modules and Their Endomorphism Rings*, The Commission on Higher Education, (RG-1-53-13-1).

Faculty of Mathematics and Natural Sciences, Institut Teknologi Bandung, Bandung 40132, Indonesia

Current address: Department of Telecommunication Engineering, Telkom University, Bandung 40257, Indonesia

Email address: khaerudin@telkomuniversity.ac.id

Faculty of Mathematics and Natural Sciences, Institut Teknologi Bandung, Bandung 40132, Indonesia

Email address: ntan@itb.ac.id

Faculty of Mathematics and Natural Sciences, Institut Teknologi Bandung, Bandung 40132, Indonesia

Email address: pudji@itb.ac.id

Contemporary Mathematics
Volume **826**, 2025
https://doi.org/10.1090/conm/826/16582

On the co-annihilating ideal graph of commutative rings

Mohd Nazim, Nadeem ur Rehman, Nazim, and Shabir Ahmad Mir

ABSTRACT. In this research article, we consider a commutative ring with unity denoted as $\mathcal{S}$ and $A(\mathcal{S})$ as the collection of all annihilating ideals of $\mathcal{S}$. We introduce a graph called the co-annihilating ideal graph of $\mathcal{S}$, denoted as $\mathbb{CAG}(\mathcal{S})$, with vertices taken from the set $A(\mathcal{S})^*$ (including all non-zero annihilating ideals of $\mathcal{S}$) and in this graph two distinct vertices $\mathcal{X}$ and $\mathcal{Y}$ are adjacent if and only if $\mathcal{X} \nsubseteq \mathcal{X}\mathcal{Y}$ and $\mathcal{Y} \nsubseteq \mathcal{X}\mathcal{Y}$. We establish the result that $\mathbb{CAG}(\mathcal{S})$ is connected with a diameter of at most two and a girth of at most four, if $\mathbb{CAG}(\mathcal{S})$ contains a cycle. Additionally, we identify rings in which the co-annihilating ideal graph takes the form of either a complete graph or a star graph. Furthermore, we undertake the task of classifying all Artinian rings $\mathcal{S}$ where the co-annihilating ideal graph $\mathbb{CAG}(\mathcal{S})$ can be matched with well-known existing graphs.

Introduction

The interaction between combinatorial objects and algebraic structures generates intriguing problems in the realm of both algebra and combinatorics. This has led to a highly active area of study focused on exploring graphs in connection with rings. Within the scope of ring theory, a multitude of research works employ combinatorial techniques to unveil valuable algebraic discoveries. In the context of this work, let's denote by $\mathcal{S}$ a unitary commutative ring that is not an integral domain. Throughout this study, $Z(\mathcal{S})$, $\mathbb{I}(\mathcal{S})$, $Nil(\mathcal{S})$, and $Min(\mathcal{S})$ stand for the sets encompassing zero-divisors, ideals, nilpotent elements, and minimal prime ideals of $\mathcal{S}$, respectively. For a subset T of $\mathcal{S}$, let $T^* = T \setminus \{0\}$. An annihilating ideal refers to an ideal with a non-zero annihilator, and $A(\mathcal{S})$ represents the set of such annihilating ideals pertaining to $\mathcal{S}$. A ring $\mathcal{S}$ is considered reduced if it lacks non-zero nilpotent elements. A ring $\mathcal{S}$ is called a decomposition ring if it can be written as $\mathcal{S} = \mathcal{S}_1 \oplus \mathcal{S}_2$, where $\mathcal{S}_1$ and $\mathcal{S}_2$ both are rings. For further clarification regarding any unclear notations or terminology in the realm of ring theory, we refer the reader to [**8**].

Consider a simple graph denoted as G, where $V(G)$ represents the set of vertices and $E(G)$ stands for the set of edges. In the context of a connected graph G, the shortest path between vertices u and v is referred to as the *distance* $d_G(u, v)$. The diameter of G, denoted as $diam(G)$, is the largest distance observed between any

2020 *Mathematics Subject Classification.* Primary 13A15, 05C10, 05C12, 05C25.

Key words and phrases. Zero-divisor graph, annihilating-ideal graph, complete graph, planar graph, genus of a graph.

pair of vertices u and v within the graph. The length of shortest cycle in G is called the *girth* of G, denoted by $gr(G)$. A vertex that shares an edge with all other vertices in G is termed a *universal vertex*. When a graph G possesses a *universal vertex* without any additional edges, it is categorized as a *star graph*. A *clique* in G is defined as a maximal complete subgraph, and the number of vertices in the largest clique of G is denoted as $\omega(G)$ (clique number). The chromatic number of G, represented as $\chi(G)$, signifies the minimum number of colors required to properly color all vertices of G, ensuring that adjacent vertices possess distinct colors. It's evident that $\chi(G)$ is always greater than or equal to $\omega(G)$. In the realm of graph terminology, K_n denotes a complete graph with n vertices, K_∞ signifies an infinite complete graph, P_n refers to a path containing n vertices, and $K_{m,n}$ represents a complete bipartite graph featuring partite sets of size m and n.

A *tree* is a type of connected graph, referred to as G, that lacks any cycles. A unicycle graph pertains to a graph G encompassing just one cycle. When it comes to *split graphs*, they are graphs in which their vertex set can be neatly divided into a *clique* and an *independent set*. If a graph G can be depicted in a way that its edges only intersect at their endpoints, it is categorized as a planar graph. Graph subdivision, on the other hand, involves generating a new graph by substituting edges with paths that do not intersect except at their endpoints. In 1930, *Kuratowski* established a relatively straightforward criterion for identifying planar graphs. As per *Kuratowski's Theorem*, a graph G is regarded as planar if and only if it doesn't include any instances of subdivided K_5 or $K_{3,3}$. An *outerplanar graph* signifies a planar graph where its vertices are positioned on the unbounded regions of the drawing. For more detailed explanations of unfamiliar notations and graph theory terminology, interested individuals can refer to sources [**20**] and [**21**]. Numerous scholarly works have addressed the association of graphs with mathematical structures like rings, groups, semigroups, and modules, as evidenced by publications such as [**1–7, 9–12, 14–17**] have explored this concept extensively. Notably, Beck [**9**] is credited as the originator of the notion of the zero-divisor graph for a commutative ring $\mathcal{S}$. In this context, the primary focus was on colorings. Beck's approach involved associating each element of the ring with a graph vertex. The exploration into the coloring of commutative rings continued with the work of Anderson and Naseer [**7**]. Additionally, Anderson and Livingston [**6**] introduced a graph denoted as $\Gamma(\mathcal{S})$, linked to the ring $\mathcal{S}$. Here, the vertices of the graph were drawn from $Z(\mathcal{S})^*$, the non-zero elements of the ring. Vertices x and y were considered adjacent if and only if their product xy equated to zero, with x and y being distinct elements of $Z(\mathcal{S})^*$. Another graph, known as the annihilating-ideal graph $\mathbb{AG}(\mathcal{S})$, was described by Behboodi and Rakeei [**10**]. This graph's vertices were elements from $A(\mathcal{S})^*$, and two vertices $\mathcal{X}$ and $\mathcal{Y}$ were linked if and only if their product $\mathcal{XY}$ equaled zero (see also [**11**]). This concept was further explored in [**10**]. Building upon these ideas, Salehifar et al. [**18**] introduced the concept of the annihilator-ideal graph $A_I(\mathcal{S})$ for a commutative ring. The vertex set of this graph was $A(\mathcal{S})^*$, and two vertices $\mathcal{X}$ and $\mathcal{Y}$ were adjacent if and only if $Ann(\mathcal{XY}) \neq Ann(\mathcal{X}) \cup Ann(\mathcal{Y})$, where $Ann(\mathcal{X}) = \{\mathcal{Y} \in \mathbb{I}(\mathcal{S}) : \mathcal{XY} = 0\}$ for $\mathcal{X} \in A(\mathcal{S})^*$. It's worth noting that $\mathbb{AG}(\mathcal{S})$ is encompassed within $A_I(\mathcal{S})$ as a subgraph.

In this paper, we introduced *the co-annihilating ideal* graph of $\mathcal{S}$, denoted by $\mathbb{CAG}(\mathcal{S})$, which is an undirected graph with vertices $A(\mathcal{S})^*$, and two distinct vertices $\mathcal{X}$ and $\mathcal{Y}$ are adjacent if and only if $\mathcal{X} \nsubseteq \mathcal{X}\mathcal{Y}$ and $\mathcal{Y} \nsubseteq \mathcal{X}\mathcal{Y}$. Each edge (path) in $\mathbb{AG}(\mathcal{S})$ is clearly a path (edge) in $\mathbb{CAG}(\mathcal{S})$. The second section contains some preliminary information as well as an examination of some basic properties of the graph $\mathbb{CAG}(\mathcal{S})$. In the third section, we determine Artinian rings $\mathcal{S}$ for which $\mathbb{CAG}(\mathcal{S})$ is isomorphic to familiar graphs like trees, unicycle graphs, split graphs, planar graphs, or an outerplanar graphs. Additionally, we address the genus of $\mathbb{CAG}(\mathcal{S})$ in the concluding part.

1. Basic properties of co-annihilating ideal graph

The essential properties of $\mathbb{CAG}(\mathcal{S})$ are examined in this section. We begin with a lemma that contains a number of $\mathbb{CAG}(\mathcal{S})$ features.

LEMMA 1.1. *Assume $\mathcal{S}$ is a commutative ring, and $\mathcal{X}, \mathcal{Y} \in A(\mathcal{S})^*$. The following assertions are valid.*

(1) *$\mathcal{X}$ and $\mathcal{Y}$ are not connected in $\mathbb{CAG}(\mathcal{S})$ if and only if $\mathcal{X} = \mathcal{X}\mathcal{Y}$ or $\mathcal{Y} = \mathcal{X}\mathcal{Y}$.*
(2) *If $\mathcal{X} \nsubseteq \mathcal{Y}$ and $\mathcal{Y} \nsubseteq \mathcal{Y}$, then $\mathcal{X}$ and $\mathcal{Y}$ are adjacent in $\mathbb{CAG}(\mathcal{S})$.*
(3) *If $\mathcal{X}$ and $\mathcal{Y}$ are connected in $\mathbb{AG}(\mathcal{S})$, this also implies a connection between them in $\mathbb{CAG}(\mathcal{S})$.*
(4) *When $\mathcal{X}$ and $\mathcal{Y}$ are linked in the context of $A_I(\mathcal{S})$, this linkage extends to a connection between them in $\mathbb{CAG}(\mathcal{S})$.*

PROOF. (1) This becomes evident by examining the definition of $\mathbb{CAG}(\mathcal{S})$.

(2) Assume that $\mathcal{X}$ and $\mathcal{Y}$ are not connected in $\mathbb{CAG}(\mathcal{S})$, then $\mathcal{X} = \mathcal{X}\mathcal{Y}$ or $\mathcal{Y} = \mathcal{X}\mathcal{Y}$ by part (1). Thus $\mathcal{X}\mathcal{Y} = \mathcal{X} \nsubseteq \mathcal{Y}$ or $\mathcal{X}\mathcal{Y} = \mathcal{Y} \nsubseteq \mathcal{X}$, leading to a contradiction.

(3) Assume that $\mathcal{X}$ and $\mathcal{Y}$ are connected in $\mathbb{AG}(\mathcal{S})$, then $\mathcal{X}\mathcal{Y} = (0)$, which implies that $\mathcal{X} \nsubseteq \mathcal{X}\mathcal{Y}$ and $\mathcal{Y} \nsubseteq \mathcal{X}\mathcal{Y}$. Hence $\mathcal{X}$ and $\mathcal{Y}$ are also connected in $\mathbb{CAG}(\mathcal{S})$.

(4) Suppose $\mathcal{X}$ and $\mathcal{Y}$ are connected in $A_I(\mathcal{S})$, then $Ann(\mathcal{X}\mathcal{Y}) \neq Ann(\mathcal{X}) \cup Ann(\mathcal{Y})$, which implies that there exists $\mathcal{K} \in \mathbb{I}(\mathcal{S})^*$ such that $\mathcal{K}\mathcal{X}\mathcal{Y} = 0$, $\mathcal{K}\mathcal{X} \neq 0$ and $\mathcal{K}\mathcal{Y} \neq 0$. Suppose on contrary that $\mathcal{X}$ and $\mathcal{Y}$ are not adjacent in $\mathbb{CAG}(\mathcal{S})$, then $\mathcal{X} = \mathcal{X}\mathcal{Y}$ or $\mathcal{Y} = \mathcal{X}\mathcal{Y}$ by part (1). Thus, $\mathcal{K}\mathcal{X} = \mathcal{K}\mathcal{X}\mathcal{Y} = 0$ or $\mathcal{K}\mathcal{Y} = \mathcal{K}\mathcal{X}\mathcal{Y} = 0$, leading to a contradiction. □

It is clear from the following example that the three graphs $\mathbb{AG}(\mathcal{S})$, $A_I(\mathcal{S})$ and $\mathbb{CAG}(\mathcal{S})$ may or may not be identical.

EXAMPLE 1.2.

(1) If $\mathcal{S} = \mathcal{F}_1 \times \mathcal{F}_2$, where $\mathcal{F}_1$, $\mathcal{F}_2$ are fields. Then $\mathbb{AG}(\mathcal{S}) = A_I(\mathcal{S}) = \mathbb{CAG}(\mathcal{S}) = K_2$.
(2) If $\mathcal{S} = \mathcal{F}_1 \times \mathcal{F}_2 \times \mathcal{F}_3$, where $\mathcal{F}_i$ for each $i = 1, 2, 3$ is a field, then $\mathbb{AG}(\mathcal{S})$ and $A_I(\mathcal{S}) = \mathbb{CAG}(\mathcal{S})$ are given in Figure 1.
(3) If $\mathcal{S} = \mathbb{Z} \times \mathbb{Z}$, where $\mathbb{Z}$ is the ring of integer, then by [**18**, Theorem 3.4], $\mathbb{AG}(\mathcal{S}) = A_I(\mathcal{S})$. Also, $2\mathbb{Z} \times (0) \sim 3\mathbb{Z} \times (0)$ is an edge of $\mathbb{CAG}(\mathcal{S})$ but not an edge of $A_I(\mathcal{S})$, which implies that $\mathbb{CAG}(\mathcal{S}) \neq A_I(\mathcal{S}) = \mathbb{AG}(\mathcal{S})$.

In the following result, we demonstrate that $\mathbb{CAG}(\mathcal{S})$ is always a connected graph and that $diam(\mathbb{CAG}(\mathcal{S})) \leq 2$. Furthermore, we show that if $\mathbb{CAG}(\mathcal{S})$ has a cycle, then $gr(\mathbb{WAG}(\mathcal{S})) \leq 4$.

PROPOSITION 1.3. *Let $\mathcal{S}$ be a commutative ring. Then $\mathbb{CAG}(\mathcal{S})$ is a connected graph with $diam(\mathbb{CAG}(\mathcal{S})) \leq 2$. Moreover, $gr(\mathbb{CAG}(\mathcal{S})) \leq 4$ if $\mathbb{CAG}(\mathcal{S})$ contains a cycle.*

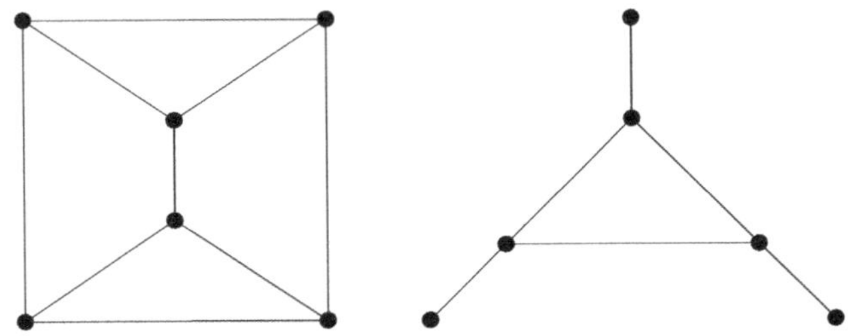

$\mathbb{CAG}(\mathcal{F}_1 \times \mathcal{F}_2 \times \mathcal{F}_3) = A_I(\mathcal{F}_1 \times \mathcal{F}_2 \times \mathcal{F}_3)$ $\mathbb{AG}(\mathcal{F}_1 \times \mathcal{F}_2 \times \mathcal{F}_3)$

FIGURE 1

PROOF. The proof is clear from [**16**, Theorem 2]. □

In the next two results, we discuss the case for which $gr(\mathbb{CAG}(\mathcal{S})) = 3$ or 4.

THEOREM 1.4. *Let $\mathcal{S}$ be a reduced commutative ring with $A(\mathcal{S})$ is an ideal of $\mathcal{S}$. Then $gr(\mathbb{CAG}(\mathcal{S})) = 3$.*

PROOF. Suppose $\mathcal{X} \in A(\mathcal{S})^*$ and $\mathcal{Y} \in Ann(\mathcal{X})^*$. Then $\mathcal{X} + \mathcal{Y} \in A(\mathcal{S})^*$, as $A(\mathcal{S})$ is an ideal of $\mathcal{S}$. Since $\mathcal{X}(\mathcal{X} + \mathcal{Y}) = \mathcal{X}^2 \neq 0$, we deduce that $\mathcal{X}$ and $\mathcal{X} + \mathcal{Y}$ are not adjacent in $\mathbb{AG}(\mathcal{S})$. If $\mathcal{X} + \mathcal{Y} = \mathcal{X}^2$, then $\mathcal{Y}(\mathcal{X} + \mathcal{Y}) = \mathcal{X}^2\mathcal{Y} = 0$. This implies that $\mathcal{Y}^2 = 0$, leading to a contradiction. Also, if $\mathcal{X} = \mathcal{X}^2$, then by utilizing Brauer's Lemma [**13**, page 172, Lemma 10.22], $\mathcal{S}$ is decomposable. This implies that $A(\mathcal{S})$ is not an ideal of $\mathcal{S}$, a contradiction. Hence $\mathcal{X} \neq \mathcal{X}^2$. Thus, $\mathcal{X} + \mathcal{Y} \sim \mathcal{X}$ is an edge of $\mathbb{CAG}(\mathcal{S})$. Similarly, one can show that $\mathcal{X} + \mathcal{Y} \sim \mathcal{Y}$ is an edge of $\mathbb{CAG}(\mathcal{S})$. Thus, $\mathcal{X} \sim (\mathcal{X} + \mathcal{Y}) \sim \mathcal{Y} \sim \mathcal{X}$ is a triangle in $\mathbb{CAG}(\mathcal{S})$. Hence $gr(\mathbb{CAG}(\mathcal{S})) = 3$. □

REMARK 1.5. If $\mathcal{S} = \mathbb{Z} \times \mathbb{Z}$, where $\mathbb{Z}$ is the ring of integer, then $\mathcal{S}$ is a reduced ring with $|Min(\mathcal{S})| = 2$ but $\mathbb{AG}(\mathcal{S}) \neq \mathbb{CAG}(\mathcal{S})$.

THEOREM 1.6. *Assume that $\mathcal{S}$ be a reduced commutative ring with $Min(\mathcal{S}) = \{\mathcal{P}_1, \mathcal{P}_2\}$. If $\mathcal{X} = \mathcal{X}\mathcal{Y}$ or $\mathcal{Y} = \mathcal{X}\mathcal{Y}$ for each $\mathcal{X}, \mathcal{Y} \in \mathcal{P}_1$ or $\mathcal{P}_2$, then $\mathbb{CAG}(\mathcal{S}) = \mathbb{AG}(\mathcal{S})$. Moreover, $gr(\mathbb{CAG}(\mathcal{S})) \in \{4, \infty\}$.*

PROOF. Since $\mathcal{S}$ is a reduced ring with $|Min(\mathcal{S})| = 2$, $\mathbb{AG}(\mathcal{S}) \cong K_{|V_1|,|V_2|}$ by [**1**, Corollary 24], where $V_1 = \{\mathcal{X} \in A(\mathcal{S})^* : \mathcal{X} \subseteq \mathcal{P}_1\}$ and $V_2 = \{\mathcal{X} \in A(\mathcal{S})^* : \mathcal{X} \subseteq \mathcal{P}_2\}$. The proof is complete if we show that no two vertices of V_1 or V_2 are adjacent in $\mathbb{CAG}(\mathcal{S})$. Suppose on contrary that there exist $\mathcal{X}, \mathcal{Y} \in V_1$ such that $\mathcal{X} \sim \mathcal{Y}$ is an edge of $\mathbb{CAG}(\mathcal{S})$. That is $\mathcal{X} \nsubseteq \mathcal{X}\mathcal{Y}$ and $\mathcal{Y} \nsubseteq \mathcal{X}\mathcal{Y}$, which is a contradiction to our assumption. Hence $\mathbb{CAG}(\mathcal{S}) = \mathbb{AG}(\mathcal{S}) \cong K_{|V_1|,|V_2|}$ and $gr(\mathbb{CAG}(\mathcal{S})) \in \{4, \infty\}$. □

Now, we discuss about the star co-anninilating ideal graph and complete co-annihilating ideal graph for an Artinian commutative ring. We start with the following lemma which is useful for proving the remaining results of this section.

LEMMA 1.7. *Let $\mathcal{S}$ be a commutative ring. Then subgraph in $\mathbb{CAG}(\mathcal{S})$ induced by non-zero nilpotent ideals of $\mathcal{S}$ can be characterized as a complete subgraph.*

PROOF. Let $\mathcal{X}$ and $\mathcal{Y}$ be two distinct non-zero nilpotent ideals of $\mathcal{S}$. We have to show that $\mathcal{X}$ and $\mathcal{Y}$ are connected in $\mathbb{CAG}(\mathcal{S})$. Suppose on contrary that $\mathcal{X} \nsim \mathcal{Y}$. Then $\mathcal{X} = \mathcal{X}\mathcal{Y}$ or $\mathcal{Y} = \mathcal{X}\mathcal{Y}$ by part (1) of Lemma 1.1. Without loss of generality, we suppose that $\mathcal{X} = \mathcal{X}\mathcal{Y}$. Since $\mathcal{Y}$ is nilpotent, there exists a positive integer m such that $\mathcal{Y}^m = 0$. Suppose $\mathcal{X}\mathcal{Y}^k \neq 0$ for each $1 \leq k < m$. Then $\mathcal{X}\mathcal{Y}^m = \mathcal{X}\mathcal{Y} \cdot \mathcal{Y}^{m-1} = \mathcal{X}\mathcal{Y}^{m-1} \neq 0$, leading to a contradiction. Now, assume that there exists the least positive integer k such that $\mathcal{X}\mathcal{Y}^k = 0$ for some $1 \leq k < m$. Then $\mathcal{X}\mathcal{Y}^k = \mathcal{X}\mathcal{Y} \cdot \mathcal{Y}^{k-1} = \mathcal{X}\mathcal{Y}^{k-1} \neq 0$, again a contradiction. Hence $\mathcal{X}$ and $\mathcal{Y}$ are adjacent in $\mathbb{CAG}(\mathcal{S})$. □

LEMMA 1.8. *Let $\mathcal{S}$ be a commutative decomposable ring. Then $\mathbb{CAG}(\mathcal{S})$ is a star graph if and only if $\mathcal{S} = \mathcal{F} \times D$, where $\mathcal{F}$ represents a field and D stands for an integral domain such that $\mathcal{X} = \mathcal{X}\mathcal{Y}$ or $\mathcal{Y} = \mathcal{X}\mathcal{Y}$ for every $\mathcal{X}, \mathcal{Y} \in \mathbb{I}(D)^*$.*

PROOF. Suppose $\mathbb{CAG}(\mathcal{S})$ is a star graph. Since $\mathcal{S}$ is decomposable, $\mathcal{S} = \mathcal{S}_1 \times \mathcal{S}_2$, where $\mathcal{S}_1$ and $\mathcal{S}_2$ are commutative rings. Suppose $\mathcal{S}_1$ is not an integral domain. Then there exist $\mathcal{X}_1, \mathcal{X}_2 \in A(\mathcal{S})^*$ such that $\mathcal{X}_1\mathcal{X}_2 = 0$. Since $\mathcal{S}_1 \times (0) \sim \mathcal{X}_1 \times \mathcal{S}_2 \sim \mathcal{X}_2 \times (0) \sim (0) \times \mathcal{S}_2 \sim \mathcal{S}_1 \times (0)$ is a cycle in $\mathbb{CAG}(\mathcal{S})$, $\mathbb{CAG}(\mathcal{S})$ is not a star graph, leading to a contradiction. As a result, $\mathcal{S}_1$ qualifies as an integral domain. Likewise, it can be demonstrated that $\mathcal{S}_2$ also possesses the characteristics of an integral domain.

Suppose that there exist $\mathcal{Y}_1, \mathcal{Y}_2 \in \mathbb{I}(\mathcal{S}_1)^*$ such that $\mathcal{Y}_1 \neq \mathcal{Y}_1\mathcal{Y}_2$ and $\mathcal{Y}_2 \neq \mathcal{Y}_1\mathcal{Y}_2$. Then $\mathcal{Y}_1 \times (0) \sim \mathcal{Y}_2 \times (0) \times (0) \times \mathcal{S}_2 \sim \mathcal{Y}_1 \times (0)$ is a cycle in $\mathbb{CAG}(\mathcal{S})$, a contradiction.

Now, assume that $\mathcal{S}_2$ is not a field with non-zero proper ideal $\mathcal{K}$. Then $\mathcal{S}_1 \times (0) \sim (0) \times \mathcal{K} \sim \mathcal{S}_1 \times \mathcal{K} \sim (0) \times \mathcal{S}_2 \sim \mathcal{S}_1 \times (0)$ is a cycle in $\mathbb{CAG}(\mathcal{S})$, leading to a contradiction.

The converse is clear. □

THEOREM 1.9. *Let $\mathcal{S}$ be an Artinian commutative ring. Then $\mathbb{CAG}(\mathcal{S})$ is a star graph if and only if $\mathbb{CAG}(\mathcal{S}) \cong K_2$.*

PROOF. Assume that $\mathbb{CAG}(\mathcal{S})$ takes the form of a star graph. In the scenario where $\mathcal{S}$ is a local, it follows from Lemma 1.7 that $\mathbb{CAG}(\mathcal{S})$ transforms into a complete graph. Taking into account our current assumption that $\mathbb{CAG}(\mathcal{S})$ is star graph, $\mathbb{CAG}(\mathcal{S}) \cong K_2$. Also, if $\mathcal{S}$ is a non-local Artinain ring, then $\mathcal{S} \cong \mathcal{S}_1 \times \mathcal{S}_2 \times \cdots \times \mathcal{S}_n$, where $\mathcal{S}_i$ is local Artinian ring and $n \geq 2$. Suppose $n \geq 3$. Consider $\mathcal{X}_1 = \mathcal{S}_1 \times (0) \times (0)$, $\mathcal{X}_2 = (0) \times \mathcal{S}_2 \times (0)$, $\mathcal{X}_3 = (0) \times (0) \times \mathcal{S}_3 \in A(\mathcal{S})^*$. Since $\mathcal{X}_i\mathcal{X}_j = 0$ for each i, j, then $\mathcal{X}_1 \sim \mathcal{X}_2 \sim \mathcal{X}_3 \sim \mathcal{X}_1$ is a cycle in $\mathbb{CAG}(\mathcal{S})$, a contradiction. Hence $n = 2$. Thus, by Lemma 1.8, $\mathcal{S} \cong \mathcal{F}_1 \times \mathcal{F}_2$, where $\mathcal{F}_1$ and $\mathcal{F}_2$ are fields. This implies that $\mathbb{CAG}(\mathcal{S}) \cong K_2$.

The converse is clear. □

Next, we'll explore the completeness property of the co-annihilating ideal graph when dealing with an Artinian ring.

THEOREM 1.10. *Let $\mathcal{S}$ be an Artinian commutative ring. Then the graph $\mathbb{CAG}(\mathcal{S})$ is a complete graph if and only if one of the subsequent condition is satisfied.*

(1) *$\mathcal{S}$ is a local ring.*

(2) *$\mathcal{S} \cong \mathcal{F}_1 \times \mathcal{F}_2$, where $\mathcal{F}_1$, $\mathcal{F}_2$ are fields.*

PROOF. Suppose $\mathbb{CAG}(\mathcal{S})$ is a complete graph. Since $\mathcal{S}$ is Artinian, $\mathcal{S} \cong \mathcal{S}_1 \times \mathcal{S}_2 \times \cdots \times \mathcal{S}_n$, where $\mathcal{S}_i$ is a local Artinian ring for each i and $n \geq 1$. Suppose $n \geq 3$. Consider $a = \mathcal{S}_1 \times (0) \times \cdots \times (0)$ and $b = \mathcal{S}_1 \times \mathcal{S}_2 \times (0) \times \cdots \times (0) \in A(\mathcal{S})^*$. Since $ab = b$, a is not adjacent to b by part (1) of Lemma 1.1, a contradiction. Hence $n \leq 2$. If $n = 1$, we are done. For $n = 2$, suppose $\mathcal{S}_1$ is not a field with the maximal ideal $\Im_1 \neq 0$. Consider $x = \mathcal{S}_1 \times (0)$ and $y = \Im_1 \times (0) \in A(\mathcal{S})^*$. Since $xy = y$, x and y are not linked in $\mathbb{CAG}(\mathcal{S})$, leading to a contradiction. Hence $\mathcal{S}_1$ is a filed. Likewise, it can be demonstrated that $\mathcal{S}_2$ forms a field.

Conversely, if $\mathcal{S}$ is local then by Lemma 1.7, $\mathbb{CAG}(\mathcal{S})$ is a complete graph. Also, if $\mathcal{S} \cong \mathcal{F}_1 \times \mathcal{F}_2$, where $\mathcal{F}_1$, $\mathcal{F}_2$ are fields, then $\mathbb{CAG}(\mathcal{S}) \cong K_2$, which is a complete graph. □

COROLLARY 1.11. *Let $\mathcal{S}$ be a non-local Artinian commutative ring. Then $\mathbb{CAG}(\mathcal{S})$ is a star graph if and only if $\mathbb{CAG}(\mathcal{S})$ is a complete graph.*

PROOF. It is clear from Theorems 1.9 and 1.10. □

2. Co-annihilating ideal graph belongs to some category of specialized graphs

In this section, we categorize Artinian rings for which the co-annihilating ideal graph exhibits isomorphism with familiar graphs such as a unicycle graph, a tree, a split graph, a planar graph, or an outerplanar graph.

THEOREM 2.1. *Let $\mathcal{S}$ be an Artinian commutative ring. Then $\mathbb{CAG}(\mathcal{S})$ is cyclic if and only if one of the following holds:*

(1) *$\mathcal{S} \cong \mathcal{S}_1 \times \mathcal{F}$, where $(\mathcal{S}_1, \Im)$ is local Artinian ring with $\Im$ is the only non-zero proper ideal of $\mathcal{S}_1$ and $\mathcal{F}$ is a field.*
(2) *$\mathcal{S}$ is a local ring with exactly three non-zero proper ideals.*

PROOF. Assume that $\mathbb{CAG}(\mathcal{S})$ is a unicycle graph. Since $\mathcal{S}$ is Artinian, $\mathcal{S} \cong \mathcal{S}_1 \times \mathcal{S}_2 \times \cdots \times \mathcal{S}_n$, where $\mathcal{S}_i$ is a local Artinian ring for each i and $n \geq 1$. Suppose $n \geq 3$. Consider $a_1 = \mathcal{S}_1 \times (0) \times (0) \times \cdots \times (0)$, $a_2 = (0) \times \mathcal{S}_2 \times (0) \times \cdots \times (0)$, $a_3 = (0) \times (0) \times \mathcal{S}_3 \times (0) \times \cdots \times (0)$, $b_1 = \mathcal{S}_1 \times \mathcal{S}_2 \times (0) \times \cdots \times (0)$, $b_2 = \mathcal{S}_1 \times (0) \times \mathcal{S}_3 \times (0) \times \cdots \times (0)$, $b_3 = (0) \times \mathcal{S}_2 \times \mathcal{S}_3 \times (0) \times \cdots \times (0) \in A(\mathcal{S})^*$. Since $a_1a_2 = a_2a_3 = a_3a_1 = 0$, $b_1 \nsubseteq b_1b_2$ and $b_2 \nsubseteq b_1b_2$, $b_2 \nsubseteq b_2b_3$ and $b_3 \nsubseteq b_2b_3$, $b_1 \nsubseteq b_1b_3$ and $b_3 \nsubseteq b_1b_3$, then $a_1 \sim a_2 \sim a_3 \sim a_1$ and $b_1 \sim b_2 \sim b_3 \sim b_1$ are two distinct cycles in $\mathbb{CAG}(\mathcal{S})$, it would lead to a contradiction. Hence, $n \leq 2$.

Assuming that $n = 2$, we have $\mathcal{S}_1$ and $\mathcal{S}_2$ as non-field structures, each possessing non-zero maximal ideals $\Im_1$ and $\Im_2$, respectively. Consider $x_1 = \mathcal{S}_1 \times (0)$, $x_2 = \Im_1 \times \Im_2$, $x_3 = (0) \times \mathcal{S}_2$, $x_4 = \Im_1 \times (0) \in A(\mathcal{S})^*$. Since $x_1 \nsubseteq x_1x_2$ and $x_2 \nsubseteq x_1x_2$, $x_2 \nsubseteq x_2x_3$ and $x_3 \nsubseteq x_2x_3$, $x_1x_3 = 0$, $x_3x_4 = 0$, $x_2 \nsubseteq x_2x_4$ and $x_4 \nsubseteq x_2x_4$, then $x_1 \sim x_2 \sim x_3 \sim x_1$ and $x_2 \sim x_3 \sim x_4 \sim x_2$ are two distinct cycles within $\mathbb{CAG}(\mathcal{S})$, it would lead to a contradiction. Consequently, either $\mathcal{S}_1$ or $\mathcal{S}_2$ must be a field.

If $\mathcal{S}_1$ and $\mathcal{S}_2$ both are fields, then $\mathbb{CAG}(\mathcal{S}) \cong K_2$, a contradiction. Thus, $\mathcal{S} \cong \mathcal{S}_1 \times \mathcal{F}$, where $\mathcal{S}_1$ is a local Artinian ring with the maximal ideal $\Im \neq 0$ and $\mathcal{F}$ is a field.

Let $\mathcal{X}$ be a non-zero proper ideal of $\mathcal{S}_1$ such that $\mathcal{X} \neq \Im$. Consider $y_1 = \Im \times (0)$, $y_2 = \mathcal{X} \times (0)$, $y_3 = \Im \times \mathcal{F}$, $y_4 = \mathcal{X} \times \mathcal{F} \in A(\mathcal{S})^*$. Since $y_1 \nsubseteq y_1y_2$ and $y_2 \nsubseteq y_1y_2$, $y_2 \nsubseteq y_2y_3$ and $y_3 \nsubseteq y_2y_3$, $y_1y_3 = (0)$, $y_3 \nsubseteq y_3y_4$ and $y_4 \nsubseteq y_3y_4$, $y_2 \nsubseteq y_2y_4$ and $y_4 \nsubseteq y_2y_4$, then $y_1 \sim y_2 \sim y_3 \sim y_1$ and $y_2 \sim y_3 \sim y_4 \sim y_2$ are two different cycles in $\mathbb{CAG}(\mathcal{S})$, a contradiction. Hence $\mathcal{S}_1$ has only one non-zero proper ideal which is $\Im_1$.

Now, Let us assume that $n = 1$. In such a situation, $\mathcal{S}$ can be identified as a local Artinian ring. As a result of Theorem 1.10, the graph $\mathbb{CAG}(\mathcal{S})$ achieves completeness. This ultimately implies that the size of $A(\mathcal{S})^*$ is 3.

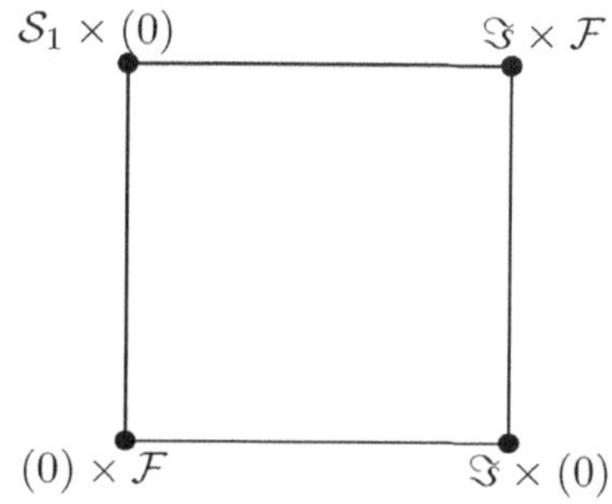

FIGURE 2. The graph $\mathbb{CAG}(\mathcal{S}_1 \times \mathcal{F})$ where $\Im$ is the only non-zero proper ideal of $(\mathcal{S}_1, \Im)$.

The converse follows from Theorem 1.10 and Figure 2. □

THEOREM 2.2. *Let $\mathcal{S}$ be an Artinian commutative ring. Then $\mathbb{CAG}(\mathcal{S})$ is tree if and only if one of the following conditions is satisfied:*

(1) *$\mathcal{S} \cong \mathcal{F}_1 \times \mathcal{F}_2$, where both $\mathcal{F}_1$ and $\mathcal{F}_2$ are fields.*
(2) *$\mathcal{S}$ is a local ring with at most two non-zero proper ideals.*

PROOF. Suppose $\mathbb{CAG}(\mathcal{S})$ is a tree. Since $\mathcal{S}$ is Artinian, $\mathcal{S} \cong \mathcal{S}_1 \times \mathcal{S}_2 \times \cdots \times \mathcal{S}_n$, where $\mathcal{S}_i$ is a local Artinian ring for each i and $n \geq 1$. Suppose $n \geq 3$. Consider $a_1 = \mathcal{S}_1 \times (0) \times (0) \times \cdots \times (0)$, $a_2 = (0) \times \mathcal{S}_2 \times (0) \times \cdots \times (0)$, $a_3 = (0) \times (0) \times \mathcal{S}_3 \times (0) \times \cdots \times (0) \in A(\mathcal{S})^*$. Since $a_1a_2 = a_2a_3 = a_3a_1 = 0$, then $a_1 \sim a_2 \sim a_3 \sim a_1$ is a cycle in $\mathbb{CAG}(\mathcal{S})$, a contradiction. Hence $n \leq 2$.

Assuming that $n = 2$ and $\mathcal{S}_1$ is a non-field with a non-zero maximal ideal $\Im_1$. Consider $x_1 = \mathcal{S}_1 \times (0)$, $x_2 = \Im_1 \times \mathcal{S}_2$, $x_3 = \Im_1 \times (0)$, $x_4 = (0) \times \mathcal{S}_2 \in A(\mathcal{S})^*$. Since $x_1 \nsubseteq x_1x_2$ and $x_2 \nsubseteq x_1x_2$, $x_2 \nsubseteq x_2x_3$ and $x_3 \nsubseteq x_2x_3$, $x_3x_4 = 0$, $x_1x_4 = 0$, then $x_1 \sim x_2 \sim x_3 \sim x_4 \sim x_1$ is a cycle in $\mathbb{CAG}(\mathcal{S})$, a contradiction. Hence $\mathcal{S}_1$ is a field. Similarly, one can show that $\mathcal{S}_2$ is a field.

Now, consider the case where $n = 1$. Then $\mathcal{S}$ takes the form of a local Artinian ring, and as a result, according to Theorem 1.10, $\mathbb{CAG}(\mathcal{S})$ is complete. Consequently, it follows that $|A(\mathcal{S})^*| \leq 2$.

The converse follows from Theorem 1.10 and the fact that $\mathbb{CAG}(\mathcal{F}_1 \times \mathcal{F}_2) \cong K_2$, where $\mathcal{F}_1$ and $\mathcal{F}_2$ are fields. □

THEOREM 2.3 ([**20**]). *Suppose we have a connected graph G. In that case, G qualifies as a split graph if and only if it does not contains any induced subgraph that is isomorphic to $2K_2$, C_4, C_5.*

THEOREM 2.4. *Let $\mathcal{S}$ be an Artinian commutative ring. Then $\mathbb{CAG}(\mathcal{S})$ is a split graph if and only if one of the following conditions is satisfied:*

(1) *$\mathcal{S} \cong \mathcal{F}_1 \times \mathcal{F}_2$, where both $\mathcal{F}_1$ and $\mathcal{F}_2$ are fields.*
(2) *$\mathcal{S}$ is a local ring with at most three non-zero proper ideals.*

PROOF. Suppose $\mathbb{CAG}(\mathcal{S})$ is a split graph. Since $\mathcal{S}$ is Artinian, $\mathcal{S} \cong \mathcal{S}_1 \times \mathcal{S}_2 \times \cdots \times \mathcal{S}_n$, where $\mathcal{S}_i$ is a local Artinian ring for each i and $n \geq 1$. Suppose $n \geq 3$. Consider $a_1 = \mathcal{S}_1 \times (0) \times (0) \times \cdots \times (0)$, $a_2 = (0) \times \mathcal{S}_2 \times (0) \times \cdots \times (0)$, $a_3 = \mathcal{S}_1 \times (0) \times \mathcal{S}_3 \times (0) \times \cdots \times (0)$, $a_4 = (0) \times \mathcal{S}_2 \times \mathcal{S}_3 \times (0) \times \cdots \times (0) \in A(\mathcal{S})^*$. Since

$a_1a_2 = 0$, $a_2a_3 = 0$, $a_3 \not\subseteq a_3a_4$ and $a_4 \not\subseteq a_3a_4$, $a_1a_4 = 0$, then $a_1 \sim a_2 \sim a_3 \sim a_4 \sim a_1$ is C_4 in $\mathbb{CAG}(\mathcal{S})$, a contradiction. Hence $n \leq 2$. Assuming that $n = 2$ and $\mathcal{S}_1$ is a non-field with a non-zero maximal ideal $\Im_1$. Consider $x_1 = \mathcal{S}_1 \times (0)$, $x_2 = \Im_1 \times \mathcal{S}_2$, $x_3 = \Im_1 \times (0)$, $x_4 = (0) \times \mathcal{S}_2 \in A(\mathcal{S})^*$. Since $x_1 \not\subseteq x_1x_2$ and $x_2 \not\subseteq x_1x_2$, $x_2 \not\subseteq x_2x_3$ and $x_3 \not\subseteq x_2x_3$, $x_3x_4 = 0$, $x_1x_4 = 0$, then $x_1 \sim x_2 \sim x_3 \sim x_4 \sim x_1$ is C_4 in $\mathbb{CAG}(\mathcal{S})$, a contradiction. Hence $\mathcal{S}_1$ is a field. Similarly, one can show that $\mathcal{S}_2$ is a field.

Now, consider the case where $n = 1$. Then $\mathcal{S}$ takes the form of a local Artinian ring, and as a result, according to Theorem 1.10, $\mathbb{CAG}(\mathcal{S})$ is complete. Consequently, it follows that $|A(\mathcal{S})^*| \leq 3$.

The converse follows from Theorem 1.10 and the fact that $\mathbb{CAG}(\mathcal{F}_1 \times \mathcal{F}_2) \cong K_2$. □

Next, our objective is to classify Artinian rings for which the co-annihilating ideal graph can be represented as a planar graph. The following result due to Kuratowski is useful for the characterization of co-annihilating ideal graph as a planar graph.

THEOREM 2.5 ([**21**]). *A graph G is planar if and only if it contains no subdivision of K_5 or $K_{3,3}$.*

THEOREM 2.6. *Let $\mathcal{S}$ be an Artinian commutative ring. Then $\mathbb{CAG}(\mathcal{S})$ is a planar graph if and only if one of the following conditions is satisfied:*

(1) *$\mathcal{S} \cong \mathcal{F}_1 \times \mathcal{F}_2 \times \mathcal{F}_3$, where $\mathcal{F}_1$, $\mathcal{F}_2$ and $\mathcal{F}_3$ represent fields.*
(2) *$\mathcal{S} \cong \mathcal{F}_1 \times \mathcal{F}_2$, where both $\mathcal{F}_1$ and $\mathcal{F}_2$ are fields.*
(3) *$\mathcal{S} \cong \mathcal{S}_1 \times \mathcal{F}$, where $(\mathcal{S}_1, \Im)$ is a local Artinian ring with $\Im$ is the only non-zero proper ideal of $\mathcal{S}_1$ and $\mathcal{F}$ is a field.*
(4) *$\mathcal{S}$ is a local ring with at most four non-zero proper ideals.*

PROOF. Suppose $\mathbb{CAG}(\mathcal{S})$ is a planar graph. Since $\mathcal{S}$ is Artinian, $\mathcal{S} \cong \mathcal{S}_1 \times \mathcal{S}_2 \times \cdots \times \mathcal{S}_n$, where $\mathcal{S}_i$ is a local Artinian ring for each i and $n \geq 1$. Suppose $n \geq 4$. Consider $x_1 = \mathcal{S}_1 \times (0) \times (0) \times \cdots \times (0)$, $x_2 = \mathcal{S}_1 \times \mathcal{S}_2 \times (0) \times \cdots \times (0)$, $x_3 = (0) \times \mathcal{S}_2 \times (0) \times \cdots \times (0)$, $y_1 = (0) \times (0) \times \mathcal{S}_3 \times (0) \times \cdots \times (0)$, $y_2 = (0) \times (0) \times (0) \times \mathcal{S}_4 \times (0) \times \cdots \times (0)$, $y_3 = (0) \times (0) \times \mathcal{S}_3 \times \mathcal{S}_4 \times (0) \times \cdots \times (0) \in A(\mathcal{S})^*$. Since $x_iy_j = 0$ for each i, j, then $K_{3,3}$ is an induced subgraph of $\mathbb{CAG}(\mathcal{S})$, a contradiction by Theorem 2.5. Hence $n \leq 3$.

First, suppose $n = 3$ and $\mathcal{S}_1$ is not a field with the maximal ideal $\Im_1 \neq 0$. Consider $a_1 = \mathcal{S}_1 \times (0) \times (0)$, $a_2 = \Im_1 \times (0) \times (0)$, $a_3 = \Im_1 \times \mathcal{S}_2 \times (0)$, $b_1 = \Im_1 \times \mathcal{S}_2 \times \mathcal{S}_3$, $b_2 = \Im_1 \times (0) \times \mathcal{S}_3$, $b_3 = (0) \times (0) \times \mathcal{S}_3 \in A(\mathcal{S})^*$. Since $a_1 \not\subseteq a_1b_1$ and $b_1 \not\subseteq a_1b_1$, $a_1 \not\subseteq a_1b_2$ and $b_2 \not\subseteq a_1b_2$, $a_1b_3 = 0$, $a_2 \not\subseteq a_2b_1$ and $b_1 \not\subseteq a_2b_1$, $a_2 \not\subseteq a_2b_2$ and $b_2 \not\subseteq a_2b_2$, $a_2b_3 = 0$, $a_3 \not\subseteq a_3b_1$ and $b_1 \not\subseteq a_3b_1$, $a_3 \not\subseteq a_3b_2$ and $b_2 \not\subseteq a_3b_2$, $a_3b_3 = 0$, then $K_{3,3}$ is an induced subgraph of $\mathbb{CAG}(\mathcal{S})$, a contradiction by Theorem 2.5. Hence $\mathcal{S}_1$ is a field. Similarly, one can show that $\mathcal{S}_2$ and $\mathcal{S}_3$ are fields.

Assuming that $n = 2$, we have $\mathcal{S}_1$ and $\mathcal{S}_2$ as non-field structures, each possessing non-zero maximal ideals $\Im_1$ and $\Im_2$, respectively. Consider $u_1 = \mathcal{S}_1 \times (0)$, $u_2 = \Im_1 \times (0)$, $u_3 = \mathcal{S}_1 \times \Im_2$, $v_1 = \Im_1 \times \mathcal{S}_2$, $v_2 = (0) \times \Im_2$, $v_3 = \Im_1 \times \Im_2 \in A(\mathcal{S})^*$. Since $u_1 \not\subseteq u_1v_1$ and $v_1 \not\subseteq u_1v_1$, $u_1v_2 = 0$, $u_1 \not\subseteq u_1v_3$ and $v_3 \not\subseteq u_1v_3$, $u_2 \not\subseteq u_2v_1$ and $v_1 \not\subseteq u_2v_1$, $u_2v_2 = 0$, $u_2 \not\subseteq u_2v_3$ and $v_3 \not\subseteq u_2v_3$, $u_3 \not\subseteq u_3v_1$ and $v_1 \not\subseteq u_3v_1$, $u_3 \not\subseteq u_3v_2$ and $v_2 \not\subseteq u_3v_2$, $u_3 \not\subseteq u_3v_3$ and $v_3 \not\subseteq u_3v_3$, then $K_{3,3}$ is an induced subgraph of $\mathbb{CAG}(\mathcal{S})$, a contradiction by Theorem 2.5. Hence at least one of $\mathcal{S}_1$ or $\mathcal{S}_2$ is a field. If $\mathcal{S}_1$ and $\mathcal{S}_2$ both are fields, then we are done. Thus, $\mathcal{S} \cong \mathcal{S}_1 \times \mathcal{F}$, where $\mathcal{S}_1$ is a local Artinian ring with the maximal ideal $\Im \neq 0$ and $\mathcal{F}$ is a field.

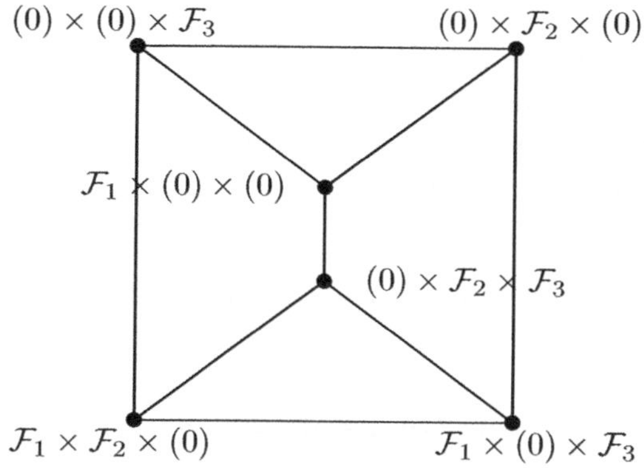

FIGURE 3. The graph $\mathbb{CAG}(\mathcal{F}_1 \times \mathcal{F}_2 \times \mathcal{F}_3)$

Let $\mathcal{X}$ be a non-zero proper ideal of $\mathcal{S}_1$ such that $\mathcal{X} \neq \Im$. Consider $p_1 = \mathcal{S}_1 \times (0)$, $p_2 = \Im_1 \times (0)$, $p_3 = \mathcal{X} \times (0)$, $q_1 = \Im_1 \times \mathcal{F}$, $q_2 = \mathcal{X} \times \mathcal{F}$, $q_3 = (0) \times \mathcal{F} \in A(\mathcal{S})^*$. Since $p_1 \nsubseteq p_1q_1$ and $q_1 \nsubseteq p_1q_1$, $p_1 \nsubseteq p_1q_2$ and $q_2 \nsubseteq p_1q_2$, $p_1q_3 = 0$, $p_2q_1 = 0$, $p_2 \nsubseteq p_2q_2$ and $q_2 \nsubseteq p_2q_2$, $p_2q_3 = 0$, $p_3 \nsubseteq p_3q_1$ and $q_1 \nsubseteq p_3q_1$, $p_3 \nsubseteq p_3q_2$ and $q_2 \nsubseteq p_3q_2$, $p_3q_3 = 0$, then $K_{3,3}$ is an induced subgraph of $\mathbb{CAG}(\mathcal{S})$, a contradiction by Theorem 2.5. Hence $\mathcal{S}_1$ has the unique non-zero proper ideal which is $\Im_1$.

Finally, consider the case where $n = 1$. Then $\mathcal{S}$ takes the form of a local Artinian ring, and as a result, according to Theorem 1.10, $\mathbb{CAG}(\mathcal{S})$ is complete. Consequently, it follows that $|A(\mathcal{S})^*| \leq 4$.

Conversely, if $\mathcal{S} \cong \mathcal{F}_1 \times \mathcal{F}_2 \times \mathcal{F}_3$, where $\mathcal{F}_1$, $\mathcal{F}_2$ and $\mathcal{F}_3$ are fields, then the planar embedding of $\mathbb{CAG}(\mathcal{S})$ is given in Figure 3. If $\mathcal{S} \cong \mathcal{F}_1 \times \mathcal{F}_2$, where $\mathcal{F}_1$ and $\mathcal{F}_2$ are fields, then $\mathbb{CAG}(\mathcal{S}) \cong K_2$, which is a planar graph. If $\mathcal{S} \cong \mathcal{S}_1 \times \mathcal{F}$, where $(\mathcal{S}_1, \Im)$ is a local Artinian ring with $\Im$ is the only non-zero proper ideal of $\mathcal{S}_1$ and $\mathcal{F}$ is a field, then the planar embedding of $\mathbb{CAG}(\mathcal{S})$ is shown in Figure 2. Also, if $\mathcal{S}$ is a local Artinian ring with at most four non-zero proper ideals, then the result follows from Theorems 1.10 and 2.6. □

THEOREM 2.7 ([**21**]). *A graph G is outerplanar if and only if it does not contain a subdivision of* K_4 *or* $K_{2,3}$.

COROLLARY 2.8. *Let* $\mathcal{S}$ *be an Artinian commutative ring. Then* $\mathbb{CAG}(\mathcal{S})$ *is an outerplanar graph if and only if one of the following conditions is satisfied:*

(1) $\mathcal{S} \cong \mathcal{F}_1 \times \mathcal{F}_2$, *where both* $\mathcal{F}_1$ *and* $\mathcal{F}_2$ *are fields.*
(2) $\mathcal{S} \cong \mathcal{S}_1 \times \mathcal{F}$, *where* $(\mathcal{S}_1, \Im)$ *is a local Artinian ring with* $\Im$ *is the only non-zero proper ideal of* $\mathcal{S}_1$ *and* $\mathcal{F}$ *is a field.*
(3) $\mathcal{S}$ *is a local ring with at most three non-zero proper ideals.*

PROOF. Suppose that $\mathbb{CAG}(\mathcal{S})$ is an outerplanar garph. Since every outerplanar graph is a planar graph, thus from Theorem 2.6, $\mathcal{S}$ is isomorphic to one of the ring from (1)- (4) in Theorem 2.6. If $\mathcal{S} \cong \mathcal{F}_1 \times \mathcal{F}_2 \times \mathcal{F}_3$, where $\mathcal{F}_1$, $\mathcal{F}_2$ and $\mathcal{F}_3$ are fields, then by Figure 3, $\mathbb{CAG}(\mathcal{S})$ is not an outerplanar graph a contradiction.
If $\mathcal{S} \cong \mathcal{F}_1 \times \mathcal{F}_2$, where both $\mathcal{F}_1$ and $\mathcal{F}_2$ are fields, then we are done. If $\mathcal{S} \cong \mathcal{S}_1 \times \mathcal{F}$, where $(\mathcal{S}_1, \Im)$ is a local Artinian ring with $\Im$ is the only non-zero proper ideal of $\mathcal{S}_1$ and $\mathcal{F}$ is a field, then again we are done.

Finally, if $\mathcal{S}$ is a local ring with at most four non-zero proper ideals, then by Theorem 1.10, $\mathbb{CAG}(\mathcal{S})$ contains K_4, which is a contradiction by Theorem 2.7. Thus, $\mathcal{S}$ has at most three non-zero proper ideals.
Conversely, if $\mathcal{S} \cong \mathcal{F}_1 \times \mathcal{F}_2$, where $\mathcal{F}_1$ and $\mathcal{F}_2$ are fields, then $\mathbb{CAG}(\mathcal{S}) \cong K_2$, which is an outerplanar graph by Theorem 2.7. If $\mathcal{S} \cong \mathcal{S}_1 \times \mathcal{F}$, where $(\mathcal{S}_1, \Im)$ is a local Artinian ring with $\Im$ is the only non-zero proper ideal of $\mathcal{S}_1$ and $\mathcal{F}$ is a field, then the outerplanar embedding of $\mathbb{CAG}(\mathcal{S})$ is shown in Figure 2. Also, if $\mathcal{S}$ is a local Artinian ring with at most three non-zero proper ideals, then the result follows from Theorems 1.10 and 2.7. □

3. Genus of $\mathbb{CAG}(\mathcal{S})$

In this section, we classify Artinian rings $\mathcal{S}$ with genus one co-annihilating-ideal graph. The *genus* of a graph G, denoted as $\gamma(G)$, is defined as the minimum integer k for which the graph can be drawn on a sphere having k handles (i.e., an oriented surface of genus k) without any edge crossings. A planar graph has genus 0 because it can be drawn on a sphere without self-crossing. The following results focus on aspects related to the genus of complete and complete bipartite graphs.

LEMMA 3.1 ([**21**]). *$\gamma(K_m) = \lceil \frac{(m-3)(m-4)}{12} \rceil$ if $m \geq 3$. In particular, $\gamma(K_m) = 1$ if $m = 5, 6, 7$.*

LEMMA 3.2 ([**21**]). *$\gamma(K_{n,m}) = \lceil \frac{(n-2)(m-2)}{4} \rceil$ if $n, m \geq 2$. In particular, $\gamma(K_{4,4}) = \gamma(K_{3,m}) = 1$ if $m = 3, 4, 5, 6$. Also $\gamma(K_{5,4}) = \gamma(K_{6,4}) = \gamma(K_{m,3}) = 2$ if $m = 7, 8, 9, 10$.*

PROPOSITION 3.3. *Let $(\mathcal{S}, \Im)$ be a local Artinian commutative ring. Then $\gamma(\mathbb{CAG}(\mathcal{S})) = 1$ if and only if $\mathcal{S}$ has at least five and at most seven non-zero proper ideals.*

PROOF. Since $\mathcal{S}$ is local, $\mathbb{CAG}(\mathcal{S})$ is a complete graph by Theorem 1.10. Hence $\gamma(\mathbb{CAG}(\mathcal{S})) = 1$ if and only if $5 \leq |A(\mathcal{S})^*| \leq 7$ by Lemma 3.1. □

THEOREM 3.4. *Let $\mathcal{S} = \mathcal{F}_1 \times \mathcal{F}_2 \times \cdots \times \mathcal{F}_n$ be an Artinian commutative ring, where $\mathcal{F}_i$ is a field for each i and $n \geq 2$. Then $\gamma(\mathbb{CAG}(\mathcal{S})) \neq 1$.*

PROOF. Suppose that $n \geq 4$. Consider $x_1 = \mathcal{F}_1 \times (0) \times (0) \times \cdots \times (0)$, $x_2 = (0) \times \mathcal{F}_2 \times (0) \times \cdots \times (0)$, $x_3 = (0) \times \mathcal{F}_2 \times \times \mathcal{F}_3 \times (0) \times \cdots \times (0)$, $x_4 = (0) \times \mathcal{F}_2 \times \mathcal{F}_3 \times \mathcal{F}_4 \times (0) \times \cdots \times (0)$, $x_5 = \mathcal{F}_1 \times (0) \times \mathcal{F}_3 \times \mathcal{F}_4 \times (0) \times \cdots \times (0)$, $x_6 = \mathcal{F}_1 \times (0) \times \mathcal{F}_3 \times (0) \times \cdots \times (0)$, $x_7 = \mathcal{F}_1 \times (0) \times (0) \times \mathcal{F}_4 \times (0) \times \cdots \times (0)$, $x_8 = (0) \times (0) \times \mathcal{F}_3 \times (0) \times \cdots \times (0)$, $x_9 = (0) \times \mathcal{F}_2 \times (0) \times \mathcal{F}_4 \times (0) \times \cdots \times (0)$, $x_{10} = (0) \times (0) \times \mathcal{F}_3 \times \mathcal{F}_4 \times (0) \times \cdots \times (0)$, $x_{11} = \mathcal{F}_1 \times \mathcal{F}_2 \times (0) \times \cdots \times (0) \in A(\mathcal{S})^*$. Since $x_1x_2 = 0$, $x_1x_3 = 0$, $x_1x_4 = 0$, $x_1x_8 = 0$, $x_1x_9 = 0$, $x_1x_{10} = 0$, $x_2x_5 = 0$, $x_2x_6 = 0$, $x_3 \nsubseteq x_3x_5$ and $x_5 \nsubseteq x_3x_5$, $x_3 \nsubseteq x_3x_6$ and $x_6 \nsubseteq x_3x_6$, $x_4 \nsubseteq x_4x_5$ and $x_5 \nsubseteq x_4x_5$, $x_4 \nsubseteq x_4x_6$ and $x_6 \nsubseteq x_4x_6$, $x_7x_8 = 0$, $x_7 \nsubseteq x_7x_9$ and $x_9 \nsubseteq x_7x_9$, $x_7 \nsubseteq x_7x_{10}$ and $x_{10} \nsubseteq x_7x_{10}$, $x_8x_{11} = 0$, $x_9 \nsubseteq x_9x_{11}$ and $x_{11} \nsubseteq x_9x_{11}$, $x_{10}x_{11} = 0$, then G' is a subgraph of $\mathbb{CAG}(\mathcal{S})$ as shown in Figure 4, which implies that $\gamma(\mathbb{CAG}(\mathcal{S})) \geq 2$.

Also, if $n = 2$ or $n = 3$, then by Theorem 2.6, $\gamma(\mathbb{CAG}(\mathcal{S})) = 0$. Hence $\gamma(\mathbb{CAG}(\mathcal{S})) \neq 1$.

□

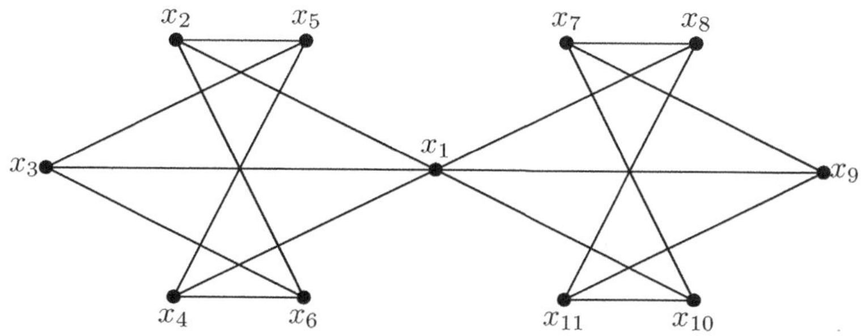

FIGURE 4. The graph G'

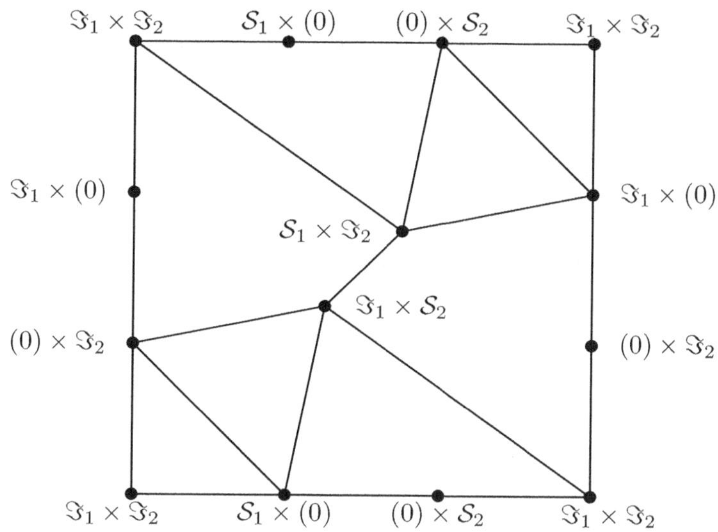

FIGURE 5. Toroidal embedding of $\mathbb{CAG}(\mathcal{S}_1 \times \mathcal{S}_2)$, with $(\mathcal{S}_1, \Im_1)$ and $(\mathcal{S}_2, \Im_2)$

THEOREM 3.5. *Let $\mathcal{S} = \mathcal{S}_1 \times \mathcal{S}_2 \times \cdots \times \mathcal{S}_n$ be an Artinian commutative ring, where $(\mathcal{S}_i, \Im_i)$ is a local Artinian ring with $\Im_i \neq 0$ for each i and $n \geq 2$. Then $\gamma(\mathbb{CAG}(\mathcal{S})) = 1$ if and only if $n = 2$ and $\Im_i$ is the only non-zero proper ideal of $\mathcal{S}_i$ for $i = 1, 2$.*

PROOF. Assume that $\gamma(\mathbb{CAG}(\mathcal{S})) = 1$. Suppose $n \geq 3$. Consider $x_1 = \mathcal{S}_1 \times (0) \times (0) \times \cdots \times (0)$, $x_2 = \Im_1 \times (0) \times (0) \times \cdots \times (0)$, $x_3 = (0) \times \mathcal{S}_2 \times (0) \times \cdots \times (0)$, $x_4 = (0) \times \Im_2 \times (0) \times \cdots \times (0)$, $y_1 = (0) \times (0) \times \mathcal{S}_3 \times (0) \times \cdots \times (0)$, $y_2 = (0) \times (0) \times \Im_3 \times (0) \times \cdots \times (0)$, $y_3 = \Im_1 \times \Im_2 \times (0) \times \cdots \times (0)$, $y_4 = \Im_1 \times (0) \times \Im_3 \times (0) \times \cdots \times (0)$, $y_5 = (0) \times \Im_2 \times \Im_3 \times (0) \times \cdots \times (0) \in A(\mathcal{S})^*$. Since $x_1y_1 = 0$, $x_1y_2 = 0$, $x_1 \nsubseteq x_1y_3$ and $y_3 \nsubseteq x_1y_3$, $x_1 \nsubseteq x_1y_4$ and $y_4 \nsubseteq x_1y_4$, $x_1y_5 = 0$, $x_2y_1 = 0$, $x_2y_2 = 0$, $x_2 \nsubseteq x_2y_3$ and $y_3 \nsubseteq x_2y_3$, $x_2 \nsubseteq x_2y_4$ and $y_4 \nsubseteq x_2y_4$, $x_2y_5 = 0$, $x_3y_1 = 0$, $x_3y_2 = 0$, $x_3 \nsubseteq x_3y_3$ and $y_3 \nsubseteq x_3y_3$, $x_3y_4 = 0$, $x_3 \nsubseteq x_3y_5$ and $y_5 \nsubseteq x_3y_5$, $x_4y_1 = 0$, $x_4y_2 = 0$, $x_4 \nsubseteq x_4y_3$ and $y_3 \nsubseteq x_4y_3$, $x_4y_4 = 0$, $x_4 \nsubseteq x_4y_5$ and $y_5 \nsubseteq x_4y_5$, then $K_{4,5}$ is an induced subgraph of $\mathbb{CAG}(\mathcal{S})$, a contradiction. Hence $n = 2$.

Let $\mathcal{X}$ be a non-zero proper ideal of $\mathcal{S}_1$ such that $\mathcal{X} \neq \Im_1$. Consider $a_1 = \mathcal{S}_1 \times (0)$, $a_2 = \Im_1 \times (0)$, $a_3 = \mathcal{X} \times (0)$, $a_4 = \mathcal{S}_1 \times \Im_2$, $b_1 = (0) \times \mathcal{S}_2$, $b_2 = (0) \times \Im_2$, $b_3 = \Im_1 \times \Im_2$, $b_4 = \mathcal{X} \times \Im_2$, $b_5 = \Im_1 \times \mathcal{S}_2 \in A(\mathcal{S})^*$. Since $a_1b_1 = 0$, $a_1b_2 = 0$, $a_1 \nsubseteq a_1b_3$ and $b_3 \nsubseteq a_1b_3$, $a_1 \nsubseteq a_1b_4$ and $b_4 \nsubseteq a_1b_4$, $a_1 \nsubseteq a_1b_5$ and $b_5 \nsubseteq a_1b_5$, $a_2b_1 = 0$, $a_2b_2 = 0$, $a_2 \nsubseteq a_2b_3$ and $b_3 \nsubseteq a_2b_3$, $a_2 \nsubseteq a_2b_4$ and $b_4 \nsubseteq a_2b_4$, $a_2 \nsubseteq a_2b_5$ and $b_5 \nsubseteq a_2b_5$, $a_3b_1 = 0$, $a_3b_2 = 0$, $a_3 \nsubseteq a_3b_3$ and $b_3 \nsubseteq a_3b_3$, $a_3 \nsubseteq a_3b_4$ and $b_4 \nsubseteq a_3b_4$, $a_3 \nsubseteq a_3b_5$ and $b_5 \nsubseteq a_3b_5$, $a_4 \nsubseteq a_4b_1$ and $b_1 \nsubseteq a_4b_1$, $a_4 \nsubseteq a_4b_2$ and $b_2 \nsubseteq a_4b_2$, $a_4 \nsubseteq a_4b_3$ and $b_3 \nsubseteq a_4b_3$, $a_4 \nsubseteq a_4b_4$ and $b_4 \nsubseteq a_4b_4$, $a_4 \nsubseteq a_4b_5$ and $b_5 \nsubseteq a_4b_5$, then $K_{4,5}$ is an induced subgraph of $\mathbb{CAG}(\mathcal{S})$, a contradiction. Consequently, $\mathcal{S}_1$ possesses a unique non-zero proper ideal, namely $\Im_1$, and similarly, it can be demonstrated that $\mathcal{S}_2$ has a unique non-zero proper ideal, which is $\Im_2$.

The converse follows from Figure 5. □

THEOREM 3.6. *Let $\mathcal{S} = \mathcal{S}_1 \times \mathcal{S}_2 \times \cdots \times \mathcal{S}_n \times \mathcal{F}_1 \times \mathcal{F}_2 \times \cdots \times \mathcal{F}_m$ be an Artinian commutative ring, where $(\mathcal{S}_i, \Im_i)$ is a local Artinian ring with $\Im_i \neq 0$ for each i, $\mathcal{F}_j$ is a field for each j and $n, m \geq 1$. Then $\gamma(\mathbb{CAG}(\mathcal{S})) = 1$ if and only if one of the following holds:*

(1) *$n = 1$, $m = 2$ and $\Im_1$ is the only non-zero proper ideal of $\mathcal{S}_1$.*
(2) *$n = m = 1$, $\Im_1$, $\Im_1^2$ and $\Im_1^3$ are the only non-zero proper ideals of $\mathcal{S}_1$.*
(3) *$n = m = 1$, $\Im_1$ and $\Im_1^2$ are the only non-zero proper ideals of $\mathcal{S}_1$.*

PROOF. Assume that $\gamma(\mathbb{CAG}(\mathcal{S})) = 1$. Suppose $n \geq 2$. Consider $a_1 = \mathcal{S}_1 \times (0) \times (0) \times \cdots \times (0)$, $a_2 = \Im_1 \times (0) \times (0) \times \cdots \times (0)$, $a_3 = (0) \times \mathcal{S}_2 \times (0) \times \cdots \times (0)$, $a_4 = (0) \times \Im_2 \times (0) \times \cdots \times (0)$, $b_1 = (0) \times (0) \times \cdots \times (0) \times \mathcal{F}_1 \times (0) \times \cdots \times (0)$, $b_2 = \Im_1 \times (0) \times \cdots \times (0) \times \mathcal{F}_1 \times (0) \times \cdots \times (0)$, $b_3 = (0) \times \Im_2 \times (0) \times \cdots \times (0) \times \mathcal{F}_1 \times (0) \times \cdots \times (0)$, $b_4 = \Im_1 \times \Im_2 \times (0) \times \cdots \times (0)$, $b_5 = \Im_1 \times \Im_2 \times (0) \times \cdots \times (0) \times \mathcal{F}_1 \times (0) \times \cdots \times (0) \in A(\mathcal{S})^*$. Since $a_1b_1 = 0$, $a_1 \nsubseteq a_1b_2$ and $b_2 \nsubseteq a_1b_2$, $a_1b_3 = 0$, $a_1 \nsubseteq a_1b_4$ and $b_4 \nsubseteq a_1b_4$, $a_1 \nsubseteq a_1b_5$ and $b_5 \nsubseteq a_1b_5$, $a_2b_1 = 0$, $a_2 \nsubseteq a_2b_2$ and $b_2 \nsubseteq a_2b_2$, $a_2b_3 = 0$, $a_2 \nsubseteq a_2b_4$ and $b_4 \nsubseteq a_2b_4$, $a_2 \nsubseteq a_2b_5$ and $b_5 \nsubseteq a_2b_5$, $a_3b_1 = 0$, $a_3b_2 = 0$, $a_3 \nsubseteq a_3b_3$ and $b_3 \nsubseteq a_3b_3$, $a_3 \nsubseteq a_3b_4$ and $b_4 \nsubseteq a_3b_4$, $a_3 \nsubseteq a_3b_5$ and $b_5 \nsubseteq a_3b_5$, $a_4b_1 = 0$, $a_4b_2 = 0$, $a_4 \nsubseteq a_4b_3$ and $b_3 \nsubseteq a_4b_3$, $a_4 \nsubseteq a_4b_4$ and $b_4 \nsubseteq a_4b_4$, $a_4 \nsubseteq a_4b_5$ and $b_5 \nsubseteq a_4b_5$, then $K_{4,5}$ is an induced subgraph of $\mathbb{CAG}(\mathcal{S})$, a contradiction. Hence $n = 1$.

Suppose $m \geq 3$. Consider $x_1 = \mathcal{S}_1 \times (0) \times (0) \times \cdots \times (0)$, $x_2 = \Im_1 \times (0) \times (0) \times \cdots \times (0)$, $x_3 = (0) \times \mathcal{F}_1 \times (0) \times \cdots \times (0)$, $x_4 = \mathcal{S}_1 \times \mathcal{F}_1 \times (0) \times \cdots \times (0)$, $y_1 = (0) \times (0) \times \mathcal{F}_2 \times (0) \times \cdots \times (0)$, $y_2 = (0) \times (0) \times (0) \times \times \mathcal{F}_3 \times (0) \times \cdots \times (0)$, $y_3 = (0) \times (0) \times \mathcal{F}_2 \times \times \mathcal{F}_3 \times (0) \times \cdots \times (0)$, $y_4 = \Im_1 \times (0) \times \mathcal{F}_2 \times (0) \times \cdots \times (0)$, $y_5 = \Im_1 \times (0) \times (0) \times \mathcal{F}_3 \times (0) \times \cdots \times (0) \in A(\mathcal{S})^*$. Since $x_1y_1 = 0$, $x_1y_2 = 0$, $x_1y_3 = 0$, $x_1 \nsubseteq x_1y_4$ and $y_4 \nsubseteq x_1y_4$, $x_1 \nsubseteq x_1y_5$ and $y_5 \nsubseteq x_1y_5$, $x_2y_1 = 0$, $x_2y_2 = 0$, $x_2y_3 = 0$, $x_2 \nsubseteq x_2y_4$ and $y_4 \nsubseteq x_2y_4$, $x_2 \nsubseteq x_2y_5$ and $y_5 \nsubseteq x_2y_5$, $x_3y_1 = 0$, $x_3y_2 = 0$, $x_3y_3 = 0$, $x_3 \nsubseteq x_3y_4$ and $y_4 \nsubseteq x_3y_4$, $x_3 \nsubseteq x_3y_5$ and $y_5 \nsubseteq x_3y_5$, $x_4y_1 = 0$, $x_4y_2 = 0$, $x_4y_3 = 0$, $x_4y_4 = 0$, $x_4y_5 = 0$, then $\mathbb{CAG}(\mathcal{S})$ contains $K_{4,5}$ as a subgraph, a contradiction. Hence $m \leq 2$.

Suppose $m = 2$. Let $\mathcal{L}$ be a non-zero proper ideal of $\mathcal{S}_1$ such that $\mathcal{L} \neq \Im_1$. Consider $\alpha_1 = \mathcal{S}_1 \times (0) \times (0)$, $\alpha_2 = \Im_1 \times (0) \times (0)$, $\alpha_3 = \mathcal{L} \times (0) \times (0)$, $\beta_1 = (0) \times \mathcal{F}_1 \times (0)$, $\beta_2 = (0) \times \mathcal{F}_1 \times \mathcal{F}_2$, $\beta_3 = (0) \times (0) \times \mathcal{F}_2$, $\beta_4 = \Im_1 \times \mathcal{F}_1 \times (0)$, $\beta_5 = \Im_1 \times (0) \times \mathcal{F}_2$, $\beta_6 = \Im_1 \times \mathcal{F}_1 \times \mathcal{F}_2$, $\beta_7 = \mathcal{L} \times \mathcal{F}_1 \times (0) \in A(\mathcal{S})^*$. Since $\alpha_1\beta_1 = 0$, $\alpha_1\beta_2 = 0$, $\alpha_1\beta_3 = 0$, $\alpha_1 \nsubseteq \alpha_1\beta_4$ and $\beta_4 \nsubseteq \alpha_1\beta_4$, $\alpha_1 \nsubseteq \alpha_1\beta_5$ and $\beta_5 \nsubseteq \alpha_1\beta_5$, $\alpha_1 \nsubseteq \alpha_1\beta_6$ and $\beta_6 \nsubseteq \alpha_1\beta_6$, $\alpha_1 \nsubseteq \alpha_1\beta_7$ and $\beta_7 \nsubseteq \alpha_1\beta_7$, $\alpha_2\beta_1 = 0$, $\alpha_2\beta_2 = 0$, $\alpha_2\beta_3 = 0$, $\alpha_2 \nsubseteq \alpha_2\beta_4$ and $\beta_4 \nsubseteq \alpha_2\beta_4$, $\alpha_2 \nsubseteq \alpha_2\beta_5$ and $\beta_5 \nsubseteq \alpha_2\beta_5$, $\alpha_2 \nsubseteq \alpha_2\beta_6$ and

$\beta_6 \not\subseteq \alpha_2\beta_6$, $\alpha_2 \not\subseteq \alpha_2\beta_7$ and $\beta_7 \not\subseteq \alpha_2\beta_7$, $\alpha_3\beta_1 = 0$, $\alpha_3\beta_2 = 0$, $\alpha_3\beta_3 = 0$, $\alpha_3 \not\subseteq \alpha_3\beta_4$ and $\beta_4 \not\subseteq \alpha_3\beta_4$, $\alpha_3 \not\subseteq \alpha_3\beta_5$ and $\beta_5 \not\subseteq \alpha_3\beta_5$, $\alpha_3 \not\subseteq \alpha_3\beta_6$ and $\beta_6 \not\subseteq \alpha_3\beta_6$, $\alpha_3 \not\subseteq \alpha_3\beta_7$ and $\beta_7 \not\subseteq \alpha_3\beta_7$, then $K_{3,7}$ is an induced subgraph of $\mathbb{CAG}(\mathcal{S})$, a contradiction. Hence $\Im_1$ is the only non-zero proper ideal of $\mathcal{S}_1$.

Now, Suppose $m = 1$. Let η be the nilpotency index of $\Im_1$. Suppose $\eta \geq 5$. Consider $u_1 = \mathcal{S}_1 \times (0)$, $u_2 = \Im_1 \times (0)$, $u_3 = \Im_1^{\eta-1} \times (0)$, $u_4 = \Im_1^{\eta-2} \times (0)$, $u_5 = \Im_1^{\eta-3} \times (0)$, $v_1 = \Im_1 \times \mathcal{F}_1$, $v_2 = \Im_1^{\eta-1} \times \mathcal{F}_1$, $v_3 = \Im_1^{\eta-2} \times \mathcal{F}_1$, $v_4 = (0) \times \mathcal{F}_1 \in A(\mathcal{S})^*$. Since $u_1 \not\subseteq u_1v_1$ and $v_1 \not\subseteq u_1v_1$, $u_1 \not\subseteq u_1v_2$ and $v_2 \not\subseteq u_1v_2$, $u_1 \not\subseteq u_1v_3$ and $v_3 \not\subseteq u_1v_3$, $u_1v_4 = 0$, $u_2 \not\subseteq u_2v_1$ and $v_1 \not\subseteq u_2v_1$, $u_2v_2 = 0$, $u_2 \not\subseteq u_2v_3$ and $v_3 \not\subseteq u_2v_3$, $u_2v_4 = 0$, $u_3v_1 = 0$, $u_3v_2 = 0$, $u_3v_3 = 0$, $u_3v_4 = 0$, $u_4 \not\subseteq u_4v_1$ and $v_1 \not\subseteq u_4v_1$, $u_4v_2 = 0$, $u_4v_3 = 0$, $u_4v_4 = 0$, $u_5 \not\subseteq u_5v_1$ and $v_1 \not\subseteq u_5v_1$, $u_5v_2 = 0$, $u_5v_3 = 0$, $u_5v_4 = 0$, then $K_{5,4}$ is an induced subgraph of $\mathbb{CAG}(\mathcal{S})$, a contradiction. Hence $\eta \leq 4$. The following three cases occur:

CASE (i) ($\eta = 4$). Let $\mathcal{X}$ be a non-zero proper ideal of $\mathcal{S}_1$ such that $\mathcal{X} \neq \Im_1, \Im_1^2, \Im_1^3$. Consider $r_1 = \mathcal{S}_1 \times (0)$, $r_2 = \Im_1 \times (0)$, $r_3 = \Im_1^2 \times (0)$, $r_4 = \Im_1^3 \times (0)$, $r_5 = \mathcal{X} \times (0)$, $s_1 = (0) \times \mathcal{F}_1$, $s_2 = \Im_1 \times \mathcal{F}_1$, $s_3 = \Im_1^2 \times \mathcal{F}_1$, $s_4 = \Im_1^3 \times \mathcal{F}_1 \in A(\mathcal{S})^*$. Since $r_1s_1 = 0$, $r_1 \not\subseteq r_1s_2$ and $s_2 \not\subseteq r_1s_2$, $r_1 \not\subseteq r_1s_3$ and $s_3 \not\subseteq r_1s_3$, $r_1 \not\subseteq r_1s_4$ and $s_4 \not\subseteq r_1s_4$, $r_2s_1 = 0$, $r_2 \not\subseteq r_2s_2$ and $s_2 \not\subseteq r_2s_2$, $r_2 \not\subseteq r_2s_3$ and $s_3 \not\subseteq r_2s_3$, $r_2s_4 = 0$, $r_3s_1 = 0$, $r_3 \not\subseteq r_3s_2$ and $s_2 \not\subseteq r_3s_2$, $r_3s_3 = 0$, $r_3s_4 = 0$, $r_4s_1 = 0$, $r_4s_2 = 0$, $r_4s_3 = 0$, $r_4s_4 = 0$, $r_5s_1 = 0$, $r_5 \not\subseteq r_5s_2$ and $s_2 \not\subseteq r_5s_2$, $r_5 \not\subseteq r_5s_3$ and $s_3 \not\subseteq r_5s_3$, $r_5 \not\subseteq r_5s_4$ and $s_4 \not\subseteq r_5s_4$, then $\mathbb{CAG}(\mathcal{S})$ contain $K_{5,4}$ as an induced subgraph, a contradiction. Consequently, the only non-zero proper ideals in $\mathcal{S}_1$ are $\Im_1$, $\Im_1^2$ and $\Im_1^3$.

CASE (ii) ($\eta = 3$). Let $\mathcal{Y}$ be a non-zero proper ideal of $\mathcal{S}_1$ such that $\mathcal{Y} \neq \Im_1, \Im_1^2$. Then by [**19**, Proposition 2.7], there exist three non-zero proper ideals $\mathcal{Y}_1, \mathcal{Y}_2, \mathcal{Y}_3$ of $\mathcal{S}_1$ such that $\mathcal{Y}_1, \mathcal{Y}_2, \mathcal{Y}_3 \neq \Im_1, \Im_1^2$. Consider $w_1 = \mathcal{S}_1 \times (0)$, $w_2 = \Im_1 \times (0)$, $w_3 = \Im_1^2 \times (0)$, $w_4 = \mathcal{Y}_1 \times (0)$, $z_1 = (0) \times \mathcal{F}_1$, $z_2 = \Im_1 \times \mathcal{F}_1$, $z_3 = \Im_1^2 \times \mathcal{F}_1$, $z_4 = \mathcal{Y}_1 \times \mathcal{F}_1$, $z_5 = \mathcal{Y}_2 \times \mathcal{F}_1 \in A(\mathcal{S})^*$. Since $w_1z_1 = 0$, $w_1 \not\subseteq w_1z_2$ and $z_2 \not\subseteq w_1z_2$, $w_1 \not\subseteq w_1z_3$ and $z_3 \not\subseteq w_1z_3$, $w_1 \not\subseteq w_1z_4$ and $z_4 \not\subseteq w_1z_4$, $w_1 \not\subseteq w_1z_5$ and $z_5 \not\subseteq w_1z_5$, $w_2z_1 = 0$, $w_2 \not\subseteq w_2z_2$ and $z_2 \not\subseteq w_2z_2$, $w_2z_3 = 0$, $w_2 \not\subseteq w_2z_4$ and $z_4 \not\subseteq w_2z_4$, $w_2 \not\subseteq w_2z_5$ and $z_5 \not\subseteq w_2z_5$, $w_3z_1 = 0$, $w_3z_2 = 0$, $w_3z_3 = 0$, $w_3 \not\subseteq w_3z_4$ and $z_4 \not\subseteq w_3z_4$, $w_3 \not\subseteq w_3z_5$ and $z_5 \not\subseteq w_3z_5$, $w_4z_1 = 0$, $w_4 \not\subseteq w_4z_2$ and $z_2 \not\subseteq w_4z_2$, $w_4 \not\subseteq w_4z_3$ and $z_3 \not\subseteq w_4z_3$, $w_4 \not\subseteq w_4z_4$ and $z_4 \not\subseteq w_4z_4$, $w_4 \not\subseteq w_4z_5$ and $z_5 \not\subseteq w_4z_5$, then $K_{4,5}$ is an induced subgraph of $\mathbb{CAG}(\mathcal{S})$, a contradiction. Consequently, the only non-zero proper ideals in $\mathcal{S}_1$ are $\Im_1$ and $\Im_1^2$.

CASE (iii) ($\eta = 2$). Let $\mathcal{K}$ be the non-zero proper ideal of $\mathcal{S}_1$ such that $\mathcal{K} \neq \Im_1$. Then by [**19**, Proposition 2.7], there exist three different ideals $\mathcal{K}_1$, $\mathcal{K}_2$ and $\mathcal{K}_3$ of $\mathcal{S}_1$ such that $\mathcal{K}_1, \mathcal{K}_2, \mathcal{K}_3 \neq \Im_1$. Consider $p_1 = \mathcal{S}_1 \times (0)$, $p_2 = \Im_1 \times (0)$, $p_3 = \mathcal{K}_1 \times (0)$, $p_4 = \mathcal{K}_2 \times (0)$, $p_5 = \mathcal{K}_3 \times (0)$, $q_1 = (0) \times \mathcal{F}_1$, $q_2 = \Im_1 \times \mathcal{F}_1$, $q_3 = \mathcal{K}_1 \times \mathcal{F}_1$, $q_4 = \mathcal{K}_2 \times \mathcal{F}_1 \in A(\mathcal{S})$. Since $p_1q_1 = 0$, $p_1 \not\subseteq p_1q_2$ and $q_2 \not\subseteq p_1q_2$, $p_1 \not\subseteq p_1q_3$ and $q_3 \not\subseteq p_1q_3$, $p_1 \not\subseteq p_1q_4$ and $q_4 \not\subseteq p_1q_4$, $p_2q_1 = 0$, $p_2q_2 = 0$, $p_2 \not\subseteq p_2q_3$ and $q_3 \not\subseteq p_2q_3$, $p_2 \not\subseteq p_2q_4$ and $q_4 \not\subseteq p_2q_4$, $p_3q_1 = 0$, $p_3 \not\subseteq p_3q_2$ and $q_2 \not\subseteq p_3q_2$, $p_3 \not\subseteq p_3q_3$ and $q_3 \not\subseteq p_3q_3$, $p_3 \not\subseteq p_3q_4$ and $q_4 \not\subseteq p_3q_4$, $p_4q_1 = 0$, $p_4 \not\subseteq p_4q_2$ and $q_2 \not\subseteq p_4q_2$, $p_4 \not\subseteq p_4q_3$ and $q_3 \not\subseteq p_4q_3$, $p_4 \not\subseteq p_4q_4$ and $q_4 \not\subseteq p_4q_4$, $p_5q_1 = 0$, $p_5 \not\subseteq p_5q_2$ and $q_2 \not\subseteq p_5q_2$, $p_5 \not\subseteq p_5q_3$ and $q_3 \not\subseteq p_5q_3$, $p_5 \not\subseteq p_5q_4$ and $q_4 \not\subseteq p_5q_4$, then $K_{4,5}$ is an induced subgraph of $\mathbb{CAG}(\mathcal{S})$, a contradiction. Hence $\mathcal{S}$ has the unique non-zero proper ideal which is $\Im_1$. Thus, by Theorem 2.6, $\gamma(\mathbb{CAG}(\mathcal{S})) = 0$, again a contradiction.

The converse follows from Figures 6, 7 and 8. □

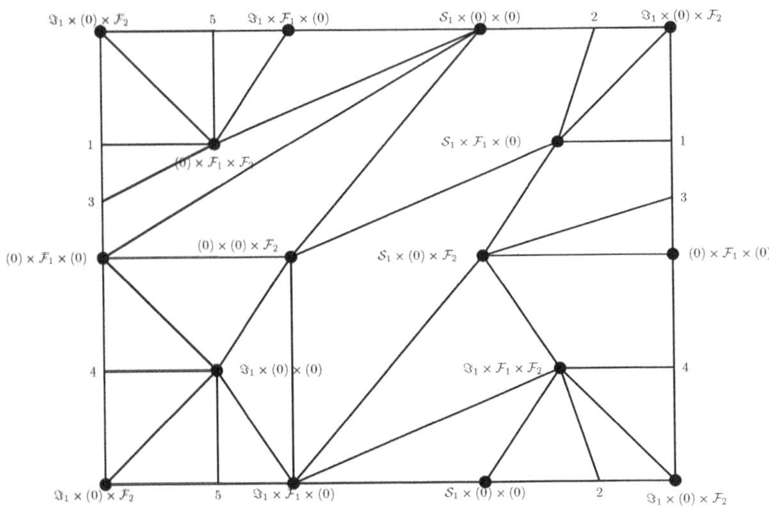

FIGURE 6. Toroidal embedding of $\mathbb{CAG}(\mathcal{S}_1 \times \mathcal{F}_1 \times \mathcal{F}_2)$ where $\Im_1$ is the only non-zero proper ideal of $\mathcal{S}_1$

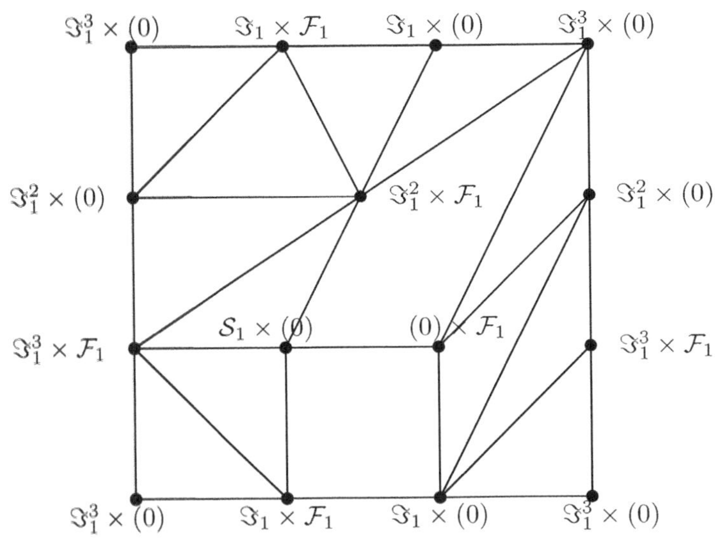

FIGURE 7. Toroidal embedding of $\mathbb{CAG}(\mathcal{S}_1 \times \mathcal{F}_1)$ where $\Im_1$, $\Im_1^2$ and $\Im_1^3$ are the only non-zero proper ideals of $\mathcal{S}_1$

4. Crosscap of $\mathbb{CAG}(\mathcal{S})$

In this section, we characterize the Artinian rings $\mathcal{S}$ for which $\mathbb{CAG}(\mathcal{S})$ has crosscap one.

Consider a non-negative integer denoted as k. Let us use N_k to represent a sphere with k crosscaps attached to it. For a specific non-negative integer value of k, it can be asserted that every connected compact surface is homeomorphic to N_k. The least positive integer k such that G may be embedded in N_k is the *crosscap*

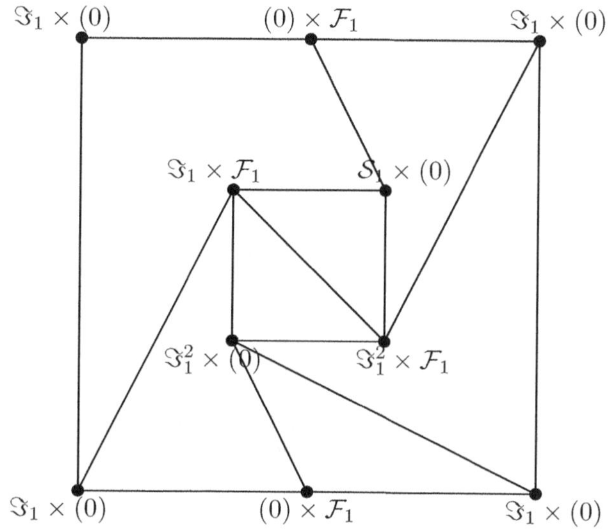

FIGURE 8. Toroidal embedding of $\mathbb{CAG}(\mathcal{S}_1 \times \mathcal{F}_1)$, where $\Im_1$ and $\Im_1^2$ are the only non-zero proper ideals of $\mathcal{S}_1$

number of G, denoted by $\overline{\gamma}(G)$. It is obvious that $\overline{\gamma}(G) \leq \overline{\gamma}(\tilde{G})$, where $\tilde{G}$ is the reduction of G, and $\overline{\gamma}(H) \leq \overline{\gamma}(G)$ for every subgraph H of G. The crosscap of the complete graph and complete bipartite graphs are given in the following results, which are useful for proving the results of this section.

LEMMA 4.1 ([**21**]). *Let $m \geq 3$. Then*

$$\overline{\gamma}(K_m) = \begin{cases} \left\lceil \frac{(m-3)(m-4)}{6} \right\rceil & \text{if } m \geq 3 \text{ and } m \neq 7; \\ 3 & \text{if } m = 7. \end{cases}$$

LEMMA 4.2 ([**21**]). *Let $n, m \geq 2$. Then*

$$\overline{\gamma}(K_{m,n}) = \left\lceil \frac{(m-2)(n-2)}{2} \right\rceil.$$

THEOREM 4.3. *let $\mathcal{S}$ be an Artinian commutative ring. Then $\overline{\gamma}(\mathbb{CAG}(\mathcal{S})) = 1$ if and only if one of the following holds:*

(1) *$\mathcal{S} \cong \mathcal{S}_1 \times \mathcal{S}_2$, where $(\mathcal{S}_i, \Im_i)$ is a local Artinain ring with $\Im_i$ is the only non-zero proper ideal of $\mathcal{S}_i$ for each $i = 1, 2$.*
(2) *$\mathcal{S} \cong \mathcal{S}_1 \times \mathcal{F}$, where $(\mathcal{S}_1, \Im_1)$ is a local Artinian ring with $\Im_1$ and $\Im_1^2$ are the only non-zero proper ideals of $\mathcal{S}_1$ and $\mathcal{F}$ is a field.*
(3) *$\mathcal{S}$ is a local ring with at least five and at most six non-zero proper ideals.*

PROOF. Suppose that $\overline{\gamma}(\mathbb{CAG}(\mathcal{S})) = 1$. Since $\mathcal{S}$ is Artinian, $\mathcal{S} \cong \mathcal{S}_1 \times \mathcal{S}_2 \times \cdots \times \mathcal{S}_n$, where $\mathcal{S}_i$ is a local Artinian ring for each i and $n \geq 1$. Suppose $n \geq 4$. Consider $x_1 = \mathcal{S}_1 \times (0) \times (0) \times \cdots \times (0)$, $x_2 = \mathcal{S}_1 \times \mathcal{S}_2 \times (0) \times \cdots \times (0)$, $x_3 = \mathcal{S}_1 \times (0) \times \mathcal{S}_3 \times (0) \times \cdots \times (0)$, $x_4 = \mathcal{S}_1 \times (0) \times (0) \times \mathcal{S}_4 \times (0) \times \cdots \times (0)$, $y_1 = (0) \times \mathcal{S}_2 \times \mathcal{S}_3 \times (0) \times \cdots \times (0)$, $y_2 = (0) \times \mathcal{S}_2 \times (0) \times \mathcal{S}_4 \times (0) \times \cdots \times (0)$, $y_3 = (0) \times (0) \times \mathcal{S}_3 \times \mathcal{S}_4 \times (0) \times \cdots \times (0)$, $y_4 = (0) \times \mathcal{S}_2 \times \mathcal{S}_3 \times \mathcal{S}_4 \times (0) \times \cdots \times (0) \in A(\mathcal{S})^*$. Since $x_1 y_i = 0$ for each $1 \leq i \leq 4$, $x_i \nsubseteq x_i y_j$ and $y_j \nsubseteq x_i y_j$ for each $2 \leq i \leq 4$ and

$1 \leq j \leq 4$, then $K_{4,4}$ is a subgraph of $\mathbb{CAG}(\mathcal{S})$. Thus, $\overline{\gamma}(\mathbb{CAG}(\mathcal{S})) > 1$, by Lemma 4.2, a contradiction. Hence $n \leq 3$. The following cases occur:

CASE (i) ($n = 3$). If $\mathcal{S}_i$ is a field for each $i = 1, 2, 3$, then $\mathbb{CAG}(\mathcal{S})$ is a planar graph by Theorem 2.6, which implies that $\overline{\gamma}(\mathbb{CAG}(\mathcal{S})) = 0$, a contradiction. Hence at least one of $\mathcal{S}_i$ say $\mathcal{S}_1$ is not a field with the maximal ideal $\Im_1 \neq 0$. Consider $a_1 = \mathcal{S}_1 \times (0) \times (0)$, $a_2 = \Im_1 \times (0) \times (0)$, $a_3 = \mathcal{S}_1 \times \mathcal{S}_2 \times (0)$, $a_4 = \Im_1 \times \mathcal{S}_2 \times (0)$, $b_1 = (0) \times (0) \times \mathcal{S}_3$, $b_2 = \Im_1 \times (0) \times \mathcal{S}_3$, $b_3 = \Im_1 \times \mathcal{S}_2 \times \mathcal{S}_3$, $b_4 = (0) \times \mathcal{S}_2 \times \mathcal{S}_3 \in A(\mathcal{S})^*$. Since $a_1b_1 = 0$, $a_1 \nsubseteq a_1b_2$ and $b_2 \nsubseteq a_1b_2$, $a_1 \nsubseteq a_1b_3$ and $b_3 \nsubseteq a_1b_3$, $a_1b_4 = 0$, $a_2b_1 = 0$, $a_2 \nsubseteq a_2b_2$ and $b_2 \nsubseteq a_2b_2$, $a_2 \nsubseteq a_2b_3$ and $b_3 \nsubseteq a_2b_3$, $a_2b_4 = 0$, $a_3b_1 = 0$, $a_3 \nsubseteq a_3b_2$ and $b_2 \nsubseteq a_3b_2$, $a_3 \nsubseteq a_3b_3$ and $b_3 \nsubseteq a_3b_3$, $a_3 \nsubseteq a_3b_4$ and $b_4 \nsubseteq a_3b_4$, $a_4b_1 = 0$, $a_4 \nsubseteq a_4b_2$ and $b_2 \nsubseteq a_4b_2$, $a_4 \nsubseteq a_4b_3$ and $b_3 \nsubseteq a_4b_3$, $a_4 \nsubseteq a_4b_4$ and $b_4 \nsubseteq a_4b_4$, then $K_{4,4}$ is a subgraph of $\mathbb{CAG}(\mathcal{S})$, a contradiction by Lemma 4.2.

CASE (ii) ($n = 2$). If $\mathcal{S}_1$ and $\mathcal{S}_2$ both are fields, then $\mathbb{CAG}(\mathcal{S}) \cong K_2$ by Theorem 2.6, which implies that $\overline{\gamma}(\mathbb{CAG}(\mathcal{S})) = 0$. Hence at least one of $\mathcal{S}_1$ or $\mathcal{S}_2$ is not a field.

Suppose $\mathcal{S}_1$ and $\mathcal{S}_2$ both are not field with maximal ideals $\Im_1 \neq 0$ and $\Im_2 \neq 0$, respectively. Let $\mathcal{X}$ be the non-zero proper ideal of $\mathcal{S}_1$ such the $\Im_1 \neq \mathcal{X}$. Consider $u_1 = \mathcal{S}_1 \times (0)$, $u_2 = \Im_1 \times (0)$, $u_3 = \mathcal{X} \times (0)$, $u_4 = \mathcal{S}_1 \times \Im_2$, $v_1 = (0) \times \mathcal{S}_2$, $v_2 = (0) \times \Im_2$, $v_3 = \Im_1 \times \Im_2$, $v_4 = \mathcal{X} \times \Im_2 \in A(\mathcal{S})^*$. Since $u_1v_1 = 0$, $u_1v_2 = 0$, $u_1 \nsubseteq u_1v_3$ and $v_3 \nsubseteq u_1v_3$, $u_1 \nsubseteq u_1v_4$ and $v_4 \nsubseteq u_1v_4$, $u_2v_1 = 0$, $u_2v_2 = 0$, $u_2 \nsubseteq u_2v_3$ and $v_3 \nsubseteq u_2v_3$, $u_2 \nsubseteq u_2v_4$ and $v_4 \nsubseteq u_2v_4$, $u_3v_2 = 0$, $u_3 \nsubseteq u_3v_3$ and $v_3 \nsubseteq u_3v_3$, $u_3 \nsubseteq u_3v_4$ and $v_4 \nsubseteq u_3v_4$, $u_4 \nsubseteq u_4v_1$ and $v_1 \nsubseteq u_4v_1$, $u_4 \nsubseteq u_4v_2$ and $v_2 \nsubseteq u_4v_2$, $u_4 \nsubseteq u_4v_3$ and $v_3 \nsubseteq u_4v_3$, $u_4 \nsubseteq u_4v_4$ and $v_4 \nsubseteq u_4v_4$, then $K_{4,4}$ is a subgraph of $\mathbb{CAG}(\mathcal{S})$, a contradiction by Lemma 4.2. Hence $\mathcal{S}_1$ has the unique non-zero proper ideal which is $\Im_1$. Similarly, one can show that $\mathcal{S}_2$ has the unique non-zero proper ideal which is $\Im_2$.

Now, suppose that $\mathcal{S} \cong \mathcal{S}_1 \times \mathcal{F}$, where $\mathcal{S}_1$ is a local Artinain ring with the maximal ideal $\Im_1 \neq 0$ and $\mathcal{F}$ is a field. Let η be the nilpotency index of $\Im_1$. Suppose $\eta \geq 4$. Consider $w_1 = \mathcal{S}_1 \times (0)$, $w_2 = \Im_1 \times (0)$, $w_3 = \Im_1^{\eta-1} \times (0)$, $w_4 = \Im_1^{\eta-2} \times (0)$, $z_1 = (0) \times \mathcal{F}$, $z_2 = \Im_1 \times \mathcal{F}$, $z_3 = \Im_1^{\eta-1} \times \mathcal{F}$, $z_4 = \Im_1^{\eta-2} \times \mathcal{F} \in A(\mathcal{S})^*$. Since $w_1z_1 = 0$, $w_1 \nsubseteq w_1z_2$ and $z_2 \nsubseteq w_1z_2$, $w_1 \nsubseteq w_1z_3$ and $z_3 \nsubseteq w_1z_3$, $w_1 \nsubseteq w_1z_4$ and $z_4 \nsubseteq w_1z_4$, $w_2z_1 = 0$, $w_2 \nsubseteq w_2z_2$ and $z_2 \nsubseteq w_2z_2$, $w_2z_3 = 0$, $w_2 \nsubseteq w_2z_4$ and $z_4 \nsubseteq w_2z_4$, $w_3z_1 = 0$, $w_3z_2 = 0$, $w_3z_3 = 0$, $w_3z_4 = 0$, $w_4z_1 = 0$, $w_4 \nsubseteq w_4z_2$ and $z_2 \nsubseteq w_4z_2$, $w_4z_3 = 0$, $w_4z_4 = 0$, then $K_{4,4}$ is a subgraph of $\mathbb{CAG}(\mathcal{S})$, a contradiction by Lemma 4.2. Hence $\eta \leq 3$. The following subcases occur:

SUBCASE (a) ($\eta = 3$). Let $\mathcal{Y}$ be the non-zero proper ideal of $\mathcal{S}_1$ such that $\mathcal{Y} \neq \Im_1, \Im_1^2$. Consider $r_1 = \mathcal{S}_1 \times (0)$, $r_2 = \Im_1 \times (0)$, $r_3 = \Im_1^2 \times (0)$, $r_4 = \mathcal{Y} \times (0)$, $s_1 = (0) \times \mathcal{F}$, $s_2 = \Im_1 \times \mathcal{F}$, $s_3 = \Im_1^2 \times \mathcal{F}$, $s_4 = \mathcal{Y} \times \mathcal{F} \in A(\mathcal{S})^*$. Since $r_1s_1 = 0$, $r_1 \nsubseteq r_1s_2$ and $s_2 \nsubseteq r_1s_2$, $r_1 \nsubseteq r_1s_3$ and $s_3 \nsubseteq r_1s_3$, $r_1 \nsubseteq r_1s_4$ and $s_4 \nsubseteq r_1s_4$, $r_2s_1 = 0$, $r_2 \nsubseteq r_2s_2$ and $s_2 \nsubseteq r_2s_2$, $r_2s_3 = 0$, $r_2 \nsubseteq r_2s_4$ and $s_4 \nsubseteq r_2s_4$, $r_3s_1 = 0$, $r_3s_2 = 0$, $r_3s_3 = 0$, $r_3 \nsubseteq r_3s_4$ and $s_4 \nsubseteq r_3s_4$, $r_4s_1 = 0$, $r_4 \nsubseteq r_4s_2$ and $s_2 \nsubseteq r_4s_2$, $r_4 \nsubseteq r_4s_3$ and $s_3 \nsubseteq r_4s_3$, $r_4 \nsubseteq r_4s_4$ and $s_4 \nsubseteq r_4s_4$, then $K_{4,4}$ is a subgraph of $\mathbb{CAG}(\mathcal{S})$, a contradiction by Lemma 4.2. Hence $\mathcal{S}$ has only two non-zero proper ideals which are $\Im_1$ and $\Im_1^2$.

SUBCASE (b) ($\eta = 2$). Let $\mathcal{K}$ be the non-zero proper ideal of $\mathcal{S}_1$ such that $\mathcal{K} \neq \Im_1$. Then by [**19**, Proposition 2.7], there exist three different ideals $\mathcal{K}_1$, $\mathcal{K}_2$ and $\mathcal{K}_3$ of $\mathcal{S}_1$ such that $\mathcal{K}_1, \mathcal{K}_2, \mathcal{K}_3 \neq \Im_1$. Consider $p_1 = \mathcal{S}_1 \times (0)$, $p_2 = \Im_1 \times (0)$,

$p_3 = \mathcal{K}_1 \times (0)$, $p_4 = \mathcal{K}_2 \times (0)$, $q_1 = (0) \times \mathcal{F}_1$, $q_2 = \Im_1 \times \mathcal{F}_1$, $q_3 = \mathcal{K}_1 \times \mathcal{F}_1$, $q_4 = \mathcal{K}_2 \times \mathcal{F}_1 \in \mathbb{CAG}(\mathcal{S})$. Since $p_1q_1 = 0$, $p_1 \nsubseteq p_1q_2$ and $q_2 \nsubseteq p_1q_2$, $p_1 \nsubseteq p_1q_3$ and $q_3 \nsubseteq p_1q_3$, $p_1 \nsubseteq p_1q_4$ and $q_4 \nsubseteq p_1q_4$, $p_2q_1 = 0$, $p_2q_2 = 0$, $p_2 \nsubseteq p_2q_3$ and $q_3 \nsubseteq p_2q_3$, $p_2 \nsubseteq p_2q_4$ and $q_4 \nsubseteq p_2q_4$, $p_3q_1 = 0$, $p_3 \nsubseteq p_3q_2$ and $q_2 \nsubseteq p_3q_2$, $p_3 \nsubseteq p_3q_3$ and $q_3 \nsubseteq p_3q_3$, $p_3 \nsubseteq p_3q_4$ and $q_4 \nsubseteq p_3q_4$, $p_4q_1 = 0$, $p_4 \nsubseteq p_4q_2$ and $q_2 \nsubseteq p_4q_2$, $p_4 \nsubseteq p_4q_3$ and $q_3 \nsubseteq p_4q_3$, $p_4 \nsubseteq p_4q_4$ and $q_4 \nsubseteq p_4q_4$, then $K_{4,4}$ is a subgraph of $\mathbb{CAG}(\mathcal{S})$, a contradiction. Hence $\mathcal{S}$ has the unique non-zero proper ideal which is $\Im_1$. Thus, by Theorem 2.6, $\overline{\gamma}(\mathbb{CAG}(\mathcal{S})) = 0$, again a contradiction.

CASE (iii) ($n = 1$). Then $\mathcal{S}$ is a local Artinian ring. Thus $\mathbb{CAG}(\mathcal{S})$ is complete by Theorem 1.10. Hence $5 \leq |A(\mathcal{S})^*| \leq 6$ by Lemma 4.1.

Conversely, if $\mathcal{S} \cong \mathcal{S}_1 \times \mathcal{S}_2$, where $(\mathcal{S}_i, \Im_i)$ is a local Artinain ring with $\Im_i$ is the only non-zero proper ideal of $\mathcal{S}_i$ for each $i = 1, 2$, then the vertex set of $\mathbb{CAG}(\mathcal{S})$ is given by $\{\mathcal{X}_1 = \mathcal{S}_1 \times (0), \mathcal{X}_2 = \mathcal{S}_1 \times \Im_2, \mathcal{X}_3 = (0) \times \mathcal{S}_2, \mathcal{X}_4 = (0) \times \Im_2, \mathcal{X}_5 = \Im_1 \times \Im_2, \mathcal{X}_6 = \Im_1 \times (0), \mathcal{X}_7 = \Im_1 \times \mathcal{S}_2\}$. Hence the projective embedding of $\mathbb{CAG}(\mathcal{S})$ is shown in Figure 9. Also, if $\mathcal{S} \cong \mathcal{S}_1 \times \mathcal{F}$, where $(\mathcal{S}_1, \Im_1)$ is a local Artinian ring with $\Im_1$ and $\Im_1^2$ are the only non-zero proper ideals of $\mathcal{S}_1$ and $\mathcal{F}$ is a field, then vertex set of $\mathbb{CAG}(\mathcal{S})$ is given by $\{\mathcal{Y}_1 = \Im_1 \times (0), \mathcal{Y}_2 = \Im_1^2 \times (0), \mathcal{Y}_3 = \Im_1 \times \mathcal{F}, \mathcal{Y}_4 = \Im_1^2 \times \mathcal{F}, \mathcal{Y}_5 = \mathcal{S}_1 \times (0), \mathcal{Y}_6 = (0) \times \mathcal{F}\}$. Hence the projective embedding of $\mathbb{CAG}(\mathcal{S})$ is shown in Figure 10. Finally, if $\mathcal{S}$ is a local ring with $5 \leq |A(\mathcal{S})^*| \leq 6$, then the result holds by Theorem 1.10 and Lemma 4.1. □

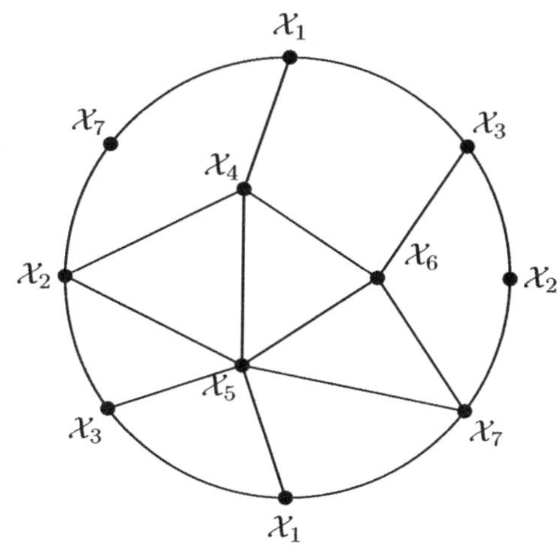

FIGURE 9. Projective embedding of $\mathbb{CAG}(\mathcal{S}_1 \times \mathcal{S}_2)$

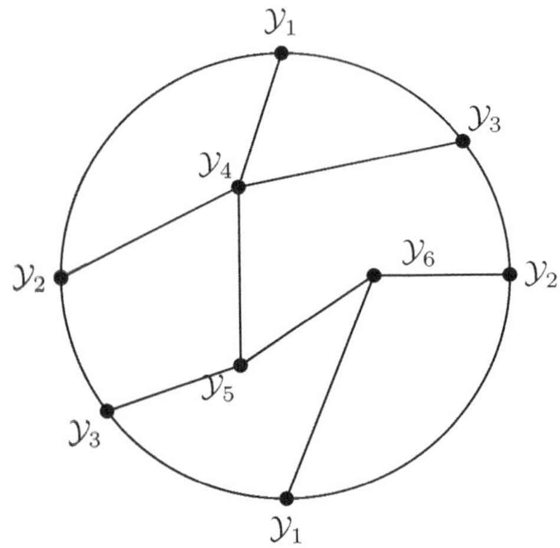

FIGURE 10. Projective embedding of $\mathbb{CAG}(\mathcal{S}_1 \times \mathcal{F})$

References

[1] G. Aalipour, S. Akbari, R. Nikandish, M.J. Nikmehr, and F. Shaveisi, *On the coloring of the annihilating-ideal graph of a commutative ring*, Discrete Math. **312** (2012), no. 17, 2620–2626, DOI 10.1016/j.disc.2011.10.020. MR2935413

[2] G. Aalipour, S. Akbari, R. Nikandish, M. J. Nikmehr, and F. Shaveisi, *Minimal prime ideals and cycles in annihilating-ideal graphs*, Rocky Mountain J. Math. **43** (2013), no. 5, 1415–1425, DOI 10.1216/RMJ-2013-43-5-1415. MR3127828

[3] M. Ahrari, Sh. A. Safari Sabet, and B. Amini, *On the girth of the annihilating-ideal graph of a commutative ring*, J. Linear Topol. Algebra **4** (2015), no. 3, 209–216. MR3567332

[4] S. Akbari and A. Mohammadian, *On the zero-divisor graph of a commutative ring*, J. Algebra **274** (2004), no. 2, 847–855, DOI 10.1016/S0021-8693(03)00435-6. MR2043378

[5] S. Akbari and A. Mohammadian, *Zero-divisor graphs of non-commutative rings*, J. Algebra **296** (2006), no. 2, 462–479, DOI 10.1016/j.jalgebra.2005.07.007. MR2201052

[6] David F. Anderson and Philip S. Livingston, *The zero-divisor graph of a commutative ring*, J. Algebra **217** (1999), no. 2, 434–447, DOI 10.1006/jabr.1998.7840. MR1700509

[7] D. D. Anderson and M. Naseer, *Beck's coloring of a commutative ring*, J. Algebra **159** (1993), no. 2, 500–514, DOI 10.1006/jabr.1993.1171. MR1231228

[8] M. F. Atiyah and I. G. Macdonald, *Introduction to commutative algebra*, Addison-Wesley Publishing Co., Reading, Mass.-London-Don Mills, Ont., 1969. MR242802

[9] István Beck, *Coloring of commutative rings*, J. Algebra **116** (1988), no. 1, 208–226, DOI 10.1016/0021-8693(88)90202-5. MR944156

[10] M. Behboodi and Z. Rakeei, *The annihilating-ideal graph of commutative rings I*, J. Algebra Appl. **10** (2011), no. 4, 727–739, DOI 10.1142/S0219498811004896. MR2834112

[11] M. Behboodi and Z. Rakeei, *The annihilating-ideal graph of commutative rings II*, J. Algebra Appl. **10** (2011), no. 4, 741–753, DOI 10.1142/S0219498811004902. MR2834113

[12] F. R. DeMeyer, T. McKenzie, and K. Schneider, *The zero-divisor graph of a commutative semigroup*, Semigroup Forum **65** (2002), no. 2, 206–214, DOI 10.1007/s002330010128. MR1911724

[13] T. Y. Lam, *A first course in noncommutative rings*, Graduate Texts in Mathematics, vol. 131, Springer-Verlag, New York, 1991, DOI 10.1007/978-1-4684-0406-7. MR1125071

[14] Mohd Nazim and Nadeem ur Rehman, *On the essential annihilating-ideal graph of commutative rings*, Ars Math. Contemp. **22** (2022), no. 3, Paper No. 5, 16. MR4481392

[15] Mohd Nazim, Nadeem ur Rehman, and Shabir Ahmad Mir, *Some properties of the essential annihilating-ideal graph of commutative rings*, Commun. Comb. Optim. **8** (2023), no. 4, 715–724. MR4602147

[16] M. J. Nikmehr and S. M. Hosseini, *More on the annihilator-ideal graph of a commutative ring*, J. Algebra Appl. **18** (2019), no. 8, 1950160, 14, DOI 10.1142/S0219498819501603. MR3977821

[17] Shane P. Redmond, *The zero-divisor graph of a non-commutative ring*, Commutative rings, Nova Sci. Publ., Hauppauge, NY, 2002, pp. 39–47. MR2037657

[18] Sepideh Salehifar, Kazem Khashyarmanesh, and Mojgan Afkhami, *On the annihilator-ideal graph of commutative rings*, Ric. Mat. **66** (2017), no. 2, 431–447, DOI 10.1007/s11587-016-0311-y. MR3715912

[19] K. Selvakumar, P. Subbulakshmi, and Jafar Amjadi, *On the genus of the graph associated to a commutative ring*, Discrete Math. Algorithms Appl. **9** (2017), no. 5, 1750058, 11, DOI 10.1142/S1793830917500586. MR3719737

[20] Douglas B. West, *Introduction to graph theory*, Prentice Hall, Inc., Upper Saddle River, NJ, 1996. MR1367739

[21] Arthur T. White, *Graphs, groups and surfaces*, North-Holland Mathematics Studies, No. 8, North-Holland Publishing Co., Amsterdam-London; American Elsevier Publishing Co., Inc., New York, 1973. MR340026

School of Computational Sciences, Faculty of Science and Technology, JSPM University, Pune-412207, India
Email address: mnazim1882@gmail.com

Department of Mathematics, Aligarh Muslim University, Aligarh-202002, India
Email address: nu.rehman.mm@amu.ac.in

Department of Mathematics, Aligarh Muslim University, Aligarh-202002, India
Email address: nrsansari03@gmail.com

Department of Mathematics, Aligarh Muslim University, Aligarh-202002, India
Email address: mirshabir967@gmail.com

Contemporary Mathematics
Volume **826**, 2025
https://doi.org/10.1090/conm/826/16583

Normally torsion-freeness and normality criteria for monomial ideals

Mehrdad Nasernejad, Veronica Crispin Quiñonez, and Jonathan Toledo

ABSTRACT. In this paper, we focus on the associated primes of powers of monomial ideals and asymptotic behavior properties such as normally torsion-freeness, normality, the strong persistence property, and the persistence property. In particular, we introduce the concept of monomial ideals of well-nearly normally torsion-free type, and show that these ideals are normal. After that, we present some results on the existence of embedded associated prime ideals in the associated primes set of powers of monomial ideals. Further, we employ them in investigating the edge and cover ideals of cones of graphs. Next, we present counterexamples to several questions concerning the relations between relevant algebraic properties of the edge ideals of clutters and complement clutters. We conclude by providing counterexamples to questions on the possible connections between normally torsion-freeness and normality of monomial ideals under polarization.

1. Introduction and preliminaries

Let R be a commutative Noetherian ring and I be an ideal of R. A well-known result of Brodmann [**7**] showed that the sequence $\{\mathrm{Ass}_R(R/I^k)\}_{k\geq 1}$ of associated prime ideals is stationary for large k. In other words, there exists a positive integer k_0 such that $\mathrm{Ass}_R(R/I^k) = \mathrm{Ass}_R(R/I^{k_0})$ for all $k \geq k_0$. The minimal such k_0 is called the *index of stability* of I, and $\mathrm{Ass}_R(R/I^{k_0})$ is called the *stable set* of associated prime ideals of I, which is denoted by $\mathrm{Ass}^{\infty}(I)$. In the context of Brodmann's result, one can ask whether it is true that $\mathrm{Ass}_R(R/I^k) \subseteq \mathrm{Ass}_R(R/I^{k+1})$ for all $k \geq 1$? McAdam [**20**] presented an example which says, in general, the question has a negative answer. We say that an ideal I of R satisfies the *persistence property* if $\mathrm{Ass}_R(R/I^k) \subseteq \mathrm{Ass}_R(R/I^{k+1})$ for all $k \geq 1$. Along this argument, an ideal I of R has the *strong persistence property* if $(I^{k+1} :_R I) = I^k$ for all $k \geq 1$. Ratliff showed in [**36**] that $(I^{k+1} :_R I) = I^k$ for all large k. Note that the strong persistence property implies the persistence property (for example see [**23**, Proposition 2.9]), however, the converse does not hold (see [**21**, Example 2.18]). In the sequel, the persistence property will often be shortened as by PP and the strong equivalent by SPP.

2020 *Mathematics Subject Classification.* Primary 13B25, 13F20, 05C25, 05E40.

Key words and phrases. Normally torsion-free monomial ideals, Normal monomial ideals, Embedded associated primes, Clutters, Polarization.

The first author is the corresponding author.

The second author was partially supported by the Lundström-Åman Foundation.

Let now I be a monomial ideal in a polynomial ring $R = K[x_1, \ldots, x_n]$ over a field K. In general, monomial ideals do not satisfy PP, see [**14**] for a counterexample. In order to further investigate PP for monomial ideals, their role in connecting commutative algebra with combinatorics comes to benefit, for instance, through combinatorial objects such as (hyper)graphs, simplicial complexes and posets. In particular, combinatorial methods have been applied to find classes of monomial ideals possessing PP (see [**24**]) as well as counterexamples in the monomial case (see [**16**] and [**22**]).

One of the innovators in this area is Villarreal, who in [**44**] introduced the concept of the *edge ideal* associated to a simple finite graph G, which is denoted by $I(G)$ and is generated by the monomials x_ix_j for every edge $\{x_j, x_j\}$ of G. Further, the Alexander dual of $I(G)$ is called the *cover ideal* of G and is denoted by $I(G)^\vee$. Next, path ideals of graphs were first introduced by Conca and De Negri [**8**] in the context of monomial ideals of linear type. So, we have the following subquestion:

Which classes of monomial ideals have the persistence property?

The answer is affirmative, when I is the cover ideal of a perfect graph (see [**9**, Corollary 5.11]) or the edge ideal of a finite graph, simple or with loops (see [**21**, Theorem 2.15] and [**37**, Corollary 2], respectively). Moreover, any polymatroidal ideal has PP (see [**15**, Proposition 3.3]).

There are two more concepts related to PP. The first one was presented in [**37**]: an ideal I in a commutative Noetherian ring R has the *symbolic strong persistence property* if $(I^{(k+1)} :_R I^{(1)}) = I^{(k)}$ for all $k \geq 1$, where $I^{(k)} = \bigcap_{\mathfrak{p}\in \mathrm{Min}(I)}(I^kR_\mathfrak{p} \cap R)$ denotes the k-th symbolic power of I. In the same paper the authors show that this property follows from SPP. However, little is known about other classes of monomial ideals which satisfy the symbolic SPP, see [**29**].

The second related concept is the following: an ideal I in a commutative Noetherian ring R is called *normally torsion-free* if $\mathrm{Ass}(R/I^k) \subseteq \mathrm{Ass}(R/I)$ for all $k \geq 1$. Examples of normally torsion-free monomial ideals appear from graph theory. Already in 1994, it was proved in [**41**] that a finite simple graph G is bipartite if and only if its edge ideal is normally torsion-free. Later it was shown in [**10**] that the cover ideals of bipartite graphs are normally torsion-free. Another example of the connection to graph theory is that normally torsion-free square-free monomial ideals are closely related to Mengerian hypergraphs. Although normally torsion-free square-free monomial ideals have been studied in [**17**, **30**, **42**], little is known for the normally torsion-free monomial ideals which are not square-free, for example, see [**34**].

We now turn our attention to the notion of the normality of ideals. An element $f \in R$ is *integral* over an ideal $I \subset R$, if there exists an equation

$$f^k + c_1f^{k-1} + \cdots + c_{k-1}f + c_k = 0 \text{ with } c_i \in I^i.$$

The *integral closure* of I is $\overline{I}$, the set of the elements in R which are integral over I. The ideal I is called *integrally closed*, if $I = \overline{I}$, and I is said to be *normal* if all powers of I are integrally closed, see [**43**] for more information. If I is a monomial ideal in a polynomial ring R, then $\overline{I}$ is the monomial ideal generated by all monomials $u \in R$ for which there exists an integer k such that $u^k \in I^k$, refer to [**13**, Theorem 1.4.2].

The last related definition for a monomial ideal in R is as follows: I is called *nearly normally torsion-free* if there exist a positive integer k and a monomial prime ideal $\mathfrak{p}$ such that $\mathrm{Ass}_R(R/I^m) = \mathrm{Min}(I)$ for all $1 \leq m \leq k$, and $\mathrm{Ass}_R(R/I^m) \subseteq$

$\mathrm{Min}(I) \cup \{\mathfrak{p}\}$ for all $m \geq k+1$, where $\mathrm{Min}(I)$ denotes the set of minimal prime ideals of I, see [**5**, Definition 2.1]. For some classes of nearly normally torsion-free monomial ideals, see [**26, 32**].

One important implication is that a normal monomial ideal has SPP by [**35**, Theorem 6.2]. This result is used in [**28**], where SPP is shown through normality in the cover ideals of the following imperfect graphs: (1) cycle graphs of odd orders, (2) wheel graphs of even orders, and (3) Helm graphs of odd orders with greater than or equal to 5. In particular, more information related to the normality of the cover ideals of imperfect graphs can be found in [**3, 4, 26**]. Moreover, based on [**13**, Theorem 1.4.6], every normally torsion-free square-free monomial ideal is normal. Hence, in the case of square-free monomial ideal, we have: normally torsion-freeness ⇒ normality ⇒ strong persistence property ⇒ persistence property.

The main aim of this work is to continue these research lines and identify the other classes of monomial ideals that satisfy PP, SPP, normality, and normally torsion-freeness. In particular, we present some criteria for normally torsion-freeness and normality of monomial ideals. In addition, we obtain some results related to the appearance of the graded maximal ideal in the associated primes of powers of monomial ideals.

Throughout this paper, $R = K[x_1, \dots, x_n]$ stands for a polynomial ring over a field K, unless stated otherwise. Also, $\mathcal{G}(I)$ denotes the unique minimal set of monomial generators of a monomial ideal $I \subset R$. The *support* of a monomial $u \in R$, denoted by $\mathrm{supp}(u)$, is the set of variables that divide u. For a monomial ideal I, we set $\mathrm{supp}(I) = \bigcup_{u \in \mathcal{G}(I)} \mathrm{supp}(u)$.

The structure of this paper is outlined as follows: In Section 2, we establish a connection between the presence of a prime monomial ideal in the associated prime set of powers of a monomial ideal and the cardinality of a maximal independent set of the monomial ideal, as demonstrated in Corollary 2.3. Subsequently, our attention shifts to examining the normality of monomial ideals within a subset of nearly normally torsion-free square-free monomial ideals known as well-nearly normally torsion-free type. In particular, we present the following theorem:

Theorem 2.12. Let $I \subset R = K[x_1, \dots, x_n]$ be of well-nearly normally torsion-free type. Then I is normal, and so has (symbolic) (strong) persistence property.

Section 3 is concerned with the existence of embedded associated prime ideals in the associated primes set of powers of monomial ideals, see Propositions 3.1 and 3.2. In particular, we show that these consequences can be employed in studying the edge and cover ideals of cones of graphs as follows:

Lemma 3.5. Let G be a connected graph and $H := C(G)$ be its cone. Then the following statements hold:

(i) If $(x_i : i \in V(G)) \in \mathrm{Ass}(J(G)^t)$ for some $t \geq 1$, then $(x_j : j \in V(H)) \in \mathrm{Ass}(J(H)^{t+1})$, where $J(G)$ and $J(H)$ denote the cover ideals of G and H, respectively.

(ii) If $(x_i : i \in V(G)) \in \mathrm{Ass}(I(G)^t)$ for some $t \geq 2$, then $(x_j : j \in V(H)) \in \mathrm{Ass}(I(H)^t)$, where $I(G)$ and $I(H)$ denote the edge ideals of G and H, respectively.

Now, let I be a normally torsion-free square-free monomial ideal with $\mathcal{G}(I) = \{u_1, \dots, u_m\}$. Let us delete one of the members of $\mathcal{G}(I)$, say u_m, and add a new square-free monomial, say v. There is no guarantee that the new square-free monomial ideal $(u_1, \ldots, u_{m-1}) + (v)$ is still normally torsion-free. We close this section with giving a result, which says that under certain conditions the square-free monomial ideal $(u_1, \ldots, u_{m-1}) + (v)$ is normally torsion-free as well, refer to Theorem 3.8.

Section 4 is devoted to comparing some algebraic properties of the edge ideals of clutters and complement clutters, see Question 4.2. Section 5 deals with the normally torsion-freeness and normality of monomial ideals under the polarization operator, refer to Question 5.2.

2. Normality of nearly normally torsion-free monomial ideals

In this section, we explore the relationship between the presence of a prime monomial ideal in the associated prime set of powers of a monomial ideal and the size of a maximal independent set of the monomial ideal, as elucidated in Corollary 2.3. Following this, we investigate the normality of monomial ideals within a specific class comprising nearly normally torsion-free square-free monomial ideals, as outlined in Theorem 2.12. Additionally, we will utilize the following proposition, which serves as an extension of [**12**, Theorem 3.5] and [**39**, Theorem 3.4], in the proof of Proposition 2.2.

To capture Proposition 2.2 and Corollary 2.3, we need to recollect some definitions. Assume that I is a square-free monomial ideal and $\Gamma \subseteq \mathcal{G}(I)$, where $\mathcal{G}(I)$ denotes the unique minimal set of monomial generators of the monomial ideal I. We say that Γ is an *independent* set in I if $\gcd(f, g) = 1$ for each $f, g \in \Gamma$ with $f \neq g$. We denote the maximum cardinality of an independent set in I by $\beta_1(I)$. Furthermore, if I is a monomial ideal, then the *deletion* of x_i from I with $1 \leq i \leq n$, denoted by $I \setminus x_i$, is obtained by setting $x_i = 0$ in every minimal generator of I, that is, we delete every minimal generator such as $u \in \mathcal{G}(I)$ with $x_i \mid u$, refer to [**12**, page 303] for definitions of deletion and independent set.

PROPOSITION 2.1. *Let $I \subset R = K[x_1, \ldots, x_n]$ be a monomial ideal, t a positive integer, $\mathfrak{p}$ a prime monomial ideal in R, and $y_1, \ldots, y_s$ be distinct variables in R such that, for each $i = 1, \ldots, s$, $\mathfrak{p} \setminus y_i \notin \mathrm{Ass}(R/(I \setminus y_i)^t)$, where $I \setminus y_i$ denotes the deletion of y_i from I. Then $\mathfrak{p} \in \mathrm{Ass}(R/I^t)$ if and only if $\mathfrak{p} \in \mathrm{Ass}(R/(I^t : \prod_{i=1}^s y_i))$.*

PROOF. ($\Leftarrow$) Let $\mathfrak{p} \in \mathrm{Ass}(R/(I^t : \prod_{i=1}^s y_i))$. In view of [**45**, Exercise 2.1.62], we have the following short exact sequence

$$0 \longrightarrow R/(I^t : \prod_{i=1}^{s} y_i) \xrightarrow{\psi} R/I^t \xrightarrow{\phi} R/(I^t, \prod_{i=1}^{s} y_i) \longrightarrow 0,$$

where $\psi(\overline{r}) = \overline{r} \prod_{i=1}^s y_i$ and $\phi(\overline{r}) = \overline{r}$. It follows also from [**40**, Exercise 9.42] that

$$\text{(1)} \quad \mathrm{Ass}(R/(I^t : \prod_{i=1}^{s} y_i)) \subseteq \mathrm{Ass}(R/I^t) \subseteq \mathrm{Ass}(R/(I^t : \prod_{i=1}^{s} y_i)) \cup \mathrm{Ass}(R/(I^t, \prod_{i=1}^{s} y_i)).$$

By virtue of $\mathfrak{p} \in \mathrm{Ass}(R/(I^t : \prod_{i=1}^s y_i))$, we deduce that $\mathfrak{p} \in \mathrm{Ass}(R/I^t)$.

($\Rightarrow$) Assume that $\mathfrak{p} \in \mathrm{Ass}(R/I^t)$. To show $\mathfrak{p} \in \mathrm{Ass}(R/(I^t : \prod_{i=1}^s y_i))$, we use induction on s. Let $J := I \setminus y_1$. Since the generators of J^t are precisely the generators of I^t that are not divisible by y_1, we get $(I^t, y_1) = (J^t, y_1)$. One can deduce from our assumption that $\mathfrak{p} \setminus y_1 \notin \mathrm{Ass}(R/J^t)$. Also, it follows from [**18**, Theorem 5.5] that $\mathfrak{p} \in \mathrm{Ass}(R/(J^t, y_1))$ if and only if $\mathfrak{p} = (\mathfrak{p}_1, y_1)$, where

$\mathfrak{p}_1 \in \mathrm{Ass}(R/J^t)$. This implies that $\mathfrak{p} \notin \mathrm{Ass}(R/(J^t, y_1))$, and so $\mathfrak{p} \notin \mathrm{Ass}(R/(I^t, y_1))$. By virtue of (1), we have $\mathrm{Ass}(R/I^t) \subseteq \mathrm{Ass}(R/(I^t : y_1)) \cup \mathrm{Ass}(R/(I^t, y_1))$, and hence $\mathfrak{p} \in \mathrm{Ass}(R/(I^t : y_1))$. Consequently, the claim is true for the case in which $s = 1$. Now, assume that the assertion has been shown for a product of $s-1$ variables, and that $\mathfrak{p} \in \mathrm{Ass}(R/I^t)$. Set $M := \prod_{i=1}^{s-1} y_i$. It follows from the induction that $\mathfrak{p} \in \mathrm{Ass}(R/(I^t : M))$. Moreover,

$$(2) \qquad \mathrm{Ass}(R/(I^t : M)) \subseteq \mathrm{Ass}(R/((I^t : M) : y_s)) \cup \mathrm{Ass}(R/((I^t : M), y_s)).$$

Let $L := I \setminus y_s$. We show that $((I^t : M), y_s) = ((L^t : M), y_s)$. Since $L \subseteq I$, it is easy to see that $((L^t : M), y_s) \subseteq ((I^t : M), y_s)$. To establish the reverse inclusion, pick a monomial u in $((I^t : M), y_s)$. If $y_s \mid u$, then $u \in ((L^t : M), y_s)$, and the proof is over. If $y_s \nmid u$, then $u \in (I^t : M)$, and hence $uM \in I^t$. This yields that there exists a monomial $f \in \mathcal{G}(I^t)$ such that $f \mid uM$. On account of $y_s \nmid u$ and $y_s \nmid M$, this gives that $y_s \nmid f$, and so $f \in L^t$. Hence, $u \in (L^t : M)$, and thus $((I^t : M), y_s) \subseteq ((L^t : M), y_s)$. The assumption implies that $\mathfrak{p} \setminus y_s \notin \mathrm{Ass}(R/L^t)$. As $\mathrm{Ass}(R/(L^t : M)) \subseteq \mathrm{Ass}(R/L^t)$, one obtains $\mathfrak{p} \setminus y_s \notin \mathrm{Ass}(R/(L^t : M))$. In light of [**18**, Theorem 5.5], we have $\mathfrak{p} \in \mathrm{Ass}(R/((L^t : M), y_s))$ if and only if $\mathfrak{p} = (\mathfrak{p}_1, y_s)$, where $\mathfrak{p}_1 \in \mathrm{Ass}(R/(L^t : M))$. This leads to $\mathfrak{p} \notin \mathrm{Ass}(R/((L^t : M), y_s))$, and so $\mathfrak{p} \notin \mathrm{Ass}(R/((I^t : M), y_s))$. From (2), this yields that $\mathfrak{p} \in \mathrm{Ass}(R/((I^t : M) : y_s)) = \mathrm{Ass}(R/(I^t : \prod_{i=1}^{s} y_i))$. This completes the inductive step, and thus the claim has been proved by induction. □

We are ready to state the first main result of this section.

PROPOSITION 2.2. *Let $I \subset R$ be a monomial ideal, ℓ a positive integer, v a square-free monomial in R with $v \in I^\ell$, and $\mathfrak{p}$ a prime monomial ideal such that $\mathfrak{p} \setminus x_i \notin \mathrm{Ass}(R/(I \setminus x_i)^s)$ for all $x_i \in \mathrm{supp}(v)$ and some positive integer s. If $\mathfrak{p} \in \mathrm{Ass}(R/I^s)$, then $s > \ell$.*

PROOF. Let $\mathfrak{p} \in \mathrm{Ass}(R/I^s)$. Without loss of generality, assume that $v = \prod_{i=1}^{r} x_i$. It follows from Proposition 2.1 that $\mathfrak{p} \in \mathrm{Ass}(R/(I^s : \prod_{i=1}^{r} x_i))$. If $s \leq \ell$, then we deduce that $(I^s : \prod_{i=1}^{r} x_i) = R$ since $v \in I^\ell$, which contradicts the fact that $\mathfrak{p} \in \mathrm{Ass}(R/(I^s : \prod_{i=1}^{r} x_i))$. This yields that $s > \ell$, as claimed. □

COROLLARY 2.3. *Let $I \subset R$ be a square-free monomial ideal, $\mathfrak{p} \subset R$ be a prime monomial ideal, and $\{u_1, \ldots, u_{\beta_1(I)}\}$ be a maximal independent set of minimal generators of I such that $\mathfrak{p} \setminus x_i \notin \mathrm{Ass}(R/(I \setminus x_i)^s)$ for all $x_i \in \mathrm{supp}(\prod_{i=1}^{\beta_1(I)} u_i)$ and some positive integer s. If $\mathfrak{p} \in \mathrm{Ass}(R/I^s)$, then $s \geq \beta_1(I) + 1$.*

In view of Corollary 2.3, one may ask the following.

QUESTION 2.4. Let $I \subset R$ be a square-free monomial ideal, $\mathfrak{p}$ a prime monomial ideal in R, and $\{u_1, \ldots, u_{\beta_1(I)}\}$ be a maximal independent set of minimal generators of I such that $\mathfrak{p} \setminus x_i \notin \mathrm{Ass}(R/(I \setminus x_i)^t)$ for all $x_i \in \bigcup_{i=1}^{\beta_1(I)} \mathrm{supp}(u_i)$ and some positive integer t.

(i) If $t \geq \beta_1(I) + 1$, then can we deduce that $\mathfrak{p} \in \mathrm{Ass}(R/I^t)$?
(ii) If $\mathfrak{p} \in \mathrm{Ass}(R/I^t)$, then is it possible the inequality $t \geq \beta_1(I) + 1$ is strict?

Before presenting the counterexample in the upcoming example, we first revisit the following result.

PROPOSITION 2.5 ([**31**, Proposition 3.3]). *Suppose that I and J are two normally torsion-free square-free monomial ideals in R such that* $\mathrm{supp}(I) \cap \mathrm{supp}(J) = \emptyset$. *Then $I \cap J = IJ$ is normally torsion-free.*

EXAMPLE 2.6. Let $I \subset K[x_1, \ldots, x_{11}]$ be the following square-free monomial ideal

$$\begin{aligned} I &= (x_1, x_2) \cap (x_2, x_3) \cap (x_3, x_4) \cap (x_4, x_5) \cap (x_5, x_1) \cap (x_1, x_7) \cap (x_2, x_8) \cap (x_3, x_9) \\ &\quad \cap (x_4, x_{10}) \cap (x_5, x_{11}) \cap (x_6, x_1) \cap (x_6, x_2) \cap (x_6, x_3) \cap (x_6, x_4) \cap (x_6, x_5) \\ &= (x_1x_2x_3x_4x_5, x_1x_2x_4x_6x_9x_{11}, x_1x_3x_4x_6x_8x_{11}, x_1x_2x_3x_4x_6x_{11}, x_1x_3x_5x_6x_8x_{10}, \\ &\quad x_2x_3x_5x_6x_7x_{10}, x_1x_2x_3x_5x_6x_{10}, x_2x_4x_5x_6x_7x_9, x_1x_2x_4x_5x_6x_9, x_1x_3x_4x_5x_6x_8, \\ &\quad x_2x_3x_4x_5x_6x_7). \end{aligned}$$

In fact, I is the cover ideal of the below Helm graph G.

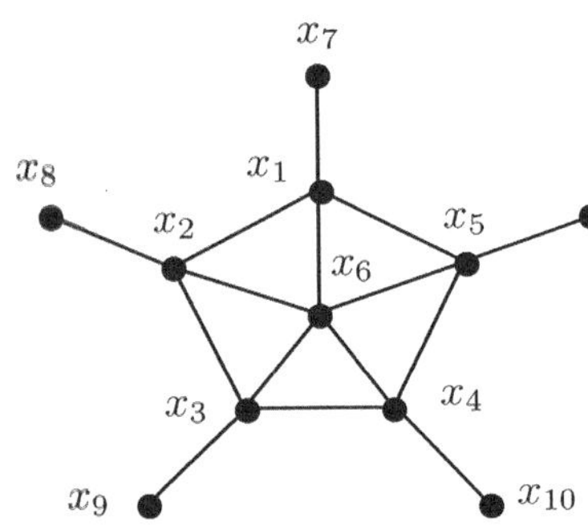

G

It is not hard to see that $\{u_1 = x_1x_2x_3x_4x_5\}$ is a maximal independent set of minimal generators of I, and so $\beta_1(I) = 1$. Using *Macaulay2* [**11**], we can derive that

$$\mathfrak{p} := (x_1, x_2, x_3, x_4, x_5, x_6) \in \mathrm{Ass}(R/I^3) \setminus (\mathrm{Ass}(R/I) \cup \mathrm{Ass}(R/I^2)).$$

On the other hand, direct computations give the following table:

$I \setminus x_1 = (x_2) \cap (x_3, x_4) \cap (x_5) \cap (x_7) \cap (x_3, x_9) \cap (x_4, x_{10}) \cap (x_6)$
$I \setminus x_2 = (x_1) \cap (x_3) \cap (x_8) \cap (x_4, x_5) \cap (x_4, x_{10}) \cap (x_5, x_{11}) \cap (x_6)$
$I \setminus x_3 = (x_2) \cap (x_4) \cap (x_5, x_1) \cap (x_1, x_7) \cap (x_9) \cap (x_5, x_{11}) \cap (x_6)$
$I \setminus x_4 = (x_1, x_2) \cap (x_3) \cap (x_5) \cap (x_1, x_7) \cap (x_2, x_8) \cap (x_{10}) \cap (x_6)$
$I \setminus x_5 = (x_2, x_3) \cap (x_4) \cap (x_1) \cap (x_2, x_8) \cap (x_3, x_9) \cap (x_{11}) \cap (x_6)$

We verify that $I \setminus x_1$ is normally torsion-free. Since $(x_3, x_4) \cap (x_3, x_9) \cap (x_4, x_{10})$ can be viewed as the cover ideal of the path graph P with $V(P) = \{x_3, x_4, x_9, x_{10}\}$ and $E(P) = \{\{x_3, x_4\}, \{x_3, x_9\}, \{x_4, x_{10}\}\}$, and by this fact that the cover ideal of any bipartite graph is normally torsion-free, we deduce that $(x_3, x_4) \cap (x_3, x_9) \cap (x_4, x_{10})$ is normally torsion-free. Now, Proposition 2.5 implies that $I \setminus x_1$ is normally torsion-free. A similar argument can be used to show that $I \setminus x_i$ is normally torsion-free for all $i = 2, \ldots, 5$. In particular, we can derive that $\mathfrak{p} \setminus x_i \notin \mathrm{Ass}((I \setminus x_i)^2)$ for all $i = 1, \ldots, 5$. Now, for $t = 2 = \beta_1(I) + 1$, we have $\mathfrak{p} \notin \mathrm{Ass}(R/I^2)$, that is, the converse of Proposition 2.2 is not true in general. This refutes (i). On the other hand, it follows from $\mathfrak{p} \in \mathrm{Ass}(R/I^3) \setminus \mathrm{Ass}(R/I^2)$ that we must have $t > 2$. Consequently, the inequality $t \geq \beta_1(I) + 1$ in Proposition 2.2 can be strict, which is a positive answer to (ii).

Along with Question 2.4, we can ask the following.

QUESTION 2.7. Let I be a nearly normally torsion-free square-free monomial ideal in R such that there exist some $m \geq 1$ and some prime monomial ideal $\mathfrak{q}$ such that $\mathrm{Ass}(I^t) = \mathrm{Min}(I)$ for all $1 \leq t \leq m$, and $\mathrm{Ass}(I^t) \subseteq \mathrm{Min}(I) \cup \{\mathfrak{q}\}$ for all $t \geq m+1$. Let also $\{u_1, \ldots, u_{\beta_1(I)}\}$ be a maximal independent set of minimal generators of I. If $\mathfrak{q} \in \mathrm{Ass}(R/I^s)$ for some $s \geq m+1$, then can we deduce that $s \geq \beta_1(I) + 1$?

Next, we describe an ideal that provides a negative answer.

EXAMPLE 2.8. Let $I \subset K[x_1, \ldots, x_7]$ be the following square-free monomial ideal

$$\begin{aligned} I &= (x_1, x_2, x_3) \cap (x_2, x_3, x_4) \cap (x_3, x_4, x_5) \cap (x_4, x_5, x_6) \cap (x_5, x_6, x_7) \\ &\quad \cap (x_6, x_7, x_1) \cap (x_7, x_1, x_2) \\ &= (x_3x_6x_7, x_3x_5x_7, x_2x_5x_7, x_3x_4x_7, x_2x_4x_7, x_1x_4x_7, x_2x_5x_6, x_2x_4x_6, x_1x_4x_6, \\ &\quad x_2x_3x_6, x_1x_3x_6, x_1x_4x_5, x_1x_3x_5, x_1x_2x_5). \end{aligned}$$

It is a common observation that the ideal I dominates the odd cycle graph C_7 with

$$E(C_7) = \{\{x_i, x_{i+1}\} : i = 1, \ldots, 6\} \cup \{\{x_1, x_7\}\}.$$

From [**31**, Theorem 4.3], and in view of the proof of [**33**, Lemma 3.2], we can deduce that I is nearly normally torsion-free such that

$$\mathrm{Ass}(I^s) = \mathrm{Min}(I) \cup \{(x_1, x_2, x_3, x_4, x_5, x_6, x_7)\} \quad \text{for all} \quad s \geq 2.$$

In addition, one can easily check that $\{u_1 = x_1x_2x_5, u_2 = x_3x_6x_7\}$ is a maximal independent set of minimal generators of I, and so $\beta_1(I) = 2$. With the notation of Question 2.7, we detect that $m = 1$, $(x_1, x_2, x_3, x_4, x_5, x_6, x_7) \in \mathrm{Ass}(I^2)$, while $2 \ngeq \beta_1(I) + 1$.

In the rest of this section, we focus on a class of nearly normally torsion-free square-free monomial ideals and show that every ideal in this class is normal. We start with Definition 2.9.

DEFINITION 2.9. Let $I \subset R$ be a square-free monomial ideal. Then we say that I is of *well-nearly normally torsion-free type* if there exist some $\ell, m \geq 1$ and some prime monomial ideal $\mathfrak{q}$ such that $\mathrm{Ass}(I^t) = \mathrm{Min}(I)$ for all $1 \leq t \leq m$, and $\mathrm{Ass}(I^t) \subseteq \mathrm{Min}(I) \cup \{\mathfrak{q}\}$ for all $t \geq m+1$ such that if $\mathfrak{q} \in \mathrm{Ass}(I^s)$ for some $s \geq m+1$, then $I^s = \bigcap_{\mathfrak{p} \in \mathrm{Min}(I)} \mathfrak{p}^s \cap \mathfrak{q}^{\ell s}$ is a minimal primary decomposition of I^s.

We demonstrate a large class of monomial ideals of well-nearly normally torsion-free type.

EXAMPLE 2.10. Suppose that C_{2n+1} is an odd cycle graph with the edge set

$$E(C_{2n+1}) = \{\{x_i, x_{i+1}\} : i = 1, \ldots, 2n\} \cup \{\{x_1, x_{2n+1}\}\}.$$

It has already been proved in [**19**, Lemma 3] that, for all $s \geq 2$, we have

$$J(C_{2n+1})^s = \bigcap_{i=1}^{2n+1} (x_i, x_{i+1})^s \cap \mathfrak{m}^{s(n+1)},$$

where $\mathfrak{m} = (x_1, \dots, x_{2n+1})$ is the maximal ideal in $R = K[x_1, \dots, x_{2n+1}]$. Notice that $\mathfrak{q} = \mathfrak{m}$ and $\ell = n + 1$. In other words, the cover ideals of odd cycle graphs are of well-nearly normally torsion-free type.

Next, we are going to give a square-free monomial ideal which is nearly normally torsion-free, but is not of well-nearly normally torsion-free type.

EXAMPLE 2.11. Let $S = K[x_1, x_2, x_3, x_4, x_5]$ and $R = S[x_6]$. Also, let $I \subset R$ be the following square-free monomial ideal

$$\begin{aligned} I =&(x_1x_3x_5, x_1x_3x_4, x_1x_2x_4, x_2x_4x_5x_6, x_2x_3x_5x_6) \\ =&(x_1, x_2) \cap (x_2, x_3) \cap (x_3, x_4) \cap (x_4, x_5) \cap (x_5, x_1) \cap (x_1, x_6). \end{aligned}$$

We claim that I is nearly normally torsion-free and also is normal. One can easily see that I is the cover ideal of the following graph G.

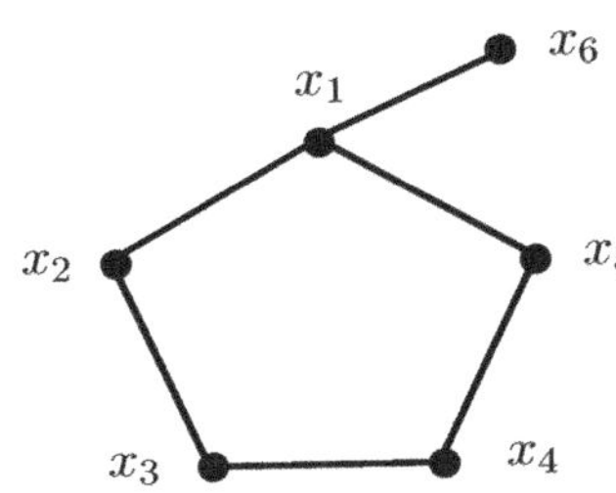

G

It follows from [**28**, Theorem 2.5] that $\mathrm{Ass}_R(R/J(G)^s) = \mathrm{Ass}_S(S/J(C_5)^s) \cup \{(x_1, x_6)\}$ for all $s \geq 1$, where $J(G)$ (respectively, $J(C_5)$) denotes the cover ideal of the graph G (respectively, odd cycle subgraph C_5). In addition, we can deduce from [**28**, Proposition 3.6] that $\mathrm{Ass}_S(S/J(C_5)^s) = \mathrm{Ass}_S(S/J(C_5)) \cup \{(x_1, x_2, x_3, x_4, x_5)\}$ for all $s \geq 2$. We therefore have, for all $s \geq 2$,

$$\begin{aligned} &\mathrm{Ass}_R(R/I^s) = \mathrm{Ass}_R(R/J(G)^s) \\ &= \{(x_1, x_2), (x_2, x_3), (x_3, x_4), (x_4, x_5), (x_5, x_1), (x_1, x_6), (x_1, x_2, x_3, x_4, x_5)\}. \end{aligned}$$

This implies that I is nearly normally torsion-free. We now turn our attention to show the normality of I. It follows from [**2**, Theorem 1.10] that $J(C_5)$ is normal, and so I is normal according to [**1**, Theorem 2.7]. In particular, [**35**, Theorem 6.2] yields that I satisfies the strong persistence property. Here, our aim is to verify that, for all $t \geq 2$ and $\ell \geq 1$, I^t cannot be presented as below

(3)
$$I^t = (x_1,x_2)^t \cap (x_2,x_3)^t \cap (x_3,x_4)^t \cap (x_4,x_5)^t \cap (x_5,x_1)^t \cap (x_1,x_6)^t \cap (x_1,x_2,x_3,x_4,x_5)^{\ell t}.$$

On the contrary, assume that I^t can be presented in the above format for some $t \geq 2$ and $\ell \geq 1$. The evaluation of x_1, x_4 at 0 in both sides of the equality (3), we get $(x_2x_3x_5x_6)^tR = (x_2x_3x_5x_6)^tR \cap (x_2, x_3, x_5)^{\ell t}$, and so $(x_2x_3x_5x_6)^t \in (x_2, x_3, x_5)^{\ell t}$. So, there exists a monomial $u = x_2^a x_3^b x_5^c \in \mathcal{G}((x_2, x_3, x_5)^{\ell t})$ such that

$u \mid (x_2x_3x_5x_6)^t$, where $a, b, c \geq 0$ and $a + b + c = \ell t$. From $u \mid (x_2x_3x_5x_6)^t$, we obtain $a \leq t$, $b \leq t$, and $c \leq t$. This implies that $\ell t = a + b + c \leq 3t$, and hence $\ell \leq 3$. It can be checked that

$$x_1^{t-1}x_2x_3^{t-1}x_4x_5^{t-1}x_6 \in \bigcap_{i=1}^{4}(x_i, x_{i+1})^t \cap (x_5, x_1)^t \cap (x_1, x_6)^t \cap (x_1, x_2, x_3, x_4, x_5)^{\ell t}.$$

We will show that $x_1^{t-1}x_2x_3^{t-1}x_4x_5^{t-1}x_6 \notin I^t$. Assume the contrary, then there are nonnegative integers a_i, where $i = 1, \ldots, 5$, with $a_1 + a_2 + a_3 + a_4 + a_5 = t$ such that

$$(x_1x_3x_5)^{a_1}(x_1x_3x_4)^{a_2}(x_1x_2x_4)^{a_3}(x_2x_4x_5x_6)^{a_4}(x_2x_3x_5x_6)^{a_5} \mid x_1^{t-1}x_2x_3^{t-1}x_4x_5^{t-1}x_6.$$

The total degree on the right is $3t$, while the total degree of the divisor is $3t+a_4+a_5$, hence, $a_1 + a_2 + a_3 = t$. Comparison of the degree of x_1 on each side leads to a contradiction. Consequently, I is not of well-nearly normally torsion-free type.

We are in a position to state the second main result of this section in the subsequent theorem.

THEOREM 2.12. *Let $I \subset R = K[x_1, \ldots, x_n]$ be of well-nearly normally torsion-free type. Then the following statements hold:*

(i) *I is normal.*
(ii) *I has the strong persistence property.*
(iii) *I has the persistence property.*
(iv) *I has the symbolic strong persistence property.*

PROOF. (i) To show the normality of I, one has to prove that $I^s = \overline{I^s}$ for all $s \geq 1$. To do this, fix $s \geq 1$. Since $I^s \subseteq \overline{I^s}$, it is enough for us to verify that $\overline{I^s} \subseteq I^s$. Let u be an arbitrary monomial in $\overline{I^s}$. It follows at once from [**13**, Theorem 1.4.2] that $u^z \in I^{sz}$ for some $z \geq 1$. Because I is of well-nearly normally torsion-free type, this implies that there exist some $\ell, m \geq 1$ and some prime monomial ideal $\mathfrak{q}$ such that $\text{Ass}(I^t) = \text{Min}(I)$ for all $1 \leq t \leq m$, and $\text{Ass}(I^t) \subseteq \text{Min}(I) \cup \{\mathfrak{q}\}$ for all $t \geq m+1$ such that if $\mathfrak{q} \in \text{Ass}(I^k)$ for some $k \geq m+1$, then $I^k = \bigcap_{\mathfrak{p}\in\text{Min}(I)} \mathfrak{p}^k \cap \mathfrak{q}^{k\ell}$ is a minimal primary decomposition of I^k. We can deduce from [**45**, Proposition 4.3.25] that $I^{(s)} = \bigcap_{\mathfrak{p}\in\text{Min}(I)} \mathfrak{p}^s$, where $I^{(s)}$ denotes the s-th symbolic power of I. We thus get $I^s = I^{(s)} = \bigcap_{\mathfrak{p}\in\text{Min}(I)} \mathfrak{p}^s$ or $I^s = \bigcap_{\mathfrak{p}\in\text{Min}(I)} \mathfrak{p}^s \cap \mathfrak{q}^{s\ell}$. This implies that $u^z \in \mathfrak{p}^{sz}$ for all $\mathfrak{p} \in \text{Min}(I)$, and so $u \in \overline{\mathfrak{p}^s}$ for all $\mathfrak{p} \in \text{Min}(I)$. In addition, it follows from [**13**, Theorem 1.4.6] that every prime monomial ideal is normal. This gives that $u \in \mathfrak{p}^s$ for all $\mathfrak{p} \in \text{Min}(I)$. Hence, $u \in \bigcap_{\mathfrak{p}\in\text{Min}(I)} \mathfrak{p}^s$. If $u^z \in \mathfrak{q}^{sz\ell}$, then $u \in \overline{\mathfrak{q}^{s\ell}}$, and [**13**, Theorem 1.4.6] yields that $u \in \mathfrak{q}^{s\ell}$. Therefore, we obtain $u \in I^s$, and so I is normal, as required.

(ii) Since normality implies SPP, the claim follows readily from (i).

(iii)-(iv) Because SPP implies PP and symbolic PP, both claims are derived from (ii). □

3. An argument on the existence of embedded associated prime ideals

In this section, our aim is to give some results which are related to the existence of embedded associated prime ideals in the associated primes set of powers of monomial ideals. In particular, we show that these consequences can be employed in studying the edge and cover ideals of cones of graphs. We begin by stating the first main result of this section.

PROPOSITION 3.1. *Let $I \subset R = K[x_1, \dots, x_n, x_{n+1}]$ be a monomial ideal, where $\gcd(u, x_{n+1}) = 1$ for all $u \in \mathcal{G}(I)$, and $\mathfrak{m}$ the ideal generated by the variables. Further, let $h = fg$ with $1 \neq f \in R, g \in \mathcal{G}(I)$ and $\gcd(h, x_{n+1}) = 1$, and $L := x_{n+1}I + hR$. If $(x_1, \dots, x_n) \in \operatorname{Ass}(I^t)$ for some $t \geq 1$, then $\mathfrak{m} \in \operatorname{Ass}(L^{t+1})$.*

PROOF. Let $(x_1, \dots, x_n) \in \operatorname{Ass}(I^t)$ for some $t \geq 1$. Hence, there exists some monomial v such that $(I^t : v) = (x_1, \dots, x_n)$. Since $x_{n+1} \notin \operatorname{supp}(I)$, we can assume $\gcd(v, x_{n+1}) = 1$ according to [**18**, Lemma 2.1]. We claim that $\mathfrak{m} = (L^{t+1} : x_{n+1}^t vh)$. To see this, we consider the following equalities:

$$\begin{aligned}(L^{t+1} : x_{n+1}^t vh) =& (\sum_{\alpha=0}^{t+1} x_{n+1}^{\alpha} I^{\alpha} h^{t+1-\alpha} : x_{n+1}^t vh)\\ =& (x_{n+1}^{t+1} I^{t+1} : x_{n+1}^t vh) + (x_{n+1}^t I^t h : x_{n+1}^t vh)\\ &+ \sum_{\alpha=0}^{t-1} (x_{n+1}^{\alpha} I^{\alpha} h^{t+1-\alpha} : x_{n+1}^t vh)\\ =& x_{n+1}(I^{t+1} : vh) + (I^t : v) + \sum_{\alpha=0}^{t-1} (I^{\alpha} h^{t-\alpha} : v).\end{aligned}$$

Due to $h \in I$, we get $(I^{\alpha} h^{t-\alpha} : v) \subseteq (I^t : v)$ for all $\alpha = 0, \dots, t-1$, and so $\sum_{\alpha=0}^{t-1} (I^{\alpha} h^{t-\alpha} : v) \subseteq (I^t : v)$. In addition, since $x_i v \in I^t$ for all $i = 1, \dots, n$, $g \in \mathcal{G}(I)$, and $f \neq 1$, this implies that $vh \in I^{t+1}$. Consequently, we have $(I^{t+1} : vh) = R$, and so $(L^{t+1} : x_{n+1}^t vh) = x_{n+1}R + (x_1, \dots, x_n) = \mathfrak{m}$, that is, $\mathfrak{m} \in \operatorname{Ass}(L^{t+1})$, as desired. □

We are ready to give the second main result.

PROPOSITION 3.2. *Assume that R, I, and $\mathfrak{m}$ are as in Proposition 3.1. Moreover, let $I \neq \mathfrak{q} = (x_1, \dots, x_n)$ and $L := I + x_{n+1}\mathfrak{q}$. Suppose that $(I^t : v) = \mathfrak{q}$ for some $t \geq 1$ and a monomial $v \in R$, such that if $x_\lambda v = f_1 \cdots f_t M$, where $1 \leq \lambda \leq n$ and $M \in R, f_1, \dots, f_t \in \mathcal{G}(I)$ are monomials, then $Mf_j/x_\lambda \neq 1$ for some $1 \leq j \leq t$. Then $\mathfrak{m} \in \operatorname{Ass}(L^t)$.*

PROOF. First note that, based on [**18**, Lemma 2.1], we can assume $\operatorname{supp}(v) \subseteq \operatorname{supp}(I)$. In particular, from $x_{n+1} \notin \operatorname{supp}(I)$, we have $\gcd(v, x_{n+1}) = 1$. It follows from our assumption that there exists some $1 \leq \lambda \leq n$ such that if $x_\lambda v = f_1 \cdots f_t M$ for some monomials $f_1, \dots, f_t \in \mathcal{G}(I)$ and monomial $M \in R$, then $Mf_j/x_\lambda \neq 1$ for some $1 \leq j \leq t$. Since $(I^t : v) = \mathfrak{q}$, this implies that $v \notin I^t$, and so $x_\lambda \nmid M$. Without loss of generality, assume that $f_1 = x_\lambda h$ for some monomial $h \in R$, and hence we can write $v = (f_2 \cdots f_t)hM$ with $hM \neq 1$. This gives rise to $v \in I^{t-1}\mathfrak{q}$. In light of the binomial expansion theorem, we have the following equalities

$$\begin{aligned}(L^t : v) =& (\sum_{\alpha=0}^{t} x_{n+1}^{t-\alpha} \mathfrak{q}^{t-\alpha} I^{\alpha} : v)\\ =& (I^t : v) + \sum_{\alpha=0}^{t-2} x_{n+1}^{t-\alpha} (\mathfrak{q}^{t-\alpha} I^{\alpha} : v) + x_{n+1}(I^{t-1}\mathfrak{q} : v)\\ =& \mathfrak{q} + x_{n+1}R = \mathfrak{m}.\end{aligned}$$

This implies that $\mathfrak{m} \in \operatorname{Ass}(L^t)$, and the proof is finished. □

Here, we provide an example, which shows that the converse of Propositions 3.1 and 3.2, respectively, is not true in general.

EXAMPLE 3.3. (i) In $R = K[x_1, \ldots, x_8]$, consider the square-free monomial ideal

$$L := (x_2x_5x_6x_8, x_1x_5x_6x_8, x_3x_4x_6x_8, x_1x_4x_6x_8, x_2x_3x_6x_8, x_3x_4x_5x_8, x_2x_4x_5x_8, x_1x_3x_5x_8, x_1x_2x_4x_8, x_1x_2x_3x_8, x_1x_2x_3x_4, x_1x_4x_5x_7x_8).$$

Now, set $h := fg$, where $g = x_1x_2x_3$ and $f = x_4$, and

$$I := (x_2x_5x_6, x_1x_5x_6, x_3x_4x_6, x_1x_4x_6, x_2x_3x_6, x_3x_4x_5, x_2x_4x_5, x_1x_3x_5, x_1x_2x_4, x_1x_2x_3, x_1x_4x_5x_7).$$

It is easy to see that $L = x_8I + hR$ satisfies the conditions in Proposition 3.1. Using *Macaulay2* [**11**], we can deduce that

$$(x_1, x_2, x_3, x_4, x_5, x_6, x_7, x_8) \in \operatorname{Ass}(L^4), \text{ but } (x_1, x_2, x_3, x_4, x_5, x_6, x_7) \notin \operatorname{Ass}(I^3).$$

(ii) Consider the ideal $I = (x_1x_2, x_2x_3, x_3x_4, x_4x_5, x_5x_1) \in K[x_1, \ldots, x_6]$ and let $L = I + x_6(x_1, x_2, x_3, x_4, x_5)$. Using *Macaulay2* [**11**], we find that $\mathfrak{m} \in \operatorname{Ass}(L^2)$, while $(x_1, x_2, x_3, x_4, x_5) \notin \operatorname{Ass}(I^2)$.

In what follows, we present some applications of Propositions 3.1 and 3.2.

DEFINITION 3.4 ([**45**, Definition 10.5.4]). The *cone* $C(G)$, over the graph G, is obtained by adding a new vertex w to G and joining every vertex of G to w.

LEMMA 3.5. *Let G be a connected graph and $H := C(G)$ be its cone. Then the following statements hold:*

(i) *If $(x_i : i \in V(G)) \in \operatorname{Ass}(J(G)^t)$ for some $t \geq 1$, then $(x_j : j \in V(H)) \in \operatorname{Ass}(J(H)^{t+1})$, where $J(G)$ and $J(H)$ denote the cover ideals of G and H, respectively.*

(ii) *If $(x_i : i \in V(G)) \in \operatorname{Ass}(I(G)^t)$ for some $t \geq 2$, then $(x_j : j \in V(H)) \in \operatorname{Ass}(I(H)^t)$, where $I(G)$ and $I(H)$ denote the edge ideals of G and H, respectively.*

PROOF. (i) Assume that the cone $H = C(G)$ is obtained by adding the new vertex w to G and joining every vertex of G to w. Then one can easily see that

$$J(H) = J(G) \cap (x_w, \prod_{i \in V(G)} x_i) = x_wJ(G) + \prod_{i \in V(G)} x_iR,$$

where $R = K[x_i : i \in V(H)]$. Since G is connected, we get $\prod_{i \in V(G)} x_i \notin \mathcal{G}(J(G))$. Now, the claim is a direct consequence of Proposition 3.1.

(ii) It is routine to check that $I(H) = I(G) + x_w(x_i : i \in V(G))$. Suppose that $(x_i : i \in V(G)) \in \operatorname{Ass}(I(G)^t)$ for some $t \geq 2$, say $(I(G)^t : v) = (x_i : i \in V(G))$ for some monomial $v \in R$. Then there exist monomials $f_1, \ldots, f_t \in \mathcal{G}(I(G))$ and a monomial $M \in R$ such that $x_1v = f_1 \cdots f_tM$. Since $x_1 \nmid M$, without loss of generality, assume that $x_1 \mid f_1$. Due to $f_1 \in \mathcal{G}(I(G))$ and G being connected, there exists some positive integer $1 \neq r \in V(G)$ such that $f_1 = x_1x_r$, and so $f_1M/x_1 = x_rM \neq 1$. As $I(G) \neq (x_i : i \in V(G))$, the claim can be deduced immediately from Proposition 3.2, and the proof is finished. □

As an immediate consequence of Lemma 3.5, we have the following result.

COROLLARY 3.6. *Let K_n denote the complete graph of order n. Then the following statements hold:*

(i) $(x_1, \ldots, x_n) \in \mathrm{Ass}(J(K_n)^t)$ *for all* $t \geq n-1$ *and* $n \geq 2$.
(ii) $(x_1, \ldots, x_n) \in \mathrm{Ass}(I(K_n)^t)$ *for all* $t \geq 2$ *and* $n \geq 3$.

PROOF. (i) By virtue of K_n being the cone of K_{n-1} and Lemma 3.5, we can establish by induction on n that $(x_1, \ldots, x_n) \in \mathrm{Ass}(J(K_n)^{n-1})$. On the other hand, according to [**2**, Corollary 1.7], we conclude that $J(K_n)$ is a normal square-free monomial ideal, and so has the persistence property. This gives rise to $(x_1, \ldots, x_n) \in \mathrm{Ass}(J(K_n)^t)$ for all $t \geq n-1$ and $n \geq 2$, as required.

(ii) First note that $(x_1, x_2, x_3) \in \mathrm{Ass}(I(K_3)^2)$. By an argument similar to part (i), we obtain $(x_1, \ldots, x_n) \in \mathrm{Ass}(I(K_n)^2)$. In addition, [**45**, Theorem 7.7.14] yields that $I(K_n)$ has the persistence property, and consequently $(x_1, \ldots, x_n) \in \mathrm{Ass}(I(K_n)^t)$ for all $t \geq 2$ and $n \geq 3$, as claimed. □

We present the following example, which is related to Lemma 3.5(i).

EXAMPLE 3.7. Let $G = C_{2n-1}$ denote the odd cycle graph of order $2n-1$, $n \geq 2$, with the vertex set $V(C_{2n-1}) = \{1, \ldots, 2n-1\}$ and the edge set

$$E(C_{2n-1}) = \{\{i, i+1\} : i = 1, \ldots, 2n-2\} \cup \{2n-1, 1\}.$$

Then the cone $H = C(C_{2n-1})$ over the graph C_{2n-1} can be obtained by adding the new vertex $2n$ and joining every vertex of C_{2n-1} to $2n$. It is obvious that $H = W_{2n}$ is the wheel graph of order $2n$. On the other hand, we know from [**28**, Propositions 3.6 and 3.10] that $(x_1, \ldots, x_{2n-1}) \in \mathrm{Ass}(J(C_{2n-1})^s)$ and $(x_1, \ldots, x_{2n-1}, x_{2n}) \in \mathrm{Ass}(J(W_{2n})^{s+1})$, where $s \geq 2$.

Can we inquire if every nearly normally torsion-free monomial ideal is a normal ideal in general? The answer is no. For example, consider the monomial ideal $I = (x^5, x^4y, xy^4, y^5z)$ in $R = K[x, y, z]$. According to [**25**, Example 2.8], we have $\mathrm{Min}(I) = \{(x, y), (x, z)\}$ and $\mathrm{Ass}(R/I^s) = \mathrm{Min}(I) \cup \{(x, y, z)\}$ for all $s \geq 2$. This means that I is nearly normally torsion-free. Now, set $u := x^4y^6$. It is easy to check that $u^2 = x^8y^{12} = (x^5)(xy^4)(xy^4)(xy^4) \in I^4$, while $u \notin I^2$. Hence, I^2 is not integrally closed, and so I is not normal.

Suppose that I is a normally torsion-free square-free monomial ideal with $\mathcal{G}(I) = \{u_1, \ldots, u_m\}$. Assume that we delete one of the members of $\mathcal{G}(I)$, say u_m, and add a new square-free monomial, say v. There is no guarantee that the new square-free monomial ideal $(u_1, \ldots, u_{m-1}) + (v)$ is still normally torsion-free. To see a counterexample, let C_9 denote the odd cycle graph of order 9 with the edge set $E(C_9) = \{\{i, i+1\} \colon i = 1, \ldots, 8\} \cup \{\{9, 1\}\}$. Assume that $L := DI(C_9)$ is the dominating ideal of C_9 in the polynomial ring $R = K[x_1, \ldots, x_9]$. Based on [**25**, Example 2.2], L is normally torsion-free. In particular, $x_3x_6x_9 \in \mathcal{G}(L)$. Now, let J be the square-free monomial ideal with $\mathcal{G}(J) = (\mathcal{G}(L) \setminus \{x_3x_6x_9\}) \cup \{x_3x_6x_9x_5\}$. Using *Macaulay2* [**11**], we obtain $\# \mathrm{Ass}(J) = 13$ while $\# \mathrm{Ass}(J^2) = 18$. This means that J is not normally torsion-free.

In the following result, we show that under certain conditions the square-free monomial ideal $(u_1, \ldots, u_{m-1}) + (v)$ is normally torsion-free as well.

THEOREM 3.8. *Let I be a normally torsion-free square-free monomial ideal in a polynomial ring $R = K[x_1, \ldots, x_n]$ over a field K with $\mathcal{G}(I) = \{u_1, \ldots, u_m\}$.*

Let $L := (u_1, \ldots, u_{m-1})S + fx_{n+1}\cdots x_{n+t}S$, *where* $S = R[x_{n+1}, \ldots, x_{n+t}]$, f *a square-free monomial with* $f \mid u_m$, $fx_{n+1}\cdots x_{n+t} \notin \mathfrak{p}^2$ *for any* $\mathfrak{p} \in \mathrm{Min}(L)$, *and* $(u_1, \ldots, u_{m-1})S$ *is normally torsion-free. Then the following statements hold:*

(i) L *is normally torsion-free.*
(ii) L *is nearly normally torsion-free.*
(iii) L *is normal.*
(iv) L *has the strong persistence property.*
(v) L *has the persistence property.*
(vi) L *has the symbolic strong persistence property.*

PROOF. (i) To simplify the notation, put $v := fx_{n+1}\cdots x_{n+t}$. Suppose, on the contrary, that L is not normally torsion-free. Let ℓ be minimal such that L^ℓ has embedded prime ideals. Due to L is a square-free monomial ideal, this gives that $\mathrm{Ass}(L) = \mathrm{Min}(L)$, and so $\ell \geq 2$. Pick an arbitrary element $\mathfrak{p} \in \mathrm{Min}(L)$. We want to show that $(L^\ell : v) = L^{\ell-1}$. It follows immediately from $v \in L$ that $L^{\ell-1} \subseteq (L^\ell : v)$. Hence, it is sufficient for us to verify the reverse inclusion. To do this, take a monomial h in $(L^\ell : v)$. We thus get $hv \in L^\ell$, and hence there exist monomials $g_1, \ldots, g_\ell \in \mathcal{G}(L)$ and some monomial f' in S such that $hv = g_1 \cdots g_\ell f'$. One can conclude readily from $\mathfrak{p} \in \mathrm{Min}(L)$ that $hv \in \mathfrak{p}^\ell$. In addition, we deduce from the assumption $v \notin \mathfrak{p}^2$ that $\mathfrak{p}$ contains exactly one variable that divides v. Consequently, $h \in \mathfrak{p}^{\ell-1}$. Since $\mathfrak{p}$ is arbitrary, we get $h \in \bigcap_{\mathfrak{p}\in\mathrm{Min}(L)} \mathfrak{p}^{\ell-1}$. It follows from [**45**, Proposition 4.3.25] that $h \in L^{(\ell-1)}$. Thanks to ℓ being minimal such that L^ℓ has embedded prime ideals, this leads to $L^{\ell-1}$ has no embedded prime ideals, and so $L^{(\ell-1)} = L^{\ell-1}$. We therefore have $h \in L^{\ell-1}$. Accordingly, one has $(L^\ell : v) \subseteq L^{\ell-1}$, and so $(L^\ell : v) = L^{\ell-1}$. Suppose that $\mathfrak{q}$ is an embedded prime ideal of L^ℓ. This implies that $\mathfrak{q} = (L^\ell : z)$ for some monomial $z \in S$. We claim that $\mathfrak{q} \setminus x_i \notin \mathrm{Ass}((L \setminus x_i)^\ell)$ for any $x_i \in \mathrm{supp}(f)$. On the contrary, assume that $\mathfrak{q} \setminus x_c \in \mathrm{Ass}((L \setminus x_c)^\ell)$ for some $x_c \in \mathrm{supp}(f)$. Since I is normally torsion-free, we can derive from [**38**, Theorem 3.21] that $I \setminus x_c$ is normally torsion-free. Observe that the hypothesis $\mathrm{supp}(f) \subseteq \mathrm{supp}(u_m)$ yields that $I \setminus x_i = L \setminus x_i$ for any $x_i \in \mathrm{supp}(f)$. This gives that $I \setminus x_c = L \setminus x_c$, and so $L \setminus x_c$ is normally torsion-free. We thus get $\mathfrak{q} \setminus x_c \in \mathrm{Min}(L \setminus x_c)$, and hence $\mathfrak{q} \in \mathrm{Min}(L)$. This leads to a contradiction. Consequently, one has $\mathfrak{q} \setminus x_i \notin \mathrm{Ass}((L \setminus x_i)^\ell)$ for any $x_i \in \mathrm{supp}(f)$. In view of [**39**, Corollary 4.5], we get $x_i \mid z$ for any $x_i \in \mathrm{supp}(f)$, that is, $f \mid z$. Here, our aim is to demonstrate that $\mathfrak{q} \setminus x_\lambda \notin \mathrm{Ass}((L \setminus x_\lambda)^\ell)$ for any $n+1 \leq \lambda \leq n+t$. Assume, for the sake of contradiction, that $\mathfrak{q} \setminus x_\lambda \in \mathrm{Ass}((L \setminus x_\lambda)^\ell)$ for some $n+1 \leq \lambda \leq n+t$. Since $L \setminus x_\lambda = (u_1, \ldots, u_{m-1})S$ and $(u_1, \ldots, u_{m-1})S$ is normally torsion-free, we have $L \setminus x_\lambda$ is normally torsion-free, and so $\mathfrak{q} \setminus x_\lambda \in \mathrm{Min}(L \setminus x_\lambda)$. This yields that $\mathfrak{q} \in \mathrm{Min}(L)$, which is a contradiction. Accordingly, we have $\mathfrak{q} \setminus x_\lambda \notin \mathrm{Ass}((L \setminus x_\lambda)^\ell)$ for any $\lambda = n+1, \ldots, n+t$. It follows again from [**39**, Corollary 4.5] that $x_\lambda \mid z$ for any $\lambda = n+1, \ldots, n+t$, that is, $x_{n+1}\cdots x_{n+t} \mid z$. This gives rise to $z = vM$ for some monomial M. Hence, $\mathfrak{q} = (L^\ell : z) = (L^{\ell-1} : M)$, and so $\mathfrak{q} \in \mathrm{Ass}(L^{\ell-1})$, which contradicts the fact that $L^{\ell-1}$ has no embedded prime ideals. This implies that L is normally torsion-free, as desired.

(ii) It is obvious from the definition of nearly normally torsion-freeness and (i).

(iii) Follows from (i).

Statements (iv)-(vi) are shown exactly as in the proof of Theorem 2.12. □

The following is an immediate consequence of Theorem 3.8.

COROLLARY 3.9. *Let I be a normally torsion-free square-free monomial ideal in $R = K[x_1, \ldots, x_n]$ with $\mathcal{G}(I) = \{u_1, \ldots, u_m\}$. Let $L := (u_1, \ldots, u_{m-1}, gu_m)S$, where $S = R[x_{n+1}, \ldots, x_{n+t}]$, g be a square-free monomial in S with* $\operatorname{supp}(g) \subseteq \{x_{n+1}, \ldots, x_{n+t}\}$, *$gu_m \notin \mathfrak{p}^2$ for any $\mathfrak{p} \in \operatorname{Min}(L)$, and $(u_1, \ldots, u_{m-1})S$ be normally torsion-free. Then the following statements hold:*

(i) *L is normally torsion-free.*
(ii) *L is nearly normally torsion-free.*
(iii) *L is normal.*
(iv) *L has the strong persistence property.*
(v) *L has the persistence property.*
(vi) *L has the symbolic strong persistence property.*

We close this section with a pair of open questions.

QUESTION 3.10. Let $I \subset R = K[x_1, \ldots, x_n]$ be a square-free monomial ideal and $\mathfrak{q}$ be a prime monomial ideal in R. Let $L := I \cap (\mathfrak{q}, x_r)$ such that $x_r \notin \operatorname{supp}(\mathfrak{q}) \cup \operatorname{supp}(I)$ and $\operatorname{Ass}(L) = \operatorname{Ass}(I) \cup \{(\mathfrak{q}, x_r)\}$. If $\mathfrak{m} = (x_1, \ldots, x_n) \in \operatorname{Ass}(L^s)$ for some $s \geq 2$, then can we deduce that $\mathfrak{m} \setminus \{x_r\} \in \operatorname{Ass}(I^s)$ or $\mathfrak{m} \setminus \{x_r\} \in \operatorname{Ass}((I \cap \mathfrak{q})^s)$?

QUESTION 3.11. Let $I \subset R$ be a square-free monomial ideal and $\mathfrak{q}$ be a prime monomial ideal in R. Let $L := I \cap (\mathfrak{q}, x_r)$ such that $x_r \notin \operatorname{supp}(\mathfrak{q}) \cup \operatorname{supp}(I)$ and $\operatorname{Ass}(L) = \operatorname{Ass}(I) \cup \{(\mathfrak{q}, x_r)\}$. If $\mathfrak{p} = (L^s : x_r^\theta v)$ for some $s \geq 2$, $\theta \geq 0$, and monomial v with $x_r \nmid v$, then can we deduce that $\mathfrak{p} \setminus x_r \in \operatorname{Ass}(I^\theta (I \cap \mathfrak{q})^{s-\theta})$?

REMARK 3.12. (i) With the notation from Question 3.10, it is possible that we simultaneously have, for some $s \geq 2$, $\mathfrak{m} = (x_1, \ldots, x_n) \in \operatorname{Ass}(L^s)$ and $\mathfrak{m} \setminus \{x_r\} \in \operatorname{Ass}(I^\theta (I \cap \mathfrak{q})^{s-\theta})$ for all $0 \leq \theta \leq s$. To see an example, consider the following square-free monomial ideal in $R = K[x_1, x_2, x_3, x_4, x_5, x_6]$,

$$L = (x_3x_4, x_2x_4, x_3x_5x_6, x_1x_5x_6, x_1x_3x_6, x_1x_4x_5, x_2x_3x_5, x_1x_2x_5, x_1x_2x_3).$$

Using *Macaulay2* [**11**], we obtain

$$\operatorname{Ass}(L) = \{(x_1, x_2, x_3), (x_1, x_3, x_4), (x_1, x_4, x_5), (x_2, x_3, x_5), (x_3, x_4, x_5), (x_2, x_4, x_6)\}.$$

Let $\mathfrak{q} := (x_2, x_4)$, $s = 3$, $x_r = x_6$, $\mathfrak{m} = (x_1, x_2, x_3, x_4, x_5, x_6)$, and

$$\operatorname{Ass}(I) = \{(x_1, x_2, x_3), (x_1, x_3, x_4), (x_1, x_4, x_5), (x_2, x_3, x_5), (x_3, x_4, x_5)\}.$$

As $x_1x_3 \in I \setminus (I \cap \mathfrak{q})$, we get $I \cap \mathfrak{q} \neq I$. Using *Macaulay2* [**11**], we deduce that $\mathfrak{m} \in \operatorname{Ass}(L^3)$, $\mathfrak{m} \setminus \{x_6\} \in \operatorname{Ass}((I \cap \mathfrak{q})^3)$, $\mathfrak{m} \setminus \{x_6\} \in \operatorname{Ass}(I^2(I \cap \mathfrak{q}))$, $\mathfrak{m} \setminus \{x_6\} \in \operatorname{Ass}(I(I \cap \mathfrak{q})^2)$, and $\mathfrak{m} \setminus \{x_6\} \in \operatorname{Ass}(I^3)$.

(ii) With the notation from Question 3.10, it is possible that we have, for some $s \geq 2$, $\mathfrak{m} = (x_1, \ldots, x_n) \in \operatorname{Ass}(L^s)$ and $\mathfrak{m} \setminus \{x_r\} \notin \operatorname{Ass}(I^\theta (I \cap \mathfrak{q})^{s-\theta})$ for some $0 \leq \theta \leq s$. For instance, in $R = K[x_1, \ldots, x_6]$, let

$$\begin{aligned} L &= (x_3x_6, x_2x_5, x_3x_4, x_2x_4, x_1x_4, x_2x_3, x_1x_3) \\ &= (x_1, x_2, x_3) \cap (x_2, x_3, x_4) \cap (x_3, x_4, x_5) \cap (x_1, x_2, x_4, x_6). \end{aligned}$$

Assume that $\mathfrak{q} := (x_1, x_2, x_4)$, $s = 2$, $x_r = x_6$, and $\mathfrak{m} = (x_1, x_2, x_3, x_4, x_5, x_6)$, and

$$\operatorname{Ass}(I) = \{(x_1, x_2, x_3), (x_2, x_3, x_4), (x_3, x_4, x_5)\}.$$

Using *Macaulay2* [**11**], we can conclude that $\mathfrak{m} \in \operatorname{Ass}(L^2)$, $\mathfrak{m} \setminus \{x_6\} \notin \operatorname{Ass}(I(I \cap \mathfrak{q}))$, and $\mathfrak{m} \setminus \{x_6\} \in \operatorname{Ass}((I \cap \mathfrak{q})^2)$, but $\mathfrak{m} \setminus \{x_6\} \notin \operatorname{Ass}(I^2)$.

4. A comparison between clutters and complement clutters

Our aim in this section is to compare some algebraic properties of the edge ideals of clutters and complement clutters. In particular, our motivation comes back to Question 6.3 in [**29**]. Firstly, we need to recall the notion of a clutter. A *clutter* (or *simple hypergraph*) $\mathcal{C}$ with vertex set $X = \{x_1, \ldots, x_n\}$ is a family of subsets of X, called *edges*, none of which is included in another. The edge ideal of a clutter, $I(\mathcal{C})$, is defined similarly as for a graph and is generated by the products of all the vertices in an edge, for every edge of $\mathcal{C}$. The cover ideal of a clutter can be defined as the ideal of all monomials M such that given any edge e of $\mathcal{C}$ there is some variable x_i such that $x_i \in e$ and $x_i | M$. Note that the vertices of these clutters become the variables of the ring to which the edge ideal and the cover ideal are allocated. Given a clutter $\mathcal{C}$ on $\{x_1, \ldots, x_n\}$ with edges $e_1, \ldots, e_r$, we define the *complement clutter*, denoted by $\mathcal{C}^c$, as the clutter whose edges are $\{x_1, \ldots, x_n\} \setminus e_i$ for each $i = 1, \ldots, r$. In particular, we have $(\mathcal{C}^c)^c = \mathcal{C}$.

QUESTION 4.1 ([**29**, Question 6.3]). Does $I(\mathcal{C})$ have the strong persistence property if and only if $I(\mathcal{C}^c)$ have the strong persistence property?

The authors give an example of a clutter $\mathcal{C}$, where neither $I(\mathcal{C})$ nor $I(\mathcal{C}^c)$ possesses SPP. Inspired by that, we formulate the following.

QUESTION 4.2.

(i) Does $I(\mathcal{C})$ have the persistence property if and only if $I(\mathcal{C}^c)$ has the persistence property?
(ii) Is $I(\mathcal{C})$ normal if and only if $I(\mathcal{C}^c)$ is normal?
(iii) Is $I(\mathcal{C})$ normally torsion-free if and only if $I(\mathcal{C}^c)$ is normally torsion-free?

We will provide an example to give a negative answer to the first two questions, which in turn will be a new counterexample to Question 4.1.

Answer. (i)-(iii) We construct an example showing that there is a clutter $\mathcal{C}$ such that $I(\mathcal{C})$ is normal (hence, has SPP and PP) but for which $I(\mathcal{C}^c)$ does not have PP (hence, does not have SPP and is not normal). Consider the clutter

$$\begin{aligned}\mathcal{C} := \{&\{x_2, x_3, x_6\}, \{x_1, x_4, x_5, x_7\}, \{x_4, x_5, x_6, x_7\}, \{x_1, x_3, x_6, x_7\}, \{x_3, x_5, x_6, x_7\},\\ &\{x_1, x_3, x_4, x_7\}, \{x_1, x_2, x_6, x_7\}, \{x_1, x_2, x_5, x_7\}, \{x_2, x_4, x_6, x_7\}, \{x_2, x_3, x_5, x_7\},\\ &\{x_2, x_3, x_4, x_7\}\}.\end{aligned}$$

Its edge ideal is $I(\mathcal{C}) = (x_2x_3x_6, x_1x_4x_5x_7, x_4x_5x_6x_7, x_1x_3x_6x_7, x_3x_5x_6x_7, x_1x_3x_4x_7, x_1x_2x_6x_7, x_1x_2x_5x_7, x_2x_4x_6x_7, x_2x_3x_5x_7, x_2x_3x_4x_7)$. On the other hand, the edge ideal of the complement clutter $\mathcal{C}^c$ is given by $I(\mathcal{C}^c) = (x_1x_4x_5x_7, x_2x_3x_6, x_1x_2x_3, x_2x_4x_5, x_1x_2x_4, x_2x_5x_6, x_3x_4x_5, x_3x_4x_6, x_1x_3x_5, x_1x_4x_6, x_1x_5x_6)$.

Our first aim is to show that $I(\mathcal{C})$ is normal. To do this, let $A := x_3x_4(x_1, x_2)$ and $B := (x_1x_2, x_1x_4, x_2x_3)$. Since (x_1, x_2) is normal, we can deduce from [**2**, Remark 1.2] that A is normal. Now, let $P = (V(P), E(P))$ be the path graph with $V(P) = \{x_1, x_2, x_3, x_4\}$ and $E(P) = \{\{x_4, x_1\}, \{x_1, x_2\}, \{x_2, x_3\}\}$. Then B is the edge ideal of P. Due to P being a bipartite graph, it follows that B is normal. In addition, we have $A+B = B$, and so $A+B$ is normal. From [**32**, Theorem 3.1], we can derive that $C := A + x_5B = (x_1x_3x_4, x_2x_3x_4, x_1x_2x_5, x_1x_4x_5, x_2x_3x_5)$ is normal. Let $D := (x_4x_5, x_1x_3, x_3x_5, x_2x_4, x_1x_2)$ be the edge ideal of $G = (V(G), E(G))$, the odd cycle graph with the vertex set $V(G) = \{x_1, x_2, x_3, x_4, x_5\}$ and the edge set

$$E(G) = \{\{x_1, x_3\}, \{x_3, x_5\}, \{x_5, x_4\}, \{x_4, x_2\}, \{x_2, x_1\}\}.$$

On account of D being the edge ideal of G, we conclude from [**45**, Corollary 10.5.9] that D is normal. Moreover, $C + D = D$, and hence $C + D$ is normal. It follows again from [**32**, Theorem 3.1] that

$$F := C + x_6D = (x_1x_3x_4, x_2x_3x_4, x_1x_2x_5, x_1x_4x_5, x_2x_3x_5, x_4x_5x_6, x_1x_3x_6, \\ x_3x_5x_6, x_2x_4x_6, x_1x_2x_6),$$

is normal. It is straightforward to check that $I(\mathcal{C}) = (x_2x_3x_6)+x_7F$. Since $(x_2x_3x_6)$ is normal, it is enough to prove that $L := (x_2x_3x_6) + F$ is normal. To see this, let $H = (V(H), E(H))$ be the graph with the vertex set $V(H) = \{x_1, x_2, x_3, x_4, x_5\}$ and the edge set $E(H) = \{\{x_2, x_3\}, \{x_1, x_3\}, \{x_3, x_5\}, \{x_5, x_4\}, \{x_4, x_2\}, \{x_2, x_1\}\}$. Then it follows from [**45**, Corollary 10.5.9] that $M := I(H)$ is normal. Moreover, note that $L = C + x_6M$. In light of $C + M = M$, we obtain $C + M$ to be normal. We can now deduce from [**32**, Theorem 3.1] that $L = C + x_6M$ is normal. We therefore get $I(\mathcal{C})$ to be normal and, hence, to have SPP and PP.

We now turn our attention to explore $I(\mathcal{C}^c)$. Using *Macaulay2* [**11**], we find that

$$\mathfrak{m} = (x_1, x_2, x_3, x_4, x_5, x_6, x_7) \in \mathrm{Ass}(I(\mathcal{C}^c)^2) \setminus \mathrm{Ass}(I(\mathcal{C}^c)^3).$$

This shows that $I(\mathcal{C}^c)$ does not satisfy PP, and, therefore, neither SPP nor normality, as claimed. Hence, the answer is negative to each one of (i)-(iii).

(iv) In the following we give a counterexample. Consider the clutter

$$\mathcal{C} = \{\{x_1, x_2\}, \{x_1, x_3, x_6\}, \{x_4, x_5, x_6\}\},$$

and its edge ideal $I(\mathcal{C}) = (x_1x_2, x_1x_3x_6, x_4x_5x_6)$. In addition, we have $I(\mathcal{C}^c) = (x_3x_4x_5x_6, x_1x_2x_3, x_2x_4x_5)$.

Our strategy to show the normally torsion-freeness of $I(\mathcal{C})$ is to employ Theorem 3.7 in [**39**]. For this purpose, we first show that $x_1x_2 \in \mathfrak{p} \setminus \mathfrak{p}^2$ for all $\mathfrak{p} \in \mathrm{Min}(I(\mathcal{C}))$, which is straightforward to check, since

$$\mathrm{Ass}(I(\mathcal{C})) = \mathrm{Min}(I(\mathcal{C})) = \{(x_1, x_4), (x_1, x_5), (x_1, x_6), (x_2, x_6), (x_2, x_3, x_4), (x_2, x_3, x_5)\}.$$

Next, it is obvious that $I(\mathcal{C}) \setminus x_1 = (x_4x_5x_6)$ is normally torsion-free. Furthermore, [**38**, Lemmas 2.5 and 3.12] implies that $I(\mathcal{C}) \setminus x_2 = (x_1x_3x_6, x_4x_5x_6) = x_6(x_1x_3, x_4x_5)$ is normally torsion-free. Consequently, since $I(\mathcal{C}) \setminus x_1$ and $I(\mathcal{C}) \setminus x_2$ are normally torsion-free, we deduce from [**39**, Theorem 3.7] that $I(\mathcal{C})$ is normally torsion-free.

We now deal with the clutter $I(\mathcal{C}^c)$. Using *Macaulay2* [**11**], we detect that

$$(x_2, x_3, x_4) \in \mathrm{Ass}(I(\mathcal{C}^c)^2) \setminus \mathrm{Ass}(I(\mathcal{C}^c)).$$

This means that $I(\mathcal{C}^c)$ is not normally torsion-free, which finishes our argument.

5. Normally torsion-freeness and normality under polarization

We continue our argument with some questions which are devoted to the normally torsion-freeness and normality of monomial ideals under the polarization operator. In the sequel, NTF will denote normally torsion-free.

DEFINITION 5.1. The process of *polarization* replaces a power x_i^t by a product of t new variables $x_{(i,1)} \cdots x_{(i,t)}$. We call $x_{(i,j)}$ a *shadow* of x_i. We will use $\widetilde{I^t}$ to denote the polarization of I^t, will use S_t for the new polynomial ring in this polarization, and will use $\widetilde{w}$ to denote the polarization in S_t of a monomial w in $R = K[x_1, \dots, x_n]$. The *depolarization* of an ideal in S_t is formed by setting $x_{(i,j)} = x_i$ for all i, j.

In general, the polarization operator is more complicated in comparison with other monomial operators, and we pose the following questions with respect to it.

QUESTION 5.2. Let I be a non-square-free monomial ideal in the polynomial ring $K[x_1, \dots, x_n]$. Then can we deduce that

(i) if $\widetilde{I}$ has the strong persistence property, then I has the strong persistence property?
(ii) if $\widetilde{I}$ has the persistence property, then I has the persistence property?
(iii) if $\widetilde{I}$ is normal, then I is normal?
(iv) I is normally torsion-free if and only if $\widetilde{I}$ is normally torsion-free? ([**38**, Question 3.23])

Before we continue, there is a need to define the expansion operator from [**6**], which will be employed as a criterion further on.

Let $R = K[x_1, \dots, x_n]$. For a fixed ordered n-tuple $(i_1, \dots, i_n)$ of positive integers, we consider the polynomial ring

$$R^{(i_1,\dots,i_n)} = K[x_{11}, \dots, x_{1i_1}, x_{21}, \dots, x_{2i_2}, \dots, x_{n1}, \dots, x_{ni_n}].$$

Let $\mathfrak{p}_j$ be the monomial prime ideal $(x_{j1}, x_{j2}, \dots, x_{ji_j}) \subseteq R^{(i_1,\dots,i_n)}$ for all $j = 1, \dots, n$. Attached to each monomial ideal $I \subset R$ with a set of monomial generators $\{\mathbf{x}^{\mathbf{a}_1}, \dots, \mathbf{x}^{\mathbf{a}_m}\}$, where $\mathbf{x}^{\mathbf{a}_i} = x_1{}^{a_i(1)} \cdots x_n{}^{a_i(n)}$ and $a_i(j)$ denotes the j-th component of the vector $\mathbf{a}_i = (a_i(1), \dots, a_i(n))$ for all $i = 1, \dots, m$. We define the *expansion of I with respect to the n-tuple* $(i_1, \dots, i_n)$, denoted by $I^{(i_1,\dots,i_n)}$, to be the monomial ideal

$$I^{(i_1,\dots,i_n)} = \sum_{i=1}^{m} \mathfrak{p}_1^{a_i(1)} \cdots \mathfrak{p}_n^{a_i(n)} \subseteq R^{(i_1,\dots,i_n)}.$$

We simply write R^* and I^*, respectively, rather than $R^{(i_1,\dots,i_n)}$ and $I^{(i_1,\dots,i_n)}$.

For example, consider $R = K[x_1, x_2, x_3]$ and the ordered 3-tuple $(1, 3, 2)$. Then we have $\mathfrak{p}_1 = (x_{11})$, $\mathfrak{p}_2 = (x_{21}, x_{22}, x_{23})$, and $\mathfrak{p}_3 = (x_{31}, x_{32})$. So for the monomial ideal $I = (x_1x_2, x_3^2)$, the ideal $I^* \subseteq K[x_{11}, x_{21}, x_{22}, x_{23}, x_{31}, x_{32}]$ is $\mathfrak{p}_1\mathfrak{p}_2 + \mathfrak{p}_3^2$, namely,

$$I^* = (x_{11}x_{21}, x_{11}x_{22}, x_{11}x_{23}, x_{31}^2, x_{31}x_{32}, x_{32}^2).$$

Answer. (i)-(iii) Consider the non-square-free monomial ideal

$$I = (x_1x_2^2x_3, x_2x_3^2x_4, x_3x_4^2x_5, x_4x_5^2x_1, x_5x_1^2x_2) \subset R = K[x_1, \dots, x_5].$$

By virtue of $(I^2 : I) \neq I$ and $\mathfrak{m} = (x_1, x_2, x_3, x_4, x_5) \in \mathrm{Ass}_R(R/I) \setminus \mathrm{Ass}_R(R/I^2)$, we deduce that I does not satisfy PP and, hence, neither SPP nor normality. On the other hand, it has already been proved in [**35**, Example 6.8] that $\widetilde{I}$ is normal, and, hence, has SPP and PP.

(iv) Consider the monomial ideal $I = (x_1, x_2) \cap (x_1, x_2, x_3)^2 \cap (x_1, x_2, x_4)^4$ in $R = K[x_1, x_2, x_3, x_4]$. We will show that I is normally torsion-free, while $\widetilde{I}$ is not. Consider $S = K[x_1, x_3, x_4]$ and the ordered 3-tuple $(2, 1, 1)$. Then by considering

$\mathfrak{p}_1 = (x_1, x_2)$, $\mathfrak{p}_2 = (x_3)$, and $\mathfrak{p}_3 = (x_4)$, for the monomial ideal $F = (x_1) \cap (x_1, x_3)^2 \cap (x_1, x_4)^4 \subset S$, we detect that the ideal $F^* \subset R$ is $\mathfrak{p}_1 \cap (\mathfrak{p}_1, \mathfrak{p}_2)^2 \cap (\mathfrak{p}_1, \mathfrak{p}_3)^4$, namely,

$$F^* = (x_1, x_2) \cap (x_1, x_2, x_3)^2 \cap (x_1, x_2, x_4)^4 = I.$$

On account of [**27**, Lemma 3.5], we know that F^* is NTF if and only if F is NTF. Hence, we will prove the latter. To do this, set $J := (x_1^3, x_1^2x_4, x_1x_4^2, x_3x_4^3)$. Then it is easy to see that $F = x_1 J$. In light of [**38**, Lemma 3.12], it is enough to show that J is NTF. Note that $J = (x_1, x_3) \cap (x_1, x_4)^3$. In what follows, we show that $J^s = (x_1, x_3)^s \cap (x_1, x_4)^{3s}$ for all $s \geq 1$. To see this, fix $s \geq 1$. It is straightforward to check that $J^s \subseteq (x_1, x_3)^s \cap (x_1, x_4)^{3s}$. Hence, one has to prove the reverse inclusion. Take a monomial $f \in \mathcal{G}((x_1, x_3)^s \cap (x_1, x_4)^{3s})$. According to [**13**, Proposition 1.2.1], we get $f = \operatorname{lcm}(u, v)$ for some $u \in \mathcal{G}((x_1, x_3)^s)$ and $v \in \mathcal{G}((x_1, x_4)^{3s})$. Hence, there exist integers $0 \leq \lambda \leq s$ and $0 \leq \theta \leq 3s$ such that $u = x_1^{\lambda} x_3^{s-\lambda}$ and $v = x_1^{\theta} x_4^{3s-\theta}$. This implies that $f = x_1^{\max\{\lambda,\theta\}} x_3^{s-\lambda} x_4^{3s-\theta}$. Here, one may consider the following cases.

CASE 1. $0 \leq \theta \leq \lambda$. Then $\max\{\lambda, \theta\} = \lambda$, and it is easy to see that

$$(x_1x_4^2)^{\lambda}(x_3x_4^3)^{s-\lambda} \mid f.$$

CASE 2. $\lambda \leq \theta \leq 2\lambda$. Then $\max\{\lambda, \theta\} = \theta$, which implies that

$$(x_1^2x_4)^{\theta-\lambda}(x_1x_4^2)^{2\lambda-\theta}(x_3x_4^3)^{s-\lambda} \mid f.$$

CASE 3. $2\lambda \leq \theta \leq 3\lambda$. Then $\max\{\lambda, \theta\} = \theta$, and we can deduce that

$$(x_1^3)^{\theta-2\lambda}(x_1^2x_4)^{3\lambda-\theta}(x_3x_4^3)^{s-\lambda} \mid f.$$

CASE 4. $3\lambda \leq \theta \leq 3s$. Then $\max\{\lambda, \theta\} = \theta$. It is not hard to check that

$$(x_1x_4^2)^{\theta}(x_3x_4^3)^{s-\theta} \mid f.$$

Consequently, we conclude that $f \in J^s$, as required. In particular, we get

$$\operatorname{Ass}(J^s) = \operatorname{Ass}(J) = \{(x_1, x_3), (x_1, x_4)\}.$$

This yields that J is NTF. We therefore obtain I to be normally torsion-free. On the other hand, using *Macaulay2* [**11**], we find that the polarization of I is

$$\begin{aligned} L := \widetilde{I} = (&z_{(0,0)}z_{(0,1)}z_{(0,2)}z_{(0,3)}, z_{(0,0)}z_{(0,1)}z_{(0,2)}z_{(1,0)}, z_{(0,0)}z_{(0,1)}z_{(1,0)}z_{(1,1)}, \\ &z_{(0,0)}z_{(1,0)}z_{(1,1)}z_{(1,2)}, z_{(1,0)}z_{(1,1)}z_{(1,2)}z_{(1,3)}, z_{(0,0)}z_{(0,1)}z_{(0,2)}z_{(3,0)}, \\ &z_{(0,0)}z_{(0,1)}z_{(1,0)}z_{(3,0)}, z_{(0,0)}z_{(1,0)}z_{(1,1)}z_{(3,0)}, z_{(1,0)}z_{(1,1)}z_{(1,2)}z_{(3,0)}, \\ &z_{(0,0)}z_{(0,1)}z_{(3,0)}z_{(3,1)}, z_{(0,0)}z_{(1,0)}z_{(3,0)}z_{(3,1)}, z_{(1,0)}z_{(1,1)}z_{(3,0)}z_{(3,1)}, \\ &z_{(0,0)}z_{(2,0)}z_{(3,0)}z_{(3,1)}z_{(3,2)}, z_{(1,0)}z_{(2,0)}z_{(3,0)}z_{(3,1)}z_{(3,2)}), \end{aligned}$$

in $S = K[z_{(0,0)}, z_{(0,1)}, z_{(0,2)}, z_{(0,3)}, z_{(1,0)}, z_{(1,1)}, z_{(1,2)}, z_{(1,3)}, z_{(2,0)}, z_{(3,0)}, z_{(3,1)}, z_{(3,2)}]$. Using *Macaulay2* [**11**], we detect that

$$\begin{aligned} &\#\operatorname{Ass}(L) = 19, \#\operatorname{Ass}(L^2) = 51, \#\operatorname{Ass}(L^3) = 85, \#\operatorname{Ass}(L^4) = 126, \\ &\#\operatorname{Ass}(L^5) = \#\operatorname{Ass}(L^6) = \#\operatorname{Ass}(L^7) = 128. \end{aligned}$$

This means that $\widetilde{I}$ is not normally torsion-free.

Now, our aim is to present a monomial ideal M, which is not normally torsion-free, while $\widetilde{M}$ is. To do this, let $M = (x_1^3x_2, x_1x_3^3, x_2^4, x_1^4x_3)$ in $S = K[x_1, x_2, x_3]$. It is a straightforward operation to check that $(x_1, x_2, x_3) \in \operatorname{Ass}(S/M^2) \setminus \operatorname{Ass}(S/M)$.

This shows that M is NTF. For convenience of notation, the polarization of M is denoted by $\widetilde{M} = (x_1x_2x_3x_5, x_1x_5x_6x_7, x_5x_6x_7x_8, x_1x_2x_3x_4x_9) \subset K[x_1, \ldots, x_9]$. Using *Macaulay2* [**11**], we deduce that

$$\begin{aligned}\widetilde{M} = &(x_1, x_5) \cap (x_1, x_6) \cap (x_1, x_7) \cap (x_1, x_8) \cap (x_2, x_5) \cap (x_2, x_6) \cap (x_2, x_7) \cap (x_3, x_5)\\ &\cap (x_3, x_6) \cap (x_3, x_7) \cap (x_4, x_5) \cap (x_5, x_9).\end{aligned}$$

Now, consider the following graph G.

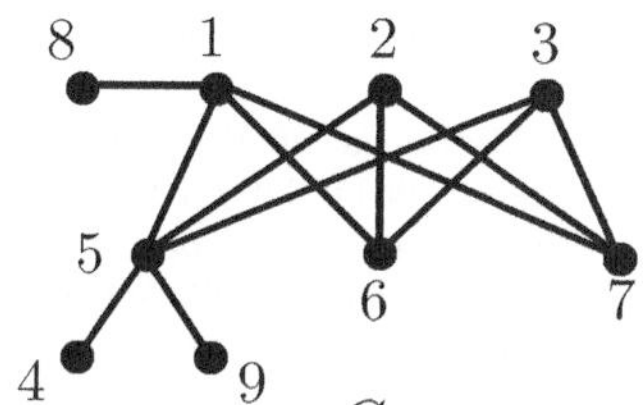

G

It is straightforward to detect that $\widetilde{M}$ is the cover ideal of the graph G. Since G has no odd cycle subgraph, this implies that G is bipartite. It follows immediately that the cover ideal of G is NTF, and so $\widetilde{M}$ is normally torsion-free, as claimed.

Acknowledgments

First, the authors are deeply grateful to the anonymous referee for careful reading of the manuscript, and for valuable suggestions which led to several improvements in the quality of this paper. In addition, this paper was prepared when the first author visited the Department of Mathematics of Uppsala University in 2023; in particular, he would like to thank Uppsala University for its hospitality.

References

[1] Ibrahim Al-Ayyoub, Mehrdad Nasernejad, Kazem Khashyarmanesh, Leslie G. Roberts, and Veronica Crispin Quiñonez, *Results on the normality of square-free monomial ideals and cover ideals under some graph operations*, Math. Scand. **127** (2021), no. 3, 441–457, DOI 10.7146/math.scand.a-128963. MR4400795

[2] Ibrahim Al-Ayyoub, Mehrdad Nasernejad, and Leslie G. Roberts, *Normality of cover ideals of graphs and normality under some operations*, Results Math. **74** (2019), no. 4, Paper No. 140, 26, DOI 10.1007/s00025-019-1068-x. MR3986727

[3] Ibrahim Al-Ayyoub, Mehrdad Nasernejad, and Leslie G. Roberts, *On the normality and associated primes of cover ideals of a class of imperfect graphs*, Math. Rep. (Bucur.) **26(76)** (2024), no. 1, 59–69, DOI 10.59277/mrar.2024.26.76.1.59. MR4717998

[4] Ibrahim Al-Ayyoub, Mehrdad Nasernejad, Kazem Khashyarmanesh, and Leslie G. Roberts, *On the strong persistence property and normality of cover ideals of theta graphs*, Comm. Algebra **51** (2023), no. 9, 3782–3791, DOI 10.1080/00927872.2023.2188413. MR4605241

[5] Claudia Andrei-Ciobanu, *Nearly normally torsionfree ideals*, Combinatorial structures in algebra and geometry, Springer Proc. Math. Stat., vol. 331, Springer, Cham, [2020] ©2020, pp. 1–13, DOI 10.1007/978-3-030-52111-0_1. MR4143236

[6] Shamila Bayati and Jürgen Herzog, *Expansions of monomial ideals and multigraded modules*, Rocky Mountain J. Math. **44** (2014), no. 6, 1781–1804, DOI 10.1216/RMJ-2014-44-6-1781. MR3310948

[7] M. Brodmann, *Asymptotic stability of* $\mathrm{Ass}(M/I^nM)$, Proc. Amer. Math. Soc. **74** (1979), no. 1, 16–18, DOI 10.2307/2042097. MR521865

[8] Aldo Conca and Emanuela De Negri, *M-sequences, graph ideals, and ladder ideals of linear type*, J. Algebra **211** (1999), no. 2, 599–624, DOI 10.1006/jabr.1998.7740. MR1666661

[9] Christopher A. Francisco, Huy Tài Hà, and Adam Van Tuyl, *Colorings of hypergraphs, perfect graphs, and associated primes of powers of monomial ideals*, J. Algebra **331** (2011), 224–242, DOI 10.1016/j.jalgebra.2010.10.025. MR2774655

[10] Isidoro Gitler, Enrique Reyes, and Rafael H. Villarreal, *Blowup algebras of ideals of vertex covers of bipartite graphs*, Algebraic structures and their representations, Contemp. Math., vol. 376, Amer. Math. Soc., Providence, RI, 2005, pp. 273–279, DOI 10.1090/conm/376/06963. MR2147027

[11] D. R. Grayson and M. E. Stillman, Macaulay2, a software system for research in algebraic geometry, `http://www.math.uiuc.edu/Macaulay2/`.

[12] Huy Tài Hà and Susan Morey, *Embedded associated primes of powers of square-free monomial ideals*, J. Pure Appl. Algebra **214** (2010), no. 4, 301–308, DOI 10.1016/j.jpaa.2009.05.002. MR2558739

[13] Jürgen Herzog and Takayuki Hibi, *Monomial ideals*, Graduate Texts in Mathematics, vol. 260, Springer-Verlag London, Ltd., London, 2011, DOI 10.1007/978-0-85729-106-6. MR2724673

[14] Jürgen Herzog and Takayuki Hibi, *The depth of powers of an ideal*, J. Algebra **291** (2005), no. 2, 534–550, DOI 10.1016/j.jalgebra.2005.04.007. MR2163482

[15] Jürgen Herzog, Asia Rauf, and Marius Vladoiu, *The stable set of associated prime ideals of a polymatroidal ideal*, J. Algebraic Combin. **37** (2013), no. 2, 289–312, DOI 10.1007/s10801-012-0367-z. MR3011344

[16] Tomá Kaiser, Matěj Stehlík, and Riste Škrekovski, *Replication in critical graphs and the persistence of monomial ideals*, J. Combin. Theory Ser. A **123** (2014), 239–251, DOI 10.1016/j.jcta.2013.12.005. MR3157809

[17] Kazem Khashyarmanesh and Mehrdad Nasernejad, *A note on the Alexander dual of path ideals of rooted trees*, Comm. Algebra **46** (2018), no. 1, 283–289, DOI 10.1080/00927872.2017.1324867. MR3764863

[18] K. Khashyarmanesh and M. Nasernejad, *Some results on the associated primes of monomial ideals*, Southeast Asian Bull. Math. **39** (2015), no. 3, 439–451. MR3410220

[19] Kazem Khashyarmanesh, Mehrdad Nasernejad, and Jonathan Toledo, *Symbolic strong persistence property under monomial operations and strong persistence property of cover ideals*, Bull. Math. Soc. Sci. Math. Roumanie (N.S.) **64(112)** (2021), no. 2, 105–131. MR4275135

[20] Stephen McAdam, *Asymptotic prime divisors*, Lecture Notes in Mathematics, vol. 1023, Springer-Verlag, Berlin, 1983, DOI 10.1007/BFb0071575. MR722609

[21] José Martínez-Bernal, Susan Morey, and Rafael H. Villarreal, *Associated primes of powers of edge ideals*, Collect. Math. **63** (2012), no. 3, 361–374, DOI 10.1007/s13348-011-0045-9. MR2957976

[22] Susan Morey and Rafael H. Villarreal, *Edge ideals: algebraic and combinatorial properties*, Progress in commutative algebra 1, de Gruyter, Berlin, 2012, pp. 85–126. MR2932582

[23] M. Nasernejad, *Asymptotic behaviour of associated primes of monomial ideals with combinatorial applications*, J. Algebra Relat. Top. **2** (2014), 15–25.

[24] Mehrdad Nasernejad, *Persistence property for some classes of monomial ideals of a polynomial ring*, J. Algebra Appl. **16** (2017), no. 6, 1750105, 17, DOI 10.1142/S0219498817501055. MR3635135

[25] Mehrdad Nasernejad, *Some counterexamples for (strong) persistence property and (nearly) normally torsion-freeness*, Ex. Countex. **3** (2023), Paper No. 100111, 4, DOI 10.1016/j.exco.2023.100111. MR4593069

[26] Mehrdad Nasernejad, Somayeh Bandari, and Leslie G. Roberts, *Normality and associated primes of closed neighborhood ideals and dominating ideals*, J. Algebra Appl. **24** (2025), no. 1, Paper No. 2550009, 19, DOI 10.1142/S0219498825500094. MR4834003

[27] Mehrdad Nasernejad and Kazem Khashyarmanesh, *On the Alexander dual of the path ideals of rooted and unrooted trees*, Comm. Algebra **45** (2017), no. 5, 1853–1864, DOI 10.1080/00927872.2016.1226855. MR3582830

[28] Mehrdad Nasernejad, Kazem Khashyarmanesh, and Ibrahim Al-Ayyoub, *Associated primes of powers of cover ideals under graph operations*, Comm. Algebra **47** (2019), no. 5, 1985–1996, DOI 10.1080/00927872.2018.1527920. MR3977715

[29] Mehrdad Nasernejad, Kazem Khashyarmanesh, Leslie G. Roberts, and Jonathan Toledo, *The strong persistence property and symbolic strong persistence property*, Czechoslovak Math. J. **72(147)** (2022), no. 1, 209–237, DOI 10.21136/CMJ.2021.0407-20. MR4389116

[30] Mehrdad Nasernejad, Veronica Crispin Quiñonez, and Winfried Hochstättler, *On the normally torsion-freeness of square-free monomial ideals*, J. Algebra Appl. **24** (2025), no. 12, Paper No. 2550283, 15, DOI 10.1142/S0219498825502834. MR4909040

[31] Mehrdad Nasernejad and Ayesha Asloob Qureshi, *Algebraic implications of neighborhood hypergraphs and their transversal hypergraphs*, Comm. Algebra **52** (2024), no. 6, 2328–2345, DOI 10.1080/00927872.2023.2300760. MR4728947

[32] Mehrdad Nasernejad, Ayesha Asloob Qureshi, Somayeh Bandari, and Aslı Musapaşaoğlu, *Dominating ideals and closed neighborhood ideals of graphs*, Mediterr. J. Math. **19** (2022), no. 4, Paper No. 152, 18, DOI 10.1007/s00009-022-02107-1. MR4440844

[33] Mehrdad Nasernejad, Ayesha Asloob Qureshi, Kazem Khashyarmanesh, and Leslie G. Roberts, *Classes of normally and nearly normally torsion-free monomial ideals*, Comm. Algebra **50** (2022), no. 9, 3715–3733, DOI 10.1080/00927872.2022.2042546. MR4442467

[34] Anda Olteanu, *Normally torsion-free lexsegment ideals*, Algebra Colloq. **22** (2015), no. 1, 23–34, DOI 10.1142/S1005386715000048. MR3296754

[35] Saeed Rajaee, Mehrdad Nasernejad, and Ibrahim Al-Ayyoub, *Superficial ideals for monomial ideals*, J. Algebra Appl. **17** (2018), no. 6, 1850102, 28, DOI 10.1142/S0219498818501025. MR3805713

[36] L. J. Ratliff Jr., *On prime divisors of I^n, n large*, Michigan Math. J. **23** (1976), no. 4, 337–352 (1977). MR457421

[37] Enrique Reyes and Jonathan Toledo, *On the strong persistence property for monomial ideals*, Bull. Math. Soc. Sci. Math. Roumanie (N.S.) **60(108)** (2017), no. 3, 293–305. MR3701891

[38] Mirsadegh Sayedsadeghi and Mehrdad Nasernejad, *Normally torsion-freeness of monomial ideals under monomial operators*, Comm. Algebra **46** (2018), no. 12, 5447–5459, DOI 10.1080/00927872.2018.1469029. MR3923772

[39] Mirsadegh Sayedsadeghi, Mehrdad Nasernejad, and Ayesha Asloob Qureshi, *On the embedded associated primes of monomial ideals*, Rocky Mountain J. Math. **52** (2022), no. 1, 275–287, DOI 10.1216/rmj.2022.52.275. MR4409930

[40] R. Y. Sharp, *Steps in commutative algebra*, London Mathematical Society Student Texts, vol. 19, Cambridge University Press, Cambridge, 1990. MR1070568

[41] Aron Simis, Wolmer V. Vasconcelos, and Rafael H. Villarreal, *On the ideal theory of graphs*, J. Algebra **167** (1994), no. 2, 389–416, DOI 10.1006/jabr.1994.1192. MR1283294

[42] Seth Sullivant, *Combinatorial symbolic powers*, J. Algebra **319** (2008), no. 1, 115–142, DOI 10.1016/j.jalgebra.2007.09.024. MR2378064

[43] Craig Huneke and Irena Swanson, *Integral closure of ideals, rings, and modules*, London Mathematical Society Lecture Note Series, vol. 336, Cambridge University Press, Cambridge, 2006. MR2266432

[44] Rafael H. Villarreal, *Cohen-Macaulay graphs*, Manuscripta Math. **66** (1990), no. 3, 277–293, DOI 10.1007/BF02568497. MR1031197

[45] Rafael H. Villarreal, *Monomial algebras*, 2nd ed., Monographs and Research Notes in Mathematics, CRC Press, Boca Raton, FL, 2015. MR3362802

Univ. Artois, UR 2462, Laboratoire de Mathématique de Lens (LML), F-62300 Lens, France

Email address: m_nasernejad@yahoo.com

ORCID: 0000-0003-1073-1934

Department of Mathematics, Uppsala University, S-751 06, Uppsala, Sweden

Email address: veronica.crispin@math.uu.se

ORCID: 0000-0002-2146-7051

Tecnológico Nacional de México, Instituto Tecnológico Del Valle de Etla, Abasolo S/N, Barrio Del Agua Buena, Santiago Suchilquitongo, 68230, Oaxaca, México

Email address: jonathan.tt@itvalletla.edu.mx

ORCID: 0000-0003-3274-1367

Contemporary Mathematics
Volume **826**, 2025
https://doi.org/10.1090/conm/826/16584

The separativity problem in terms of varieties and diagonal reduction

Pere Ara, Ken Goodearl, Pace P. Nielsen, Enrique Pardo, and Francesc Perera

ABSTRACT. We provide two new formulations of the separativity problem.

First, it is known that separativity (and strong separativity) in von Neumann regular (and exchange) rings is tightly connected to unit-regularity of certain kinds of elements. By refining this information, we characterize separative regular rings in terms of a special type of inner inverse operation, which is defined via a single identity. This shows that the separative regular rings form a subvariety of the regular rings. The separativity problem reduces to the question of whether every element of the form $(1 - aa')bac(1 - a'a)$ in a regular ring is unit-regular, where a' is an inner inverse for a.

Second, it is known that separativity in exchange rings is equivalent to regular matrices over corner rings being reducible to diagonal matrices via elementary row and column operations. We show that for 2×2 invertible matrices, four elementary operations are sufficient, and in general also necessary. Dropping the invertibility hypothesis, but specializing to separative regular rings, we show that three elementary row operations together with three elementary column operations are sufficient, and again in general also necessary. The separativity problem can subsequently be reframed in terms of the explicit number of operations needed to diagonally reduce.

2020 *Mathematics Subject Classification.* Primary 16E50; Secondary 15A20, 16E20, 16U40, 16U60.

Key words and phrases. Diagonally reduce, exchange rings, inner inverses, regular rings, separativity.

The first, fourth and fifth authors were partially supported by the Spanish State Research Agency (grant No. PID2020-113047GB-I00/AEI/10.13039/501100011033). The first and fifth authors were partially supported by the Comissionat per Universitats i Recerca de la Generalitat de Catalunya (grant No. 2017-SGR-1725) and by the Spanish State Research Agency through the Severo Ochoa and María de Maeztu Program for Centers and Units of Excellence in R&D (CEX2020-001084-M). The fourth author was partially supported by PAI III grant FQM-298 of the Junta de Andalucía, and by the grant "Operator Theory: an interdisciplinary approach", reference ProyExcel 00780, a project financed in the 2021 call for Grants for Excellence Projects, under a competitive bidding regime, aimed at entities qualified as Agents of the Andalusian Knowledge System, in the scope of the Plan Andaluz de Investigación, Desarrollo e Innovación (PAIDI 2020), Consejería de Universidad, Investigación e Innovación of the Junta de Andalucía. This work was partially supported by a grant from the Simons Foundation (#963435 to Pace P. Nielsen).

1. Introduction

In 1954, Kaplansky [**15**] posed three problems for abelian groups, which are:

- If $A \lesssim B$ and $B \lesssim A$ (that is, each is isomorphic to a submodule of the other), then is $A \cong B$?
- If $A \oplus A \cong B \oplus B$, then is $A \cong B$?
- If C is finitely generated and $A \oplus C \cong B \oplus C$, then is $A \cong B$?

These questions have since become a standard against which to judge one's understanding of a category of modules over a ring R. Depending on the qualities of the ring R, and on the characteristics of the module category under consideration, these questions can range from "trivial" to "important unsolved problem."

When R is a (von Neumann) regular ring, and we focus on the category $\mathrm{FP}(R)$ of finitely generated projective right R-modules, it is sometimes possible to characterize when Kaplansky's problems have positive answers. For example, the third question has a positive answer in $\mathrm{FP}(R)$ for exactly those regular rings that are unit-regular; this was first proven true by Handelman, see [**13**, Theorem 2]. Unit-regular rings are also the regular rings with stable rank 1, as shown independently by Fuchs [**8**, Corollary 1 and Theorem 4] and Kaplansky (unpublished). For additional related results and history see [**11**, Chapter 4].

One can significantly strengthen the second hypothesis from Kaplansky's third test problem, by asking for any two modules $A, B \in \mathscr{C}$, where $\mathscr{C}$ is some category of modules:

$$\text{If } A \oplus A \cong A \oplus B \cong B \oplus B, \text{ then is } A \cong B? \tag{1.1}$$

(This strengthens the second hypothesis of the third test question by combining the cases when $C = A$ and $C = B$. Interestingly, this is also a specialization of the second test question.) When the answer to (1.1) is positive, then following the literature one says that $\mathscr{C}$ is *separative*. Moreover, in the case when $\mathscr{C} = \mathrm{FP}(R)$, one says that the ring R is *separative*.

Since separativity depends on isomorphisms between modules and not equalities, then instead of working in the (large) category $\mathrm{FP}(R)$, it is occasionally more convenient to work with $V(R)$, the set of *isomorphism types* of modules in $\mathrm{FP}(R)$. The set $V(R)$ is a monoid, by defining for any two modules $A, B \in \mathrm{FP}(R)$ the addition

$$[A] + [B] = [A \oplus B].$$

Separativity of R is equivalent to the assertion that $2[A] = [A] + [B] = 2[B]$ implies $[A] = [B]$, in $V(R)$.

The initial study of separativity for regular rings, and more generally for exchange rings, began with the papers [**1**] and [**2**]. Since that time, many other papers have explored this topic. Separativity has a number of important and restrictive consequences in regular rings; see [**2**, Theorem 6.1] for five examples. For each of those five properties, it is an open question whether every regular ring has that property. Thus, a major open problem in this area, which motivates much of this paper, is the following.

SEPARATIVITY PROBLEM. *Is every regular (or exchange) ring separative?*

This problem has now been open for over 30 years. Outside the class of exchange rings, separativity can easily fail, as for the first Weyl algebra and for the coordinate ring of the 2-sphere; see [**12**, Section 2]. However, perhaps the most instructive parallel for us is in the category of finite rank, torsion-free abelian groups (abstractly these are just the subgroups of finite-dimensional vector spaces over the rational field $\mathbb{Q}$). This category is well-documented as the home of pathology in module theory (see [**9, 10**]). Despite this, examples of failure of separative cancellation are rare, and tied up with nontrivial number theory; see [**24**] for details, as well as a concrete construction of a large class of explicit examples where separativity fails. The separativity problem has numerous equivalent reformulations; for a recent example in terms of products of idempotents see [**14**].

In this paper we continue developing the theory of separative regular rings. Specifically, after a brief review of facts about inner inverses in Section 2, we explain in Section 3 how to define separative regular (and separative exchange) rings in new ways in terms of regularity and unit-regularity. A key result is the following ring-theoretic reformulation of separativity.

Separativity via unit-regularity (Corollary 3.25 with Lemma 4.5). *A regular ring R, with an inner inverse operation $'$, is separative if and only if for any idempotent $e \in R$, and for any $a, b, c \in eRe$, the element*

$$x := (1 - aa')bac(1 - a'a) + (1 - e)$$

is unit-regular in R.

We also prove a similar, but more complicated, characterization of separativity in exchange rings. Moreover, the methods apply to other cancellation properties besides separativity, as multiple examples demonstrate.

A consequence of the previous theorem is that the separativity problem is equivalent to determining whether or not every such element x in a regular ring is unit-regular (where we may even take $e = 1$, by passing to a corner ring). Another consequence of this work is a description of separative regular rings as a subvariety of regular rings. This is worked out explicitly in Theorem 4.6. Interestingly, the separative exchange rings also form a variety; see Theorem 5.3 and its proof. A third consequence is a simpler diagonal reduction process for 2×2 regular matrices over separative exchange rings. In the last couple of sections we undertake the task of minimizing the number of elementary operations needed to diagonally reduce such matrices. Ultimately, we prove the following two other main results:

Diagonal reduction of invertible matrices (Theorem 6.11). *Over a separative exchange ring, every invertible 2×2 matrix can be diagonally reduced using four elementary operations.*

Generally, four elementary operations are also necessary. Removing the invertibility hypothesis raises the number of needed operations, even when we restrict to regular rings. In that case, we have:

Diagonal reduction over separative regular rings (Theorem 7.5). *Over a separative regular ring, every 2×2 matrix can be diagonally reduced using just three row operations together with three column operations. Nothing less on either side can work in general.*

Readers may find Theorem 6.2 and Proposition 6.6 of independent interest. The first concerns connections between stable rank 1 and diagonal reduction. The second is a statement about completable unimodular rows in exchange rings.

2. Inner inverses

Throughout this paper, let R be a unital ring. Recall that an element $a \in R$ is (von Neumann) *regular* if it has an *inner inverse* $b \in R$ (also called a *quasi-inverse*), which satisfies $aba = a$. Inner inverses are in general not unique; for example, any element of R is an inner inverse for 0.

When every element of a ring R is regular, then R is called a *regular ring*. These rings are a central fixture of modern noncommutative ring theory. The reader is directed to [**11**] for a more thorough study of these rings. In case R is regular, then we may as well fix some inner inverse $a' \in R$ for each element $a \in R$. We can then view $'$ as a unary operation on R, called an *inner inverse operation*. The category of regular rings with a distinguished inner inverse operation is a variety, in the universal algebra sense, satisfying the axioms of rings together with the identity

$$aa'a = a.$$

In general, inner inverse operations on a given ring are not unique. In fact, there are many potential ways to improve a given inner inverse. For example, let $(R, ')$ be a regular ring (with a distinguished inner inverse operation), and let $a \in R$. Fixing

$$a^\dagger := a'aa',$$

then a quick computation shows that

$$aa^\dagger a = a \quad \text{and} \quad a^\dagger a a^\dagger = a^\dagger.$$

Thus, $a^\dagger$ is an inner inverse for a, and in addition a is an inner inverse for $a^\dagger$. In this case, we say that a and $a^\dagger$ form a pair of *reflexive inverses*. We can view † as a unary operation on R, called a *reflexive inverse operation*.

Since every regular ring has a reflexive inverse operation, it is tempting to simply declare that the initial operation $'$ is a reflexive inverse operation, and indeed one can do so. But why stop there? In fact, there is an infinite sequence of further refinements that can be made to inner inverse operations, as shown in [**6, 22, 25**]. In general, however, there is no operation satisfying all of those improvements simultaneously. This is, of course, no problem if one only needs finitely many improvements.

Yet, there are often good reasons to not even assume that $'$ is a reflexive inverse operation. To see why this is the case, we first recall another standard definition. Given a ring R, then an element $a \in R$ is *unit-regular* in the ring R when there is some unit inner inverse. If every element of R has a unit inner inverse, then R is called a *unit-regular ring*.

Put another way, if R is a unit-regular ring, then we may force $a' \in \mathrm{U}(R)$ for each $a \in R$, in which case we say that $'$ is a *unit inner inverse operation*. It is interesting to observe that these types of inner inverses can also be described equationally. First, note that if $a \in \mathrm{U}(R)$ and if a' is *any* inner inverse for a, then from $aa'a = a$, we get $a' = a^{-1}$. Thus, $'$ is a unit inner inverse operation if and only if

$$a'a'' = 1 \quad \text{and} \quad a''a' = 1.$$

Surprisingly, these two identities can be expressed by a single identity; namely,

$$a'a''a''a' = 1,$$

since this forces a' to be both left and right invertible and hence a unit.

When given a unit-regular ring $(R, ')$ where $'$ is merely an inner inverse operation, then in general it is impossible to find a derived operation that is a unit inner inverse operation. Indeed, [**23**, Proposition 2.15] tells us that there is no such derived operation that is *uniformly* defined over the rings $\mathbb{M}_n(F)$, for F a field, as the integer $n \geq 1$ varies. Thus, there is no such derived operation at all on $\prod_{n=1}^{\infty} \mathbb{M}_n(F)$.

We now have two potential routes for improving a given inner inverse. We may force it to be reflexive, and when working in a unit-regular ring we may force it to be a unit. Unfortunately, these routes rarely are compatible.

LEMMA 2.1. *Let $(R, ')$ be a regular ring. Then $'$ is both a reflexive inverse operation and a unit inner inverse operation if and only if $R = 0$.*

PROOF. Any reflexive inverse $0'$ for 0 satisfies

$$0' = 0'00' = 0,$$

and this can be a unit only in the zero ring. Conversely, in the zero ring, there is only one inner inverse operation, and it satisfies the stated properties. □

For the following discussion, it is convenient to view ring elements in a slightly different way. Recall that there is a natural isomorphism $R \cong \operatorname{End}(R_R)$, where an element $a \in R$ corresponds to the endomorphism "left multiplication by a." Hereafter, we will freely identity elements of R with their left multiplication action on the free right R-module R_R. Thus, by $\operatorname{im}(a)$ we simply mean the right ideal aR, and by $\ker(a)$ we mean the right annihilator in R. Note that we are privileging endomorphisms of the *right* R-module R_R over endomorphisms of the left R-module ${}_RR$, since from the beginning we made a choice to work in the category $\operatorname{FP}(R)$ of finitely generated projective *right* R-modules.

We can classify regular and unit-regular elements according to their behavior as endomorphisms, as follows. (Note that these classifications are part of a more general situation, which appears in the solutions to Exercises 4.14A$_1$ and 4.14C in [**18**].) An element $a \in R$ is regular if and only if $\operatorname{im}(a) \subseteq^{\oplus} R_R$. Letting a' be any inner inverse, then aa' is an idempotent, and

$$\operatorname{im}(a) = aR = aa'R \subseteq^{\oplus} R_R.$$

Also, $a'a$ is an idempotent, and

$$\ker(a) = \operatorname{ann}_r(a) = (1 - a'a)R \subseteq^{\oplus} R_R.$$

Continuing to assume that a is regular, then it is unit-regular in R if and only if

$$\ker(a) \cong \operatorname{coker}(a) := R/aR,$$

or in other words

$$(1 - a'a)R \cong (1 - aa')R.$$

The complementary idempotents aa' and $1 - aa'$, as well as their opposites $a'a$ and $1 - a'a$, thus play an important role.

Following [**17**, Proposition 21.20], two idempotents $e, f \in R$ are said to be *isomorphic* (also, sometimes called *equivalent*) when $eR \cong fR$, or equivalently $Re \cong Rf$. Thus, the two idempotents $a'a$ and aa' are always isomorphic (via left multiplication by a), but the complementary idempotents $1 - a'a$ and $1 - aa'$ are isomorphic exactly when a is unit-regular.

An element $b \in R$ is an inner inverse to a if and only if (left multiplication by) b takes the summand $\mathrm{im}(a)$ isomorphically to a direct sum complement of $\ker(a)$ in R_R and left multiplication by a is the inverse isomorphism. Nonuniqueness of inner inverses follows from the fact that b can act arbitrarily on a direct sum complement of $\mathrm{im}(a)$. To be a reflexive inverse, $\ker(b)$ must be a direct sum complement of $\mathrm{im}(a)$. In other words, $a, b \in R$ are a pair of reflexive inverses if and only if

- $R_R = \mathrm{im}(b) \oplus \ker(a) = \mathrm{im}(a) \oplus \ker(b)$ and
- left multiplications by a and b yield inverse isomorphisms between $\mathrm{im}(b)$ and $\mathrm{im}(a)$.

3. Defining separativity in a new way

The topic at hand is separative cancellation (1.1). However, this is a complicated condition, involving many isomorphisms. Instead of immediately characterizing that situation in terms of regularity, we first begin with the following two simpler forms of cancellation.

DEFINITION 3.1. A ring R is *strongly separative* when

$$A \oplus A \cong A \oplus B \;\Rightarrow\; A \cong B,$$

for any $A, B \in \mathrm{FP}(R)$. As noted in [**2**, Lemma 5.1], this condition is equivalent to

$$A \oplus C \oplus C \cong B \oplus C \;\Rightarrow\; A \oplus C \cong B, \tag{3.2}$$

for any $A, B, C \in \mathrm{FP}(R)$. In case (3.2) is only assumed to hold when $R_R \cong A \oplus C \oplus C$, we shall say that R is *core strongly separative.*

Any strongly separative ring is core strongly separative, but the converse can fail.

EXAMPLE 3.3. There exists a core strongly separative ring that is not strongly separative. Moreover, we can force the ring to be regular.

PROOF. Let M be the commutative (additive) monoid generated by two elements u and x, subject to the two relations

$$u + x = u \quad \text{and} \quad x + x = x.$$

Any element of this monoid is thus uniquely represented as one of the elements in

$$\{x\} \cup \{ku \,:\, k \in \mathbb{Z}_{\geq 0}\}.$$

(Define a homomorphism $M \to \mathbb{N}$ by sending $x \mapsto 0$ and $u \mapsto 1$; this shows that the multiples of u are distinct, and none are equal to x except perhaps 0. Similarly, if N is the two element commutative monoid $\{0, \infty\}$ subject to the rule $\infty + z = \infty$ for any $z \in N$, then the map $M \to N$ sending $x \mapsto \infty$ and $u \mapsto \infty$ shows that x is nonzero. Alternatively, a normal form theorem, like Bergman's diamond lemma, works.)

Given any $y, z \in M$, we see (by case analysis, or using the monoid N above) that

$$y + z = 0 \;\Rightarrow y = z = 0.$$

This means that M is *conical.* A similarly easy computation by cases shows that M is a *refinement* monoid as defined in [**2**]. Thus, by [**4**, Theorem B], together with the paragraph that follows that theorem, we know that M is isomorphic to the monoid $V(R)$, for some (unital) regular ring R.

The ring R is not strongly separative, since on the monoid side $2x = x$ does not imply $x = 0$. However, R is core strongly separative, which we see as follows. The isomorphism class $[R_R]$ can be made to correspond to u (from the paragraph following [**4**, Theorem B]). Suppose that

$$u = a + c + c = b + c$$

for some $a, b, c \in M$. If $c = 0$, then $a + c = b$ and we are done. If $c = x$, then we must have $a = b = u$, and again $a + c = b$. Finally, if $c = ku$ for some integer $k \geq 1$, then we have $u = a + 2c = a + 2ku$, which is impossible. $\square$

Core strong separativity is defined in terms of modules. However, we can reexpress it ring theoretically. To do so, we first introduce an important class of elements in rings, in the following lemma–definition.

LEMMA 3.4. *Let R be a ring, and let $x \in R$ be a regular element with inner inverse $x' \in R$. The following are equivalent:*

(1) $xR \lesssim^{\oplus} R/xR$, *or in other words* $\operatorname{im}(x)$ *is isomorphic to a direct summand of* $\operatorname{coker}(x)$.
(2) *If* $xR \oplus Y = R_R$ *for some submodule* $Y \subseteq R_R$, *then* $xR \lesssim^{\oplus} Y$.
(3) $xx'R \lesssim^{\oplus} (1 - xx')R$.
(4) $Rxx' \lesssim^{\oplus} R(1 - xx')$.

Any regular element satisfying these conditions is called a **right repeater**.

PROOF. The regularity hypothesis on x guarantees that $xR \subseteq^{\oplus} R_R$. Every direct summand complement of xR in R_R is isomorphic to R/xR. Also, $(1-xx')R$ is one of those direct sum complements. As the relation $\lesssim^{\oplus}$ is an isomorphism invariant, this proves the equivalence of the first three conditions. Finally, from the left-right symmetry of isomorphic idempotents, [**17**, Proposition 21.20], we see that (3) $\Leftrightarrow$ (4), since both conditions assert that the idempotent xx' is isomorphic to a subidempotent of $(1 - xx')$. $\square$

An important class of examples of right repeaters, which we suggest that the reader keep in mind, are the following. Given an integer $n \geq 1$ and a field F, suppose that $R = \mathbb{M}_n(F)$. Then a matrix $x \in R$ is a right repeater exactly when $\operatorname{rank}(x) \leq n/2$.

Returning to the situation where R is an arbitrary ring, assume that the premise of core strong separativity, $R_R \cong A \oplus C \oplus C \cong B \oplus C$, is instantiated by some $A, B, C \in \mathrm{FP}(R)$. After replacing these modules with isomorphic copies, and renaming as necessary, we may assume that there are internal direct sum decompositions $R = A \oplus C \oplus C_1 = B \oplus C_2$, where $C \cong C_1 \cong C_2$. Letting $x \in R$ be any endomorphism that annihilates B, and that sends C_2 isomorphically to C_1, we see that x is a right repeater.

Conversely, let $x \in R$ be any right repeater. By Lemma 3.4(3), $(1-xx')R = A \oplus C$ with $C \cong C_1 := xR$. Putting $B := (1 - x'x)R$ and $C_2 := x'xR$, we now have

$$R_R = A \oplus C \oplus C_1 = B \oplus C_2,$$

with $C \cong C_1 \cong C_2$. Thus right repeaters are exactly what is needed to instantiate the premise of the implication defining core strong separativity. Moreover, the conclusion of that implication is the isomorphism $(1 - xx')R = A \oplus C \cong B = (1 - x'x)R$. So, the conclusion holds if and only if the right repeater is unit-regular. In other words:

PROPOSITION 3.5. *A ring is core strongly separative if and only if all right repeaters in the ring are unit-regular.*

We will do better than this, by showing that unit-regularity of a special subset of the right repeaters suffices. We will work in an arbitrary ring, where some elements of R may fail to have any inner inverse, but we will still find it convenient to fix a unary operation $'$ on R such that $aa'a = a$ whenever a happens to be regular. For simplicity, we refer to $'$ as an inner inverse operation on the regular elements.

PROPOSITION 3.6. *Let R be a ring, and fix an inner inverse operation $'$ on the regular elements of R. If $a \in R$ is regular, and if $b \in R$ is such that $(1 - aa')baa'$ is regular, then*

$$x := (1 - aa')ba$$

is a right repeater.

PROOF. Assuming that a is regular, then $aR = aa'R$ and hence

$$xR = (1 - aa')baR = (1 - aa')baa'R.$$

Assuming $(1 - aa')baa'$ is regular, then we see that $xR \subseteq^{\oplus} R_R$. Thus, x is regular. Since $xR \subseteq^{\oplus} (1 - aa')R$, we may fix an inner inverse $x^{\dagger}$ for x, possibly different from x', such that $aa'R \subseteq^{\oplus} (1 - xx^{\dagger})R$.

Now, $xx^{\dagger} = (1 - aa')bax^{\dagger}$ is an idempotent. Hence, $ax^{\dagger}xx^{\dagger}(1 - aa')b$ is an isomorphic idempotent. (Here we are using the easy fact, called the "flip-and-double trick," that if cd is an idempotent, then $dcdc$ is also an idempotent isomorphic to cd. Take $c := (1 - aa')b$ and $d := ax^{\dagger}$.) Therefore, xR is isomorphic to the summand

$$ax^{\dagger}xx^{\dagger}(1 - aa')bR \subseteq^{\oplus} aR = aa'R \subseteq^{\oplus} (1 - xx^{\dagger})R. \qquad \square$$

REMARK 3.7. With access to x and to the parameters a and b, the previous proof explicitly constructs an isomorphism, using the flip-and-double trick, from $\operatorname{im}(x)$ to a summand in a complement of $\operatorname{im}(x)$.

We are now ready to limit the right repeaters needed to establish core strong separativity.

THEOREM 3.8. *Let R be a ring, and fix a distinguished inner inverse operation $'$ on the set of regular elements of R. Then R is core strongly separative if and only if the right repeaters from* Proposition 3.6 *are all unit-regular.*

PROOF. By Proposition 3.5, it suffices to show the "if" part. Assume every right repeater of that special form is unit-regular. Also, assume $R_R = A \oplus C \oplus C_1 = B \oplus C_2$ for some right ideals $A, B, C, C_1, C_2 \subseteq R_R$, with $C \cong C_1 \cong C_2$.

Let $a \in R$ be any endomorphism with kernel B that sends C_2 isomorphically to C. Notice that $(1 - aa')R$ is a direct sum complement of $\operatorname{im}(a)$, as is $A \oplus C_1$, and so $(1 - aa')R \cong A \oplus C_1$. Write $(1 - aa')R = A' \oplus C_1'$ with $A \cong A'$ and $C_1 \cong C_1'$.

Let b be any endomorphism with kernel $(1 - aa')R$, that sends C isomorphically to C_1'. Then it holds that $b = (1 - aa')baa'$. Also b is regular, having an image that is a direct summand.

Now, putting $x := (1 - aa')ba$, we see that this is an endomorphism with kernel B that sends C_2 isomorphically to C_1'. Since $A' \oplus C$ is a direct sum complement of C_1' in R_R, then

$$A \oplus C \cong A' \oplus C \cong \operatorname{coker}(x) \cong \ker(x) = B,$$

where the third isomorphism follows from the unit-regularity of x. $\square$

COROLLARY 3.9. *A regular ring* $(R, ')$ *is core strongly separative if and only if* $(1 - aa')ba$ *is unit-regular, for every* $a, b \in R$.

Another way to think about this corollary is that core strongly separative regular rings have special inner inverse operations, where every element of the form $(1 - aa')ba$ has a unit inner inverse.

Having dealt with the special case of *core* strong separativity, let us now handle the non-core case, which is the more natural class. We will find it convenient to restrict this classification to the class of exchange rings (which, importantly, contains the regular rings). A ring R is an *exchange ring* when the right R-module R_R satisfies the finite exchange property in direct sums of modules. Alternative characterizations for exchange rings appear throughout the literature, but see [**21**] for an introduction to this topic.

Lemma 3.10 describes a nice characterization of strong separativity (and separativity) in exchange rings, which in retrospect partly explains the "core" restriction.

LEMMA 3.10. *Let R be an exchange ring. The following hold:*

(1) *R is strongly separative if and only if*

$$A \oplus C \oplus C \cong B \oplus C \lesssim R_R \Rightarrow A \oplus C \cong B,$$

for all $A, B, C \in \mathrm{FP}(R)$.

(2) *R is separative if and only if*

$$A \oplus C \oplus C \cong B \oplus C \oplus C \lesssim R_R \Rightarrow A \oplus C \cong B \oplus C,$$

for all $A, B, C \in \mathrm{FP}(R)$.

PROOF. Part (2) is exactly Proposition 2.8 and Corollary 2.9 from [**2**]. (There is a minor typo in the proof, where the reference to part (i) of Lemma 2.7 should be to part (ii).) Part (1) follows from the same argument, *mutatis mutandis*. □

THEOREM 3.11. *An exchange ring R is strongly separative if and only if eRe is core strongly separative for each idempotent $e \in R$.*

PROOF. ($\Rightarrow$): As pointed out on page 125 of [**2**], strong separativity in exchange rings passes to corner rings. Thus, these corners are also core strongly separative.

($\Leftarrow$): Assume all corners are core strongly separative. Also assume that

$$A \oplus C \oplus C \cong B \oplus C \lesssim R_R.$$

After replacing these modules with isomorphic copies, we may assume there are internal direct sum decompositions

$$A \oplus C \oplus C_1 = B \oplus C_2 = X \subseteq^{\oplus} R_R \tag{3.12}$$

where $C \cong C_1 \cong C_2$. Write $X = eR$ for some idempotent $e \in R$.

We need a (very) little Morita theory; we include the details for completeness. If $fR \subseteq^{\oplus} eR$ for some idempotent $f \in R$, then we can replace f by efe, which is another idempotent with $efeR = fR$. Thus, given summands $f_1R, f_2R \subseteq^{\oplus} eR$, for some idempotents $f_1, f_2 \in R$, then

$$f_1R \cong f_2R \text{ if and only if } ef_1eR \cong ef_2eR.$$

The latter isomorphism is evidenced by left multiplications via two (reflexive inverse) elements of the corner ring eRe, by [**17**, Proposition 21.20(2)].

There is a natural correspondence between the right ideals of R contained in eR and the right ideals of eRe; see [**17**, Theorem 21.11(1)]. The argument from the previous paragraph shows that this correspondence respects isomorphisms on cyclic projective modules. Applying this correspondence to (3.12), we have

$$Ae \oplus Ce \oplus C_1e = Be \oplus C_2e = eRe,$$

with $Ce \cong C_1e \cong C_2e$. By the core strong separativity of eRe, we have $Ae \oplus Ce \cong Be$. Applying the correspondence again, $A \oplus C \cong B$. □

In order to translate the previous theorem into the language of right repeaters, we first need to understand how inner inverses behave in corner rings. Given any ring R, suppose that $a \in R$ is regular, with an inner inverse a'. Further, suppose that $e \in R$ is an idempotent, and that $a \in eRe$. In particular, $eae = a$ and hence $a' = (eae)'$. Now, a' might not belong to the corner ring eRe, but we do have

$$a(ea'e)a = aa'a = a.$$

Thus, a is not only regular in R, but also in the corner ring eRe, with an inner inverse $ea'e \in eRe$. Of course, conversely, if $a \in eRe$ is regular in eRe, then it is also regular in the overring R.

In other words, given any inner inverse operation $'$ on the regular elements of a ring R, there is a derived inner inverse operation on all of the regular elements of the corner ring eRe, defined by $a^\dagger := ea'e$. With this understanding in place, we can now classify strongly separative rings in terms of regularity and unit-regularity.

THEOREM 3.13. *Let R be a ring with a distinguished inner inverse operation $'$ on its regular elements. Then R is strongly separative if and only if for each idempotent $e \in R$, for each regular element $a \in eRe$, and for each $b \in eRe$ with $(1 - aa')baa'$ regular, then $(1 - aa')ba$ is unit-regular in eRe.*

PROOF. ($\Leftarrow$): Let $e \in R$ be idempotent, and let $a, b \in eRe$ with a regular and with $(1-aa')baa'$ regular. Since $(1-aa')baa'eaa' = (1-aa')baa'aa' = (1-aa')baa'$, we have

$$(1 - aa')ba(ea'e)R = (1 - aa')baa'R \subseteq^{\oplus} R_R$$

Moreover, using the notation $a^\dagger = ea'e$, then since $a, b \in eRe$ we have

$$(1 - aa')ba(ea'e) = (e - a(ea'e))ba(ea'e) = (e - aa^\dagger)baa^\dagger \in eRe,$$

and hence it is regular in that corner ring. Clearly, $(1 - aa')ba = (e - aa^\dagger)ba$.

Thus, as a and b range over the elements of eRe, subject to the fact that a is regular and $(1 - aa')baa'$ is regular, then the quantity $(1 - aa')ba$ ranges over the special right repeaters of Proposition 3.6, but in $(eRe, {}^\dagger)$. As all of these elements are unit-regular by assumption, then by Theorem 3.8 every corner ring of R is core strongly separative. Theorem 3.11 then implies that R is strongly separative.

($\Rightarrow$): Assume R is strongly separative. Then by Theorem 3.11, every corner ring is core strongly separative, and hence all right repeaters in those corner rings are unit-regular (in that corner). The computation from the first paragraph of this proof, together with Proposition 3.6, shows that the elements we are considering are indeed right repeaters in the respective corner rings. □

COROLLARY 3.14. *Let R be a regular ring. Then R is strongly separative if and only if for each idempotent $e \in R$, and for all $a, b \in eRe$, then $(1 - aa')ba$ is unit-regular in eRe.*

Left repeaters are defined in the obvious way, by using a left-right symmetric version of Lemma 3.4. The previous work in this section can now be repeated for left repeaters, with very little changes to the proofs. In regular rings, elements of the form $ac(1-a'a)$ are always left repeaters, and a regular ring is core strongly separative if and only if these special left repeaters are all unit-regular. Using the bullet points at the very end of Section 2, one can show that left repeaters and right repeaters come in reflexive inverse pairs.

In matrix rings over fields, the left repeaters and right repeaters are exactly the same elements; namely, those whose rank is no more than half the number of rows. In the ring of column-finite matrices over a field, these classes of elements are different, as

$$E_{2,1}+E_{4,2}+E_{6,3}+\cdots=\begin{bmatrix} 0 & & & & \\ 1 & 0 & & & \\ & 0 & 0 & & \\ & 1 & 0 & 0 & \\ & & & & \ddots \end{bmatrix}$$

is a right repeater but not a left repeater.

Having dealt with strong separativity, we now can tackle separativity.

DEFINITION 3.15. A ring R is *separative* when

$$A\oplus A\cong A\oplus B\cong B\oplus B \ \Rightarrow\ A\cong B,$$

for any $A,B\in \mathrm{FP}(R)$. As noted on page 115 of [**2**], this condition is equivalent for exchange rings to

$$A\oplus C\oplus C\cong B\oplus C\oplus C \ \Rightarrow\ A\oplus C\cong B\oplus C, \tag{3.16}$$

for any $A,B,C\in \mathrm{FP}(R)$. In case (3.16) is only assumed to hold when $R_R\cong A\oplus C\oplus C$, we shall say that R is *core separative.*

As we have no examples of exchange rings that are not separative, we do not know whether the two classes of rings in Definition 3.15 are distinct for exchange rings. However, we do have:

EXAMPLE 3.17. There is a core separative ring R, together with some modules $A,B,C\in \mathrm{FP}(R)$ satisfying $A\oplus C\oplus C\cong B\oplus C\oplus C$, and yet $A\oplus C\not\cong B\oplus C$.

PROOF. From page 114 of [**2**], there is a commutative (additive) monoid M that is conical, has the refinement property, but is not separative (where separativity is a direct translation of (3.16) to refinement monoids via [**2**, Lemma 2.1]). Let $M'=M\sqcup\{\infty\}$, which is still conical, has the refinement property, and fails separativity.

From the paragraph after Theorem 3.4 in [**7**], we know that $M'\cong V(R)$, for some ring R, where ∞ corresponds to $[R_R]$. To show that R is a core separative ring, it suffices to show that the corresponding condition in M' holds, with $[R_R]$ replaced by ∞. Thus, suppose that $\infty=a+c+c=b+c+c$ for some $a,b,c\in M'$. Then either $c=\infty$ or $a=b=\infty$. In any case, $a+c=b+c$. □

Just as right repeaters instantiate the premise of (the implication defining) core strong separativity, the following elements instantiate the premise of core separativity.

DEFINITION 3.18. Let R be a ring, and let $x \in R$. We say that x is a (two-sided) *repeater* if it is both a left and right repeater.

Describing a special collection of repeaters in general rings is slightly more complicated than it was for right repeaters. This is done as follows:

PROPOSITION 3.19. *Let R be a ring, and fix an inner inverse operation $'$ on its regular elements. Assume that $a \in R$ is regular, and that $c \in R$ is such that $a'ac(1-a'a)$ is regular. Setting $y := ac(1-a'a)$, further suppose that $b \in R$ is such that $(1-yy')byy'$ is regular. Then*

$$x := (1-yy')bac(1-a'a)$$

is a repeater.

PROOF. By the left-right symmetric version of Proposition 3.6, we know that y is a left repeater, and hence regular. Thus, by a second application of Proposition 3.6, we know that $x = (1-yy')by$ is a right repeater. It just remains to show that x is a left repeater.

Notice that $a'aR \subseteq \ker(x)$. By the "flip-and-double trick" (mentioned in the proof of Proposition 3.6, but this time with $c := x'(1-yy')ba$ and $d := a'ac(1-a'a)$, so that $cd = x'x$), we then know that

$$x'xR \cong a'ac(1-a'a)x'xx'(1-yy')baR \subseteq^{\oplus} \ker(x) = (1-x'x)R. \qquad \square$$

We can modify Proposition 3.5 and Theorem 3.8 to work for core separativity, by essentially the same proofs.

THEOREM 3.20. *Let R be a ring, and fix a distinguished inner inverse operation $'$ on the set of regular elements of R. Then R is core separative if and only if all repeaters in R are unit-regular, if and only if the repeaters from* Proposition 3.19 *are all unit-regular.*

REMARK 3.21. The referee has pointed out that the unit-regularity of repeaters in a separative regular ring follows from [**2**, Proposition 6.2]. In fact, one can show that in a general ring R, any right repeater a satisfies $\mathrm{ann}_\ell(a)R = R(1-a)R = R$, and hence all repeaters satisfy the condition $(*)$ of [**2**, Proposition 6.2].

Lemma 3.22 will help us give a simpler set of repeaters in regular rings. It is unclear whether a similar result holds for exchange rings.

LEMMA 3.22. *If $(R, ')$ is a regular ring, then any right multiple of a right repeater is a right repeater. Similarly, left multiples of left repeaters are left repeaters.*

Consequently, for any $a, b, c \in R$, then $(1-aa')bac(1-a'a)$ is a repeater.

PROOF. Assume that $x \in R$ is a right repeater, so xR is isomorphic to a summand in $(1-xx')R$. Let $y \in R$ be arbitrary. Now, since xy is regular, we know that $xyR \subseteq^{\oplus} R_R$. Thus, $xyR \subseteq^{\oplus} xR$. Hence, xyR is isomorphic to a summand in $(1-xx')R$. As $(1-xx')R$ is a direct summand in a complement of xyR, we have xyR is isomorphic to a summand in a complement of xyR.

The second sentence of the lemma follows by symmetry.

Since $(1-aa')bac(1-a'a)$ is a right multiple of the right repeater $(1-aa')ba$, it is also a right repeater. By symmetry, it is also a left repeater. $\square$

COROLLARY 3.23. *A regular ring R is core separative if and only if the element $(1-aa')bac(1-a'a)$ is unit-regular, for every $a, b, c \in R$.*

PROOF. The repeaters in the statement of Theorem 3.20 can be replaced with the simpler ones from Lemma 3.22, using essentially the same proof. □

Our work with corner rings also generalizes from the strongly separative case to the separative case. In particular, (again, by modifying proofs in trivial ways, as needed) we have:

THEOREM 3.24. *An exchange ring is separative if and only if all corner rings are core separative.*

We thus arrive at a new characterization of separative regular rings.

COROLLARY 3.25. *A regular ring $(R, ')$ is separative if and only if for every idempotent $e \in R$, and for every $a, b, c \in eRe$, the element*

$$(1 - aa')bac(1 - a'a)$$

is unit-regular in eRe.

By passing to a corner ring, if necessary, we thus see that the separativity problem is equivalent to deciding whether or not every element in a regular ring of the form

$$(1 - aa')bac(1 - a'a)$$

is unit-regular.

4. Viewing algebras as varieties

To begin this section we quickly review the notion of a *variety*, in the sense of universal algebra. A variety is a collection of algebras defined using two pieces of information. First, we need a set of operations, called the *signature*. For instance, the signature of rings is usually given by $\{+, \cdot, -, 0, 1\}$, where $+, \cdot$ are binary operations, $-$ is a unary operation, and $0, 1$ are nullary operations (i.e., constants). Second, we need a set of universal identities satisfied under these operations. The standard identities for rings, each universally quantified over the variables, are:

$$\begin{aligned}
x + y &= y + x && \text{(commutative law for addition)}\\
(x + y) + z &= x + (y + z) && \text{(associative law for addition)}\\
x + 0 &= x && \text{(0 is a right additive identity)}\\
x + (-x) &= 0 && \text{(negation yields a right additive inverse)}\\
(x \cdot y) \cdot z &= x \cdot (y \cdot z) && \text{(associative law for multiplication)}\\
x \cdot 1 &= x && \text{(1 is a right multiplicative identity)}\\
1 \cdot x &= x && \text{(1 is a left multiplicative identity)}\\
x \cdot (y + z) &= (x \cdot y) + (x \cdot z) && \text{(left distributivity of multiplication over addition)}\\
(y + z) \cdot x &= (y \cdot x) + (z \cdot x) && \text{(right distributivity of multiplication over addition)}
\end{aligned}$$

The list of identities will sometimes vary, since different combinations of identities will give the same consequences. For instance, we didn't include the fact 0 is a left additive identity because this follows from commutativity of addition in conjunction with 0 being a right additive identity. Also, the signature of rings will sometimes vary, as we can (for instance) avoid referring to 0 if we like. It is often more important to pick a convenient representation, rather than attempting to minimize the axioms and signature.

The variety of rings is the collection of 6-tuples $(R, +, \cdot, -, 0, 1)$, where R is a set, and the other symbols are operations on R satisfying the identities above. We will denote this variety as **Ring**.

A variety forms a category, where a morphism is a map that respects each of the operations. For instance, a morphism in the variety of rings $(R, +, \cdot, -, 0, 1) \to (S, +, \cdot, -, 0, 1)$ is a map $f\colon R \to S$ where

$$f(x+y)=f(x)+f(y),\ f(x\cdot y)=f(x)\cdot f(y),\ f(-x)=-f(x),\ f(0)=0, \text{ and } f(1)=1.$$

We refer to f as a **Ring**-homomorphism when emphasizing its connection to the variety.

Beginning students in algebra sometimes learn a definition of rings where, say, the axiom of "right multiplicative identity" is given as $\exists y\, \forall x\, (x \cdot y = x)$. However, when defining varieties we are not allowed to use existential quantification, nor implications, nor other Boolean connectives. Thus, varieties are sometimes referred to as *equational classes*. Fortunately, the axioms of rings are amenable to being rewritten as pure identities, after adjoining a few additional operations, such as the nullary operation 1.

The universal formula

$$x \cdot y = 0 \Rightarrow (x = 0 \vee y = 0),$$

which asserts that there are no nonzero zero-divisors, is not a pure identity. Moreover, unlike for rings, the class of domains cannot be made into a variety, due to:

THEOREM 4.1 (Birkhoff's theorem; see [**5**, Theorem 9.6.1]). *Given any signature, a collection of algebras under that signature forms a variety if and only if the collection is closed under homomorphic images, subalgebras, and direct products.*

Powerful results like Birkhoff's theorem make the study of varieties especially interesting. Another such result, which we will make use of in this paper, is the following:

THEOREM 4.2 ([**5**, Theorem 9.3.7]). *Given any variety and given any set X, there is a free algebra on the set X in the variety.*

With this short introduction out of the way, we now connect these ideas to the work in Section 3. We can view the class of regular rings with a distinguished inner inverse operation as a variety, denoted **Reg**, by using the same signature as for rings, except that we adjoin an extra unary operation $'$. The identities are the same as for rings, except that we adjoin one extra identity

$$aa'a = a,$$

which forces a' to be an inner inverse for a. Unless context suggests otherwise, other varieties in this section are understood to include the signatures and identities of **Reg**. Whenever possible, we will use exactly the same signature, and there is a special name for this situation.

DEFINITION 4.3. Given a variety $\mathbf{V}$, a *subvariety* is another variety with the same signature as $\mathbf{V}$ that is subject to the identities of $\mathbf{V}$, but there may also be additional identities.

For example, by adjoining the one new identity

$$a'aa' = a'$$

we have a subvariety (of **Reg**) where the distinguished inner inverse operation always yields a reflexive inverse. Similarly, unit regular rings (or, more properly, regular rings with a distinguished unit inner inverse) form a subvariety by adjoining the one new identity

$$a'a''a''a' = 1.$$

Thinking of the symbols a and b as parameters, we can put $x := (1 - aa')ba$. Then the core strongly separative regular rings form a subvariety by adjoining the one new identity

$$x'x''x''x' = 1,$$

which forces the special right repeaters of Corollary 3.9 to be unit-regular. Similarly, putting

$$x := (1 - aa')bac(1 - a'a),$$

and asserting $x'x''x''x' = 1$, we have defined the core separative regular rings as a subvariety, according to Corollary 3.23.

Our next goal is to explain how to similarly varietize the strongly separative, as well as separative, regular rings. We need a uniform method for referring to idempotents. This is done, as follows:

LEMMA 4.4. *Given a regular ring $(R, ')$, the set of idempotents of R equals the set of elements of the form $e := aa' + a(1 - aa')$ for some $a \in R$.*

PROOF. First, since $aa'a = a$, we have $(1 - aa')a = 0$. Thus

$$e^2 = aa'aa' + aa'a(1-aa') + a(1-aa')aa' + a(1-aa')a(1-aa') = aa' + a(1-aa') = e.$$

Conversely, let $f \in R$ be an arbitrary idempotent. Taking $a = f$, then

$$e = ff' + f(1 - ff') = ff' + f - ff' = f.$$

So the expression represents all idempotents. □

We also must describe unit-regularity in corner rings. The following result of Lam and Murray from [**20**] is exactly what is needed to transfer that information back to the original ring.

LEMMA 4.5. *Let R be a ring, let $e \in R$ be an idempotent, and let $x \in eRe$. Then x is unit-regular in eRe if and only if $x + (1 - e)$ is unit-regular in R.*

With this lemma in hand, we are now ready to prove:

THEOREM 4.6. *The class of strongly separative regular rings forms a subvariety of* **Reg**, *seen by adjoining the identity*

$$x_e'x_e''x_e''x_e' = 1,$$

where

$$x_e := (1 - eae(eae)')ebeae + (1 - e)$$

is an expression in the three parameters a, b, e, where e is an idempotent parameter. If instead we use the four parameter expression

$$x_e := (1 - eae(eae)')ebeaece(1 - (eae)'eae) + (1 - e),$$

this describes the class of separative regular rings as a subvariety of **Reg**.

PROOF. A ring is strongly separative (or separative) exactly when all x_e are unit-regular in R, by using Lemma 4.5 together with Corollary 3.14 (respectively, together with Corollary 3.25). The identity is exactly what is needed to assert unit-regularity of x_e in a regular ring. □

To end this section, we mention three useful applications of these varieties. First, they point us towards a concrete (but somewhat complicated) regular ring that could help to answer the separativity problem.

THEOREM 4.7. *If there is a nonseparative regular ring, then the free* **Reg**-*algebra on three (or more) generators is such a ring.*

PROOF. Assume there is a nonseparative regular ring $(R, ')$. After passing to a corner if necessary, then by Corollary 3.25 there are three elements $a, b, c \in R$ such that the repeater $x := (1 - aa')bac(1 - a'a)$ is not unit-regular in R.

Let S be the free **Reg**-algebra on three generators p, q, r. Then there is a unique **Reg**-homomorphism $\varphi\colon S \to R$ determined by sending p, q, r respectively to a, b, c. Fixing $y := (1-pp')qpr(1-p'p) \in S$, then $\varphi(y) = x$. Now y is not unit-regular in S, since x is not unit-regular in R. But y is a repeater in S, so S is not separative. □

Second, they bring uniformity and explicitness to constructions. For example, it was known by combining [**1**, Theorem 3.4] and [**3**, Theorem 2.10] that:

PROPOSITION 4.8. *An exchange ring R is separative if and only if for each idempotent $e \in R$, then each regular matrix in $\mathbb{M}_2(eRe)$ can be reduced to a diagonal matrix by a finite number of elementary row and column operations over the ring eRe.*

We show, in Section 7, that for a (core) separative regular ring R, then each matrix $A \in \mathbb{M}_2(R)$ needs only three row operations and three column operations to diagonally reduce. Moreover, these operations are explicitly definable via regular ring expressions, treating the entries of A as parameters, as long as the inner inverse operation satisfies the new identity used to varietize the core separative regular rings. More precisely, we only need the unit-regularity of exactly one repeater, easily defined in terms of the entries of A. Similarly, the 2×1 matrices over a (core) strongly separative ring can be explicitly diagonally reduced using the unit-regularity of only one right repeater.

Third, these varieties can be used to strengthen results in the literature, while giving added evidence that there are nonseparative regular rings. Here is one example. Fix a field F, and let $n \in \mathbb{Z}_{\geq 1}$ be a parameter that we allow to vary. Recall that in Section 2 we mentioned the fact, appearing in [**23**, Proposition 2.15], that there is no uniformly defined derived operation that is a unit inner inverse operation for all of the rings $\mathbb{M}_n(F)$, in terms of a fixed (but arbitrary) inner inverse operation $'$ on $\mathbb{M}_n(F)$. The proof shows more; there is no way to diagonally reduce all 2×1 matrices uniformly over these rings. But, as we mentioned at the end of the previous paragraph, there is a uniform way to diagonally reduce 2×1 matrices in the variety of core strongly separative regular rings. Thus, there is no uniform way to view $(\mathbb{M}_n(F), ')$ as an element of that variety under some derived operation.

In other words, given access to a distinguished (but arbitrary) inner inverse operation $'$ on $\mathbb{M}_n(F)$, and given a matrix $x \in \mathbb{M}_n(F)$ that we know satisfies $\operatorname{rank}(x) \leq n/2$ because we also have access to an explicit isomorphism from xR to a summand of $(1 - xx')R$, then we cannot uniformly define an isomorphism

$(1-x'x)R \to (1-xx')R$ from this information. The separativity problem reduces to the question of whether we could define such an isomorphism uniformly when given just one additional small piece of information; namely, that $(1-x'x)R$ contains an isomorphic copy of xR. Frankly, this seems unlikely.

There are other properties of separative regular rings, besides the cancellation condition

$$A \oplus 2C \cong B \oplus 2C \;\Rightarrow\; A \oplus C \cong B \oplus C,$$

that allow for varietization. Indeed, as noted on pages 398–399 of [**23**], one can varietize the class of rings where 2×2 matrices are reducible to diagonal matrices, by introducing sixteen new diagonal reduction operations. Then, after adjoining an inner inverse operation and after forcing the diagonal reduction identities to hold in corner rings, one has an alternative varietal definition of separative regular rings. Our work in Section 7 describes those sixteen operations using derived expressions.

5. Varieties of exchange rings

Much of the work from the previous section likewise allows us to define a variety of separative exchange rings. We first make exchange rings a variety, **Exch**. One way this can be done is by using the description of exchange rings found in [**16**], as follows. Adjoin to the signature of rings a unary operation ^{s}, and adjoin to the identities of rings the equality

$$(1 - a^s a)(1 + (1-a)a^s) = 0. \tag{5.1}$$

The element a^s is called a *suitabilizer* for a, and the theory of such elements was first studied in [**16**], although suitabilizers had appeared implicitly in previous papers. Think of the suitabilizer map as a replacement for the inner inverse operation in regular rings.

Next, we need to show that idempotents and regular elements can be uniformly described in exchange rings.

LEMMA 5.2. *Let R be an exchange ring with suitabilizer map ^{s}.*

(1) *The set of idempotents of R equals the set of elements of the form*

$$e := (1 + a^s(1-a))a$$

for some $a \in R$.

(2) *The set of regular elements of R equals the set of elements of the form*

$$p = p(a,b) := a(1 + (ba)^s(1-ba))ba$$

for some $a, b \in R$.

PROOF. (1) The complement of the idempotent from [**16**, Proposition 3.3(C)] is exactly e. Moreover, given an idempotent $f \in R$, then taking $a = f$ we immediately get $e = f$.

(2) First, if we set $g := (1 + (ba)^s(1-ba))ba$, then this is an idempotent by part (1). Now, taking

$$q = q(a,b) := (1 + (ba)^s(1-ba))b$$

we find

$$pqp = ag^3 = ag = p.$$

Thus p is a regular element with inner inverse q. Conversely, suppose c is a regular element of R with an inner inverse d. Taking $a = c$ and $b = d$, then $p = c$, so it represents all regular elements. □

We are now ready to generalize Theorem 4.6 from regular rings to exchange rings. As the resulting varietal descriptions are quite complicated, and not subvarieties of **Exch**, we will only give the full description in the proof.

THEOREM 5.3. *Strongly separative exchange rings form a variety extending* **Exch**, *via two new unary operations with two new identities. The same is true for separative exchange rings.*

PROOF. We will only handle the case of strongly separative exchange rings, as similar arguments work for separative exchange rings.

First, let $(R, {}^s)$ be any exchange ring. For any two elements $a, b \in R$ define $p(a,b)$ and $q(a,b)$ as in the Lemma 5.2. Given any regular element $x \in R$, then we can fix (once and for all) elements $a_x, b_x \in R$ such that $p(a_x, b_x) = x$. We may then define an inner inverse operation $'$ on the regular elements of R by the rule $x' = q(a_x, b_x)$. Now, as we run over all pairs of elements $a, b \in R$, not only does $p(a,b)$ run over all regular elements, but $q(a,b)$ runs over enough inner inverses to cover all of the inner inverses under the operation $'$.

Next, suppose that $c \in (1 - xx')Rxx'$ is regular. Then

$$p((1 - xx')cxx', c') = p(c, c') = c.$$

On the other hand, for any $a, b \in R$, we have $p((1 - xx')axx', b) \in (1 - xx')Rxx'$. So, the elements of the form $p((1 - xx')axx', b)$ represent all regular elements of $(1 - xx')Rxx'$.

Putting this all together, given parameters a, b, c, d, then

$$z := p\Big((1 - p(a,b)q(a,b))cp(a,b)q(a,b), d\Big)p(a,b)$$

is a right repeater, by Proposition 3.6. Moreover, as these four parameters range across the elements of a ring, the expression ranges across all the special (and possibly more) right repeaters needed in Theorem 3.8 under some fixed inner inverse operation on that ring. We can assert that z is unit-regular by adjoining two unary operations u, v, and by adding the two identities $zu(z)z = z$ and $u(a)v(a)u(a) = 1$.

We have thus described how the core strongly separative exchange rings form a varietal extension of **Exch**. To remove the "core" restriction, first introduce an idempotent parameter e. Then, replace each of the other four parameters a, b, c, d in z by the corresponding corner elements eae, ebe, ece, ede; call the resulting expression z_e. Finally, by Lemma 4.5 together with Theorem 3.13 we just need to assert that $z_e + (1 - e)$ is unit-regular. This is done as in the previous paragraph, by introducing two new unary operations and two new identities, but replacing z by $z_e + (1 - e)$. □

We end this section with a couple of questions regarding whether or not these varietal descriptions could be simplified further.

QUESTION 5.4. Can the regular elements of an algebra in **Exch** be represented by an expression using only one parameter?

QUESTION 5.5. Can the varieties in this section be defined without new operations?

6. Minimizing the number of elementary operations

The observation that separative exchange rings are GE-rings, by [**3**, Theorem 2.8], lets us diagonally reduce using elementary matrices, rather than using arbitrary units (which may be quite complicated). The proof that separative exchange rings are GE-rings requires a large number of elementary transformations. We will focus on the case of 2×2 matrices, where this number can be significantly reduced, and in fact minimized. An important fact that often lets us reduce this number, and which will be used repeatedly, is the normalizing equality

$$\begin{bmatrix} d_1 & 0 \\ 0 & d_2 \end{bmatrix} \begin{bmatrix} 1 & x \\ 0 & 1 \end{bmatrix} = \begin{bmatrix} 1 & d_1 x d_2^{-1} \\ 0 & 1 \end{bmatrix} \begin{bmatrix} d_1 & 0 \\ 0 & d_2 \end{bmatrix} \tag{6.1}$$

for invertible diagonal matrices. Of course, there is a similar equality involving lower-triangular 2×2 elementary matrices.

The rings where only a small number of elementary operations are needed to diagonally reduce 2×2 matrices can be completely characterized. Recall that a ring R has *stable rank* 1 when for any $a, b \in R$, if $aR + bR = R$, then there is some $x \in R$ with $a + bx \in \mathrm{U}(R)$. The following theorem, tying stable rank 1 to diagonal reduction, appears to be new.

THEOREM 6.2. *A ring R has stable rank 1 if and only if every matrix in $\mathrm{GL}_2(R)$ can be reduced to a diagonal matrix by at most three elementary operations.*

PROOF. Throughout this proof, let $X \in \mathrm{GL}_2(R)$ be an arbitrary invertible matrix.

($\Rightarrow$): Assume R has stable rank 1. Write the first row of X as $[a, b]$ for some $a, b \in R$. As XX^{-1} is the identity matrix, we have $aR + bR = R$. Thus, from the definition of stable rank 1, we may fix some $x \in R$ with $a + bx \in \mathrm{U}(R)$. This means that after one elementary column operation, we can force the northwest corner of X to be a unit. Using that unit, we can finish the diagonal reduction after two more elementary operations.

($\Leftarrow$): When X can be diagonally reduced by elementary operations, we claim that the reduction can be accomplished by using only column operations, but the same number. To see why, suppose that M_1 and M_2 are finite products of elementary 2×2 matrices, and that $D := M_1 X M_2$ is diagonal. Then, $X = M_1^{-1} D M_2^{-1}$. By using the normalizing identity (6.1) finitely many times, we have $M_1 D = D M_3$ for some matrix M_3 that is also a product of elementary matrices, with the same number of elementary matrices as in M_1. Rearranging, we obtain $M_1^{-1} D = D M_3^{-1}$. Thus, $X = D M_3^{-1} M_2^{-1}$, and so $D = X M_2 M_3$.

Now, assume every matrix in $\mathrm{GL}_2(R)$ can be diagonally reduced by at most three elementary operations. The identity operation is an elementary operation, so we can use exactly three operations. Further, note that a product of two upper-triangular elementary matrices is again elementary, since

$$\begin{bmatrix} 1 & p \\ 0 & 1 \end{bmatrix} \begin{bmatrix} 1 & q \\ 0 & 1 \end{bmatrix} = \begin{bmatrix} 1 & p+q \\ 0 & 1 \end{bmatrix}.$$

The same is true for lower-triangular elementary matrices. So, without loss of generality, we may assume that the diagonal reduction operations are represented by alternatingly upper-triangular and lower-triangular matrices (by combining like types as necessary).

We first show that if $[a,b]$ is the top row of an invertible matrix X, and if $ay+bz=1$ for some $y,z\in R$, then $[a,b]$ is the top row of a new invertible matrix $Y\in \mathrm{GL}_2(R)$, where the left column of Y^{-1} is $\left[\begin{smallmatrix} y\\ z\end{smallmatrix}\right]$. Let M_1 be a product of three elementary matrices such that $XM_1=\left[\begin{smallmatrix} v&0\\ 0&w\end{smallmatrix}\right]$ for some $v,w\in R$. Any diagonal matrix in $\mathrm{GL}_2(R)$ must have unit entries along the diagonal, and so $v,w\in \mathrm{U}(R)$. Now,

$$1=\begin{bmatrix} a & b\end{bmatrix}\begin{bmatrix} y\\ z\end{bmatrix}=\begin{bmatrix} a & b\end{bmatrix}M_1M_1^{-1}\begin{bmatrix} y\\ z\end{bmatrix}=\begin{bmatrix} v & 0\end{bmatrix}M_1^{-1}\begin{bmatrix} y\\ z\end{bmatrix}.$$

Thus,

$$M_1^{-1}\begin{bmatrix} y\\ z\end{bmatrix}=\begin{bmatrix} v^{-1}\\ t\end{bmatrix}$$

for some $t\in R$. Fix $M_2=\left[\begin{smallmatrix} 1&0\\ -tv&1\end{smallmatrix}\right]$, so $M_2M_1^{-1}\left[\begin{smallmatrix} y\\ z\end{smallmatrix}\right]=\left[\begin{smallmatrix} v^{-1}\\ 0\end{smallmatrix}\right]$. Note that $[v,0]M_2=[v,0]$. Fixing Y to be $\left[\begin{smallmatrix} v&0\\ 0&1\end{smallmatrix}\right]M_2M_1^{-1}$, then the top row of Y and the left column of Y^{-1} are as desired.

Finally, we show that given $a,b\in R$ with $aR+bR=R$, then there is some element $x\in R$ with $a+bx\in \mathrm{U}(R)$. Fix any two elements $r,s\in R$ such that $ar+bs=1$. Since $ar+(bs)1=1$, the previous paragraph implies that there is an invertible matrix whose top row is $[a,\,bs]$.

Since $bs=(1-ar)$, we have $(bs)^2=1-2ar+arar$. Thus,

$$a(2r-rar)+(bs)(bs)=2ar-arar+(1-2ar+arar)=1.$$

By another application of the work above, there is some matrix $Y\in \mathrm{GL}_2(R)$ whose first row is $[a,bs]$, and where the left column of Y^{-1} is $\left[\begin{smallmatrix} 2r-rar\\ bs\end{smallmatrix}\right]$. There are now two cases to consider.

First, suppose that

$$Y\begin{bmatrix} 1 & 0\\ p_1 & 1\end{bmatrix}\begin{bmatrix} 1 & p_2\\ 0 & 1\end{bmatrix}\begin{bmatrix} 1 & 0\\ p_3 & 1\end{bmatrix}=\begin{bmatrix} v & 0\\ 0 & w\end{bmatrix}$$

for some $p_1,p_2,p_3,v,w\in R$. Comparing the upper right corners of both sides, we have $bs+(a+bsp_1)p_2=0$. With this equality in hand, then comparing the upper left corners we have $a+bsp_1=v\in \mathrm{U}(R)$. Hence, we may take $x=sp_1$.

Second, suppose that

$$Y\begin{bmatrix} 1 & p_1\\ 0 & 1\end{bmatrix}\begin{bmatrix} 1 & 0\\ p_2 & 1\end{bmatrix}\begin{bmatrix} 1 & p_3\\ 0 & 1\end{bmatrix}=\begin{bmatrix} v & 0\\ 0 & w\end{bmatrix}$$

for some $p_1,p_2,p_3,v,w\in R$. Taking inverses of both sides, then the left-right dual of the computation from the previous paragraph shows that $(2r-rar)-p_1bs=v^{-1}\in \mathrm{U}(R)$. So,

$$v^{-1}=r+r(1-ar)-p_1bs=r+(r-p_1)bs.$$

Applying the left-right symmetric version of [**18**, (1.25)] to the equality $ar+(bs)=1$, then there is some $z\in R$ with $a+bsz\in \mathrm{U}(R)$. Hence, we may take $x=sz$. □

Over a nonzero ring, the matrix $\left[\begin{smallmatrix} 0&1\\ 1&0\end{smallmatrix}\right]$ cannot be diagonally reduced using fewer than three elementary operations. So, this provides a natural lower bound on how many elementary operations are generally needed. Of course, individual matrices may be diagonally reduced by fewer than three elementary operations.

There are separative regular rings that do not have stable rank 1, such as the ring of column-finite matrices over a field. So, we have shown that for separative exchange rings, at least four elementary operations are necessary, in general, to diagonally reduce 2×2 invertible matrices. Our next goal is to show that this number of operations is optimal. We begin with the following powerhouse lemma about exchange rings.

LEMMA 6.3. *Let R be an exchange ring, and let $a, b \in R$. If $aR + bR = R$, then there exists an idempotent $e \in aR$ with $1 - e \in bR$. Moreover, given any idempotent $f \in aR$, we can choose e (depending on f) so that $f \in eR$. In particular, if $aR \subseteq^{\oplus} R_R$, then we can choose e so that $eR = aR$.*

PROOF. Assume $f \in aR$ is an idempotent. Slightly modifying the terminology of [**26**, p. 440], let us say that a module P has property $(N)_{\mathrm{finite}}$ when given any module decomposition $P = \sum_{i \in I} N_i$ with $|I| < \infty$, then there exist submodules $N_i' \subseteq N_i$ with $P = \bigoplus_{i \in I} N_i'$. From [**21**, Proposition 2.9], we know that a projective module has the finite exchange property if and only if it has property $(N)_{\mathrm{finite}}$. In particular, R_R has this property.

Note that [**26**, Lemma 2.5] is still true with $(N)_{\mathrm{finite}}$ in place of (N) after limiting the index set I to be finite, using the same proof. Consequently, since $R = fR \oplus (1 - f)R = aR + bR$, then there exist right ideals $A' \subseteq aR$ and $B' \subseteq bR$ such that $R = fR \oplus A' \oplus B'$.

Letting $e \in R$ be the idempotent with $eR = fR \oplus A'$ and $(1 - e)R = B'$, then $e \in aR$ and $1 - e \in bR$. Clearly, $f \in eR$.

To prove the last sentence, suppose that $aR \subseteq^{\oplus} R_R$. Then there is some idempotent $f \in aR$ such that $aR = fR$. In this case $A' = 0$, and so $eR = fR \oplus A' = fR = aR$. $\square$

Given any matrix $A := \left[\begin{smallmatrix} a & b \\ c & d \end{smallmatrix}\right] \in \mathrm{GL}_2(R)$, the top row $[a, b]$ is *right unimodular*, meaning

$$aR + bR = R.$$

In the opposite direction, if a right unimodular row is the top row of an invertible matrix, we say that it is *completable.* It is easy to find examples of right unimodular rows that are not completable. For instance, in the ring $\mathrm{CFM}_\omega(F)$ of $\omega \times \omega$ column-finite matrices over a field F, the right unimodular row generated by

(6.4)

$$\begin{bmatrix} 1 & & & & & & \\ 0 & 0 & & & & & \\ & 1 & 0 & & & & \\ & 0 & 0 & 0 & & & \\ & & 1 & 0 & 0 & & \\ & & 0 & 0 & 0 & 0 & \\ & & & & & & \ddots \end{bmatrix} \quad \text{and} \quad \begin{bmatrix} 0 & & & & & & \\ 1 & 0 & & & & & \\ & 0 & 0 & & & & \\ & 1 & 0 & 0 & & & \\ & & 0 & 0 & 0 & & \\ & & 1 & 0 & 0 & 0 & \\ & & & & & & \ddots \end{bmatrix}$$

is not completable. Before we give the proof, let us recall some facts about completability.

Given $m, n \in \mathbb{Z}_{>0}$, identify R^m with the set of column matrices of size $m \times 1$, and do the same for R^n. Then each right R-module homomorphism $R^n \to R^m$ corresponds to left multiplication by an $m \times n$ matrix. In particular, by taking $m = 1$, we have a correspondence between rows of length n and their left multiplication action $R^n \to R$. Clearly, any such row is right unimodular if and only if the corresponding left multiplication map is surjective.

In that case, from projectivity write $R^n \cong P \oplus R$, where P is the kernel of the left multiplication action. The given right unimodular row is completable if and only if $P \cong R^{n-1}$; see [**19**, Corollary 4.9] for the proof. Specializing to the case when $n = 2$, we have:

COROLLARY 6.5. *Let R be a ring. Every 1×2 right unimodular row is completable if and only if R is self-cancellative (meaning that for any $P \in \mathrm{FP}(R)$, then $R^2 \cong P \oplus R$ implies $P \cong R$).*

Taking $R = \mathrm{CFM}_\omega(F)$, then there is an obvious isomorphism $R^2 \to R$ given by sending the rows of the first coordinate of R^2 to the odd indexed rows in the output and similarly sending the rows of the second coordinate of R^2 to the even indexed rows in the output. This isomorphism is exactly left multiplication by the row generated by (6.4), which explains how we found that row in the first place. Lack of completability boils down to the fact that the kernel of the homomorphism is zero, and hence it is not isomorphic to R.

Proposition 6.6 gives us a strong way to complete matrices over exchange rings.

PROPOSITION 6.6. *Let R be an exchange ring, and let $[a, b] \in R^2$ be a completable row. Then there exist $c, d \in R$ such that $A := \left[\begin{smallmatrix} a & b \\ c & d \end{smallmatrix}\right] \in \mathrm{GL}_2(R)$, with $RcR = R$ and c regular.*

PROOF. As $[a, b]$ is completable, fix some matrix $A := \left[\begin{smallmatrix} a & b \\ * & * \end{smallmatrix}\right] \in \mathrm{GL}_2(R)$. Moreover, as $[a, b]$ is right unimodular, then by Lemma 6.3 we can find an idempotent $e \in aR$ with $1 - e \in bR$. Write $e = ar$ and $1 - e = bs$ for some $r, s \in R$.

Let $E_1 := \left[\begin{smallmatrix} 1 & -rb \\ 0 & 1 \end{smallmatrix}\right]$. We find

$$AE_1 = \begin{bmatrix} a & (1-e)b \\ * & z \end{bmatrix},$$

for some $z \in R$. The upper-right corner of AE_1 is $(1 - e)b = bsb$, which is regular with inner inverse s. In particular $R(1 - e)b \subseteq^{\oplus} {}_RR$.

Notice that $R(1 - e)b + Rz = R$, since AE_1 is (left-)invertible. Using Lemma 6.3 a second time, fix an idempotent $f \in R(1 - e)b$ with $1 - f \in Rz$ and with $R(1 - e)b = Rf$. In particular,

$$(1-e)b(1-f) = 0. \tag{6.7}$$

Write $f = t(1 - e)b$ and $1 - f = uz$, for some $t, u \in R$.

Let $E_2 := \left[\begin{smallmatrix} 1 & 0 \\ -zt & 1 \end{smallmatrix}\right]$. We find

$$E_2AE_1 = \begin{bmatrix} a & (1-e)b \\ c_1 & z(1-f) \end{bmatrix},$$

for some $c_1 \in R$. The lower-right corner of E_2AE_1 is $z(1-f) = zuz$, which is regular with inner inverse u. By using Lemma 6.3 a third time, we can fix an idempotent $g \in z(1 - f)R$ so that $1 - g \in c_1R$ and $gR = z(1 - f)R$. Write $g = z(1 - f)v$ and $1 - g = c_1w$, for some $v, w \in R$.

Let $E_3 := \left[\begin{smallmatrix} 1 & 0 \\ -(1-f)vc_1+(1-f)v & 1 \end{smallmatrix}\right]$. From (6.7), the top rows of E_2AE_1 and $E_2AE_1E_3$ agree, and we find

$$E_2AE_1E_3 = \begin{bmatrix} a & (1-e)b \\ (1-g)c_1 + g & z(1-f) \end{bmatrix}.$$

Let $c_2 := (1-g)c_1+g$ be the lower left corner entry of $E_2AE_1E_3$. Since $gc_2 = g$, and $(1-g)c_2w = (1-g)$, we see that $Rc_2R = R$. However, c_2 may not yet be regular. From [**3**, Corollary 2.2], we may fix an idempotent $p \in c_2R$ with $RpR = R$. Now, by a fourth use of Lemma 6.3, fix an idempotent $h \in c_2R$ with $1-h \in z(1-f)R$ and with $p \in hR$. Write $h = c_2x$ and $1-h = z(1-f)y$, for some $x, y \in R$.

Let $E_4 := \left[\begin{smallmatrix} 1 & 0 \\ -(1-f)yc_2 & 1 \end{smallmatrix}\right]$. We find

$$E_2AE_1E_3E_4 = \begin{bmatrix} a & (1-e)b \\ hc_2 & z(1-f) \end{bmatrix}.$$

Letting $c := hc_2 = c_2xc_2$, we see that c is regular, with inner inverse x. Moreover,

$$RcR \supseteq RcxR = RhR \supseteq RpR = R.$$

Finally, multiply on the right by E_1^{-1} to replace $(1-e)b$ with b but leave a and c unchanged. We see that

$$E_2AE_1E_3E_4E_1^{-1} = \begin{bmatrix} a & b \\ c & * \end{bmatrix} \in \mathrm{GL}_2(R),$$

which is a matrix of the form we want. □

Next, we will show that over separative exchange rings, matrices of the type given in Proposition 6.6 can be diagonally reduced using four elementary operations.

PROPOSITION 6.8. *Let R be a separative exchange ring. Assume that $a, b, c, d \in R$ are chosen such that $A := \left[\begin{smallmatrix} a & b \\ c & d \end{smallmatrix}\right] \in \mathrm{GL}_2(R)$, with $RcR = R$ and c regular. Then there exist upper-triangular elementary matrices U_1, U_2 and lower-triangular elementary matrices L_1, L_2 such that $U_2L_1AU_1L_2$ is diagonal.*

PROOF. We will break the proof into four successive steps. In the first step we will construct an upper-triangular matrix U_1 so that $AU_1 = \left[\begin{smallmatrix} a' & b' \\ c' & d' \end{smallmatrix}\right]$ has new special properties. Throughout this proof primes are never inner inverse operations.

STEP 1. We force $d' \in Rd$, with $d'R = pR$ for some idempotent $p \in R$ that additionally satisfies the equality $R(1-p)R = R$; moreover, we force $Rb'R = R$ with b' regular.

We break this step into three stages. First, since $cR \subseteq^{\oplus} R_R$, then by Lemma 6.3 there exists an idempotent $e \in cR$ with $1-e \in dR$ and with $cR = eR$. Write $e = cr$ and $1-e = ds$, for some $r, s \in R$. Note that r is an inner inverse for c.

Setting $U_{1,1} := \left[\begin{smallmatrix} 1 & -rd \\ 0 & 1 \end{smallmatrix}\right]$, then we have

$$AU_{1,1} = \begin{bmatrix} a & b - ard \\ c & (1-e)d \end{bmatrix}.$$

Notice that $d' := (1-e)d = dsd \in Rd$ is regular, with inner inverse s. Taking $p := ds = 1-e$, we have $d'R = pR$. Moreover,

$$R(1-p)R = ReR = RcR = R.$$

For the second and third stages we will not change the bottom row of $AU_{1,1}$, and thus not undo any of these facts about d'. Before starting the second stage, we will perform an auxiliary calculation. (Many of the ideas we will use already appear in the proof of Proposition 6.6.) Notice that $Ra+Rc=R$, and so by Lemma 6.3 we may fix an idempotent $f\in Rc$ with $1-f\in Ra$ and with $Rc=Rf$. In particular,

$$c(1-f)=0. \tag{6.9}$$

Write $f=tc$, for some $t\in R$. Letting $V:=\left[\begin{smallmatrix}1 & -ate\\ 0 & 1\end{smallmatrix}\right]$, then the top row of $VAU_{1,1}$ is

$$[a(1-f), b-ard].$$

In particular, this row is right unimodular.

We are now ready for the second stage. Using the exchange property, fix an idempotent $g\in a(1-f)R$ with $1-g\in(b-ard)R$. Write $g=a(1-f)v$, for some $v\in R$.

Let $U_{1,2}:=E_{1,1}+E_{2,2}+[-(1-f)v(b-ard)+(1-f)v]E_{1,2}$. By (6.9) we see that the bottom rows of $AU_{1,1}$ and $AU_{1,1}U_{1,2}$ are the same, and we find

$$AU_{1,1}U_{1,2}=\begin{bmatrix} a & g+(1-g)(b-ard)\\ c & (1-e)d\end{bmatrix}.$$

Let $z\in R$ denote the upper-right entry of $AU_{1,1}U_{1,2}$. Then z is full (i.e., the two-sided ideal it generates is the entire ring) because $g=gz\in RzR$ and also $1-g\in(1-g)(b-ard)R\subseteq RzR$. The matrix $U_{1,2}$ plays exactly the same role that E_3 did in the proof of Proposition 6.6. Analogously, we can construct another upper-triangular elementary matrix $U_{1,3}$, playing a role similar to that of E_4 from that other proof, such that right multiplication by $U_{1,3}$ again leaves the bottom row of $AU_{1,1}U_{1,2}$ alone, but replaces the upper-right entry with a full element that is also regular. The reader is invited to work out the details.

Letting $U_1:=U_{1,1}U_{1,2}U_{1,3}$, we see that U_1 is an upper-triangular elementary matrix. The matrix $AU_1=\left[\begin{smallmatrix}a' & b'\\ c' & d'\end{smallmatrix}\right]$ now has the claimed properties for Step 1. In Step 2 we will construct a lower-triangular elementary matrix L_1 such that $L_1AU_1=\left[\begin{smallmatrix}a'' & b''\\ c'' & d''\end{smallmatrix}\right]$ satisfies a new special property.

STEP 2. We force d'' to be a unit.

We break this step into two stages. First, by exactly the same idea as in the first stage of the first step, we can find a lower-triangular matrix $L_{1,1}$ so that the lower-right corner of $L_{1,1}AU_1$, which we can call d_1, satisfies $d_1\in d'R$, and $Rd_1=Rq$ for some idempotent $q\in R$ with $R(1-q)R=R$. Since d_1 is regular, we have $d_1R=p'R$ for some idempotent $p'\in R$. But since

$$p'R=d_1R\subseteq d'R=pR$$

we have $R(1-p')\supseteq R(1-p)$, and hence $R(1-p')R=R$. (In other words, the new element d_1 continues to have some of those special properties we forced on d' in Step 1.)

By the argument in the second and third paragraphs of the proof of [**3**, Theorem 2.8], which is the only place we use (a special case of) separativity, we see that d_1 is unit-regular. (Another version of this same argument appears as [**2**, Theorem 6.2].) Write $d_1=uh$ for some unit $u\in\mathrm{U}(R)$ and some idempotent $h\in R$.

For the second stage, using the exchange property fix an idempotent $e'\in Rd_1=Rh$ such that $1-e'\in Rb'$. Applying another lower-triangular elementary matrix $L_{1,2}$, we can force the lower-right corner to become $d'':=uh+u(1-h)(1-e')$. Since d'' is a unit if and only if $u^{-1}d''$ is a unit, it suffices to see that

$$h+(1-h)(1-e')=1-(1-h)e'$$

is a unit. This is clear because $(1-h)e'$ squares to zero, since $e' \in Rh$. Taking $L_1 := L_{1,2}L_{1,1}$, then Step 2 is complete.

STEPS 3 AND 4. Force the nondiagonal entries to be zero.

Using the unit entry, we can easily accomplish this task with an elementary column operation followed by an elementary row operation. □

In the previous proof, we used two row and two column operations. Rewriting the resulting expression, we have

$$A = \begin{bmatrix} 1 & 0 \\ q_1 & 1 \end{bmatrix} \begin{bmatrix} 1 & q_2 \\ 0 & 1 \end{bmatrix} \begin{bmatrix} d_1 & 0 \\ 0 & d_2 \end{bmatrix} \begin{bmatrix} 1 & 0 \\ q_3 & 1 \end{bmatrix} \begin{bmatrix} 1 & q_4 \\ 0 & 1 \end{bmatrix} \tag{6.10}$$

for some elements $q_1, q_2, q_3, q_4 \in R$. The placement of the diagonal matrix is irrelevant, because we have the normalizing equality (6.1), as well as its dual for lower-triangular matrices. Thus, in hindsight, we could have proved Proposition 6.8 using only row operations, or using only column operations. Exploiting this fact, let us now prove:

THEOREM 6.11. *Let R be a separative exchange ring. Any element of* $\mathrm{GL}_2(R)$ *can be diagonally reduced using four elementary operations.*

PROOF. Given $A \in \mathrm{GL}_2(R)$, let $[a, b]$ be the top row of A. By Proposition 6.6, it is also the top row of another matrix $A' \in \mathrm{GL}_2(R)$, where the bottom-left entry of A' is full and regular. Thus, putting (6.1) together with Proposition 6.8, then there exist two lower-triangular elementary matrices L_1, L_2 and two upper-triangular elementary matrices U_1, U_2, such that $A'U_1L_1U_2L_2$ is diagonal.

This means that $[a, b]U_1L_1U_2L_2 = [u, 0]$ for some unit $u \in R$. But then

$$[a, b]U_1L_1 = [u, 0]L_2^{-1}U_2^{-1} = [u, *].$$

Hence, AU_1L_1 has a unit entry in its upper-left corner. Thus, we can fully reduce to a diagonal matrix using just two more elementary operations. □

Looking at the proof of Theorem 6.11, we see that in fact any element of $\mathrm{GL}_2(R)$ can be diagonally reduced using four elementary (column) operations, effected by starting with an upper-triangular matrix. By symmetry considerations, we could just have easily have done the same thing starting with a lower-triangular matrix.

7. Diagonal reduction of more matrices

It has long been known that every regular matrix A over a separative exchange ring admits a diagonal reduction, by [**1**, Theorem 3.1]. This means that there are invertible matrices P, Q such that $D := PAQ$ is diagonal. If A is 2×2, then so are P and Q, and from the results of the previous section we know that each can be expressed as products of at most four elementary matrices and a diagonal invertible matrix, the "diagonal part." Thus, since we can incorporate both diagonal parts into D, we see that four row operations and four column operations are sufficient to diagonally reduce A. But are they necessary?

It is straightforward to come up with examples showing that we must use at least three (elementary) row operations and at least three (elementary) column operations, as follows. Let $[a, b]$ be any right unimodular row over a separative regular ring R that is not completable. (We gave an explicit example of such a row, generated by the matrices in (6.4).)

First, we claim that using only column operations, it is impossible to transform this row to the form $[x, 0]$. Assuming otherwise, then x would be right invertible, say $xy = 1$ for some $y \in R$. Then the row $[x, 0]$ is completable to $\left[\begin{smallmatrix} x & 0 \\ 1-yx & y \end{smallmatrix}\right]$ with inverse $\left[\begin{smallmatrix} y & 1-yx \\ 0 & x \end{smallmatrix}\right]$. Hence, the original row $[a, b]$ would have been completable.

Next, let U and L be arbitrary upper-triangular and lower-triangular (respectively) elementary matrices. Thus, taking $A_1 := \left[\begin{smallmatrix} a & b \\ 0 & 0 \end{smallmatrix}\right]$, then the top row of LUA_1 cannot be put in the form $[x, 0]$ using any number of column operations. Similarly, taking $A_2 := \left[\begin{smallmatrix} 0 & 0 \\ a & b \end{smallmatrix}\right]$, then ULA_2 cannot be made diagonal using any number of column operations. Combining this information, the matrix

$$\begin{bmatrix} (a,0) & (b,0) \\ (0,a) & (0,b) \end{bmatrix} \in \mathbb{M}_2(R \times R)$$

cannot be made diagonal using only two row operations and any number of column operations. By working over the opposite ring R^{op}, we can reverse the roles of the column and row operations. The matrix

$$\begin{bmatrix} (a,0,a^{\mathrm{op}},0) & (b,0,0,a^{\mathrm{op}}) \\ (0,a,b^{\mathrm{op}},0) & (0,b,0,b^{\mathrm{op}}) \end{bmatrix} \in \mathbb{M}_2(R \times R \times R^{\mathrm{op}} \times R^{\mathrm{op}})$$

needs at least three column operations and at least three row operations to diagonally reduce.

Our next goal is to show that over separative *regular* rings, we can reduce to a diagonal matrix using only three row operations and three column operations, thus improving on the naive bound of four row and four column operations (mentioned in the first paragraph of this section). We are unsure whether this result applies to regular 2×2 matrices over separative exchange rings. As we will focus on regular rings, many of the arguments are simpler than those that appeared in the previous section. We will make essential use of repeaters.

Proposition 7.1 describes a series of elementary operations we can use to first produce a unit-regular entry. In order to simplify notation, let $\mathscr{U}$ be the subgroup of GL_2 consisting of upper-triangular elementary matrices, and similarly let $\mathscr{L}$ be the subgroup of lower-triangular elementary matrices.

PROPOSITION 7.1. *Let R be a regular ring, and let $A \in \mathbb{M}_2(R)$ be an arbitrary 2×2 matrix. Then there is a matrix in $\mathscr{U}\mathscr{L}A\mathscr{U}\mathscr{L}$ whose upper left corner entry, which we will call z, is a repeater. In particular, if R is (core) separative, then z is unit-regular.*

PROOF. Write $A = \left[\begin{smallmatrix} a & b \\ * & * \end{smallmatrix}\right]$ for some $a, b \in R$. Letting $U_1 := \left[\begin{smallmatrix} 1 & -a'-a'b \\ 0 & 1 \end{smallmatrix}\right]$ and letting $L_1 := \left[\begin{smallmatrix} 1 & 0 \\ a & 1 \end{smallmatrix}\right]$, then $x := (1 - aa')ba$ is the upper left corner of AU_1L_1. This is a right repeater.

Repeat the previous work, but now use row operations instead of column operations. Fix $AU_1L_1 = \left[\begin{smallmatrix} x & * \\ y & * \end{smallmatrix}\right]$, for some $y \in R$. Letting $L_2 := \left[\begin{smallmatrix} 1 & 0 \\ -x'-yx' & 1 \end{smallmatrix}\right]$ and $U_2 = \left[\begin{smallmatrix} 1 & x \\ 0 & 1 \end{smallmatrix}\right]$, then $z := xy(1 - x'x)$ is the upper left corner of $U_2L_2AU_1L_1$. This is visually a left repeater. It also remains a right repeater, by Lemma 3.22. Thus, it is a repeater. $\square$

Note that the previous proposition and proof work just as well for matrices in $\mathbb{M}_n(R)$ with $n > 2$, by restricting attention to the northwest 2×2 corner of such matrices. The proof also shows that using just column operations, we can force the northwest corner to be a right repeater.

With the existence of a unit-regular entry, it is straightforward to diagonally reduce. First, there is one other result that we will need, generalizing Lemma 6.3.

LEMMA 7.2. *Let R be an exchange ring, and let $a, b \in R$. If $aR + bR = pR$ for some idempotent $p \in R$, then there exist orthogonal idempotents $e \in aR$ and $f \in bR$ with $e + f = p$. Moreover, given any idempotent $q \in aR$, we can choose e (depending on q) so that $q \in eR$. In particular, if $aR \subseteq^{\oplus} R_R$, then we can choose e so that $eR = aR$.*

PROOF. We have

$$aR + bR + (1-p)R = R_R = X \oplus qR \oplus (1-p)R,$$

where X is any direct summand complement of qR in pR. By [**26**, Lemma 2.5], with $(N)_{\text{finite}}$ in place of (N), we have

$$R_R = qR \oplus (1-p)R \oplus A' \oplus B' \oplus C', \tag{7.3}$$

for some submodules $A' \subseteq aR$ and $B' \subseteq bR$ and $C' \subseteq (1-p)R$.

Clearly $C' = 0$, since $(1-p)R$ is a direct sum complement to $C' \subseteq (1-p)R$. Also, using the inclusion $qR \oplus A' \oplus B' \subseteq aR + bR = pR$, since pR is a direct summand complement to $(1-p)R$, we must have from (7.3) that $qR \oplus A' \oplus B' = pR$.

Write $p = e + f$ for orthogonal idempotents e, f such that $eR = qR \oplus A'$ and $fR = B'$. This is a decomposition into orthogonal (sub)idempotents, as desired.

The last sentence is proved exactly as before. □

PROPOSITION 7.4. *Let R be a regular ring, and let $A := \left[\begin{smallmatrix} a & b \\ c & d \end{smallmatrix}\right] \in \mathbb{M}_2(R)$, where $a \in R$ is unit-regular. Then there is an element of $\mathscr{L}\mathscr{U} A \mathscr{L}\mathscr{U}$ that is diagonal.*

PROOF. As a is unit-regular, write $a = eu$ for an idempotent $e \in R$ and a unit $u \in R$. Write $aR + bR = pR$ for some idempotent $p \in R$. Applying the strong form of Lemma 7.2, we can find a decomposition $p = ar + bs$ into orthogonal subidempotents, for some $r, s \in R$, but also guaranteeing that $arR = aR$. In other words, $eurR = euR$, so by [**18**, Ex. 21.4] we may replace r by a unit.

Letting $L_1 := \left[\begin{smallmatrix} 1 & 0 \\ sr^{-1} & 1 \end{smallmatrix}\right]$, then the top row of AL_1 is $[pr^{-1}, b]$. Letting $U_{1,1} := \left[\begin{smallmatrix} 1 & -rb \\ 0 & 1 \end{smallmatrix}\right]$, then the upper right entry of $AL_1U_{1,1}$ is zero. Letting $D := \left[\begin{smallmatrix} r & 0 \\ 0 & 1 \end{smallmatrix}\right]$, we can write

$$AL_1U_{1,1}D = \begin{bmatrix} p & 0 \\ x & y \end{bmatrix}$$

for some $x, y \in R$. Multiplying by D simplifies later computations, and we will explain how to remove it later.

Let $h := x(1-p)[x(1-p)]'$, which is an idempotent. Setting

$$U_{1,2} := \begin{bmatrix} 1 & -(1-p)[x(1-p)]'y \\ 0 & 1 \end{bmatrix},$$

then

$$AL_1U_{1,1}DU_{1,2} = \begin{bmatrix} p & 0 \\ x & (1-h)y \end{bmatrix}.$$

Next, fix an idempotent q such that $Rp + Rx(1-p) = Rq$. From the left-right symmetric version of Lemma 7.2, we can fix $v, w \in R$ so that $q = vp + wx(1-p)$ is an orthogonal sum of idempotents where $Rp = Rvp$. Again applying [**18**, Ex. 21.4] we may assume v is a unit.

From orthogonality, we know $vpwx(1-p) = 0$. Since v is a unit, $pwx(1-p) = 0$. In other words, $(1-p)wx(1-p) = wx(1-p)$. Thus, without losing any generality, we might as well have fixed $w \in (1-p)R$.

Now, letting $U_2 := \left[\begin{smallmatrix} 1 & v^{-1}wh \\ 0 & 1 \end{smallmatrix}\right]$, we find

$$U_2AL_1U_{1,1}DU_{1,2} = \begin{bmatrix} p + v^{-1}whx & 0 \\ x & (1-h)y \end{bmatrix}.$$

Set $z := p + v^{-1}whx$, which is the upper left corner entry of this matrix. We compute

$$\begin{aligned} vz = vp + whx &= vp + whx(1-p) + whxp \\ = vp + wx(1-p) + (1-p)whxp &= (1 + (1-p)whxp)q. \end{aligned}$$

Thus, $[1 + (1-p)whxp]^{-1}vz = q$. Since

$$Rx \subseteq Rx(1-p) + Rxp \subseteq Rx(1-p) + Rp = Rq,$$

we can write $x = tq$ for some $t \in R$. Therefore, by taking

$$L_2 := \begin{bmatrix} 1 & 0 \\ -t[1 + (1-p)whxp]^{-1}v & 1 \end{bmatrix},$$

then $L_2U_2AL_1U_{1,1}DU_{1,2}$ is diagonal.

Of course, multiplying on the right by D^{-1} keeps it diagonal. As $U_1 := U_{1,1}DU_{1,2}D^{-1} \in \mathscr{U}$, we are done. □

Putting Proposition 7.1 together with Proposition 7.4, and using the second paragraph of this section, we now know:

THEOREM 7.5. *Over a separative regular ring, every 2×2 matrix can be diagonally reduced using just three row operations and three column operations. Nothing less on either side can work in general.*

More precisely, given any matrix $A \in \mathbb{M}_2(R)$, when R is a core separative regular ring, then

$$\mathscr{L}\mathscr{U}\mathscr{L}A\mathscr{U}\mathscr{L}\mathscr{U}$$

contains a diagonal element. By a symmetric construction (or, alternatively, noting that conjugation by $\left[\begin{smallmatrix} 0 & 1 \\ 1 & 0 \end{smallmatrix}\right]$ is an automorphism of $\mathbb{M}_2(R)$), we similarly know that

$$\mathscr{U}\mathscr{L}\mathscr{U}A\mathscr{L}\mathscr{U}\mathscr{L}$$

always contains a diagonal element.

Analyzing the proof, we similarly see that for any 1×2 matrix A over a core strongly separative ring, then $A\mathscr{U}\mathscr{L}\mathscr{U}$ contains a diagonal row matrix $[* \;\; 0]$.

The separativity problem can now be restated as Question 7.6:

QUESTION 7.6. Does there exist a regular ring, where some 2×2 matrix over that ring cannot be diagonally reduced using just three row operations and three column operations?

Acknowledgments

We thank Kevin O'Meara for posing problems and contributing to many conversations that led to much of this paper. We also thank T. Y. Lam and the anonymous referee for recommendations which led to improvements in the paper. The third author thanks the Mathematics Department of the Universitat Autonóma de Barcelona for their hospitality during his visit.

References

[1] P. Ara, K. R. Goodearl, K. C. O'Meara, and E. Pardo, *Diagonalization of matrices over regular rings*, Linear Algebra Appl. **265** (1997), 147–163. MR1466896

[2] P. Ara, K. R. Goodearl, K. C. O'Meara, and E. Pardo, *Separative cancellation for projective modules over exchange rings*, Israel J. Math. **105** (1998), 105–137, DOI 10.1007/BF02780325. MR1639739

[3] P. Ara, K. R. Goodearl, K. C. O'Meara, and R. Raphael, *K_1 of separative exchange rings and C^*-algebras with real rank zero*, Pacific J. Math. **195** (2000), no. 2, 261–275, DOI 10.2140/pjm.2000.195.261. MR1782176

[4] Pere Ara, Joan Bosa, and Enrique Pardo, *The realization problem for finitely generated refinement monoids*, Selecta Math. (N.S.) **26** (2020), no. 3, Paper No. 33, 63, DOI 10.1007/s00029-020-00559-5. MR4103463

[5] George M. Bergman, *An invitation to general algebra and universal constructions*, 2nd ed., Universitext, Springer, Cham, 2015, DOI 10.1007/978-3-319-11478-1. MR3309721

[6] George M. Bergman, *Strong inner inverses in endomorphism rings of vector spaces*, Publ. Mat. **62** (2018), no. 1, 253–284, DOI 10.5565/PUBLMAT6211812. MR3738191

[7] George M. Bergman and Warren Dicks, *Universal derivations and universal ring constructions*, Pacific J. Math. **79** (1978), no. 2, 293–337. MR531320

[8] L. Fuchs, *On a substitution property of modules*, Monatsh. Math. **75** (1971), 198–204, DOI 10.1007/BF01299099. MR296096

[9] László Fuchs, *Infinite abelian groups. Vol. I*, Pure and Applied Mathematics, Vol. 36, Academic Press, New York-London, 1970. MR255673

[10] László Fuchs, *Infinite abelian groups. Vol. II*, Pure and Applied Mathematics, Vol. 36-II, Academic Press, New York-London, 1973. MR349869

[11] K. R. Goodearl, *Von Neumann regular rings*, 2nd ed., Robert E. Krieger Publishing Co., Inc., Malabar, FL, 1991. MR1150975

[12] K. R. Goodearl, *Von Neumann regular rings and direct sum decomposition problems*, Abelian groups and modules (Padova, 1994), Math. Appl., vol. 343, Kluwer Acad. Publ., Dordrecht, 1995, pp. 249–255. MR1378203

[13] David Handelman, *Perspectivity and cancellation in regular rings*, J. Algebra **48** (1977), no. 1, 1–16, DOI 10.1016/0021-8693(77)90289-7. MR447329

[14] S. K. Jain and A. Leroy, *Some remarks on the decomposition into products of idempotents*, preprint (2024).

[15] Irving Kaplansky, *Infinite abelian groups*, University of Michigan Press, Ann Arbor, MI, 1954. MR65561

[16] Dinesh Khurana, T. Y. Lam, and Pace P. Nielsen, *Exchange elements in rings, and the equation $XA - BX = I$*, Trans. Amer. Math. Soc. **369** (2017), no. 1, 495–516, DOI 10.1090/tran6652. MR3557782

[17] T. Y. Lam, *A first course in noncommutative rings*, 2nd ed., Graduate Texts in Mathematics, vol. 131, Springer-Verlag, New York, 2001, DOI 10.1007/978-1-4419-8616-0. MR1838439

[18] T. Y. Lam, *Exercises in classical ring theory*, 2nd ed., Problem Books in Mathematics, Springer-Verlag, New York, 2003. MR2003255

[19] T. Y. Lam, *Serre's problem on projective modules*, Springer Monographs in Mathematics, Springer-Verlag, Berlin, 2006, DOI 10.1007/978-3-540-34575-6. MR2235330

[20] T. Y. Lam and Will Murray, *Unit regular elements in corner rings*, Bull. Hong Kong Math. Soc. **1** (1997), no. 1, 61–65. MR1466836

[21] W. K. Nicholson, *Lifting idempotents and exchange rings*, Trans. Amer. Math. Soc. **229** (1977), 269–278, DOI 10.2307/1998510. MR439876

[22] Pace P. Nielsen and Janez Šter, *Connections between unit-regularity, regularity, cleanness, and strong cleanness of elements and rings*, Trans. Amer. Math. Soc. **370** (2018), no. 3, 1759–1782, DOI 10.1090/tran/7080. MR3739190

[23] K. C. O'Meara and R. M. Raphael, *Uniform diagonalisation of matrices over regular rings*, Algebra Universalis **45** (2001), no. 4, 383–405, DOI 10.1007/s000120050221. MR1816975

[24] K. C. O'Meara and C. Vinsonhaler, *Separative cancellation and multiple isomorphism in torsion-free abelian groups*, J. Algebra **221** (1999), no. 2, 536–550, DOI 10.1006/jabr.1999.7990. MR1728395

[25] Kevin C. O'Meara, *Another multiplicative property of a regular ring that has no multiplicative proof*, Semigroup Forum **101** (2020), no. 3, 769–778, DOI 10.1007/s00233-020-10086-6. MR4182154

[26] Josef Stock, *On rings whose projective modules have the exchange property*, J. Algebra **103** (1986), no. 2, 437–453, DOI 10.1016/0021-8693(86)90145-6. MR864422

Departament de Matemàtiques, Universitat Autònoma de Barcelona, 08193 Bellaterra, Barcelona, Spain, and Centre de Recerca Matemàtica, Edifici Cc, Campus de Bellaterra, 08193 Cerdanyola del Vallès, Barcelona, Spain

Email address: `pere.ara@uab.cat`

URL: `https://www.crm.cat/person/77/ara-pere/`

Department of Mathematics, University of California, Santa Barbara, California 93106

Email address: `goodearl@math.ucsb.edu`

URL: `https://web.math.ucsb.edu/~goodearl/`

Department of Mathematics, Brigham Young University, Provo, Utah 84602

Email address: `pace@math.byu.edu`

URL: `https://mathdept.byu.edu/~pace/`

Departamento de Matemáticas, Facultad de Ciencias, Universidad de Cádiz, Campus de Puerto Real, 11510 Puerto Real (Cádiz), Spain

Email address: `enrique.pardo@uca.es`

URL: `https://sites.google.com/gm.uca.es/webofenriquepardo/inicio`

Departament de Matemàtiques, Universitat Autònoma de Barcelona, 08193 Bellaterra, Barcelona, Spain, and Centre de Recerca Matemàtica, Edifici Cc, Campus de Bellaterra, 08193 Cerdanyola del Vallès, Barcelona, Spain

Email address: `francesc.perera@uab.cat`

URL: `https://mat.uab.cat/web/perera`

Contemporary Mathematics
Volume **826**, 2025
https://doi.org/10.1090/conm/826/16585

Revisiting constructions of codes over Galois rings through q-ary lifts

Frédérique Oggier

ABSTRACT. We revisit known constructions of self-dual codes over Galois rings through the view point of q-ary lifts from the finite field $\mathbb{F}_{p^r}$, p a prime. We show that known results about the lifts of codes over $\mathbb{Z}_{p^m}$ can be easily extended and thus this approach provides a valid alternative way to study codes over Galois rings. We conclude with some open questions.

1. Introduction

Codes over the ring $\mathbb{Z}_{p^m}$ of integers modulo p^m, for p a prime, and over the ring of p-adic integers, have been well studied. The structure of cyclic codes over such rings is presented e.g., in [**2**, **5**]. Duality and self-dual codes are considered in e.g., [**3**], while their weight enumerators are investigated in e.g., [**4**]. There is a natural common approach to these works, which consists in starting with codes over $\mathbb{Z}_p$ and then to lift them to $\mathbb{Z}_{p^m}$ for $m \geq 2$.

In [**8**], the question of a similar lift starting from the finite field $\mathbb{F}_{p^r}$ instead is positively answered, and cyclic codes over $\mathbb{F}_{p^r}$ are then lifted to cyclic codes over the Galois ring $GR(p^m, r)$ for $m \geq 2$. Codes over Galois rings also have their own studies, see e.g., [**7**] for constructions of self-dual codes, and reference therein for earlier works on codes over Galois rings.

This leads to the natural question of how results about lifts of codes over $\mathbb{Z}_{p^m}$ for $m \geq 2$ can be extended to lifts of codes over the Galois ring $GR(p^m, r)$ for $m \geq 2$. In Section 2, we recall an iterative construction of self-dual codes over Galois rings, and motivate an alternative lift approach. In Section 3, we show that known lift results about self-duality and Hamming distance are easily extended from $\mathbb{Z}_{p^m}$ to $GR(p^m, r)$ for $m \geq 2$, thus providing an alternative approach to study codes over Galois rings. Finally in Section 4, we suggest some open research directions. We recall below some context and background about codes over the rings of interest.

Let p be a prime, and m, r be positive integers. Let $\mathbb{Z}_{p^m}$ denote the ring of integers modulo p^m and $\mathbb{Z}_{p^\infty}$ be the ring of p-adic integers. Let $\mathbb{F}_{p^r}$ denote the finite field of order p^r. A Galois ring is constructed as $GR(p^m, r) \simeq \mathbb{Z}_{p^m}[x]/f(x)$ where $f(x)$ is a monic basic irreducible polynomial in $\mathbb{Z}_{p^m}[x]$ of degree r, meaning that f is not a zero divisor (it is regular), and that both f and $\bar{f} = \mu(f)$, the image of f under the projection $\mu : \mathbb{Z}_{p^m}[x] \to \mathbb{Z}_p[x]$, are irreducible [**1**, Sections 6.1, 6.2]. We have $GR(p, r) \simeq \mathbb{F}_{p^r}$ and $GR(p^m, 1) \simeq \mathbb{Z}_{p^m}$.

Every Galois ring $GR(p^m,r)$ contains a primitive p^r-1th root of unity ζ, and we have $GR(p^m,r)\simeq \mathbb{Z}_{p^m}[x]/(h(x)) = \mathbb{Z}_{p^m}[\zeta]$ where $h(x)|x^{p^r-1}-1$ for $h(x)$ irreducible in $\mathbb{Z}_{p^m}[x]$ and $\zeta = x+(h(x))$. The polynomial $h(x)$ is determined by the integral version of Hensel's Lemma, from a primitive polynomial in $\mathbb{Z}_p[x]$ of degree r, that is from the minimal polynomial over the finite field $\mathbb{F}_p$ of a primitive element of $\mathbb{F}_{p^r}$ [**1**, Section 6.2].

A linear code $\mathcal{C}$ of length n over R, where R is either $GR(p^m,r)$ or $\mathbb{Z}_{p^\infty}$, is a submodule of R^n. Let $\{m_0,m_1,\dots,m_s\}$ be a set of nonnegative integers such that $m_i<m_j$ for $i<j$. After a suitable permutation of the coordinates, a non-zero linear code $\mathcal{C}$ over R can be described by a generator matrix G of the form

$$G=\begin{bmatrix} p^{m_0}I_{k_0} & p^{m_0}A_{0,1} & p^{m_0}A_{0,2} & p^{m_0}A_{0,3} & \dots & p^{m_0}A_{0,s} & p^{m_0}A_{0,s+1}\\ 0 & p^{m_1}I_{k_1} & p^{m_1}A_{1,2} & \dots & \dots & \dots & p^{m_1}A_{1,s+1}\\ \vdots & \vdots & \vdots & \vdots & \vdots & \vdots & \vdots\\ 0 & 0 & 0 & 0 & \dots & p^{m_s}I_{k_s} & p^{m_s}A_{s,s+1}\end{bmatrix} \tag{1.1}$$

where the columns are grouped into blocks of size $k_0,k_1,\dots,k_{s+1}$ such that $\sum_{i=0}^{s+1}k_i = n$, and where for $R=GR(p^m,r)$, $m_i=i$ and $s=m-1$ (see [**7**] for $R=GR(p^m,r)$). The matrix G is said to be in systematic or standard form.

DEFINITION 1.1. The rank of a code $\mathcal{C}$, denoted by rank($\mathcal{C}$), is defined to be the number of nonzero rows of its generator matrix G in systematic form.

A matrix in systematic form of the form (1.1), or a code $\mathcal{C}$ having such a generator matrix, is said to be of type

$$(p^{m_0})^{k_0}(p^{m_1})^{k_1}(p^{m_2})^{k_2}\cdots(p^{m_s})^{k_s}.$$

When $R=GR(p^m,r)$, $\mathcal{C}$ is thus of type $(p^0)^{k_0}(p^1)^{k_1}(p^2)^{k_2}\cdots(p^{m-1})^{k_{m-1}}$ and $\mathcal{C}$ contains p^k codewords, where

$$k=r\Big(\sum_{i=0}^{m}(m-i)k_i\Big). \tag{1.2}$$

In particular, when $r=1$, $R=\mathbb{Z}_{p^m}$, and [**3**]

$$k=\sum_{i=0}^{m}(m-i)k_i.$$

This is because

$$\mathcal{C}=\{(u_0,\dots,u_s)G,(u_0,\dots,u_s)\in GR(p^m,r)^{s+1}\}$$

but if the elements of the jth row of G are divisible by p^i, then we can restrict u_{j-1} to $GR(p^{m-i},r)$.

Under the usual inner product $\cdot$ on R^n, the dual $\mathcal{C}^\perp$ of $\mathcal{C}$ is defined as

$$\mathcal{C}^\perp=\{\mathbf{x}\in R^n,\mathbf{x}\cdot\mathbf{y}=0 \text{ for all } \mathbf{y}\in\mathcal{C}\}.$$

A code $\mathcal{C}$ is said to be self-dual if $\mathcal{C}=\mathcal{C}^\perp$.

2. Motivation

We start by reviewing a method to build self-dual codes over $GR(p^m, r)$, presented in [7]. Proposition 3.4 from [7] starts with an $e \times n$ generator matrix G_0 of a self-dual code, and extends it into an $(e+1) \times (n+2)$ generator matrix G of a new self-dual code. Assuming there exists $c \in GR(p^m, r)$ such that $c^2 = -1$ (from [7, Lemma 3.1], such a c exists if and only if either $p \equiv 1 \pmod 4$ with any r or $p \equiv -1 \pmod 4$ with r even), and given G_0, a generator matrix of a self-dual code $\mathcal{C}_0$ over $GR(p^m, r)$ with rows $\mathbf{r}_1, \dots \mathbf{r}_e$, let $\mathbf{x}$ be a vector in $GR(p^m, r)^n$ with $\mathbf{x} \cdot \mathbf{x} = -1$ and set $y_i = \mathbf{x} \cdot \mathbf{r}_i$ for $1 \le i \le e$. Then

$$G = \left[\begin{array}{cc|c} 1 & 0 & \mathbf{x} \\ \hline -y_1 & cy_1 & \\ \vdots & \vdots & G_0 \\ -y_e & cy_e & \end{array}\right] \tag{2.1}$$

generates a self-dual code $\mathcal{C}$ over $GR(p^m, r)$ of length $n+2$.

Here is a simple observation about the above construction.

LEMMA 2.1. *Under the above assumptions, start with*

$$G_0 = [1, c], c \in GR(p^m, r), c^2 = -1.$$

If $\mathbf{x}$ *is chosen in* $\mathbb{Z}_{p^m}^2$*, then the systematic form of*

$$G_1 = \begin{bmatrix} 1 & 0 & x_1 & x_2 \\ -y_1 & cy_1 & 1 & c \end{bmatrix}$$

is

$$\begin{bmatrix} 1 & 0 & x_1 & x_2 \\ 0 & 1 & x_2 & -x_1 \end{bmatrix}.$$

In particular, the systematic form has coefficients in $\mathbb{Z}_{p^m}$.

PROOF. Indeed, to obtain the systematic form for G_1, start by adding y_1 times the first row to the second row:

$$\begin{bmatrix} 1 & 0 & x_1 & x_2 \\ 0 & cy_1 & 1 + x_1 y_1 & c + x_2 y_1 \end{bmatrix}.$$

The first row lives in $\mathbb{Z}_{p^m}^4$ by assumption on $\mathbf{x}$. By construction, $y_1 = \mathbf{x} \cdot [1, c] = x_1 + cx_2$. It thus has inverse $y_1^{-1} = -x_1 + x_2 c$, since

$$(x_1 + cx_2)(-x_1 + x_2 c) = -x_1^2 + x_2^2 c^2 = -x_1^2 - x_2^2 = -\mathbf{x} \cdot \mathbf{x} = 1$$

recalling that $c^2 = -1$ (this also shows that y_1 cannot be 0, otherwise $x_1 = -cx_2$ and it is not possible that $x_1^2 + x_2^2 = -1$).

Since y_1 is a unit, and $c^2 = -1$ so c is a unit too, it is then enough to multiply the second row by $c^{-1} y_1^{-1} = (-c) y_1^{-1}$ to get the systematic form:

$$\begin{bmatrix} 1 & 0 & x_1 & x_2 \\ 0 & 1 & -cy_1^{-1}(1 + x_1 y_1) & -cy_1^{-1}(c + x_2 y_1) \end{bmatrix}.$$

But then, recalling that $y_1^{-1} = -x_1 + x_2 c$, we get

$$-cy_1^{-1}(1 + x_1 y_1) = -c(-x_1 + x_2 c) - cx_1 = -c^2 x_2 = x_2 \in \mathbb{Z}_{p^m},$$

and

$$-cy_1^{-1}(c + x_2 y_1) = y_1^{-1} - cx_2 = -x_1 + x_2 c - cx_2 = -x_1 \in \mathbb{Z}_{p^m}.$$

This shows that the systematic form of G_1 is

$$\begin{bmatrix} 1 & 0 & x_1 & x_2 \\ 0 & 1 & x_2 & -x_1 \end{bmatrix}.$$

□

Self-dual codes $\mathcal{C}_{4,1}, \ldots, \mathcal{C}_{4,10}$ of length 4 over $GR(3^2, 2)$ presented in Table 2 of [**7**] all use $\mathbf{x} \in \mathbb{Z}_{3^2}^2$.

EXAMPLE 2.2. For example, the last code in Table 2 of [**7**] is given by

$$\begin{bmatrix} 1 & 0 & 8 & 4 \\ 5w+6 & 8w+4 & 1 & w+1 \end{bmatrix}$$

for w such that $w^2 = 7w + 7$. A generator in systematic form is

$$\begin{bmatrix} 1 & 0 & 8 & 4 \\ 0 & 1 & 4 & 1 \end{bmatrix}.$$

Recall from Section 1 that $GR(p^m, r) \simeq \mathbb{Z}_{p^m}[x]/(h(x)) = \mathbb{Z}_{p^m}[\zeta]$ where $h(x)|x^{p^m-1} - 1$ for $h(x)$ irreducible in $\mathbb{Z}_{p^m}[x]$. For $p = 3$, $m = 2$, we have

$$x^8 - 1 = (x^2 - 1)(x^2 + 1)(x^4 + 1) \in \mathbb{Z}[x]$$
$$x^4 + 1 = (x^2 + 4x - 1)(x^2 - 4x - 1) \in \mathbb{Z}_{3^2}[x],$$

so we may take ζ such that $\zeta^2 = -4\zeta + 1$. Then $\zeta^4 = -1$ and $\zeta = \zeta_8$ is an 8th root of unity in $\mathbb{Z}_{3^2}[\zeta]$, so

$$GR(3^2, 2) \simeq \mathbb{Z}_{3^2}[w] \simeq \mathbb{Z}_{3^2}[\zeta],$$

for w such that $w^2 = 7w + 7$. This isomorphism sends w to $\zeta^2 - 1 = -4\zeta$, since $(\zeta^2 - 1)^2 - 7(\zeta^2 - 1) = 7$.

We then project $GR(3^2, 2)$ to $GR(3, 2) \simeq \mathbb{F}_{3^2} \simeq \mathbb{F}_3(\bar{\zeta})$, for $\bar{\zeta}$ such that $\bar{\zeta}^2 = -\bar{\zeta} + 1$, which maps ζ to $\bar{\zeta}$.

PROPOSITION 2.3.

(1) *The codes* $\mathcal{C}_{4,1}, \ldots, \mathcal{C}_{4,10}$ *over* $GR(3^2, 2)$ [**7**, Table 2] *constructed from* (2.1) *with* $G_0 = [1, w + 1]$ *and parameters* $[-y_1, cy_1, \mathbf{x}]$, *namely*

$$\begin{aligned}
\mathcal{C}_{4,1} &= [4w + 3, w + 5, [1, 5]] \\
\mathcal{C}_{4,2} &= [7w + 5, 2w, [2, 2]] \\
\mathcal{C}_{4,3} &= [2w, 2w + 4, [2, 7]] \\
\mathcal{C}_{4,4} &= [8w + 4, 4w + 3, [4, 1]] \\
\mathcal{C}_{4,5} &= [w + 6, 4w + 5, [4, 8]] \\
\mathcal{C}_{4,6} &= [8w + 3, 5w + 4, [5, 1]] \\
\mathcal{C}_{4,7} &= [w + 5, 5w + 6, [5, 8]] \\
\mathcal{C}_{4,8} &= [7w, 7w + 5, [7, 2]] \\
\mathcal{C}_{4,9} &= [2w + 4, 7w, [7, 7]] \\
\mathcal{C}_{4,10} &= [5w + 6, 8w + 4, [8, 4]]
\end{aligned}$$

reduce to four self-dual codes over $\mathbb{F}_3$, *described by* $[I_2|A]$ *where*

$$A = \begin{bmatrix} x_1 & x_2 \\ x_2 & -x_1 \end{bmatrix}, x_1, x_2 \in \{1, 2\}.$$

(2) *Keeping the notations of the above computations, the codes* $\mathcal{C}_{6,1},\ldots,\mathcal{C}_{6,10}$ *over* $GR(3^2,2)$ [**7**, Table 3] *constructed from* (2.1) *with*

$$G_0 = \begin{bmatrix} 1 & 0 & 1 & 4 \\ 5w+4 & w+6 & 1 & w+1 \end{bmatrix}$$

and parameters $[-y_1, cy_1, -y_2, cy_2, \mathbf{x}]$, *namely*

$$\begin{aligned}
\mathcal{C}_{6,1} &= [4w+4, 4, w+3, 7w+8, [0,1,w+4,w+7]] \\
\mathcal{C}_{6,2} &= [4w+4, 4, 4w+3, w+5, [0,1,w+7,w+4]] \\
\mathcal{C}_{6,3} &= [5w+5, 5, 6w+6, 6, [0,1,2w+2,5w+5]] \\
\mathcal{C}_{6,4} &= [5w+5, 5, 6, 3w+3, [0,1,2w+5,5w+2]] \\
\mathcal{C}_{6,5} &= [5w+5, 5, 3w+6, 6w, [0,1,2w+8,5w+8]] \\
\mathcal{C}_{6,6} &= [4w+4, 4, w+6, 4w+5, [0,1,4w+1,7w+1]] \\
\mathcal{C}_{6,7} &= [4w+4, 4, 4w+6, 7w+2, [0,1,4w+4,7w+7]] \\
\mathcal{C}_{6,8} &= [4w+4, 4, 7w+6, w+8, [0,1,4w+7,7w+4]]
\end{aligned}$$

$$\begin{aligned}
\mathcal{C}_{6,9} &= [5w+5, 5, 0, 0, [0,1,5w+2,2w+5]] \\
\mathcal{C}_{6,10} &= [5w+5, 5, 3w, 3w+6, [0,1,5w+5,2w+2]]
\end{aligned}$$

reduce to two codes over $\mathbb{F}_{3^2}$, *described by*

$$\begin{aligned}
&[2\bar{\zeta}+1, 1, 2\bar{\zeta}, 2\bar{\zeta}+2, [0,1,2\bar{\zeta}+1, 2\bar{\zeta}+1]] \\
&[\bar{\zeta}+2, 2, 0, 0, [0,1,\bar{\zeta}+2,\bar{\zeta}+2]].
\end{aligned}$$

PROOF.

(1) Using Lemma 2.1, all these codes have a systematic generator matrix of the form

$$\begin{bmatrix} 1 & 0 & x_1 & x_2 \\ 0 & 1 & x_2 & -x_1 \end{bmatrix}$$

where, modulo 3, $[x_1, x_2] \pmod 3 \in \{[1,2],[2,2],[2,1],[1,1]\}$. It is clear that the first row is orthogonal to the second one, and since $1^2+x_1^2+x_2^2 = 0 \pmod 3$ for $x_1, x_2 \in \{1, 2 \equiv -1 \pmod 3\}$, the code is included inside its dual, and it is further self-dual since it is has the same dimension (namely 2) as its dual.

(2) The proof that these codes reduce to two codes over $\mathbb{F}_{3^2}$ is a direct computation, which just involves mapping w to $\bar{\zeta}$. From (2.1), the corresponding projected matrices in $\mathbb{F}_{3^2}$ are

$$G' = \left[\begin{array}{cc|cccc} 1 & 0 & 0 & 1 & 2\bar{\zeta}+1 & 2\bar{\zeta}+1 \\ \hline 2\bar{\zeta}+1 & 1 & 1 & 0 & 1 & 4 \\ 2\bar{\zeta} & 2\bar{\zeta}+2 & 5\bar{\zeta}^2-1 & \bar{\zeta}^2+5 & 1 & \bar{\zeta}^2 \end{array}\right],$$

$$G'' = \left[\begin{array}{cc|cccc} 1 & 0 & 0 & 1 & \bar{\zeta}+2 & \bar{\zeta}+2 \\ \hline \bar{\zeta}+2 & 2 & 1 & 0 & 1 & 4 \\ 0 & 0 & 5\bar{\zeta}^2-1 & \bar{\zeta}^2+5 & 1 & \bar{\zeta}^2 \end{array}\right].$$

A direct computation shows that $G'(G')^T = G''(G'')^T = \mathbf{0}$.

□

Proposition 2.3 suggests the following change of view point: instead of the iterative construction proposed in [**7**], the families of codes $\mathcal{C}_{4,1},\dots,\mathcal{C}_{4,10}$ and $\mathcal{C}_{6,1},\dots,\mathcal{C}_{6,10}$ could be regarded as q-ary lifts of two self-dual codes over $\mathbb{F}_{3^2}$, respectively p-ary lifts of four self-dual codes over $\mathbb{F}_3$. The p-ary lift of codes over $\mathbb{Z}_{p^r}$ has been well studied in the context of codes over p-adic integers (see e.g. [**3**]), while q-ary lifts have been recently introduced [**8**], as we discuss next.

3. q-ary lifts

For $e<m$, define the map $\Psi_{m,e}: GR(p^m,r)\simeq \mathbb{Z}_{p^m}[\zeta]\to GR(p^e,r)\simeq\mathbb{Z}_{p^e}[\bar{\zeta}]$, which sends $\sum_{i=0}^{r-1}a_i\zeta^i$, $a_i=\sum_{j=0}^{m-1}a_{ij}p^j\in\mathbb{Z}_{p^m}$, to

$$\Psi_{m,e}\Big(\sum_{i=0}^{r-1}a_i\zeta^i\Big)=\sum_{i=0}^{e-1}\bar{a}_i\bar{\zeta}^i. \tag{3.1}$$

The coefficient $\bar{a}_i=\sum_{j=0}^{e-1}a_{ij}p^j$ is the projection of $a_i \pmod{p^e}$, and ζ is a primitive p^r-1th root of unity. More precisely, a minimal polynomial in $\mathbb{F}_p[x]$ for a primitive p^r-1th root of unity of the form $x^r+c_{r-1}x^{r-1}+\dots+c_0$ may be lifted using Hensel's Lemma to $x^r+(p^k-p+c_{r-1})x^{r-1}+\dots+(p^k-p)+c_0\in\mathbb{Z}_{p^k}$, for $k=e,m$. Indeed $p^k-p+c_j<p^k$ since $c_j<p$, and conversely, the polynomial $x^r+(p^m-p+c_{r-1})x^{r-1}+\dots+(p^m-p)+c_0\in\mathbb{Z}_{p^m}$ with root ζ is projected to $x^r+(p^e-p+c_{r-1})x^{r-1}+\dots+(p^e-p)+c_0\in\mathbb{Z}_{p^e}$ with root $\bar{\zeta}$ and both are projected to $x^r+c_{r-1}x^{r-1}+\dots+c_0\in\mathbb{Z}_p[x]$. It is clear that $\Psi_{m,e}(\sum_{i=0}^{r-1}a_i\zeta^i)+\Psi_{m,e}(\sum_{i=0}^{r-1}b_i\zeta^i)=\Psi_{m,e}(\sum_{i=0}^{r-1}(a_i+b_i)\zeta^i)$. Multiplicativity follows from the polynomial equations satisfied by ζ and $\bar{\zeta}$ as just discussed.

When $r=1$ and $m=\infty$, this is the map Ψ_e defined in [**3**, (4)]. Since codes over $\mathbb{F}_p$ can be lifted to $\mathbb{Z}_{p^m}$ and $\mathbb{Z}_{p^\infty}$ (see e.g. [**2, 3**]), it is natural to ask about the lifting of codes over $\mathbb{F}_{p^r}$, a question asked in [**8**], which led to proving the lift of cyclic codes, and to the following projective system:

$$\begin{array}{ccccccc}
\mathbb{F}_{p^{rs}}\simeq GR(p,rs) & \longleftarrow & GR(p^2,rs) & \longleftarrow & GR(p^3,rs) & \longleftarrow & \cdots\\
| & & | & & | & & \\
\mathbb{F}_{p^r}\simeq GR(p,r) & \longleftarrow & GR(p^2,r) & \longleftarrow & GR(p^3,r) & \longleftarrow & \cdots\\
| & & | & & | & & \\
\mathbb{F}_p\simeq\mathbb{Z}_p\simeq GR(p,1) & \longleftarrow & \mathbb{Z}_{p^2} & \longleftarrow & \mathbb{Z}_{p^3} & \longleftarrow & \cdots
\end{array}$$

which motivates the introduction of the map (3.1). This map $\Psi_{m,e}$ is extended componentwise to matrices. If $\Psi_{m,e}(A)=B$ for some matrices A,B, then the inverse image of B is denoted by $\Psi_{m,e}^{-1}(B)$ and comprises all matrices A' such that $\Psi_{m,e}(A')=B$, which includes A.

Definition 3.1. Let $1\le e_1\le e_2$ be integers. A code $\mathcal{C}_1$ over $GR(p^{e_1},r)$ lifts to a code $\mathcal{C}_2$ over $GR(p^{e_2},r)$, denoted by $\mathcal{C}_1\prec\mathcal{C}_2$, if (1) the rank of $\mathcal{C}_1$ is equal to the rank of $\mathcal{C}_2$ and (2) $\mathcal{C}_2$ has a generator matrix G_2 such that $\Psi_{e_2,e_1}(G_2)$ is a generator matrix of $\mathcal{C}_1$.

The case $r=1$ is found in [**3**, Definition 1]. In [**8**], lifts of cyclic codes are done via their generator polynomials.

PROPOSITION 3.2. *Suppose $p \neq 2$. A self-dual code of type $(1)^k$ over $GR(p^{e_1}, r)$ lifts to a self-dual code $\mathcal{C}_2$ of type $(1)^k$ over $GR(p^{e_2}, r)$.*

PROOF. We will prove that such a lift is always possible from $GR(p^m, r)$ to $GR(p^{m+1}, r)$ and thus iteratively from $GR(p^{e_1}, r)$ to $GR(p^{e_2}, r)$. Let $\mathcal{C}_m, \mathcal{C}_{m+1}$ be type $(1)^k$ codes over $GR(p^m, r)$, $GR(p^{m+1}, r)$ respectively. They then have generator matrices in systematic form given by $G_m = [I|A_m], G_{m+1} = [I|A_{m+1}]$. The code $\mathcal{C}_m$, $\mathcal{C}_{m+1}$ are self-dual if and only if

$$I + A_m A_m^T \equiv 0 \pmod{(p^m)}, I + A_{m+1} A_{m+1}^T \equiv 0 \pmod{(p^{m+1})}.$$

Suppose $\mathcal{C}_{m+1}$ is a lift of $\mathcal{C}_m$. Then by definition

$$\Psi_{m+1,m}(G_{m+1}) = G_m, \Psi_{m+1,m}(A_{m+1}) = A_m.$$

A general form for A_{m+1} was found in [**3**, Theorem 3.20] for codes over $\mathbb{Z}_{p^\infty}$, assuming $p \neq 2$, which we may adjust here as

$$A_{m+1} = \left(\tfrac{p+3}{2} I + \tfrac{p+1}{2} \Psi_{m+1,m}^{-1}(A_m) \Psi_{m+1,m}^{-1}(A_m^T) + p^m K_m\right) \Psi_{m+1,m}^{-1}(A_m) \tag{3.2}$$

for K_m skew symmetric. Indeed, $\Psi_{m+1,m}(A_{m+1}) \equiv (\frac{p+3}{2} I + \frac{p+1}{2}(-I))A_m = A_m \pmod{(p^m)}$. Next, set $B = \frac{p+3}{2} I + \frac{p+1}{2} \Psi_{m+1,m}^{-1}(A_m A_m^T)$ and compute

$$A_{m+1} A_{m+1}^T \equiv (B + p^m K_m) \Psi_{m+1,m}^{-1}(A_m A_m^T)(B^T - p^m K_m) \pmod{(p^{m+1})}$$

since K_m is skew symmetric. First, the term in p^{2m} is $0 \pmod{(p^{m+1})}$. Then, the term in p^m is $p^m K_m (\Psi_{m+1,m}^{-1}(A_m A_m^T) B^T - B \Psi_{m+1,m}^{-1}(A_m A_m^T)) K_m$. Since $I + A_m A_m^T \equiv 0 \pmod{(p^m)} \iff A_m A_m^T = -I + p^m S$ for some matrix S with coefficients in $GR(p^m, r)$ which is thus symmetric, we have that

$$B = \tfrac{p+3}{2} I + \tfrac{p+1}{2}(-I + p^m S) = I + \tfrac{p+1}{2} p^m S$$

is symmetric. Then $\Psi_{m+1,m}^{-1}(A_m A_m^T) B - B \Psi_{m+1,m}^{-1}(A_m A_m^T) = 0$.

We are thus left to compute $B \Psi_{m+1,m}^{-1}(A_m A_m^T) B$:

$$\begin{aligned}
&B\Psi_{m+1,m}^{-1}((A_m A_m^T)B \\
&= B(-I + \tfrac{p+1}{2} p^m S)B \\
&= -B^2 + \tfrac{p+1}{2} p^m BSB \\
&\equiv -(I + p^m S) + p^m (I + \tfrac{p+1}{2} p^m S) S (I + \tfrac{p+1}{2} p^m S) \pmod{(p^{m+1})} \\
&\equiv -I \pmod{(p^{m+1})}
\end{aligned}$$

as desired. □

EXAMPLE 3.3. Consider the code $\mathcal{C}_4$ over $GR(3^2, 2)$ from [**7**], with $w^2 = 7w + 7$, generated by $(1, 0, 1, 4), (5w+4, w+6, 1, w+1)$. Applying again the iterative method (2.1) to $\mathcal{C}_4$, with

$$G_0 = \begin{bmatrix} 1 & 0 & 1 & 4 \\ 5w+4 & w+6 & 1 & w+1 \end{bmatrix}$$

and parameters $\mathbf{x} = [0, 1, w+1, w+1]$ yields the code $\mathcal{C}_6$ over $GR(3^2, 2)$, generated by

$$\left[\begin{array}{cc|cccc} 1 & 0 & 0 & 1 & w+1 & w+1 \\ \hline 4w+4 & 4 & 1 & 0 & 1 & 4 \\ 7w+3 & 4w+2 & 5w+4 & w+6 & 1 & w+1 \end{array}\right].$$

In systematic form, the generator matrix becomes

$$\begin{bmatrix} 1 & 0 & 0 & 1 & w+1 & w+1 \\ 0 & 1 & 0 & 7w+4 & 5w+3 & w+5 \\ 0 & 0 & 1 & 4w+7 & 7w+2 & 5w+6 \end{bmatrix},$$

or, expressed in terms of ζ such that $\zeta^2 = -4\zeta + 1$ instead of w,

$$\begin{bmatrix} 1 & 0 & 0 & 1 & 5\zeta+1 & 5\zeta+1 \\ 0 & 1 & 0 & 8\zeta+4 & 7\zeta+3 & 5\zeta+5 \\ 0 & 0 & 1 & 2\zeta+7 & 8\zeta+2 & 7\zeta+6 \end{bmatrix}.$$

Projected on $GR(3,2) \simeq \mathbb{F}_{3^2}$, the generator matrix becomes

$$\begin{bmatrix} 1 & 0 & 0 & 1 & 2\bar{\zeta}+1 & 2\bar{\zeta}+1 \\ 0 & 1 & 0 & 2\bar{\zeta}+1 & \bar{\zeta} & 2\bar{\zeta}+2 \\ 0 & 0 & 1 & 2\bar{\zeta}+1 & 2\bar{\zeta}+2 & \bar{\zeta} \end{bmatrix}.$$

Conversely, from (3.2), we have

$$A_2 = \left(3I + 2\Psi_{2,1}^{-1}(A_1A_1^T) + 3K_1\right)\Psi_{2,1}^{-1}(A_1)$$

for K_1 a skew symmetric matrix. Take e.g.

$$K_1 = \begin{bmatrix} 0 & 6\zeta & 6\zeta \\ 3\zeta & 0 & 6\zeta \\ 3\zeta & 3\zeta & 0 \end{bmatrix}, \begin{bmatrix} 1 & 2\bar{\zeta}+1 & 2\bar{\zeta}+1 \\ 2\bar{\zeta}+1 & \bar{\zeta} & 2\bar{\zeta}+2 \\ 2\bar{\zeta}+1 & 2\bar{\zeta}+2 & \bar{\zeta} \end{bmatrix} \in \Psi_{2,1}^{-1}(A_1A_1^T)$$

to get

$$A_2 = \begin{bmatrix} 3\zeta+4 & 2\zeta+1 & 2\zeta+1 \\ 2\zeta+1 & \zeta+3 & 5\zeta+2 \\ 2\zeta+1 & 5\zeta+2 & \zeta+3 \end{bmatrix} \equiv \begin{bmatrix} 1 & 5\zeta+1 & 5\zeta+1 \\ 8\zeta+4 & 7\zeta+3 & 5\zeta+5 \\ 2\zeta+7 & 8\zeta+2 & 7\zeta+6 \end{bmatrix} \pmod{(3)}.$$

Recall that the Hamming distance between two vectors count the number of coordinates in which they differ, the Hamming weight of a vector counts the number of its non-zero coordinates, and that the minimum Hamming distance of a code is the smallest Hamming distance between distinct codewords in the code.

DEFINITION 3.4. If R is finite, $\mathcal{C}$ is said to be MDS (maximum distance separable) if its minimum Hamming distance $d_H(\mathcal{C})$ satisfies

$$d_H(\mathcal{C}) = n - \log_{|R|}(|\mathcal{C}|) + 1.$$

The next result is a slight variation of [**4**, Theorem 2.5.].

PROPOSITION 3.5. *Suppose that $\mathcal{C}_1$ is a code of type $(1)^k$ over $GR(p^{e_1},r)$ which lifts to a code $\mathcal{C}_2$ of type $(1)^k$ over $GR(p^{e_2},r)$. Then*

$$d_H(\mathcal{C}_1) = d_H(\mathcal{C}_2).$$

PROOF. We will prove that lifting such a code does not change the minimum Hamming distance, that is $d_H(\mathcal{C}_m) = d_H(\mathcal{C}_{m+1}) = d$ for $\mathcal{C}_m, \mathcal{C}_{m+1}$ two type $(1)^k$ codes over $GR(p^m,r)$, $GR(p^{m+1},r)$ respectively, with parity check matrices H_m, H_{m+1}. First observe that for a codeword $\mathbf{v}$ of weight d in $\mathcal{C}_m$, then $p\mathbf{v}$ is a codeword of $\mathcal{C}_{m+1}$ of weight d, since multiplication by p does not introduce new terms which are 0, and $p\mathbf{v} \in \mathcal{C}_{m+1} \iff pH_{m+1}\mathbf{v} \equiv 0 \pmod{(p^{m+1})} \iff pH_m\mathbf{v} \equiv 0 \pmod{(p^{m+1})} \iff H_m\mathbf{v} \equiv 0 \pmod{(p^m)}$. Thus $d(\mathcal{C}_{m+1}) \leq d_H(\mathcal{C}_m)$. Suppose there exists $\mathbf{w} \in \mathcal{C}_{m+1}$ which is not $\mathbf{0}$ of weight strictly less than d. Then $\Psi_{m+1,m}(\mathbf{w})$ is a vector of weight strictly less than d in $\mathcal{C}_m$, which is only possible if $\Psi_{m+1,m}(\mathbf{w}) = \mathbf{0}$, that is $\mathbf{w} = p^m\mathbf{u}$. But then $p^m\mathbf{u} \in \mathcal{C}_{m+1} \iff p^m H_{m+1}\mathbf{u} \equiv 0 \pmod{(p^{m+1})} \iff p^{m-1}H_m\mathbf{v} \equiv 0 \pmod{(p^m)}$ yielding a codeword in $\mathcal{C}_m$, of weight strictly less than d, a contradiction. □

EXAMPLE 3.6. A code over $\mathbb{F}_3$ given by $[I_2|A]$ where

$$A = \begin{bmatrix} x_1 & x_2 \\ x_2 & -x_1 \end{bmatrix}, x_1, x_2 \in \{1,2\}$$

is self-dual, and it is clear that its parity check matrix satisfies that there are three of its columns which are linearly dependent (say column 1, 2 and 3), but no two columns are linearly dependent so its minimum Hamming distance is 3. As a result, codes $\mathcal{C}_{4,1}, \ldots, \mathcal{C}_{4,10}$ over $R = GR(3^2, 2)$ from Proposition 2.3 all have minimum Hamming distance of 3. These codes are also MDS since

$$3 = n - \log_{|R|}(|\mathcal{C}|) = 4 - \log_{81}(81^2) + 1.$$

We next give an example of construction of lift of a cyclic code as per [**8**], and revisit it with the current approach.

EXAMPLE 3.7. From Example 2.2, recall that

$$GR(3^2, 2) \simeq \mathbb{Z}_{3^2}[w] \simeq \mathbb{Z}_{3^2}[\zeta],$$

for w such that $w^2 = 7w + 7$ and ζ an 8th root of unity such that $\zeta^2 = -4\zeta + 1$, and the isomorphism sends w to $\zeta^2 - 1 = -4\zeta$.

Consider the projection $GR(3^2, 2) \to GR(3, 2) \simeq \mathbb{F}_9 \simeq \mathbb{F}_3(\bar{\zeta})$, where $\bar{\zeta}$ satisfies $\bar{\zeta}^2 = -4\bar{\zeta} + 1 = -\bar{\zeta} + 1$ and is a primitive 8th root of unity.

In $\mathbb{F}_9[x] \simeq \mathbb{F}_3(\bar{\zeta})[x]$, where $\bar{\zeta}^4 = -1$,

$$x^4 - 1 = (x-1)(x+1)(x^2+1), x^2 + 1 = (x - \bar{\zeta}^2)(x + \bar{\zeta}^2) = x^2 - \bar{\zeta}^4.$$

and in $GR(3^2, 2)[x]$, this factorization is lifted as

$$x^4 - 1 = (x-1)(x+1)(x^2+1), x^2 + 1 = (x - \zeta^2)(x + \zeta^2) = x^2 - \zeta^4.$$

Accordingly, $\bar{g}(x) = (x-1)(x - \bar{\zeta}^2)$ generates a cyclic code $\mathcal{C}_1$ over $\mathbb{F}_9$, which lifts to a cyclic code $\mathcal{C}_2$ over $GR(3^2, 2)$, generated by $g(x) = (x-1)(x - \zeta^2)$.

From the view point of generator matrices, a generator matrix G_1 for $\mathcal{C}_1$ is given by

$$\begin{bmatrix} \bar{\zeta}^2 & -(\bar{\zeta}^2+1) & 1 & 0 \\ 0 & \bar{\zeta}^2 & -(\bar{\zeta}^2+1) & 1 \end{bmatrix}.$$

Similarly a generator matrix G_2 for the cyclic code $\mathcal{C}_2$ over $GR(3^2, 2)$ is

$$\begin{bmatrix} \zeta^2 & -(\zeta^2+1) & 1 & 0 \\ 0 & \zeta^2 & -(\zeta^2+1) & 1 \end{bmatrix}$$

and $\Psi_{2,1}(G_2) = G_1$.

The systematic form of G_1 is

$$\begin{bmatrix} 1 & 0 & \bar{\zeta} - 1 & -\bar{\zeta} \\ 0 & 1 & \bar{\zeta} + 1 & -\bar{\zeta} + 1 \end{bmatrix}.$$

Looking at the columns of the corresponding parity check matrix H_1 shows that $d_H(\mathcal{C}_1) = 3$, and correspondingly $d_H(\mathcal{C}_2) = 3$.

4. Concluding Remarks

In Definition 1.1, the rank of a code $\mathcal{C}$ is defined to be the number of nonzero rows of its generator matrix G in systematic form. In this context, e.g., [**3**, **4**], a code is said to be MDR (maximum distance separable with respect to rank) if its Hamming distance is $n-k+1$.

There is another definition of being maximum distance separable with respect to rank, which relates to the rank distance [**6**].

Definition 4.1. Let $x = (x_1, \dots, x_n)$ be a vector in $\mathbb{F}_{q^m}^n$. Its rank weight is the maximal number of linearly independent coordinates x_i over $\mathbb{F}_q$. For $\mathcal{C}$ an $[n, k]$ linear code over $\mathbb{F}_{q^m}$, its rank distance is the minimum over the rank weight of all its nonzero codewords.

A code is then MRD (maximum rank distance) if its rank distance is $n-k+1$.

A first open question would be to look at codes over Galois rings using the latter definition of rank.

We note that in [**4**], formulas are given to compute the weight enumerator for lifted codes over $\mathbb{Z}_{p^e}$. Generalizing these formulas to the lift of Galois rings would be a natural second open question.

References

[1] Gilberto Bini and Flaminio Flamini, *Finite commutative rings and their applications*, The Kluwer International Series in Engineering and Computer Science, vol. 680, Kluwer Academic Publishers, Boston, MA, 2002. With a foreword by Dieter Jungnickel, DOI 10.1007/978-1-4615-0957-8. MR1919698

[2] A. R. Calderbank and N. J. A. Sloane, *Modular and p-adic cyclic codes*, Des. Codes Cryptogr. **6** (1995), no. 1, 21–35, DOI 10.1007/BF01390768. MR1334198

[3] Steven T. Dougherty and Young Ho Park, *Codes over the p-adic integers*, Des. Codes Cryptogr. **39** (2006), no. 1, 65–80, DOI 10.1007/s10623-005-2542-x. MR2201384

[4] Steven T. Dougherty, Sun Young Kim, and Young Ho Park, *Lifted codes and their weight enumerators*, Discrete Math. **305** (2005), no. 1-3, 123–135, DOI 10.1016/j.disc.2005.08.004. MR2186686

[5] Steven T. Dougherty and Young Ho Park, *On modular cyclic codes*, Finite Fields Appl. **13** (2007), no. 1, 31–57, DOI 10.1016/j.ffa.2005.06.004. MR2284665

[6] È. M. Gabidulin, *Theory of codes with maximum rank distance* (Russian), Problemy Peredachi Informatsii **21** (1985), no. 1, 3–16. MR791529

[7] Jon-Lark Kim and Yoonjin Lee, *Construction of MDS self-dual codes over Galois rings*, Des. Codes Cryptogr. **45** (2007), no. 2, 247–258, DOI 10.1007/s10623-007-9117-y. MR2341887

[8] Young Ho Park, *The q-adic liftings of codes over finite fields*, Korean J. Math. **26** (2018), no. 3, 537–544, DOI 10.11568/kjm.2018.26.3.537. MR3865924

Nanyang Technological University, 639798 Singapore

Current address: School of Mathematics, the University of Birmingham, B15 2TT, United Kingdom

Email address: `f.e.oggier@bham.ac.uk`

Contemporary Mathematics
Volume **826**, 2025
https://doi.org/10.1090/conm/826/16586

Twisted group algebras of Abelian groups: a survey

A. Duarte and C. Polcino Milies

Abstract. We survey results on the structure of twisted group algebras of Abelian groups; in particular, we describe a method to fully determine their structure. We also consider how to extend the notion of essential idempotent to these algebras and give criteria for their existence.

1. Introduction

The theory of projective representations of finite groups over the complex number field was created and developed by I. Schur in [**37**, **40**, **41**]. For results in this area see [**12**, Section VII.51], [**8**] or [**23**]. The problem of determining all projective representations of a finite group G involves central extensions of certain groups and the study of group algebras of these kind of extensions leads very naturally to the concept a twisted group algebra. For an interesting exposition of the work of Schur, see [**11**, Section IV.5].

In Section 2 of this paper we state basic facts regarding twisted group algebras and include some commentaries on the origin of this concept. In Section 3 we recall the definition of essential idempotents in group algebras and state some of its applications. This definition of this kind of idempotents strongly depends on the idea of idempotent determined by a subgroup, which is widely used in the study of group algebras. In Section 4 we see how idempotents from subgroups can be defined in twisted group algebras of Abelian groups and we show how they give a method to fully describe the structure of these algebras. We also give an example to illustrate how to apply this method.

In Section 5 we give criteria for the existence of essential idempotents in twisted group algebras. Since these results have not been published elsewhere, proofs of our statements are included.

2. Basic facts

We introduce the definition of a twisted group algebra following the same line of ideas of Schur, as mentioned above.

Definition 2.1. Let N and G be two groups. A group Γ is called an **extension** of N by G if there exists a short exact sequence

$$1 \to N \xrightarrow{i} \Gamma \xrightarrow{\pi} G \to 1.$$

This work was supported by FAPESP, Proc. No 2020/16594-0.

If $i(N)$ is contained in the center of Γ then Γ is said to be a **central extension** of G by N.

Identifying N with its image $i(N) \subset \Gamma$ we have that $\Gamma/N \cong G$. Let $\overline{G} = \{\overline{g} \mid g \in G\}$ denote a set of representatives of cosets of G in Γ. Then, every element $x \in \Gamma$ can be written as $x = n\overline{g}$ with $n \in N$ and $\overline{g} \in \overline{G}$.

Consequently, given a commutative ring R, every element in the group algebra $R\Gamma$ can be written in the form $\alpha = \sum_{n,\overline{g}} a_{n\overline{g}} n\overline{g}$, with $a_{n\overline{g}} \in R$, hence also in the form $\alpha = \sum_{\overline{g}\in\overline{G}} \gamma_{\overline{g}}\overline{g}$, with $\gamma_{\overline{g}} \in RN$.

Thus, the group algebra $R\Gamma$ is very similar to the group algebra $(RN)G^*$ of the group $G^* = \Gamma/N$ with coefficients in the ring RN, except for the way elements are multiplied. In fact, since elements in RN are central, we have

$$(\gamma_{\overline{g}}\overline{g})(\delta_{\overline{h}}\overline{h}) = (\gamma_{\overline{g}}\delta_{\overline{h}})(\overline{g}\overline{h}).$$

The product $\overline{g}\overline{h}$ is certainly in the class of $\overline{gh}$ but the representative chosen may not be precisely this element, so we have

$$\overline{g}\overline{h} = t(g,h)\overline{gh} \quad \text{with} \quad t(g,h) \in N.$$

Inspired by this fact, we can formulate a more general concept.

DEFINITION 2.2. Let G be a group and R a commutative ring whose set of invertible elements we denote by $\mathcal{U}(R)$. A map $t : G \times G \to \mathcal{U}(R)$ is called a **twisting** or a **factor set** if, for $x, y, z \in G$ we have that

$$t(g,h).t(gh,\ell) = t(h,\ell).t(g,h\ell).$$

If also

$$t(g,1) = t(1,g) = 1,$$

the twisting is called **normalized**.

Consider a set of symbols $\overline{G} = \{\overline{g} \mid g \in G\}$. The **twisted group algebra** of G over R with twisting t, denoted R^tG, is the set of finite sums

$$R^tG = \Big\{\sum_{g\in G} a_g\overline{g} \mid a_g \in R\Big\}$$

where addition is defined componentwise and multiplication is given by the following rules

$$\bar{x}.\bar{y} = t(x,y)\overline{xy} \qquad \text{for all } x, y \in G,$$
$$\bar{x}\lambda = \lambda\bar{x} \qquad \text{for all } x \in G \text{ and } \lambda \in R,$$

extended linearly.

An element $g \in G$ is called **regular** if $t(g,h) = t(h,g)$ whenever $h \in C_G(g)$, the centralizer of g in G. We shall denote by G_0 the set of all regular elements of G, i.e.

$$G_0 = \{g \in G \mid t(g,h) = t(h,g),\ \forall h \in C_G(g)\}. \tag{1}$$

There is a close connection between factor sets and 2-cocycles as used in cohomology, actually both concepts coincide (see, for example [**21**, p. 83]).

Several results in this area can be proved via cohomological concepts or using projective representation theory, but in the present paper we shall use only classical ring theory.

As it is the case for group algebras, twisted group algebras can be characterized by a universal property.

PROPOSITION 2.3. *Let G be a group and R a commutative ring. Given a ring A containing R and a mapping $f : G \to A$ such that $f(g)f(h) = t(g,h)f(gh)$, there exists a ring homomorphism $f^* : R^tG \to A$ such that $f = f^* \circ i$, where $i : G \to R^tG$ is the map such that $i(g) = \overline{g}$, for all $g \in G$.*

The proof is basically the same as in the group algebra case [**36**, Proposition 3.2.7].

The next elementary result is of interest to establish a connection to coding theory.

PROPOSITION 2.4. *Let $C = \langle g \rangle$ be a cyclic group of order n, R a commutative ring and λ an invertible element in R. Let $R^{t_\lambda}C$ be the corresponding twisted group algebra. Then*

$$R^{t_\lambda}C \cong \frac{R[X]}{(X^n - \lambda)}.$$

PROOF. Consider the linear map $\phi : R^{t_\lambda}C \to \frac{R[X]}{(X^n-\lambda)}$ such that $\phi(\overline{g^i}) = X^i, 0 \leq i \leq n-1$ (here, as usual, to simplify notations we write X^i to actually denote the class of this element in the factor ring).

If $i + j < n$ then $\overline{g^i}\ \overline{g^j} = \overline{g^{i+j}}$ while $X^iX^j = X^{i+j}$ and, if $i + j \geq n$, writing $i + j = n + r$ then $\overline{g^i g^j} = \lambda\overline{g^r}$ and it follows easily that also $X^iX^j = \lambda X^r$.

Hence, ϕ is a ring isomorphism. □

DEFINITION 2.5. Let λ be an element of a finite field $\mathbb{F}$. A linear code $\mathcal{C} \subset \mathbb{F}^n$ is called a **λ-constacyclic code** if for every vector $(a_0, a_1, \ldots, a_{n-2}, a_{n-1})$ in the code, we have that also the vector $(\lambda a_{n-1}, a_0, a_1, \ldots, a_{n-2})$ is in the code.

Given a polynomial $f \in \mathbb{F}[X]$ we shall denote by $[f]$ its equivalence class in the factor ring $R[X]/(X^n - \lambda)$. The map $\varphi : \mathcal{C} \to R[X]/(X^n - \lambda)$ given by

$$(a_0, a_1, \ldots, a_{n-2}, a_{n-1}) \in \mathbb{F}[X] \quad \mapsto \quad [a_0 + a_1X + \ldots + a_{n-2}X^{n-2} + a_{n-1}X^{n-1}],$$

is a linear isomorphism and it is easy to see that $\mathcal{C}$ is constacyclic if and only if $\varphi(\mathcal{C})$ is an ideal of $R[X]/(X^n - \lambda)$.

In view of Proposition 2.4 it is clear that a code is constacyclic if and only if it can be constructed as an ideal of a twisted group algebra. This approach has been taken in several recent works; see, for example, [**9**], [**10**], [**14**], [**22**], [**42**].

3. Essential idempotents in group algebras

It is well-known that a *cyclic code* can be realized as an ideal in the group algebra of a cyclic group. Inspired by this fact, S.D. Berman [**2**] in 1967 and MacWilliams [**29**] in 1970 introduced independently the notion of an Abelian code: one such code, over a field $\mathbb{F}$ is an ideal of the group algebra $\mathbb{F}A$ of a finite Abelian group A. It is then natural to further extend this definition. A **group code** over a field $\mathbb{F}$ is an ideal of the group algebra $\mathbb{F}G$ of a finite group G. Codes of this type have special algebraic properties which prove useful in decodification (see [**13**] and [**19**]) and many of the classical codes can be realized in this way, as shown in [**24**] and [**25**].

We shall restrict ourselves to the case when $char(\mathbb{F}) \not| \, |G|$ when, according to Maschke's Theorem [**36**, Theorem 3.4.7], the group algebra is semisimple. In this case, $\mathbb{F}G$ can be written as a (unique) direct sum of a finite number of two-sided ideals $\{A_i\}_{1\leq i\leq r}$, called the *simple components* of $\mathbb{F}G$, which are simple algebras, and also a direct sum (in more than one way) of minimal left ideals.

This implies that there exists a unique family $e_1, \ldots, e_r$ of orthogonal idempotents in $\mathbb{F}G$ which are central, such that none of them can be written as a sum of two non-zero central idempotents, and such that $\sum_{i=1}^{r} e_i = 1$. These are called the *primitive central idempotents* of $\mathbb{F}G$. Moreover, every two-side ideal I of $\mathbb{F}G$ is generated by an idempotent element which is a sum of some of the primitive idempotents, called its *constituents*.

There is a natural way to exhibit idempotents of $\mathbb{F}G$ constructed from subgroups of G. If H is a subgroup, then the element

$$\widehat{H} = \sum_{h\in H} h,$$

is an idempotent and it is central if and only if H is normal in G.

As shown in [**36**, Proposition 3.6.7] we have that $(\mathbb{F}G)(1 - \widehat{H}) = \Delta(G, H)$, the kernel of the natural projection $\pi : \mathbb{F}G \to \mathbb{F}[G/H]$ and it is easy to see that $(FG)\widehat{H} \cong \mathbb{F}[G/H]$.

If H is a normal subgroup of G, then $\widehat{H}$ is a central idempotent and, as such, a sum of its constituents. Let e be a primitive central idempotent of $\mathbb{F}G$. Then:

- If e is not a constituent of $\widehat{H}$ we have $e\widehat{H} = 0$.
- If e is a constituent of $\widehat{H}$ we have $e\widehat{H} = e$.

In this last case, we have that $(\mathbb{F}G)e \subset (\mathbb{F}G)\widehat{H}$ and thus the minimal code $(\mathbb{F}G)e$ is a repetition code. We are interested, of course, in codes which are not of this type.

DEFINITION 3.1. A primitive idempotent e in the group algebra $\mathbb{F}G$, is called an **essential idempotent** if $e\widehat{H} = 0$, for every subgroup $H \neq (1)$ in G. A minimal ideal of $\mathbb{F}G$ is called an **essential ideal** if it is generated by an essential idempotent and **non essential** otherwise.

Notice that, if e is a central idempotent, then the map $\pi : G \to Ge$, given by $\pi(g) = ge$ is a group epimorphism. We can use this map to characterize essential idempotents.

PROPOSITION 3.2. *Let $e \in \mathbb{F}G$ be a primitive central idempotent. Then e is essential if and only if the map $\pi : G \to Ge$, is a group isomorphism.*

As an immediate consequence, we have the following.

COROLLARY 3.3 ([**5**, Corollary 2.4]). *If G is Abelian and $\mathbb{F}G$ contains an essential idempotent, then G is cyclic. Hence, if G is Abelian but not cyclic, all the minimal ideals of $\mathbb{F}G$ give repetition codes*

On the other hand, it can be shown that group algebras of cyclic codes always contain essential idempotents.

THEOREM 3.4 ([**18**, Theorem 9]). *Let $C_n = \langle g\rangle$ be cyclic group of order n a and let e_0 be the sum of all essential idempotents in $\mathbb{F}_qC_n$ (where $\mathbb{F}_q$ denotes the field with q elements). Then:*

(i) *$dim(F_qC_n)e_0 = \varphi(n)$ where φ denotes Euler's Totient function.*

(ii) *There exist precisely* $\varphi(n)/m$ *essential idempotents in* F_qC_n.

This kind of idempotents can be used to prove the following.

THEOREM 3.5 ([**5**, Theorem 3.3]). *Every minimal ideal in the semisimple group algebra* $\mathbb{F}A$ *of a finite Abelian group* A *is permutation equivalent to a minimal ideal in the group algebra* $\mathbb{F}C$ *of a cyclic group* C *of the same order.*

Results similar to those above have been obtained in the case of group algebras of finite nilpotent groups [**17**].

Essential idempotents play a role in the generation of binary simplex codes and also of codes of constant weight (see [**5**] and [**6**]).

4. Structure of twisted group algebras of Abelian groups

The number of simple components of a twisted group algebra, in the semisimple case, was computed by W.R. Reynolds [**38**], [**39**]. Results on radicals of twisted group algebras were given by D.S Passman [**34**].

Explicit descriptions of idempotents in special cases were given by G. Karpilovsky [**23**, Chapter1], P. Grover and A.K. Bhandari [**20**], T.Zh. Mollov [**30**] and A.J.van Zanten [**44**].

As seen in the previous section, idempotents of the form $\widehat{H} = \sum_{h\in H} h$ where H is a subgroup of a group G are very useful for the study of group algebras. Due to the special character of the multiplication in a twisted group algebra it can be readily seen that elements of this type are not idempotents in this structure.

We begin by extending this idea to the case when the subgroup is cyclic, which will suffice for our study. The proofs of the results quoted in this section can be found in [**16**].

We shall need some well-known facts.

Let $C = \langle g\rangle$ be a cyclic group of order n and let λ be an invertible element in R. Then, the map $t_\lambda : C\times C \to \mathcal{U}(R)$ given by

$$t_\lambda(g^i, g^j) = \begin{cases} 1 & \text{if } i+j<n,\\ \lambda & \text{if } i+j\geq n.\end{cases} \tag{2}$$

is a twisting.

THEOREM 4.1. *Let* $C = \langle g\rangle$ *be a cyclic group of order* n *and let* R^tC *be its twisted group algebra over a commutative ring* R. *Set*

$$\lambda = \prod_{\ell=1}^{n-1} t(g, g^\ell). \tag{3}$$

Then $R^tC \cong R^{t_\lambda}C$ *where* t_λ *is as above.*

A proof using ideas from cohomology is given in [**21**, p. 80]. A direct proof can be seen in [**16**, Theorem 2.1], which also shows that R^tC and $R^{t_\lambda}C$ are actually the same as sets, with the same operations, though constructed from different bases.

Since it is easily seen that $R^{t_\lambda}C$ is commutative the result above gives the following.

COROLLARY 4.2. *The twisted group algebra of a cyclic group over a commutative ring is commutative.*

Twistings for Abelian groups can be exibited in a similar way.

Given a finite Abelian group A, written as a direct product $A = C_{m_1} \times \cdots \times C_{m_s}$, where $C_{m_i} = \langle g_i \rangle$ is cyclic of order m_i, and invertible elements $\lambda_i \in R$, $1 \le i \le s$, set

$$t_{\lambda_i}(g_i^j, g_i^k) = \begin{cases} 1, & \text{for } j + k < m_i, \\ \lambda_i, & \text{for } j + k \ge m_i, \end{cases}$$

which is a twisting of $C_{m_i} = \langle g_i \rangle$ over R.

We denote by t_Λ the twisting of A defined as follows. Given $a = g_1^{i_1} \cdots g_s^{i_s}$, $b = g_1^{j_1} \cdots g_s^{j_s} \in A$ we set:

$$t_\Lambda(a,b) = t_\Lambda(g_1^{i_1} \cdots g_s^{i_s}, g_1^{j_1} \cdots g_s^{j_s}) = \prod_{k=1}^{s} t_{\lambda_k}(g_k^{i_k}, g_k^{j_k}). \tag{4}$$

where $\Lambda = (\lambda_1, \ldots, \lambda_s)$.

THEOREM 4.3. *Let t be a twisting of A over $\mathbb{F}$ such that $R^t A$ is commutative. Then, there exists a twisting t_Λ as defined in (4) such that $R^t A = R^{t_\Lambda} A$.*

Conversely, a twisted group algebra of the form $R^{t_\Lambda} A$ is commutative.

We are now ready to introduce subgroup idempotents. We remark that, in the present case, they depend not only on the given subgroup but also on certain elements in a field of coefficients.

PROPOSITION 4.4. *Let $C = \langle g \rangle$ be a cyclic group of order n and $t = t_\lambda$, with λ in a field $\mathbb{F}$, a twisting of C over $\mathbb{F}$. Given a root α of $X^n - \lambda$ in field $\mathbb{K}$ containing $\mathbb{F}$, we set*

$$\widehat{C}_\alpha = \frac{1}{n} \sum_{j=0}^{n-1} \alpha^{-j} \bar{g}^j.$$

Then, $\widehat{C}_\alpha$ is an idempotent of the twisted group algebra $\mathbb{K}^{t_\lambda} C$.

Moreover, if $\beta \ne \alpha$ is another root of $X^n - \lambda$, then $\widehat{C}_\alpha \widehat{C}_\beta = 0$.

All the primitive idempotents of a twisted group algebra of a cyclic group can be obtained in this way.

LEMMA 4.5. *Let $\mathbb{K}^t C$ be the twisted group algebra of a cyclic group $C = \langle g \rangle$, of order n, and $\mathbb{K}$ a splitting field of the polynomial $X^n - \lambda$ over $\mathbb{F}$ such that $char(K) \nmid |G|$ where λ is as given in formula (2). Let $\{\alpha_i\}_{1 \le i \le n}$ be the set of all roots of the polynomial $X^n - \lambda$ in $\mathbb{K}$. Then*

$$\{\widehat{C}_{\alpha_i} \mid 1 \le i \le n\}, \tag{5}$$

is the set of all primitive idempotents of $\mathbb{F}^t C$.

As before, this result can be extended to finite Abelian groups.

THEOREM 4.6. *Let A be a finite Abelian group written as a direct product $A = C_{m_1} \times \cdots \times C_{m_s}$, where $C_{m_i} = \langle g_i \rangle$ is cyclic of order m_i, and $\mathbb{F}$ a finite field. Assume that the twisted group algebra $\mathbb{F}^{t_\Lambda} A$ is endowed with a twisting t_Λ as defined in (4), with $\lambda_i \in \mathbb{F}$, $1 \le i \le s$.*

Let $\mathbb{K}$ be the splitting field of the polynomial $f = \prod_{i=1}^{t} (X^{m_i} - \lambda_i)$; let $\mathcal{R}_i = \{\alpha_{ij} \mid 1 \le j \le m_i\}$ be the set of all roots of the polynomial $X^{m_i} - \lambda_i$, in $\mathbb{K}$, $1 \le$

$i \leq m_i$, and set $\mathcal{R} = \prod_{i=1}^{s} \mathcal{R}_i$. For each subset of roots $\alpha = (\alpha_{1j_1}, \ldots, \alpha_{sj_s}) \in \mathcal{R}$, we set:

$$e_\alpha = \widehat{(C_{m_1})}_{\alpha_{1j_1}} \cdots \widehat{(C_{m_s})}_{\alpha_{sj_s}}.$$

Then

$$\{e_\alpha \mid \alpha \in \mathcal{R}\},$$

is the set of primitive idempotents of $\mathbb{K}^{t_\Lambda} A$.

The case of a finite Abelian group can be reduced to the previous one due to a simple remark.

LEMMA 4.7. *Let A be a finite Abelian group, R a commutative ring, $R^t A$ the corresponding twisted group algebra and $A_0 = \{a \in A \mid t(a,h) = t(h,a), \ \forall h \in A\}$ the set of regular elements of A. supp($\widehat{H}_\beta$) $= H \subset G_0$, by Lemma 4.2 Then, the center of $\mathbb{F}^t A$ is the twisted group algebra*

$$\mathcal{Z}(\mathbb{F}^t A) \ = \ \mathbb{F}^{t_0} A_0.$$

PROOF. It is easy to see that $\mathbb{F}^{t_0} A_0 \subseteq \mathcal{Z}(\mathbb{F}^t A)$. Conversely, if $\gamma = \sum_{a \in A} x_a \overline{a} \in \mathcal{Z}(\mathbb{F}^t A)$ and $x_b \neq 0$ for some $b \in A$, then $t(b,a) = t(a,b)$ for all $a \in A$, which implies that $\mathcal{Z}(\mathbb{F}^t A) \subseteq \mathbb{F}^{t_0} A_0$, as required. □

Since the central primitive idempotents of a semisimple algebra are also the primitive idempotents of its center, to find the primitive idempotents of a twisted group algebra of the form $\mathbb{K}^t A$, using the lemma above, all we need is to determine A_0, its set of regular elements and then use Theorem 4.6.

Finally, the idempotents of the original twisted group algebra $\mathbb{F}^t A$ can be obtained via the process known as Galois descent.

With the notations above, let $\mathcal{G} = Gal(\mathbb{K}, \mathbb{F})$ be the Galois group of $\mathbb{K}$ over $\mathbb{F}$. For $\sigma \in \mathcal{G}$ and $\alpha = (\alpha_{1i_1}, \ldots, \alpha_{si_s}) \in \mathcal{R}$, we set

$$\sigma \cdot \alpha = (\alpha_{1i_1}^\sigma, \ldots, \alpha_{si_s}^\sigma).$$

Thus, $\mathcal{G}$ acts on $\mathcal{R}$.

We also define an action of $\mathcal{G}$ on $\mathbb{K}^{t_\Lambda} A$ setting:

$$\sigma\Big(\sum_{a \in A} b_a \overline{a}\Big) = \sum_{a \in A} b_a^\sigma \overline{a} \quad \sigma \in \mathcal{G}.$$

If $\sigma \in \mathcal{G}$, then

$$\begin{aligned} \sigma(e_\alpha) &= \sigma \cdot \Big(\widehat{(C_{m_1})}_{\alpha_{1i_1}} \cdots \widehat{(C_{m_s})}_{\alpha_{si_s}}\Big) \\ &= \sigma\widehat{(C_{m_1})}_{\alpha_{1i_1}} \cdots \sigma\widehat{(C_{m_s})}_{\alpha_{si_s}} \\ &= \widehat{(C_{m_1})}_{\alpha_{1i_1}^\sigma} \cdots \widehat{(C_{m_s})}_{\alpha_{si_s}^\sigma} \\ &= e_{\sigma \cdot \alpha}. \end{aligned}$$

Hence, for $\alpha \in \mathcal{R}$ and $\sigma \in \mathcal{G}$, we get $\sigma(e_\alpha) = e_{\sigma\cdot\alpha}$. Since $e_{\sigma\cdot\alpha}$ is also a primitive central idempotent, we have $\mathcal{G}$ acting on $\{e_\alpha \,|\, \alpha \in \mathcal{R}\}$. We denote

$$S_\alpha = \{\sigma \cdot \alpha \,|\, \sigma \in \mathcal{G}\}.$$

PROPOSITION 4.8. *If $\alpha \in \mathcal{R}$, then $\widetilde{e_\alpha} = \sum_{\beta \in S_\alpha} e_\beta$ is a primitive idempotent of $\mathbb{F}^{t_\Lambda} A$. In addition, every primitive idempotent of $\mathbb{F}^{t_\Lambda} A$ is of form $\widetilde{e_\alpha}$ for some $\alpha \in \mathcal{R}$.*

Let

$$St(\alpha) = \{\sigma \in \mathcal{G} \mid \sigma \cdot \alpha = \alpha\},$$

denote the *stabilizer* of α in $\mathcal{G}$.

For an idempotent $e_\alpha = \sum_{a \in A} k_a a$ we can obtain the expression of the corresponding $\widetilde{e_\alpha}$ in $\mathbb{F}^t A$ computing:

$$\begin{aligned}\widetilde{e_\alpha} &= \frac{1}{|St(\alpha)|}\sum_{\sigma\in\mathcal{G}}\Big(\sum_{a\in A} k_a^\sigma \overline{a}\Big)\\ &= \frac{1}{|St(\alpha)|}\sum_{a\in A}\Big(\sum_{\sigma\in\mathcal{G}} k_a^\sigma\Big)\overline{a}\\ &= \frac{1}{|St(\alpha)|}\sum_{a\in A} tr_{\mathbb{K}|\mathbb{F}}(k_a)\overline{a}.\end{aligned}$$

It is also possible to fully describe the simple components of this algebra.

Let $\mathcal{R}$ be as constructed in Theorem 4.6 and take $\widetilde{e_{\alpha_i}}$, for some $\alpha_i \in \mathcal{R}$, a primitive idempotent of $\mathbb{F}\,^t A$. Notice that, since $\mathbb{F}\ tA_0$ is a commutative algebra, its simple component $\mathbb{F}\ tA_0\widetilde{e_{\alpha_i}}$ is a field.

THEOREM 4.9. *With the notation above, we have:*
(i) $\mathbb{F}^t A\, e_i \cong M_d(\mathbb{F}^t A_0\, e_i)$, *where* $d = \sqrt{[A : A_0]}$.
(ii) $\mathbb{F}^t A_0\, \widetilde{e_{\alpha_i}} \cong \mathbb{F}(\alpha_i)$ *and* $[\mathbb{F}^t A_0\, \widetilde{e_{\alpha_i}} : \mathbb{F}] = |S_\alpha|$.

An example.

Let $\mathbb{F} = \mathbb{F}_5$ be the finite field with 5 elements and set $A = \langle x\rangle \times \langle y\rangle \times \langle z\rangle$ where x and y are of order 2 and z is of order 3.

We define $t : A \times A \to \mathcal{U}(\mathbb{F}_5)$ as follows:

$$t(x^{i_1}y^{j_1}z^{k_1}, x^{i_2}y^{j_2}z^{k_2}) = (-1)^{j_1 i_2}\, t_\lambda(z^{k_1}, z^{k_2}).$$

where t_λ is as in equation (2) and we take $\lambda = -1$.

Since $o(x) = o(y) = 2$ and the multiplicative order of -1 in $\mathcal{U}(\mathbb{F}_5)$ is also equal to 2, it follows immediately that the map $(x_{i_1} j^{j_1}, x^{i_2}y^{j_2}) \mapsto (-1)^{i_2 j_1}$ is a twisting for $\langle x\rangle \times \langle y\rangle$ and clearly t_λ is a twisting for $\langle z\rangle$, so the map t is a twisting of A over $\mathbb{F}$.

Notice that an element $a = x^{i_1}y^{j_1}z^{k_1}$ is in A_0 if and only if $t(a,b) = t(b,a)$ for all $b = x^{i_2}y^{j_2}z^{k_2} \in A$. Using the formula for t we get $(-1)^{j_1 i_2} t_\lambda(z^{k_1}, z^{k_2} = (-1)^{j_2 i_1} t_\lambda(z^{k_2}, z^{k_1})$ hence $(-1)^{j_1 i_2} = (-1)^{j_2 i_1}$, so we must have $i_1 = j_1 = 0$. Thus $A_0 = \langle z\rangle$.

Let $\mathbb{K}$ be a splitting field for $X^3 + 1 = (X+1)(X^2 + 4X + 1)$ over $\mathbb{F}$. If α is one of the roots of $(X^2 + 4X + 1)$ the the other is $\beta = 1 - \alpha$ and, according to Proposition 4.4 the idempotents of $\mathbb{K}\,^t A_0$, which are also the idempotents of $\mathbb{K}\,^t A$ are:

$$\begin{aligned}e_{-1} &= 2\,(\overline{1} - \overline{z} + \overline{z^2}),\\ e_\alpha &= 2\,(\overline{1} + (1+\alpha)\overline{z} - \alpha\overline{z^2}),\\ e_{1-\alpha} &= 2\,(\overline{1} + \alpha\overline{z} + (1-\alpha)\overline{z^2}),\end{aligned}$$

Computing stabilizers we get $St(-1) = \mathcal{G}$, $St(\alpha) = \{id\} = St(1+\alpha)$. Then, the central primitive idempotents of $\mathbb{F}^t A$:

$$\widetilde{e_{-1}} = 2\,(\overline{1} - \overline{z} + \overline{z^2}),$$

and

$$\begin{aligned}\widetilde{e_\alpha} &= \frac{2}{|St(\alpha)|}\left(Tr_{\mathbb{K}/\mathbb{F}_5}(1)\overline{1} + Tr_{\mathbb{K}/\mathbb{F}_5}(1+\alpha)\overline{z} - Tr_{\mathbb{K}/\mathbb{F}_5}(\alpha)\overline{z^2}\right)\\ &= 2\left(2\overline{1} + (1-\alpha+\alpha)\overline{z} - (\alpha+1-\alpha)\overline{z^2}\right)\\ &= -\overline{1} + 2\overline{z} + 3\overline{z^2}.\end{aligned}$$

Finally, the simple components of the center are

$$\mathbb{F}_5^{t-1}\langle z\rangle\widetilde{e_{-1}} \cong \mathbb{F}_5 \quad \text{and} \quad \mathbb{F}_5^{t-1}\langle z\rangle\widetilde{e_\alpha} \cong \mathbb{F}_{25}.$$

Since $[A : A_0] = 4$, the simple components of $\mathbb{F}_5^t A$ are:

$$\mathbb{F}_5^t A\widetilde{e_{-1}} \cong M_2(\mathbb{F}_5) \quad \text{and} \quad \mathbb{F}_5^t A\widetilde{e_\alpha} \cong M_2(\mathbb{F}_{25}).$$

5. Essential idempotents

We now consider the concept of essential idempotent in the context of twisted group algebras of Abelian groups following [**15**]. To do so, we need to extend the construction of idempotents given in Section 4.

PROPOSITION 5.1. *Let t be a twisting of an Abelian group A over a field $\mathbb{F}$ and let A_0 be the set of t-regular elements of A. Let H be a subgroup of A_0 and assume that for each element $h \in H$ there exists $\beta_h \in \mathbb{F}^*$, such that $t(h,k) = \beta_h\beta_k\beta_{hk}^{-1}$, for all $h, k \in H$. Then*

$$\widehat{H}_\beta = \frac{1}{|H|}\sum_{h\in H}\beta_h^{-1}\overline{h}$$

is a central idempotent of $\mathbb{F}^{\,t}A$.

PROOF. Since $supp(\widehat{H}_\beta) = H \subset A_0$, it follows from Lemma 4.7 that $\widehat{H}$ is central.

For an element $k \in H$ we compute:

$$\begin{aligned}\overline{k}\widehat{H}_\beta &= \frac{\overline{k}}{|H|}\sum_{h\in H}\beta_h^{-1}\,\overline{h} = \sum_{h\in H}\beta_h^{-1}\,t(k,h)\,\overline{kh}\\ &= \sum_{h\in H}\beta_k\beta_{kh}^{-1}\,\overline{kh} = \beta_k\widehat{H}_\beta.\end{aligned}$$

Thus

$$\widehat{H}_\beta^2 = \frac{1}{|H|}\sum_{k\in H}\beta_k^{-1}\overline{k}\widehat{H}_\beta = \frac{1}{|H|}\sum_{k\in H}\beta_k^{-1}\beta_k\widehat{H}_\beta = \widehat{H}_\beta$$

showing that $\widehat{H}_\beta$ is an idempotent. □

DEFINITION 5.2. With the notations above, let H be a subgroup of A_0 and $\beta : H \to \mathbb{F}^*$ a map denoted as $\beta(h) = \beta_h$, for all $h \in H$. We say that the (H, β) is ***t*-essential pair**, if $t(h,k) = \beta_h\beta_k\beta_{hk}^{-1}$, for all $h, k \in H$.

Suppose that (H, β) is a t-essential pair and let τ be a left transversal of H in G. Set

$$\mathcal{B} = \{\overline{g}\widehat{H}_\beta \,|\, g \in \tau\} \;\subset\; I = \mathbb{F}^{\,t}G\widehat{H}_\beta.$$

Given $y \in A$, we can write $y = gh$ with $g \in \tau$ and $h \in H$ so

$$\overline{y}\widehat{H}_\beta = t(g,h)^{-1}\overline{g}\overline{h}\widehat{H}_\beta = [t(g,h)^{-1}\beta_h]\overline{g}\widehat{H}_\beta.$$

This shows that $\mathcal{B} = \{\bar{g}\widehat{H}_\beta \mid g \in \tau\}$ is a set of generators of the ideal $I = \mathbb{F}^tG\widehat{H}_\beta$.

As the elements of $\mathcal{B}$ have disjoint support they also are linearly independent. So, we have proved the following.

PROPOSITION 5.3. *Let* (H,β) *be a* t*-essential pair and let* τ *be a left transversal of* H *in* G. *Then,*

$$\mathcal{B} = \{\bar{g}\widehat{H}_\beta \mid g \in \tau\}$$

is a basis of $\mathbb{F}^tG\,\widehat{H}_\beta$ *as a vector space over* $\mathbb{F}$.

An element $\alpha \in \mathbb{F}^tG\,\widehat{H}_\beta$ is thus of the form

$$\alpha = \sum_{g\in\tau} k_g\bar{g}\widehat{H}_\beta = \frac{1}{|H|}\sum_{g\in\tau}\sum_{h\in H} k_g\beta_h^{-1}\bar{g}\bar{h} = \frac{1}{|H|}\sum_{g\in\tau}\sum_{h\in H} k_g\beta_g\beta_{gh}^{-1}\bar{g}\bar{h}$$

So, if we consider the map $T : \mathbb{F}^tG \to \mathbb{F}^tG$ defined on the elements of the basis $\overline{G}$ of $\mathbb{F}^tG$ by $T(\overline{g}) = \beta_g\beta_{gh}^{-1}$ then

$$\alpha = T\left(\frac{1}{|H|}\sum_{g\in\tau}\sum_{h\in H} k_g\overline{gh}\right).$$

Since T is a monomial transformation, this shows that the code defined by $\mathbb{F}^tG\,\widehat{H}_\beta$ is equivalent to a repetition code.

As in the case of group algebras, we have that if $e \in \mathbb{F}^tG$ is a primitive central idempotent, then

- If e is not a constituent of $\widehat{H}_\beta$ we have $e\widehat{H}_\beta = 0$.
- If e is a constituent of $\widehat{H}_\beta$ we have $e\widehat{H}_\beta = e$.

To avoid minimal ideals which define codes equivalent to repetition codes we introduce the following.

DEFINITION 5.4. Let $e \in \mathbb{F}^tA$ be a primitive central idempotent. We say that e is an **essential idempotent** if $e\widehat{H}_\beta = 0$, for all t-essential pair (H,β) with $H \neq \{1\}$.

A minimal ideal of $\mathbb{F}^tG$ is called an **essential ideal** if it is generated by an essential idempotent and **non essential** otherwise.

DEFINITION 5.5. Let e be a primitive central idempotent of $\mathbb{F}^tA$. The subgroup

$$K_e = \{g \in A \mid \bar{g}\,e = \beta_g\,e, \text{ for some } \beta_g \in \mathbb{F}\} \tag{6}$$

of A is called the **kernel** of e.

A straightforward computation shows that $K_e \subset A_0$.

Also, if $g,h \in K_e$, then

$$\bar{g}\bar{h}e = \bar{g}(\bar{h}e) = \bar{g}(\beta_h e) = \beta_h\bar{g}e = \beta_g\beta_h e,$$
$$\bar{g}\bar{h}e = t(g,h)\,\overline{gh}\,e = t(g,h)\beta_{gh}e,$$

which implies

$$\beta_g\beta_h = t(g,h)\beta_{gh},$$

whenever $g,h \in K$. This shows that $t(g,h) = \beta_g\beta_h\beta_{gh}^{-1}$, for all $g,h \in K$. and we are done. This argument shows that (K_e,β) is an t-essential pair.

Thus, we can consider the idempotent

$$\widehat{K_{e\beta}} = \frac{1}{|K_e|} \sum_{k \in K_e} \beta_k^{-1} \bar{k} \in \mathbb{F}^t A \tag{7}$$

Note that $\widehat{K_{e\beta}}\, e = \frac{1}{|K_e|} \sum_{k \in K_e} \beta_k^{-1} \bar{k} e = \frac{1}{|K_e|} \sum_{k \in K_e} \beta_k^{-1} \beta_k e = e$.

THEOREM 5.6. *A primitive central idempotent $e \in \mathbb{F}^t G$ is an essential idempotent if, and only if, $K_e = \{1\}$.*

PROOF. Assume that $K = K_e \neq \{1\}$. As shown above, (K, β) is a t-essential pair. Then $e\widehat{K}_\beta = e$, which implies that e is not an essential idempotent.

Conversely, assume that $K_e = \{1\}$, and suppose that e is not an essential idempotent. Then, there exists a t-essential pair (H, β), $H \neq \{1\}$, such that $e\widehat{H}_\beta \neq 0$. Since e is a primitive central idempotent and $\widehat{H}_\beta$ is central, we must have $e\widehat{H}_\beta = e$. But this implies that for all $h \in H$ we have

$$\bar{h}e = \bar{h}e\widehat{H}_\beta = e\bar{h}\widehat{H}_\beta = e\beta_h\widehat{H}_\beta = \beta_h e,$$

so $h \in K_e$, for all $h \in H$, a contradiction. Hence e is essential. □

Set $\Gamma_e = \{a\bar{g}e \,|\, a \in \mathbb{F}^*, g \in A_0\}$ and denote $\mathcal{U}(\mathbb{F}^t A)$ the group of units of $\mathbb{F}^t A$. Notice that Γ_e is a central multiplicative subgroup of $\mathcal{U}(\mathbb{F}^t A)$ containing the subgroup $\mathbb{F}^* e$. We define

$$\pi : A_0 \to \frac{\Gamma_e}{\mathbb{F}^* e} \quad \text{by } \pi(g) = [\bar{g}e], \tag{8}$$

where $[\bar{g}e]$ denotes the class of $\bar{g}e$ in $\Gamma_e/\mathbb{F}^* e$.

PROPOSITION 5.7. *A primitive central idempotent $e \in \mathbb{F}^t A$ is essential if, and only if π is a group isomorphism.*

PROOF. Since $supp(e) \in A_0$, we have $K_e \subset A_0$. Then, $g \in ker(\pi)$ if and only if $[\bar{g}e] = [e]$ and so $\bar{g}e = \beta_g e$ for some $\beta_g \in \mathbb{F}^*$. It follows that $g \in ker(\pi)$ if and only if $g \in K_e$. □

COROLLARY 5.8. *Let A be a finite abelian group. If $\mathbb{F}^t G$ contains an essential idempotent, then A_0 is cyclic.*

PROOF. Since e is a primitive central idempotent of $\mathbb{F}^t A$ it is also a primitive idempotent of $\mathcal{Z}(\mathbb{F}^t A) = \mathbb{F}^{t_0} A_0$. Then, $\mathbb{F}^t A_0 e$ is a field. As Γ_e is a group contained in $(\mathbb{F}^{t_0} G_0 e)^*$, it follows that Γ_e is cyclic. By the Proposition above A_0 is an isomorphism to $\Gamma_e/\mathbb{F}^* e$ which is cyclic. Hence G_0 is cyclic. □

We now investigate when $\mathbb{F}^{t_0} A_0$ contains essential idempotents for A_0 cyclic. We shall need two easy lemmas.

LEMMA 5.9. *Let p be a prime integer with $gcd(p, q) = 1$. The polynomial $X^p - \lambda \in \mathbb{F}[X]$ has a root in $\mathbb{F}$ if and only if $X^p - \lambda$ is reducible in $\mathbb{F}[X]$.*

PROOF. The "if" part is clear. Conversely, the proof of [**26**, Theorem 3.75], shows that $X^p - \lambda \in \mathbb{F}[X]$ has a root in $\mathbb{F}$. □

LEMMA 5.10. *Let p be a prime integer with $gcd(p, q) = 1$ and assume that $X^p - \lambda \in \mathbb{F}[X]$ is a reducible polynomial. Then, either $X^p - \lambda$ has a unique root in $\mathbb{F}$ or it splits in $\mathbb{F}$ as a product of distinct linear factors.*

PROOF. Assume that $X^p - \lambda$ has two distinct roots a and b in $\mathbb{F}$. Then, $ab^{-1} \neq 1$ and $(X)^p - 1$ has a root in $\mathbb{F}$.

If θ is a primitive root of $X^p - 1$, then $X^p - \lambda = \prod_{i=0}^{p-1}(X - a\theta^i)$ splits in $\mathbb{F}$ into distinct linear factors. □

Let C be the cyclic group of order n and $t = t_\lambda$ for a non zero element λ of $\mathbb{F} = \mathbb{F}_q$ a twisting.

PROPOSITION 5.11. *Let p be a prime integer with $gcd(p, q) = 1$ such that p divides n. If $X^p - \lambda$ splits in $\mathbb{F}$ into distinct linear factors, then $\mathbb{F}^tC$ does not contain essential idempotents.*

PROOF. Let $C = \langle g \rangle$, set $H = \langle h \rangle$, where $h = g^{n/p}$ and write $X^p - \lambda = \prod_{i=1}^{p}(X - b_i)$, $b_i \in \mathbb{F}$. For each index $i = 1, \ldots, p$, set $\beta_i : H \to \mathbb{F}^*$, defined by $\beta_i(h) = b_i$. Note that, for all i, we have that (H, β_i) is an t-essential pair and $\{\widehat{H}_{\beta_i}\}$ is a complete set of primitive central idempotents of $\mathbb{F}^tC$.

Assume, by the way of contradiction, that e is a essential idempotent in $\mathbb{F}^tC$. Since (H, β_i) are t-essential pairs, we have $e\widehat{H}_{\beta_i} = 0$ for all $i = 1, \ldots, p$. But this implies

$$e = e \cdot 1 = e \cdot \sum_{i=1}^{p} \widehat{H}_{\beta_i} = \sum_{i=1}^{p} e\widehat{H}_{\beta_i} = 0,$$

a contradiction. □

Let A be an Abelian group such that its subgroup of t-regular elements $A_0 = \langle g \rangle$ is cyclic of order n_0. We know from Proposition 4.1 that $\mathbb{F}^tA_0$ is equal to $\mathbb{F}^{t_\lambda}A_0$ for some $\lambda \in \mathbb{F}^*$.

Since A_0 is cyclic, it can be written as a direct product $A_0 = C_1 \times \cdots \times C_r$ with $C_i = \langle g_i \rangle$ is a cyclic group of order $p_i^{n_i}$, where p_i is a prime integer and n_i ia a positive integer, $1 \leq i \leq r$.

Assume that each polynomial $X^{p_i} - \lambda$ which is reducible in $\mathbb{F}[X]$ contains a unique root in $\mathbb{F}$, $1 \leq i \leq r$ and set

$$\mathcal{E} = \{1 \leq i \leq r \mid X^{p_i} - \lambda \text{ is reducible in } \mathbb{F}[X]\}.$$

If $i \in \mathcal{E}$, then there exists a unique $b_i \in \mathbb{F}$ such that $b_i^{p_i} = \lambda$. In this case set $h_i = g_i^{p_i^{n_i-1}}$, $H_i = \langle h_i \rangle$ and $\beta_i(h_i^j) = b_i^j$.

Then, (H_i, β_i) is a t-essential pair for all $i \in \mathcal{E}$. Set

$$e_0 = \prod_{i \in \mathcal{E}} (1 - \widehat{H_{i\beta_i}}),$$

which is a central idempotent of $\mathbb{F}^tG$.

Since $\widehat{H_{i\beta_i}} = \frac{1}{p_i} \sum_{j=0}^{p-1} \beta_i^{-j} \overline{h_i}^j$, for all $i \in \mathcal{E}$, the coefficient of $\prod_{i \in \mathcal{E}} \overline{h_i}$ in the expression of e_0 is $(-1)^{|\mathcal{E}|} \prod_{i \in \mathcal{E}} \frac{\beta_i^{-1}}{p_i} \neq 0$. Hence, e_0 is nonzero.

PROPOSITION 5.12. *Assume that A_0 is cyclic group. Then, a primitive central idempotent $e \in \mathbb{F}^tA$ is essential if and only if $ee_0 = e$.*

PROOF. Assume that e is essential. Then, $e\widehat{H_{i\beta_i}} = 0$ for all $i \in \mathcal{E}$, which implies that $ee_0 = \prod_{i \in \mathcal{E}} e(1 - \widehat{H_{i\beta_i}}) = \prod_{i \in \mathcal{E}} e = e$.

Conversely, assume that $ee_0 = e$ but e is not essential. Then there exists some t-essential pair (H, β) such that $e\widehat{H}_\beta = e$. Write $H = \langle h \rangle$; then $\beta_{h^i}\beta_{h^i} =$

$t(h^i, h^j)\beta_{h^{i+j}}$, which implies that $\beta_h^{|H|} = \lambda$. For each p_i dividing $|H|$, we obtain $(\beta_h^{|H|/p_i})^{p_i} = \lambda$, so $\beta_h^{|H|/p_i} = b_i$ and thus $i \in \mathcal{E}$. Since $\widehat{H_{i\beta_i}}\widehat{H}_\beta = \widehat{H}_\beta$, we have $e\widehat{H_{i\beta_i}} = e$, which means that $e(1 - \widehat{H_{i\beta_i}}) = 0$ and so $ee_0 = 0$, a contradiction. □

Notice that the proposition above implies that e_0 is precisely the sum of all essential idempotents in $\mathbb{F}^tA$.

We are now ready to state the state following:

THEOREM 5.13 ([**15**]). *With the notation above the twisted group algebra $\mathbb{F}^tA$ contains an essential idempotent if and only if the following contiditions hold:*
(i) *A_0 is cyclic;*
(ii) *$|A_0|$ and $(q-1)/m$ are relatively prime and m is the multiplicative order of λ, where λ is such that $t|_{A_0\times A_0} = t_\lambda$.*

We finish this section with two examples.

EXAMPLE 5.14. Let $\mathbb{F}$ be the field $\mathbb{F}_7$, $A = C_8 = \langle g\rangle$ and let t be the twisting of A over $\mathbb{F}$ defined by

$$t(g^i, g^j) = \begin{cases} 1, & i+j < n \\ 3, & i+j \geq n. \end{cases}$$

As $X^8 - 3$ has no roots in $\mathbb{F}$, we see that there is no t-essential pair (H, β) with $H \neq \{1\}$, so every primitive central idempotent of $\mathbb{F}^{t_3}G$ is essential.

EXAMPLE 5.15. Consider the case when $\mathbb{F} = \mathbb{F}_3$, $A = C_2 \times C_4 = \langle x\rangle \times \langle y\rangle$ and set

$$t(x^iy^j, x^ky^\ell) = (-1)^{i\ell}.$$

Then, $A_0 = \langle y^2\rangle$ has order 2. Since 2 divides $|A_0|$ and the roots of $X^2 - 1$ are ± 1, we have that A_0 does not satisfy the condition in Theorem 5.13, so this algebra contains no essential idempotents.

References

[1] Elwyn R. Berlekamp, *Algebraic coding theory*, McGraw-Hill Book Co., New York-Toronto-London, 1968. MR238597

[2] S. D. Berman, *Semisimple cyclic and abelian codes II*, Kibernetika, **3** (1967), 17-23.

[3] Swati Bhardwaj and Madhu Raka, *Two dimensional constacyclic codes of arbitrary length over finite fields*, Indian J. Pure Appl. Math. **53** (2022), no. 1, 49–61, DOI 10.1007/s13226-021-00087-8. MR4387375

[4] Victor Bovdi, *Structure of normal twisted group rings*, Publ. Math. Debrecen **51** (1997), no. 3-4, 279–293, DOI 10.5486/pmd.1997.1873. MR1485224

[5] Gladys Chalom, Raul A. Ferraz, and C. Polcino Milies, *Essential idempotents and simplex codes*, J. Algebra Comb. Discrete Struct. Appl. **4** (2017), no. 2, 181–188, DOI 10.13069/jacodesmath.284931. MR3601349

[6] Gladys Chalom, Raul A. Ferraz, and César Polcino Milies, *Essential idempotents and codes of constant weight*, São Paulo J. Math. Sci. **11** (2017), no. 2, 253–260, DOI 10.1007/s40863-017-0074-8. MR3716690

[7] S. B. Conlon, *Twisted group algebras and their representations*, J. Austral. Math. Soc. **4** (1964), 152–173. MR168663

[8] Tania-Luminiţa Costache, *On irreducible projective representations of finite groups*, Surv. Math. Appl. **4** (2009), 191–214. MR2606502

[9] Javier de la Cruz and Ricardo Villanueva-Polanco, *Public key cryptography based on twisted dihedral group algebras*, Adv. Math. Commun. **18** (2024), no. 3, 857–877, DOI 10.3934/amc.2022031. MR4729290

[10] Javier De La Cruz and Wolfgang Willems, *Twisted group codes*, IEEE Trans. Inform. Theory **67** (2021), no. 8, 5178–5184, DOI 10.1109/TIT.2021.3089003. MR4306326

[11] Charles W. Curtis, *Pioneers of representation theory: Frobenius, Burnside, Schur, and Brauer*, History of Mathematics, vol. 15, American Mathematical Society, Providence, RI; London Mathematical Society, London, 1999, DOI 10.1090/hmath/015. MR1715145

[12] Charles W. Curtis and Irving Reiner, *Representation theory of finite groups and associative algebras*, Pure and Applied Mathematics, Vol. XI, Interscience Publishers (a division of John Wiley & Sons, Inc.), New York-London, 1962. MR144979

[13] N. Doan, S. A. Hashemi, M. Mondelli and W. J. Gross, *Decoding Reed-Muller Codes With Successive Codeword Permutations*, IEEE Transactions on Communications, **70**, 11 (2022), 7134-7145, DOI 10.1109/TCOMM.2022.3211101.

[14] S. T. Dougherty, Serap Şahinkaya, and Bahattin Yıldız, *Skew G-codes*, J. Algebra Appl. **22** (2023), no. 2, Paper No. 2350056, 19, DOI 10.1142/S0219498823500561. MR4541642

[15] André Duarte, *Projective essential idempotents*, IEEE Trans. Inform. Theory **70** (2024), no. 8, 5566–5572, DOI 10.1109/tit.2024.3395545. MR4773197

[16] André Duarte, Raul Antonio Ferraz, and César Polcino Milies, *Twisted group algebras of Abelian groups*, Finite Fields Appl. **95** (2024), Paper No. 102386, 14, DOI 10.1016/j.ffa.2024.102386. MR4708496

[17] A. Duarte, A. Pereira, and C. Polcino Milies, *Nilpotent group codes*, J. Algebra Appl. **23** (2024), no. 3, Paper No. 2450060, 14, DOI 10.1142/S0219498824500609. MR4688828

[18] Raul A. Ferraz and C. Polcino Milies, *Essential idempotents in group algebras and coding theory*, Indian J. Pure Appl. Math. **52** (2021), no. 3, 747–760, DOI 10.1007/s13226-021-00187-5. MR4356278

[19] G. David Forney Jr., *On decoding* BCH *codes*, IEEE Trans. Inform. Theory **IT-11** (1965), 549–557, DOI 10.1109/tit.1965.1053825. MR189923

[20] Pooja Grover and Ashwani K. Bhandari, *Explicit determination of certain minimal constabelian codes*, Finite Fields Appl. **18** (2012), no. 6, 1037–1060, DOI 10.1016/j.ffa.2012.09.004. MR3019183

[21] Nathan Jacobson, *Lectures in abstract algebra. Vol III: Theory of fields and Galois theory*, D. Van Nostrand Co., Inc., Princeton, N.J.-Toronto, Ont.-London-New York, 1964. MR172871

[22] Yan Jia, *On quasi-twisted codes over finite fields*, Finite Fields Appl. **18** (2012), no. 2, 237–257, DOI 10.1016/j.ffa.2011.08.001. MR2890550

[23] Gregory Karpilovsky, *Group representations. Vol. 3*, North-Holland Mathematics Studies, vol. 180, North-Holland Publishing Co., Amsterdam, 1994. MR1280715

[24] A. V. Kelarev and P. Solé, *Error-correcting codes as ideals in group rings*, Abelian groups, rings and modules (Perth, 2000), Contemp. Math., vol. 273, Amer. Math. Soc., Providence, RI, 2001, pp. 11–18, DOI 10.1090/conm/273/04419. MR1817147

[25] Peter Landrock and Olaf Manz, *Classical codes as ideals in group algebras*, Des. Codes Cryptogr. **2** (1992), no. 3, 273–285, DOI 10.1007/BF00141972. MR1181735

[26] Rudolf Lidl and Harald Niederreiter, *Introduction to finite fields and their applications*, Cambridge University Press, Cambridge, 1986. MR860948

[27] K. J. Horadam and A. A. I. Perera, *Codes from cocycles*, Applied algebra, algebraic algorithms and error-correcting codes (Toulouse, 1997), Lecture Notes in Comput. Sci., vol. 1255, Springer, Berlin, 1997, pp. 151–163, DOI 10.1007/3-540-63163-1_12. MR1634112

[28] Garry Hughes, *Constacyclic codes, cocycles and a $u+v|u-v$ construction*, IEEE Trans. Inform. Theory **46** (2000), no. 2, 674–680, DOI 10.1109/18.825841. MR1748996

[29] F. J. MacWilliams, *Binary codes which are ideals in the group algebra of an Abelian group*, Bell System Tech. J. **49** (1970), 987–1011, DOI 10.1002/j.1538-7305.1970.tb01812.x. MR265054

[30] Todor Zh. Mollov and Nako A. Nachev, *Minimal idempotents of twisted group algebras of cyclic 2-groups*, Southeast Asian Bull. Math. **26** (2002), no. 4, 593–601, DOI 10.1007/s100120200062. MR2047489

[31] Todor Zh. Mollov and Nako A. Nachev, *On the commutative twisted group algebras*, Comm. Algebra **35** (2007), no. 10, 3064–3070, DOI 10.1080/00927870701405074. MR2356142

[32] M. D. G. Olvera, L. A. L. Ramosand B. Torecillas Jover, *Secure group communications using twisted group rings*, Mathematics, **10** (2022), 1–16, DOI 10.3390/math10162845.

[33] D. S. Passman, *On the semisimplicity of twisted group algebras*, Proc. Amer. Math. Soc. **25** (1970), 161–166, DOI 10.2307/2036548. MR263939

[34] D. S. Passman, *Radicals of twisted group rings*, Proc. London Math. Soc. (3) **20** (1970), 409–437, DOI 10.1112/plms/s3-20.3.409. MR271246

[35] Donald S. Passman, *Infinite crossed products*, Pure and Applied Mathematics, vol. 135, Academic Press, Inc., Boston, MA, 1989. MR979094

[36] César Polcino Milies and Sudarshan K. Sehgal, *An introduction to group rings*, Algebra and Applications, vol. 1, Kluwer Academic Publishers, Dordrecht, 2002, DOI 10.1007/978-94-010-0405-3. MR1896125

[37] J. Schur, *Über die Darstellung der endlichen Gruppen durch gebrochen lineare Substitutionen* (German), J. Reine Angew. Math. **127** (1904), 20–50, DOI 10.1515/crll.1904.127.20. MR1580631

[38] W. F. Reynolds, *Projective representations of finite groups in cyclotomic fields*, Illinois J. Math. **9** (1965), 191–198. MR174631

[39] William F. Reynolds, *Twisted group algebras over arbitrary fields*, Illinois J. Math. **15** (1971), 91–103. MR279209

[40] J. Schur, *Untersuchungen über die Darstellung der endlichen Gruppen durch gebrochene lineare Substitutionen* (German), J. Reine Angew. Math. **132** (1907), 85–137, DOI 10.1515/crll.1907.132.85. MR1580715

[41] J. Schur, *Über die Darstellung der symmetrischen und der alternierenden Gruppe durch gebrochene lineare Substitutionen* (German), J. Reine Angew. Math. **139** (1911), 155–250, DOI 10.1515/crll.1911.139.155. MR1580818

[42] Minjia Shi and Yiping Zhang, *Quasi-twisted codes with constacyclic constituent codes*, Finite Fields Appl. **39** (2016), 159–178, DOI 10.1016/j.ffa.2016.01.010. MR3475547

[43] R.P. Slessor, *Performance of Codes based on Crossed Product Algebras*, PhD Thesis, U. of Southampton, 2011.

[44] A. J. van Zanten, *Primitive idempotent tables of cyclic and constacyclic codes*, Des. Codes Cryptogr. **87** (2019), no. 6, 1199–1225, DOI 10.1007/s10623-018-0495-0. MR3947343

[45] Rongsheng Wu and Minjia Shi, *A modified Gilbert-Varshamov bound for self-dual quasi-twisted codes of index four*, Finite Fields Appl. **62** (2020), 101627, 11, DOI 10.1016/j.ffa.2019.101627. MR4053143

Instituto de Ciências Matemáticas e de Computação, Universidade de São Paulo, Caixa Postal 668, CEP 13560-970, Brazil
Email address: andreduarte@icmc.usp.br

Instituto de Matemática e Estatística, Universidade de São Paulo, Caixa Postal 66.281, CEP 05314-970, Brazil
Email address: polcino@ime.usp.br

Contemporary Mathematics
Volume **826**, 2025
https://doi.org/10.1090/conm/826/16587

Residue structures

Louis H. Rowen

ABSTRACT. We consider residue structures R/G where G is a subgroup, not necessarily an ideal, of a ring R. Special instances include Krasner's construction of quotient hyperfields, and Petit's construction of nonassociative algebras. The residue construction, treated formally, satisfies the Noether isomorphism theorems, and is cast in a broader categorical setting which includes categorical products, sums, and various extensions.

1. Introduction and preliminaries

The residue group $\mathcal{A}/G$ of a group $\mathcal{A}$ modulo a normal subgroup G is one of the most familiar setups in algebra, lying at the foundation of group theory. The cosets form the residue group, as learned in a beginning course in abstract algebra.

Analogously, given a ring R and a subgroup G of $(R, +)$, one might ask about the structure of R/G. In order for the cosets to be a ring, G must be an ideal of R. In this paper we consider what happens when we drop this extra assumption on G. Krasner [**17, 19**] achieved considerable success in field theory while using this approach, and the multiplicative version also can be applied to recover various "semiassociative algebras," including an example of Dickson described in [**26**].

Krasner's residue construction takes us to the power set $\mathcal{P}(R)$ of a set R, in which some but not all of the algebraic properties of R are preserved. In particular, additive inverses are lost, drawing us from ring theory to semiring theory, and beyond. Our objective is to lay out Krasner's construction in its general structural framework, which extends the usual theory of residue structures.

Although the relevant functors are provided, tying the original ring to the Krasner residue, the residue construction encounters some categorical hurdles, which we circumvent by considering the more inclusive setting of "pairs."

Connections are given to some recent work in nonassociative algebras ([**5, 27**]), and general linear algebra ([**3**], [**4**]).

1.1. Underlying algebraic structures.

2020 *Mathematics Subject Classification.* Primary 16Y20, 16Y60, 16W99; Secondary 14T10, 20N20.

Key words and phrases. bimagma, hyperfield, hypergroup, hypermonoid, Krasner, linear algebra, magma, semiassociative algebra, pair, residue, semiring, subgroup.

The research of the author was supported by the ISF grant 1994/20 and the Anshel Peffer Chair.

First we review some definitions. As in [**24**], the theory is presented for magmas, in order to include more varied algebraic constructions. $\mathbb{N}^*$ denotes the positive natural numbers, and $\mathbb{N} = \mathbb{N}^* \cup 0$.

DEFINITION 1.1.

(i) A **magma** $(\mathcal{M}, *, \iota)$ is a set $\mathcal{M}$ with a binary operation $\mathcal{M} \times \mathcal{M} \to \mathcal{M}$, not necessarily associative, where the operation is denoted by concatenation. In this paper a magma always has a neutral element, denoted ι, i.e., $\iota * b = b * \iota = b$ for all $b \in \mathcal{M}$.[1]

(ii) A **monoid** is a magma with an associative operation, not necessarily commutative.

(iii) A submonoid $\mathcal{N}$ of a monoid $(\mathcal{M}, \cdot, 1)$ is **normal** if, for all $b_i \in \mathcal{M}$,
  (a) Any two cosets $b_1\mathcal{N}$, $b_2\mathcal{N}$ are either disjoint or equal.
  (b) $(b_1\mathcal{N})(b_2\mathcal{N}) = (b_1 b_2)\mathcal{N}$.
  (This generalizes the notion of a normal subgroup of a group.) We also say $(\mathcal{M}, \cdot, 1)$ **normalizes** $\mathcal{N}$.

If the operation of a monoid is denoted as "+" (resp. "·") then the neutral element is denoted as 0 (resp. 1). We often delete · from the notation, using concatenation. Our point of departure is the following basic structure result based on [**17**].

LEMMA 1.2.

(i) *If G is a normal submonoid of a monoid $\mathcal{T}$, then there is the* ***residue monoid*** $\bar{\mathcal{T}} := \mathcal{T}/G$*, the set of cosets, which has the operation*

$$\bar{a}_1 \bar{a}_2 = \overline{a_1 a_2},$$

writing $\bar{a}_i$ for $a_i G$.

(ii) *$\bar{\mathcal{T}}$ is a group if and only if for every $a \in \mathcal{T}$ there is $a' \in \mathcal{T}$ such that $aa' \in G$.*

PROOF. (i) Just as for the familiar case of groups. Namely $(a_1G)(a_2G) = (a_1Ga_2)G = (a_1a_2)GG = (a_1a_2)G$. The neutral element is $\bar{\iota} = \iota G$. Note that distinct cosets still are disjoint.

(ii) $aa' \in G$ if and only if $\bar{a}\bar{a}' = \overline{aa'} = \bar{\iota}$. □

1.2. Semirings and semifields.

Our interest in this paper lies in ring-like structures.

DEFINITION 1.3.

(i) A **pre-semiring** $(\mathcal{S}, \cdot, +, 1)$ ([**10**], called an nd-semiring in [**16**]) is both a multiplicative monoid $(\mathcal{S}, \cdot, 1)$ and a commutative additive monoid $(\mathcal{S}, +)$ (but not necessarily satisfying distributivity). We can always adjoin a 0 element that is additively neutral and also is multiplicatively absorbing, so we assume that any pre-semiring also has such a 0.

(ii) A **semiring** [**8, 11**] is an associative pre-semiring $\mathcal{A}$ that satisfies all the properties of a ring (including distributivity of multiplication over addition), except possibly negation.

[1]In [**24**] these magmas are called *unital*. The element ι is called e in [**24**], but we reserve e for another purpose in Remark 3.4. If a magma $\mathcal{M}$ did not already contain a neutral element ι, we could adjoin it formally by declaring the operations $\iota * b = b * \iota = b$ for $b \in \mathcal{M}$.

(iii) A semiring $\mathcal{A}$ is a **semifield** if $\mathcal{S}^\times := \mathcal{S} \setminus \{0\}$ is a group (not necessarily abelian).

EXAMPLE 1.4. Here is the best known example of a semifield which is not a ring, which plays a major role in tropical algebra. If $(\mathcal{G}, +)$ is an ordered abelian monoid, the **max-plus algebra** on $\mathcal{A} := \mathcal{G} \cup \{-\infty\}$ is given by defining multiplication to be the original addition on $\mathcal{G}$, with $-\infty$ additively absorbing, and addition to be the maximum. Thus the zero element of $\mathcal{A}$ is $-\infty$.

Sometimes we need some distributivity, associativity, and commutativity.

DEFINITION 1.5. The **distributive nucleus** $\mathcal{N}(\mathcal{A})$ of a pre-semiring $\mathcal{A}$ is the subset $\mathcal{N}$ which associates and distributes with all elements of $\mathcal{A}$, to wit, for all $a \in \mathcal{N}$ and $b_i \in \mathcal{A}$:

(i) $a(b_1b_2) = (ab_1)b_2, \qquad (b_1a)b_2 = b_1(ab_2), \qquad (b_1b_2)a = b_1(b_2a),$

(ii) $a(b_1 + b_2) = ab_1 + ab_2, \qquad (b_1 + b_2)a = b_1a + b_2a.$

The **distributive nuclear center** $\mathcal{Z}(\mathcal{A})$ is the subset $\mathcal{Z} \subseteq \mathcal{N}$ also satisfying $ab = ba$ for all $a \in \mathcal{Z}$, $b \in \mathcal{A}$.

LEMMA 1.6. *The distributive nucleus $\mathcal{N} = \mathcal{N}(\mathcal{A})$ and the distributive nuclear center $\mathcal{Z} = \mathcal{Z}(\mathcal{A})$ of a pre-semiring $\mathcal{A}$ are multiplicative submonoids of $\mathcal{A}$.*

PROOF. Suppose $a_i \in \mathcal{N}$ and $b_i \in \mathcal{A}$. Then

$$(a_1a_2)(b_1b_2) = a_1(a_2(b_1b_2)) = a_1((a_2b_1)b_2) = (a_1(a_2b_1))b_2 = (a_1a_2)b_1)b_2.$$

$$(b_1(a_1a_2))b_2 = ((b_1a_1)a_2)b_2 = (b_1a_1)(a_2b_2) = b_1(a_1(a_2b_2)) = b_1((a_1a_2)b_2).$$

$$\begin{aligned}(a_1a_2)(b_1 + b_2) &= a_1(a_2(b_1 + b_2)) = a_1((a_2b_1 + a_2b_2) \\ &= a_1(a_2b_1) + a_1(a_2b_2) = (a_1a_2)b_1 + (a_1a_2)b_2.\end{aligned} \tag{1.1}$$

If $a_i \in \mathcal{Z}$ then

$$(a_1a_2)b_1 = a_1(a_2b_1) = a_1(b_1a_2) = (a_1b_1)a_2 = (b_1a_1)a_2 = b_1(a_1a_2).$$

□

REMARK 1.7. In contrast to the familiar cases in distributive nonassociative algebras, the distributive nucleus of a pre-semiring need not be a subring, since it need not be closed under addition! This often is the case with hypermagmas, such as the power set of the phase hyperfield of Example 2.4 below.

1.2.1. *Digression: The structure of semifields.*

Pre-semirings and semirings (as well as magmas and monoids) readily fit into the framework of universal algebra [**15**, Chapter 2], in which homomorphisms are defined via congruences. Although the theory is quite elegant, one finds it difficult to deal explicitly with congruences, so, in [**13, 31, 33**], multiplication was given the traditional role of addition in order to utilize group theory.

DEFINITION 1.8. A **kernel** of a semifield $\mathcal{S}$ is a normal subgroup K of $\mathcal{S}^\times$ which is **convex** in the sense that if $a_1, a_2 \in K$ and $r_1, r_2 \in \mathcal{S}$ with $r_1 + r_2 = 1$, then $r_1a_1 + r_2a_2 \in K$.

Other equivalent criteria are given in [**13**, Theorem 3.2]. Thus, the category of semifields in universal algebra, is isomorphic to the category of lattice-ordered abelian groups, where the kernels correspond to the convex subgroups, cf. [**33**]. We ask here, "What happens when one drops the convexity condition on a kernel?"

In order to frame the answer more broadly, we bring in rather inclusive algebraic concepts.

1.3. Sets with a two-sided monoid action.

The notion of an algebraic structure having a designated subgroup was considered formally in [**16**], and developed in [**3**], from which we extract the essence, also taking [**24**] into account.

DEFINITION 1.9. Let $\mathcal{T}$ be a monoid with a neutral element 1.

(i) A **left $\mathcal{T}$-set** is a set $\mathcal{A}$ together with a (left) $\mathcal{T}$-action $\mathcal{T} \times \mathcal{A} \to \mathcal{A}$ (denoted as concatenation), for which, for all $a_i \in \mathcal{T}$, $b \in \mathcal{A}$,
  (a) $1b = b$.
  (b) $(a_1a_2)b = a_1(a_2b)$.

(ii) A **right $\mathcal{T}$-set** is defined analogously.

(iii) A **$\mathcal{T}$-biset** is a left and right $\mathcal{T}$-set $\mathcal{A}$, for which $(a_1b)a_2 = a_1(ba_2)$ for all $a_i \in \mathcal{T}$ and $b \in \mathcal{A}$.

(iv) A **$\mathcal{T}$-bimagma** is a magma $(\mathcal{A}, *, \iota)$, which also is a $\mathcal{T}$-biset satisfying
  (a) ι is absorbing, i.e. $a\iota = \iota = \iota a$, for all $a \in \mathcal{T}$.
  (b) The action is left and right **distributive** over $\mathcal{T}$, in the sense that

$$a(b_1 * b_2) = ab_1 * ab_2, \quad a(b_1 * b_2) = b_1a * b_2a, \text{for all } a \in \mathcal{T},\ b_i \in \mathcal{A}.$$

We make an important restriction.

DEFINITION 1.10.

(i) A $\mathcal{T}$-set $\mathcal{A}$ is **weakly admissible** if $\mathcal{T} \subseteq \mathcal{A}$. Then we can define $\mathcal{T}_0 = \mathcal{T} \cup \{\iota\}$, and declare $\iota\mathcal{A} = \mathcal{A}\iota = \iota$. This makes $\mathcal{T}_0$ a monoid, and $\mathcal{A}$ a weakly admissible $\mathcal{T}_0$-set.

A formalism to make a $\mathcal{T}$-bimagma weakly admissible is given in [**30**, Lemma 2.5].

DEFINITION 1.11.

(i) A **$\mathcal{T}$-bimodule** (called *mosaic* in [**24**]) is an abelian monoid which is a $\mathcal{T}$-set. (Note in this case that the actions $\mathcal{T} \times \mathcal{A} \to \mathcal{A}$ and $\mathcal{A} \times \mathcal{T} \to \mathcal{A}$ are linear over $\mathcal{A}$)

(ii) A **$\mathcal{T}$-pre-semiring** is a pre-semiring $\mathcal{A}$ which is a unital, weakly admissible $\mathcal{T}$-bimodule with $\mathcal{T}$ in the distributive nucleus of $\mathcal{A}$.

(iii) A **$\mathcal{T}$-semiring** is a $\mathcal{T}$-pre-semiring which is a semiring.

EXAMPLE 1.12.

(i) $\mathcal{A}$ is an associative ring with a multiplicative submonoid $\mathcal{T}$ (our main example). In particular, $\mathcal{A}$ can be a graded algebra over a field F, where $\mathcal{T}$ is comprised of the homogeneous elements. Special cases:
  (a) (Most pertinent to this paper) The **Ore polynomial extension** $F[x; \sigma, \delta]$ where σ is an automorphism of the field F and δ is a derivation of F. Here $\mathcal{T}$ is the monoid of monomials. $F[x; \sigma, \delta]$ has the usual definition of addition of polynomials, but multiplication is given by $xa = \sigma(a)x + \delta(a)$, extended associatively and distributively.
  (b) $\mathcal{A}$ is the group algebra $F[G]$, with $\mathcal{T} = G$.
  (c) $\mathcal{A}$ is a symbol algebra $\langle x, y\rangle$, where $x_n = \alpha \in F$ and $y^n = \beta \in F$, and $yx = \mu xy$ where μ is a primitive root of 1; $\mathcal{T} = \cup_{0 \le i,j \le n-1} F x^i y^j$ is the monoid of products of its generators.

(ii) $\mathcal{A}$ is an algebra over a field F, and $\mathcal{T} = F^\times$.
(iii) $\mathcal{A}$ is an associative ring with a left ideal $\mathcal{T}$.
(iv) $\mathcal{A}$ is a monoid with a unit element, and $\mathcal{T}$ is its set of monoid endomorphisms.
(v) $\mathcal{A}$ is a field, and $\mathcal{T}$ is its group of automorphisms.

Note that (i)–(iii) are weakly admissible. Here is another major example originally studied by Marty [**21**].

2. Hypermagmas

We follow the general treatment of Nakamura and Reyes [**24**]. Define $\mathcal{P}^\sharp(\mathcal{H}) := \mathcal{P}(\mathcal{H}) \setminus \emptyset$.

DEFINITION 2.1 ([**21**], [**24**]).

(i) (i) A **hypermagma** $(\mathcal{H}, \boxdot, \iota)$ is a set $\mathcal{H}$ with a binary "hyperoperation"[2]

$$\boxdot : \mathcal{H} \times \mathcal{H} \to \mathcal{P}^\sharp(\mathcal{H}),$$

with neutral element ι (i.e., $\iota \boxdot a = a \boxdot \iota = \{a\}$), whose natural extension to $\mathcal{P}^\sharp(\mathcal{H})$ obtained by defining $S_1 \cdot S_2 := \cup_{s_i \in S_i} s_1 \boxdot s_2$ for $S_1, S_2 \in \mathcal{P}^\sharp(\mathcal{H})$, makes $\mathcal{P}^\sharp(\mathcal{H})$ a weakly admissible left and right $\mathcal{H}$-bimagma, viewing $\mathcal{H} \subseteq \mathcal{P}^\sharp(\mathcal{H})$ by identifying a with $\{a\}$.

(ii) A **sub-hypermagma** $\mathcal{H}'$ of $\mathcal{H}$ satisfies $a_1 \boxdot a_2 \subseteq \mathcal{P}(\mathcal{H}')$ for $a_1, a_2 \in \mathcal{H}'$.

(iii) A **hyperinverse** (if it exists) of an element a in a hypermagma $(\mathcal{H}, \boxdot, \iota)$ is an element denoted a^{-1}, for which $\iota \in a \boxdot a^{-1}$ and $\iota \in a^{-1} \boxdot a$.

(ii) A **hypermonoid**[3] is a hypermagma $\mathcal{H}$ which is abelian and associative in the sense that $a_1 \boxdot a_2 = a_2 \boxdot a_1$ and $(a_1 \boxdot a_2) \boxdot a_3 = a_1 \boxdot (a_2 \boxdot a_3)$, $\forall a_i \in \mathcal{H}$.

We follow the customary notation for hypermonoids, using $\boxplus$ instead of $\boxdot$, called **hyperaddition**, and 0 for ι.

(a) Then a hyperinverse of an element a is called a **hypernegative** and denoted as $-a$, i.e., $0 \in (-a) \boxplus a$ and $0 \in a \boxplus (-a)$.

(b) A **hypergroup** is a hypermonoid $(\mathcal{H}, \boxplus, 0)$ for which every element a has a unique hypernegative, and which is **reversible** in the following sense:
$a_3 \in a_1 \boxplus a_2$ iff $a_2 \in a_3 \boxplus (-a_1)$.

(c) A **hypersemiring** (resp. **hyperring**) is a hypermonoid (resp. hypergroup) $\mathcal{H}$, which is also a multiplicative monoid, $(\mathcal{H}, \cdot)$, with $\cdot$ distributing over hyperaddition.

(d) A **hyperfield** is a hyperring $\mathcal{H}$ for which $(\mathcal{H} \setminus \{0\}, \cdot)$ is a group.

(iii) A $\mathcal{T}$-**hypermonoid** $(\mathcal{H}, \boxplus, 0)$ is a hypermonoid which also is a $\mathcal{T}$-set with the $\mathcal{T}$-action distributing over $\boxplus$, $a(\boxplus a_i) = \boxplus(a a_i)$ and $(\boxplus a_i)a = \boxplus(a_i a)$ for every $a \in \mathcal{T}$, $a_i \in \mathcal{H}$.

---

[2][**24**] permits the product of two elements to be the empty set $\emptyset$, called ∞ in [**24**], with $\emptyset \cdot S = S \cdot \emptyset = \emptyset$.

[3]In [**24**] this is called "canonical." Other authors such as [**20**] use the term "canonical" for what we call "reversible."

LEMMA 2.2.

(i) *Any hypermonoid* $(\mathcal{H}, \boxplus, 0)$ *gives rise to a monoid* $\mathcal{P}^\sharp(\mathcal{H})$, *with the operation given by*

$$S_1 + S_2 = \cup\{s_1 \boxplus s_2 : s_i \in S_i\}.$$

(ii) *If furthermore* $\mathcal{H}$ *is a weakly admissible* $\mathcal{T}$*-hypersemiring then* $\mathcal{P}^\sharp(\mathcal{A})$ *is a weakly admissible* $\mathcal{T}$*-pre-semiring,*[4] *with* $\mathcal{T}$ *in the distributive nucleus, when we identify the elements* $a \in \mathcal{H}$ *with the singletons* $\{a\} \in \mathcal{P}^\sharp(\mathcal{A})$, *and define the operations elementwise, i.e., for* $S \in \mathcal{P}^\sharp(\mathcal{A})$:

(a) $aS = \{as : s \in S\}$, $Sa = \{sa : s \in S\}$,

(b) $S_1S_2 = \{s_1s_2 : s_i \in S_i\}$.

(iii) *Any hypersemiring* $\mathcal{H}$ *is an* $\mathcal{H}$*-hypersemiring where we define* $aa_1 = a \cdot a_1$.

PROOF. (i) Associativity is checked elementwise.
(ii) The axioms of Definition 1.9 are verified elementwise.
(iii) $\mathcal{P}^\sharp(\mathcal{H})$ is an $\mathcal{H}$-set by definition. □

One can easily present the noncommutative version, with $\boxdot$ as the hyperoperation; then hyperinverses are required to satisfy $(a_1 \boxdot a_2)^{-1} = (a_2^{-1}) \boxdot (a_1^{-1})$.

LEMMA 2.3. *Any hypermagma* $(\mathcal{H}, \boxdot, \iota)$ *gives rise to a* $\mathcal{H}$*-bimagma* $\mathcal{P}^\sharp(\mathcal{H})$, *with the operation* $\cdot$ *given by*

$$S_1 \cdot S_2 = \cup\{s_1 \boxdot s_2 : s_i \in S_i\}.$$

The operation $\cdot$ *is associative on* $\mathcal{P}^\sharp(\mathcal{H})$ *if* $\boxdot$ *is associative on* $\mathcal{H}$.

2.1. Examples of hyperfields. Here are some major examples of hyperfields, recalled from [**2, 19, 32**]. A large collection is given in [**23**, §2], and an assortment of other related examples is found in [**24**].

EXAMPLE 2.4.

(i) The **Krasner hyperfield** [**19**] is the set $\mathcal{H} := \{0, 1\}$, with the usual multiplication law, and with hyperaddition defined by $x \boxplus 0 = 0 \boxplus x = x$ for all x, and $1 \boxplus 1 = \{0, 1\}$.

(ii) The **hyperfield of signs** $\mathcal{H} := \{1, 0, -1\}$, with the usual multiplication law, and hyperaddition defined by $1 \boxplus 1 = 1$, $-1 \boxplus -1 = -1$, $x \boxplus 0 = 0 \boxplus x = x$ for all x, and $1 \boxplus -1 = -1 \boxplus 1 = \{0, 1, -1\}$.

(iii) The **phase hyperfield**. Let S^1 denote the complex unit circle, and take $\mathcal{H} = S^1 \cup \{0\}$. Nonzero points a and b are **antipodes** if $a = -b$. Multiplication is defined as usual (so corresponds on S^1 to addition of angles). We denote an open arc of less than 180 degrees connecting two distinct points a, b of the unit circle by $(a\,b)$. The hypersum is given, for $a, b \neq 0$, by

$$a \boxplus b = \begin{cases} (a\,b) \text{ if } a \neq b \text{ and } a \neq -b; \\ \{-a, 0, a\} \text{ if } a = -b, \\ \{a\} \text{ if } a = b. \end{cases}$$

[4] In general $\mathcal{P}^\sharp(\mathcal{H})$ is not distributive, cf. Remark 3.4(i),(ii), for example, but does satisfy $S(S_1 * S_2) \subseteq SS_1 * SS_2$ and $(S_1 * S_2)S \subseteq S_1S * S_2S$ for all $S, S_i \in \mathcal{P}^\sharp(\mathcal{H})$, cf. [**22**, Proposition 1.1].

(iv) The **tropical hyperfield** consists of the set $\mathcal{H} = \mathbb{R} \cup \{-\infty\}$, with $-\infty$ as the zero element and 0 as the unit, equipped with the addition $a \boxplus b = \{a\}$ if $a > b$, $a \boxplus b = \{b\}$ if $a < b$, and $a \boxplus a = [-\infty, a]$.

The **signed tropical hyperfield** [**2**, Example 3.18] $\mathcal{H}$ is the union of two disjoint copies of $\mathbb{R}$, the first one being identified with $\mathbb{R}$, and denoted by $\mathbb{R}$, the second one being denoted by $(-)\mathbb{R}$, with a zero element $-\infty$ adjoined. The signed tropical hyperfield is described in [**2**, Example 5.20], and is otherwise known as the "real tropical numbers," or the "tropical real hyperfield," see [**32**].

3. Krasner's residue construction

Now we are equipped to return to the question, "What can be said when we take the residues with respect to a normal subgroup?" The answer is intriguing. The pertinence of hyperfields is Krasner's lovely construction [**19**], known in the literature as the **quotient hyperfield**, done here more generally.

3.1. Quotient hypermagmas.

DEFINITION 3.1. Suppose $\mathcal{T}$ is a monoid, and $\mathcal{M}$ is a $\mathcal{T}$-hypermagma, normalizing a submonoid G of $\mathcal{T}$. Define the **quotient hypermagma** $\mathcal{H} = \mathcal{M}/G$ to have the operation $\boxdot : \mathcal{H} \times \mathcal{H} \to \mathcal{P}(\mathcal{H})$ given by

$$S_1G \boxdot S_2G = \{cG : c \in S_1G \boxdot S_2G\},$$

where S_iG denotes $\{sG : s \in S\}$.

In particular, if $\mathcal{M}$ is a $\mathcal{T}$-magma then the operation $\boxdot : \mathcal{H} \times \mathcal{H} \to \mathcal{P}(\mathcal{H})$ is given by

$$b_1G \boxdot b_2G = \{cG : c \in b_1G * b_2G\},$$

for $b_i \in \mathcal{M}$. The definition has been given in greater generality in order to be able to repeat the construction, cf. Theorem 4.6 below.

LEMMA 3.2. *Suppose $\mathcal{T}$ is a monoid and $\mathcal{M}$ is a weakly admissible $\mathcal{T}$-magma, which normalizes a submonoid G of $\mathcal{T}$. Then G is a normal submonoid of $\mathcal{T}$, so we have a monoid $\bar{\mathcal{T}} := \mathcal{T}/G$.*

(i) *The quotient hypermagma $\mathcal{H} = \mathcal{M}/G$ is indeed a hypermagma, under the hyperoperation $(a_1G) \boxdot (a_2G) = \{aG : a \in (a_1G) \cdot (a_2G)\}$.*
(ii) *If $\mathcal{M}$ is a weakly admissible $\mathcal{T}$-module, then $\mathcal{H} := \mathcal{M}/G$ is a $\bar{\mathcal{T}}$-hypermodule, under the hyperaddition $(a_1G) \boxplus (a_2G) = \{aG : a \in a_1G + a_2G)\}$ and the action $\bar{a}(bG) = (ab)G$.*

PROOF. (i) Clearly $(a_1G) \boxdot (a_2G) = (a_1G) \boxdot (a_2G) = (a_1a_2)G = a_1(a_2G)$.

(ii) Since $\mathcal{M}$ normalizes G, its subset $\mathcal{T}$ normalizes G, so G is a normal submonoid of $\mathcal{T}$, and $\bar{\mathcal{T}}$ becomes a monoid under the operation $\overline{a_1}\,\overline{a_2} = \overline{a_1a_2}$. The hyperaddition matches the hyperoperation of (i). □

EXAMPLE 3.3.

(i) Suppose F is a field with a subgroup G. Define the familiar **quotient hyperfield** $\mathcal{H} = F/G$ to have hyperaddition as in (i) and the usual residue multiplication $(b_1G)(b_2G) = (b_1b_2)G$ for $b_i \in F$.

(ii) (Projectivization) Similarly to (i), suppose that A is an algebra over a field F, define $P = A/F^\times$, with hyperaddition as in Lemma 3.2 and the usual residue multiplication $(b_1F^\times)(b_2F^\times) = (b_1b_2)F^\times$ for $b_i \in A$.

REMARK 3.4. Suppose that the hypernegative -1 exists in the quotient hypermonoid $\mathcal{H} = \mathcal{M}/G$. Define $e = 1 \boxplus (-1) \in \mathcal{P}^\sharp(\mathcal{H})$ (so $0 \in e$).

(i) $e = \{g_1 - g_2 : g_i = G\}$, and $ee = \{(g_1 - g_2)g_3 \boxplus (g_4 - g_5)g_6 : g_i = G\}$.

(ii) $e \boxplus e = \{(g_1 - g_2) - (g_4 - g_5) : g_i = G\}$.

(iii) Thus e encapsulates the question in [**23**], "Under what conditions is $F \setminus G$ spanned by (or, equals) $G - G$, for a multiplicative subgroup G of a field F?"

(iv) If $G = \{\pm 1\}$, then $0 \in a \boxplus a$ for all $a \in \mathcal{H}$.

The failure of distributivity in quotient hyperfields enhances the following result.

LEMMA 3.5. *$ee = e \boxplus e$ in $\mathcal{S}/G$, for a semiring $\mathcal{S}$ with $-1 \in G$.*

PROOF. $ee \subseteq e \boxplus e$, by direct verification in Remark 3.4(i),(ii), taking $g_3 = g_6 = 1$. For the converse, note that $(g_1 - g_2)g_3 \boxplus (g_4 - g_5)g_6 = (g_1g_3 - g_2g_3) \boxplus (g_4g_6 - g_5g_6)$. □

3.1.1. *Description of major hyperfields as quotient hyperfields.*
We want to identify the different examples of Example 2.4 as quotient hyperfields. Towards this end, an **isomorphism** $f : \mathcal{H} \to \mathcal{H}'$ of hypermonoids is defined as a 1:1 and onto map satisfying $f(a_1 \boxplus a_2) = f(a_1) \boxplus f(a_2)$ for all $a_i \in \mathcal{H}$.

EXAMPLE 3.6.

(i) The Krasner hyperfield is isomorphic to the quotient hyperfield $F/F^\times$, for any field F.

(ii) The hyperfield of signs is isomorphic to the quotient hyperfield $K/K_{>0}$ for each linearly ordered field $(K, \le)$, where $K_{>0}$ is the group of positive elements of K.

(iii) The phase hyperfield is isomorphic to the quotient hyperfield $\mathbb{C}/\mathbb{R}_{>0}$.

(iv) The tropical hyperfield is isomorphic to the quotient hyperfield F/G, where F denotes a field with a surjective non-archimedean valuation $v : F \to \mathbb{R} \cup \{+\infty\}$, and $G := \{f \in F : v(f) = 0\}$.

(v) The signed tropical hyperfield is isomorphic to the quotient hyperfield K/G where K is a linearly ordered non-archimedean field with a surjective valuation $\mathrm{val} : K \to \mathbb{R} \cup \{+\infty\}$, and

$$G := \{f \in K : f > 0, \ \mathrm{val}\, f = 0\}.$$

REMARK 3.7. Krasner's original use [**17, 18**] of quotient hyperfields was in valuation theory, in which he succeeded in approximating complete fields of characteristic $p > 0$ by fields of characteristic 0. This aspect is developed further in [**20**].

REMARK 3.8. If $\mathcal{A}$ is a pre-semiring, then $\mathcal{A}$ is a bimagma over $\mathcal{N} := \mathcal{N}(\mathcal{A})$ (which is a monoid, by Lemma 1.6). Thus we can mod out by any normal multiplicative subgroup of $\mathcal{N}$, as done in §3.1.

3.2. Multiplicative Krasner residues.

One can take hypermultiplication instead of hyperaddition, done for magmas in some interesting cases in [**24**, §2.2, Examples 2.15 and 2.16]. We go in a slightly different direction, and take a unified approach.

3.2.1. *m-Hypersemirings and nonassociative algebras.*

DEFINITION 3.9.

(i) An **m-hypermagma** is a hypermagma $(\mathcal{H}, \boxdot, 1)$, where $1 \in \mathcal{H}$ is the unit element.

(ii) An **m-hypersemiring** is an m-hypermagma in which $(\mathcal{H}, +)$ also is an additive monoid. (Thus the addition extended elementwise to $\mathcal{P}^\sharp(\mathcal{H})$ makes $\mathcal{P}^\sharp(\mathcal{H})$ a pre-semiring.)

In any pre-semiring $\mathcal{S}$ with commutative addition, reversing the roles of addition and multiplication, we factor out a subgroup of $(\mathcal{S}, +)$. As in Definition 3.1, we have

DEFINITION 3.10. Suppose $\mathcal{A}$ is an m-hypersemiring, and $(G, +)$ is an additive subgroup of $\mathcal{A}$. Then we define an m-hypersemiring $\mathcal{H} = \mathcal{A}/G$, by defining coset addition as usual, and hypermultiplication

$$(S_1 + G) \boxdot (S_2 + G) = \{(s_1 + g_1) \boxdot (s_2 + g_2) \ + G : s_i \in S_i, g_i \in G.\}$$

LEMMA 3.11. *Notation as in Definition* 3.10, *Suppose $\mathcal{A}$ is an m-hypersemiring, and L is a left ideal of $\mathcal{A}$. Then $\mathcal{H} = \mathcal{A}/L$ is an m-hypersemiring, by defining coset addition as usual, and hypermultiplication*

$$(S_1 + L) \boxdot (S_2 + L) = \{(s_1 + b_1) \boxdot s_2 + L : s_i \in S_i, b_1 \in L.\}$$

PROOF. The formula follows since $(s_1 + b_1) \boxdot L \subseteq L$. □

EXAMPLE 3.12. Suppose R is a ring and L is a left ideal. Then R/L is an m-hyperring, seen by defining coset addition as usual, and hypermultiplication

$$(r_1 + L) \boxdot (r_2 + L) = \{(r_1 + b)r_2 + L : b \in L\}.$$

3.2.2. *Lie Krasner residues.*

In Definition 3.9(ii) we did not require the operation $\boxdot$ to be associative, or even distributive, so we have the following special case of Definition 3.10.

DEFINITION 3.13. Suppose $(G, 0)$ is an additive sub-pre-semiring of a pre-semiring $\mathcal{S}$. Define hypermultiplication

$$(r_1 + G) \boxdot (r_2 + G) = \{(r_1 r_2 + r_1 g_1 + g_2 r_2) + G : g_1, g_2 \in G.\},$$

where $r_i \in \mathcal{S}$, $g_i \in G$:

This enables us to define hyperLie algebras.

LEMMA 3.14. *If L_1 is a Lie subalgebra of a Lie algebra L, then the residue module L/L_1 has the Lie hyperbracket*

$$[a_1 + L_1, a_2 + L_1]_\diamond = \{[a_1, a_2] + [a_1, y_2] + [a_2, y_1] + L_1 : a_i \in L, \ y_i \in L_1\},$$

which satisfies the following properties, for all $a_i \in L$:

(i) $0 \in [a_1 + L_1, a_1 + L_1]_\diamond$,

(ii) $0 \in [a_1 + L_1, a_2 + L_1]_\diamond + [a_2 + L_1, a_1 + L_1]_\diamond$,

(iii)

$$\begin{aligned}0 \in [[a_1 + L_1, a_2 + L_1]_\diamond]_\diamond, a_3 + L_1]_\diamond + [[a_2 + L_1, \ a_3 + L_1]_\diamond]_\diamond, a_1 + L_1]_\diamond \\ + [[a_3 + L_1, a_1 + L_1]_\diamond]_\diamond, a_2 + L_1]_\diamond\end{aligned} \tag{3.1}$$

(iv) *If $0 \in \sum_i a_i + L_1$ then $0 \in \sum_i [a_i + L_1, a + L_1]_\diamond$ for all $a \in \mathcal{L}$.*

PROOF. (i) $0 = [a_1, a_1] \in [a_1 + L_1, a_1 + L_1]_\diamond$.

(ii) $0 = [a_1, a_2] + [a_2, a_1] \in [a_1 + L_1, a_2 + L_1]_\diamond + [a_2 + L_1, a_1 + L_1]_\diamond$.

(iii) $0 = [[a_1, a_2]], a_3] + [[a_2, a_3]], a_1] + [[a_3, a_1]], a_2] \in [[a_1 + L_1, a_2 + L_1]_\diamond]_\diamond, a_3 + L_1]_\diamond + [[a_2 + L_1, a_3 + L_1]_\diamond]_\diamond, a_1 + L_1]_\diamond + [[a_3 + L_1, a_1 + L_1]_\diamond]_\diamond, a_2 + L_1]_\diamond$.

(iv) If $y = \sum_i a_i \in L_1$ then writing $0 = (a_1 - y) + \sum_{i>1} a_i$, we get $0 \in [0, a] \in \sum_i [a_i + L_1, a + L_1]_\diamond$ for all $a \in \mathcal{L}$. □

3.3. Nonassociative algebras arising from multiplicative Krasner residues.

REMARK 3.15. One obtains intriguing results specializing m-hyperstructures to classical structures, when one has a canonical way of choosing an element in a coset. For instance, let us tie in Example 3.12 with a nonassociative algebra construction which dates back at least to Dickson [**9**], and given explicitly for Ore extensions in Petit [**25**, § 1]:

Given $f \in R := F[x; \sigma, \delta]$ of degree n, cf. Example 1.12, define the left R-module $R_n = R/Rf$, viewed as the Ore polynomials of degree $< n$, writing $[g]$ for the residue of g. Given $[g], [h] \in R$, define $[g][h]$ to be the residue $[gh]$ of gh, i.e., the unique Ore polynomial of R_n which can be written in the form $gh - qf$ for $q \in R$. This defines a nonassociative algebra R_f which actually is the associative algebra R/Rf when $Rf \triangleleft R$.

More generally, one has:

PROPOSITION 3.16. *Suppose R is a filtered algebra (such as an Ore extension), which has a division algorithm for $f \in R$, in the sense that given any $g \in R$ we can write $g = qf + r$ where* $\deg r < \deg f$, *with q, r uniquely determined by g. Then one defines a distributive, nonassociative multiplication on $R_f := R/Rf$ by taking $[g_1][g_2] = [r]$ where $g_1 g_2 = qf + r$ as above.*

PROOF. We need to verify distributivity, i.e., $[[g_1]([g_2] + [g_3])] = [[g_1][g_2]] + [[g_1][g_3]]$ Write $g_1 g_i = q_i f + r_i$ where $\deg r_i < \deg f$. Then $\deg(r_2 + r_3) \le \deg f$ and

$$g_1(g_2 + g_3) = g_1 g_2 + g_1 g_3 = (q_2 + q_3)f + (r_2 + r_3).$$

□

THEOREM 3.17. *Suppose, as in Proposition* 3.16, *that R is a filtered algebra and $L = Rf$ is a left ideal of R, with* $\deg f = n$. *There is a natural (as explained in the proof) bijection Φ from the m-hypersemiring $\mathcal{H} = \mathcal{A}/L$ to R_f sending $g + L$ to the polynomial $\hat{g}$ of lowest degree in $g + L$.*

PROOF. Since $\deg \hat{g} < n$, we have $\Phi : \mathcal{H} \to R_f$, clearly a module homomorphism. If $\Phi(g_1) = \Phi(g_2)$ then $g_1 - g_2 \in L$, of degree $< n$, so $g_1 - g_2 = 0$. On the other hand, if $\deg g < n$ then $\hat{g} = g = \Phi(Lg)$. □

Pumpluen [**26, 27**] and Brown-Pumpluen [**6**] have shown that R_f is a nonassociative version of associative cyclic central simple algebras. Petit algebras are seen in [**5, 28**] to be part of a wider class of "semiassociative algebras," namely algebras A whose nucleus contains an étale F-algebra K, over which A is minimally faithful as a K, K-bimodule, and Blachar, Haile, Matzri, Rein, Vishne [**5**] develop an extensive theory of these algebras.

4. Categorical properties of the residue construction

The residue construction is surprisingly tractable with regard to the Noether structure theory. First we look at magma homomorphisms via universal algebra.

DEFINITION 4.1.

(i) A **magma homomorphism** is a map $f : \mathcal{M} \to \mathcal{M}'$ satisfying $f(b_1 * b_2) = f(b_1) * f(b_2)$ for all $b_i \in \mathcal{M}$.
(ii) A **surjection** of magmas is an onto magma homomorphism.
(iii) An **ismomorphism** of magmas is a 1:1 surjection.

REMARK 4.2. If $f : \mathcal{M} \to \mathcal{M}'$ is a homomorphism, with $\mathcal{M}$ a weakly admissible $\mathcal{T}$-bimagma, then we can view $f(\mathcal{M})$ as a weakly admissible bimagma over $f(\mathcal{T})$ in the obvious way.

4.0.1. *$\subseteq$-morphisms.*
Morphisms are more sophisticated at the level of hypermagmas.

DEFINITION 4.3. Let $(\mathcal{H}, \boxdot)$ (resp. $(\mathcal{H}', \boxdot)$) be a hypermagma. A **$\subseteq$-morphism**, (analogous to "colax morphism" in [**24**]) is a map $f : \mathcal{H} \to \mathcal{H}'$ satisfying the following condition:

(i) $f(a_1 \boxdot a_2) \subseteq f(a_1) \boxdot f(a_2), \quad \forall a_1, a_2 \in \mathcal{H}.$

A **$\subseteq$-morphism** $f : (\mathcal{H}, \boxplus) \to (\mathcal{H}', \boxplus)$ of $\mathcal{T}$-hypersemirings is a multiplicative map $f : \mathcal{H} \to \mathcal{H}'$, i.e., satisfying $f(a_1 a_2) = f(a_1) f(a_2)$ for $a_i \in \mathcal{H}$, which also is a $\subseteq$-morphism.

REMARK 4.4. A nonassociative algebra is an m-hypersemiring in the obvious way, that $a_1 \boxdot a_2$ is just the product. Then a $\subseteq$-morphism is simply an algebra homomorphism.

There is a weaker version of morphism for hypermagmas, which we note for reference.

DEFINITION 4.5. A **weak morphism**[5] $f : (\mathcal{H}, \boxdot) \to (\mathcal{H}', \boxdot)$ of hypermagmas satisfies $\iota \in a_1 \boxdot a_2$ implies $\iota \in f(a_1) \boxdot f(a_2)$ for $a_i \in \mathcal{H}$.

4.0.2. *Noether-type isomorphism theorems.*

THEOREM 4.6. *Suppose that $\mathcal{M}$ is a $\mathcal{T}$-bimagma, and $\mathcal{M}$ normalizes a submonoid G of $\mathcal{T}$.*

(i) *If $f : \mathcal{M} \to \mathcal{M}'$ is a homomorphism of magmas then there is a $\subseteq$-morphism $\bar{f} : \mathcal{M}/G \to f(\mathcal{M})/f(G)$ given by*
$$\bar{f}(bG) = \{b'G : \ b' \in \mathcal{M}, \ f(b)f(G) = f(b')f(G)\}.$$
(ii) *If $\mathcal{M}$ also normalizes a submonoid $G_1 \supseteq G$ of $\mathcal{T}$, then*
$$\mathcal{M}/G_1 \cong (\mathcal{M}/G)/(G_1/G).$$
(iii) *Suppose $\mathcal{M}$ is a $\mathcal{T}$-pre-semiring, and $f : \mathcal{M} \to \mathcal{M}'$ is a pre-semiring homomorphism. Then $f(\mathcal{M})$ is a $f(\mathcal{T})$-pre-semiring, and for a multiplicative submonoid G of $\mathcal{T}$ which is normal in $\mathcal{A}$, the induced map $\bar{f} : \mathcal{M}/G \to f(\mathcal{M})/f(G)$ is a $\subseteq$-morphism of $\mathcal{T}$-hypersemirings.*

[5]This is the standard definition for hypergroups, but could be vacuous for hypermagmas lacking hypernegatives.

(iv) *If G_i are normal submonoids of $\mathcal{M}_i$, then $(\prod \mathcal{M}_i)/(\prod G_i) \cong \prod(\mathcal{M}_i/G_i)$.*

PROOF. (i)

$$\begin{aligned}\bar{f}(b_1Gb_2G) &= \{b'G : f(b_1b_2)f(G) = f(b')f(G)\}\\ &= \{b'G : f(b_1)f(b_2)f(G) = f(b')f(G)\}\\ &\subseteq \{b_1'G : f(b_1)f(G) = f(b_1')f(G)\}\{b_2'G : f(b_2)f(G) = f(b_2')f(G)\}\\ &= \bar{f}(b_1G)\bar{f}(b_2G).\end{aligned} \tag{4.1}$$

(ii) The cosets of G match up with the cosets of G_1, i.e., $(aG)\bar{G}_1$ with aG_1.
(iii) $\bar{f}$ is obviously multiplicative, so apply (i). □

(iv) They are isomorphic as multiplicative monoids, and

$$\begin{aligned}(a_i) \boxplus (a_i') &= \left\{(a_i) + (g_i) + (a_i') + \prod G_i : (g_i) \in \prod G_i\right\}\\ &= \{(a_i + g_i + a_i') + \prod G_i) : g_i \in G_i\} = (a_i \boxplus a_i').\end{aligned} \tag{4.2}$$

These results are enhanced by noting that the category of magmas, with magma homomomorphisms, is a faithful subcategory of the category of hypermagmas, with $\subseteq$-morphisms.

4.0.3. *Noether-type properties of the multiplicative Krasner construction.*

As in §3.2.1 we can reverse addition and multiplication, and get the multiplicative version of Theorem 4.6, which is

THEOREM 4.7. *Define m-$\subseteq$-morphism as in Definition 4.3, but with addition and multiplication reversed. Suppose $\mathcal{A}$ is an m-hypersemiring, and $(G, +)$ is an additive subgroup of $\mathcal{A}$.*

(i) *If $G_1 \supseteq G$ is another additive subgroup of $\mathcal{A}$, then*
$$\mathcal{A}/G_1 \cong (\mathcal{A}/G)/(G_1/G).$$
(ii) *Suppose $f : \mathcal{A} \to \mathcal{A}'$ is a pre-semiring homomorphism. Then $f(\mathcal{A})$ is a pre-semiring, and for an additive subgroup of $\mathcal{A}$, the induced map $\bar{f}$: $\mathcal{A}/G \to f(\mathcal{A})/f(G)$ is an m-$\subseteq$-morphism of hypersemirings.*
(iii) *If G_i are additive subgroups of $\mathcal{A}_i$ then $(\prod \mathcal{A}_i)/(\prod G_i) \cong \prod(\mathcal{A}_i/G_i)$.*

REMARK 4.8. An intriguing way of interpreting the multiplicative Krasner residue is that the product of two elements belongs to the product of the cosets, but we do not know which one it is, much as in quantum computing. Then one could choose the product and perhaps obtain a semiassociative algebra. So we can get better functorial results by reversing the direction of Theorem 3.17:

THEOREM 4.9. *Notation as in Theorem 3.17, the m-hypersemiring $\mathcal{H} = \mathcal{A}/L$ is terminal in the sense that if, as a module, $\mathcal{A}/L$ has a given structure as a nonassociative algera A, the identity map is an m-$\subseteq$-morphism.*

PROOF. $a_1 \cdot a_2 \in a_1 \boxdot a_2$. □

The same idea could be applied to Example 3.3(ii), which is "terminal" with respect to the skew matrix algebras of [**5**, §6] in the following sense (where $e_{ij}e_{k\ell} = \delta_{jk}c_{ijl}e_{il}$, with each $0 \neq c_{ijl} \in F$[6]):

---

[6][**5**, §6] permits some $c_{ijl} = 0$; to handle this possibility one must consider monoids with 0, and broaden the definition of residue monoid.

PROPOSITION 4.10. *Suppose $A = \oplus Fe_{ij}$ is a skew matrix algebra. Then the map $A \to M_n(F)/F^\times$ (sending $e_{ij} \mapsto e_{ij}F^\times$) is an m-$\subseteq$-morphism.*

PROOF. As with Theorem 4.9. □

One may be led to study m-hyperstructures in this light, but very little has been done yet in this direction.

5. Pairs

Although seen to be quite pervasive, the process of modding out a normal submonoid has some categorical drawbacks. Even such a basic construction as polynomials does not work for hyperfields, since lack of "double distributivity" ruins associativity of multiplication of polynomials. A similar argument shows that the Krasner residue construction cannot be functorial with respect to tensor products unless one has considerable collapsing (as in [**24**]).

Furthermore, utilizing Lemma 2.2 is precarious since $|\mathcal{P}^\sharp(\mathcal{H})| = 2^{|\mathcal{P}^\sharp(H)|}$. In particular $|\mathcal{P}^\sharp(\mathcal{H}_1) \times \mathcal{P}^\sharp(\mathcal{H}_2)| = 2^{|\mathcal{P}^\sharp(H_1)|+|\mathcal{P}^\sharp(H_2)|}$ whereas

$$|\mathcal{P}^\sharp(\mathcal{H}_1 \times \mathcal{H}_2)| = 2^{|\mathcal{P}^\sharp(H_1)||\mathcal{P}^\sharp(H_2)|}.$$

Thus, $\mathcal{P}^\sharp(\mathcal{H}_1) \times \mathcal{P}^\sharp(\mathcal{H}_2) \subset \mathcal{P}^\sharp(\mathcal{H}_1 \times \mathcal{H}_2)$. So one is led to a broader view of $\mathcal{T}$-bimagmas which includes the residue constructions. This was done in [**29**], and generalized in [**3**, **16**]. The pre-semiring version is given here. The main challenge is to find a replacement for 0, which plays a secondary role in residue structures to the sets $\{b : 0 \in bG\}$. We fix a submonoid $\mathcal{T} \subseteq \mathcal{Z}(\mathcal{A})$.

DEFINITION 5.1.

(i) A **pair** $(\mathcal{A}, \mathcal{A}_0)$ is a $\mathcal{T}$-module[7] $\mathcal{A}$ together with a subset $\mathcal{A}_0$ containing 0, satisfying $ab_0, b_0a \in \mathcal{A}_0$ for all $a \in \mathcal{T}$, $b_0 \in \mathcal{A}_0$.
(ii) A **pre-semiring pair** is a pair for which $(\mathcal{A}, \mathcal{A}_0)$ is a $\mathcal{T}$-semiring.
(iii) The **hyperpair** $(\mathcal{A}, \mathcal{A}_0)$ of a hypermonoid $\mathcal{H}$ is $(\mathcal{A}, \mathcal{A}_0)$, where $\mathcal{A} = \mathcal{P}^\sharp(\mathcal{H})$, $\mathcal{A}_0 = \{S \in \mathcal{A} : 0 \in S\}$ and $\mathcal{T} = \mathcal{H} \setminus \{0\}$. (More generally, we could take $\mathcal{H}$ to be a $\mathcal{T}$-hypermonoid where $\mathcal{T}$ is a given monoid.)

5.1. Surpassing relations and negation maps.

There are two other ingredients for pairs which generalize hyperfield theory.

5.1.1. *Surpassing relations.*

DEFINITION 5.2. A **surpassing relation** on a weakly admissible pair $(\mathcal{A}, \mathcal{A}_0)$, denoted $\preceq$, is a pre-order satisfying the following:

(1) $a_1 \preceq a_2$ for $a_1, a_2 \in \mathcal{T}$ implies $a_1 = a_2$.
(2) $b \preceq b + c$ for all $b \in \mathcal{A}$ and $c \in \mathcal{A}_0$.
(3) $b \preceq 0$ for $b \in \mathcal{A}$ implies $b = 0$.

EXAMPLE 5.3. Here are two main examples of surpassing relations.

(i) (Compare with [**1**, Definition 2.8]) Define $\preceq_0$ by $b_1 \preceq_0 b_2$ when $b_2 = b_1 + c$ for some $c \in \mathcal{A}_0$. This is the surpassing relation used mainly in supertropical mathematics [**14**], in [**29**], and in [**3**].
(ii) On a hyperpair $(\mathcal{P}^\sharp(\mathcal{H}), \mathcal{P}^\sharp(\mathcal{H})_0)$, cf., Definition 5.1(iii), we define $\preceq_\subseteq$ on $\mathcal{P}^\sharp(\mathcal{H})$ by set inclusion. This surpassing relation fits in well with residue hypermagmas, and is the main application in this note.

[7]More generally, one could take $\mathcal{T}$-magmas, with $\iota \in \mathcal{A}_0$.

A pre-semiring pair $(\mathcal{A}, \mathcal{A}_0)$ is **$\preceq$-distributive** if $a(b_1 + b_2) \preceq ab_1 + ab_2$ and $(b_1 + b_2)a \preceq b_1 a + b_2 a$ for all $a \in \mathcal{T}$ and $b_i \in \mathcal{A}$.

PROPOSITION 5.4. *In Example* 3.12, *R/L, is $\preceq_{\subseteq}$-distributive.*

PROOF.

$$\begin{aligned}(r_1 + L) \boxdot ((r_2 + L) + (r_3 + L)) &= (r_1 + L) \boxdot (r_2 + r_3 + L) \\ &= \{(r_1 + a)(r_2 + r_3) + L : a \in L\} \\ &\preceq_{\subseteq} \{(r_1 + a)r_2 + L : a \in L\} + \{(r_1 + a')r_3 + L : a' \in L\} \\ &= (r_1 + L) \boxdot (r_2 + L) + (r_1 + L) \boxdot (r_3 + L).\end{aligned} \tag{5.1}$$

□

5.1.2. *Negation maps.*

DEFINITION 5.5.

(1) A weakly admissible pair $(\mathcal{A}, \mathcal{A}_0)$ satisfies **weak Property N** when
   (i) there is an element $1^\dagger \in \mathcal{T}$ such that $e := 1 + 1^\dagger = 1^\dagger + 1 \in \mathcal{A}_0$, and $1^\dagger b = b1^\dagger$ for each $b \in \mathcal{A}$.
   (ii) $1 + 1' = e$ for each $1' \in \mathcal{T}$ such that $1 + 1' \in \mathcal{A}_0$. (Thus e is uniquely defined.)
(2) When $(1^\dagger)^\dagger = 1$, the map $a \mapsto (-)a := a1^\dagger$ is called a **negation map**. The pair $(\mathcal{A}, \mathcal{A}_0)$ is **uniquely negated** if $a + a' \in \mathcal{A}_0$ implies $a' = (-)a$, for $a \in \mathcal{T}$.

REMARK 5.6. There is a doubling procedure given in [**3**, §4.4, esp. Lemma 4.20] which embeds a pair into a uniquely negated pair.

5.1.3. *Function pairs.*

DEFINITION 5.7. Write X^S for the functions from a set S to a set X, and $X^{(S)}$ for the functions with finite support. Write X^S_{mon} for those functions whose support is a singleton.

EXAMPLE 5.8.

(i) If $(\mathcal{A}, \mathcal{A}_0)$ is a pair and S is a monoid then $\mathcal{T}^S_{mon}$ is a submonoid of $\mathcal{Z}(\mathcal{A}^S)$, and $(\mathcal{A}^S, \mathcal{A}^S_0)$ and $(\mathcal{A}^{(S)}, \mathcal{A}^{(S)}_0)$ are pairs with respect to $\mathcal{T}^S_{mon}$, whose unit element is the function sending 1 to 1 and all other elements to 0.
(ii) If the pair $(\mathcal{A}, \mathcal{A}_0)$ has a surpassing relation $\preceq$ then $(\mathcal{A}^S, \mathcal{A}^S_0)$ has a surpassing relation $\preceq^S$ given by $f \preceq g$ if $f(s) \preceq g(s)$ for each $s \in S$.
(iii) If the pair $(\mathcal{A}, \mathcal{A}_0)$ has a negation map, then $(\mathcal{A}^S, \mathcal{A}^S_0)$ has a negation map given by $(-)f : s \mapsto (-)f(s)$.
(iv) When $\mathcal{A}$ is a pre-semiring, $A^{(S)}$ becomes a pre-semiring under the convolution product $fg : s \to \sum_{u+v=s} f(u)g(v)$. $A^{(S)}$ is a semiring when A is a semiring.
(v) Write $X[\lambda]$ for $X^{\mathbb{N}}$, where $f(\lambda)$ denotes the function sending n to $f(n)$, for $n \in \mathbb{N}$. when $\mathcal{A}$ is a semiring pair, $(\mathcal{A}[\lambda], \mathcal{A}_0[\lambda])$ is a semiring pair over $\mathcal{T}[\lambda]_{mon}$, the monomials with coefficients in $\mathcal{T}$.

5.2. Common examples of pairs.

EXAMPLE 5.9.

(i) Any $\mathcal{T}$-bimagma $\mathcal{A}$ defines a pair, by taking $\mathcal{A}_0 = \{0\}$.

(ii) Take the hyperpair $(\mathcal{P}^\sharp(\mathcal{H}), \mathcal{P}^\sharp(\mathcal{H})_0)$ of Definition 5.1(iii) when $\mathcal{H}$ is a hypersemiring, which is a $\preceq_\subseteq$-distributive pre-semiring pair, in view of the footnote to Lemma 2.2(ii).

When $\mathcal{H}$ is a hyperring, the hypernegation makes $(\mathcal{A}, \mathcal{A}_0)$ uniquely negated, and $e = 1 \boxplus -1$.

(iii) **Supertropical pairs**, based on [**14**], are described in [**3**, Example 4.2], and are fundamental to tropical algebra. They satisfy $a_1 + a_2 \in \mathcal{A}_0$ if and only if $a_1 = a_2$, so $(\mathcal{A}, \mathcal{A}_0)$ is uniquely negated, with $e = 1 + 1$.

(iv) ([**3**, Example 4.1]) For an arbitrary monoid $\mathcal{T}$, take $\mathcal{A} = \mathcal{T}_0 \cup \{\infty\}$, where $a_1 + a_2 = \infty$ and $a\infty = \infty a = \infty = a + \infty = \infty + a$ for all a, a_i in $\mathcal{T}$. $\mathcal{A}_0 = \{0, \infty\}$. $(\mathcal{A}, \mathcal{A}_0)$ has the negation map $a \mapsto a$, but is not uniquely negated. This pair plays a key role as a terminal object in [**24**]. If instead we declared $a + a = a$ then we still would have Property N, but without a negation map.

5.3. Morphisms of pairs.

Other examples of pairs are given in [**3**, §4] and [**30**, §3.1]. Here is the general version, for pairs, of a weak morphism.

DEFINITION 5.10. A **weak morphism** $f : (\mathcal{A}, \mathcal{A}_0) \to (\mathcal{A}', \mathcal{A}'_0)$ of pairs satisfies $a_1 + a_2 \in \mathcal{A}_0$ implies $f(a_1) + f(a_2) \in \mathcal{A}'_0$ for all $a_i \in \mathcal{T}$.

A **weak morphism** of pre-semiring pairs is a weak morphism which also is a multiplicative map.

Better, we can insert the surpassing relation into our categories.

DEFINITION 5.11. Let $(\mathcal{A}, +, 0)$ (resp. $(\mathcal{A}', +, 0')$) be weakly admissible bimodules over a monoid $\mathcal{T}$ (resp. $\mathcal{T}'$).

(i) When $\mathcal{A}'$ has a surpassing relation $\preceq$, a **$\preceq$-map** is a map $f : \mathcal{A} \to \mathcal{A}'$ satisfying the following conditions:
- $f(0) = 0'$.
- $f(\mathcal{T}) \subseteq \mathcal{T}'$.

(ii) In this case, a **$\preceq$-morphism** $f : (\mathcal{A}, \mathcal{A}_0) \to (\mathcal{A}', \mathcal{A}'_0)$ satisfies the following conditions:
- $f(b_1) \preceq f(b_2), \quad \forall b_1 \preceq b_2 \in \mathcal{A}$.
- $f(ab) = f(a)f(b), \;\; f(ba) = f(b)f(a)$, for $a \in \mathcal{T}$, $b \in \mathcal{A}$.
- $f(b_1 + b_2) \preceq f(b_1) + f(b_2), \quad \forall b_1, b_2 \in \mathcal{A}$.

LEMMA 5.12 (As in [**2**, Lemma 2.10]). *Every $\preceq$-morphism of a pair is a weak morphism.*

LEMMA 5.13. *Any $\subseteq$-morphism $f : \mathcal{H} \to \mathcal{H}'$ of $\mathcal{T}$-bimagmas extends to a $\preceq$-morphism $\hat{f} : \mathcal{P}^\sharp(\mathcal{H}) \to \mathcal{P}^\sharp(\mathcal{H}')$ given by $\hat{f}(S) = \{f(a) : a \in S\}$, with $\hat{f}(\mathcal{P}^\sharp(\mathcal{H}))_0 \subseteq \mathcal{P}^\sharp(\mathcal{H}')_0$.*

PROOF. Clearly if $S_1 \subseteq S_2$ then $f(S_1) \subseteq f(S_2)$, and

$$\begin{aligned} f(S_1 \boxplus S_2) &= \{f(a) : a \in S_1 \boxplus S_2\} = \{f(a_1 \boxplus a_2) : a_i \in S_i\} \\ &\subseteq \{f(a_1) \boxplus f(a_2) : a_i \in S_i\} = f(S_1) \boxplus f(S_2) \end{aligned} \tag{5.2}$$

If $0 \in S$ then $0 = f(0) \in f(S)$. □

6. Categorical constructions

DEFINITION 6.1. We list the basic categories.

- $\mathcal{T}$-**Hmon** is the category of $\mathcal{T}$-hypermonoids, with $\subseteq$-morphisms.
- $\mathcal{T}$-**Pr** is the category of $\mathcal{T}$-pairs with weak morphisms.
- $\mathcal{T}$-**Hmod** is the category of $\mathcal{T}$-hypermodules (where $\mathcal{T}$ can vary, and the morphism acts like a monoid homomorphism on $\mathcal{T}$), with weak morphisms.
- $\mathcal{T}$-**Hsring** is the full subcategory of $\mathcal{T}$-**Hmod** whose objects are $\mathcal{T}$-hypersemirings.
- $\mathcal{T}$-**SringPr** is the subcategory of $\mathcal{T}$-**Pr** whose objects are pre-semiring $\mathcal{T}$-pairs, with weak morphisms of pre-semirings.
- $\mathcal{T}$-$\mathbf{Pr}_{\preceq}$ is the category of $\mathcal{T}$-pairs with $\preceq$-morphisms.
- $\mathcal{T}$-$\mathbf{Hmod}_{\subseteq}$ is the category of $\mathcal{T}$-hypermodules (where $\mathcal{T}$ can vary, and the morphism acts like a monoid homomorphism on $\mathcal{T}$), with $\subseteq$-morphisms.
- $\mathcal{T}$-$\mathbf{Hsring}_{\subseteq}$ is the full subcategory of $\mathcal{T}$-**Hmod** whose objects are $\mathcal{T}$-hypersemirings.
- **presrngPr** is the category of pre-semiring pairs with weak morphisms.
- For any category **C**, the name $\mathbf{C}_{\preceq}$, means the subcategory whose objects have a surpassing relation $\preceq$, and the morphisms are $\preceq$-morphisms.

One reason to consider pairs is that they are preserved under the main constructions of category theory.

REMARK 6.2.

(i) $\mathcal{T}$-**Hmod** resp. $\mathcal{T}$-$\mathbf{Hmod}_{\subseteq}$, both are closed under products and sums, which respect weak morphisms (resp. $\preceq$-morphisms).
(ii) The residue construction respects products and sums in $\mathcal{T}$-**Hmod**. Namely, in the notation of Definition 3.1, $\oplus(\mathcal{M}_i/G_i) \cong (\oplus\mathcal{M}_i)/\oplus G_i)$, under the correspondence $(b_iG_i) \mapsto (b_i)(\oplus G_i)$.
(iii) $\mathcal{T}$-**Hsring** is closed under products and sums.
(iv) The direct sum of $\mathcal{T}$-pairs $(\mathcal{A}_i, \mathcal{A}_{i0})$ is the $\mathcal{T}$-pair $\oplus(\mathcal{A}_i, \mathcal{A}_{i0})$, under the diagonal action $a(b_i) = (ab_i)$. Hence, $\mathcal{T}$-**Pr** is closed under sums (and likewise products), as is its subcategory $\mathcal{T}$-**SringPr**.

REMARK 6.3. The direct product of $\mathcal{T}_i$-pairs $(\mathcal{A}_i, \mathcal{A}_{i0})$ is the $\prod \mathcal{T}_i$-pair $\prod(\mathcal{A}_i, \mathcal{A}_{i0})$, viewed componentwise. Thus, **presrngPr** is closed under products. If $(\mathcal{A}, \mathcal{A}_0)$ is a pre-semiring pair, then

(i) the function construction of Example 5.8(iv) yields a pre-semiring pair.
(ii) the matrix pair $(M_n(\mathcal{A}), M_n(\mathcal{A}_0))$ exists over $\cup_{i,j}\mathcal{T}e_{i,j} \cup \{1\}$, where $e_{i,j}$ are the usual matrix units.

In this way, **presrngPr** is closed under polynomial extensions, power series extensions, Laurent polynomial extensions, and matrix extensions.

As shown in [**30**], tensor products of $\mathcal{T}$-pairs exist, following the lines of [**7**], but taking the $\mathcal{T}$-action into account, as follows.

DEFINITION 6.4. One defines the **free** $\mathcal{T}$-**module** $(\mathcal{F}(X), *)$ on a set X. Let $\mathcal{M}_1$ be a right $\mathcal{T}$-module and $\mathcal{M}_2$ be a left $\mathcal{T}$-module, and $\mathcal{F}_i = \mathcal{F}(M_i), i = 1, 2$. Define the $\mathcal{T}$-**tensor product module** $\mathcal{M}_1 \otimes_{\mathcal{T}} \mathcal{M}_2$ to be the module $(\mathcal{F}_1 \oplus \mathcal{F}_2)/\mathrm{Cong}$, where Cong is the congruence generated by all

$$\left((v_1 * w_1, x_2), (v_1, x_2) * (w_1, x_2))\right), \quad \left((x_1, v_2 * w_2), ((x_1, v_2) * (x_1, w_2))\right), \tag{6.1}$$

$$\left((x_1 a, x_2), (x_1, a x_2)\right), \qquad \forall x_i \in \mathcal{M}_i, v_i, w_i \in \mathcal{F}_i,\ a \in \mathcal{T}. \tag{6.2}$$

If $(\mathcal{M}_i, \mathcal{M}_{i0})$ are pairs, then we define

$$(\mathcal{M}_1 \otimes_{\mathcal{T}} \mathcal{M}_2)_0 = \mathcal{M}_1 \otimes_{\mathcal{T}} \mathcal{M}_{20} + \mathcal{M}_{10} \otimes_{\mathcal{T}} \mathcal{M}_2,$$

to obtain a pair. There are some rather subtle issues, which are treated in [**30**], especially the tensor product of (weak) morphisms.

6.1. Functors connecting our categories.

We get pre-semirings from hypersemirings by passing to the power set.

THEOREM 6.5.

(i) *The Krasner residue map induces a faithful functor from $\mathcal{T}$-**Pr** to $\mathcal{T}$-**Hmon**, which restricts to a faithful functor from $\mathcal{T}$-**SringPr** to $\mathcal{T}$-**Hsring**$_\subseteq$.*
(ii) *There is a fully faithful functor from $\mathcal{T}$-**Hmon** to $\mathcal{T}$-**Pr**, by passing to the power set, which restricts to a faithful functor from $\mathcal{T}$-**Hsring** to $\mathcal{T}$-**SringPr**.*
(iii) *For each category **C** listed above with surpassing relation $\preceq$, the identity map induces faithful functor from $\preceq$-**C** to **C**.*

PROOF. In all of these, one checks that the morphisms match.
(i) By Lemma 3.2. The Krasner residue of a $\mathcal{T}$-pair is a $\mathcal{T}$-pair.
(ii) By Example 5.9(ii) and Lemma 2.2.
(iii) By Lemma 5.12. □

7. Recent results about pairs

Many well-established theories can be extended in the language of pairs. We quote some results from [**3**], for instance.

7.1. Matrix theory.

We start with a $\mathcal{T}$-pair $(\mathcal{A}, \mathcal{A}_0)$ which we assume has a negation map (by doubling, if necessary, cf. Remark 5.6), with $\mathcal{T}$ commutative. Zeilberger [**34**] has recast proofs of matrix theory in combinatorial terms, which are amenable to be formulated in terms of pairs. We say that a matrix $A = (a_{i,j})$ is **tangible** if each of its entries is in $\mathcal{T}_0$.

For example, for a tangible $n \times n$ matrix A we define the $(-)$-determinant

$$|A|_+ = \sum_{\pi \in S_n \text{ even}} a_\pi, \qquad |A|_- = \sum_{\pi \in S_n \text{ odd}} a_\pi, \qquad |A| = |A|_+ - |A|_-. \tag{7.1}$$

THEOREM 7.1 (Generalized Laplace identity, [**3**, Theorem L]). *Laplace's well-known identity*

$$|A| = \sum_{j=1}^{n} (-)^{i+j} a'_{i,j} a_{i,j},$$

for any i, holds over a pair with a negation map, where (a_{ij}) is a square matrix, and (a'_{ij}) the associated comatrix.

More generally, fix $I = \{i_1, \dots, i_m\} \subset \{1, 2, \dots, n\}$, *and for any set* $J = \{j_1, \dots, j_m\} \subset \{1, 2, \dots, n\}$, *write* $(-)^J$ *for* $(-)^{j_1+\cdots+j_m}$, $(a_{I,J})$ *for the* $m \times m$ *minor* $(a_{i,j} : i \in I, j \in J)$, *and* $(a'_{I,J})$ *for the* $(-)$*-determinant of the* $(n-m)\times(n-m)$ *minor obtained by deleting all rows from* I *and all columns from* J. *Then*

$$|A| = \sum_{J:|J|=m} (-)^I(-)^J a'_{I,J}|a_{I,J}|. \tag{7.2}$$

A version of the Cayley-Hamilton theorem over commutative semring pairs is given in [**3**, Theorem K].

7.2. Linear algebra over pairs.

DEFINITION 7.2.

(i) A **vector pair** over $(\mathcal{A}, \mathcal{A}_0)$ is a pair $(\mathcal{V}, \mathcal{V}_0) := (\mathcal{A}, \mathcal{A}_0)^{(S)}$, the set of elements in $\mathcal{A}^S$ with finite support, taken over an index set S. In this discussion we take $S = \{1, \dots, n\}$, so $\mathcal{V} = \mathcal{A}^S$.

(ii) A set of vectors $\{\mathbf{v}_i \in \mathcal{V} : i \in I\}$ is **dependent** if $\sum_{i\in I'} a_i\mathbf{v}_i \in \mathcal{V}_0$ for some nonempty finite subset $I' \subseteq I$ and $a_i \in \mathcal{T}$.

(iii) A vector is **tangible** if each of its entries is in $\mathcal{T}_0$.

A matrix A is **singular** if $|A| \in \mathcal{A}_0$. As in the classical theory, one takes the matrix $A = (v_{ij})$ whose rows are the $\mathbf{v}_i = (v_{i1}, \dots, v_{in})$. Here are some of the main ways of defining the rank of A :

(1) (The row rank) The maximal number of independent rows of A;
(2) (The column rank) The maximal number of independent columns of A;
(3) (The submatrix rank) The size of the largest nonsingular square submatrix of A.

Over a Krasner residue hyperfield, [**4**] notes that the submatrix rank is less than or equal to the row rank and the column rank, and in the other direction, the rows of an $m \times n$ matrix are dependent if $m > n$.

These assertions hold in several situations, but not all, as investigated in [**3**]. Furthermore, the sign hyperfield is a counterexample to equality of the submatrix rank with the column rank, cf. [**3**, Example 8.3].

Acknowledgments

Thanks to the referee for helpful suggestions.

References

[1] Marianne Akian, Stéphane Gaubert, and Alexander Guterman, *Tropical Cramer determinants revisited*, Tropical and idempotent mathematics and applications, Contemp. Math., vol. 616, Amer. Math. Soc., Providence, RI, 2014, pp. 1–45, DOI 10.1090/conm/616/12324. MR3221324

[2] Marianne Akian, Stephane Gaubert, and Louis Rowen, *Semiring systems arising from hyperrings*, J. Pure Appl. Algebra **228** (2024), no. 6, Paper No. 107584, 47, DOI 10.1016/j.jpaa.2023.107584. MR4682509

[3] M. Akian, S. Gaubert, and L.-Rowen, *Linear algebra over pairs*, `arXiv:2310.05257v2` [math.RA], 2025.

[4] M. Baker and T. Zhang *Baker, Matthew and Zhang, Tianyi, On some notions of rank for matrices over tracts*, `arXiv:1601.01204v4`, 2017.

[5] Guy Blachar, Darrell Haile, Eliyahu Matzri, Edan Rein, and Uzi Vishne, *Semiassociative algebras over a field*, J. Algebra **649** (2024), 35–84, DOI 10.1016/j.jalgebra.2024.03.006. MR4725768

[6] C. Brown and S. Pumplün, *Nonassociative cyclic extensions of fields and central simple algebras*, J. Pure Appl. Algebra **223** (2019), no. 6, 2401–2412, DOI 10.1016/j.jpaa.2018.08.018. MR3906555

[7] Guillermo Cortiñas, Christian Haesemeyer, Mark E. Walker, and Charles Weibel, *Toric varieties, monoid schemes and cdh descent*, J. Reine Angew. Math. **698** (2015), 1–54, DOI 10.1515/crelle-2012-0123. MR3294649

[8] A. Almeida Costa, *Sur la théorie générale des demi-anneaux* (French), Publ. Math. Debrecen **10** (1963), 14–29, DOI 10.5486/pmd.1963.10.1-4.03. MR168608

[9] L. E. Dickson, *Linear algebras with associativity not assumed*, Duke Math. J. **1** (1935), no. 2, 113–125, DOI 10.1215/S0012-7094-35-00112-0. MR1545870

[10] L. Gatto and L. Rowen, *Lie pairs*, Communications in Mathematics 32 (2024), Issue 2 (Special issue: CIMPA schools "Nonassociative Algebras and related topics, Brazil'2023" and "Current Trends in Algebra, Philippines'2024") - DOI 10.46298/cm.12413 `arXiv:2309.03867`.

[11] Jonathan S. Golan, *Semirings and their applications*, Kluwer Academic Publishers, Dordrecht, 1999. Updated and expanded version of *The theory of semirings, with applications to mathematics and theoretical computer science* [Longman Sci. Tech., Harlow, 1992; MR1163371 (93b:16085)], DOI 10.1007/978-94-015-9333-5. MR1746739

[12] Joachim Hilgert and Karl H. Hofmann, *Semigroups in Lie groups, semialgebras in Lie algebras*, Trans. Amer. Math. Soc. **288** (1985), no. 2, 481–504, DOI 10.2307/1999949. MR776389

[13] H. C. Hutchins and H. J. Weinert, *Homomorphisms and kernels of semifields*, Period. Math. Hungar. **21** (1990), no. 2, 113–152, DOI 10.1007/BF01946851. MR1070951

[14] Zur Izhakian and Louis Rowen, *Supertropical algebra*, Adv. Math. **225** (2010), no. 4, 2222–2286, DOI 10.1016/j.aim.2010.04.007. MR2680203

[15] Nathan Jacobson, *Basic algebra. II*, W. H. Freeman and Co., San Francisco, CA, 1980. MR571884

[16] Jaiung Jun, Kalina Mincheva, and Louis Rowen, *$\mathcal{T}$-semiring pairs*, Kybernetika (Prague) **58** (2022), no. 5, 733–759, DOI 10.14736/kyb-2022-5-0733. MR4538623

[17] Marc Krasner, *Approximation des corps valués complets de caractéristique $p \neq 0$ par ceux de caractéristique* 0 (French), Colloque d'algèbre supérieure, tenu à Bruxelles du 19 au 22 décembre 1956, Centre Belge de Recherches Mathématiques, Établissements Ceuterick, Louvain, 1957, pp. 129–206. MR106218

[18] Marc Krasner, *Approximation des corps valués complets de caractéristique $p \neq 0$ par ceux de caractéristique* 0 (French), Colloque d'algèbre supérieure, tenu à Bruxelles du 19 au 22 décembre 1956, Centre Belge de Recherches Mathématiques, Établissements Ceuterick, Louvain, 1957, pp. 129–206. MR106218

[19] Marc Krasner, *A class of hyperrings and hyperfields*, Internat. J. Math. Math. Sci. **6** (1983), no. 2, 307–311, DOI 10.1155/S0161171283000265. MR701303

[20] Katarzyna Kuhlmann, Alessandro Linzi, and Hanna Stojałowska, *Orderings and valuations in hyperfields*, J. Algebra **611** (2022), 399–421, DOI 10.1016/j.jalgebra.2022.08.006. MR4474641

[21] F. Marty, *Rôle de la notion de hypergroupe dans i' tude de groupes non abeliens*, Comptes Rendus Acad. Sci. Paris 201, (1935), 636–638.

[22] Ch. G. Massouros, *Methods of constructing hyperfields*, Internat. J. Math. Math. Sci. **8** (1985), no. 4, 725–728, DOI 10.1155/S0161171285000813. MR821630

[23] C. Massouros and G. Massouros, *On the Borderline of Fields and Hyperfields*, Mathematics **11**, (2023), 1289, DOI 10.3390/math11061289.

[24] S. Nakamura, and M. L. Reyes, *Categories of hypermagmas, hypergroups, and related hyperstructures*, `arXiv:2304.09273`, 2025.

[25] J.-C. Petit *Sur certains quasi-corps généralisant un type d'anneau quotient*, Séminaire Dubreil. Algèbre et théorie des nombres, 20(2) (1966), 1–18.

[26] S. Pumplün, *Algebras whose right nucleus is a central simple algebra*, J. Pure Appl. Algebra **222** (2018), no. 9, 2773–2783, DOI 10.1016/j.jpaa.2017.10.019. MR3783018

[27] C. Brown and S. Pumplün, *Nonassociative cyclic extensions of fields and central simple algebras*, J. Pure Appl. Algebra **223** (2019), no. 6, 2401–2412, DOI 10.1016/j.jpaa.2018.08.018. MR3906555

[28] S. Pumplün, *Nonassociative cyclic algebras and the semiassociative Brauer monoid*, Rend. Circ. Mat. Palermo (2) **73** (2024), no. 8, 3253–3275, DOI 10.1007/s12215-024-01105-4. MR4835195

[29] Louis Halle Rowen, *Algebras with a negation map*, Eur. J. Math. **8** (2022), no. 1, 62–138, DOI 10.1007/s40879-021-00499-0. MR4389491

[30] L. H. Rowen, *Tensor products of pairs*, Preprint (2024).

[31] E. M. Vechtomov and D. V. Chuprakov, *The principal kernels of semifields of continuous positive functions* (Russian, with English and Russian summaries), Fundam. Prikl. Mat. **14** (2008), no. 4, 87–107, DOI 10.1007/s10958-009-9688-4; English transl., J. Math. Sci. (N.Y.) **163** (2009), no. 5, 500–514. MR2482035

[32] O. Viro, *Hyperfields for Tropical Geometry I. Hyperfields and dequantization*, arXiv:1006.3034, 2010.

[33] H. J. Weinert and R. Wiegandt, *On the structure of semifields and lattice-ordered groups*, Period. Math. Hungar. **32** (1996), no. 1-2, 129–147, DOI 10.1007/BF01879738. MR1407915

[34] Doron Zeilberger, *A combinatorial approach to matrix algebra*, Discrete Math. **56** (1985), no. 1, 61–72, DOI 10.1016/0012-365X(85)90192-X. MR808086

Department of Mathematics, Bar-Ilan University, Ramat-Gan 52900, Israel
Email address: rowen@math.biu.ac.il

Contemporary Mathematics
Volume **826**, 2025
https://doi.org/10.1090/conm/826/16588

Weights with maximal symmetry and failures of the MacWilliams identities

Jay A. Wood

In memoriam:
Melvin G. Rothenberg, 1934–2023
Erna B. Yackel, 1939–2022
Robert J. Zimmer, 1947–2023

ABSTRACT. This paper examines the w-weight enumerators of weights w with maximal symmetry over finite chain rings and matrix rings over finite fields. In many cases, including the homogeneous weight, the MacWilliams identities for w-weight enumerators fail because there exist two linear codes with the same w-weight enumerator whose dual codes have different w-weight enumerators.

1. Introduction

The MacWilliams identities [**17**] reveal a relationship between the Hamming weight enumerator (hwe) of a linear code C over a finite field $\mathbb{F}_q$ and the Hamming weight enumerator of its dual code $C^{\perp}$:

$$\operatorname{hwe}_{C^{\perp}}(X,Y) = \frac{1}{|C|}\operatorname{hwe}_C(X+(q-1)Y, X-Y).$$

One way to try to generalize this result is to use any integer-valued weight w on a finite ring R with 1. The homogeneous weight, suitably normalized, is an example. Assume $w(0)=0$ and $w(r)>0$ for $r \neq 0$. Denote the maximum value of w by $w_{\max}$. For a vector $x=(x_1,x_2,\ldots,x_n)\in R^n$, set $w(x)=\sum_{i=1}^n w(x_i)$. Define the w-weight enumerator (wwe) of a left R-linear code $C \subseteq R^n$ by

$$\operatorname{wwe}_C(X,Y) = \sum_{c\in C} X^{n w_{\max}-w(c)} Y^{w(c)}.$$

Do MacWilliams identities hold for w-weight enumerators?

This paper will show that MacWilliams identities seldom hold for weights having maximal symmetry when the ring is a finite chain ring or a matrix ring over a finite field. For example, suppose $R=\mathbb{Z}/4\mathbb{Z}$ and w is an integer-valued weight on R with maximal symmetry (i.e., $w(1)=w(3)$ for this ring). Then the only weights for which the MacWilliams identities hold are multiples of the Hamming weight and multiples of the homogeneous weight (the Lee weight for this ring),

2020 *Mathematics Subject Classification.* Primary 94B05.

Key words and phrases. Dual code, finite Frobenius rings, homogeneous weight, MacWilliams identities, partition enumerator, counter-examples.

Corollary 9.6. Similarly, let $R = M_{2\times 2}(\mathbb{F}_q)$ and w be an integer-valued weight with maximal symmetry (i.e., the value of $w(r)$ depends only on the rank of r). Then the only weights for which the MacWilliams identities hold are multiples of the Hamming weight (any q) and multiples of the homogeneous weight ($q = 2$ only), Theorem 19.5. More generally, the homogeneous weight on $R = M_{k\times k}(\mathbb{F}_q)$, $k \geq 2$, satisfies the MacWilliams identities if and only if $k = q = 2$, Theorem 19.4.

REMARK 1.1. There are good reasons for concentrating on chain rings and matrix rings. After finite fields, they are among the simplest examples of finite rings, and they are sufficiently concrete to allow for detailed analysis. Because of this, they are natural testing grounds for conjectures involving linear codes over finite rings.

The reason the MacWilliams identities often fail is that there exist R-linear codes $C, D \subseteq R^n$ for some n such that $\mathrm{wwe}_C = \mathrm{wwe}_D$ but $\mathrm{wwe}_{C^\perp} \neq \mathrm{wwe}_{D^\perp}$. This claim presents two challenges:

- to construct linear codes $C, D \subseteq R^n$ with $\mathrm{wwe}_C = \mathrm{wwe}_D$ in such a way that it is then possible ...
- to detect differences $\mathrm{wwe}_{C^\perp} \neq \mathrm{wwe}_{D^\perp}$ in the w-weight enumerators of their dual codes.

A weight w on a finite ring R with 1 has maximal symmetry when $w(uru') = w(r)$ for all $r \in R$ and units u, u' in R. Any left linear code $C \subseteq R^n$ can be viewed as the image of an injective homomorphism $\Lambda : M \to R^n$ of left R-modules, for some finite left R-module M. The group of units of R acts on M on the left, and the maximal symmetry hypothesis implies that $x \mapsto w(x\Lambda)$ is constant on each orbit of this group action. Writing $[x]$ for the orbit of $x \in M$, we see that

$$\mathrm{wwe}_C(X, Y) = \sum_{\text{orbits } [x]} |[x]| X^{nw_{\max} - w(x\Lambda)} Y^{w(x\Lambda)}. \tag{1.2}$$

In essence, the choice of homomorphism Λ determines the 'orbit weights' $\omega(x) = w(x\Lambda)$ assigned to each orbit $[x]$ in M.

In order to find another linear code $D \subseteq R^n$ with $\mathrm{wwe}_C = \mathrm{wwe}_D$, we permute the weights assigned to orbits while keeping the sum in (1.2) unchanged. For example, in $R = M_{2\times 2}(\mathbb{F}_2)$ there is one orbit of size 6 consisting of matrices of rank 2 and three orbits, each of size 3, of matrices of rank 1. We then seek linear codes whose orbit weights behave as follows:

orbit	$[x_1]$	$[x_2]$	$[x_3]$	$[y]$
size of orbit	3	3	3	6
$w(x\Lambda)$	a	b	b	c
$w(x\Lambda')$	a	c	c	b

Section 16 shows that constructions of this type can be carried out for all matrix rings $M_{k\times k}(\mathbb{F}_q)$ over finite fields.

When R is a finite chain ring, all the nonzero orbits have different sizes, so the permutation idea does not work. However, one can use different modules as the domains of the defining homomorphisms. For example, the ring $\mathbb{Z}/8\mathbb{Z}$ has three modules of size 8: $M = \mathbb{Z}/8\mathbb{Z}$ itself, $\mathbb{Z}/4\mathbb{Z} \oplus \mathbb{Z}/2\mathbb{Z}$, and $M' = \mathbb{Z}/2\mathbb{Z} \oplus \mathbb{Z}/2\mathbb{Z} \oplus \mathbb{Z}/2\mathbb{Z}$. Considering M and M', the orbits of M are $\{1, 3, 5, 7\}$, $\{2, 6\}$, $\{4\}$, and $\{0\}$, while the orbits of M' are the 8 subsets of size 1. By choosing certain unions of orbits of M', say $\{100, 101, 110, 111\}$, $\{011, 010\}$, $\{001\}$, $\{000\}$, of the same size as the orbits of M, one can construct homomorphisms $\Lambda : M \to R^n$ and $\Lambda' : M' \to R^n$ achieving the same weights on corresponding (unions of) orbits. Section 8 details these constructions.

In order to show that $\mathrm{wwe}_{C^\perp} \neq \mathrm{wwe}_{D^\perp}$, it is enough to show that $A_j(C^\perp) \neq A_j(D^\perp)$ for some $j > 0$; here, $A_j(C^\perp)$ is the number of codewords $v \in C^\perp$ with $w(v) = j$. The easiest case to understand is when $v \in C^\perp$ has exactly one nonzero entry; such a v is called a *singleton.* When $C \subseteq R^n$ is the image of a homomorphism $\Lambda : M \to R^n$, the components of $\Lambda = (\lambda_1, \lambda_2, \ldots, \lambda_n)$ are elements $\lambda_i \in \mathrm{Hom}_R(M, R)$. By understanding how many elements $r \in R$ annihilate any given λ_i, i.e., $\lambda_i r = 0$, one can write down formulas for the contributions of singletons to $A_j(C^\perp)$, Proposition 5.2. When j is sufficiently small, only singletons can contribute to $A_j(C^\perp)$, Corollary 5.5. This technique turns out to be surprisingly effective in allowing one to prove that $A_j(C^\perp) \neq A_j(D^\perp)$ in a large number of situations.

This paper is divided into three parts. The first establishes notation and ideas that can apply to any finite ring with 1. In particular, codes will usually be linear codes over a finite ring with 1. A weight w on R will be assumed to have maximal symmetry and have positive integer values, and w will be extended additively to R^n. The second part examines the construction of linear codes and the analysis of singleton dual codewords over a finite chain ring, while the third part does the same for matrix rings over finite fields. An appendix provides a short outline of a proof of the MacWilliams identities over finite Frobenius rings using the Fourier transform and the Poisson summation formula.

Part 1. Generalities

2. Preliminaries

This section will review without proof some terminology and results from [**28**] about characters of finite abelian groups, finite Frobenius rings, additive and linear codes and their dual codes, symmetry groups, and weights.

Let A be a finite abelian group. A *character* of A is a group homomorphism $\pi : A \to (\mathbb{C}^\times, \cdot)$ from A to the multiplicative group of nonzero complex numbers. Denote by $\widehat{A} = \mathrm{Hom}_{\mathbb{Z}}(A, \mathbb{C}^\times)$ the set of all characters of A; $\widehat{A}$ is a finite abelian group under pointwise multiplication of functions. The groups A and $\widehat{A}$ are isomorphic, but not naturally so; $|\widehat{A}| = |A|$. The double character group is naturally isomorphic to the group: $\mathrm{Hom}_{\mathbb{Z}}(\widehat{A}, \mathbb{C}^\times) \cong A$; $a \in A$ corresponds to evaluation at a of $\pi \in \widehat{A}$, i.e., $\pi \mapsto \pi(a)$.

For a subgroup $B \subseteq A$, define its *annihilator* by

$$(\widehat{A} : B) = \{\pi \in \widehat{A} : \pi(b) = 1, \text{for all } b \in B\}.$$

Then $(\widehat{A} : B)$ is a subgroup of $\widehat{A}$, $(\widehat{A} : B) \cong \mathrm{Hom}_{\mathbb{Z}}(A/B, \mathbb{C}^\times)$, and $|(\widehat{A} : B)| = |A/B| = |A|/|B|$. Identifying $A \cong \mathrm{Hom}_{\mathbb{Z}}(\widehat{A}, \mathbb{C}^\times)$, we have $(A : (\widehat{A} : B)) = B$.

Throughout this paper R will denote a finite (associative) ring with 1; R may be noncommutative. The group of units (invertible elements) of R is denoted $\mathcal{U} = \mathcal{U}(R)$. The *Jacobson radical* J of R is the intersection of all maximal left ideals of R; J is itself a two-sided ideal of R. The left/right *socle* $\mathrm{soc}({}_RR)$, $\mathrm{soc}(R_R)$ of R is the left/right ideal generated by the minimal left/right ideals of R. A ring R (perhaps infinite) is *Frobenius* if ${}_RR/J \cong \mathrm{soc}({}_RR)$ and $R/J_R \cong \mathrm{soc}(R_R)$

[**15**, Theorem (16.14)]; a theorem of Honold [**13**] says that one of these isomorphisms suffices for finite rings.

Every finite ring R has an underlying additive abelian group. Its character group $\widehat{R}$ is a bimodule over R. The two scalar multiplications are written in exponential form, with $\pi \in \widehat{R}$, $r, s \in R$:

$$({}^{r}\pi)(s) = \pi(sr), \quad \pi^{r}(s) = \pi(rs).$$

A finite ring R is Frobenius if and only if $R \cong \widehat{R}$ as left (resp., right) R-modules, [**25**, Theorem 3.10]. This implies that a finite Frobenius ring admits a character χ, called a *generating character*, such that $r \mapsto {}^{r}\chi$ is an isomorphism of left R-modules (resp., $r \mapsto \chi^{r}$ is an isomorphism of right R-modules). A generating character has the property that any one-sided ideal of R that is contained in $\ker \chi$ must be the zero ideal.

An *additive code* of length n over R is an additive subgroup $C \subseteq R^n$. If $C \subseteq R^n$ is a left, resp., right, R-submodule, then C is a left (resp., right) R-*linear code*. One way to present a left R-linear code is as the image $C = \operatorname{im} \Lambda$ of a homomorphism $\Lambda : M \to R^n$ of left R-modules, and similarly for right linear codes.

We will write homomorphisms of left R-modules with inputs on the left, so that preservation of scalar multiplication is $(rx)\phi = r(x\phi)$, where $r \in R$, $x \in M$, M a left R-module, and ϕ a homomorphism of left R-modules with domain M.

Define the *standard dot product* on R^n by

$$x \cdot y = \sum_{i=1}^{n} x_i y_i \in R, \tag{2.1}$$

for $x = (x_1, x_2, \ldots, x_n), y = (y_1, y_2, \ldots, y_n) \in R^n$. Given an additive code $C \subseteq R^n$, define dual codes by

$$\begin{aligned} \mathcal{L}(C) &= \{y \in R^n : y \cdot x = 0, \text{for all } x \in C\}, \\ \mathcal{R}(C) &= \{y \in R^n : x \cdot y = 0, \text{for all } x \in C\}. \end{aligned} \tag{2.2}$$

We sometimes write $C^{\perp}$ for $\mathcal{L}(C)$ or $\mathcal{R}(C)$. When R is Frobenius, with generating character χ, also define

$$\begin{aligned} \mathfrak{L}(C) &= \{y \in R^n : \chi(y \cdot x) = 1, \text{for all } x \in C\}, \\ \mathfrak{R}(C) &= \{y \in R^n : \chi(x \cdot y) = 1, \text{for all } x \in C\}. \end{aligned} \tag{2.3}$$

Using the isomorphisms $r \mapsto {}^{r}\chi$ and $r \mapsto \chi^{r}$ of R to $\widehat{R}$, there are isomorphisms $R^n \to \widehat{R}^n$ of left, resp., right, R-modules given by $x \mapsto {}^{x}\chi$ and $x \mapsto \chi^{x}$, where ${}^{x}\chi(y) = \chi(y \cdot x)$ and $\chi^{x}(y) = \chi(x \cdot y)$, for $x, y \in R^n$. Under the isomorphism $x \mapsto {}^{x}\chi$, $\mathfrak{R}(C)$ is taken to $(\widehat{R}^n : C)$, while under the isomorphism $x \mapsto \chi^{x}$, $\mathfrak{L}(C)$ is taken to $(\widehat{R}^n : C)$.

LEMMA 2.4. *Suppose R is Frobenius and $C \subseteq R^n$ is an additive code. Then*

- $|C| \cdot |\mathfrak{L}(C)| = |C| \cdot |\mathfrak{R}(C)| = |R^n|$;
- $\mathfrak{L}(\mathfrak{R}(C)) = C = \mathfrak{R}(\mathfrak{L}(C))$.

REMARK 2.5. Note that $\mathcal{R}(C) \subseteq \mathfrak{R}(C)$ and $\mathcal{L}(C) \subseteq \mathfrak{L}(C)$. In general these containments will be proper. However, if C is a left R-linear code, then $\mathcal{R}(C) = \mathfrak{R}(C)$. Similarly, $\mathcal{L}(C) = \mathfrak{L}(C)$ if C is right R-linear.

A *weight* on R is a function $w : R \to \mathbb{C}$ from R to the complex numbers $\mathbb{C}$ with $w(0) = 0$. In most of this paper we will study weights having positive integer values, except for $w(0) = 0$. A weight w will be extended additively to R^n, so that $w(v) = \sum_{i=1}^n w(v_i) \in \mathbb{C}$, where $v = (v_1, v_2, \ldots, v_n) \in R^n$.

Every weight w on R has two *symmetry groups*, left and right:

$$\begin{aligned} G_{\text{lt}}(w) &= \{u \in \mathcal{U} : w(ur) = w(r) \text{ for all } r \in R\}, \\ G_{\text{rt}}(w) &= \{u \in \mathcal{U} : w(ru) = w(r) \text{ for all } r \in R\}. \end{aligned} \tag{2.6}$$

A weight w has *maximal symmetry* when $G_{\text{lt}}(w) = G_{\text{rt}}(w) = \mathcal{U}$.

EXAMPLE 2.7. The most well-known weight on R is the *Hamming weight* H, defined by $\text{H}(0) = 0$ and $\text{H}(r) = 1$ for $r \neq 0$. The Hamming weight has maximal symmetry.

EXAMPLE 2.8. Another well-known weight on R having maximal symmetry is the *homogeneous weight* $\text{W} : R \to \mathbb{R}$. The homogeneous weight was first introduced in [**5**] over integer residue rings and generalized to all finite rings and modules in [**14**] and [**10**]. A homogeneous weight is characterized by the choice of a real number $\zeta > 0$ and the following properties [**10**]:

- $\text{W}(0) = 0$;
- $G_{\text{lt}}(\text{W}) = \mathcal{U}$; and
- $\sum_{x \in Rr} \text{W}(x) = \zeta |Rr|$ for nonzero principal left ideals $Rr \subseteq R$.

The last property says that all nonzero left principal ideals of R have the same average weight ζ. In fact, the average weight property holds for all nonzero left ideals of R if and only if R is Frobenius [**10**, Corollary 1.6].

When $R = \mathbb{F}_q$, all the nonzero elements are units, so $\text{W}(u) = \text{W}(1)$ for all units u. Thus W is a constant multiple (namely, $\text{W}(1)$) times the Hamming weight. Note that $\zeta = (q-1)\text{W}(1)/q$ over $\mathbb{F}_q$.

Greferath and Schmidt [**10**, Theorem 1.3] prove that homogeneous weights exist on any R by giving an explicit formula for W in terms of ζ and the Möbius function μ (see [**22**, §5.5]) of the poset of principal left ideals of R; namely:

$$\text{W}(r) = \zeta \left(1 - \frac{\mu(0, Rr)}{|\mathcal{U}r|}\right), \quad r \in R. \tag{2.9}$$

This formula implies that all the values of W are rational multiples of ζ. By choosing ζ appropriately, one can produce a homogeneous weight on R with integer values. Another consequence of the formula is that any two homogeneous weights on R are scalar multiples of each other: if W and W' are homogeneous weights on R with average weights ζ and ζ', respectively, then $\text{W}' = (\zeta'/\zeta)\text{W}$.

EXAMPLE 2.10. Let $R = M_{k\times k}(\mathbb{F}_q)$ be the ring of $k \times k$ matrices over the finite field $\mathbb{F}_q$. The *rank weight* R is defined by $\text{R}(r) = \text{rk}(r)$, the usual rank of the matrix $r \in R$. The rank weight has maximal symmetry.

3. MacWilliams identities

In her 1962 doctoral dissertation, Florence Jessie MacWilliams gave a formula relating the Hamming weight enumerator of a linear code over a finite field to the Hamming weight enumerator of its dual code [**16, 17**]. In this section, we will describe this work of MacWilliams as well as some of its generalizations.

For any linear code $C \subseteq R^n$ over a finite ring R, the *Hamming weight enumerator* is the following homogeneous polynomial of degree n:

$$\operatorname{hwe}_C(X,Y) = \sum_{x \in C} X^{n-\mathrm{H}(x)} Y^{\mathrm{H}(x)}, \tag{3.1}$$

where H is the Hamming weight, as in Example 2.7. The formula relating hwe_C and $\operatorname{hwe}_{C^\perp}$ is quoted next.

THEOREM 3.2 (MacWilliams identities [**16, 17**]). *If $C \subseteq \mathbb{F}_q^n$ is a linear code over the finite field $\mathbb{F}_q$, then*

$$\operatorname{hwe}_{C^\perp}(X,Y) = \frac{1}{|C|}\operatorname{hwe}_C(X+(q-1)Y, X-Y).$$

REMARK 3.3. Note in particular that the formula for $\operatorname{hwe}_{C^\perp}$ depends only on hwe_C and not on a more detailed knowledge of the code C. By applying the MacWilliams identities to $C^\perp$ and $C = (C^\perp)^\perp$, the roles of C and $C^\perp$ can be reversed.

We isolate one consequence of the MacWilliams identities.

COROLLARY 3.4. *If C and D are two linear codes over $\mathbb{F}_q$ with* $\operatorname{hwe}_C = \operatorname{hwe}_D$, *then* $\operatorname{hwe}_{C^\perp} = \operatorname{hwe}_{D^\perp}$.

The MacWilliams identities for the Hamming weight enumerator can be generalized in several ways. One way is to generalize the algebraic structure of the codes. There are versions of the MacWilliams identities with the Hamming weight enumerator for additive codes over finite abelian groups [**6**], as well as for left (or right) linear codes over a finite Frobenius ring [**25**, Theorem 8.3]. In the latter, one replaces q with $|R|$ and $C^\perp$ with $\mathcal{R}(C)$ (with $\mathcal{L}(C)$ if C is right linear).

Another way to generalize the MacWilliams identities is to generalize the enumerator. There are two broad ways of doing this, stemming from two interpretations of the exponents in (3.1). Following Gluesing-Luerssen [**8**], one of the generalizations will be called *partition enumerators*; the other will be called *w-weight enumerators*. These enumerators will be defined below, and the Hamming weight enumerator will be an example of both. While most of the following material can be formulated for additive codes over finite abelian groups, the discussion here will be restricted to linear codes over finite rings.

Suppose S is a finite set. A *partition* of S is a collection $\mathcal{P} = \{P_i\}$ of nonempty subsets of S such that the subsets are pairwise disjoint and cover S, i.e., $S = \biguplus_i P_i$. The subsets P_i are called the *blocks* of the partition.

Suppose a finite ring R has a partition $\mathcal{P} = \{P_i\}_{i=1}^m$. Define counting functions $n_i : R^n \to \mathbb{N}$, $i = 1, 2, \ldots, m$, by $n_i(x) = |\{j : x_j \in P_i\}|$, for $x = (x_1, x_2, \ldots, x_n) \in R^n$. The counting functions count how many entries of x belong to each block of the partition. For a linear code $C \subseteq R^n$ define the *partition enumerator* associated to C and the partition $\mathcal{P}$ to be the following homogeneous polynomial of degree n in the variables $Z_1, Z_2, \ldots, Z_m$:

$$\operatorname{pe}_C^{\mathcal{P}}(Z_1, \ldots, Z_m) = \sum_{x \in C} \prod_{i=1}^m Z_i^{n_i(x)}. \tag{3.5}$$

Examples of such partition enumerators include:

- the *complete enumerator* (ce) using the singleton partition $\mathcal{P} = \{\{r\}\}_{r \in R}$;

- a *symmetrized enumerator* (se) based on a partition consisting of the orbits of a group action on R;
- the Hamming (weight) enumerator based on the partition with blocks $\{0\}$ and the set difference $R - \{0\}$.

While the literature refers to the examples above as weight enumerators, the first two do not involve weights, so we will use the shorter names indicated.

Suppose R has two partitions $\mathcal{P} = \{P_i\}_{i=1}^{m}$ and $\mathcal{Q} = \{Q_j\}_{j=1}^{m'}$. Also suppose $\mathcal{P}$ is a *refinement* of $\mathcal{Q}$, i.e., each block P_i is contained in some (unique) block Q_j; write $j = f(i)$. Write the partition enumerators of a linear code $C \subseteq R^n$, using variables Z_i, $i = 1, 2, \ldots, m$, for $\mathcal{P}$, and $\mathcal{Z}_j$, $j = 1, 2, \ldots, m'$, for $\mathcal{Q}$:

$$\mathrm{pe}_C^{\mathcal{P}}(Z_1, \ldots, Z_m) \quad \text{and} \quad \mathrm{pe}_C^{\mathcal{Q}}(\mathcal{Z}_1, \ldots, \mathcal{Z}_{m'}).$$

The specialization of variables $Z_i \rightsquigarrow \mathcal{Z}_{f(i)}$ allows us to write the $\mathcal{Q}$-enumerator in terms of the $\mathcal{P}$-enumerator:

$$\mathrm{pe}_C^{\mathcal{Q}}(\mathcal{Z}_1, \ldots, \mathcal{Z}_{m'}) = \mathrm{pe}_C^{\mathcal{P}}(Z_1, \ldots, Z_m)\big|_{Z_i \rightsquigarrow \mathcal{Z}_{f(i)}}. \tag{3.6}$$

The MacWilliams identities are known to generalize to the complete enumerator and certain symmetrized enumerators, over finite fields [**18**] and finite Frobenius rings [**25**]. The MacWilliams identities generalize to so-called reflexive partition enumerators over finite Frobenius rings; see [**8**] for details. For the symmetrized enumerator, see Theorem 11.6. A short review of the main arguments used for proving the MacWilliams identities over finite Frobenius rings is in Appendix A.

For the other type of enumerator, suppose R is a finite ring with 1, and w is a weight on R with positive integer values (except $w(0) = 0$). Denote the largest value of w by $w_{\max}$. For any left R-linear code $C \subseteq R^n$, define the *w-weight enumerator* of C by

$$\mathrm{wwe}_C(X, Y) = \sum_{x \in C} X^{n w_{\max} - w(x)} Y^{w(x)}. \tag{3.7}$$

The w-weight enumerator is a homogeneous polynomial of degree $n w_{\max}$ in X and Y. Different codewords in C may have the same weight. Collecting terms in (3.7) leads to

$$\mathrm{wwe}_C(X, Y) = \sum_{j=0}^{n w_{\max}} A_j^w(C) X^{n w_{\max} - j} Y^j, \tag{3.8}$$

where $A_j^w(C)$ is the number of codewords of C having weight j:

$$A_j^w(C) = |\{x \in C : w(x) = j\}|. \tag{3.9}$$

We write $A_j(C)$ when w is clear from context. To save space in examples in later sections we will often write wwe with $X = 1$ and $Y = t$, so that $\mathrm{wwe}_C = \sum_j A_j(C) t^j$. When $w = \mathrm{H}$, the Hamming weight, we recover the Hamming weight enumerator hwe.

REMARK 3.10. A disadvantage of using the notation $\mathrm{wwe}_C = \sum_j A_j(C) t^j$ is that information about the length n of the code is lost. Of course, if the length of C is known, then the homogeneous form (3.7) of wwe is easily recovered. For example, suppose C is a linear code of length n, and let D be the linear code of length $n+1$ obtained by appending a zero to each codeword of C. Since $w(0) = 0$, there are no changes in the weights of the codewords, so that $A_j(C) = A_j(D)$ for all j. However, $\mathrm{wwe}_D(X, Y) = X\, \mathrm{wwe}_C(X, Y)$.

The partition enumerators and the w-weight enumerators are related. Given a weight w on R, let $\mathcal{Q}$ be the partition of R into the orbits $\operatorname{orb}(r)$ of $G_{\mathrm{lt}}(w)$ acting on R on the left, and let $\mathcal{P}$ be the complete partition of R; $\mathcal{P}$ is a refinement of every partition, hence a refinement of $\mathcal{Q}$. Use variables Z_r, $r \in R$, for $\mathcal{P}$, and $\mathcal{Z}_{\operatorname{orb}(r)}$ for $\mathcal{Q}$. Then the specialization of variables $\mathcal{Z}_{\operatorname{orb}(r)} \rightsquigarrow X^{w_{\max}-w(r)}Y^{w(r)}$ allows us to write wwe_C in terms of $\mathrm{se}_C^{\mathcal{Q}}$ for any R-linear code $C \subseteq R^n$:

$$\mathrm{wwe}_C(X,Y) = \mathrm{se}_C^{\mathcal{Q}}(\mathcal{Z}_{\operatorname{orb}(r)})\big|_{\mathcal{Z}_{\operatorname{orb}(r)} \rightsquigarrow X^{w_{\max}-w(r)}Y^{w(r)}}\,. \tag{3.11}$$

This specialization is well-defined by the definition of the symmetry group $G_{\mathrm{lt}}(w)$: the value of w is constant on every left orbit of $G_{\mathrm{lt}}(w)$. Section 11 gives details of this situation over finite chain rings.

One way to view the MacWilliams identities is in terms of the diagram in Figure 1 below. In the diagram, the map $\mathcal{R}$ sends a left R-linear code of size

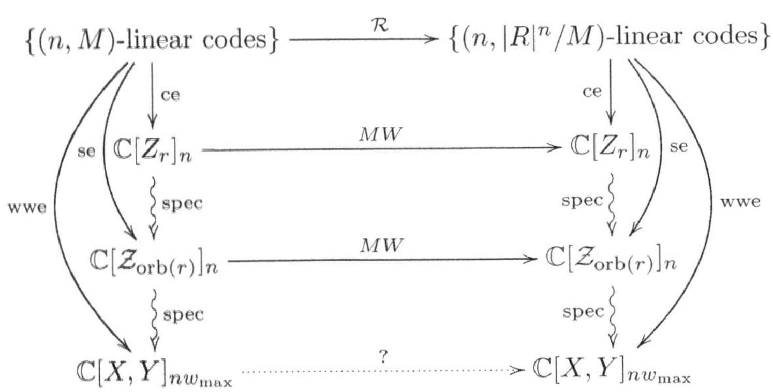

FIGURE 1. Relations among enumerators

M to its right R-linear dual code $\mathcal{R}(C)$. Under favorable circumstances (e.g., R Frobenius), the dual code has size $|\mathcal{R}(C)| = |R|^n/|C|$. The vertical maps ce, se, wwe associate to a linear code its complete enumerator, $\mathcal{Q}$-symmetrized enumerator, and w-weight enumerator, respectively. The other vertical maps (both called spec) are the specializations of variables described in (3.6) and (3.11).

Because the MacWilliams identities hold for ce and se, the horizontal maps MW are the MacWilliams transforms that provide the linear changes of variables. The solid arrows in the diagram commute.

The big question under study in this paper is whether there is a horizontal map '?' that makes the diagram commute for the w-weight enumerator. If such a map exists, then the following property holds: if $\mathrm{wwe}_C = \mathrm{wwe}_D$, then $\mathrm{wwe}_{\mathcal{R}(C)} = \mathrm{wwe}_{\mathcal{R}(D)}$. We refer to this property by saying *the weight w respects duality.* To formalize:

DEFINITION 3.12. A weight w on a finite ring R *respects duality* if $\mathrm{wwe}_C = \mathrm{wwe}_D$ implies $\mathrm{wwe}_{\mathcal{R}(C)} = \mathrm{wwe}_{\mathcal{R}(D)}$ for all left R-linear codes $C, D \subseteq R^n$, $n \geq 1$.

Corollary 3.4 says that the Hamming weight on $\mathbb{F}_q$ respects duality. If a weight w does not respect duality, then the MacWilliams identities cannot hold for wwe.

As an example, let us see what happens when a weight is multiplied by a positive constant. Suppose $w : R \to \mathbb{Z}$ is a weight on R, and let $\tilde{w} = cw$, with c a positive integer. Denote the w-weight enumerators for w and $\tilde{w}$ by wwe and wwe^c respectively.

LEMMA 3.13. *Let $w : R \to \mathbb{Z}$ be a weight on R, and let $\tilde{w} = cw$, with c a positive integer. Then*

$$\mathrm{wwe}_C^c(X, Y) = \mathrm{wwe}_C(X^c, Y^c).$$

PROOF. For any element $r \in R$, we have

$$X^{\tilde{w}_{\max} - \tilde{w}(r)} Y^{\tilde{w}(r)} = (X^c)^{w_{\max} - w(r)} (Y^c)^{w(r)}. \qquad \square$$

PROPOSITION 3.14. *Let $w : R \to \mathbb{Z}$ be a weight on R, and let $\tilde{w} = cw$, with c a positive integer. Then $\tilde{w}$ respects duality if and only if w respects duality.*

Even if a weight w does not respect duality (and hence wwe does not satisfy the MacWilliams identities), it is still possible to determine both wwe_C and $\mathrm{wwe}_{\mathcal{R}(C)}$ by calculating se_C, using the MacWilliams identities for se to find $\mathrm{se}_{\mathcal{R}(C)}$, and then specializing variables to get wwe_C and $\mathrm{wwe}_{\mathcal{R}(C)}$. But one cannot go directly from wwe_C to $\mathrm{wwe}_{\mathcal{R}(C)}$.

The main objective of this paper is to show that it is rare for a weight having maximal symmetry to respect duality, at least over finite chain rings or matrix rings over finite fields. In addition to Theorem 3.2 for the Hamming weight enumerator and its generalization to finite Frobenius rings, the MacWilliams identities for w-weight enumerators are known to hold for the Lee weight on $\mathbb{Z}/4\mathbb{Z}$ [**12**] (and see Theorem 6.5) and the homogeneous weight on the matrix ring $M_{2\times 2}(\mathbb{F}_2)$, Theorem 12.12. The MacWilliams identities for w-weight enumerators are known to fail for the Rosenbloom-Tsfasman weight on matrices [**7**], the Lee weight on $\mathbb{Z}/m\mathbb{Z}$, $m \geq 5$ [**1**], and the homogeneous weight on $\mathbb{Z}/m\mathbb{Z}$ for composite $m \geq 6$ [**30**]. In all cases, the failure is proved by showing that the weight does not respect duality: there exist linear codes C and D with $\mathrm{wwe}_C = \mathrm{wwe}_D$, yet $\mathrm{wwe}_{\mathcal{R}(C)} \neq \mathrm{wwe}_{\mathcal{R}(D)}$, by virtue of $A_j^w(\mathcal{R}(C)) \neq A_j^w(\mathcal{R}(D))$ for some j.

The hypothesis that w has maximal symmetry is important. There are results about 'Lee weights' of different types that can be valid because they secretly tap into the Hamming weight or the Lee weight on $\mathbb{Z}/4\mathbb{Z}$; cf., [**31**].

4. Linear codes via multiplicity functions

In later sections, linear codes will be presented as images of homomorphisms of left R-modules. In turn, the homomorphisms will be described in terms of multiplicity functions. In this section, the use of multiplicity functions to describe linear codes will be summarized briefly. Addititonal information can be found in [**26**, §3] or [**28**, §7].

As throughout this paper, let R be a finite ring with 1. Suppose M is a finite unital left R-module; *unital* means $1x = x$ for all $x \in M$. We call any homomorphism $\lambda : M \to R$ of left R-modules a *linear functional* on M. Define $M^\sharp = \mathrm{Hom}_R(M, R)$ to be the set of all left linear functionals on M. We will write inputs to linear functionals on the left, so that $(rx)\lambda = r(x\lambda)$ for $r \in R$, $x \in M$, and $\lambda \in M^\sharp$; $M^\sharp$ is a right R-module, with λr given by $x(\lambda r) = (x\lambda)r$ for $r \in R$, $x \in M$, and $\lambda \in M^\sharp$. The left R-module M admits a left action by the group of units $\mathcal{U}$ of R using left scalar multiplication. Denote the orbit of $x \in M$ by $\mathrm{orb}(x)$

or by $[x]$. Similarly, the right R-module $M^\sharp$ admits a right $\mathcal{U}$-action, with orbits denoted $\operatorname{orb}(\lambda)$ or $[\lambda]$.

A left R-linear code of length n *parametrized* by M is the image $C = \operatorname{im}\Lambda$ of a homomorphism $\Lambda : M \to R^n$ of left R-modules. The module M is the *information module* of the linear code C. Denote the components of Λ by $\Lambda = (\lambda_1, \lambda_2, \dots, \lambda_n)$, with each $\lambda_i : M \to R$ being a linear functional on M. We call the λ_i the *coordinate functionals* of the linear code C.

Suppose w is a weight on R. Then the weight w and a parametrized code C given by $\Lambda : M \to R^n$ define a *weight function* $W_\Lambda : M \to \mathbb{C}$ by $W_\Lambda(x) = w(x\Lambda) = \sum_{i=1}^n w(x\lambda_i)$, $x \in M$.

LEMMA 4.1. *Suppose w is a weight on R with symmetry groups* (2.6), *and suppose C is an R-linear code parametrized by $\Lambda : M \to R^n$. Then*

(1) *the weight function $W_\Lambda : M \to \mathbb{C}$ is constant on each left $G_{\mathrm{lt}}(w)$-orbit $\operatorname{orb}(x) \subseteq M$;*
(2) *$W_{\Lambda'} = W_\Lambda$, if $\Lambda' = (\lambda_{\tau(1)}u_1, \dots, \lambda_{\tau(n)}u_n)$, where τ is a permutation of $\{1, 2, \dots, n\}$ and $u_1, \dots, u_n \in G_{\mathrm{rt}}(w)$.*

PROOF. These follow directly from (2.6). □

REMARK 4.2. If the weight w has nonnegative integer values, the weight function W_Λ determines the w-weight enumerator of the linear code C:

$$\mathrm{wwe}_C = \frac{1}{|\ker \Lambda|} \sum_{x \in M} t^{W_\Lambda(x)}.$$

When Λ is injective, the w-weight enumerator can be written in terms of the sizes of orbits:

$$\mathrm{wwe}_C = \sum_{[x] \subseteq M} |[x]| t^{W_\Lambda(x)}. \tag{4.3}$$

Lemma 4.1 shows that the weight function W_Λ depends only on the numbers of coordinate functionals belonging to different $G_{\mathrm{rt}}(w)$-orbits in $M^\sharp$. We formalize this observation next. Write $F(X, Y)$ for the set of all functions from X to Y.

Given an information module M, define the orbit spaces:

$$\mathcal{O} = G_{\mathrm{lt}}(w) \backslash M, \quad \mathcal{O}^\sharp = M^\sharp / G_{\mathrm{rt}}(w).$$

Lemma 4.1 shows that W_Λ depends only on a function $\eta \in F(\mathcal{O}^\sharp, \mathbb{N})$, where $\eta([\lambda])$ is the number of coordinate functionals that belong to the orbit $\operatorname{orb}(\lambda) = [\lambda]$. (In [**21**, §3.6], Peterson and Weldon call η the *modular representation* of the linear code C.)

The weight w induces an additive map $W : F(\mathcal{O}^\sharp, \mathbb{N}) \to F(\mathcal{O}, \mathbb{C})$. For $\eta \in F(\mathcal{O}^\sharp, \mathbb{N})$, define $\omega = W(\eta)$, the *list of orbit weights*, by

$$\omega([x]) = \sum_{[\lambda] \in \mathcal{O}^\sharp} w(x\lambda)\eta([\lambda]). \tag{4.4}$$

By (2.6), the value of $w(x\lambda)$ is well-defined. If w has values in $\mathbb{Z}$ or $\mathbb{Q}$, then W has values in $F(\mathcal{O}, \mathbb{Z})$ or $F(\mathcal{O}, \mathbb{Q})$, accordingly. In these latter cases, tensoring with $\mathbb{Q}$ yields a linear transformation $W : F(\mathcal{O}^\sharp, \mathbb{Q}) \to F(\mathcal{O}, \mathbb{Q})$.

By ordering the elements of $\mathcal{O}$ and $\mathcal{O}^\sharp$, one can define a matrix W whose rows are indexed by $\mathcal{O}$, whose columns are indexed by $\mathcal{O}^\sharp$, and whose entry at position $([x],[\lambda])$ is the well-defined value $w(x\lambda)$:

$$W_{[x],[\lambda]} = w(x\lambda), \quad [x] \in \mathcal{O}, [\lambda] \in \mathcal{O}^\sharp. \tag{4.5}$$

Treating η and ω as column vectors, (4.4) is just matrix multiplication: $\omega = W\eta$.

Any element $\eta \in F(\mathcal{O}^\sharp, \mathbb{N})$, called a *multiplicity function*, determines an R-linear code as the image of the homomorphism $\Lambda_\eta : M \to R^N$ of left R-modules given by sending $x \in M$ to the N-tuple $(\ldots, x\lambda, \ldots)$, where a representative of each orbit $[\lambda] \in \mathcal{O}^\sharp$ is repeated $\eta([\lambda])$ times; $N = \sum_{[\lambda]\in\mathcal{O}^\sharp} \eta([\lambda])$. Said another way, treat elements $\lambda \in M^\sharp$ as columns of a generator matrix, with representatives of $[\lambda]$ repeated $\eta([\lambda])$ times. The resulting linear code C_η is well-defined up to monomial equivalence; its list $\omega = W\eta$ of orbit weights is well-defined. Using ω, one can write down the w-weight enumerator of C_η, as in Remark 4.2:

$$\mathrm{wwe}_{C_\eta} = \sum_{x\in M} t^{w(x\Lambda_\eta)} = \sum_{[x]\in\mathcal{O}} |[x]| t^{\omega([x])},$$

assuming that Λ_η is injective. (If Λ_η is not injective, we must divide by $|\ker \Lambda_\eta|$.)

REMARK 4.6. The definition of $W : F(\mathcal{O}^\sharp, \mathbb{N}) \to F(\mathcal{O}, \mathbb{C})$ is valid whether or not $w(0) = 0$. When $w(0) = 0$, then $W(\eta)([0]) = \omega([0]) = 0$ for any $\eta \in F(\mathcal{O}^\sharp, \mathbb{N})$. Also, when $w(0) = 0$, $W([\delta_{[0]}]) = 0$, where $\delta_{[0]}$ is the indicator function of the orbit $[0] \in \mathcal{O}^\sharp$ of the zero-functional.

More formally, define

$$F_0(\mathcal{O}^\sharp, \mathbb{N}) = \{\eta \in F(\mathcal{O}^\sharp, \mathbb{N}) : \eta([0]) = 0\},$$
$$F_0(\mathcal{O}, \mathbb{C}) = \{\omega \in F(\mathcal{O}, \mathbb{C}) : \omega([0]) = 0\}.$$

When $w(0) = 0$, the image of W is contained in $F_0(\mathcal{O}, \mathbb{C})$. We denote the restriction of W to $F_0(\mathcal{O}^\sharp, \mathbb{N})$ by $W_0 : F_0(\mathcal{O}^\sharp, \mathbb{N}) \to F_0(\mathcal{O}, \mathbb{C})$; W_0 is an additive map. The map W_0 is represented by a matrix whose rows are indexed by the nonzero orbits $[x] \in \mathcal{O}$, whose columns are indexed by the nonzero orbits $[\lambda] \in \mathcal{O}^\sharp$, and with entries given as in (4.5). When $w(0) = 0$, the map W can never be injective (because $\delta_{[0]} \in \ker W$), but the map W_0 often is injective. When w has values in $\mathbb{Q}$ and $w(0) = 0$, the linear transformation $W_0 : F_0(\mathcal{O}^\sharp, \mathbb{Q}) \to F_0(\mathcal{O}, \mathbb{Q})$ is often invertible. This will be an important tool in later sections.

We conclude this section with a short discussion of the effective length of codes. The next lemma is a variant of [**26**, (6.1)].

LEMMA 4.7. *Suppose $C \subseteq R^n$ is an R-linear code. Then*

$$\sum_{c\in C} w(c) = \sum_{i=1}^{n} |\ker \lambda_i| \sum_{b\in \operatorname{im} \lambda_i} w(b). \tag{4.8}$$

PROOF. Write $\lambda_1, \ldots, \lambda_n$ for the coordinate functionals of C. Then,

$$\sum_{c\in C} w(c) = \sum_{c\in C}\sum_{i=1}^{n} w(c_i) = \sum_{c\in C}\sum_{i=1}^{n} w(c\lambda_i).$$

Interchange the finite sums, and use that λ_i is a homomorphism, so $\sum_{c\in C} w(c\lambda_i) = |\ker \lambda_i| \sum_{b\in \operatorname{im}\lambda_i} w(b)$. □

There are situations where (4.8) simplifies. Define a weight w on R to be *egalitarian* if there exists a constant γ such that, for any nonzero left ideal $B \subseteq R$, $\sum_{b\in B} w(b) = \gamma|B|$. This definition is due to [**14**]. The homogeneous weight on a finite Frobenius ring is an example of an egalitarian weight; see Example 2.8.

Define the *effective length* of a linear code C to be $\operatorname{efflng}(C) = |\{i : \lambda_i \neq 0\}|$. If C is given by a generator matrix, the effective length counts the number of nonzero columns of the generator matrix.

PROPOSITION 4.9. *Suppose $C \subseteq R^n$ is an R-linear code and w is an egalitarian weight on R. Then*

$$\sum_{c\in C} w(c) = \gamma|C| \operatorname{efflng}(C).$$

PROOF. Use Lemma 4.7. For $\lambda_i \neq 0$, set $B = \operatorname{im}\lambda_i$ and note that $|C| = |\ker\lambda_i| \cdot |\operatorname{im}\lambda_i|$. □

COROLLARY 4.10. *Let w be an egalitarian weight on R. Suppose $C, D \subseteq R^n$ are two R-linear codes. If $\operatorname{wwe}_C = \operatorname{wwe}_D$, then the effective lengths of C and D are equal.*

PROOF. The hypothesis means that $A_j^w(C) = A_j^w(D)$ for all j. But

$$|C| = \sum_{j=0}^{nw_{\max}} A_j^w(C) \quad \text{and} \quad \sum_{c\in C} w(c) = \sum_{j=0}^{nw_{\max}} jA_j^w(C),$$

so $|C| = |D|$ and $\sum_{c\in C} w(c) = \sum_{d\in D} w(d)$. Apply Proposition 4.9. □

5. Singletons in dual codes

In this section, we describe some general results on the contributions to $\operatorname{wwe}_{C^\perp}$ coming from singleton vectors.

As usual, suppose R is a finite ring with 1 and $C \subseteq R^n$ is a left R-linear code. Its dual code $C^\perp$ is the right dual code $\mathcal{R}(C)$ of (2.2). We assume w is an integer-valued weight on R with $w(r) > 0$ for $r \neq 0$ and $w(0) = 0$. Let $\mathring{w} = \min\{w(r) : r \neq 0\}$, so that $\mathring{w} > 0$.

We say that a vector $v \in R^n$ is a *singleton* if v has exactly one nonzero entry. Given a vector $v \in R^n$, recall that the weight of the vector is $w(v) = \sum_{j=1}^n w(v_j)$. The smallest possible nonzero weight of a vector is $\mathring{w}$, which is attained by any singleton whose nonzero entry r has $w(r) = \mathring{w}$.

We want to write down the contributions of singletons to the w-weight enumerator of a linear code, especially to a dual code. As in (3.9), recall that $A_j(C) = |\{x \in C : w(x) = j\}|$. To track the contributions of singletons we write

$$A_j^{\text{sing}}(C) = |\{x \in C : x \text{ is a singleton and } w(x) = j\}|.$$

Of course, $A_j^{\text{sing}}(C) \leq A_j(C)$. Equality will be addressed in Corollary 5.5 below.

For any $\lambda \in M^\sharp$ and positive integer j, define $\operatorname{ann}_{\text{rt}}(\lambda, j) = \{r \in R : \lambda r = 0 \text{ and } w(r) = j\}$, the set of elements in R of weight j that annihilate λ.

LEMMA 5.1. *Suppose w has maximal symmetry. For any $\lambda \in M^\sharp$ and $u \in \mathcal{U}$, $\operatorname{ann}_{\text{rt}}(\lambda u, j) = u^{-1} \operatorname{ann}_{\text{rt}}(\lambda, j)$. In particular, $|\operatorname{ann}_{\text{rt}}(\lambda u, j)| = |\operatorname{ann}_{\text{rt}}(\lambda, j)|$ for any $u \in \mathcal{U}$, $\lambda \in M^\sharp$.*

PROOF. Suppose $w(r) = j$. By maximal symmetry, $w(ur) = w(r) = j$ for all $u \in \mathcal{U}$. Because $(\lambda u)(u^{-1}r) = \lambda r$, we see that $r \in \operatorname{ann}_{\mathrm{rt}}(\lambda, j)$ if and only if $u^{-1}r \in \operatorname{ann}_{\mathrm{rt}}(\lambda u, j)$. □

PROPOSITION 5.2. *Assume w is an integer-valued weight on R with maximal symmetry. If C is a linear code determined by a multiplicity function η, then, for any positive integer j,*

$$A_j^{\mathrm{sing}}(C^\perp) = \sum_{[\lambda]\in\mathcal{O}^\sharp} |\operatorname{ann}_{\mathrm{rt}}(\lambda, j)|\eta([\lambda]).$$

PROOF. Suppose C has coordinate functionals $\lambda_1, \dots, \lambda_n$. Let v be a singleton vector with nonzero entry $r \in R$ appearing in position i. Then, using the form of the standard product (2.1), $v \in C^\perp$ if and only if $\lambda_i r = 0$. Thus

$$A_j^{\mathrm{sing}}(C^\perp) = \sum_{i=1}^{n} |\operatorname{ann}_{\mathrm{rt}}(\lambda_i, j)|,$$

which reduces to the stated formula because of Lemma 5.1. □

In later sections, formulas for $|\operatorname{ann}_{\mathrm{rt}}(\lambda, j)|$ will be very specific, depending on the nature of the ring R.

REMARK 5.3. The analysis of dual codewords is later sections depends almost entirely on using Proposition 5.2 to determine $A_j^{\mathrm{sing}}(C^\perp)$. In turn, Proposition 5.2 works because the dot product of a singleton with nonzero entry r in position j with another vector x is just $x_j r$. If one were to use a different inner product, the expression could be more complicated: $\sum_i x_i \alpha_{i,j} r$, where $\alpha_{i,j}$ is the inner product of the standard singleton basis vectors e_i and e_j. If there is only one term in the sum (so that $(\alpha_{i,j})$ is an invertible monomial matrix), the analysis of singleton dual codewords stays the same. Otherwise, a whole new approach would be needed.

LEMMA 5.4. *Suppose $v \in R^n$ has weight $w(v)$ satisfying $\mathring{w} \le w(v) < 2\mathring{w}$. Then v must be a singleton.*

PROOF. Suppose v has at least two nonzero entries, say in positions j_1, j_2. Then $w(v) \ge w(v_{j_1}) + w(v_{j_2}) \ge 2\mathring{w}$. □

COROLLARY 5.5. *If $\mathring{w} \le d < 2\mathring{w}$, then $A_d(C) = A_d^{\mathrm{sing}}(C)$.*

In later sections, Corollary 5.5 will be applied mostly to dual codes, in tandem with Proposition 5.2.

REMARK 5.6. In order that $A_j^{\mathrm{sing}}(C)$ be nonzero, it is necessary that $j = w(r)$ for some $r \in R$.

It is possible that $A_j(C) = A_j^{\mathrm{sing}}(C)$ even when $j \ge 2\mathring{w}$. For example: when $j = w(r)$ is not equal to a linear combination of the form $\sum_{s:w(s)<j} c_s w(s)$ with c_s being nonnegative integers.

Part 2. Finite chain rings

6. Definitions and a positive result

A finite ring R with 1 is a *chain ring* if its left ideals form a chain under set inclusion. In particular, R has a unique maximal left ideal, denoted $\mathfrak{m}$, so that R is a local ring. Examples of chain rings include finite fields, $\mathbb{Z}/p^m\mathbb{Z}$ with p prime, Galois rings, $\mathbb{F}_q[X]/(X^m)$; cf., [**19**]. Every finite chain ring is Frobenius [**24**, Lemma 14].

From [**4**, Lemma 1] we know that $\mathfrak{m}$ is a principal ideal, say $\mathfrak{m} = R\theta = \theta R$, that $\theta^m = 0$ for some (smallest) $m \geq 1$, and that every left or right ideal of R is a two-sided ideal of the form $R\theta^j = \theta^j R$, $j = 0, 1, \ldots, m$. In particular, $\mathfrak{m}$ is a two-sided ideal, so that $R/\mathfrak{m}$ is a finite field, say $R/\mathfrak{m} \cong \mathbb{F}_q$, of order q, a prime power. Write (θ^j) for $R\theta^j = \theta^j R$. Thus, all the ideals of R are displayed here:

$$R = (\theta^0) \supset (\theta) \supset (\theta^2) \supset \cdots \supset (\theta^{m-1}) \supset (\theta^m) = (0). \tag{6.1}$$

Each quotient $(\theta^j)/(\theta^{j+1})$ is a one-dimensional vector space over $R/\mathfrak{m}$, with basis element $\theta^j + (\theta^{j+1})$. It follows that

$$|(\theta^j)| = q^{m-j}, \quad j = 0, 1, \ldots, m. \tag{6.2}$$

In particular, $|R| = q^m$. The group of units of R, denoted $\mathcal{U} = \mathcal{U}(R)$, equals the set difference $R - \mathfrak{m}$. The group $\mathcal{U}$ acts of R on the left and on the right by multiplication. The orbits of the actions are exactly the set differences $\text{orb}(\theta^j) = (\theta^j) - (\theta^{j+1})$, which have size

$$|\text{orb}(\theta^j)| = q^{m-j-1}(q-1), \quad \text{for } j < m. \tag{6.3}$$

In particular, the left orbits of $\mathcal{U}$ equal the right orbits: $\mathcal{U}\theta^j = \theta^j\mathcal{U}$.

From (6.1), we see that every element $r \in R$ has the form $r = u\theta^j$ where u is a unit of R and j is uniquely determined by r (the largest i such that $r \in (\theta^i)$). Note that the annihilator of (θ^j) is (θ^{m-j}).

Let w be a weight on R with positive integer values for $r \neq 0$ in R. Assume that w has maximal symmetry, so that $w(ur) = w(ru) = w(r)$ for all $r \in R$ and units $u \in \mathcal{U}$. This means that w is constant on the $\mathcal{U}$-orbits $\text{orb}(\theta^j) = (\theta^j) - (\theta^{j+1})$. Define w_j as the common value of w on $\text{orb}(\theta^j)$, so that $w_j = w(u\theta^j) = w(\theta^j u)$ for all units $u \in \mathcal{U}$. Then $w_0, w_1, \ldots, w_{m-1}$ are positive integers, and $w_m = 0$.

EXAMPLE 6.4. Choosing $\zeta = q - 1$, we see from Example 2.8 and (6.1) that the homogeneous weight w on a chain ring R has the following integer values:

$$\mathrm{w}(r) = \begin{cases} 0, & r = 0, \\ q, & r \in (\theta^{m-1}) - (0), \\ q-1, & r \in R - (\theta^{m-1}). \end{cases}$$

Then $\mathrm{w}_0 = \cdots = \mathrm{w}_{m-2} = q - 1$, $\mathrm{w}_{m-1} = q$, and $\mathrm{w}_m = 0$.

Do the MacWilliams identities hold for the homogeneous weight enumerator over a finite chain ring R? We will see that the answers depend on q and m.

When $m = 1$, then $\theta = 0$, so that R is a finite field $\mathbb{F}_q$. As we saw in Example 2.8, the homogeneous weight w on $\mathbb{F}_q$ equals a multiple of the Hamming weight. By Theorems 3.2 and Lemma 3.13, the homogeneous weight over finite fields respects duality.

For $m \geq 2$, there is one special case ($m = q = 2$, Theorem 6.5, below) where the MacWilliams identities hold for the homogeneous weight. In the remaining cases, we will see in Theorem 10.2 that the homogeneous weight does not respect duality.

These results apply to the chain rings $\mathbb{Z}/p^m\mathbb{Z}$, so that the MacWilliams identities hold for the homogeneous weight over $\mathbb{Z}/p\mathbb{Z}$, p prime, and over $\mathbb{Z}/4\mathbb{Z}$ [**12**, Equation (9)], but not for other prime powers. More generally, over $\mathbb{Z}/m\mathbb{Z}$, m not a prime power (so that $\mathbb{Z}/m\mathbb{Z}$ is not a chain ring), the homogeneous weight does not respect duality [**30**, Theorem 6.2].

THEOREM 6.5. *The MacWilliams identities hold for the homogeneous weight enumerator over a finite chain ring R with $q = 2$ and $m = 2$. If $C \subseteq R^n$ is a linear code and $C^\perp$ is its dual code, then*

$$\mathrm{howe}_{C^\perp}(X, Y) = \frac{1}{|C|}\,\mathrm{howe}_C(X + Y, X - Y).$$

PROOF. Appendix A outlines of a proof of the MacWilliams identities over finite Frobenius rings and describes the Fourier transform. Here, we provide details relevant to the chain rings appearing in this theorem.

We know that $|R| = 4$, with $R = \{0, 1, \theta, 1+\theta\}$. The values of the homogeneous weight, with $\zeta = 1$, are:

r	0	1	θ	$1+\theta$
$\mathrm{w}(r)$	0	1	2	1

The additive group of R could be a cyclic group of order 4 (in which case $\theta = 2$, in order that (θ) be a maximal ideal) or a Klein 4-group. In either case, there exists a generating character χ of R with the following values. (What is crucial is that $\chi(\theta) = -1$.)

	r	0	1	θ	$1+\theta$
cyclic	$\chi(r)$	1	i	-1	$-i$
Klein	$\chi(r)$	1	1	-1	-1

Define $f : R \to \mathbb{C}[X, Y]$ by $f(r) = X^{2-\mathrm{w}(r)}Y^{\mathrm{w}(r)}$. For either choice of χ, the Fourier transform $\widehat{f}$ of (A.2) is the same:

r	0	1	θ	$1+\theta$
$f(r)$	X^2	XY	Y^2	XY
$\widehat{f}(r)$	$(X+Y)^2$	$X^2 - Y^2$	$(X-Y)^2$	$X^2 - Y^2$

Note that the values of $\widehat{f}$ have the form $\widehat{f}(r) = (X + Y)^{2-\mathrm{w}(r)}(X - Y)^{\mathrm{w}(r)}$. I.e., $\widehat{f}(r) = f(r)|_{X \leftarrow X+Y, Y \leftarrow X-Y}$. The rest of the argument in Appendix A now carries through. □

7. Modules over a chain ring

We continue to assume that R is a finite chain ring with maximal ideal $\mathfrak{m} = (\theta)$ such that $R/\mathfrak{m} \cong \mathbb{F}_q$ and $\theta^m = 0$, for some integer $m \geq 2$. By forming quotients from (6.1), we define cyclic R-modules $Z_k = R/(\theta^k)$, of order q^k, $k = 1, 2, \ldots, m$. The module Z_1 is the unique simple R-module. Also define semisimple modules $S_k = Z_1 \oplus \cdots \oplus Z_1$ with k summands, $k = 2, 3, \ldots, m$; $|S_k| = q^k$.

In the next two sections, we will construct two families of linear codes with the same w-weight enumerators. Here is a brief sketch of the construction. We will build examples using Z_k and S_k, $2 \leq k \leq m$, as the underlying information modules. On Z_k, choose a multiplicity function. The weights of elements will be constant along nonzero orbits $\mathcal{O}$, which have sizes $q^k - q^{k-1}, q^{k-1} - q^{k-2}, \ldots, q^2 - q, q - 1$. We then need to build a linear code based on S_k with the same w-weight enumerator. This entails choosing subsets of S_k with sizes matching the sizes $q^k - q^{k-1}, q^{k-1} - q^{k-2}, \ldots, q^2 - q, q - 1$. We do this by fixing a filtration of S_k using linear subspaces. We can then solve for a multiplicity function for S_k. If the original multiplicity function on Z_k calls for one functional from each class, then the multiplicity function on S_k is reasonably nice. We then calculate the common weight enumerator for these multiplicity functions.

In this section, we study the orbit structure of Z_k as well as the subsets arising from a filtration of S_k. In Section 8, the multiplicity functions are described and analyzed.

In order to define linear codes over Z_k and S_k, let us examine their linear functionals. Recall first that the left linear functionals of R itself are given by right multiplications by elements of R. That is, $R^\sharp \cong R$ as right R-modules, with $r \in R$ corresponding to the left linear functional $\rho_r \in R^\sharp$ defined by $r'\rho_r = r'r$, $r' \in R$.

If $\lambda \in Z_k^\sharp$, then the composition with the natural quotient map must equal ρ_r for some $r \in R$.

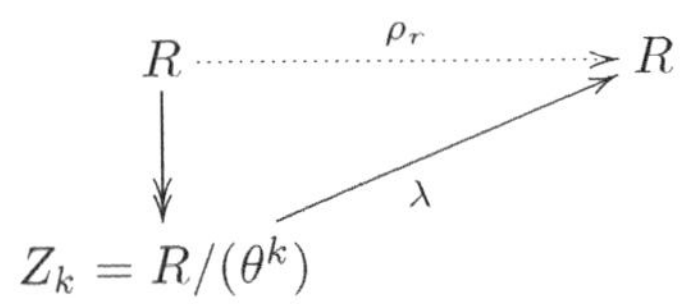

Conversely, $\rho_r : R \to R$ factors through Z_k if and only if $(\theta^k) \subseteq \ker \rho_r$; i.e., if and only if $\theta^k r = 0$. This occurs when $r \in (\theta^{m-k})$. Thus $Z_k^\sharp \cong (\theta^{m-k})$ as right R-modules. In particular, $Z_1^\sharp \cong (\theta^{m-1})$ as right R-modules.

As for $S_k = Z_1 \oplus \cdots \oplus Z_1$, we have $S_k^\sharp \cong (\theta^{m-1}) \oplus \cdots \oplus (\theta^{m-1})$ as right R-modules. For $s = (s_1, \ldots, s_k) \in S_k$ and $\mu = \langle \mu_1\theta^{m-1}, \ldots, \mu_k\theta^{m-1}\rangle \in S_k^\sharp$, $s\mu = \sum_{i=1}^k s_i\mu_i\theta^{m-1} \in R$. Both S_k and $S_k^\sharp$ are k-dimensional vector spaces over $R/\mathfrak{m} \cong \mathbb{F}_q$.

In order to exploit maximal symmetry in Lemma 4.1, we want to understand the orbit structures of the left actions of $\mathcal{U}$ on Z_k and S_k, and, to a lesser extent, the orbit structures of the right actions of $\mathcal{U}$ on $Z_k^\sharp$ and $S_k^\sharp$.

Because (θ^k) is a two-sided ideal of R, Z_k is itself a chain ring. Its left R-submodules are the same as its left ideals:

$$Z_k \supset R\theta \supset R\theta^2 \supset \cdots \supset R\theta^{k-1} \supset \{0\}. \tag{7.1}$$

Note that (7.1) can be viewed as a filtration of Z_k by R-submodules.

LEMMA 7.2. *The orbits of the left action of $\mathcal{U}$ on Z_k are:*

$$\mathrm{orb}(1) = \mathrm{orb}(\theta^0), \mathrm{orb}(\theta), \ldots, \mathrm{orb}(\theta^{k-1}), \{0\} = \mathrm{orb}(0) = \mathrm{orb}(\theta^k).$$

The sizes of the orbits are: $|\mathrm{orb}(0)| = 1$ *and*

$$|\mathrm{orb}(\theta^i)| = q^{k-i-1}(q-1), \quad i = 0, 1, \ldots, k-1.$$

PROOF. Apply (6.3) to the chain ring Z_k. □

Similar to Lemma 7.2, the orbits of the right action of $\mathcal{U}$ on $Z_k^\sharp$ are $\mathrm{orb}(\theta^{m-k})$, $\mathrm{orb}(\theta^{m-k+1}), \ldots, \mathrm{orb}(\theta^{m-1}), \{0\} = \mathrm{orb}(\theta^m)$.

Because S_k is a vector space over $\mathbb{F}_q \cong R/\mathfrak{m}$, the action of $\mathcal{U}$ on S_k reduces to the action of the multiplicative group $\mathbb{F}_q^\times$. The $\mathcal{U}$-orbits are $\{0\}$ and $L - \{0\}$, for every 1-dimensional subspace $L \subseteq S_k$. The same structure applies to the dual vector space $S_k^\sharp$; the $\mathcal{U}$-orbits are $\{0\}$ and the nonzero elements of 1-dimensional subspaces.

For later use in Section 8, we will identify subsets (say, $\mathcal{S}_1, \ldots, \mathcal{S}_k$) of S_k that will be the counterparts to the orbits $\operatorname{orb}(1), \ldots, \operatorname{orb}(\theta^{k-1})$ of Z_k (in reverse order). In particular, we want two features:

- each $\mathcal{S}_i$ is a union of $\mathcal{U}$-orbits in S_k;
- $|\mathcal{S}_i| = |\operatorname{orb}(\theta^{k-i})| = q^{i-1}(q-1)$ for $i = 1, \ldots, k$.

To define the $\mathcal{S}_i$, we define a filtration on S_k. Recall that S_k is a k-dimensional vector space over $R/\mathfrak{m} \cong \mathbb{F}_q$. Write elements of $S_k = Z_1 \oplus \cdots \oplus Z_1$ as row vectors of length k over $\mathbb{F}_q$. (Row vectors will be written as $(x_1, \ldots, x_k)$, while column vectors will be written as $\langle x_1, \ldots, x_k \rangle$.) Define vector subspaces of S_k: $V_0 = \{0\}$ and $V_i = \{(0, \ldots, 0, s_{k-i+1}, \ldots, s_k) \in S_k : s_j \in \mathbb{F}_q\}$, for $i = 1, 2, \ldots, k$. In V_i, the first $k - i$ entries are zero; the last i entries vary over $\mathbb{F}_q$. Then, $\dim_{\mathbb{F}_q} V_i = i$ for $i = 0, 1, \ldots, k$, so that $|V_i| = q^i$, and

$$S_k = V_k \supset V_{k-1} \supset \cdots \supset V_1 \supset V_0 = \{0\}. \tag{7.3}$$

Set $\mathcal{S}_i = V_i - V_{i-1}$, for $i = 1, \ldots, k$; then $|\mathcal{S}_i| = q^{i-1}(q-1)$. Set $\mathcal{S}_0 = \{0\}$. Because the V_i are vector subspaces, the $\mathcal{S}_i$ are unions of $\mathcal{U}$-orbits.

We also want to understand the linear functionals on S_k in terms of the filtration (7.3). To that end, we examine the dual filtration of $S_k^\sharp$ defined by annihilators $\mathcal{V}_i = \operatorname{ann}(V_i) = \{\mu \in S_k^\sharp : V_i\mu = 0\}$. Then, $\dim_{\mathbb{F}_q} \mathcal{V}_i = k - \dim_{\mathbb{F}_q} V_i = k - i$, $|\mathcal{V}_i| = q^{k-i}$, and

$$\{0\} = \mathcal{V}_k \subset \mathcal{V}_{k-1} \subset \cdots \subset \mathcal{V}_1 \subset \mathcal{V}_0 = S_k^\sharp. \tag{7.4}$$

If we view elements of $S_k^\sharp$ as column vectors $\mu = \langle \mu_1\theta^{m-1}, \ldots, \mu_k\theta^{m-1} \rangle$, then $\mathcal{V}_i$ consists of those μ whose last i entries equal zero. The zero entries of $\mathcal{V}_i$ align with the nonzero entries of elements of V_i.

The set differences $\mathcal{V}_i - \mathcal{V}_{i+1}$ consist of all $\mu = \langle \mu_1\theta^{m-1}, \ldots, \mu_k\theta^{m-1} \rangle$ with $\mu_{k-i+1} = \cdots = \mu_k = 0$, $\mu_{k-i} \neq 0$, and $\mu_1, \ldots, \mu_{k-i-1} \in \mathbb{F}_q$. The $\mathcal{U}$-orbit of μ is the set of all nonzero scalar multiplies of μ. In each orbit there is exactly one 'normalized' representative with $\mu_{k-i} = 1$. For $i = 0, 1, \ldots, k-1$, let B_i be the subset of $\mathcal{V}_i - \mathcal{V}_{i+1}$ consisting of all the normalized representatives; i.e.,

$$B_i = \{\mu \in S_k : \mu_{k-i} = 1, \mu_{k-i+1} = \cdots = \mu_k = 0\}; \tag{7.5}$$

as a special case, set $B_k = \{0\}$. Then $|B_i| = q^{k-i-1}$, except $|B_k| = 1$.

LEMMA 7.6. *Let $s \in \mathcal{S}_i$. Then, for $j = 0, 1, \ldots, k-1$,*

$$|B_j \cap \operatorname{ann}(s)| = \begin{cases} q^{k-j-1}, & i \leq j \leq k-1, \\ 0, & j = i-1, \\ q^{k-j-2}, & 0 \leq j \leq i-2. \end{cases}$$

PROOF. The case $\mathcal{S}_0 = \{0\}$ has $\operatorname{ann}(0) = S_k^\sharp$. Then $|B_j \cap \operatorname{ann}(0)| = |B_j| = q^{k-j-1}$ for all $0 \leq j \leq k-1$.

Now let $1 \leq i \leq k$. The element $s \in \mathcal{S}_i = V_i - V_{i-1}$ is nonzero and has the form $s = (0, \ldots, 0, s_{k-i+1}, \ldots, s_k)$, with $s_{k-i+1} \neq 0$. Any $\mu \in \mathcal{V}_{i-1} - \mathcal{V}_i$ has the form $\mu = \langle \mu_1\theta^{m-1}, \ldots, \mu_{k-i+1}\theta^{m-1}, 0, \ldots, 0 \rangle$ with $\mu_{k-i+1} \neq 0$. Thus $s\mu = s_{k-i+1}\mu_{k-i+1}\theta^{m-1} \neq 0$, so that $|B_{i-1} \cap \operatorname{ann}(s)| = 0$. For $i \leq j \leq k-1$, use the definition of $\mathcal{V}_i$ to see that $\mathcal{V}_j \subseteq \mathcal{V}_i = \operatorname{ann}(V_i) \subseteq \operatorname{ann}(s)$. So $B_j \subseteq \operatorname{ann}(s)$ and $|B_j \cap \operatorname{ann}(s)| = |B_j| = q^{k-j-1}$.

Because $s \neq 0$, $\operatorname{ann}(s)$ is a vector subspace of $S_k^\sharp$ with $\dim_{\mathbb{F}_q} \operatorname{ann}(s) = k-1$. By dimension counting, $\dim(\mathcal{V}_j \cap \operatorname{ann}(s))$ equals $k-j-1$ or $k-j$. The case $\dim(\mathcal{V}_j \cap \operatorname{ann}(s)) = k-j$ occurs when $\mathcal{V}_j \subseteq \operatorname{ann}(s)$. This is the case $i \leq j \leq k-1$ above. When $j = i-1$, $\mathcal{V}_{i-1} \cap \operatorname{ann}(s) = \mathcal{V}_i$, so that $|B_{i-1} \cap \operatorname{ann}(s)| = 0$, as we saw above. Finally, let $0 \leq j \leq i-2$. Then $\dim(\mathcal{V}_j \cap \operatorname{ann}(s)) = k-j-1$ and $\dim(\mathcal{V}_{j+1} \cap \operatorname{ann}(s)) = k-j-2$. This implies $|(\mathcal{V}_j - \mathcal{V}_{j+1}) \cap \operatorname{ann}(s)| = |\mathcal{V}_j \cap \operatorname{ann}(s)| - |\mathcal{V}_{j+1} \cap \operatorname{ann}(s)| = q^{k-j-2}(q-1)$. Taking normalized representatives implies $|B_j \cap \operatorname{ann}(s)| = q^{k-j-2}$. □

8. Two families of linear codes with the same wwe

We continue to assume that R is a finite chain ring with maximal ideal $\mathfrak{m} = (\theta)$ such that $R/\mathfrak{m} \cong \mathbb{F}_q$ and $\theta^m = 0$, for some integer $m \geq 2$. Let w be an integer-valued weight on R with maximal symmetry. Denote the common value of w on $\operatorname{orb}(\theta^i)$ by $w_i > 0$, $i = 0, 1, \ldots, m-1$, and $w_m = w(0) = 0$.

To begin, we use $w_0, \ldots, w_{m-1}$ to define several numerical quantities. For $i = 0, 1, \ldots, m-1$, define

$$a_i = \sum_{j=0}^{i} q^j w_{m-j-1}. \tag{8.1}$$

Also define, for $k = 2, 3, \ldots, m$,

$$\Delta_k = kq^{k-1} w_{m-1} - \sum_{i=0}^{k-1} q^{k-i-1} a_i. \tag{8.2}$$

Recall the cyclic R-module $Z_k = R/(\theta^k)$ and the semisimple R-module $S_k = Z_1 \oplus \cdots \oplus Z_1$ (k summands) from Section 6. We will construct R-linear codes with Z_k and S_k as their underlying information modules. The linear codes will be images of homomorphisms $\Lambda : Z_k \to R^n$ and $\Gamma : S_k \to R^n$ of left R-modules. As explained in Section 4, the linear codes are determined by their multiplicity functions.

Definition 8.3. Define a left R-linear code C_k parametrized by Z_k by using the linear functionals given by right multiplication by each of $\theta^{m-k}, \ldots, \theta^{m-1}$, each repeated $q^{k-1} w_{m-1}$ times, and the zero functional, repeated $\max\{0, -\Delta_k\}$ times; cf., (8.2).

Equivalently, C_k has a generator matrix of size $1 \times (kq^{k-1} w_{m-1} + \max\{0, -\Delta_k\})$, with entries $\theta^{m-k}, \ldots, \theta^{m-1}$, each repeated $q^{k-1} w_{m-1}$ times, plus entries of 0, repeated $\max\{0, -\Delta_k\}$ times. Thus, 0 does not appear if $\Delta_k \geq 0$, and 0 appears $-\Delta_k$ times when $\Delta_k < 0$.

Proposition 8.4. *The linear code C_k of Definition 8.3 has length $kq^{k-1} w_{m-1} + \max\{0, -\Delta_k\}$. Its weight function W_Λ has values*

$$W_\Lambda(\theta^i) = q^{k-1} w_{m-1}(w_{m-k+i} + \cdots + w_{m-1}),$$

for $i = 0, 1, \ldots, k-1$, and the w-weight enumerator of C_k is

$$\begin{aligned} \mathrm{wwe}_{C_k} &= 1 + \sum_{i=0}^{k-1} |\operatorname{orb}(\theta^i)| t^{W_\Lambda(\theta^i)} \\ &= 1 + \sum_{i=0}^{k-1} q^{k-i-1}(q-1) t^{q^{k-1} w_{m-1} \sum_{j=i}^{k-1} w_{m-k+j}}. \end{aligned}$$

PROOF. The formula for the length follows directly from Definition 8.3. In calculating $W_\Lambda(\theta^i)$, remember that $w_m = w(\theta^m) = w(0) = 0$.

$$\begin{aligned} W_\Lambda(\theta^i) &= \sum_{j=m-k}^{m-1} q^{k-1} w_{m-1} w(\theta^i \theta^j) \\ &= q^{k-1} w_{m-1} \left(w_{i+m-k} + \cdots + w_{m-1} \right). \end{aligned}$$

The functional in $Z_k^\sharp$ given by right multiplication by θ^{m-k} is injective, so that Λ is also injective. Then wwe_{C_k} follows from (4.3). □

We now want to define linear codes D_k parametrized by $\Gamma : S_k \to R^n$ such that $\mathrm{wwe}_{C_k} = \mathrm{wwe}_{D_k}$. The form of wwe_{C_k} was determined by (4.3), in particular by the sizes of the orbits $\mathrm{orb}(\theta^i)$ and the value of W_Λ on those orbits. In order to be able to match terms in the equation $\mathrm{wwe}_{C_k} = \mathrm{wwe}_{D_k}$, we make use of the subsets $\mathcal{S}_1, \ldots, \mathcal{S}_k$ of S_k defined following (7.3). Also recall the sets B_i defined in (7.5). We will design Γ so that W_Γ is constant on each $\mathcal{S}_i$, with value equal to the value of W_Λ on $\mathrm{orb}(\theta^{k-i})$.

DEFINITION 8.5. Define a left R-linear code D_k parametrized by $\Gamma : S_k \to R^n$ by using the linear functionals in $\cup_i B_i$, with each $\mu \in B_i$ repeated a_i times (cf., (8.1)), $i = 0, 1, \ldots, k-1$, and the zero-functional in B_k repeated $\max\{\Delta_k, 0\}$ times (cf., (8.2)).

Likewise, D_k has a generator matrix of size $k \times (\sum_{i=0}^{k-1} q^{k-i-1} a_i + \max\{\Delta_k, 0\})$, with columns given by $\mu \in \cup_i B_i$, with each $\mu \in B_i$ repeated a_i times, $i = 0, 1, \ldots, k$, the zero-column in B_k repeated $\max\{\Delta_k, 0\}$ times. Thus, zero-columns do not appear if $\Delta_k \leq 0$, and the zero-column appears Δ_k times when $\Delta_k > 0$.

We now express the values of W_Γ in terms of $a_0, a_1, \ldots, a_{k-1}$.

PROPOSITION 8.6. *The weight function $W_\Gamma : S_k \to \mathbb{Z}$ is constant on each $\mathcal{S}_i$, with $W_\Gamma(\mathcal{S}_0) = 0$ and, for $i = 1, \ldots, k$,*

$$W_\Gamma(\mathcal{S}_i) = \left(q^{k-i} a_{i-1} + \sum_{j=0}^{i-2} q^{k-j-2}(q-1) a_j \right) w_{m-1}.$$

PROOF. If $i = 0$, then $\mathcal{S}_0 = \{0\}$, so that $W_\Gamma(\mathcal{S}_0) = 0$. Now let $i = 1, \ldots, k$, and let $s \in \mathcal{S}_i$. Using Lemma 7.6, we see that

$$\begin{aligned} W_\Gamma(s) &= \sum_{j=0}^{k-1} a_j \sum_{\mu \in B_j} w(s\mu) = \sum_{j=0}^{i-1} a_j \sum_{\mu \in B_j} w(s\mu) \\ &= \left(q^{k-i} a_{i-1} + \sum_{j=0}^{i-2} q^{k-j-2}(q-1) a_j \right) w_{m-1}. \end{aligned}$$

This formula depends only on i, so W_Γ is constant on each $\mathcal{S}_i$. □

In order that $\mathrm{wwe}_{C_k} = \mathrm{wwe}_{D_k}$, we need $W_\Lambda(\mathrm{orb}(\theta^{k-i})) = W_\Gamma(\mathcal{S}_i)$, for $i = 1, \ldots, k$. That is, canceling a common factor, we need

$$q^{k-1}(w_{m-i} + \cdots + w_{m-1}) = q^{k-i} a_{i-1} + \sum_{j=0}^{i-2} q^{k-j-2}(q-1) a_j, \tag{8.7}$$

for $i = 1, \ldots, k$. The number of additive terms on each side of the equation is i. Starting with $i = 1$ and working upwards, we solve a triangular system recursively for $a_0, a_1, \ldots, a_{k-1}$.

LEMMA 8.8. *The solutions of* (8.7) *are*

$$a_i = w_{m-1} + qw_{m-2} + \cdots + q^i w_{m-i-1},$$

for $i = 0, 1, \ldots, k-1$. *This formula matches* (8.1).

PROOF. Exercise, by induction. The terms simplify by telescoping. □

THEOREM 8.9. *For each* $k = 2, 3, \ldots, m$, *the codes* C_k *and* D_k *have the same length and satisfy* $\mathrm{wwe}_{C_k} = \mathrm{wwe}_{D_k}$.

PROOF. The definition of Δ_k in (8.2) guarantees that the codes have the same length. The a_i of (8.1) were defined so that Lemma 8.8 holds. Thus $W_\Gamma(\mathcal{S}_i) = W_\Lambda(\mathrm{orb}(\theta^{k-i}))$, for $i = 1, \ldots, k$. Because $|\mathrm{orb}(\theta^{k-i})| = |\mathcal{S}_i|$, the equality of the w-weight enumerators follows from (4.3). □

REMARK 8.10. It is possible to generalize the constructions of C_k and D_k by allowing more general expressions for the multiplicities of the linear functionals appearing in C_k. In Definition 8.3, $\theta^{m-k}, \ldots, \theta^{m-1}$ could be repeated $q^{k-1}w_{m-1}b_{m-k}$, $\ldots, q^{k-1}w_{m-1}b_{m-1}$ times, respectively. One can then mimic (8.7) and Lemma 8.8 to solve for the multiplicities a_i used in defining D_k. One must be careful in choosing the b-variables in order that the a-variables come out nonnegative. Sufficiently large b-variables should work. The factors of $q^{k-1}w_{m-1}$ are present so that the a-variables are integers. The present work does not need this level of generality.

In later sections, we will use the formula below for Δ_k, which is expressed in terms of some numerical quantities defined next.

DEFINITION 8.11. For $i = 1, 2, \ldots, m$, define $\epsilon_i = w_i - w_{i-1}$. For example, $\epsilon_1 = w_1 - w_0$, while $\epsilon_m = -w_{m-1}$ (as $w_m = 0$). For $i = 1, \ldots, m-2$, define $\epsilon'_i = \epsilon_i$, and define $\epsilon'_{m-1} = (q-1)w_{m-1} - qw_{m-2} = q\epsilon_{m-1} + \epsilon_m$.

Also define these polynomial expressions in q: $p_0 = 0$, and

$$p_i = 1 + 2q + 3q^2 + \cdots + iq^{i-1}, \quad i = 1, 2, \ldots. \tag{8.12}$$

Then for $i = 1, 2, \ldots$, the p_i are positive and satisfy the formulas

$$p_i - qp_{i-1} = 1 + q + q^2 + \cdots + q^{i-1}. \tag{8.13}$$

PROPOSITION 8.14. *For* $k = 2, 3, \ldots, m$,

$$\Delta_k = p_{k-1}\epsilon'_{m-1} + \sum_{j=2}^{k-1} q^j p_{k-j}\epsilon'_{m-j}.$$

PROOF. Start with (8.2), and replace a_i using (8.1):

$$\Delta_k = kq^{k-1}w_{m-1} - \sum_{i=0}^{k-1} q^{k-i-1} \sum_{j=0}^{i} q^j w_{m-1-j}.$$

Interchange the order of summation and use (8.13):

$$\begin{aligned}
\Delta_k &= kq^{k-1}w_{m-1} - \sum_{j=0}^{k-1}\sum_{i=j}^{k-1} q^{k-i+j-1}w_{m-1-j} \\
&= kq^{k-1}w_{m-1} - \sum_{j=0}^{k-1} q^j(p_{k-j} - qp_{k-j-1})w_{m-1-j} \\
&= kq^{k-1}w_{m-1} - \sum_{j=0}^{k-1} q^j p_{k-j}w_{m-1-j} + \sum_{j=0}^{k-1} q^{j+1}p_{k-j-1}w_{m-1-j}.
\end{aligned}$$

Re-index the last summation with $\ell = j+1$ (the $\ell = k$ term vanishes), separate some initial terms, and combine the rest using Definition 8.11:

$$\begin{aligned}
\Delta_k &= kq^{k-1}w_{m-1} - \sum_{j=0}^{k-1} q^j p_{k-j}w_{m-1-j} + \sum_{\ell=1}^{k} q^\ell p_{k-\ell}w_{m-\ell} \\
&= kq^{k-1}w_{m-1} - p_k w_{m-1} + qp_{k-1}\epsilon_{m-1} + \sum_{j=2}^{k-1} q^j p_{k-j}\epsilon_{m-j}.
\end{aligned}$$

Simplify the coefficient of w_{m-1} and again use Definition 8.11:

$$\begin{aligned}
\Delta_k &= p_{k-1}\epsilon_m + qp_{k-1}\epsilon_{m-1} + \sum_{j=2}^{k-1} q^j p_{k-j}\epsilon_{m-j} \\
&= p_{k-1}\epsilon'_{m-1} + \sum_{j=2}^{k-1} q^j p_{k-j}\epsilon'_{m-j}.
\end{aligned}$$

□

9. Analysis of dual codewords of low weight

Our ultimate objective is to prove that for some $k = 2, 3, \ldots, m$, the codes C_k and D_k of Theorem 8.9 have dual codes with different weight enumerators: $\mathrm{wwe}_{C_k^\perp} \neq \mathrm{wwe}_{D_k^\perp}$. We will try to do this in the most direct way—by showing that $C_k^\perp$ and $D_k^\perp$ have different numbers of codewords of the smallest possible weight. With that in mind, let's develop some notation.

Recall that we are assuming the chain ring R is equipped with an integer-valued weight w of maximal symmetry. The common value on $\mathrm{orb}(\theta^i)$ is denoted $w_i > 0$, and $w_m = w(0) = 0$. Let $\mathring{w} = \min\{w_i : i = 0, 1, \ldots, m-1\}$, so that $\mathring{w} > 0$. Define $\mathring{I} = \{i : w_i = \mathring{w}\}$, the set of exponents of θ that achieve the minimum value of the weight; $\mathring{I}$ is nonempty.

We now turn our attention to the linear codes C_k and D_k of Theorem 8.9 and codewords of weight $d < 2\mathring{w}$ in their dual codes. All such codewords must be singletons by Lemma 5.4. We will abuse notation slightly by using the phrase 'singleton θ^i' to mean a singleton whose nonzero entry is a unit multiple of θ^i.

LEMMA 9.1. *Suppose an integer d satisfies $\mathring{w} \leq d < 2\mathring{w}$. Let $I_d = \{i : w_i = d\}$. If $0 \in I_d$, then, for $k = 2, 3, \ldots, m$,*

$$A_d(C_k^\perp) - A_d(D_k^\perp) = -|\mathrm{orb}(1)|\Delta_k - \sum_{\substack{i\in I_d \\ 0<i<k}} (k-i)q^{k-1}w_{m-1}|\mathrm{orb}(\theta^i)|.$$

If $0 \notin I_d$ and I_d is nonempty, then, for $k = 2, 3, \ldots, m$,

$$A_d(C_k^\perp) - A_d(D_k^\perp) = -\sum_{\substack{i \in I_d \\ i<k}} (k-i) q^{k-1} w_{m-1} |\mathrm{orb}(\theta^i)|.$$

PROOF. Nonzero contributions to $A_d(C_k^\perp) - A_d(D_k^\perp)$ are made by singletons of weight d in $C_k^\perp$ or $D_k^\perp$. The nonzero entry of a singleton of weight d must be a unit multiple of θ^i with $i \in I_d$. In order for a singleton to belong to a dual code, its nonzero entry—located in position j, say—must annihilate column j of the generator matrix of the primal code.

If $0 \in I_d$, a singleton 1 annihilates only zero-columns. (Remember that the nonzero entry of a singleton 1 is a unit.) The number of zero-columns is determined by Δ_k: if $\Delta_k > 0$, then D_k has Δ_k zero-columns; if $\Delta_k < 0$, then C_k has $-\Delta_k$ zero-colums. The net contribution to $A_d(C_k^\perp) - A_d(D_k^\perp)$ is $-|\mathrm{orb}(1)|\Delta_k$, the number of units times the number of zero-columns.

If $0 < i \in I_d$, then a singleton θ^i annihilates all the columns of D_k. Such singletons contribute $-|\mathrm{orb}(\theta^i)| \cdot \mathrm{length}(D_k)$ to $A_d(C_k^\perp) - A_d(D_k^\perp)$. On the other hand, when $i \geq k$, a singleton θ^i annihilates all columns of C_k. Such singletons contribute $|\mathrm{orb}(\theta^i)| \cdot \mathrm{length}(C_k)$ to $A_d(C_k^\perp) - A_d(D_k^\perp)$. When $i < k$, a singleton θ^i annihilates all columns of C_k except those with entries $\theta^{m-k}, \ldots, \theta^{m-i-1}$. Such singletons contribute $|\mathrm{orb}(\theta^i)|(\mathrm{length}(C_k) - (k-i)q^{k-1}w_{m-1})$ to $A_d(C_k^\perp) - A_d(D_k^\perp)$. Because C_k and D_k have the same length, the total contribution of such singletons is 0 when $i \geq k$ and $-(k-i)q^{k-1}w_{m-1}|\mathrm{orb}(\theta^i)|$ when $i < k$. Summing over $i \in I_d$ completes the proof. □

Our main interest is dual codewords of weight $\mathring{w}$. For $k = 2, 3, \ldots, m$, define

$$\delta_k = A_{\mathring{w}}(C_k^\perp) - A_{\mathring{w}}(D_k^\perp). \tag{9.2}$$

Our aim is to show, whenever possible, for a given weight w, that $\delta_k \neq 0$ for some k. We restate Lemma 9.1 for the case where $d = \mathring{w}$.

LEMMA 9.3. *Fix $k = 2, 3, \ldots, m$. If $0 \in \mathring{I}$, then*

$$\delta_k = -|\mathrm{orb}(1)|\Delta_k - \sum_{\substack{i \in \mathring{I} \\ 0<i<k}} (k-i) q^{k-1} w_{m-1} |\mathrm{orb}(\theta^i)|.$$

If $0 \notin \mathring{I}$, then

$$\delta_k = -\sum_{\substack{i \in \mathring{I} \\ i<k}} (k-i) q^{k-1} w_{m-1} |\mathrm{orb}(\theta^i)|.$$

We draw three corollaries, using notation from Definition 8.11.

COROLLARY 9.4. *If $0 \notin \mathring{I}$, then $\delta_k < 0$ for all $k = 1 + \min \mathring{I}, \ldots, m$. Weights w on R with $w_0 > \mathring{w}$ do not respect duality.*

COROLLARY 9.5. *Suppose $\mathring{I} = \{0\}$. Then the following hold:*

1. *If $m \geq 3$ and j, $1 \leq j \leq m-1$, is the largest index with $\epsilon'_j \neq 0$, then $\delta_{m-j+1} \neq 0$.*
2. *If $m = 2$ and $\epsilon'_1 = (q-1)w_1 - qw_0 \neq 0$, then $\delta_2 \neq 0$.*
3. *If $m = 2$, $\epsilon'_1 = (q-1)w_1 - qw_0 = 0$, and $q > 2$, then $A_{w_1}(C_2^\perp) < A_{w_1}(D_2^\perp)$.*

Weights w on R with $w_0 < w_i$ for all $i = 1, 2, \ldots, m-1$ do not respect duality, except when $m = 2$, $q = 2$, and $w_1 = 2w_0$; cf., Theorem 6.5.

PROOF. Suppose $m \geq 3$. By the hypothesis on $\mathring{I}$, $\epsilon'_1 = w_1 - w_0 > 0$, so there exists a maximal index j, $1 \leq j \leq m-1$, with $\epsilon'_j \neq 0$. Thus, $\epsilon'_\ell = 0$ for $\ell = j+1, \ldots, m-1$. From Proposition 8.14, $\Delta_{m-j+1} = q^{m-j}\epsilon'_j \neq 0$. From Lemma 9.3 and the hypothesis, $\delta_{m-j+1} = -|\mathrm{orb}(1)|\Delta_{m-j+1} \neq 0$.

When $m = 2$, $\epsilon'_1 = (q-1)w_1 - qw_0$, not $w_1 - w_0$. If $\epsilon'_1 \neq 0$, the proof proceeds as above: $\Delta_2 \neq 0$, and $\delta_2 \neq 0$.

When $m = 2$ and $\epsilon'_1 = 0$, then $\Delta_2 = \delta_2 = 0$. However, $\epsilon'_1 = 0$ means that $(q-1)w_1 = qw_0$, i.e., $w_1 = (q/(q-1))w_0 > w_0$. But note, for integers $q \geq 2$, that $q/(q-1) \leq 2$ with equality holding if and only if $q = 2$. Assuming $q > 2$, we have $\mathring{w} = w_0 < w_1 < 2\mathring{w}$. By Lemma 9.1 applied to $d = w_1$ and $0 \notin I_d$, we see that $A_{w_1}(C_2^\perp) < A_{w_1}(D_2^\perp)$.

When $m = 2$, $q = 2$, and $w_1 = 2w_0$, we are in the situation of Theorem 6.5, where the MacWilliams identities hold. □

COROLLARY 9.6. *Let R be a finite chain ring with $m = 2$. Then every weight w on R having maximal symmetry does not respect duality, except for multiples of the Hamming weight (any q) or the homogeneous weight ($q = 2$ only).*

PROOF. Because $m = 2$, there are only w_0 and w_1. If $w_0 > w_1$, Corollary 9.4 implies w does not respect duality. If $w_0 = w_1$, w is a multiple of the Hamming weight, and the MacWilliams identities hold [**25**, Theorem 8.3]. If $w_0 < w_1$, then Corollary 9.5 applies: w does not respect duality, except for multiples of the homogeneous weight if $q = 2$. □

10. Weak monotonicity and final arguments

When $\{0\} \subsetneq \mathring{I}$, the formula for δ_k in Lemma 9.3 is difficult to exploit systematically; the combinatorics can be formidable. (But not always: see Example 11.9.) In order to make progress, we will assume that the weight w on the chain ring R is *weakly monotone*; i.e., we assume

$$\mathring{w} = w_0 \leq w_1 \leq \cdots \leq w_{m-2} \leq w_{m-1}. \tag{10.1}$$

This hypothesis implies that $\epsilon_i \geq 0$ for $i = 1, 2, \ldots, m-1$ in Definition 8.11. However, $\epsilon_m = -w_{m-1} < 0$. The weakly monotone hypothesis allows us to state our main result.

THEOREM 10.2. *Let R be a finite chain ring with a weakly monotone weight w. Then w does not respect duality, except when*

- *w is a multiple of the Hamming weight, or*
- *$m = 2$, $q = 2$, and $w_1 = 2w_0$.*

Theorem 10.2 will follow from Theorem 10.10, which we will prove after we establish some technical lemmas.

Equalities are possible in (10.1). Given a weakly monotone weight w, define j_0 to be the largest index such that

$$\mathring{w} = w_0 = \cdots = w_{j_0} < w_{j_0+1}. \tag{10.3}$$

Similarly, define j_1 to be the smallest index such that

$$w_{j_1-1} < w_{j_1} = \cdots = w_{m-1}. \tag{10.4}$$

There are three situations to highlight, depending upon how many nonzero values w takes.

- If w has only one nonzero value, then $w_0 = \cdots = w_{m-1}$, so that $j_0 = m-1$ and $j_1 = 0$. The weight w is a multiple of the Hamming weight.
- If w has exactly two nonzero values, then $w_0 = \cdots = w_{j_0} < w_{j_0+1} = \cdots = w_{m-1}$, so that $j_1 = j_0+1$. The homogeneous weight is an example of this, with $j_0 = m-2$, $j_1 = m-1$.
- If w has three or more values, define j_2 so that
$$\cdots w_{j_2-1} < w_{j_2} = \cdots = w_{j_1-1} < w_{j_1} = \cdots = w_{m-1}.$$
Said another way, j_1 is the largest index less than m with $\epsilon_{j_1} > 0$, and j_2 is the second-largest index less than m with $\epsilon_{j_2} > 0$.

The weight w has a least two nonzero values if and only if $j_1 > 0$. In that case, $j_0 < j_1 \leq m-1$, with

$$\epsilon_{j_1} > 0 \text{ and } \epsilon_{j_1+1} = \cdots = \epsilon_{m-1} = 0. \tag{10.5}$$

If w has at least three values, then $j_0 < j_2 < j_1 \leq m-1$ and, in addition to (10.5), we have

$$\epsilon_{j_2} > 0 \text{ and } \epsilon_{j_2+1} = \cdots = \epsilon_{j_1-1} = 0. \tag{10.6}$$

The key to our analysis is a simplified expression for δ_k; cf., (8.12).

LEMMA 10.7. *Suppose* $\{0,1,\ldots,j_0\} = \mathring{I}$, *with* $j_0 \geq 1$. *If k is an integer,* $2 \leq k \leq j_0+1$, *then*

$$\delta_k = -q^m(q-1)\sum_{j=1}^{k-1} q^{j-1} p_{k-j}\epsilon_{m-j}.$$

If k is an integer, $j_0+2 \leq k \leq m$, *then*

$$\begin{aligned}\delta_k = {} & -((k-j_0-1)q^{m+k-j_0-2} - q^{m+k-j_0-3} - \cdots - q^{m-1})\epsilon_m \\ & -q^m(q-1)\sum_{j=1}^{k-1} q^{j-1} p_{k-j}\epsilon_{m-j}.\end{aligned}$$

PROOF. For any integer k, $2 \leq k \leq m$, Lemma 9.3 and (6.3) imply

$$\begin{aligned}\delta_k &= -|\mathrm{orb}(1)|\Delta_k - \sum_{i=1}^{\min\{j_0,k-1\}} (k-i)q^{k-1}w_{m-1}|\mathrm{orb}(\theta^i)| \\ &= -q^{m-1}(q-1)\Delta_k - \sum_{i=1}^{\min\{j_0,k-1\}} (k-i)q^{m+k-i-2}(q-1)w_{m-1}.\end{aligned}$$

In the formula for Δ_k given in Proposition 8.14, use $\epsilon'_{m-1} = q\epsilon_{m-1} + \epsilon_m = q\epsilon_{m-1} - w_{m-1}$. Because of telescoping sums, the w_{m-1}-terms cancel completely when $2 \leq k \leq j_0+1$ or cancel partially when $j_0+2 \leq k \leq m$. The terms that remain are as stated. □

In the case of $j_0+2 \leq k \leq m$, note that the ϵ_m-term is positive, as $-\epsilon_m = w_{m-1} > 0$ and the numerical sum is positive. The other terms are nonpositive. The balance between the terms appears to be problematic. When $2 \leq k \leq j_0+1$, $\delta_k \leq 0$.

We will also need information about how δ_k changes when $k \geq j_0+2$.

LEMMA 10.8. *If $k \geq j_0 + 2$ and $k \geq m - j_1$, then*

$$\delta_{k+1} - \delta_k = -q^{m+k-j_0-2}(q-1) \times \left\{ (k-j_0)\epsilon_m + q^{j_0+1} \sum_{i=m-j_1}^{k} (k-i+1)\epsilon_{m-i} \right\}.$$

PROOF. Because $k \geq j_0+2$, the second formula in Lemma 10.7 applies to both δ_k and δ_{k+1}. All but the highest order terms cancel, leaving

$$\delta_{k+1} - \delta_k = -(k-j_0)q^{m+k-j_0-2}(q-1)\epsilon_m - q^{m+k-1}(q-1)\sum_{i=1}^{k}(k-i+1)\epsilon_{m-i}.$$

Because $j_1 \geq m-k$, (10.5) implies that terms vanish in the summation for $i = 1, 2, \ldots, m - j_1 - 1$. □

COROLLARY 10.9. *Suppose $j_0 + j_1 = m - 1$. If $k \geq j_0 + 2$, then*

$$\delta_{k+1} - \delta_k = -q^{m+k-j_0-2}(q-1)(k-j_0)\left(\epsilon_m + q^{j_0+1}\epsilon_{j_1}\right) - q^{m+k-1}(q-1) \sum_{i=m-j_1+1}^{k} (k-i+1)\epsilon_{m-i}.$$

PROOF. Note first that $j_0 + j_1 = m - 1$ implies $j_0 + 2 = m - j_1 + 1$. Thus, if $k \geq j_0 + 2$, then $k \geq m - j_1$ is automatic. Apply Lemma 10.8 and notice that the term inside the summation with $i = m - j_1$ is $(k - m + j_1 + 1)\epsilon_{j_1} = (k - j_0)\epsilon_{j_1}$. □

The next theorem gives a more detailed description of the claims in Theorem 10.2.

THEOREM 10.10. *Let R be a finite chain ring with a weakly monotone weight w. Let j_0 and j_1 be as defined in (10.3) and (10.4). Then the following statements hold.*

(1) *If $j_0 = m - 1$, then w is a multiple of the Hamming weight.*
(2) *If $j_0 = 0$, then Corollary 9.5 applies.*

In the following statements, assume $1 \leq j_0 < m - 1$.

(3) *If $j_0 + j_1 \geq m$, then $\delta_k < 0$ for $m - j_1 + 1 \leq k \leq j_0 + 1$.*
(4) *If $j_0 + j_1 \leq m - 2$, then $\delta_k > 0$ for $j_0 + 2 \leq k \leq m - j_1$.*
(5) *If $j_0 + j_1 = m - 1$ and $w_{m-1} \neq q^{j_0+1}\epsilon_{j_1}$, then $\delta_{j_0+2} \neq 0$.*
(6) *Suppose $j_0 + j_1 = m - 1$ and $w_{m-1} = q^{j_0+1}\epsilon_{j_1}$. If w has at least three nonzero values, then $\delta_k < 0$ for $k \geq m - j_2 + 1$.*
(7) *Suppose $j_0 + j_1 = m - 1$, $w_{m-1} = q^{j_0+1}\epsilon_{j_1}$, and w has two nonzero values. Then $j_1 = j_0 + 1$, $m = 2j_0 + 2$ is even, and*

$$A_{w_{m-1}}(C_k^{\perp}) - A_{w_{m-1}}(D_k^{\perp}) < 0$$

for $k > j_1$.

REMARK 10.11. The length of the 'run' $w_{j_1} = \cdots = w_{m-1}$ is $m - j_1$. The length of the 'run' $w_0 = \cdots = w_{j_0}$ is $j_0 + 1$. Their difference is

$$(m - j_1) - (j_0 + 1) = \begin{cases} +, & j_0 + j_1 \leq m - 2, \\ 0, & j_0 + j_1 = m - 1, \\ -, & j_0 + j_1 \geq m. \end{cases}$$

PROOF OF THEOREM 10.10. The first two claims were explained after (10.4). If $1 \leq j_0 < m - 1$ and $j_0 + j_1 \geq m$, then any k satisfying $m - j_1 + 1 \leq k \leq j_0 + 1$ has $\delta_k < 0$ by Lemma 10.7 and (10.5). Namely, the first formula in Lemma 10.7 applies, and $\epsilon_{j_1} > 0$ appears in the formula because $j_1 \geq m - k + 1$.

If $1 \leq j_0 < m-1$ and $j_0+j_1 \leq m-2$, then any k satisfying $j_0+2 \leq k \leq m-j_1$ has $\delta_k > 0$. Now the second formula in Lemma 10.7 applies, and only the (positive) ϵ_m-term survives, because $j_1 < m - k + 1$.

If $1 \leq j_0 < m - 1$ and $j_0 + j_1 = m - 1$, then $j_0 + 2 = m - j_1 + 1$. Setting $k = j_0 + 2$, we see from the second formula of Lemma 10.7 that

$$\begin{aligned} \delta_{j_0+2} &= -(q^m - q^{m-1})\epsilon_m - q^m(q-1)q^{j_0}\epsilon_{j_1} \\ &= -q^{m-1}(q-1)\left(\epsilon_m + q^{j_0+1}\epsilon_{j_1}\right). \end{aligned}$$

As $\epsilon_m = -w_{m-1}$, we see that $w_{m-1} \neq q^{j_0+1}\epsilon_{j_1}$ implies $\delta_{j_0+2} \neq 0$.

Next, suppose $1 \leq j_0 < m - 1$, $j_0 + j_1 = m - 1$, and $w_{m-1} = q^{j_0+1}\epsilon_{j_1}$. We just saw that this implies $\delta_{j_0+2} = 0$. Applying Corollary 10.9, first with $k = j_0 + 2$, and then recursively, we see, for any $k \geq j_0 + 2$, that δ_k has the form:

$$\delta_k = - \sum_{i=m-j_1+1}^{k-1} c(k)_i \epsilon_{m-i},$$

where each $c(k)_i$ is a positive integer depending on k. Because each $\epsilon_{m-i} \geq 0$, each $\delta_k \leq 0$, and $\delta_k < 0$ if at least one $\epsilon_{m-i} > 0$ in the interval of summation. Remember (10.6). If $k \geq m - j_2 + 1$, then $m - j_2 + 1 > m - j_1 + 1 = j_0 + 2$, $m - (k-1) \leq j_2$, and $\epsilon_{j_2} > 0$ appears in the expression for δ_k. Thus $\delta_k < 0$.

Finally, suppose $1 \leq j_0 < m - 1$, $j_0 + j_1 = m - 1$, $w_{m-1} = q^{j_0+1}\epsilon_{j_1}$, and w has exactly two nonzero values. That means that $j_1 = j_0 + 1$, so that $m = 2j_0 + 2$. In addition, we must have $\epsilon_{j_1} = w_{m-1} - w_0$. Using this in the equation $w_{m-1} = q^{j_0+1}\epsilon_{j_1}$ yields $q^{j_0+1}w_0 = (q^{j_0+1} - 1)w_{m-1}$.

Because the coefficients in this last equation are relatively prime, there exists a positive integer s such that $w_0 = (q^{j_0+1} - 1)s$ and $w_{m-1} = q^{j_0+1}s$. Calculate: $2w_0 - w_{m-1} = (q^{j_0+1} - 2)s > 0$, because $j_0 \geq 1$ and $q \geq 2$. (But see Remark 10.12.) Thus $\mathring{w} = w_0 < w_{m-1} < 2\mathring{w}$. Using the second formula of Lemma 9.1, with $d = w_{m-1}$ and $0 \notin I_d$, we see that $A_{w_{m-1}}(C_k^\perp) - A_{w_{m-1}}(D_k^\perp) < 0$ for any $k > j_1 = \min I_d$. □

REMARK 10.12. In general, for positive integers $q \geq 2$, $q^{j_0+1} - 2 \geq 0$, with equality if and only if $j_0 = 0$ and $q = 2$. Equality, again, points to Theorem 6.5.

EXAMPLE 10.13. When $m = 3$, the only situation not covered by Corollaries 9.4, 9.5, or Theorem 10.2 is when $w_0 = w_2 < w_1$. In this case, $\epsilon_1' = w_1 - w_0 > 0$ and $\epsilon_2' = (q-1)w_2 - qw_1 < 0$. Then Proposition 8.14 and Lemma 9.3 imply that

$$\delta_2 = -|\mathrm{orb}(1)|\Delta_2 = -q^2(q-1)((q-1)w_2 - qw_1) > 0.$$

Thus, this w does not respect duality. We conclude that, for $m = 3$, the only weights that respect duality are multiples of the Hamming weight.

11. Symmetrized enumerators and examples

In this section, there are details of the MacWilliams identities for the symmetrized enumerator for the action of the full group $\mathcal{U}$ of units on a finite chain ring R, followed by several examples. The details supplement the general outline provided in Appendix A.

Suppose R is a finite chain ring with $\mathfrak{m} = (\theta)$, $R/\mathfrak{m} \cong \mathbb{F}_q$, and $\theta^m = 0$. As we have seen earlier, the group of units $\mathcal{U}$ acts on R on the left, with orbits $\operatorname{orb}(1), \operatorname{orb}(\theta), \ldots, \operatorname{orb}(\theta^{m-1}), \operatorname{orb}(\theta^m) = \{0\}$. The dual action of $\mathcal{U}$ on $R^\sharp \cong R_R$ has the same orbit structure.

For an element $r \in R$, define $\nu(r)$ via $r \in \operatorname{orb}(\theta^{\nu(r)})$; i.e., $\nu(r)$ is the exponent i of θ such that $r = u\theta^i$ for some unit $u \in \mathcal{U}$. Define $f : R \to \mathbb{C}[Z_0, \ldots, Z_m]$ by $f(r) = Z_{\nu(r)}$, $r \in R$. Then define $F : R^n \to \mathbb{C}[Z_0, \ldots, Z_m]$ by

$$F(x) = \prod_{\ell=1}^{n} f(x_\ell) = \prod_{\ell=1}^{n} Z_{\nu(x_\ell)}, \quad x = (x_1, x_2, \ldots, x_n) \in R^n. \tag{11.1}$$

The *symmetrized enumerator* of an additive code $C \subseteq R^n$ is the following element $\mathrm{se}_C \in \mathbb{C}[Z_0, \ldots, Z_m]$:

$$\mathrm{se}_C = \mathrm{se}_C(Z_0, \ldots, Z_m) = \sum_{x \in C} F(x) = \sum_{x \in C} \prod_{\ell=1}^{n} Z_{\nu(x_\ell)}.$$

Any finite chain ring R is a Frobenius ring, with a generating character χ. We will use the following properties of χ [**25**]:

(1) for any nonzero ideal I of R, $\sum_{r \in I} \chi(r) = 0$;
(2) every $\pi \in \widehat{R}$ has the form $\pi = \chi^r$ for some unique $r \in R$;
(3) $\chi^r(0) = 1$ for all $r \in R$.

We first calculate the sum of a character over the orbits $\operatorname{orb}(\theta^i)$.

LEMMA 11.2. *Suppose $i, j = 0, 1, \ldots, m$ and $r \in \operatorname{orb}(\theta^j)$. Then*

$$\sum_{s \in \operatorname{orb}(\theta^i)} \chi^r(s) = \begin{cases} 0, & i + j \leq m - 2, \\ -q^{m-i-1}, & i + j = m - 1, \\ q^{m-i-1}(q-1), & i + j \geq m, i < m, \\ 1, & i + j \geq m, i = m. \end{cases}$$

PROOF. When $i = m$, $\theta^m = 0$. Then $\sum_{s \in \operatorname{orb}(\theta^m)} \chi^r(s) = 1$, as $\chi^r(0) = 1$. For $i = 0, 1, \ldots, m-1$, $\operatorname{orb}(\theta^i) = (\theta^i) - (\theta^{i+1})$, so, for χ itself,

$$\begin{aligned} \sum_{s \in \operatorname{orb}(\theta^i)} \chi(s) &= \sum_{s \in (\theta^i)} \chi(s) - \sum_{s \in (\theta^{i+1})} \chi(s) \\ &= \begin{cases} 0, & i = 0, 1, \ldots, m-2, \\ -1, & i = m-1, \end{cases} \end{aligned} \tag{11.3}$$

using property (1) of χ.

Now suppose $r \in \operatorname{orb}(\theta^j)$, so that $r = u\theta^j$, $u \in \mathcal{U}$. Left multiplication by r maps $\operatorname{orb}(\theta^i)$ onto $\operatorname{orb}(\theta^{i+j})$, with each element in $\operatorname{orb}(\theta^{i+j})$ being hit $|\operatorname{orb}(\theta^i)|/|\operatorname{orb}(\theta^{i+j})|$

times. (Because $\mathcal{U}\theta^i = \theta^i\mathcal{U}$, units can be moved across powers of θ; for any unit u, $u\theta^i = \theta^i u'$ for some unit u'.) This implies that

$$\sum_{s\in\mathrm{orb}(\theta^i)} \chi^r(s) = \sum_{s\in\mathrm{orb}(\theta^i)} \chi(rs) = \frac{|\mathrm{orb}(\theta^i)|}{|\mathrm{orb}(\theta^{i+j})|} \sum_{t\in\mathrm{orb}(\theta^{i+j})} \chi(t).$$

Using (6.3) and (11.3), we get the stated result. □

REMARK 11.4. The formulas in Lemma 11.2 depend only on the orbit of r, not r itself. This is a general feature of character sums over the blocks of a partition coming from a group action [**8**, Theorem 2.6].

Define the *generalized Kravchuk matrix* K by

$$K_{ij} = \sum_{s\in\mathrm{orb}(\theta^i)} \chi^r(s), \quad r \in \mathrm{orb}(\theta^j).$$

We calculate the Fourier transform of f and F as in (A.2).

LEMMA 11.5. *For any $r \in R$, the Fourier transform of f is*

$$\widehat{f}(r) = \sum_{i=0}^{m} Z_i K_{ij}, \quad r \in \mathrm{orb}(\theta^j).$$

The Fourier transform of F is

$$\widehat{F}(r_1, r_2, \ldots, r_n) = \prod_{\ell=1}^{n} \widehat{f}(r_\ell) = \prod_{\ell=1}^{n} \left(\sum_{i=0}^{m} Z_i K_{i,\nu(r_\ell)} \right).$$

PROOF. The identity $\widehat{F}(r_1, r_2, \ldots, r_n) = \prod_{\ell=1}^{n} \widehat{f}(r_\ell)$ is well-known, [**25**, Proposition A.5], so it is enough to calculate $\widehat{f}(r)$:

$$\begin{aligned} \widehat{f}(r) &= \sum_{s\in R} \chi(rs) f(s) = \sum_{s\in R} \chi^r(s) Z_{\nu(s)} \\ &= \sum_{i=0}^{m} \sum_{s\in\mathrm{orb}(\theta^i)} \chi^r(s) Z_i = \sum_{i=0}^{m} Z_i \sum_{s\in\mathrm{orb}(\theta^i)} \chi^r(s). \end{aligned}$$ □

We now have the MacWilliams identities for the symmetrized enumerator over a finite chain ring; cf., [**25**, Theorem 8.4], [**8**, Theorem 3.5].

THEOREM 11.6. *Suppose R is a finite chain ring and $C \subseteq R^n$ is a left R-linear code. Then, using $C^\perp = \mathcal{R}(C)$,*

$$\mathrm{se}_{C^\perp}(Z_0, \ldots, Z_m) = \frac{1}{|C|} \mathrm{se}_C(\mathcal{Z}_0, \ldots, \mathcal{Z}_m)\Big|_{\mathcal{Z}_j = \sum_{i=0}^{m} Z_i K_{ij}}.$$

PROOF. Follow the outline in Appendix A, apply Lemma 11.5, and note that $\mathcal{R}(C) = \mathfrak{R}(C)$ for left R-linear codes. □

A version of this theorem, valid for the partition determined by the homogeneous weight, appears in [**20**, Theorem 2.1]

EXAMPLE 11.7. Let $R = \mathbb{Z}/8\mathbb{Z}$. Then $\mathcal{U} = \{1,3,5,7\}$. The $\mathcal{U}$-orbits are $\mathrm{orb}(1) = \mathcal{U}$, $\mathrm{orb}(2) = \{2,6\}$, $\mathrm{orb}(4) = \{4\}$, and $\mathrm{orb}(0) = \{0\}$. The generalized Kravchuk matrix is

$$K = \begin{bmatrix} 0 & 0 & -4 & 4 \\ 0 & -2 & 2 & 2 \\ -1 & 1 & 1 & 1 \\ 1 & 1 & 1 & 1 \end{bmatrix}.$$

Let w be the homogeneous weight, so that $\mathrm{w}_0 = \mathrm{w}_1 = 1$, $\mathrm{w}_2 = 2$, and $\mathrm{w}_3 = 0$. The linear codes C_3 and D_3 of Theorem 8.9 have the following codewords, with multiplicities listed above the horizontal line, telling how many times the given entry is repeated.

8	8	8
0	0	0
1	2	4
2	4	0
3	6	4
4	0	0
5	2	4
6	4	0
7	6	4

2	2	2	2	4	4	8
0	0	0	0	0	0	0
4	4	4	4	0	0	0
0	0	4	4	4	4	0
4	4	0	0	4	4	0
0	4	0	4	0	4	4
4	0	4	0	0	4	4
0	4	4	0	4	0	4
4	0	0	4	4	0	4

Then the symmetrized enumerators of the codes are:

$$\begin{aligned} \mathrm{se}_{C_3} &= 4Z_0^8 Z_1^8 Z_2^8 + 2Z_1^8 Z_2^8 Z_3^8 + Z_2^8 Z_3^{16} + Z_3^{24}, \\ \mathrm{se}_{D_3} &= 4Z_2^{16} Z_3^8 + 2Z_2^{12} Z_3^{12} + Z_2^8 Z_3^{16} + Z_3^{24}. \end{aligned}$$

One can compute the symmetrized enumerators of the dual codes (via Theorem 11.6 and SageMath [**23**], say), but the results have too many terms to include here.

Specializing $Z_i \rightsquigarrow t^{\mathrm{w}_i}$ (and taking a Taylor expansion for the dual codes) yields the homogeneous weight enumerators:

$$\begin{aligned} \mathrm{howe}_{C_3} &= 1 + t^{16} + 2t^{24} + 4t^{32}, \\ \mathrm{howe}_{D_3} &= 1 + t^{16} + 2t^{24} + 4t^{32}, \\ \mathrm{howe}_{C_3^\perp} &= 1 + 16t + 1848t^2 + 60400t^3 + \cdots, \\ \mathrm{howe}_{D_3^\perp} &= 1 + 48t + 1832t^2 + 64656t^3 + \cdots. \end{aligned}$$

The computed value $\delta_3 = A_1(C_3^\perp) - A_1(D_3^\perp) = -32$ matches the value given in the second formula of Lemma 10.7.

EXAMPLE 11.8. Still use $R = \mathbb{Z}/8\mathbb{Z}$, but change the weight to $w_0 = 1, w_1 = w_2 = 2, w_3 = 0$. The multiplicities of both codes (call them $C_{3'}$ and $D_{3'}$) change accordingly:

8	8	8	6
0	0	0	0
1	2	4	0
2	4	0	0
3	6	4	0
4	0	0	0
5	2	4	0
6	4	0	0
7	6	4	0

2	2	2	2	6	6	10
0	0	0	0	0	0	0
4	4	4	4	0	0	0
0	0	4	4	4	4	0
4	4	0	0	4	4	0
0	4	0	4	0	4	4
4	0	4	0	0	4	4
0	4	4	0	4	0	4
4	0	0	4	4	0	4

Now the symmetrized enumerators are:

$$\begin{aligned} \mathrm{se}_{C_{3'}} &= 4Z_0^8 Z_1^8 Z_2^8 Z_3^6 + 2Z_1^8 Z_2^8 Z_3^{14} + Z_2^8 Z_3^{22} + Z_3^{30}, \\ \mathrm{se}_{D_{3'}} &= 4Z_2^{20} Z_3^{10} + 2Z_2^{16} Z_3^{14} + Z_2^8 Z_3^{22} + Z_3^{30}. \end{aligned}$$

The w-weight enumerators are

$$\begin{aligned} \mathrm{wwe}_{C_{3'}} &= 1 + t^{16} + 2t^{32} + 4t^{40}, \\ \mathrm{wwe}_{D_{3'}} &= 1 + t^{16} + 2t^{32} + 4t^{40}, \\ \mathrm{wwe}_{C_{3'}^{\perp}} &= 1 + 24t + 1074t^2 + 36584t^3 + \cdots, \\ \mathrm{wwe}_{D_{3'}^{\perp}} &= 1 \qquad\quad + 1354t^2 + 34304t^3 + \cdots. \end{aligned}$$

For this w, $\mathring{I} = \{0\}$. Lemma 9.3 implies that $\delta_{3'} = -4\Delta_{3'} = 24$, which matches the computed value.

EXAMPLE 11.9. For yet another example, still use $R = \mathbb{Z}/8\mathbb{Z}$, but change the weight to $w_0 = 1, w_1 = 2, w_2 = 1, w_3 = 0$, which lies outside the scope of the main results given in previous sections; cf., Example 10.13. The multiplicities of both codes (call them $C_{3''}$ and $D_{3''}$) change accordingly:

4	4	4	11
0	0	0	0
1	2	4	0
2	4	0	0
3	6	4	0
4	0	0	0
5	2	4	0
6	4	0	0
7	6	4	0

1	1	1	1	5	5	9
0	0	0	0	0	0	0
4	4	4	4	0	0	0
0	0	4	4	4	4	0
4	4	0	0	4	4	0
0	4	0	4	0	4	4
4	0	4	0	0	4	4
0	4	4	0	4	0	4
4	0	0	4	4	0	4

Now the symmetrized enumerators are:

$$\begin{aligned} \mathrm{se}_{C_{3''}} &= 4Z_0^4 Z_1^4 Z_2^4 Z_3^{11} + 2Z_1^4 Z_2^4 Z_3^{15} + Z_2^4 Z_3^{19} + Z_3^{23}, \\ \mathrm{se}_{D_{3''}} &= 4Z_2^{16} Z_3^7 + 2Z_2^{12} Z_3^{11} + Z_2^4 Z_3^{19} + Z_3^{23}. \end{aligned}$$

The w-weight enumerators are

$$\begin{aligned} \mathrm{wwe}_{C_{3''}} &= 1 + t^4 + 2t^{12} + 4t^{16}, \\ \mathrm{wwe}_{D_{3''}} &= 1 + t^4 + 2t^{12} + 4t^{16}, \\ \mathrm{wwe}_{C_{3''}^{\perp}} &= 1 + 63t + 2111t^2 + 51635t^3 + \cdots, \\ \mathrm{wwe}_{D_{3''}^{\perp}} &= 1 + 23t + 1195t^2 + 38431t^3 + \cdots. \end{aligned}$$

In this example, $0 \in \mathring{I}$, and Lemma 9.3 implies that $\delta_{3''} = 44 - 4 = 40$, which matches the computed value. The weight w does not satisfy the hypotheses of Corollaries 9.4, 9.5 or Theorem 10.2; nonetheless, we see that w does not respect duality. Example 10.13 uses $k = 2$ to reach the same conclusion.

REMARK 11.10. Although outside the main thrust of this paper, one could consider using two weights over a finite ring R. (I thank Marcus Greferath for this idea.) One weight, say w, is used for a left linear code $C \subseteq R^n$, and the other weight, say w', is used for the dual code $\mathcal{R}(C) \subseteq R^n$. We then say that the weights w, w' *respect duality* when, for all left linear codes $C, D \subseteq R^n$, if $\mathrm{wwe}_C = \mathrm{wwe}_D$, then $\mathrm{wwe}'_{\mathcal{R}(C)} = \mathrm{wwe}'_{\mathcal{R}(D)}$.

The next example suggests that nothing will be gained by this generalization of using two weights.

EXAMPLE 11.11. Let $R = \mathbb{Z}/4\mathbb{Z}$, so that $m = 2$. Let w, w' be two weights on R having maximal symmetry. Write $w_0 = w(\pm 1)$, $w_1 = w(2)$, $w_2 = w(0) = 0$, $w'_0 = w'(\pm 1)$, $w'_1 = w'(2)$, and $w'_2 = w'(0) = 0$.

The codewords of the linear codes C_2 and D_2 of Theorem 8.9 are as follows. Multiplicities of entries are listed above the line. There are two cases, depending on the sign of $\Delta_2 = w_1 - 2w_0$.

- If $w_1 - 2w_0 \geq 0$:

C_2:

$2w_1$	$2w_1$
0	0
1	2
2	0
3	2

D_2:

$w_1 - 2w_0$	$w_1 + 2w_0$	w_1	w_1
0	0	0	0
0	2	0	2
0	0	2	2
0	2	2	0

Both linear codes have length $4w_1$ and the same w-weight enumerator:

$$\mathrm{wwe}_{C_2} = \mathrm{wwe}_{D_2} = 1 + t^{2w_1^2} + 2t^{2w_1(w_0+w_1)}.$$

The number of singleton r vectors in the dual codes are displayed next.

r	± 1	2
Number in $C_2^\perp$	0	$2w_1$
Number in $D_2^\perp$	$2(w_1 - 2w_0)$	$4w_1$

- If $w_1 - 2w_0 < 0$:

C_2:

$2w_0 - w_1$	$2w_1$	$2w_1$
0	0	0
0	1	2
0	2	0
0	3	2

D_2:

$w_1 + 2w_0$	w_1	w_1
0	0	0
2	0	2
0	2	2
2	2	0

Both codes have length $2w_0 + 3w_1$ and the same w-weight enumerator:

$$\mathrm{wwe}_{C_2} = \mathrm{wwe}_{D_2} = 1 + t^{2w_1^2} + 2t^{2w_1(w_0+w_1)}.$$

The number of singleton r vectors in the dual codes are displayed next.

r	± 1	2
Number in $C_2^\perp$	$2(2w_0 - w_1)$	$2w_0 + w_1$
Number in $D_2^\perp$	0	$2w_0 + 3w_1$

In both cases, the differences in the number of singleton vectors (the number for $C_2^\perp$ minus the number for $D_2^\perp$) are $2(2w_0 - w_1)$ for $r = \pm 1$ and $-2w_1$ for $r = 2$.

- If $w_1' < w_0'$, then the singleton 2 vectors yield $A_{w_1'}(C_2^\perp) - A_{w_1'}(D_2^\perp) = A_{w_1'}^{\text{sing}}(C_2^\perp) - A_{w_1'}^{\text{sing}}(D_2^\perp) = -2w_1 < 0$.
- If $w_0' = w_1'$ (so that w' is a multiple of the Hamming weight), then both types of singletons contribute. Thus, $A_{w_0'}(C_2^\perp) - A_{w_0'}(D_2^\perp) = A_{w_0'}^{\text{sing}}(C_2^\perp) - A_{w_0'}^{\text{sing}}(D_2^\perp) = 4(w_0 - w_1)$. This expression is nonzero unless the weight w is a multiple of the Hamming weight. When $w_0 = w_1$, we are in the situation of the Hamming weight, where the MacWilliams identities hold.
- If $w_0' < w_1'$, then singleton ± 1 vectors yield $A_{w_0'}(C_2^\perp) - A_{w_0'}(D_2^\perp) = A_{w_0'}^{\text{sing}}(C_2^\perp) - A_{w_0'}^{\text{sing}}(D_2^\perp) = 2(2w_0 - w_1)$. This expression is nonzero unless the weight w is a multiple of the homogeneous weight.
  - If $w_1 = 2w_0$ and $w_0' < w_1' < 2w_0'$, then $A_{w_1'}(C_2^\perp) - A_{w_1'}(D_2^\perp) = A_{w_1'}^{\text{sing}}(C_2^\perp) - A_{w_1'}^{\text{sing}}(D_2^\perp) = -2w_1 < 0$.
  - If $w_1 = 2w_0$ and $w_1' = 2w_0'$, then the MacWilliams identities hold for the homogeneous weight, Theorem 6.5.
  - If $w_1 = 2w_0$ and $w_1' > 2w_0'$, then counting doubleton vectors with entries of ± 1 yields $A_{2w_0'}(C_2^\perp) - A_{2w_0'}(D_2^\perp) = 2w_1 > 0$.

In summary, the only situations where w, w' respect duality are when both weights are multiples of the Hamming weight or both weights are multiples of the homogeneous weight. In all other situations, duality is not respected.

Part 3. Matrix rings over finite fields

12. Matrix modules, their orbits, and a positive result

We begin our study of matrix rings by describing certain matrix modules, the orbits of the group of units, the homogeneous weight, and the MacWilliams identities for $M_{2\times 2}(\mathbb{F}_2)$.

Fix integers k, m with $2 \le k \le m$. Let $R = M_{k\times k}(\mathbb{F}_q)$ be the ring of $k \times k$ matrices over a finite field $\mathbb{F}_q$, and let $M = M_{k\times m}(\mathbb{F}_q)$ be the left R-module of $k \times m$ matrices over $\mathbb{F}_q$. Both R and M are vector spaces over $\mathbb{F}_q$. The scalar multiplication of R on M is the multiplication of matrices. The group $\mathrm{GL}(k, \mathbb{F}_q)$ of invertible $k \times k$ matrices is the group of units $\mathcal{U} = \mathcal{U}(R)$ of R; $\mathcal{U}$ acts on M on the left via matrix multiplication. The size of $\mathrm{GL}(k, \mathbb{F}_q)$ is:

$$|\mathrm{GL}(k, \mathbb{F}_q)| = (q^k - 1)(q^k - q)\cdots(q^k - q^{k-1}). \tag{12.1}$$

Our first objective is to understand the cyclic left R-submodules of M and the $\mathcal{U}$-orbits in M.

Given a $k \times m$ matrix $x \in M$, denote by $\mathrm{rowsp}(x)$ the row space of x, i.e., the $\mathbb{F}_q$-linear subspace of $\mathbb{F}_q^m$ spanned by the rows of x. A left R-module is *cyclic* if it is generated by one element; i.e., it has the form Rx for some element x in the module. Denote the orbit of $x \in M$ under the action of $\mathcal{U}$ by $\mathrm{orb}(x)$ or $[x]$; denote the rank of a matrix x by $\mathrm{rk}\, x$.

LEMMA 12.2. *Let $x \in M = M_{k\times m}(\mathbb{F}_q)$. Then*

- *for $y \in M$, $y \in Rx$ if and only if* $\mathrm{rowsp}(y) \subseteq \mathrm{rowsp}(x)$;

- *for* $y \in M$, $Ry = Rx$ *if and only if* $\operatorname{rowsp}(y) = \operatorname{rowsp}(x)$ *if and only if* $\operatorname{orb}(y) = \operatorname{orb}(x)$;
- *if* $y \in \operatorname{orb}(x)$, *then* $\operatorname{rk} y = \operatorname{rk} x$.

PROOF. If $y = rx$, then the rows of y are linear combinations of the rows of x. This implies $\operatorname{rowsp}(y) \subseteq \operatorname{rowsp}(x)$. Conversely, if $\operatorname{rowsp}(y) \subseteq \operatorname{rowsp}(x)$, then each row of y is a linear combination of the rows of x, say $y_i = \sum_{j=1}^{k} r_{ij} x_j$, for some $r_{ij} \in \mathbb{F}_q$, where the rows of x and y are denoted with subscripts. Define $r \in R$ by $r = (r_{ij})$; then $y = rx$.

For the second item, apply the first item twice, symmetrically in y and x. When $\operatorname{rowsp}(y) = \operatorname{rowsp}(x)$, both x and y row reduce to the same row-reduced echelon form, which means they are in the same $\mathcal{U}$-orbit. □

Let $\mathcal{P}_M$ be the partially ordered set (poset) of all cyclic left R-submodules of $M = M_{k\times m}(\mathbb{F}_q)$, and let $\mathcal{P}_{k,m}$ be the poset of all linear subspaces of dimension at most k in $\mathbb{F}_q^m$; both partial orders are set inclusion. Define $\rho : \mathcal{P}_M \to \mathcal{P}_{k,m}$ by $\rho(Rx) = \operatorname{rowsp}(x)$; ρ is well-defined by Lemma 12.2. Conversely, given a linear subspace $V \subseteq \mathbb{F}_q^m$, define

$$\psi(V) = \{x \in M : \operatorname{rowsp}(x) \subseteq V\}.$$

PROPOSITION 12.3. *When* $\dim V \leq k$, $\psi(V)$ *is a cyclic left R-submodule of M. The map* $\rho : \mathcal{P}_M \to \mathcal{P}_{k,m}$ *is an isomorphism of posets, with inverse given by* ψ.

PROOF. Suppose $\dim V \leq k$. Choose a basis of V, and define $x \in M$ to have the chosen basis of V as its first $\dim V$ rows, followed by rows of zeros. Then $\operatorname{rowsp}(x) = V$. By Lemma 12.2, $\psi(V) = Rx$ is a cyclic module. The argument also shows that ρ is surjective. Lemma 12.2 implies that ρ is injective and preserves inclusion. □

COROLLARY 12.4. *The orbits* $\operatorname{orb}(x)$, $x \in M$, *of the left action of* $\mathcal{U} = \mathrm{GL}(k, \mathbb{F}_q)$ *on* $M = M_{k\times m}(\mathbb{F}_q)$ *are in one-to-one correspondence with the linear subspaces of dimension at most k contained in* $\mathbb{F}_q^m$. *More precisely, the orbits of matrices of rank j are in one-to-one correspondence with linear subspaces of dimension j in* $\mathbb{F}_q^m$. *The linear subspace corresponding to* $\operatorname{orb}(x)$ *is* $\operatorname{rowsp}(x)$.

There are similar results for linear functionals on M. A *linear functional* on M is a homomorphism $\lambda : M \to R$ of left R-modules; inputs will be written on the left, so that $(rx)\lambda = r(x\lambda)$ for $r \in R$ and $x \in M$. The collection of all linear functionals is denoted $M^\sharp = \operatorname{Hom}_R(M, R)$; $M^\sharp$ is a right R-module, with addition defined point-wise and λr defined by $x(\lambda r) = (x\lambda)r$, where $\lambda \in M^\sharp$, $r \in R$, $x \in M$. When $M = M_{k\times m}(\mathbb{F}_q)$, then $M^\sharp = M_{m\times k}(\mathbb{F}_q)$, with the evaluation $x\lambda \in R$, $x \in M$, $\lambda \in M^\sharp$, being matrix multiplication.

The orbits $\operatorname{orb}(\lambda)$, $\lambda \in M^\sharp$, of the right action of $\mathcal{U} = \mathrm{GL}(k, \mathbb{F}_q)$ on $M^\sharp = M_{m\times k}(\mathbb{F}_q)$ are in one-to-one correspondence with the linear subspaces of dimension at most k contained in $\mathbb{F}_q^m$. More precisely, the orbits of linear functionals of rank j are in one-to-one correspondence with linear subspaces of dimension j in $\mathbb{F}_q^m$. The linear subspace corresponding to $\operatorname{orb}(\lambda)$ is the column space $\operatorname{colsp}(\lambda)$.

REMARK 12.5. Given a linear functional $\lambda \in M^\sharp$, i.e., $\lambda : M \to R$, its kernel consists of $x \in M$ such that $\operatorname{rowsp}(x) \subseteq \operatorname{colsp}(\lambda)^\perp$. Here, for a linear subspace $Y \subseteq \mathbb{F}_q^m$, denote by $Y^\perp$ its orthogonal with respect to the standard dot product on

$\mathbb{F}_q^m$. Given $\lambda_i \in M^\sharp$, $i = 1, 2, \ldots, n$, define $\Lambda : M \to R^n$ by $x\Lambda = (x\lambda_1, \ldots, x\lambda_n) \in R^n$. Then $\ker \Lambda$ consists of $x \in M$ with $\operatorname{rowsp}(x) \subseteq \cap \operatorname{colsp}(\lambda_i)^\perp = (\operatorname{colsp}(\lambda_1) + \cdots + \operatorname{colsp}(\lambda_n))^\perp$. This latter uses the fact that $(X+Y)^\perp = X^\perp \cap Y^\perp$, for linear subspaces $X, Y \subseteq \mathbb{F}_q^m$. In particular, if $\operatorname{colsp}(\lambda_1), \ldots, \operatorname{colsp}(\lambda_n)$ span $\mathbb{F}_q^m$, then Λ is injective.

We record the number of $\mathcal{U}$-orbits and their sizes, depending on their rank. The *q-binomial coefficient* $\left[{m \atop j}\right]_q$ for $1 \le j \le m$ is defined by

$$\begin{bmatrix} m \\ j \end{bmatrix}_q = \frac{(q^m - 1)(q^{m-1} - 1) \cdots (q^{m-j+1} - 1)}{(q^j - 1)(q^{j-1} - 1) \cdots (q - 1)}.$$

For $m \ge 0$, $\left[{m \atop 0}\right]_q = 1$. If $j < 0$ or $j > m$, then $\left[{m \atop j}\right]_q = 0$.

PROPOSITION 12.6. *There is one orbit of rank 0, of size $\mathscr{S}_0 = 1$, in M. For any integer j, $1 \le j \le k$, all orbits of rank j matrices in M have the same size. The number and size $\mathscr{S}_j$ of orbits of rank j matrices in M are:*

number	*size*
$\begin{bmatrix} m \\ j \end{bmatrix}_q$	$\mathscr{S}_j = \prod_{i=0}^{j-1} (q^k - q^i)$

The ratio $\mathscr{S}_{j+1}/\mathscr{S}_j$ satisfies $\mathscr{S}_{j+1}/\mathscr{S}_j = q^k - q^j$, for $0 \le j \le k-1$.

The size of a cyclic submodule Rr depends only on $\operatorname{rk} r$: $|Rr| = q^{k \operatorname{rk} r}$.

We note that the sizes $\mathscr{S}_j$ and $|Rr|$ do not depend on m.

PROOF. It is well-known (e.g., [**22**, Theorem 3.2.6]) that the q-binomial coefficient $\left[{m \atop j}\right]_q$ counts the number of j-dimensional linear subspaces in $\mathbb{F}_q^m$, so the number of orbits follows from Corollary 12.4.

In addition to the left action of $\mathcal{U} = \mathrm{GL}(k, \mathbb{F}_q)$ on $M = M_{k \times m}(\mathbb{F}_q)$, there is also a right action of $\mathrm{GL}(m, \mathbb{F}_q)$ on M via matrix multiplication; both actions preserve rank. As matrix multiplication is associative, these two actions commute. Thus right multiplication by $P \in \mathrm{GL}(m, \mathbb{F}_q)$ maps the $\mathcal{U}$-orbit $\operatorname{orb}(x)$, $x \in M$, to $\operatorname{orb}(xP)$, and the two orbits have the same size.

Suppose the integer j satisfies $1 \le j \le k$. Choose $x_0 \in M$ to have the first j standard basis vectors (i.e., $(1, 0, \ldots, 0)$, etc.) as its first j rows, with the remaining rows being all zeros. Pick any $y \in M$ with $\operatorname{rk} y = j$. Choose a basis of $\operatorname{rowsp}(y)$, and extend it to a basis of $\mathbb{F}_q^m$. Use this basis of $\mathbb{F}_q^m$ as the rows of a matrix $P \in \mathrm{GL}(m, \mathbb{F}_q)$. Then the rows of $x_0 P$ consist of the chosen basis of $\operatorname{rowsp}(y)$, followed by $k - j$ zero-rows. Thus we have $\operatorname{rowsp}(x_0 P) = \operatorname{rowsp}(y)$, so that $\operatorname{orb}(y) = \operatorname{orb}(x_0 P)$, by Lemma 12.2. We conclude that $|\operatorname{orb}(y)| = |\operatorname{orb}(x_0 P)| = |\operatorname{orb}(x_0)|$, so that all orbits of rank j matrices have the same size.

As for the size $\mathscr{S}_j$ of an orbit of rank j matrices, it is enough to calculate $|\operatorname{orb}(x_0)|$ using $|\operatorname{orb}(x_0)| = |\mathcal{U}|/|\operatorname{stab}(x_0)|$, where $\operatorname{stab}(x_0) = \{u \in \mathcal{U} : u x_0 = x_0\}$ is the stabilizer subgroup of x_0. Then $u \in \operatorname{stab}(x_0)$ has the form

$$u = \left[\begin{array}{c|c} I_j & B \\ \hline 0 & D \end{array}\right],$$

with I_j the $j \times j$ identity matrix, B arbitrary, and D invertible. Then

$$|\operatorname{orb}(x_0)| = |\mathrm{GL}(k, \mathbb{F}_q)| / (q^{j(k-j)} |\mathrm{GL}(k-j, \mathbb{F}_q)|),$$

which simplifies as claimed.

The same argument using the right action of $\mathrm{GL}(m, \mathbb{F}_q)$ implies that the size of a cyclic submodule $Rr \subseteq M$ depends only on $\operatorname{rk} r$. Indeed, suppose $\operatorname{rk} r_1 = \operatorname{rk} r_2$. By row and column operations, there are units $u_1 \in \mathrm{GL}(k, \mathbb{F}_q)$ and $u_2 \in \mathrm{GL}(m, \mathbb{F}_q)$ such that $r_2 = u_1 r_1 u_2$. By Lemma 12.2, $Rr_2 = Rr_1 u_2$. Right multiplication by u_2 maps Rr_1 isomorphically to $Rr_1 u_2$. We conclude that $|Rr_2| = |Rr_1 u_2| = |Rr_1|$. For $j = 1, 2, \ldots, k$, let $r \in M$ be the following matrix of rank j:

$$r = \left[\begin{array}{c|c} I_j & 0 \\ \hline 0 & 0 \end{array}\right].$$

Then Rr consists of all matrices in M whose last $m - j$ columns are zero. The cyclic submodule Rr has size $|Rr| = q^{kj} = q^{k \operatorname{rk} r}$. □

We now turn our attention to the homogeneous weight on $R = M_{k\times k}(\mathbb{F}_q)$. Because of Proposition 12.3, the Möbius function for the poset of principal left ideals of R equals the Möbius function for the poset of linear subspaces of $\mathbb{F}_q^k$, which, following [**11**, (2.7)], is

$$\mu(V_1, V_2) = (-1)^c q^{\binom{c}{2}}, \quad V_1 \subseteq V_2 \subseteq \mathbb{F}_q^k, \tag{12.7}$$

where $c = \dim V_2 - \dim V_1$ is the codimension of V_1 in V_2.

Equation (2.9) yields the following formula for the homogeneous weight w on R; this formula also appears in [**14**, Proposition 7]. We write $\rho = \operatorname{rk}(r)$, $r \in R$:

$$\mathrm{w}(r) = \begin{cases} 0, & \rho = 0, \\ \zeta\left(1 - \frac{(-1)^\rho}{(q^k-1)(q^{k-1}-1)\cdots(q^{k-\rho+1}-1)}\right), & \rho > 0. \end{cases} \tag{12.8}$$

Note that $\mathrm{w}(r)$ depends only on ρ, which is consistent with w being constant on left $\mathcal{U}$-orbits. Write w_ρ for the common value of $\mathrm{w}(r)$ where $\operatorname{rk} r = \rho$. By choosing ζ appropriately, namely

$$\zeta = (q^k - 1)(q^{k-1} - 1) \cdots (q - 1)/q,$$

the homogeneous weight will have integer values.

EXAMPLE 12.9. For $M_{2\times 2}(\mathbb{F}_q)$, the homogeneous weight is

	w_0	w_1	w_2
general q	0	$q^2 - q$	$q^2 - q - 1$
$q = 2$	0	2	1
$q = 3$	0	6	5

,

with average weight $\zeta = (q^2 - 1)(q - 1)/q$ in general, so that $\zeta = 3/2$ for $q = 2$, and $\zeta = 16/3$ for $q = 3$.

EXAMPLE 12.10. For $M_{3\times 3}(\mathbb{F}_q)$, the homogeneous weight is:

	w_0	w_1	w_2	w_3
q	0	$q^5 - q^4 - q^3 + q^2$	$q^5 - q^4 - q^3 + q$	$q^5 - q^4 - q^3 + q + 1$
$q = 2$	0	12	10	11
$q = 3$	0	144	138	139

,

with average weight $\zeta = (q^3 - 1)(q^2 - 1)(q - 1)/q$ in general, so that $\zeta = 21/2$ for $q = 2$, and $\zeta = 416/3$ for $q = 3$.

LEMMA 12.11. *The homogeneous weight on $M_{k\times k}(\mathbb{F}_q)$ satisfies*

$$0 = \mathrm{w}_0 < \mathrm{w}_2 < \mathrm{w}_4 < \cdots < \zeta < \cdots < \mathrm{w}_3 < \mathrm{w}_1.$$

Moreover, $2\mathrm{w}_2 - \mathrm{w}_1 > 0$ *for all* $k \geq 2$, $q \geq 2$, except *for* $k = q = 2$, *where* $2\mathrm{w}_2 - \mathrm{w}_1 = 0$.

PROOF. The denominator in (12.8) is an increasing function of $\rho = \mathrm{rk}(r)$. This, together with the alternating sign of $(-1)^\rho$, yields the inequalities among the w_ρ. By moving ζ to the left side of (12.8), one sees that $\mathrm{w}_\rho - \zeta = -(-1)^\rho\zeta/((q^k - 1)\cdots(q^{k-\rho+1}-1))$. This implies that $\mathrm{w}_\rho - \zeta$ is positive when ρ is odd and negative when ρ is even.

For $k \geq 2$, one calculates that

$$2\mathrm{w}_2 - \mathrm{w}_1 = \zeta\frac{(q^k-2)(q^{k-1}-1)-2}{(q^k-1)(q^{k-1}-1)}.$$

Using $k \geq 2$ and $q \geq 2$, the numerator satisfies

$$(q^k-2)(q^{k-1}-1)-2 \geq (q^2-2)(q-1)-2 = q(q+1)(q-2).$$

This last expression is positive when $q > 2$ and vanishes when $q = 2$. Even for $q = 2$, the earlier inequality is strict when $k > 2$. Thus, $2\mathrm{w}_2 - \mathrm{w}_1 > 0$, except for $k = q = 2$, where $2\mathrm{w}_2 - \mathrm{w}_1 = 0$. □

Using Lemma 5.4, we see that any nonzero vector v with $\mathrm{w}(v) < \mathrm{w}_1$ must be a singleton. Any nonzero vector with $\mathrm{w}(v) = \mathrm{w}_1$ must be a singleton (or a doubleton, i.e., two nonzero entries, only for $M_{2\times 2}(\mathbb{F}_2)$).

THEOREM 12.12. *The MacWilliams identities hold for the homogeneous weight over* $R = M_{2\times 2}(\mathbb{F}_2)$. *For a linear code* $C \subseteq R^n$,

$$\mathrm{howe}_{C^\perp}(X,Y) = \frac{1}{|C|}\,\mathrm{howe}_C(X+3Y, X-Y).$$

PROOF. As in the proof of Theorem 6.5, we provide details to be used in the argument outlined in Appendix A.

From Example 12.9, we have that $\mathrm{w}_0 = 0$, $\mathrm{w}_1 = 2$, and $\mathrm{w}_2 = 1$. A generating character for R is $\chi(r) = (-1)^{\mathrm{tr}\, r}$, $r \in R$, where tr is the matrix trace. Define $f : R \to \mathbb{C}[X,Y]$ by $f(r) = X^{2-\mathrm{w}(r)}Y^{\mathrm{w}(r)}$. The value of $f(r)$ depends only on $\mathrm{rk}\, r$:

$\mathrm{rk}\, r$	0	1	2
$f(r)$	X^2	Y^2	XY

A calculation shows the Fourier transform (A.2) depends only on $\mathrm{rk}\, r$:

$$\widehat{f}(r) = \begin{cases} X^2 + 9Y^2 + 6XY = (X+3Y)^2, & \mathrm{rk}\, r = 0,\\ X^2 + Y^2 - 2XY = (X-Y)^2, & \mathrm{rk}\, r = 1,\\ X^2 - 3Y^2 + 2XY = (X+3Y)(X-Y), & \mathrm{rk}\, r = 2.\end{cases}$$

Note that $\widehat{f}(r)$ has the form of $f(r)$ with a linear substitution: $X \leftarrow X + 3Y$, $Y \leftarrow X - Y$. Applying these details, the rest of the argument in Appendix A carries through. □

There is a Gray map from $M_{2\times 2}(\mathbb{F}_2)$ equipped with the homogeneous weight to $\mathbb{F}_4^2$ equipped with the Hamming weight (of $\mathbb{F}_4$) [**2**].

13. W-matrix

In this section, we determine the W-matrix of (4.5) for a weight w on $R = M_{k\times k}(\mathbb{F}_q)$ that has maximal symmetry.

As usual, let $R = M_{k\times k}(\mathbb{F}_q)$, and suppose w is a weight on R with maximal symmetry. Suppose $r \in R$. By row and column reduction there exist units $u_1, u_2 \in \mathcal{U} = \mathrm{GL}(k, \mathbb{F}_q)$ such that

$$u_1 r u_2 = \left[\begin{array}{c|c} I_\rho & 0 \\ \hline 0 & 0 \end{array}\right],$$

where $\rho = \operatorname{rk} r$. Thus $w(r) = w(u_1 r u_2) = w(\left[\begin{smallmatrix} I_\rho & 0 \\ 0 & 0 \end{smallmatrix}\right])$, which says that the value of $w(r)$ depends only on the rank of r. Write $w_0, w_1, \ldots, w_k$ for the value of w on matrices of rank $0, 1, \ldots, k$, respectively.

REMARK 13.1. While $w(0) = 0$ is part of the definition of a weight, some of the results of this section will be more natural to state if we allow w_0 to be viewed as an indeterminate. We will proceed with w_0 as an indeterminate, and later show, in Theorem 13.16 and Corollary 13.18, how the general results are affected when we set $w_0 = 0$.

The information module M will be $M = M_{k\times m}(\mathbb{F}_q)$ with $m \geq k$. Then $\operatorname{Hom}_R(M, R) = M_{m\times k}(\mathbb{F}_q)$, achieved by right multiplication against M; i.e., the evaluation pairing $M \times \operatorname{Hom}_R(M, R) \to R$ sends $x \in M$ and $\lambda \in \operatorname{Hom}_R(M, R)$ to $x\lambda \in R$.

Because of maximal symmetry, the symmetry groups of w are $G_{\mathrm{lt}} = G_{\mathrm{rt}} = \mathrm{GL}(k, \mathbb{F}_q)$. The orbit space $\mathcal{O} = G_{\mathrm{lt}}\backslash M$ is represented by row-reduced echelon matrices of size $k \times m$, and $\mathcal{O}^\sharp = \operatorname{Hom}_R(M, R)/G_{\mathrm{rt}}$ is represented by column-reduced echelon matrices of size $m \times k$. The matrix transpose maps $\mathcal{O} \leftrightarrow \mathcal{O}^\sharp$ bijectively. The sets $\mathcal{O}$ and $\mathcal{O}^\sharp$ are partitioned by rank, and elements of $\mathcal{O}$ and $\mathcal{O}^\sharp$ correspond to linear subspaces of $\mathbb{F}_q^m$ of dimension at most k, by Corollary 12.4. (Left orbits in $\mathcal{O}$ are viewed in terms of row spaces; their transposes, right orbits in $\mathcal{O}^\sharp$, are viewed in terms of column spaces.)

The rows of W are indexed by elements $[x] \in \mathcal{O}$, ordered so that ranks go from 0 to k. Similarly, the columns of W are indexed by elements $[\lambda] \in \mathcal{O}^\sharp$, ordered to match $\mathcal{O}$ under the bijection $\mathcal{O} \leftrightarrow \mathcal{O}^\sharp$. The $[x], [\lambda]$-entry of W is simply $w(x\lambda)$, i.e., the value of w at the evaluation $x\lambda \in R$. The value $w(x\lambda)$ is well-defined by the definition of the symmetry groups. By maximal symmetry, the value $w(x\lambda)$ depends only on the rank $\operatorname{rk}(x\lambda)$. The matrix W is square of size $N_{k,m}$, the number of linear subspaces of dimension at most k in $\mathbb{F}_q^m$.

Suppose $[x] \in \mathcal{O}$ corresponds to the linear subspace $X \subseteq \mathbb{F}_q^m$ and $[\lambda] \in \mathcal{O}^\sharp$ corresponds to Y. We seek to express $\operatorname{rk}(x\lambda)$, and so $w(x\lambda)$, in terms of X and Y.

For a linear subspace $X \subseteq \mathbb{F}_q^m$, denote by $X^\perp$ its orthogonal with respect to the standard dot product on $\mathbb{F}_q^m$. Then $\dim X^\perp = m - \dim X$ and $(X^\perp)^\perp = X$, for all linear subspaces $X \subseteq \mathbb{F}_q^m$. Also, $(X \cap Y)^\perp = X^\perp + Y^\perp$, for linear subspaces $X, Y \subseteq \mathbb{F}_q^m$.

LEMMA 13.2. *For linear subspaces $X, Y \subseteq \mathbb{F}_q^m$ representing $[x] \in \mathcal{O}$ and $[\lambda] \in \mathcal{O}^\sharp$, respectively:*

(1) $\dim X - \dim(X \cap Y^\perp) = \dim Y - \dim(Y \cap X^\perp)$, *and*
(2) $\operatorname{rk}(x\lambda) = \dim X - \dim(X \cap Y^\perp) = \dim Y - \dim(Y \cap X^\perp)$.

PROOF. Consider $(X \cap Y^{\perp})^{\perp} = X^{\perp} + Y$, and compare dimensions:

$$\begin{aligned} m - \dim(X \cap Y^{\perp}) &= \dim X^{\perp} + \dim Y - \dim(X^{\perp} \cap Y) \\ &= m - \dim X + \dim Y - \dim(X^{\perp} \cap Y). \end{aligned}$$

We conclude that $\dim X - \dim(X \cap Y^{\perp}) = \dim Y - \dim(X^{\perp} \cap Y)$.

Choose as a representative x a $k \times m$ matrix whose first $\dim(X \cap Y^{\perp})$ rows form a basis for $X \cap Y^{\perp}$, whose next $\dim X - \dim(X \cap Y^{\perp})$ rows complete to a basis of X, and whose remaining rows are zeros. Choose a representative λ by reversing the roles of X and Y: its first $\dim(Y \cap X^{\perp})$ columns form a basis for $Y \cap X^{\perp}$, its next $\dim Y - \dim(Y \cap X^{\perp})$ columns complete to a basis of Y, and its remaining columns are zeros. Then $x\lambda$ has the form

$$x\lambda = \left[\begin{array}{c|c|c} 0 & 0 & 0 \\ \hline 0 & Z & 0 \\ \hline 0 & 0 & 0 \end{array}\right],$$

where Z is a square matrix of size $(\dim X - \dim(X \cap Y^{\perp})) \times (\dim Y - \dim(X^{\perp} \cap Y))$. The matrix Z is nonsingular. If not, there exists a nonzero vector v with $vZ = 0$. Then $[0|v|0]x \in Y^{\perp}$, which violates the choice of basis of $X \cap Y^{\perp}$ in the construction of x. The formula for $\operatorname{rk}(x\lambda)$ now follows. □

The matrix W is symmetric when we use the bijection $\mathcal{O} \leftrightarrow \mathcal{O}^{\sharp}$ to align the indexing.

Because some of our later results depend upon inverting W, we need to understand when the matrix W is invertible. We will be able to transform W into a block diagonal format by making use of the Möbius function of the poset $\mathcal{P}_{k,m}$ of linear subspaces of dimension at most k in $\mathbb{F}_q^m$. Versions of this block diagonal format can be found in [**9**, §4], [**28**, Theorem 9.6], and [**29**, §6].

Recall that we index the rows and columns of W by linear subspaces of dimension at most k in $\mathbb{F}_q^m$, with ranks increasing from 0 to k.

Define a matrix P, with rows and columns indexed by linear subspaces of dimension $\leq k$ in $\mathbb{F}_q^m$, using the same ordering as for W. The entry $P_{\alpha,\beta}$ is given by

$$P_{\alpha,\beta} = \begin{cases} \mu(0,\beta) = (-1)^{\dim \beta} q^{\binom{\dim \beta}{2}}, & \text{if } \beta \subseteq \alpha, \\ 0, & \text{if } \beta \not\subseteq \alpha. \end{cases} \tag{13.3}$$

Because we are ordering rows and columns so that ranks increase, we see that P is lower triangular. Its diagonal entries are $P_{\alpha,\alpha} = (-1)^{\dim \alpha} q^{\binom{\dim \alpha}{2}} \neq 0$. Thus P is invertible over $\mathbb{Q}$.

For $j = 0, 1, \ldots, k$, define an incidence matrix $\mathscr{I}_j$ over $\mathbb{Q}$, square of size $\left[{m \atop j} \right]_q$, with rows and columns indexed by linear subspaces of dimension j in $\mathbb{F}_q^m$, using the dimension j portion of the ordering used for W and P. The α, δ-entry of $\mathscr{I}_j$ is given by

$$(\mathscr{I}_j)_{\alpha,\delta} = \begin{cases} 1, & \alpha \cap \delta^{\perp} = 0, \\ 0, & \alpha \cap \delta^{\perp} \neq 0. \end{cases}$$

The incidence matrices $\mathscr{I}_j$ are invertible by [**29**, Proposition 6.7].

Our main objective in this section is to prove Theorem 13.4.

THEOREM 13.4. *For positive integers $2 \le k \le m$ and a weight w on $M_{k\times k}(\mathbb{F}_q)$ having maximal symmetry and w_0 indeterminate, we have*

$$PWP^\top = \begin{bmatrix} c_0\mathscr{I}_0 & 0 & \cdots & 0 \\ 0 & c_1\mathscr{I}_1 & \cdots & 0 \\ \vdots & & \ddots & \vdots \\ 0 & 0 & \cdots & c_k\mathscr{I}_k \end{bmatrix},$$

where, for $j = 0, 1, \ldots, k$,

$$c_j = (-1)^j q^{\binom{j}{2}} \sum_{\ell=0}^{j} (-1)^\ell q^{\binom{\ell}{2}} \begin{bmatrix} j \\ \ell \end{bmatrix}_q w_\ell. \tag{13.5}$$

Before we prove Theorem 13.4, we prove some preliminary lemmas and propositions that will be used in the proof. We begin by quoting the well-known Cauchy Binomial Theorem, e.g., [**22**, Theorem 3.2.4].

THEOREM 13.6 (Cauchy Binomial Theorem). *For a positive integer k,*

$$\prod_{i=0}^{k-1}(1+xq^i) = \sum_{j=0}^{k} q^{\binom{j}{2}} \begin{bmatrix} k \\ j \end{bmatrix}_q x^j.$$

In particular, using $x = -1$, for k positive,

$$\sum_{j=0}^{k} (-1)^j q^{\binom{j}{2}} \begin{bmatrix} k \\ j \end{bmatrix}_q = 0. \tag{13.7}$$

In a vector space, a *frame* is an ordered set of linearly independent vectors; if there are b such vectors, we call the frame a *b-frame.*

LEMMA 13.8. *Let V be a vector space over $\mathbb{F}_q$ with $\dim V = v$, and let D be a linear subspace of V with $\dim D = d$. Then*

$$|\{B \subseteq V : \dim B = b, B \cap D = 0\}| = q^{bd} \begin{bmatrix} v-d \\ b \end{bmatrix}_q.$$

PROOF. Count the number of b-frames outside of D, and divide by $|\mathrm{GL}(b, \mathbb{F}_q)|$, (12.1). Then, factoring out b factors of q^d from

$$\frac{(q^v - q^d)(q^v - q^{d+1})\cdots(q^v - q^{d+b-1})}{(q^b - 1)(q^b - q)\cdots(q^b - q^{b-1})}$$

yields the stated result. □

LEMMA 13.9. *Let V be a vector space over $\mathbb{F}_q$ with $\dim V = v$, and let D be a linear subspace of V with $\dim D = d$. For any $j = 1, 2, \ldots, d$,*

$$|\{B \subseteq V : \dim B = b, \dim(B \cap D) = j\}| = q^{(b-j)(d-j)} \begin{bmatrix} d \\ j \end{bmatrix}_q \begin{bmatrix} v-d \\ b-j \end{bmatrix}_q.$$

PROOF. The count equals the number of j-dimensional subspaces $J \subseteq D$ times the number of subspaces B of dimension b with $B \cap D = J$. The number of j-dimensional subspaces of D is $\left[\begin{smallmatrix} d \\ j \end{smallmatrix}\right]_q$. The set of b-dimensional subspaces $B \subseteq V$ with $B \cap D = J$ is in one-to-one correspondence with the set of $(b-j)$-dimensional subspaces of V/J that intersect D/J trivially. By Lemma 13.8, the number of such subspaces is $q^{(b-j)(d-j)} \left[\begin{smallmatrix} v-d \\ b-j \end{smallmatrix}\right]_q$. □

Lemma 13.9 sharpens [**29**, Lemma 6.9]; the latter's $C_1(b)$ is $q^{b(m-b)}$.

LEMMA 13.10. *Let V be a vector space over $\mathbb{F}_q$ with $\dim V = v$, and let A and D be linear subspaces of V with $\dim A = a$ and $\dim D = d$. If $\dim(A \cap D) = i$, then, for any $j = 0, 1, \ldots, i$,*

$$|\{B \subseteq A : \dim B = b, \dim(B \cap D) = j\}| = q^{(b-j)(i-j)} \begin{bmatrix} i \\ j \end{bmatrix}_q \begin{bmatrix} a-i \\ b-j \end{bmatrix}_q .$$

PROOF. Note that $B \cap D = B \cap (A \cap D)$. Apply Lemmas 13.8 and 13.9 with ambient space A and subspace $A \cap D$. □

LEMMA 13.11. *Suppose $\alpha, \delta \subseteq \mathbb{F}_q^m$ are linear subspaces. Then $\dim(\alpha \cap \delta^{\perp}) \geq \dim \alpha - \dim \delta$. If $\dim \alpha > \dim \delta$, then $\dim(\alpha \cap \delta^{\perp}) > 0$.*

PROOF. Using $\alpha + \delta^{\perp} \subseteq \mathbb{F}_q^m$, compare dimensions:

$$m \geq \dim(\alpha + \delta^{\perp}) = \dim \alpha + m - \dim \delta - \dim(\alpha \cap \delta^{\perp}),$$

from which the result follows. □

PROPOSITION 13.12. *If $\dim(\alpha \cap \delta^{\perp}) > 0$, then $(PW)_{\alpha,\delta} = 0$. If $\alpha \cap \delta^{\perp} = 0$, then, writing $a = \dim \alpha$,*

$$(PW)_{\alpha,\delta} = \sum_{\beta \subseteq \alpha} \mu(0, \beta) w_{\dim \beta} = \sum_{r=0}^{a} (-1)^r q^{\binom{r}{2}} \begin{bmatrix} a \\ r \end{bmatrix}_q w_r .$$

In particular, the matrix PW is block upper triangular.

Likewise, if $\dim(\gamma \cap \beta^{\perp}) > 0$, then $(WP^{\top})_{\beta,\gamma} = 0$. If $\gamma \cap \beta^{\perp} = 0$, then, writing $c = \dim \gamma$,

$$(WP^{\top})_{\beta,\gamma} = \sum_{\epsilon \subseteq \gamma} \mu(0, \epsilon) w_{\dim \epsilon} = \sum_{s=0}^{c} (-1)^s q^{\binom{s}{2}} \begin{bmatrix} c \\ s \end{bmatrix}_q w_s .$$

In particular, the matrix $WP^{\top}$ is block lower triangular.

PROOF. From the definition of the matrix P and Lemma 13.2,

$$(PW)_{\alpha,\delta} = \sum_{\beta \subseteq \alpha} \mu(0, \beta) w_{\dim \beta - \dim(\beta \cap \delta^{\perp})} .$$

In this formula, the subscript $\dim \beta - \dim(\beta \cap \delta^{\perp}) = \dim \delta - \dim(\delta \cap \beta^{\perp}) \leq \dim \delta - \dim(\delta \cap \alpha^{\perp}) = \dim \alpha - \dim(\alpha \cap \delta^{\perp})$, as $\beta \subseteq \alpha$, so that $\alpha^{\perp} \subseteq \beta^{\perp}$. Writing $i = \dim(\alpha \cap \delta^{\perp})$, we see that the subscript $\dim \beta - \dim(\beta \cap \delta^{\perp}) \leq a - i$. Thus

$$\begin{aligned} (PW)_{\alpha,\delta} &= \sum_{r=0}^{a-i} \sum_{\substack{\beta \subseteq \alpha \\ \dim \beta - \dim(\beta \cap \delta^{\perp}) = r}} \mu(0, \beta) w_r \\ &= \sum_{r=0}^{a-i} \sum_{j=0}^{i} \sum_{\substack{\beta \subseteq \alpha \\ \dim \beta = r+j \\ \dim(\beta \cap \delta^{\perp}) = j}} \mu(0, \beta) w_r . \end{aligned}$$

By Lemma 13.10 and (12.7), the coefficient C_r of w_r is

$$\begin{aligned} C_r &= \sum_{j=0}^{i} q^{r(i-j)} \begin{bmatrix} i \\ j \end{bmatrix}_q \begin{bmatrix} a-i \\ r \end{bmatrix}_q (-1)^{r+j} q^{\binom{r+j}{2}} \\ &= (-1)^r q^{\binom{r}{2}} q^{ir} \begin{bmatrix} a-i \\ r \end{bmatrix}_q \left(\sum_{j=0}^{i} (-1)^j q^{\binom{j}{2}} \begin{bmatrix} i \\ j \end{bmatrix}_q \right) \\ &= \begin{cases} (-1)^r q^{\binom{r}{2}} \left[{a \atop r} \right]_q, & i = 0, \\ 0, & i > 0. \end{cases} \end{aligned}$$

To simplify, we used the identity $\binom{r+j}{2} = \binom{r}{2} + jr + \binom{j}{2}$ and (13.7).

If $\dim \delta < \dim \alpha$, Lemma 13.11 implies $i = \dim(\alpha \cap \delta^\perp) > 0$. Using dimension to create blocks, we see that PW is block upper triangular.

Essentially the same arguments yield the results about $WP^\top$. □

Recall that the matrix P is lower triangular, so that $P^\top$ is upper triangular. Thus $PWP^\top$ will be both lower and upper block triangular, hence block diagonal. The exact form of $PWP^\top$ is the next result.

PROPOSITION 13.13. *If* $\dim \alpha \neq \dim \delta$, *then* $(PWP^\top)_{\alpha,\delta} = 0$. *If* $\dim \alpha = \dim \delta = a$, *then*

$$(PWP^\top)_{\alpha,\delta} = \begin{cases} (-1)^a q^{\binom{a}{2}} \sum_{j=0}^{a} (-1)^j q^{\binom{j}{2}} \left[{a \atop j} \right]_q w_j, & \alpha \cap \delta^\perp = 0, \\ 0, & \alpha \cap \delta^\perp \neq 0. \end{cases}$$

PROOF. We first show that $\alpha \cap \delta^\perp \neq 0$ implies $(PWP^\top)_{\alpha,\delta} = 0$. Assume $\alpha \cap \delta^\perp \neq 0$. For any $\gamma \subseteq \delta$, we have $\delta^\perp \subseteq \gamma^\perp$, so that $\alpha \cap \delta^\perp \subseteq \alpha \cap \gamma^\perp$. Thus $\alpha \cap \gamma^\perp \neq 0$ for all $\gamma \subseteq \delta$.

Using the definition of P, we see that

$$(PWP^\top)_{\alpha,\delta} = \sum_{\gamma \subseteq \delta} (PW)_{\alpha,\gamma} \mu(0, \gamma). \tag{13.14}$$

By Proposition 13.12, all the $(PW)_{\alpha,\gamma}$-terms in (13.14) vanish. Thus $(PWP^\top)_{\alpha,\delta} = 0$, as claimed.

Essentially the same argument using

$$(PWP^\top)_{\alpha,\delta} = \sum_{\beta \subseteq \alpha} \mu(0, \beta)(WP^\top)_{\beta,\delta}$$

shows that $\delta \cap \alpha^\perp \neq 0$ implies $(PWP^\top)_{\alpha,\delta} = 0$.

From Lemma 13.2 we have the equation $\dim \alpha - \dim \delta = \dim(\alpha \cap \delta^\perp) - \dim(\delta \cap \alpha^\perp)$. If $\dim \alpha \neq \dim \delta$, at least one of $\alpha \cap \delta^\perp$ or $\delta \cap \alpha^\perp$ is nonzero. Thus $(PWP^\top)_{\alpha,\delta} = 0$.

At last, suppose $\dim \alpha = \dim \delta$. In this situation, note that $\alpha \cap \delta^\perp = 0$ if and only if $\delta \cap \alpha^\perp = 0$. As above, if $\alpha \cap \delta^\perp \neq 0$, then $(PWP^\top)_{\alpha,\delta} = 0$. If $\alpha \cap \delta^\perp = 0$, then the equality of dimensions yields $\mathbb{F}_q^m = \alpha \oplus \delta^\perp$. For any $\gamma \subsetneq \delta$, we have $\delta^\perp \subsetneq \gamma^\perp$, so that $\alpha \cap \gamma^\perp \neq 0$. By Proposition 13.12, $(PW)_{\alpha,\gamma} = 0$ for $\gamma \subsetneq \delta$. Thus, (13.14) implies that

$$(PWP^\top)_{\alpha,\delta} = (PW)_{\alpha,\delta} \mu(0, \delta),$$

which gives the stated formula. □

PROOF OF THEOREM 13.4. The nonzero entries in Proposition 13.13 depend only on $\dim\alpha$ and equal the c_j in the statement of Theorem 13.4. Whether $\alpha\cap\delta^\perp = 0$ is marked by the incidence matrix $\mathscr{I}_j$. □

COROLLARY 13.15. *The matrix W is invertible over $\mathbb{Q}$ if and only if*

$$c_j = (-1)^j q^{\binom{j}{2}} \sum_{\ell=0}^{j} (-1)^\ell q^{\binom{\ell}{2}} \begin{bmatrix} j \\ \ell \end{bmatrix}_q w_\ell \neq 0,$$

for all $j = 0, 1, \ldots, k$.

PROOF. The matrices P, $P^\top$, and $\mathscr{I}_j$ are all invertible over $\mathbb{Q}$. □

Recall from Remark 13.1 that we have been treating w_0 as an indeterminate. When w_0 is set equal to 0, every entry of the first row and first column of the matrix W is 0. Then W cannot be invertible. Equivalently, $c_0 = 0$ in Corollary 13.15.

To get around this lack of invertibility, we make the following adjustments, as in Remark 4.6. Define a matrix W_0 with rows and columns indexed by the *nonzero* elements of $\mathcal{O}$ and $\mathcal{O}^\sharp$, respectively, ordered so that ranks go from 1 to k. The $[x], [\lambda]$-entry is again $w(x\lambda)$, with $w(0) = 0$ now. Similarly, define a matrix P_0 with rows and columns indexed by the nonzero elements of $\mathcal{O}$ and (α, β)-entry given by (13.3). Then the counterparts of Theorem 13.4 and Corollary 13.15 are the following.

THEOREM 13.16. *For positive integers $2 \leq k \leq m$ and a weight w on $M_{k\times k}(\mathbb{F}_q)$ having maximal symmetry and $w(0) = 0$, we have*

$$P_0 W_0 P_0^\top = \begin{bmatrix} c_1\mathscr{I}_1 & 0 & \cdots & 0 \\ 0 & c_2\mathscr{I}_2 & \cdots & 0 \\ \vdots & & \ddots & \vdots \\ 0 & 0 & \cdots & c_k\mathscr{I}_k \end{bmatrix},$$

where, for $j = 1, \ldots, k$,

$$c_j = (-1)^j q^{\binom{j}{2}} \sum_{\ell=1}^{j} (-1)^\ell q^{\binom{\ell}{2}} \begin{bmatrix} j \\ \ell \end{bmatrix}_q w_\ell. \tag{13.17}$$

PROOF. The relationships between the matrices P and P_0 and between W and W_0, when $w_0 = 0$, are given below. The notations row(a) and col(a) mean a row, resp., column, vector, all of whose entries are a.

$$P = \left[\begin{array}{c|c} 1 & \mathrm{row}(0) \\ \hline \mathrm{col}(1) & P_0 \end{array}\right], \quad W|_{w_0=0} = \left[\begin{array}{c|c} 0 & \mathrm{row}(0) \\ \hline \mathrm{col}(0) & W_0 \end{array}\right].$$

Then

$$(PWP^\top)|_{w_0=0} = \left[\begin{array}{c|c} 0 & \mathrm{row}(0) \\ \hline \mathrm{col}(0) & P_0W_0P_0^\top \end{array}\right],$$

and the result follows from Theorem 13.4, with $w_0 = 0$. □

COROLLARY 13.18. *The matrix W_0 is invertible over $\mathbb{Q}$ if and only if*

$$c_j = (-1)^j q^{\binom{j}{2}} \sum_{\ell=1}^{j} (-1)^\ell q^{\binom{\ell}{2}} \begin{bmatrix} j \\ \ell \end{bmatrix}_q w_\ell \neq 0,$$

for all $j = 1, \ldots, k$.

REMARK 13.19. The extension property (EP) for w holds when the W_0 map is injective (zero right null space) for all information modules. We see that EP holds if and only if all $c_j \neq 0$ for $j = 1, 2, \ldots, k$. See [**28**, Theorem 9.5]. In particular, when W_0 is invertible, then $W_0 : F_0(\mathcal{O}^\sharp, \mathbb{Q}) \to F_0(\mathcal{O}, \mathbb{Q})$ is an isomorphism, Remark 4.6.

REMARK 13.20. It is possible to generalize Propositions 13.12 and 13.13 to the context of an alphabet $A = M_{k\times\ell}(\mathbb{F}_q)$ and information module $M = M_{k\times m}(\mathbb{F}_q)$, with $m \geq \ell \geq k$. This paper does not require such a level of generality.

14. Locally constant functions

In this section, we examine a type of multiplicity function that will feature prominently in the construction of linear codes in Section 16.

As in previous sections, $R = M_{k\times k}(\mathbb{F}_q)$ and $M = M_{k\times m}(\mathbb{F}_q)$, with $k \leq m$. The ring R is equipped with a weight w having maximal symmetry and positive integer values. As in Remark 13.1, we will treat w_0 as an indeterminate. Recall that the orbit spaces $\mathcal{O}$ and $\mathcal{O}^\sharp$ are ordered by rank, from 0 to k. Write $\mathcal{O}_i$ and $\mathcal{O}_i^\sharp$ for the collections of orbits of rank equal to i. Recall that orbits are denoted by $\operatorname{orb}(x)$ or $[x]$. Given sets $B \subseteq A$, we denote the indicator function of B by $1_B : A \to \mathbb{Z}$, with

$$1_B(x) = \begin{cases} 1, & x \in B, \\ 0, & x \notin B. \end{cases}$$

The type of multiplicity function to be considered has the form $1_{\mathcal{O}_i^\sharp}$ or a linear combination of such indicator functions.

Given a multiplicity function $\eta : \mathcal{O}^\sharp \to \mathbb{N}$ (or, more generally, $\eta : \mathcal{O}^\sharp \to \mathbb{Q}$), we refer to $\omega = W\eta$ as the *list of orbit weights* of η. The list ω is a function $\omega : \mathcal{O} \to \mathbb{Q}$. We say that η, resp., ω, is *locally constant* if $\eta = \sum_{j=0}^k a_j 1_{\mathcal{O}_j^\sharp}$, resp., $\omega = \sum_{j=0}^k b_j 1_{\mathcal{O}_j}$, for rational constants a_j, b_j. Said another way, η is locally constant if $\operatorname{rk}\lambda_1 = \operatorname{rk}\lambda_2$ implies $\eta([\lambda_1]) = \eta([\lambda_2])$; i.e., η is constant on each $\mathcal{O}_j^\sharp$. Similar comments apply to ω.

Let $\eta_j = 1_{\mathcal{O}_j^\sharp}$, and $\omega_j = W\eta_j$. We show that ω_j is locally constant.

PROPOSITION 14.1. *Let $\eta_j = 1_{\mathcal{O}_j^\sharp}$. Then $\omega_j = W\eta_j$ is locally constant. If $i = \operatorname{rk} x$, then*

$$\omega_j([x]) = \sum_{d=0}^{\min\{j,m-i\}} q^{(m-i-d)(j-d)} \begin{bmatrix} m-i \\ d \end{bmatrix}_q \begin{bmatrix} i \\ j-d \end{bmatrix}_q w_{j-d}.$$

In particular, when $j = 1$, $i = \operatorname{rk} x = 1, 2, \ldots, k$, and $w_0 = 0$,

$$\omega_1([x]) = q^{m-i} \begin{bmatrix} i \\ 1 \end{bmatrix}_q w_1 = (q^{m-i} + q^{m-i+1} + \cdots + q^{m-1}) w_1,$$

is an increasing function of i.

PROOF. Because $\omega_j = W\eta_j$, we have

$$\omega_j([x]) = \sum_{[\lambda]\in\mathcal{O}^\sharp} w(x\lambda)\eta_j([\lambda]) = \sum_{[\lambda]\in\mathcal{O}_j^\sharp} w(x\lambda).$$

Using Lemma 13.2, represent $\operatorname{orb}(x) = [x]$ by a linear subspace $X \subseteq \mathbb{F}_q^m$, $\dim X = i$, and $\operatorname{orb}(\lambda) = [\lambda]$ by Y, $\dim Y = j$. Then $w(x\lambda) = w_{i-\dim(X\cap Y^\perp)} = w_{j-\dim(Y\cap X^\perp)}$. Lemma 13.9 now implies

$$\begin{aligned}\omega_j([x]) &= \sum_{d=0}^{\min\{j,m-i\}} |\{Y : \dim Y = j, \dim(Y \cap X^\perp) = d\}| w_{j-d} \\ &= \sum_{d=0}^{\min\{j,m-i\}} q^{(m-i-d)(j-d)} \begin{bmatrix} m-i \\ d \end{bmatrix}_q \begin{bmatrix} i \\ j-d \end{bmatrix}_q w_{j-d}.\end{aligned}$$

When $j = 1$, the formula simplifies as stated. □

COROLLARY 14.2. *If η is locally constant, then so is $\omega = W\eta$.*

15. $\overline{W}$-matrix

There is an 'averaged' version $\overline{W}$ of the W-matrix that will be useful in our analysis of dual codewords in Section 18. We describe $\overline{W}$ in this section. We continue to assume that the information module is $M = M_{k\times m}(\mathbb{F}_q)$ and that w_0 is indeterminate.

LEMMA 15.1. *For $i = 0, 1, \ldots, k$ and $[\lambda] \in \mathcal{O}_j^\sharp$, the value of*

$$\sum_{[x]\in\mathcal{O}_i} W_{[x],[\lambda]} = \sum_{d=0}^{i} q^{(i-d)(m-j-d)} \begin{bmatrix} m-j \\ d \end{bmatrix}_q \begin{bmatrix} j \\ i-d \end{bmatrix}_q w_{i-d}$$

depends only on i and $j = \operatorname{rk}\lambda$.

PROOF. The proof of Proposition 14.1 applies, interchanging the roles of $i = \operatorname{rk} x$ and $j = \operatorname{rk}\lambda$. □

DEFINITION 15.2. Define $\overline{W}$ to be the integer $(k+1)\times(k+1)$ matrix

$$\overline{W}_{ij} = \sum_{[x]\in\mathcal{O}_i} W_{[x],[\lambda]}, \quad [\lambda] \in \mathcal{O}_j^\sharp, \tag{15.3}$$

for $i, j = 0, 1, \ldots, k$. The definition is well-defined by Lemma 15.1.

EXAMPLE 15.4. When $k = 2$ and $m = 3$, we see that

$$\overline{W} = \begin{bmatrix} w_0 & w_0 & w_0 \\ (1+q+q^2)w_0 & (1+q)w_0 + q^2 w_1 & w_0 + (q+q^2)w_1 \\ (1+q+q^2)w_0 & w_0 + (q+q^2)w_1 & (1+q)w_1 + q^2 w_2 \end{bmatrix}.$$

We next formalize the relationship between $\overline{W}$ and W, in order to determine when $\overline{W}$ is invertible.

Recall that W is square of size $N_{k,m}$, the number of linear subspaces of dimension at most k in $\mathbb{F}_q^m$. Define B to be a $(k+1)\times N_{k,m}$ matrix. For $i = 0, 1, \ldots, k$, row i of B is the indicator function $1_{\mathcal{O}_i}$ for the collection of linear subspaces of dimension i in $\mathbb{F}_q^m$. Similarly, define an $N_{k,m}\times(k+1)$ matrix E. For $j = 0, 1, \ldots, k$, column j of E is $(1/\left[\begin{smallmatrix} m \\ j \end{smallmatrix}\right]_q)1_{\mathcal{O}_j^\sharp}$. Notice that $BE = I_{k+1}$, and EB is block diagonal, with the block indexed by rank j being $1/\left[\begin{smallmatrix} m \\ j \end{smallmatrix}\right]_q$ times the square all-one matrix of size $\left[\begin{smallmatrix} m \\ j \end{smallmatrix}\right]_q$.

LEMMA 15.5. *We have: $\overline{W} = BWE$.*

PROOF. Left multiplication by B gives the sums of Lemma 15.1. Right multiplication by E averages those (equal!) sums over $[\lambda] \in \mathcal{O}_j^\sharp$. □

Recall the P-matrix of (13.3).

LEMMA 15.6. *We have: $BPEB = BP$ and $EBP^\top E = P^\top E$.*

PROOF. The matrix BP has rows that are locally constant, and EB acts as the identity when it right multiplies matrices with rows that are locally constant. Similarly, $P^\top E$ has columns that are locally constant, and EB acts as the identity when it left multiplies matrices with columns that are locally constant. □

Our next result is the counterpart of Theorem 13.4. Set $Q_1 = BPE$ and $Q_2 = BP^\top E$. One verifies that Q_1 is lower triangular and Q_2 is upper triangular. Their i,j-entries are

$$(Q_1)_{i,j} = (-1)^j q^{\binom{j}{2}} \begin{bmatrix} m-j \\ i-j \end{bmatrix}_q, \quad (Q_2)_{i,j} = (-1)^i q^{\binom{i}{2}} \begin{bmatrix} j \\ i \end{bmatrix}_q. \tag{15.7}$$

In particular, the diagonal entries of both are $(-1)^j q^{\binom{j}{2}}$, $j = 0, 1, \ldots, k$, so that both Q_1 and Q_2 are invertible.

THEOREM 15.8. *We have $Q_1 \overline{W} Q_2 = BPWP^\top E$ and*

$$Q_1 \overline{W} Q_2 = \begin{bmatrix} c_0 & 0 & \ldots & 0 & \ldots & 0 \\ 0 & q^{m-1} c_1 & \ldots & 0 & \ldots & 0 \\ \vdots & \vdots & \ddots & \vdots & & \vdots \\ 0 & 0 & \ldots & q^{j(m-j)} c_j & \ldots & 0 \\ \vdots & \vdots & & \vdots & \ddots & \vdots \\ 0 & 0 & \ldots & 0 & \ldots & q^{k(m-k)} c_k \end{bmatrix},$$

where $c_0, c_1, \ldots, c_k$ are in (13.5).

PROOF. The first equation follows from Lemma 15.6. The second equation follows from Proposition 13.13. The factor $q^{j(m-j)}$ arises from counting the number of α of dimension j in $\mathbb{F}_q^m$ that satisfy $\alpha \cap \delta^\perp = 0$ for a fixed δ of dimension j. That count uses Lemma 13.8. □

COROLLARY 15.9. *The matrix W is invertible if and only if $\overline{W}$ is invertible.*

When $w_0 = 0$ one can define a $k \times k$ matrix $\overline{W}_0$ using (15.3), but only for $i, j = 1, 2, \ldots, k$. By defining smaller versions of B, E, Q_1, Q_2, one can prove the following results using a proof similar to that of Theorem 13.16.

THEOREM 15.10. *Suppose $w_0 = 0$. Then $Q_{0,1} \overline{W}_0 Q_{0,2} = B_0 P_0 W_0 P_0^\top E_0$ and*

$$Q_{0,1} \overline{W}_0 Q_{0,2} = \begin{bmatrix} q^{m-1} c_1 & \ldots & 0 & \ldots & 0 \\ \vdots & \ddots & \vdots & & \vdots \\ 0 & \ldots & q^{j(m-j)} c_j & \ldots & 0 \\ \vdots & & \vdots & \ddots & \vdots \\ 0 & \ldots & 0 & \ldots & q^{k(m-k)} c_k \end{bmatrix},$$

where $c_1, \ldots, c_k$ are in (13.5), *but with $w_0 = 0$.*

COROLLARY 15.11. *When $w_0 = 0$, the matrix W_0 is invertible if and only if $\overline{W}_0$ is invertible.*

16. Constructions

In this section, we construct two linear codes C and D over $R = M_{k\times k}(\mathbb{F}_q)$ with $\mathrm{wwe}_C = \mathrm{wwe}_D$, assuming that w has maximal symmetry, that $w_0 = 0$, and that the associated W_0 matrix is invertible, Corollary 13.18. Both linear codes will have information module $M = M_{k\times m}(\mathbb{F}_q)$, with $m > k$.

Here is a sketch of the main idea behind the construction. Suppose C is the image of $\Lambda : M \to R^n$. Recall that the maximal symmetry hypothesis means that W_Λ is constant on any left $\mathcal{U}$-orbit in the information module M. Suppose Λ has the property that N chosen orbits of rank s have the same value v_1 of W_Λ. Pick one orbit of rank $s+1$, and denote by v_2 its value of W_Λ. Now try to swap values on those orbits: try to find a linear code D, the image of $\Gamma : M \to R^n$, so that W_Γ has value v_2 on the N chosen orbits of rank s, value v_1 on the chosen orbit of rank $s+1$, and $W_\Gamma = W_\Lambda$ on all other orbits. If N times the size of a rank s orbit equals the size of a rank $s+1$ orbit, then $\mathrm{wwe}_C = \mathrm{wwe}_D$, provided such a D exists and provided that C and D have the same length; cf., Remark 3.10. If the codes have different lengths, we append enough zero-functionals to the shorter code so that the lengths become equal. Because $w_0 = 0$, the additional zero-functionals have no effect on the weights.

We first need a few facts about q-binomial coefficients.

LEMMA 16.1. *For integers* $0 \le s \le m$,

$$\begin{bmatrix} m \\ s \end{bmatrix}_q = \begin{bmatrix} m \\ m-s \end{bmatrix}_q; \quad \begin{bmatrix} m \\ s+1 \end{bmatrix}_q = \frac{q^{m-s}-1}{q^{s+1}-1}\begin{bmatrix} m \\ s \end{bmatrix}_q;$$

$$\begin{bmatrix} m \\ s \end{bmatrix}_q \le \begin{bmatrix} m \\ s+1 \end{bmatrix}_q, \quad 0 \le s < m/2;$$

$$\begin{bmatrix} m \\ s \end{bmatrix}_q = q^{m-s}\begin{bmatrix} m-1 \\ s-1 \end{bmatrix}_q + \begin{bmatrix} m-1 \\ s \end{bmatrix}_q, \quad 0 \le s < m;$$

$$\begin{bmatrix} m \\ s \end{bmatrix}_q = \begin{bmatrix} m-1 \\ s-1 \end{bmatrix}_q + q^s\begin{bmatrix} m-1 \\ s \end{bmatrix}_q, \quad 0 \le s < m.$$

PROOF. Most of the identities are in [**22**, Chapter 3] or its exercises. The inequality follows from the preceding identity, because the multiplying fraction is at least 1 for $s < m/2$. □

LEMMA 16.2. *Suppose integers* k, m, s *satisfy* $0 < s < k < m$. *Then*

$$q^k \le \begin{bmatrix} m \\ s \end{bmatrix}_q - \begin{bmatrix} m-1 \\ s \end{bmatrix}_q.$$

PROOF. The result is true for $s = 1$: $\left[\begin{smallmatrix} m \\ 1 \end{smallmatrix}\right]_q - \left[\begin{smallmatrix} m-1 \\ 1 \end{smallmatrix}\right]_q = q^{m-1} \ge q^k$. Now suppose $s \ge 2$, so that $2 \le s \le m-2$. By symmetry and monotonicity in Lemma 16.1, we see that $\left[\begin{smallmatrix} m-1 \\ s-1 \end{smallmatrix}\right]_q \ge \left[\begin{smallmatrix} m-1 \\ 1 \end{smallmatrix}\right]_q = q^{m-2} + q^{m-3} + \cdots + q + 1 \ge q^{m-2}$. Use Lemma 16.1 again to see that

$$\begin{bmatrix} m \\ s \end{bmatrix}_q - \begin{bmatrix} m-1 \\ s \end{bmatrix}_q = q^{m-s}\begin{bmatrix} m-1 \\ s-1 \end{bmatrix}_q \ge q^2 q^{m-2} = q^m > q^k. \quad \square$$

For the constructions, assume $k \ge 2$, $m > k$, and set $w_0 = 0$. We assume a weight w on R has maximal symmetry, and we assume the associated W_0-matrix is invertible. There will be a construction for each value of $s = 1, 2, \ldots, k-1$. The integer s determines the ranks of the orbits that will be swapped.

Fix an integer s with $1 \leq s < k < m$, and choose an orbit $\operatorname{orb}(\lambda_0) \in \mathcal{O}^\sharp$ with $\operatorname{rk}\lambda_0 = 1$. As in Corollary 12.4, this orbit corresponds to the linear subspace $L_0 = \operatorname{colsp}(\lambda_0) \subseteq \mathbb{F}_q^m$, with $\dim L_0 = 1$. As $\dim L_0^\perp = m-1$, there are $\left[{m-1 \atop s}\right]_q$ linear subspaces of dimension s contained in $L_0^\perp$, leaving $\left[{m \atop s}\right]_q - \left[{m-1 \atop s}\right]_q$ linear subspaces of dimension s in $\mathbb{F}_q^m$ that are not contained in $L_0^\perp$. By Lemma 16.2, there are at least $q^k - q^s < q^k$ linear subspaces of dimension s in $\mathbb{F}_q^m$ that are not contained in $L_0^\perp$. Similarly, using $m - s \geq 2$, there are $\left[{m-1 \atop s+1}\right]_q \geq 1$ linear subspaces of dimension $s+1$ contained in $L_0^\perp$.

Choose distinct s-dimensional linear subspaces $X_1, X_2, \ldots, X_{q^k-q^s}$ of $\mathbb{F}_q^m$ that are not contained in $L_0^\perp$. Also choose one $(s+1)$-dimensional linear subspace $Y \subseteq L_0^\perp$. By Corollary 12.4, these linear subspaces correspond to distinct orbits $[x_1] = \operatorname{orb}(x_1), \ldots, [x_{q^k-q^s}] = \operatorname{orb}(x_{q^k-q^s})$ and $[y] = \operatorname{orb}(y)$ in $\mathcal{O}$, with $\operatorname{rk} x_i = s$ and $\operatorname{rk} y = s+1$. Then $|\operatorname{orb}(y)| = \sum_{i=1}^{q^k-q^s} |\operatorname{orb}(x_i)|$, by Proposition 12.6.

We now consider several indicator functions on $\mathcal{O}^\sharp$ and find the weights they determine at the orbits $[x_1], \ldots, [x_{q^k-q^s}]$ and $[y]$.

Let $1_{[\lambda_0]} : \mathcal{O}^\sharp \to \mathbb{Z}$ be the indicator function of $[\lambda_0] = \operatorname{orb}(\lambda_0) \in \mathcal{O}^\sharp$:

$$1_{[\lambda_0]}([\lambda]) = \begin{cases} 1, & [\lambda] = [\lambda_0], \\ 0, & [\lambda] \neq [\lambda_0]. \end{cases}$$

From Lemma 13.2, the weights of $1_{[\lambda_0]}$ at the orbits $[x_i]$, $1 \leq i \leq q^k - q^s$, and $[y]$ are $(W_0 1_{[\lambda_0]})([x]) = w(x\lambda_0) = w_{\dim X - \dim(X \cap L_0^\perp)}$:

$$(W_0 1_{[\lambda_0]})([x]) = \begin{cases} w_1, & [x] = [x_i], \\ 0, & [x] = [y]. \end{cases}$$

The exact values of $W_0 1_{[\lambda_0]}$ at other inputs will not be relevant. What is crucial is that the value at $[y]$ is 0 and that the values at the $[x_i]$ are equal and positive.

Recall that $\mathcal{O}_1^\sharp = \{\operatorname{orb}(\lambda) \in \mathcal{O}^\sharp : \operatorname{rk}\lambda = 1\}$; let $1_{\mathcal{O}_1^\sharp}$ be its indicator function. The orbit weights $\omega_1 = W_0 1_{\mathcal{O}_1^\sharp}$ are found in Proposition 14.1. The indicator function $1_{\mathcal{O}_+^\sharp}$ of the set of all nonzero orbits $\mathcal{O}_+^\sharp$ is locally constant, so $W_0 1_{\mathcal{O}_+^\sharp}$ is also locally constant, Corollary 14.2. Set $\alpha_1 = (W_0 1_{\mathcal{O}_+^\sharp})(\operatorname{orb}(x))$, when $\operatorname{rk} x = 1$, and $\alpha_2 = (W_0 1_{\mathcal{O}_+^\sharp})(\operatorname{orb}(x))$ when $\operatorname{rk} x = 2$. Both α_1, α_2 are positive integers.

Define a function $\varsigma^{(s)} \in F_0(\mathcal{O}, \mathbb{Z})$ by

$$\varsigma^{(s)}(\operatorname{orb}(x)) = \begin{cases} -1, & \operatorname{orb}(x) = \operatorname{orb}(x_i), \\ 1, & \operatorname{orb}(x) = \operatorname{orb}(y), \\ 0, & \text{otherwise}. \end{cases}$$

Because we are assuming W_0 is invertible, Remark 13.19 says that $W_0 : F_0(\mathcal{O}^\sharp, \mathbb{Q}) \to F_0(\mathcal{O}, \mathbb{Q})$ is an isomorphism. Thus $W_0^{-1}\varsigma^{(s)} \in F_0(\mathcal{O}^\sharp, \mathbb{Q})$ exists, but it has rational values of both signs. To clear denominators, choose a positive integer c sufficiently large so that $\sigma^{(s)} = cw_1 W_0^{-1}\varsigma^{(s)}$ has integer values. Then $\sigma^{(s)}(\operatorname{orb}(0)) = 0$, as $\sigma^{(s)} \in F_0(\mathcal{O}^\sharp, \mathbb{Z})$, and $W_0\sigma^{(s)} = cw_1\varsigma^{(s)}$, so that $(W_0\sigma^{(s)})(\operatorname{orb}(x))$ equals 0 or $\pm cw_1$.

We collect the values of $W_0\eta$ at the $[x_i]$ and $[y]$, for various η.

(16.3)

η	$(W_0\eta)([x_i])$	$(W_0\eta)([y])$
$1_{[\lambda_0]}$	w_1	0
$1_{\mathcal{O}^\sharp_+}$	α_1	α_2
$1_{\mathcal{O}^\sharp_1}$	$(q^{m-s}+\cdots+q^{m-1})w_1$	$(q^{m-s-1}+\cdots+q^{m-1})w_1$
$\sigma^{(s)}$	$-cw_1$	cw_1

For later use, write $B_{m,s} = q^{m-s-1}+\cdots+q^{m-1}$; $B_{m,s} > 0$.

Some of the values of $\sigma^{(s)}$ will be negative. Choose a positive integer a sufficiently large so that all the values of $aw_1 1_{\mathcal{O}^\sharp_+} + \sigma^{(s)}$ are nonnegative. Choose an integer $b \geq 1$ large enough that $a(\alpha_1 - \alpha_2) < bq^{m-s-1}$. (If $\alpha_1 < \alpha_2$, then $b = 1$ suffices.) Set

$$\Delta = \sum_{\mathrm{orb}(\lambda)\in\mathcal{O}^\sharp_+} \sigma^{(s)}(\mathrm{orb}(\lambda)), \tag{16.4}$$

$$\varepsilon = aw_1 1_{\mathcal{O}^\sharp_+} + b1_{\mathcal{O}^\sharp_1} + (c + a(\alpha_2 - \alpha_1) + bq^{m-s-1})1_{[\lambda_0]}.$$

Define two multiplicity functions $\eta_C, \eta_D \in F(\mathcal{O}^\sharp, \mathbb{Q})$ by setting

$$\begin{aligned} \eta_C &= \varepsilon + \max(\Delta, 0)1_{[0]}, \\ \eta_D &= \varepsilon + \sigma^{(s)} - \min(\Delta, 0)1_{[0]}. \end{aligned} \tag{16.5}$$

THEOREM 16.6. *Let $R = M_{k\times k}(\mathbb{F}_q)$, $k \geq 2$, and $M = M_{k\times m}(\mathbb{F}_q)$, $k < m$. Let w be a weight on R with maximal symmetry, positive integer values, and $w(0) = 0$. Assume the associated W_0-matrix is invertible.*

Then, for any integer s, $1 \leq s < k$, the multiplicity functions η_C and η_D of (16.5) *have nonnegative integer values, i.e., $\eta_C, \eta_D \in F(\mathcal{O}^\sharp, \mathbb{N})$, and they define left R-linear codes C and D, respectively. The two codes have the same length. Their weights at the orbits* $\mathrm{orb}(x_i)$, $i = 1, 2, \ldots, q^k - q^s$, *and* $\mathrm{orb}(y)$ *are*

η	$(W\eta)(\mathrm{orb}(x_i))$	$(W\eta)(\mathrm{orb}(y))$
η_C	$(c + a\alpha_2 + bB_{m,s})w_1$	$(a\alpha_2 + bB_{m,s})w_1$
η_D	$(a\alpha_2 + bB_{m,s})w_1$	$(c + a\alpha_2 + bB_{m,s})w_1$

At all other orbits, their weights agree. In particular, $\mathrm{wwe}_C = \mathrm{wwe}_D$.

PROOF. All of $a, w_1, b, c, \alpha_1, \alpha_2, \max(\Delta, 0)$, and $-\min(\Delta, 0)$ are nonnegative integers. The integer $b \geq 1$ was chosen so that $a(\alpha_2 - \alpha_1) + bq^{m-s-1} > 0$, so the values of η_C are nonnegative integers. The positive integer a was chosen so that $aw_1 1_{\mathcal{O}^\sharp_+} + \sigma^{(s)}$ has nonnegative values, which implies that the values of η_D are nonnegative integers.

Both η_C and η_D contain the term $b1_{\mathcal{O}^\sharp_1}$. Because the linear subspace spanned by $\{\mathrm{colsp}(\lambda) : \lambda \in \mathcal{O}^\sharp_1\}$ is all of $\mathbb{F}_q^m$, Remark 12.5 implies that Λ_C and Λ_D are both injective. The value of Δ was chosen to satisfy $\Delta = \mathrm{efflng}(D) - \mathrm{efflng}(C)$. If $\Delta > 0$, then D is longer, and we add Δ zero-functionals to C; if $\Delta < 0$, then C is longer, and we add $-\Delta$ zero-functionals to D. Thus the codes have the same length.

The weights $(W\eta)(\mathrm{orb}(x_i))$ and $(W\eta)(\mathrm{orb}(y))$ follow from (16.3) and (16.5). Because $W\eta_D - W\eta_C = W\sigma^{(s)} = cw_1\varsigma^{(s)}$, the form of $\varsigma^{(s)}$ implies that the weights for C and D are equal at all other orbits. As noted earlier, $|\mathrm{orb}(y)| = \sum_{i=1}^{q^k-q^s} |\mathrm{orb}(x_i)|$, by Proposition 12.6. We conclude that $\mathrm{wwe}_C = \mathrm{wwe}_D$. ☐

REMARK 16.7. Here is a summary of the motivations for various parts of the construction. The integer c was chosen to clear denominators so that $\sigma^{(s)}$ would have integer values. However, $\sigma^{(s)}$ has both positive and negative values, because W has nonnegative entries and $W\sigma^{(s)} = cw_1\varsigma^{(s)}$ has mixed signs. So, a was chosen so that $aw_1 1_{\mathcal{O}_+^\sharp} + \sigma^{(s)}$ has nonnegative integer values. The function $1_{\mathcal{O}_+^\sharp}$ (resp., $1_{\mathcal{O}_1^\sharp}$) is used because its weights are the same at each $[x_i]$. The integer $b \geq 1$ was chosen so that $(W_0(aw_1 1_{\mathcal{O}_+^\sharp} + b1_{\mathcal{O}_1^\sharp}))(\mathrm{orb}(x_i)) < (W_0(aw_1 1_{\mathcal{O}_+^\sharp} + b1_{\mathcal{O}_1^\sharp}))(\mathrm{orb}(y))$ and also to guarantee that Λ_C and Λ_D are injective. Then the coefficient of $1_{[\lambda_0]}$ was chosen so that the weights interchange when $\sigma^{(s)}$ is added to η_C. The function $1_{[\lambda_0]}$ is used because it changes the weights at each $[x_i]$ in the same way but not the weight at $[y]$.

17. Degeneracies

The constructions of Theorem 16.6 made use of the invertibility of the matrix W_0. In this section, we describe a construction in the situation where W_0 is not invertible. We continue to assume information module $M = M_{k\times m}(\mathbb{F}_q)$ and weight w with maximal symmetry, positive integer values, and $w_0 = 0$.

Corollary 13.18 says that W_0 is singular when at least one

$$c_j = (-1)^j q^{\binom{j}{2}} \sum_{s=1}^{j} (-1)^s q^{\binom{s}{2}} \begin{bmatrix} j \\ s \end{bmatrix}_q w_s,$$

$j = 1, 2, \ldots, k$, vanishes. By hypothesis on w, $w_1 > 0$, so that $c_1 \neq 0$.

LEMMA 17.1. *Suppose $c_j = 0$ for some $j = 2, 3, \ldots, k$. Then any row of the matrix P_0 indexed by a linear subspace $\gamma \subseteq \mathbb{F}_q^m$ with $\dim\gamma = j$ belongs to $\ker W_0$.*

PROOF. The second part of Proposition 13.12 shows that, if $\dim\gamma = j$, then $(WP^\top)_{\beta,\gamma} = 0$. This holds for any β and any γ with $\dim\gamma = j$. This means the columns of $P^\top$ indexed by γ with $\dim\gamma = j$ belong to $\ker W$. Those columns are the same as the rows of P indexed by γ with $\dim\gamma = j$. Because $w_0 = 0$, the same analysis applies with W_0 and P_0. □

Suppose $c_j = 0$ for some $j = 2, 3, \ldots, k$. Pick a row v_γ of P_0 indexed by a linear subspace $\gamma \subseteq \mathbb{F}_q^m$ with $\dim\gamma = j$. Define two multiplicity functions $\eta_\pm \in F_0(\mathcal{O}^\sharp, \mathbb{N})$ based on the positive, resp., negative, parts of v_γ:

$$\begin{aligned} \eta_+ &= (v_\gamma + |v_\gamma|)/2 + 1_{\mathcal{O}_1^\sharp}, \\ \eta_- &= -(v_\gamma - |v_\gamma|)/2 + 1_{\mathcal{O}_1^\sharp}, \end{aligned}$$

where $|v_\gamma|$ means the vector obtained from v_γ by taking the absolute value of each entry. The terms $1_{\mathcal{O}_1^\sharp}$ are included so that the associated homomorphisms $\Lambda_{\eta_\pm}$ are injective; see Remark 12.5. Note that $\eta_+ - \eta_- = v_\gamma$, so that $W_0\eta_+ = W_0\eta_-$. Modify η_+ by setting $\eta_+([0]) = 1$; set $\eta_-([0]) = 0$.

PROPOSITION 17.2. *Let $C_\pm$ be the linear codes determined by $\eta_\pm$. Then C_+ and C_- have the same length, and* $\mathrm{wwe}_{C_+} = \mathrm{wwe}_{C_-}$.

PROOF. In (13.3), there are $\left[\begin{smallmatrix} j \\ r \end{smallmatrix}\right]_q$ nonzero terms in rank r positions of the row v_γ. Then the difference in lengths of C_+ and C_- is

$$\begin{aligned}\operatorname{length}(C_+) - \operatorname{length}(C_-) &= \sum_{\substack{r=0 \\ r \text{ even}}}^{j} q^{\binom{r}{2}} \begin{bmatrix} j \\ r \end{bmatrix}_q - \sum_{\substack{r=0 \\ r \text{ odd}}}^{j} q^{\binom{r}{2}} \begin{bmatrix} j \\ r \end{bmatrix}_q \\ &= \sum_{r=0}^{j} (-1)^r q^{\binom{r}{2}} \begin{bmatrix} j \\ r \end{bmatrix}_q = 0,\end{aligned}$$

by (13.7). Zero-functionals don't change the value of $W\eta_+$. We then have $W\eta_+ = W\eta_-$, so that the w-weight enumerators are equal. □

This same construction was used (with the Hamming weight) in [**27**, p. 703] and [**28**, p. 145].

REMARK 17.3. The swapping idea in Theorem 16.6 does not always work in the degenerate case because $\varsigma^{(s)}$ is not always in $\operatorname{im} W_0$. See Remark 18.13 for more details.

18. Analysis of singleton dual codewords

In this section, we analyze dual codewords that are singleton vectors. We will then apply this analysis to the codes constructed using (16.5). The key result, Theorem 18.12, shows how the contributions of singletons of rank i to $A^{\text{sing}}_{w_i}(D^\perp) - A^{\text{sing}}_{w_i}(C^\perp)$ depend on the parameter s used in (16.5).

Given a left R-linear code $C \subseteq R^n$, recall the right dual code $\mathcal{R}(C)$ from (2.2). We will often denote $\mathcal{R}(C)$ by $C^\perp$.

When an R-linear code C is given by a multiplicity function η, a singleton vector v belongs to $C^\perp$ when the nonzero entry r of v right-annihilates the coordinate functional λ in that position: $\lambda r = 0$. Remember that $\lambda \in \operatorname{Hom}_R(M, R) = M_{m\times k}(\mathbb{F}_q)$ and $r \in R = M_{k\times k}(\mathbb{F}_q)$.

Given a functional λ with $\operatorname{rk}\lambda = j$, we will determine how many elements $r \in R$ with $\operatorname{rk} r = i$ satisfy $\lambda r = 0$. For $\lambda \in \operatorname{Hom}_R(M, R)$, define

$$\operatorname{ann}(i, \lambda) = \{r \in R : \operatorname{rk} r = i \text{ and } \lambda r = 0\}.$$

Recall that the sizes $\mathscr{S}_j$ of $\mathcal{U}$-orbits were given in Proposition 12.6.

LEMMA 18.1. *Suppose* $\lambda \in \operatorname{Hom}_R(M, R)$. *Then the size of* $\operatorname{ann}(i, \lambda)$ *depends only on* $\operatorname{rk}\lambda$. *If* $\operatorname{rk}\lambda = j$, *then*

$$|\operatorname{ann}(i, \lambda)| = \begin{cases} \mathscr{S}_i \left[\begin{smallmatrix} k-j \\ i \end{smallmatrix}\right]_q, & i \le k - j, \\ 0, & i > k - j. \end{cases}$$

PROOF. View $M_{m\times k}(\mathbb{F}_q)$ and $M_{k\times k}(\mathbb{F}_q)$ as spaces of $\mathbb{F}_q$-linear transformations $\mathbb{F}_q^m \to \mathbb{F}_q^k$ and $\mathbb{F}_q^k \to \mathbb{F}_q^k$, respectively, with inputs written on the left. Then $\lambda r = 0$ means that $\operatorname{im}\lambda \subseteq \ker r$. Given that $\operatorname{rk}\lambda = j$ and $\operatorname{rk} r = i$, we see that a necessary condition for $\lambda r = 0$ is that $j = \dim \operatorname{im}\lambda \le \dim\ker r = k - i$.

Given $\operatorname{im}\lambda \subseteq \mathbb{F}_q^k$, the number of linear subspaces K (candidates for $\ker r$) satisfying $\operatorname{im}\lambda \subseteq K \subseteq \mathbb{F}_q^k$ and $\dim K = k - i$ equals the number of linear subspaces of dimension $k - i - j$ in $\mathbb{F}_q^k / \operatorname{im}\lambda$, a vector space of dimension $k - j$. That number is $\left[{k-j \atop k-i-j}\right]_q = \left[{k-j \atop i}\right]_q$. For a given K, there is a $\mathcal{U}$-orbit's worth of matrices r with $\ker r = K$, and hence $\operatorname{rk} r = i$. The size of that orbit is $\mathscr{S}_i$. □

We will abuse notation and write $|\operatorname{ann}(i,j)|$ for the common value $|\operatorname{ann}(i,\lambda)|$ when $\operatorname{rk}\lambda = j$.

EXAMPLE 18.2. Let $k = 2$. We display the values of $|\operatorname{ann}(i,j)|$ in a matrix with row index i and column index j, $i,j = 0,1,2$:

$$\begin{bmatrix} 1 & 1 & 1 \\ (q^2-1)(q+1) & q^2-1 & 0 \\ (q^2-1)(q^2-q) & 0 & 0 \end{bmatrix}.$$

Suppose a left R-linear code C is given by a multiplicity function $\eta : \mathcal{O}^\sharp \to \mathbb{N}$. For $j = 0,1,\ldots,k$, define

$$\overline{\eta}_j = \sum_{[\lambda]\in\mathcal{O}^\sharp_j} \eta([\lambda]).$$

Then $\overline{\eta}_j$ counts the number of coordinate functionals having rank j. Set $\overline{\eta} = \langle \overline{\eta}_0, \overline{\eta}_1, \ldots, \overline{\eta}_k\rangle \in \mathbb{N}^{k+1}$. We call the $\overline{\eta}_j$ 'rank-sums'.

PROPOSITION 18.3. *Suppose a left R-linear code C is given by a multiplicity function η. Then the contribution of singletons of rank i to $A^{\text{sing}}_{w_i}(C^\perp)$ is*

$$\sum_{j=0}^{k-i} |\operatorname{ann}(i,j)|\overline{\eta}_j.$$

In particular, the contribution of singletons of rank k to $A^{\text{sing}}_{w_k}(C^\perp)$ is $|\mathrm{GL}(k,\mathbb{F}_q)|\overline{\eta}_0$.

PROOF. There are a total of $\overline{\eta}_j$ coordinate functionals of rank j. For each one, apply Lemma 18.1. □

The larger i is, the fewer terms there are in the summation.

COROLLARY 18.4. *For a linear code C given by multiplicity function η,*

$$A^{\text{sing}}_d(C^\perp) = \sum_{i:w_i=d} \sum_{j=0}^{k-i} |\operatorname{ann}(i,j)|\overline{\eta}_j.$$

Having seen the importance of the rank-sums $\overline{\eta}_j$, we next see how they behave with respect to the $\overline{W}$-matrix. Given a function $\omega : \mathcal{O} \to \mathbb{Q}$, define

$$\overline{\omega}_i = \sum_{[x]\in\mathcal{O}_i} \omega([x]), \quad i = 0,1,\ldots,k.$$

Set $\overline{\omega} = \langle \overline{\omega}_0, \overline{\omega}_1, \ldots, \overline{\omega}_k\rangle$.

PROPOSITION 18.5. *Suppose $\eta : \mathcal{O}^\sharp \to \mathbb{Q}$. If $\omega = W\eta$, then $\overline{\omega} = \overline{W}\overline{\eta}$.*

PROOF. Sum the rows indexed by rank i elements of $\mathcal{O}$ in $\omega = W\eta$, change the order of summation, and use Lemma 15.1 and (15.3). □

Because our ultimate objective is to find linear codes C and D of the same length with $\mathrm{wwe}_C = \mathrm{wwe}_D$ and $\mathrm{wwe}_{C^\perp} \neq \mathrm{wwe}_{D^\perp}$, we now apply Proposition 18.3 and Corollary 18.4 to two linear codes C and D of the same length. Write $\delta\eta =$

$\eta_D - \eta_C$ and $\delta A_d^{\rm sing} = A_d^{\rm sing}(D^\perp) - A_d^{\rm sing}(C^\perp)$. The net contribution of rank i singletons to $\delta A_{w_i}^{\rm sing}$ then simplifies to

$$\sum_{j=0}^{k-i} |\mathrm{ann}(i,j)| \delta\overline{\eta}_j, \tag{18.6}$$

and $\delta A_d^{\rm sing} = \sum_{i:w_i=d} \sum_{j=0}^{k-i} |\mathrm{ann}(i,j)| \delta\overline{\eta}_j$.

REMARK 18.7. The last two formulas will be the main tools for showing that two dual codes have different w-weight enumerators. But the fomulas cut both ways. If all the $\delta\overline{\eta}_j$ vanish, then singletons cannot detect differences between $\mathrm{wwe}_{C^\perp}$ and $\mathrm{wwe}_{D^\perp}$. See Example 21.5.

Now consider specifically the linear codes C and D constructed by (16.5). In this case, we see that

$$\begin{aligned} \delta\eta &= \sigma^{(s)} - \Delta 1_{[0]}, \\ \overline{\eta}_D - \overline{\eta}_C &= \langle -\Delta, \overline{\sigma}_1^{(s)}, \ldots, \overline{\sigma}_k^{(s)} \rangle, \\ \overline{\omega}_D - \overline{\omega}_C &= \overline{W}\overline{\sigma}^{(s)} = \langle 0, \ldots, 0, -cw_1(q^k - q^s), cw_1, 0, \ldots, 0 \rangle, \end{aligned}$$

where the nonzero entries of $\overline{\omega}_D - \overline{\omega}_C$ are in positions s and $s+1$.

Recall from Theorem 16.6 that C and D have the same length. This is reflected in the fact that the sum of the entries of $\overline{\eta}_D - \overline{\eta}_C$ is $-\Delta + \sum_j \overline{\sigma}_j^{(s)} = 0$, from (16.4). This allows us to re-write the net contribution (18.6) of rank i singletons to $\delta A_{w_i}^{\rm sing}$ as $\sum_{j=1}^{k} (|\mathrm{ann}(i,j)| - |\mathrm{ann}(i,0)|)\, \overline{\sigma}_j^{(s)}$. To write this equation in matrix form, define a $k \times k$ matrix $\overline{\mathrm{Ann}}$ by

$$\overline{\mathrm{Ann}}_{i,j} = |\mathrm{ann}(i,j)| - |\mathrm{ann}(i,0)| \quad i,j = 1,2,\ldots,k.$$

In summary, the net contributions of singletons of rank i to $\delta A_{w_i}^{\rm sing}$ are given by the entries of

$$\overline{\mathrm{Ann}}\, \overline{\sigma}^{(s)}. \tag{18.8}$$

LEMMA 18.9. *The matrix $\overline{\mathrm{Ann}}$ is invertible over $\mathbb{Q}$.*

PROOF. Define a $(k+1) \times (k+1)$ matrix $\mathcal{A}$ by

$$\mathcal{A}_{i,j} = |\mathrm{ann}(i,j)|, \quad i,j = 0,1,2,\ldots,k.$$

Then $\mathcal{A}$ is upper 'anti-triangular' by Lemma 18.1, i.e., $\mathcal{A}_{i,j} = 0$ when $i+j > k$. Then $\det \mathcal{A} = \pm \prod_{i=0}^{k} \mathscr{S}_i$ is nonzero. Thus $\mathcal{A}$ is invertible.

Subtract the zeroth column of $\mathcal{A}$ from every other column. The resulting matrix is still invertible and has the form

$$\left[\begin{array}{c|c} 1 & \mathrm{row}(0) \\ \hline \mathrm{col}(*) & \overline{\mathrm{Ann}} \end{array}\right],$$

where $\mathrm{col}(*)$ is a column of nonzero entries. Thus $\overline{\mathrm{Ann}}$ is invertible. □

The matrices $\overline{W}_0$ and $\overline{\mathrm{Ann}}$ are both $k \times k$, so they both define $\mathbb{Q}$-linear transformations $\mathbb{Q}^k \to \mathbb{Q}^k$. We next explore how these linear transformations behave with respect to certain filtrations of $\mathbb{Q}^k$.

Define a $k \times k$ matrix T over $\mathbb{Q}$:

$$T_{i,j} = (-1)^i q^{\binom{i}{2}} \begin{bmatrix} j \\ i \end{bmatrix}_q, \quad i,j = 1,2,\ldots,k.$$

By standard conventions, $\left[\begin{smallmatrix} j \\ i \end{smallmatrix}\right]_q = 0$ when $i > j$. This implies that T is upper-triangular and invertible. (By (15.7), T is just $Q_{0,2}$.)

LEMMA 18.10. *The matrices $\overline{W}_0 T$ and $\overline{\mathrm{Ann}}\, T$ are lower triangular.*

The proof of Lemma 18.10 will utilize Lemma 18.11.

LEMMA 18.11. *For integers $0 \le i < j \le k$,*

$$\sum_{\ell=0}^{j} (-1)^\ell q^{\binom{\ell}{2}} \begin{bmatrix} j \\ \ell \end{bmatrix}_q \begin{bmatrix} k-\ell \\ i \end{bmatrix}_q = 0.$$

PROOF. We first prove the edge cases $i = 0$ or $j = k$, and then prove the remaining cases by induction on k.

Suppose $i = 0$. Then the term $\left[\begin{smallmatrix} k-\ell \\ 0 \end{smallmatrix}\right]_q = 1$ for all $\ell = 0, \dots j$. The sum reduces to (13.7), which vanishes, as $j > 0$.

Suppose $j = k$. We observe that

$$\begin{bmatrix} k \\ \ell \end{bmatrix}_q \begin{bmatrix} k-\ell \\ i \end{bmatrix}_q = \begin{bmatrix} k \\ i \end{bmatrix}_q \begin{bmatrix} k-i \\ \ell \end{bmatrix}_q,$$

and that $\left[\begin{smallmatrix} k-\ell \\ i \end{smallmatrix}\right]_q = 0$ for $\ell > k - i$. Then

$$\begin{aligned} \sum_{\ell=0}^{k} (-1)^\ell q^{\binom{\ell}{2}} \begin{bmatrix} k \\ \ell \end{bmatrix}_q \begin{bmatrix} k-\ell \\ i \end{bmatrix}_q &= \sum_{\ell=0}^{k-i} (-1)^\ell q^{\binom{\ell}{2}} \begin{bmatrix} k \\ i \end{bmatrix}_q \begin{bmatrix} k-i \\ \ell \end{bmatrix}_q \\ &= \begin{bmatrix} k \\ i \end{bmatrix}_q \sum_{\ell=0}^{k-i} (-1)^\ell q^{\binom{\ell}{2}} \begin{bmatrix} k-i \\ \ell \end{bmatrix}_q = 0, \end{aligned}$$

by (13.7), as $k - i > 0$.

We prove the remaining cases, $0 < i < j < k$, by induction on k. The first case is when $k = 3$, with $i = 1 < j = 2$. By direct calculation,

$$\sum_{\ell=0}^{2} (-1)^\ell q^{\binom{\ell}{2}} \begin{bmatrix} 2 \\ \ell \end{bmatrix}_q \begin{bmatrix} 3-\ell \\ 1 \end{bmatrix}_q = (q^2 + q + 1) - (q+1)^2 + q = 0.$$

For the induction step, suppose $0 < i < j < k + 1$. Apply an identity from Lemma 16.1, so that

$$\begin{aligned} \sum_{\ell=0}^{j} (-1)^\ell q^{\binom{\ell}{2}} \begin{bmatrix} j \\ \ell \end{bmatrix}_q \begin{bmatrix} k+1-\ell \\ i \end{bmatrix}_q &= \sum_{\ell=0}^{j} (-1)^\ell q^{\binom{\ell}{2}} \begin{bmatrix} j \\ \ell \end{bmatrix}_q \begin{bmatrix} k-\ell \\ i-1 \end{bmatrix}_q \\ &\quad + q^i \sum_{\ell=0}^{j} (-1)^\ell q^{\binom{\ell}{2}} \begin{bmatrix} j \\ \ell \end{bmatrix}_q \begin{bmatrix} k-\ell \\ i \end{bmatrix}_q = 0, \end{aligned}$$

because both sums on the right side vanish by the induction hypothesis or the edge cases. □

PROOF OF LEMMA 18.10. We need to show that the i, j-entry of each product vanishes when $i < j$. For $\overline{W}_0 T$, the result is the $w_0 = 0$ analog of Proposition 13.12, as T is $Q_{0,2}$. For $\overline{\mathrm{Ann}}\, T$,

$$(\overline{\mathrm{Ann}}\, T)_{i,j} = \sum_{\ell=1}^{j} \mathscr{S}_i \left(\begin{bmatrix} k-\ell \\ i \end{bmatrix}_q - \begin{bmatrix} k \\ i \end{bmatrix}_q \right) (-1)^\ell q^{\binom{\ell}{2}} \begin{bmatrix} j \\ \ell \end{bmatrix}_q.$$

Note that the sum does not change if we include $\ell = 0$. Applying the distributive law and (13.7), as $j > 0$, we see from Lemma 18.11 that

$$(\overline{\mathrm{Ann}}\, T)_{i,j} = \mathscr{S}_i \sum_{\ell=0}^{j} (-1)^\ell q^{\binom{\ell}{2}} \begin{bmatrix} j \\ \ell \end{bmatrix}_q \begin{bmatrix} k-\ell \\ i \end{bmatrix}_q = 0. \qquad \square$$

We now define two filtrations of $\mathbb{Q}^k$. Let $e_1 = (1, 0, \ldots, 0), \ldots, e_k = (0, \ldots, 0, 1)$ be the standard basis vectors of $\mathbb{Q}^k$. For $i = 1, 2, \ldots, k$, define linear subspaces of $\mathbb{Q}^k$:

$$\begin{aligned} \mathcal{V}_i &= \mathrm{span}\{e_i, e_{i+1}, \ldots, e_k\}, \\ \mathcal{T}_i &= \mathrm{span}\{\text{columns } i, i+1, \ldots, k \text{ of the matrix } T\}. \end{aligned}$$

Then $\dim \mathcal{V}_i = \dim \mathcal{T}_i = k - i + 1$, and

$$\begin{aligned} \mathbb{Q}^k &= \mathcal{V}_1 \supset \mathcal{V}_2 \supset \cdots \supset \mathcal{V}_k \supset \{0\}, \\ \mathbb{Q}^k &= \mathcal{T}_1 \supset \mathcal{T}_2 \supset \cdots \supset \mathcal{T}_k \supset \{0\}. \end{aligned}$$

Lemma 18.10 shows that the matrices $\overline{\mathrm{Ann}}$ and $\overline{W}_0$ (when invertible) map the $\mathcal{T}$-filtration isomorphically to the $\mathcal{V}$-filtration.

THEOREM 18.12. *Let C and D be the linear codes constructed using* (16.5) *and $\varsigma^{(s)}$ for some s, $1 \le s < k$, with $\overline{W}_0$ invertible. Then the contribution by rank i singletons to $\delta A^{\mathrm{sing}}_{w_i} = A^{\mathrm{sing}}_{w_i}(D^\perp) - A^{\mathrm{sing}}_{w_i}(C^\perp)$ is zero if $i < s$. The contribution of rank s singletons to $\delta A^{\mathrm{sing}}_{w_s}$ is nonzero.*

PROOF. The vector $\bar{\varsigma}^{(s)}$ belongs to $\mathcal{V}_s - \mathcal{V}_{s+1}$. Then $\bar{\sigma}^{(s)} = \overline{W}_0^{\,-1} \bar{\varsigma}^{(s)} \in \mathcal{T}_s - \mathcal{T}_{s+1}$. This, in turn, implies that $\overline{\mathrm{Ann}}\, \bar{\sigma}^{(s)} \in \mathcal{V}_s - \mathcal{V}_{s+1}$. By (18.8), singletons of rank i, $i < s$, make zero contribution to $\delta A^{\mathrm{sing}}_{w_i}$, while the singletons of rank s make a nonzero contribution to $\delta A^{\mathrm{sing}}_{w_s}$. $\square$

We point out that Theorem 18.12 makes no claims about the contribution of singletons of rank k to $\delta A^{\mathrm{sing}}_{w_k}$.

REMARK 18.13. If W_0 is degenerate because $c_j = 0$, then $\overline{W}_0$ is also degenerate, and $\overline{W}_0$ maps $\mathcal{T}_j$ into $\mathcal{V}_{j+1}$, and $\mathcal{T}_i$, $i < j$, will map to a proper linear subspace of $\mathcal{V}_i$. If j is the largest index such that $c_j = 0$, then $\bar{\sigma}^{(s)}$ will be in the image of $\overline{W}_0$ provided $j < s$. Recalling that $s < k$ and that $c_1 \neq 0$, we see that $\bar{\sigma}^{(s)} \in \mathrm{im}\, \overline{W}_0$ when $2 \le j < s < k$. For example, when $k = 3$, no such s can exist.

19. Main results

We are now in a position to prove that a large number of weights with maximal symmetry, including the homogeneous weight, do not respect duality.

As usual in this part of the paper, let $R = M_{k\times k}(\mathbb{F}_q)$ with $k \ge 2$. Suppose w is a weight on R that has maximal symmetry, positive integer values, and $w(0) = 0$. The value $w(r)$, $r \in R$, depends only on the rank $\mathrm{rk}\, r$ of r. Write w_i for the common value $w(r)$ when $\mathrm{rk}\, r = i$.

We will say that a weight w is *nondegenerate* if the expressions c_i of (13.5) are nonzero for all $i = 1, 2, \ldots, k$. (Note that $c_0 = 0$ because $w(0) = 0$, and $c_1 = w_1 > 0$.) If at least one of $c_2, \ldots, c_k$ vanishes, we say that w is *degenerate*.

Let $\mathring{w} = \min\{w_1, w_2, \ldots, w_k\}$; $\mathring{w}$ is a positive integer. Write $\mathring{I} = \{i : w_i = \mathring{w}\}$ for set of indices i where w_i achieves the minimum positive value $\mathring{w}$. In general, for an integer $d \geq \mathring{w}$, set $I_d = \{i : w_i = d\}$; depending on d, I_d may be empty. Of course, $\mathring{I} = I_{\mathring{w}}$ is nonempty.

THEOREM 19.1. *Assume w is a nondegenerate weight on $R = M_{k\times k}(\mathbb{F}_q)$, with maximal symmetry, positive integer values, and $w(0) = 0$. Suppose there is an integer d such that $\mathring{w} \leq d < 2\mathring{w}$, I_d is nonempty, and $k \notin I_d$. Then w does not respect duality: there exist linear codes C and D over R of the same length such that $\mathrm{wwe}_C = \mathrm{wwe}_D$ and $A_d(C^\perp) \neq A_d(D^\perp)$. In particular, if $k \notin \mathring{I}$, then w does not respect duality.*

PROOF. Suppose d satisfies the stated hypotheses, and let $s = \max I_d$; then $s < k$. Construct linear codes C and D over R using (16.5) using $s = \max I_d$. Because $s < k$, Theorem 16.6 implies that the codes have the same length and that $\mathrm{wwe}_C = \mathrm{wwe}_D$.

As for the dual codes, Corollary 5.5 says that $A_d(D^\perp) - A_d(C^\perp) = A_d^{\mathrm{sing}}(D^\perp) - A_d^{\mathrm{sing}}(C^\perp) = \delta A_d^{\mathrm{sing}}$. The only singletons that can contribute to $\delta A_d^{\mathrm{sing}}$ are those of rank i with $i \in I_d$, Corollary 18.4. Theorem 18.12 says that singletons of rank $i < s$ make zero contributions to $\delta A_d^{\mathrm{sing}}$, while singletons of rank s make a nonzero contribution to $\delta A_d^{\mathrm{sing}}$. We conclude that $A_d(D^\perp) - A_d(C^\perp) = \delta A_d^{\mathrm{sing}} \neq 0$. □

REMARK 19.2. If $k \in I_d$, the arguments given above are not conclusive. In Theorem 18.12 there is always the possibility that the contributions of singletons of rank k could cancel the contributions of singletons of lower rank $i \in I_d$. Even if $I_d = \{k\}$, it is possible that singletons of rank k make zero contributions. This happens for the homogeneous weight, for example; see Corollary 4.10 and Proposition 18.3.

COROLLARY 19.3. *Suppose a nondegenerate w satisfies $\mathring{w} < w_i < 2\mathring{w}$ for some i, $1 \leq i \leq k$. Then w does not respect duality.*

PROOF. By the definition of $\mathring{w}$, there is some index j so that $w_j = \mathring{w}$. As $w_j < w_i$, we have $i \neq j$. The index k can equal at most one of i or j. Apply Theorem 19.1 to the other one. □

We can, at long last, prove that the homogeneous weight on $M_{k\times k}(\mathbb{F}_q)$ does not respect duality, provided $k > 2$ or $q > 2$. The homogeneous weight on any finite Frobenius ring has the Extension Property [**10**, Theorem 2.5]. Since the matrix ring $M_{k\times k}(\mathbb{F}_q)$ is Frobenius, it follows that W_0 and $\overline{W}_0$ are invertible for the homogeneous weight and any information module M, Remark 13.19. This says that the homogeneous weight is nondegenerate.

THEOREM 19.4. *Let $R = M_{k\times k}(\mathbb{F}_q)$, $k \geq 2$, with the homogeneous weight w. Then w respects duality if and only if $k = 2$ and $q = 2$.*

PROOF. The 'if' portion is Theorem 12.12. For the converse, Lemma 12.11 says $\mathrm{w}_2 = \mathring{\mathrm{w}}$ is the smallest nonzero value of w, while w_1 is the largest value, with $\mathrm{w}_2 < \mathrm{w}_1 \leq 2\mathrm{w}_2$. There is equality $\mathrm{w}_1 = 2\mathrm{w}_2$ if and only if $k = 2$ and $q = 2$. For all other values of k and q, there is strict inequality: $\mathrm{w}_2 < \mathrm{w}_1 < 2\mathrm{w}_2$. Now apply Theorem 19.1 with $d = \mathrm{w}_1$. □

The last result of this section determines, for $k = 2$, all the weights with maximal symmetry (nondegenerate or not) that respect duality. The list is a short one: the Hamming weight (any q) and the homogeneous weight (only when $q = 2$).

THEOREM 19.5. *Let w be a weight on $R = M_{2\times 2}(\mathbb{F}_q)$ having maximal symmetry, positive integer values, and $w(0) = 0$. Assume w is neither a multiple of the Hamming weight ($w_1 = w_2$) nor, when $q = 2$, a multiple of the homogeneous weight ($w_1 = 2w_2$ when $q = 2$). Then w does not respect duality.*

PROOF. Using $m = k + 1 = 3$ and $s = 1$ (the only possible value of s), Example 15.4 gives the $\overline{W}_0$-matrix:

$$\overline{W}_0 = \begin{bmatrix} q^2 w_1 & (q^2+q)w_1 \\ (q^2+q)w_1 & (q+1)w_1 + q^2 w_2 \end{bmatrix}.$$

Then $\det \overline{W}_0 = q^3 w_1(-(q+1)w_1 + qw_2)$. As $w_1 > 0$ by hypothesis, $\det \overline{W}_0$ vanishes only when $-(q+1)w_1 + qw_2 = 0$.

First consider the nondegenerate case, where $-(q+1)w_1 + qw_2 \neq 0$. Use the construction of (16.5) with $m = 3$ and $s = 1$ to produce linear codes C and D with $\mathrm{wwe}_C = \mathrm{wwe}_D$. The net counts of singleton vectors in the dual codes depends only on $\bar{\sigma}$. A calculation shows that

$$\begin{aligned} \overline{W}_0 \langle (q+1)w_1 + (q^2-q)w_2, -q^2 w_1 \rangle &\\ &= (q(-(q+1)w_1 + qw_2))\langle -(q^2-q), 1\rangle \\ &= (q(-(q+1)w_1 + qw_2))\bar{\varsigma}^{(1)}. \end{aligned}$$

Thus, up to scaling, we take $\bar{\sigma}^{(1)} = \langle (q+1)w_1 + (q^2-q)w_2, -q^2 w_1\rangle$.

Using Example 18.2, we see that the matrix $\overline{\mathrm{Ann}}$ is

$$\overline{\mathrm{Ann}} = \begin{bmatrix} -q(q^2-1) & -(q^2-1)(q+1) \\ -(q^2-1)(q^2-q) & -(q^2-1)(q^2-q) \end{bmatrix}.$$

By (18.8), the net contributions to $\delta A^{\mathrm{sing}}_{w_j} = A^{\mathrm{sing}}_{w_j}(D^\perp) - A^{\mathrm{sing}}_{w_j}(C^\perp)$ by rank j singletons are given by the entries of $\overline{\mathrm{Ann}}\,\bar{\sigma}$:

$$\overline{\mathrm{Ann}}\,\bar{\sigma} = q(q-1)(q^2-1)\begin{bmatrix} (q+1)w_1 - qw_2 \\ (q^2-q-1)w_1 - (q^2-q)w_2 \end{bmatrix}. \tag{19.6}$$

The contribution for $j = 1$ is nonzero because w is nondegenerate. The contribution for $j = 2$ is nonzero, provided w is not a multiple of the homogeneous weight; see Example 12.9.

Because $k = 2$, there are just a few (nondegenerate) cases:

- If $w_1 < w_2$, then $\delta A_{w_1} = \delta A^{\mathrm{sing}}_{w_1} \neq 0$, by (19.6).
- If $w_2 < w_1$ and w is not homogeneous, then $\delta A_{w_2} = \delta A^{\mathrm{sing}}_{w_2} \neq 0$, by (19.6).
- If w is homogeneous, apply Theorem 19.4 (except when $q = 2$).
- If $w_1 = w_2$, w is a multiple of the Hamming weight.

Now suppose w is degenerate, so that $qw_2 = (q+1)w_1$. Because q and $q+1$ are relatively prime, there exists a positive integer τ such that $w_1 = q\tau$ and $w_2 = (q+1)\tau$. Then $w_1 < w_2 < 2w_1$, as $q \geq 2$. The degenerate matrix $\overline{W}_0$ becomes

$$\overline{W}_{\mathrm{degen}} = \begin{bmatrix} q^2 w_1 & q(q+1)w_1 \\ q(q+1)w_1 & (q+1)^2 w_1 \end{bmatrix}.$$

A basis for $\ker \overline{W}_{\mathrm{degen}}$ is $\langle -(q+1), q\rangle$.

Use the linear codes $C_\pm$ of Proposition 17.2. They have the same length and w-weight enumerators. Their net rank-sums $\bar{\eta}_i(C_+) - \bar{\eta}_i(C_-)$ are $1, -(q+1), q,$

respectively. Then the contributions of singletons are:

$$\overline{\text{Ann}}\begin{bmatrix} -(q+1) \\ q \end{bmatrix} = \begin{bmatrix} 0 \\ (q^2-1)(q^2-q) \end{bmatrix}.$$

Because $w_1 < w_2 < 2w_1$, Corollary 5.5 applies to w_2. Then $A_{w_2}(C_+^{\perp}) - A_{w_2}(C_-^{\perp}) = A_{w_2}^{\text{sing}}(C_+^{\perp}) - A_{w_2}^{\text{sing}}(C_-^{\perp}) = (q^2-1)(q^2-q) > 0$. □

In Section 22, the case of $R = M_{3\times 3}(\mathbb{F}_2)$ is discussed in detail.

20. Rank partition enumerators

Section 3 described various enumerators including the complete enumerator and symmetrized enumerators. In this section, we focus on a particular enumerator, the rank partition enumerator, over the matrix ring $M_{k\times k}(\mathbb{F}_q)$. The rank partition enumerator is a partition enumerator associated to rank, and it is coarser than the symmetrized enumerator associated to the group action of $\mathrm{GL}(k, \mathbb{F}_q)$ acting on $M_{k\times k}(\mathbb{F}_q)$.

On $R = M_{k\times k}(\mathbb{F}_q)$, define the *rank partition* $\mathcal{RK} = \{R_i\}_{i=0}^{k}$, with

$$R_i = \{s \in R : \operatorname{rk} s = i\}.$$

As in Section 3, define counting functions $n_i : R^n \to \mathbb{N}$, $i = 0, 1, \ldots, k$, by $n_i(x) = |\{j : x_j \in R_i\}|$, for $x = (x_1, x_2, \ldots, x_n) \in R^n$. For a linear code $C \subseteq R^n$, define the *rank partition enumerator* rpe_C associated to C to be the homogeneous polynomial of degree n in the variables $Z_0, Z_1, \ldots, Z_k$ given by

$$\mathrm{rpe}_C(Z) = \sum_{x\in C}\prod_{j=1}^{n} Z_{\operatorname{rk} x_j} = \sum_{x\in C}\prod_{i=0}^{k} Z_i^{n_i(x)}.$$

If w is a weight on R with maximal symmetry and positive integer values, then the value of $w(r)$, $r \in R$, depends only on the rank $\operatorname{rk} r$ of r. Write w_i for $w(r)$ when $\operatorname{rk} r = i$, and denote by $w_{\max}$ the largest value of w. Then the specialization of variables $Z_i \rightsquigarrow X^{w_{\max}-w_i}Y^{w_i}$ allows one to write wwe_C in terms of rpe_C:

$$\mathrm{wwe}_C(X,Y) = \mathrm{rpe}_C(Z)|_{Z_i \rightsquigarrow X^{w_{\max}-w_i}Y^{w_i}}. \tag{20.1}$$

As an example of some of the results of [**8**], we will show that the rank partition enumerator satisfies the MacWilliams identities. Then (20.1) will allow us to calculate $\mathrm{wwe}_{C^{\perp}}$ for many examples. This will be one way to illustrate the main results of Section 19 (and to prove additional results).

In order to show that the MacWilliams identities hold for the rank partition enumerator, we refer to the argument outlined in Appendix A and provide details on the relevant Fourier transforms.

It is well-known ([**25**, Example 4.4]) that $R = M_{k\times k}(\mathbb{F}_q)$ is a Frobenius ring with a generating character χ. To describe the standard generating character χ, we first recall the standard generating character θ_q of $\mathbb{F}_q$: $\theta_q(a) = \zeta^{\mathrm{Tr}_{q\to p}(a)}$, $a \in \mathbb{F}_q$, where $q = p^e$, p prime, $\zeta = \exp(2\pi i/p) \in \mathbb{C}^{\times}$, and $\mathrm{Tr}_{q\to p}$ is the absolute trace from $\mathbb{F}_q \to \mathbb{F}_p$. Then define $\chi(s) = \theta_q(\operatorname{tr} s)$, $s \in R$, where tr is the matrix trace. Because the matrix trace over $\mathbb{F}_q$ satisfies $\operatorname{tr}(rs) = \operatorname{tr}(sr)$, we see that $\chi(rs) = \chi(sr)$ for all $r, s \in R$. As is the case for all generating characters, χ has the property that $\ker\chi$ contains no nonzero left or right ideal of R, [**25**, Lemma 4.1].

LEMMA 20.2. *Each partition block R_i of $\mathcal{RK}$ is invariant under left or right multiplication by units. If $\operatorname{rk} r_1 = \operatorname{rk} r_2$, then $\sum_{s\in R_i}\chi(sr_1) = \sum_{s\in R_i}\chi(sr_2)$ for all i.*

PROOF. The rank of a matrix is invariant under multiplication by units. If $\operatorname{rk} r_1 = \operatorname{rk} r_2$, then, using row and column operations, there are units u_1, u_2 such that $r_2 = u_1 r_1 u_2$. Thus,

$$\sum_{s\in R_i} \chi(sr_2) = \sum_{s\in R_i} \chi(su_1r_1u_2) = \sum_{s\in R_i} \chi(u_2su_1r_1) = \sum_{s\in R_i} \chi(sr_1),$$

using the property $\chi(rs) = \chi(sr)$ and the bi-invariance of R_i. □

Define the *Kravchuk matrix* K of size $(k+1)\times(k+1)$ for the rank partition $\mathcal{RK}$ by

$$K_{i,j} = \sum_{s\in R_i} \chi(sr), \quad i,j = 0,1,\ldots,k,$$

where $r \in R$ has $\operatorname{rk} r = j$. This sum is well-defined by Lemma 20.2.

In order to develop an explicit formula for $K_{i,j}$, we first remark that R_i, being invariant under left multiplication by units, equals the disjoint union of the left $\mathcal{U}$-orbits it contains. We already know the number and sizes of the $\mathcal{U}$-orbits in R, Proposition 12.6. So, we turn our attention to sums of the form $\sum_{t\in\operatorname{orb}(s)} \chi(tr)$, for $r \in R$.

Let $\mathcal{P}_R$ be the poset of all principal left ideals of R under set containment; all the left ideals of R are principal. (This is the type of poset used in Proposition 12.3, with $M = R$ and $m = k$.) Fix an element $r \in R$, and define two functions $f_r, g_r : \mathcal{P}_R \to \mathbb{C}$ by

$$f_r(Rs) = \sum_{t\in Rs} \chi(tr), \quad g_r(Rs) = \sum_{t\in\operatorname{orb}(s)} \chi(tr).$$

The definition of g_r is well-defined by Lemma 12.2. We collect some facts about f_r and g_r in the next lemma. For $r \in R$, its left *annihilator* is $\operatorname{ann}_{\mathrm{lt}}(r) = \{s \in R : sr = 0\}$; the left annihilator is a left ideal of R.

LEMMA 20.3. *For any $r, s \in R$, we have*

$$f_r(Rs) = \sum_{Rt\subseteq Rs} g_r(Rt).$$

The values of f_r are

$$f_r(Rs) = \begin{cases} |Rs|, & \text{if } Rs \subseteq \operatorname{ann}_{\mathrm{lt}}(r), \\ 0, & \text{otherwise.} \end{cases}$$

PROOF. The first equality reflects the fact that the left ideal Rs is invariant under left multiplication by units, and hence Rs equals the disjoint union of the left $\mathcal{U}$-orbits it contains.

As in (2.3), denote $\mathfrak{R}(Rs) = \{r \in R : \chi(tr) = 1, \text{for all } t \in Rs\}$. We claim that $r \in \mathfrak{R}(Rs)$ if and only if $Rs \subseteq \operatorname{ann}_{\mathrm{lt}}(r)$. The 'if' direction is clear: if $Rs \subseteq \operatorname{ann}_{\mathrm{lt}}(r)$, then $tr = 0$, and hence $\chi(tr) = 1$, for all $t \in Rs$. Conversely, suppose $r \in \mathfrak{R}(Rs)$. Then $Rsr \subseteq \ker\chi$. We conclude that $Rsr = 0$, because any left ideal in $\ker\chi$ must be zero. Thus $Rs \subseteq \operatorname{ann}_{\mathrm{lt}}(r)$. The second formula now follows from Lemma A.1. □

PROPOSITION 20.4. *Let μ be the Möbius function of the poset $\mathcal{P}_R$. Then*

$$g_r(Rs) = \sum_{Rt\subseteq Rs} \mu(Rt, Rs) f_r(Rt), \quad Rs \in \mathcal{P}_R.$$

Simplifying, we have

$$g_r(Rs) = \sum_{j=0}^{\operatorname{rk} s - \operatorname{rk}(sr)} (-1)^{\operatorname{rk} s - j} q^{\binom{\operatorname{rk} s - j}{2}} q^{kj} \begin{bmatrix} \operatorname{rk} s - \operatorname{rk}(sr) \\ j \end{bmatrix}_q .$$

PROOF. The first equation comes from Möbius inversion, [**22**, Theorem 5.5.5], because of the first equation in Lemma 20.3.

By Proposition 12.3, the poset $\mathcal{P}_R$ is isomorphic to the poset $\mathcal{P}_{k,k}$ of linear subspaces of $\mathbb{F}_q^k$. This allows us to translate the equation for $g_r(Rs)$ into geometric language. Let $r \in R$ correspond to a linear subspace $Y \subseteq \mathbb{F}_q^k$, with $\dim Y = \operatorname{rk} r$. Then $|Rr| = q^{k \dim Y} = q^{k \operatorname{rk} r}$, by Proposition 12.6. Similarly, let $s, t \in R$ correspond to X, T, with $\dim X = \operatorname{rk} s$ and $\dim T = \operatorname{rk} t$. The Möbius function of $\mathcal{P}_{k,k}$ is in (12.7). The condition $Rs \subseteq \operatorname{ann}_{\mathrm{lt}}(r)$ becomes $X \subseteq Y^{\perp}$.

Using Lemma 20.3, the formula for $g_r(Rs)$ simplifies:

$$\begin{aligned} g_r(Rs) &= \sum_{Rt \subseteq Rs \cap \operatorname{ann}_{\mathrm{lt}}(r)} \mu(Rt, Rs)|Rt| \\ &= \sum_{T \subseteq X \cap Y^{\perp}} (-1)^{\dim X - \dim T} q^{\binom{\dim X - \dim T}{2}} q^{k \dim T} \\ &= \sum_{j=0}^{\dim(X \cap Y^{\perp})} (-1)^{\operatorname{rk} s - j} q^{\binom{\operatorname{rk} s - j}{2}} q^{kj} \begin{bmatrix} \dim(X \cap Y^{\perp}) \\ j \end{bmatrix}_q . \end{aligned}$$

Finally, Lemma 13.2 implies $\dim(X \cap Y^{\perp}) = \operatorname{rk} s - \operatorname{rk}(sr)$. □

To simplify notation slightly, define

$$B(i, \ell) = \sum_{j=0}^{\ell} (-1)^{i-j} q^{\binom{i-j}{2}} q^{kj} \begin{bmatrix} \ell \\ j \end{bmatrix}_q , \tag{20.5}$$

for $i = 0, 1, \ldots, k$ and $0 \le \ell \le i$. Then $\sum_{t \in \operatorname{orb}(s)} \chi(tr) = g_r(Rs) = B(\operatorname{rk} s, \operatorname{rk} s - \operatorname{rk}(sr))$. In addition, suppose there are linear subspaces $C \subseteq A \subseteq \mathbb{F}_q^k$ with $\dim A = a$ and $\dim C = c$. Then define $I(a, b, c, d) = |\{B \subseteq A : \dim B = b, \text{ and } \dim(B \cap C) = d\}|$. By Lemma 13.10 (with $D = A \cap C$),

$$I(a, b, c, d) = q^{(b-d)(c-d)} \begin{bmatrix} c \\ d \end{bmatrix}_q \begin{bmatrix} a - c \\ b - d \end{bmatrix}_q .$$

PROPOSITION 20.6. *The Kravchuk matrix K has entries*

$$K_{i,j} = \sum_{\ell=0}^{i} I(k, i, k - j, \ell) B(i, \ell),$$

for $i, j = 0, 1, \ldots, k$.

PROOF. As mentioned earlier, if $j = \operatorname{rk} r$, the sum in $K_{i,j} = \sum_{s \in R_i} \chi(sr)$ can be split up into sums over the left $\mathcal{U}$-orbits contained in R_i. The individual sums over orbits depend upon $\operatorname{rk} s$ and $\operatorname{rk}(sr)$, so we need to count the number of orbits $\operatorname{orb}(s)$ with $\operatorname{rk} s = i$ for various values of $\operatorname{rk}(sr)$.

In terms of linear subspaces, we need to count the number of linear subspaces X of $\mathbb{F}_q^k$ with $\dim X = i$ and $\dim(X \cap Y^{\perp}) = \ell$. Because $\dim Y^{\perp} = k - \operatorname{rk} r = k - j$, this count is $I(k, i, k - j, \ell)$. □

EXAMPLE 20.7. For $k = 2$, the Kravchuk matrix is:

$$K = \begin{bmatrix} 1 & 1 & 1 \\ (q^2-1)(q+1) & q^2-q-1 & -q-1 \\ (q^2-q)(q^2-1) & -q^2+q & q \end{bmatrix}.$$

Suppose $C \subseteq R^n$ is an additive code. The annihilators $\mathfrak{L}(C)$ and $\mathfrak{R}(C)$ were defined in (2.3). The MacWilliams identities for the rank partition enumerator are next.

THEOREM 20.8. *Let $R = M_{k\times k}(\mathbb{F}_q)$ with Kravchuk matrix K. If C is an additive code in R^n, then*

$$\begin{aligned} \operatorname{rpe}_{\mathfrak{L}(C)}(Z_i) &= \frac{1}{|C|} \operatorname{rpe}_C(Z_j)|_{Z_j=\sum_i K_{i,j} Z_i}, \\ \operatorname{rpe}_{\mathfrak{R}(C)}(Z_i) &= \frac{1}{|C|} \operatorname{rpe}_C(Z_j)|_{Z_j=\sum_i K_{i,j} Z_i}. \end{aligned}$$

The formulas are reversible in C and $\mathfrak{L}(C)$, resp., C and $\mathfrak{R}(C)$.

If $C \subseteq R^n$ is a left, resp., right, R-linear code, then $\mathcal{R}(C) = \mathfrak{R}(C)$, resp., $\mathcal{L}(C) = \mathfrak{L}(C)$.

PROOF. We add details to the outline provided in Appendix A. Let $V = \mathbb{C}[Z_0, Z_1, \ldots, Z_k]$, and define $f : R \to V$ by $f(s) = Z_{\operatorname{rk} s}$, $s \in R$. We calculate the Fourier transform of f, as in (A.2). Write $j = \operatorname{rk} r$.

$$\begin{aligned} \widehat{f}(r) &= \sum_{s\in R} \chi(rs) Z_{\operatorname{rk} s} = \sum_{s\in R} \chi(sr) Z_{\operatorname{rk} s} \\ &= \sum_{i=0}^{k} \sum_{s\in R_i} \chi(sr) Z_i = \sum_{i=0}^{k} K_{i,j} Z_i. \end{aligned}$$

Note that $\widehat{f}(r)$ depends only on $\operatorname{rk} r$, so that $\widehat{f}(r)$ equals $f(r)$ after applying the linear substitution $Z_j \leftarrow \sum_{i=0}^k K_{i,j} Z_i$.

To reverse roles, use Lemma 2.4 and apply the formulas to the pair $\mathfrak{L}(C)$ and $\mathfrak{R}(\mathfrak{L}(C)) = C$ and the pair $\mathfrak{R}(C)$ and $\mathfrak{L}(\mathfrak{R}(C)) = C$. For the case of linear codes, see Remark 2.5. $\square$

21. Examples

In this section, we calculate a number of examples over $R = M_{2\times 2}(\mathbb{F}_2)$. Set $m = 3$, so that the information module is $M = M_{2\times 3}(\mathbb{F}_2)$. The orbit spaces $\mathcal{O}_1$ and $\mathcal{O}_2$ have representatives in reduced row-echelon form and are ordered as follows:

$$\begin{aligned} \mathcal{O}_1 &= [\left[\begin{smallmatrix} 1&0&1\\0&0&0 \end{smallmatrix}\right], \left[\begin{smallmatrix} 0&1&0\\0&0&0 \end{smallmatrix}\right], \left[\begin{smallmatrix} 0&0&1\\0&0&0 \end{smallmatrix}\right], \left[\begin{smallmatrix} 1&1&1\\0&0&0 \end{smallmatrix}\right], \left[\begin{smallmatrix} 0&1&1\\0&0&0 \end{smallmatrix}\right], \left[\begin{smallmatrix} 1&0&0\\0&0&0 \end{smallmatrix}\right], \left[\begin{smallmatrix} 1&1&0\\0&0&0 \end{smallmatrix}\right]], \\ \mathcal{O}_2 &= [\left[\begin{smallmatrix} 1&0&0\\0&1&0 \end{smallmatrix}\right], \left[\begin{smallmatrix} 1&0&1\\0&1&0 \end{smallmatrix}\right], \left[\begin{smallmatrix} 1&0&0\\0&0&1 \end{smallmatrix}\right], \left[\begin{smallmatrix} 1&1&0\\0&0&1 \end{smallmatrix}\right], \left[\begin{smallmatrix} 1&0&0\\0&1&1 \end{smallmatrix}\right], \left[\begin{smallmatrix} 0&1&0\\0&0&1 \end{smallmatrix}\right], \left[\begin{smallmatrix} 1&0&1\\0&1&1 \end{smallmatrix}\right]]. \end{aligned} \tag{21.1}$$

The rank 1 orbits have size 3, and the rank 2 orbits have size 6, as in Proposition 12.6. The representatives of $\mathcal{O}_1^\sharp$ and $\mathcal{O}_2^\sharp$ are the transposes of the representatives of $\mathcal{O}_1$ and $\mathcal{O}_2$, using the same orderings, with rank 1 coming before rank 2.

The W_0-matrix has size 14×14, while $\overline{W}_0$ is 2×2:

$$W_0 = \begin{bmatrix} 0 & 0 & w_1 & 0 & w_1 & w_1 & w_1 & w_1 & 0 & w_1 & w_1 & w_1 & w_1 & w_1 \\ 0 & w_1 & 0 & w_1 & w_1 & 0 & w_1 & w_1 & w_1 & 0 & w_1 & w_1 & w_1 & w_1 \\ w_1 & 0 & w_1 & w_1 & w_1 & 0 & 0 & 0 & w_1 & w_1 & w_1 & w_1 & w_1 & w_1 \\ 0 & w_1 & w_1 & w_1 & 0 & w_1 & 0 & w_1 & w_1 & w_1 & w_1 & w_1 & w_1 & 0 \\ w_1 & w_1 & w_1 & 0 & 0 & 0 & w_1 & w_1 & w_1 & w_1 & w_1 & 0 & w_1 & w_1 \\ w_1 & 0 & 0 & w_1 & 0 & w_1 & w_1 & w_1 & w_1 & w_1 & w_1 & w_1 & 0 & w_1 \\ w_1 & w_1 & 0 & 0 & w_1 & w_1 & 0 & w_1 & w_1 & w_1 & 0 & w_1 & w_1 & w_1 \\ w_1 & w_1 & 0 & w_1 & w_1 & w_1 & w_1 & w_2 & w_2 & w_1 & w_1 & w_2 & w_1 & w_2 \\ 0 & w_1 & w_1 & w_1 & w_1 & w_1 & w_1 & w_2 & w_1 & w_1 & w_2 & w_2 & w_2 & w_1 \\ w_1 & 0 & w_1 & w_1 & w_1 & w_1 & w_1 & w_1 & w_1 & w_2 & w_2 & w_2 & w_1 & w_2 \\ w_1 & w_1 & w_1 & w_1 & w_1 & w_1 & 0 & w_1 & w_2 & w_2 & w_1 & w_2 & w_2 & w_1 \\ w_1 & w_1 & w_1 & w_1 & 0 & w_1 & w_1 & w_2 & w_2 & w_2 & w_2 & w_1 & w_1 & w_1 \\ w_1 & w_1 & w_1 & w_1 & w_1 & 0 & w_1 & w_1 & w_2 & w_1 & w_2 & w_1 & w_2 & w_2 \\ w_1 & w_1 & w_1 & 0 & w_1 & w_1 & w_1 & w_2 & w_1 & w_2 & w_1 & w_1 & w_2 & w_2 \end{bmatrix},$$

$$\overline{W}_0 = \begin{bmatrix} 4w_1 & 6w_1 \\ 6w_1 & 3w_1 + 4w_2 \end{bmatrix}.$$

Akin to Figure 1 on page 368, the rank partition enumerator of a linear code specializes to the w-weight enumerator of the code under the specialization $Z_0 \rightsquigarrow 1$, $Z_1 \rightsquigarrow t^{w_1}$ and $Z_2 \rightsquigarrow t^{w_2}$. Using the Kravchuk matrix K from Example 20.7, with $q = 2$:

$$K = \begin{bmatrix} 1 & 1 & 1 \\ 9 & 1 & -3 \\ 6 & -2 & 2 \end{bmatrix},$$

the MacWilliams identities for the rank partition enumerator yield the rank partition enumerator for the dual code, Theorem 20.8. Using the same specialization, the w-weight enumerator of the dual code is obtained.

In all of the examples that follow, multiplicity functions η and lists ω of orbit weights are written in terms of the ordering of $\mathcal{O}$ given in (21.1), with ranks separated by vertical lines. All calculations were performed in SageMath [**23**]. Rank partition enumerators of dual codes are not listed because they would use too much space. Only the lowest order terms in the w-weight enumerators of dual codes are displayed.

It is differences such as $A_j(D^\perp) - A_j(C^\perp)$ that ultimately matter, so we will write $\delta\overline{\eta} = \overline{\eta}(D) - \overline{\eta}(C)$. By (18.6) and Example 18.2 with $q = 2$, the net contributions of rank i singletons to $\delta A^{\text{sing}}_{w_i} = A^{\text{sing}}_{w_i}(D^\perp) - A^{\text{sing}}_{w_i}(C^\perp)$ appear as the entries in the vector

$$\begin{bmatrix} 1 & 1 & 1 \\ 9 & 3 & 0 \\ 6 & 0 & 0 \end{bmatrix} \delta\overline{\eta}. \tag{21.2}$$

Example 21.3. Set $w_1 = 2$ and $w_2 = 3$. Then $\mathring{w} = 2$, and $\mathring{w} \leq w_i < 2\mathring{w}$ for $i = 1, 2$. This is a degenerate case, as $\det W_0 = 0$. Use the codes in Proposition 17.2, with an extra rank 1 functional so that the associated homomorphisms Λ are injective; see Remark 12.5. The multiplicity functions and lists of orbit weights are:

$$\begin{aligned} \eta_+ &= \langle 1|1, 0, 0, 0, 0, 0, 0|2, 0, 0, 0, 0, 0, 0\rangle, \\ \omega_+ &= \langle 0|4, 4, 2, 4, 6, 6, 6|8, 6, 6, 6, 8, 6, 8\rangle, \\ \eta_- &= \langle 0|1, 1, 0, 0, 0, 1, 1|0, 0, 0, 0, 0, 0, 0\rangle, \\ \omega_- &= \langle 0|4, 4, 2, 4, 6, 6, 6|8, 6, 6, 6, 8, 6, 8\rangle. \end{aligned}$$

The equation $\omega_+ = \omega_-$ is a feature of the construction in Proposition 17.2. Both codes have length 4, and $\delta\overline{\eta} = \langle -1, 3, -2\rangle$. Then (21.2) implies that $\delta A_2 = \delta A^{\text{sing}}_2 =$

0, while $\delta A_3 = \delta A_3^{\text{sing}} = -6$. The enumerators are

$$\begin{aligned}
\mathrm{se}_{C_+} &= Z_0^4 + 3Z_0^3 Z_1 + 9Z_0^2 Z_1^2 + 27 Z_0 Z_1^3 + 6Z_0^2 Z_2^2 + 18 Z_0 Z_1 Z_2^2, \\
\mathrm{se}_{C_-} &= Z_0^4 + 3Z_0^3 Z_1 + 9Z_0^2 Z_1^2 + 33 Z_0 Z_1^3 + 18 Z_1^4; \\
\mathrm{wwe}_{C_+} &= 1 + 3t^2 + 9t^4 + 33t^6 + 18t^8, \\
\mathrm{wwe}_{C_-} &= 1 + 3t^2 + 9t^4 + 33t^6 + 18t^8; \\
\mathrm{wwe}_{C_+^\perp} &= 1 + 12t^2 + 6t^3 + 36t^4 + \cdots, \\
\mathrm{wwe}_{C_-^\perp} &= 1 + 12t^2 \qquad\quad + 54t^4 + \cdots .
\end{aligned}$$

EXAMPLE 21.4. Set $w_1 = 1$ and $w_2 = 2$, so the weight of a matrix equals its rank. Then $\mathring{w} = 1$, but $w_2 = 2\mathring{w}$, so Corollary 5.5 applies only to $w_1 = 1$. Use the codes given by (16.5), with ς and σ given here:

$$\begin{aligned}
\varsigma &= \langle 0, -1, -1, 0, 0, 0, 0 | 0, 0, 0, 0, 0, 0, 1 \rangle, \\
\sigma &= \langle -1, -1, -1, -1, -3, 1, -1 | 0, 0, 0, 0, 0, 2, 2 \rangle.
\end{aligned}$$

The scaling is such that $W_0 \sigma = 2\varsigma$, so that $c = 2$. One calculates that $\alpha_1 = 10 w_1 = 10$ and $\alpha_2 = 9w_1 + 4w_2 = 17$. It suffices to take $a = 3$ and $b = 0$. Then $\Delta = -3$. The multiplicity functions and lists of orbit weights are

$$\begin{aligned}
\eta_{C_2} &= \langle 0 | 3, 3, 3, 26, 3, 3, 3 | 3, 3, 3, 3, 3, 3, 3 \rangle, \\
\eta_{D_2} &= \langle 3 | 2, 2, 2, 25, 0, 4, 2 | 3, 3, 3, 3, 3, 5, 5 \rangle; \\
\omega_{C_2} &= \langle 0 | 30, 53, 53, 53, 30, 53, 30 | 74, 74, 74, 74, 74, 74, 51 \rangle, \\
\omega_{D_2} &= \langle 0 | 30, 51, 51, 53, 30, 53, 30 | 74, 74, 74, 74, 74, 74, 53 \rangle.
\end{aligned}$$

Both codes have length 65, and $\delta\overline{\eta} = \langle 3, -7, 4 \rangle$. Then, using (21.2), $\delta A_1 = \delta A_1^{\text{sing}} = 6$ and $\delta A_2^{\text{sing}} = 18$. The enumerators are

$$\begin{aligned}
\mathrm{se}_{C_2} &= Z_0^{65} + 9Z_0^{35} Z_1^{30} + 12 Z_0^{12} Z_1^{53} + 6 Z_0^{26} Z_1^{27} Z_2^{12} + 36 Z_0^3 Z_1^{50} Z_2^{12}, \\
\mathrm{se}_{D_2} &= Z_0^{65} + 9Z_0^{35} Z_1^{30} + 6 Z_0^{14} Z_1^{51} + 6 Z_0^{12} Z_1^{53} + 6 Z_0^3 Z_1^{50} Z_2^{12} \\
&\quad + 24 Z_0^5 Z_1^{46} Z_2^{14} + 6 Z_0^{28} Z_1^{21} Z_2^{16} + 6 Z_0^7 Z_1^{42} Z_2^{16}; \\
\mathrm{wwe}_{C_2} &= 1 + 9t^{30} + 6t^{51} + 12t^{53} + 36t^{74}, \\
\mathrm{wwe}_{D_2} &= 1 + 9t^{30} + 6t^{51} + 12t^{53} + 36t^{74}; \\
\mathrm{wwe}_{C_2^\perp} &= 1 + 132t + 15762t^2 + 1674894t^3 + \cdots, \\
\mathrm{wwe}_{D_2^\perp} &= 1 + 138t + 16176t^2 + 1695210t^3 + \cdots .
\end{aligned}$$

Even though $\delta A_2^{\text{sing}} = 18$, $\delta A_2 = 414$; there are dual codewords of weight 2 coming from vectors with two nonzero entries, both of rank 1, that account for the difference. This illustrates the importance of the strict inequality $\mathring{w} \le d < 2\mathring{w}$ in Corollary 5.5.

EXAMPLE 21.5. Set $w_1 = 4$ and $w_2 = 5$. Then $\mathring{w} = 4$ and $\mathring{w} \le w_i < 2\mathring{w}$, $i = 1, 2$, so Corollary 5.5 applies to both w_1 and w_2.

In this example, the codes given by (16.5) will be used; call then C_3 and D_3. In addition, two other codes C_4, D_4 will be given. They are designed so that their lists of orbit weights have just three different values. One of the values applies to all the rank 2 orbits. The other two values apply to three, resp., four, rank 1 orbits. For the code C_4, the three rank 1 orbits having the same weight are linearly independent, while for D_4 they are linearly dependent.

For C_3 and D_3, ς and σ are given here:

$$\begin{aligned}\varsigma &= \langle 0,-1,-1,0,0,0,0|0,0,0,0,0,0,1\rangle,\\ \sigma &= \langle 2,2,2,-1,3,1,2|0,0,0,0,0,-4,-4\rangle.\end{aligned}$$

The scaling is such that $W_0\sigma = 8\varsigma$, so that $c = 2$. One calculates that $\alpha_1 = 10w_1 = 40$ and $\alpha_2 = 9w_1 + 4w_2 = 56$. It suffices to take $a = 1$ and $b = 0$. Then $\Delta = 3$. The multiplicity functions and lists of orbit weights are

$$\begin{aligned}\eta_{C_3} &= \langle 3|4,4,4,22,4,4,4|4,4,4,4,4,4,4\rangle,\\ \eta_{D_3} &= \langle 0|6,6,6,21,7,5,6|4,4,4,4,4,0,0\rangle;\\ \omega_{C_3} &= \langle 0|160,232,232,232,160,232,160\\ &\qquad |296,296,296,296,296,296,224\rangle,\\ \omega_{D_3} &= \langle 0|160,224,224,232,160,232,160\\ &\qquad |296,296,296,296,296,296,232\rangle.\end{aligned}$$

Both codes have length 77, and $\delta\overline{\eta} = \langle -3,11,-8\rangle$. Then, using (21.2), $\delta A_4 = \delta A_4^{\text{sing}} = 6$ and $\delta A_5 = \delta A_5^{\text{sing}} = -18$. The enumerators are

$$\begin{aligned}\text{se}_{C_3} &= Z_0^{77} + 9Z_0^{37}Z_1^{40} + 12Z_0^{19}Z_1^{58} + 6Z_0^{25}Z_1^{36}Z_2^{16} + 36Z_0^{7}Z_1^{54}Z_2^{16},\\ \text{se}_{D_3} &= Z_0^{77} + 9Z_0^{37}Z_1^{40} + 6Z_0^{21}Z_1^{56} + 6Z_0^{19}Z_1^{58} + 6Z_0^{21}Z_1^{48}Z_2^{8}\\ &\quad + 6Z_0^{5}Z_1^{64}Z_2^{8} + 24Z_0^{6}Z_1^{59}Z_2^{12} + 6Z_0^{7}Z_1^{54}Z_2^{16};\\ \text{wwe}_{C_3} &= 1 + 9\,t^{160} + 6\,t^{224} + 12\,t^{232} + 36\,t^{296},\\ \text{wwe}_{D_3} &= 1 + 9\,t^{160} + 6\,t^{224} + 12\,t^{232} + 36\,t^{296};\\ \text{wwe}_{C_3^\perp} &= 1 + 165\,t^4 + 18\,t^5 + 21186\,t^8 + \cdots,\\ \text{wwe}_{D_3^\perp} &= 1 + 171\,t^4 \qquad\quad + 21918\,t^8 + \cdots.\end{aligned}$$

The codes C_4, D_4 have multiplicity functions and lists of orbit weights:

$$\begin{aligned}\eta_{C_4} &= \langle 0|2,3,3,1,2,3,2|2,6,2,6,6,2,6\rangle,\\ \eta_{D_4} &= \langle 0|2,2,4,2,2,2,2|6,6,2,2,6,2,6\rangle;\\ \omega_{C_4} &= \langle 0|136,144,144,136,136,144,136\\ &\qquad |192,192,192,192,192,192,192\rangle,\\ \omega_{D_4} &= \langle 0|136,144,136,136,136,144,144\\ &\qquad |192,192,192,192,192,192,192\rangle.\end{aligned}$$

Both codes have length 46, and $\delta\overline{\eta} = \langle 0,0,0\rangle$. Then, using (21.2), $\delta A_4 = \delta A_4^{\rm sing} = 0$ and $\delta A_5 = \delta A_5^{\rm sing} = 0$. The enumerators are

$$\begin{aligned}
\mathrm{se}_{C_4} &= Z_0^{46} + 12Z_0^{12}Z_1^{34} + 9Z_0^{10}Z_1^{36} + 6Z_0Z_1^{33}Z_2^{12}\\
&\qquad + 18Z_0^2Z_1^{28}Z_2^{16} + 18Z_0^3Z_1^{23}Z_2^{20},\\
\mathrm{se}_{D_4} &= Z_0^{46} + 12Z_0^{12}Z_1^{34} + 9Z_0^{10}Z_1^{36} + 36Z_0^2Z_1^{28}Z_2^{16} + 6Z_0^4Z_1^{18}Z_2^{24};\\
\mathrm{wwe}_{C_4} &= 1 + 12\,t^{136} + 9\,t^{144} + 42\,t^{192},\\
\mathrm{wwe}_{D_4} &= 1 + 12\,t^{136} + 9\,t^{144} + 42\,t^{192};\\
\mathrm{wwe}_{C_4^\perp} &= 1 + 48\,t^4 + 4059\,t^8 + 1440\,t^9 + 522\,t^{10} + 290160\,t^{12} + \cdots,\\
\mathrm{wwe}_{D_4^\perp} &= 1 + 48\,t^4 + 4059\,t^8 + 1440\,t^9 + 522\,t^{10} + 290112\,t^{12} + \cdots.
\end{aligned}$$

The calculation shows that $\delta A_{12} \neq 0$, but this would be difficult to detect theoretically.

EXAMPLE 21.6. Set $w_1 = 3$ and $w_2 = 7$. Then $\mathring{w} = 3$, but $w_2 > 2\mathring{w}$, so Corollary 5.5 applies only to $w_1 = 3$. Use the codes given by (16.5), with ς and σ given here:

$$\begin{aligned}
\varsigma &= \langle 0,-1,-1,0,0,0,0|0,0,0,0,0,0,1\rangle,\\
\sigma &= \langle -3,-3,-3,-5,-11,5,-3|0,0,0,0,0,6,6\rangle.
\end{aligned}$$

The scaling is such that $W_0\sigma = 30\varsigma$, so that $c = 10$. One calculates that $\alpha_1 = 10w_1 = 30$ and $\alpha_2 = 9w_1 + 4w_2 = 55$. It suffices to take $a = 4$ and $b = 0$. Then $\Delta = -11$. The multiplicity functions and lists of orbit weights are

$$\begin{aligned}
\eta_{C_5} &= \langle 0|12,12,12,122,12,12,12|12,12,12,12,12,12,12\rangle,\\
\eta_{D_5} &= \langle 11|9,9,9,117,1,17,9|12,12,12,12,12,18,18\rangle;\\
\omega_{C_5} &= \langle 0|360,690,690,690,360,690,360\\
&\qquad |990,990,990,990,990,990,660\rangle,\\
\omega_{D_5} &= \langle 0|360,660,660,690,360,690,360\\
&\qquad |990,990,990,990,990,990,690\rangle.
\end{aligned}$$

Both codes have length 278, and $\delta\overline{\eta} = \langle 11,-23,12\rangle$. Then, using (21.2), $\delta A_3 = \delta A_3^{\rm sing} = 30$ and $\delta A_7^{\rm sing} = 66$. The enumerators are

$$\begin{aligned}
\mathrm{se}_{C_5} &= Z_0^{278} + 9Z_0^{158}Z_1^{120} + 12Z_0^{48}Z_1^{230} + 6Z_0^{122}Z_1^{108}Z_2^{48}\\
&\qquad + 36Z_0^{12}Z_1^{218}Z_2^{48},\\
\mathrm{se}_{D_5} &= Z_0^{278} + 9Z_0^{158}Z_1^{120} + 6Z_0^{58}Z_1^{220} + 6Z_0^{48}Z_1^{230} + 6Z_0^{12}Z_1^{218}Z_2^{48}\\
&\qquad + 24Z_0^{20}Z_1^{204}Z_2^{54} + 6Z_0^{128}Z_1^{90}Z_2^{60} + 6Z_0^{28}Z_1^{190}Z_2^{60};\\
\mathrm{wwe}_{C_5} &= 1 + 9\,t^{360} + 6\,t^{660} + 12\,t^{690} + 36\,t^{990},\\
\mathrm{wwe}_{D_5} &= 1 + 9\,t^{360} + 6\,t^{660} + 12\,t^{690} + 36\,t^{990};\\
\mathrm{wwe}_{C_5^\perp} &= 1 + 582\,t^3 + 316947\,t^6 \qquad\qquad + 152382900\,t^9 + \cdots,\\
\mathrm{wwe}_{D_5^\perp} &= 1 + 612\,t^3 + 326649\,t^6 + 66\,t^7 + 154592448\,t^9 + \cdots.
\end{aligned}$$

Even though $w_2 = 7 > 2\mathring{w} = 6$, we still have $\delta A_7 = \delta A_7^{\rm sing}$. The reason is that w_2 is not an integer multiple of w_1: a vector can have weight 7 only if it is a singleton with nonzero entry of rank 2. See Remark 5.6.

22. The case of $M_{3\times 3}(\mathbb{F}_q)$

In this section, we show that most weights of maximal symmetry on $M_{3\times 3}(\mathbb{F}_q)$ do not respect duality. There is one situation that remains unsettled.

THEOREM 22.1. *Let $R = M_{3\times 3}(\mathbb{F}_q)$, and suppose w is a weight on R with maximal symmetry, positive integer values, and $w(0) = 0$. If w is nondegenerate and not a multiple of the Hamming weight, then w does not respect duality. If w is degenerate because $-(q+1)w_1 + qw_2 = 0$, then w does not respect duality.*

The situation where w is degenerate because $-(q^2+q+1)w_1+q(q^2+q+1)w_2-q^3w_3 = 0$ is unsettled.

PROOF. In order to display the matrices $\overline{W}_0$ and $\overline{\mathrm{Ann}}$ in a way that fits on the page, we name certain polynomials in q:

$$f_+ = q+1, \quad f_- = q-1, \quad f_2 = q^2+q+1.$$

The $\overline{W}_0$-matrix is

$$\overline{W}_0 = \begin{bmatrix} q^3w_1 & q^2f_+w_1 & qf_2w_1 \\ q^2f_2w_1 & qf_+^2w_1 + q^4w_2 & f_2w_1 + q^2f_2w_2 \\ qf_2w_1 & f_+w_1 + q^2f_+w_2 & f_2w_2 + q^3w_3 \end{bmatrix}.$$

Consistent with Theorem 15.10, the determinant of $\overline{W}_0$ factors as

$$\det \overline{W}_0 = -q^6w_1(-f_+w_1 + qw_2)(-f_2w_1 + qf_2w_2 - q^3w_3).$$

The annihilator matrix $\overline{\mathrm{Ann}}$ is

$$\overline{\mathrm{Ann}} = \begin{bmatrix} -q^2f_-f_2 & -qf_+f_-f_2 & -f_-f_2^2 \\ -q^2f_+^2f_-^2f_2 & -qf_+f_-^2f_2^2 & -qf_+f_-^2f_2^2 \\ -q^3f_+f_-^3f_2 & -q^3f_+f_-^3f_2 & -q^3f_+f_-^3f_2 \end{bmatrix}.$$

Suppose w is nondegenerate, i.e., $\overline{W}_0$ is invertible. We first collect the results of some calculations of (18.8) made using SageMath. The leading terms of $\overline{\mathrm{Ann}}\,\bar{\sigma}^{(s)}$, where $\bar{\sigma}^{(s)} = \overline{W}_0^{-1}\bar{\varsigma}^{(s)}$, for the swaps $\bar{\varsigma}^{(1)} = \langle -q^3+q, 1, 0\rangle$ and $\bar{\varsigma}^{(2)} = \langle 0, -q^3+q^2, 1\rangle$ are:

$$\overline{\mathrm{Ann}}\,\bar{\sigma}^{(1)} = \left\langle \frac{q^5-q^3-q^2+1}{w_1}, *, \frac{\alpha_1}{\beta_1} \right\rangle, \tag{22.2}$$

$$\overline{\mathrm{Ann}}\,\bar{\sigma}^{(2)} = \left\langle 0, \frac{q^7-q^6-q^5+q^3+q^2-q}{f_+w_1 - qw_2}, \frac{\alpha_2}{\beta_2} \right\rangle. \tag{22.3}$$

Note that the denominators are nonzero because $\overline{W}_0$ is invertible.

The term at rank 3 of $\overline{\mathrm{Ann}}\,\bar{\sigma}^{(1)}$ has numerator:

$$\begin{aligned} \alpha_1 = {} & f_+f_-^3f_2(q^6+q^5-q^4-2q^3+q+1)w_1^2 \\ & - qf_+f_-^3f_2(q^6+2q^5-3q^3-2q^2+1)w_1w_2 + q^3f_+^2f_-^4f_2^2w_2^2 \\ & + q^3f_+f_-^3f_2(q^4-2q^2-q+1)w_1w_3 - q^5f_+^2f_-^4f_2w_2w_3, \end{aligned}$$

and nonzero denominator

$$\beta_1 = w_1(-f_+w_1 + qw_2)(-f_2w_1 + qf_2w_2 - q^3w_3).$$

The term at rank 3 of $\overline{\operatorname{Ann}}\,\bar{\sigma}^{(2)}$ has numerator:

$$\begin{aligned}\alpha_2 = qf_+f_-^3f_2(q^3-q^2-q-1)w_1 \\ - q^2f_+f_-^3f_2(q^3-q-1)w_2 + q^4f_+f_-^4f_2w_3\end{aligned}$$

and nonzero denominator

$$\beta_2 = (-f_+w_1+qw_2)(-f_2w_1+qf_2w_2-q^3w_3).$$

Using SageMath, one can solve for those nonzero weights with $\alpha_1=\alpha_2=0$, namely (up to uniform scale factors):

$$w_1=1, \qquad w_2=\frac{q^3-q-1}{q^3-q}, \qquad w_3=\frac{q^5-q^4-q^3+q+1}{q^5-q^4-q^3+q^2};$$

$$w_1=1, \qquad w_2=\frac{q+1}{q}, \qquad w_3=\frac{q^2+q+1}{q^2}.$$

In the first case, one confirms that w is nondegenerate and that $w_2<w_3<w_1$. In the second case, one notes that w is degenerate (both $-f_+w_1+qw_2$ and $-f_2w_1+qf_2w_2-q^3w_3$ vanish in the factorization of $\det \overline{W}_0$) and that $w_1<w_2<w_3$.

One can also solve for those nonzero weights where the sum of the entries of $\overline{\operatorname{Ann}}\,\bar{\sigma}^{(2)}$ vanishes:

$$w_1=(q+1)w_2-qw_3, \tag{22.4}$$

or where $w_1=w_3$ and $\alpha_2=0$:

$$w_2=(q^4-q^2-q-1)w_1/(q^4-q^2-q)<w_1=w_3. \tag{22.5}$$

Now apply the results of the calculations, still assuming that w is nondegenerate.

- If $w_1<\min\{w_2,w_3\}$, then $\delta A_{w_1}=\delta A^{\text{sing}}_{w_1}\neq 0$, using (22.2).
- If $w_2<\min\{w_1,w_3\}$, then $\delta A_{w_2}=\delta A^{\text{sing}}_{w_2}\neq 0$, using (22.3).
- If $w_3<\min\{w_1,w_2\}$, then $\delta A_{w_3}=\delta A^{\text{sing}}_{w_3}\neq 0$. At least one of α_1,α_2 is nonzero (because $w_3<w_2$), so use the corresponding (22.2) or (22.3).
- If $\mathring{w}=w_1=w_2<w_3$, then $\delta A_{\mathring{w}}=\delta A^{\text{sing}}_{\mathring{w}}\neq 0$, using (22.3).
- If $\mathring{w}=w_1=w_3<w_2$, then $\delta A_{\mathring{w}}=\delta A^{\text{sing}}_{\mathring{w}}\neq 0$, using (22.3). Here, $\alpha_2\neq 0$ by (22.5).
- If $\mathring{w}=w_2=w_3<w_1$, then $\delta A_{\mathring{w}}=\delta A^{\text{sing}}_{\mathring{w}}\neq 0$, using (22.3). The sum of the rank 2 and rank 3 contributions does not vanish. If it did, (22.4) and $w_2=w_3$ would imply $w_1=w_2=w_3$. This contradicts the hypothesis that $w_2=w_3<w_1$.
- If $w_1=w_2=w_3$, then w is a multiple of the Hamming weight.

If w is degenerate with $-(q+1)w_1+qw_2=0$, then $w_1<w_2<2w_1$. The construction of Proposition 17.2, with $j=2$, has $\delta\bar{\eta}=\langle 1,-(q+1),q,0\rangle$. Then, dropping the initial term of $\delta\bar{\eta}$, we have

$$\overline{\operatorname{Ann}}\,\delta\bar{\eta}=q^3f_+f_-^2f_2\langle 0,1,f_-\rangle.$$

- If $w_1<w_3$, then $\delta A_{w_2}=\delta A^{\text{sing}}_{w_2}\neq 0$, using Corollary 5.5. This still works if $w_2=w_3$ because $1+f_-=q\neq 0$.
- If $w_3\leq w_1$, then $\delta A_{w_3}=\delta A^{\text{sing}}_{w_3}\neq 0$. □

In the situation where w is degenerate with $-(q^2+q+1)w_1 + q(q^2+q+1)w_2 - q^3 w_3 = 0$, the construction of Proposition 17.2, with $j = 3$, has $\delta\overline{\eta} = \langle 1, -(q^2+q+1), q(q^2+q+1), -q^3\rangle$. After dropping the rank 0 term, we have

$$\overline{\operatorname{Ann}}\,\delta\overline{\eta} = q^3 f_+ f_-^3 f_2\langle 0,0,1\rangle.$$

If $\mathring{w} \leq w_3 < 2\mathring{w}$, then $\delta A_{w_3} = \delta A^{\text{sing}}_{w_3} \neq 0$, and w does not respect duality.

From the degeneracy equation we have $(q^2+q+1)(qw_2 - w_1) = q^3 w_3$. Because q^3 and q^2+q+1 are relatively prime, there exists a positive integer a such that $qw_2 - w_1 = q^3 a$ and $w_3 = (q^2+q+1)a$. Similarly, $w_1 = q(w_2 - q^2 a)$, so there exists a positive integer b such that $w_1 = qb$ and $w_2 - q^2 a = b$. In all, $w_1 = qb$, $w_2 = q^2 a + b$, and $w_3 = (q^2+q+1)a$, for some positive integers a, b.

Because of Proposition 3.14, only the ratio $\rho = a/b \in \mathbb{Q}$, $\rho > 0$, matters. One can show that $\mathring{w} = w_3$ when $0 < \rho \leq q/(q^2+q+1)$ and that $\mathring{w} = w_1 < w_3 < 2w_1$ when $q/(q^2+q+1) < \rho < 2q/(q^2+q+1)$. Thus, for $0 < \rho < 2q/(q^2+q+1)$, w does not respect duality. For $\rho \geq 2q/(q^2+q+1)$, singletons alone are not enough to decide whether w respects duality.

For example, when the weight w over $\mathbb{F}_2$ has $w_1 = 2$, $w_2 = 5$, and $w_3 = 7$ (so $\rho = 1 > 4/7$), calculations like those in Example 21.5 show that $\delta A_6 \neq 0$. The dual codewords involved have three nonzero entries of rank 1. The combinatorics of dual codewords of this type can be very complicated (see [**1**, §3] for the situation over finite fields), and we will not pursue the matter further.

Appendix A. Fourier transform

This appendix will be a brief review without proof of the use of the Fourier transform and the Poisson summation formula in proving the MacWilliams identities for additive codes over a finite Frobenius ring. The main ideas go back to Gleason (see [**3**]) and can be generalized to additive codes over a finite abelian group. Details can be found [**28**].

In this appendix, R is a finite Frobenius ring with generating character χ. The annihilators $\mathfrak{L}(C)$ and $\mathfrak{R}(C)$ were defined in (2.3).

LEMMA A.1. *Suppose $C \subseteq R^n$ is an additive code. If $y \in R^n$, then*

$$\sum_{x\in C} \chi(x\cdot y) = \begin{cases} |C|, & \text{if } y \in \mathfrak{R}(C),\\ 0, & \text{otherwise,}\end{cases}$$

$$\sum_{x\in C} \chi(y\cdot x) = \begin{cases} |C|, & \text{if } y \in \mathfrak{L}(C),\\ 0, & \text{otherwise.}\end{cases}$$

Let V be a vector space over the complex numbers $\mathbb{C}$. If $f : R^n \to V$ is any function, define its *Fourier transform* $\widehat{f} : R^n \to V$ by

$$\widehat{f}(x) = \sum_{y\in R^n} \chi(x\cdot y) f(y), \quad x \in R^n. \tag{A.2}$$

There is an inversion formula:

$$f(x) = \frac{1}{|R^n|} \sum_{y\in R^n} \chi(-y\cdot x)\widehat{f}(y), \quad x \in R^n.$$

This version of the Fourier transform differs from that in [**28**] in that the isomorphism $x \mapsto {}^x\chi$ has been incorporated into the definition.

Suppose, in addition, that V is an algebra over $\mathbb{C}$. If $f_i : R \to V$ for $i = 1, 2, \ldots, n$, and $F : R^n \to V$ is given by $F(r_1, r_2, \ldots, r_n) = \prod_{i=1}^n f_i(r_i)$, then $\widehat{F}(x_1, x_2, \ldots, x_n) = \prod_{i=1}^n \widehat{f}_i(x_i)$.

THEOREM A.3 (Poisson summation formula). *Suppose R is Frobenius and $C \subseteq R^n$ is an additive code. If $f : R^n \to V$, then*

$$\sum_{y \in \mathfrak{R}(C)} f(y) = \frac{1}{|C|} \sum_{x \in C} \widehat{f}(x).$$

REMARK A.4. There is another version of the Fourier transform, with $\widehat{f}(x) = \sum_{y \in R^n} \chi(y \cdot x) f(y)$, that incorporates the isomorphism $x \mapsto \chi^x$ instead. Then the Poisson summation formula has the form

$$\sum_{y \in \mathfrak{L}(C)} f(y) = \frac{1}{|C|} \sum_{x \in C} \widehat{f}(x).$$

If χ has the property that $\chi(rs) = \chi(sr)$, $r, s \in R$, then the two versions of the Fourier transform agree and both forms of the Poisson summation formula are valid. This situation occurs over $M_{k \times k}(\mathbb{F}_q)$.

In order to prove the MacWilliams identities over a finite Frobenius ring R for a partition enumerator pe or a w-weight enumerator wwe as described in Section 3, here is the standard argument. There are generalizations of this argument in [**8**]. For a partition enumerator whose partition $\mathcal{P} = \{P_i\}_{i=1}^m$ of R has m blocks, set $V = \mathbb{C}[Z_1, \ldots, Z_m]$, with one variable for each block. Define $[r] = i$ when $r \in P_i$. For a w-weight enumerator, whose weight w has positive integer values with maximum value $w_{\max}$, set $V = \mathbb{C}[X, Y]$. Define $f : R \to V$ by

$$f(a) = \begin{cases} Z_{[a]}, & \text{for pe,} \\ X^{w_{\max} - w(a)} Y^{w(a)}, & \text{for wwe.} \end{cases}$$

On R^n, define $F : R^n \to V$ by $F(x_1, x_2, \ldots, x_n) = \prod_{i=1}^n f(x_i)$. For an additive code $C \subseteq R^n$, note that

$$\sum_{x \in C} F(x) = \begin{cases} \mathrm{pe}_C(Z_1, \ldots, Z_m), & \text{for pe,} \\ \mathrm{wwe}_C(X, Y), & \text{for wwe.} \end{cases}$$

The next steps depend on the specific ring, partition, and weight w:

- calculate the Fourier transform of $f : R \to V$;
- find $\widehat{F}$ by the product formula above;
- recognize the form of $\widehat{F}$ as an enumerator (if possible);
- apply the Poisson summation formula.

Some care must be taken to show that the form of $\widehat{F}$ is that of an enumerator. Care must also be taken to check if one can reverse the roles of the code and its annihilator. See [**8**] for details.

Acknowledgment

We thank the anonymous referee for many helpful comments and suggestions.

References

[1] N. Abdelghany and J. A. Wood, *Failure of the MacWilliams identities for the Lee weight enumerator over $\mathbb{Z}_m$, $m \geqslant 5$*, Discrete Math. **343** (2020), no. 11, 112036, 12, DOI 10.1016/j.disc.2020.112036. MR4119402

[2] A. Alahmadi, H. Sboui, P. Solé, and O. Yemen, *Cyclic codes over $M_2(\mathbb{F}_2)$*, J. Franklin Inst. **350** (2013), no. 9, 2837–2847, DOI 10.1016/j.jfranklin.2013.06.023. MR3146951

[3] E. F. Assmus Jr. and H. F. Mattson Jr., *Coding and combinatorics*, SIAM Rev. **16** (1974), 349–388, DOI 10.1137/1016056. MR359982

[4] W. E. Clark and D. A. Drake, *Finite chain rings*, Abh. Math. Sem. Univ. Hamburg **39** (1973), 147–153, DOI 10.1007/BF02992827. MR332875

[5] I. Constantinescu and V. Heise, *A metric for codes over residue class rings of integers* (Russian, with Russian summary), Problemy Peredachi Informatsii **33** (1997), no. 3, 22–28; English transl., Problems Inform. Transmission **33** (1997), no. 3, 208–213 (1998). MR1476368

[6] P. Delsarte, *Bounds for unrestricted codes, by linear programming*, Philips Res. Rep. **27** (1972), 272–289. MR314545

[7] S. T. Dougherty and M. M. Skriganov, *MacWilliams duality and the Rosenbloom-Tsfasman metric* (English, with English and Russian summaries), Mosc. Math. J. **2** (2002), no. 1, 81–97, 199, DOI 10.17323/1609-4514-2002-2-1-81-97. MR1900585

[8] H. Gluesing-Luerssen, *Fourier-reflexive partitions and MacWilliams identities for additive codes*, Des. Codes Cryptogr. **75** (2015), no. 3, 543–563, DOI 10.1007/s10623-014-9940-x. MR3336966

[9] M. Greferath, T. Honold, C. Mc Fadden, J. A. Wood, and J. Zumbrägel, *MacWilliams' extension theorem for bi-invariant weights over finite principal ideal rings*, J. Combin. Theory Ser. A **125** (2014), 177–193, DOI 10.1016/j.jcta.2014.03.005. MR3207471

[10] M. Greferath and S. E. Schmidt, *Finite-ring combinatorics and MacWilliams' equivalence theorem*, J. Combin. Theory Ser. A **92** (2000), no. 1, 17–28, DOI 10.1006/jcta.1999.3033. MR1783936

[11] P. Hall, *The Eulerian functions of a group*, Quart. J. Math. **7** (1936), 134–151.

[12] A. R. Hammons Jr., P. V. Kumar, A. R. Calderbank, N. J. A. Sloane, and P. Solé, *The $\mathbf{Z}_4$-linearity of Kerdock, Preparata, Goethals, and related codes*, IEEE Trans. Inform. Theory **40** (1994), no. 2, 301–319, DOI 10.1109/18.312154. MR1294046

[13] T. Honold, *Characterization of finite Frobenius rings*, Arch. Math. (Basel) **76** (2001), no. 6, 406–415, DOI 10.1007/PL00000451. MR1831096

[14] T. Honold and A . A. Nechaev, *Fully weighted modules and representations of codes* (Russian, with Russian summary), Problemy Peredachi Informatsii **35** (1999), no. 3, 18–39; English transl., Problems Inform. Transmission **35** (1999), no. 3, 205–223. MR1730800

[15] T. Y. Lam, *Lectures on modules and rings*, Graduate Texts in Mathematics, vol. 189, Springer-Verlag, New York, 1999, DOI 10.1007/978-1-4612-0525-8. MR1653294

[16] F. J. MacWilliams, *COMBINATORIAL PROBLEMS OF ELEMENTARY ABELIAN GROUPS*, ProQuest LLC, Ann Arbor, MI, 1962. Thesis (Ph.D.)–Radcliffe College. MR2939359

[17] J. MacWilliams, *A theorem on the distribution of weights in a systematic code*, Bell System Tech. J. **42** (1963), 79–94, DOI 10.1002/j.1538-7305.1963.tb04003.x. MR149978

[18] F. J. MacWilliams and N. J. A. Sloane, *The theory of error-correcting codes. I*, North-Holland Mathematical Library, Vol. 16, North-Holland Publishing Co., Amsterdam-New York-Oxford, 1977. MR465509

[19] B. R. McDonald, *Finite rings with identity*, Pure and Applied Mathematics, Vol. 28, Marcel Dekker, Inc., New York, 1974. MR354768

[20] M. Moeini, R. Rezaei, and K. Samei, *MacWilliams identity for linear codes over finite chain rings with respect to homogeneous weight*, Bull. Korean Math. Soc. **58** (2021), no. 5, 1163–1173, DOI 10.4134/BKMS.b200751. MR4319450

[21] W. W. Peterson and E. J. Weldon Jr., *Error-correcting codes*, 2nd ed., The M.I.T. Press, Cambridge, Mass.-London, 1972. MR0347444

[22] B. E. Sagan, *Combinatorics: the art of counting*, Graduate Studies in Mathematics, vol. 210, American Mathematical Society, Providence, RI, [2020] ©2020, DOI 10.1090/gsm/210. MR4249619

[23] The Sage Developers, *SageMath, the Sage Mathematics Software System (Version 9.5)*, 2022, `https://www.sagemath.org`.

[24] J. A. Wood, *Extension theorems for linear codes over finite rings*, Applied algebra, algebraic algorithms and error-correcting codes (Toulouse, 1997), Lecture Notes in Comput. Sci., vol. 1255, Springer, Berlin, 1997, pp. 329–340, DOI 10.1007/3-540-63163-1_26. MR1634126

[25] J. A. Wood, *Duality for modules over finite rings and applications to coding theory*, Amer. J. Math. **121** (1999), no. 3, 555–575. MR1738408

[26] J. A. Wood, *The structure of linear codes of constant weight*, Trans. Amer. Math. Soc. **354** (2002), no. 3, 1007–1026, DOI 10.1090/S0002-9947-01-02905-1. MR1867370

[27] J. A. Wood, *Code equivalence characterizes finite Frobenius rings*, Proc. Amer. Math. Soc. **136** (2008), no. 2, 699–706, DOI 10.1090/S0002-9939-07-09164-2. MR2358511

[28] J. A. Wood, *Foundations of linear codes defined over finite modules: the extension theorem and the MacWilliams identities*, Codes over rings, Ser. Coding Theory Cryptol., vol. 6, World Sci. Publ., Hackensack, NJ, 2009, pp. 124–190, DOI 10.1142/9789812837691_0004. MR2850303

[29] J. A. Wood, *Isometry groups of additive codes over finite fields*, J. Algebra Appl. **17** (2018), no. 10, 39 pages, paper 1850198. MR3866771

[30] J. A. Wood, *Homogeneous weight enumerators over integer residue rings and failures of the MacWilliams identities*, Rev. Un. Mat. Argentina **64** (2023), no. 2, 333–353. MR4771211

[31] B. Yildiz and S. Karadeniz, *Linear codes over $\mathbb{Z}_4 + u\mathbb{Z}_4$: MacWilliams identities, projections, and formally self-dual codes*, Finite Fields Appl. **27** (2014), 24–40, DOI 10.1016/j.ffa.2013.12.007. MR3168981

Department of Mathematics, Western Michigan University, 1903 W. Michigan Ave., Kalamazoo, Michigan 49008

Email address: `jay.wood@wmich.edu`

Selected Published Titles in This Series

826 **Ahmed Laghribi and Andre Leroy, Editors,** Algebraic Structures and Applications, 2025

821 **Bogdan D. Suceavă, Adara M. Blaga, Marian Ioan Munteanu, Cezar Oniciuc, Shoo Seto, Gabriel-Eduard Vîlcu, and Lihan Wang, Editors,** Recent Advances in Differential Geometry and Related Areas, 2025

820 **Sudeb Mitra, Rakesh Kumar Parmar, Saminathan Ponnusamy, and Toshiyuki Sugawa, Editors,** Geometric Function Theory and Related Topics, 2025

819 **Howard S. Cohl, Roberto S. Costas-Santos, and Robert S. Maier, Editors,** Applications and q-Extensions of Hypergeometric Functions, 2025

818 **Howard S. Cohl, Roberto S. Costas-Santos, and Robert S. Maier, Editors,** Classical Hypergeometric Functions and Generalizations, 2025

817 **Mikhail Khovanov, Joshua Sussan, and Anton Zeitlin, Editors,** From Representation Theory to Mathematical Physics and Back, 2025

816 **Luca F. Di Cerbo and Laurenţiu G. Maxim, Editors,** Geometry and Topology of Aspherical Manifolds, 2025

815 **Daniel Orr and Joshua Jeishing Wen, Editors,** Macdonald Theory and Beyond, 2025

814 **Susan Montgomery and Siu-Hung Ng, Editors,** Quantum Groups, Hopf Algebras, and Applications, 2025

813 **Thomas Creutzig, Julia Plavnik, and David Ridout, Editors,** Quantum Symmetries, 2025

812 **Ross Geoghegan, Craig R. Guilbault, and Kim Ruane, Editors,** Topology at Infinity of Discrete Groups, 2025

811 **Krishnendu Gongopadhyay and Shashikant A. Katre, Editors,** Geometry, Groups and Mathematical Philosophy, 2025

810 **Robert J. Berman and Yanir A. Rubinstein, Editors,** Convex and Complex: Perspectives on Positivity in Geometry, 2025

809 **Fabrice Baudoin and Luca Rizzi, Editors,** New Trends in Sub-Riemannian Geometry, 2025

808 **Loring W. Tu, Editor,** Group Actions and Equivariant Cohomology, 2024

807 **Galina Filipuk, Editor,** Recent Progress in Special Functions, 2024

806 **Maximilian Engel, Hildeberto Jardón-Kojakhmetov, and Cinzia Soresina, Editors,** Topics in Multiple Time Scale Dynamics, 2024

805 **Anthony Iarrobino, Pedro Macias Marques, Maria Evelina Rossi, and Jean Vallès, Editors,** Deformation of Artinian Algebras and Jordan Type, 2024

804 **Mahir Bilen Can and Jörg Feldvoss, Editors,** A Glimpse into Geometric Representation Theory, 2024

803 **Peter Gothen, Margarida Melo, and Montserrat Teixidor i Bigas, Editors,** Moduli Spaces and Vector Bundles—New Trends, 2024

802 **Ralph M. Kaufmann, Martin Markl, and Alexander A. Voronov, Editors,** Higher Structures in Topology, Geometry, and Physics, 2024

801 **K. A. Brown, T. J. Hodges, M. Vancliff, and J. J. Zhang, Editors,** Recent Advances in Noncommutative Algebra and Geometry, 2024

800 **Avinoam Mann, Louis H. Rowen, David J. Saltman, Aner Shalev, Lance W. Small, and Uzi Vishne, Editors,** Amitsur Centennial Symposium, 2024

799 **Alberto A. Condori, Elodie Pozzi, William T. Ross, and Alan A. Sola, Editors,** Recent Progress in Function Theory and Operator Theory, 2024

798 **Marat V. Markin, Igor V. Nikolaev, and Carsten Trunk, Editors,** Advances in Functional Analysis and Operator Theory, 2024

For a complete list of titles in this series, visit the
AMS Bookstore at **www.ams.org/bookstore/conmseries/**.

Printed and bound by CPI Group (UK) Ltd, Croydon, CR0 4YY
15/09/2026
14962376-0002